Springer Series in Reliability Engineering

Series Editor

Hoang Pham, Industrial and Systems Engineering, Rutgers, The State University of New Jersey, Piscataway, USA

Today's modern systems have become increasingly complex to design and build, while the demand for reliability and cost effective development continues. Reliability is one of the most important attributes in all these systems, including aerospace applications, real-time control, medical applications, defense systems, human decision-making, and home-security products. Growing international competition has increased the need for all designers, managers, practitioners, scientists and engineers to ensure a level of reliability of their product before release at the lowest cost. The interest in reliability has been growing in recent years and this trend will continue during the next decade and beyond.

The Springer Series in Reliability Engineering publishes books, monographs and edited volumes in important areas of current theoretical research development in reliability and in areas that attempt to bridge the gap between theory and application in areas of interest to practitioners in industry, laboratories, business, and government.

Now with 100 volumes!

Indexed in Scopus and EI Compendex

Interested authors should contact the series editor, Hoang Pham, Department of Industrial and Systems Engineering, Rutgers University, Piscataway, NJ 08854, USA. Email: **hopham@soe.rutgers.edu, or Anthony Doyle, Executive Editor, Springer, London. Email: anthony.doyle@springer.com**.

He Li · Ke Feng · Mohammad Yazdi ·
Hong-Zhong Huang
Editors

Data-Driven Methods for Reliability and Safety Engineering: Applications in Industrial Systems

Leveraging AI, Machine Learning, and Advanced Analytics to Enhance Risk Assessment, Decision-Making, and System Performance

Springer

Editors
He Li
School of Engineering
Liverpool John Moores University
Liverpool, UK

Mohammad Yazdi
School of Engineering
Macquarie University
Sydney, NSW, Australia

Ke Feng
School of Mechanical Engineering
Xi'an Jiaotong University
Xi'an, China

Hong-Zhong Huang
Center for System Reliability and Safety
University of Electronic Science
and Technology of China
Chengdu, China

ISSN 1614-7839 ISSN 2196-999X (electronic)
Springer Series in Reliability Engineering
ISBN 978-3-032-22872-7 ISBN 978-3-032-22873-4 (eBook)
https://doi.org/10.1007/978-3-032-22873-4

This Springer imprint is published by the registered company Springer Nature Switzerland AG
The registered company address is: Gewerbestrasse 11, 6330 Cham, Switzerland

Preface

The pursuit of reliability and safety has always been one of the most critical dimensions of engineering. From the earliest machines to today's highly integrated and technologically advanced infrastructures, humanity's progress has been shaped through innovation and its ability to manage risk and uncertainty. In aerospace systems, safety margins are razor thin; in nuclear power, the consequences of failure are profound; in industrial and manufacturing contexts, efficiency must coexist with protection. The demand for structured and scientific approaches to reliability continues to grow.

This book is written against the backdrop of that demand. Engineering systems are becoming increasingly complex, operating under uncertainty, and expected to deliver uncompromising levels of safety and performance. Traditional methods of design and assessment remain valuable, yet they require reinforcement with integrated methodologies that combine statistical rigor, computational intelligence, data fusion, and risk-informed decision-making.

The chapters collected here demonstrate such integration. They present an impressive breadth of approaches, response surface methodology, Bayesian estimation, mission reliability evaluation, risk-informed exemptions, and more, applied to real engineering contexts. What unites these contributions is their dual commitment: advancing theoretical understanding and addressing practical challenges. The book bridges the worlds of research and application in a deliberate way, offering both scientific depth and operational value.

A central theme across the book is the recognition of uncertainty as an unavoidable aspect of engineering. Instead of treating uncertainty as a nuisance to be eliminated, the contributors present frameworks for quantifying, managing, and making decisions in its presence. These frameworks are mathematically rigorous and operationally meaningful, providing tools that can be applied in industries that must function with limited test data, constrained resources, and rapidly changing environments.

The intended audience for this book extends across disciplines and professions. For researchers, it provides consolidated knowledge of recent methodologies and case studies, forming a foundation for deeper inquiry and innovation. For practitioners and engineers, it offers techniques and insights that can be implemented directly in system design, testing, and evaluation. For educators, it serves as a resource that connects theory with application, helping students understand how principles function in practice. For policy-makers and regulators, it illustrates how modern reliability and risk-informed approaches can guide decisions that balance safety, cost, and societal impact.

This project represents a collaborative effort. The contributing authors bring diverse expertise from aerospace, nuclear engineering, industrial safety, and related domains. Each chapter reflects both individual depth and collective perspective. The reviewers and colleagues who supported the development of this volume strengthened its quality through constructive feedback, and their contributions deserve equal recognition.

In closing, this book aspires to serve as a reliable reference for those engaged in ensuring the safety and resilience of complex systems. Whether the reader is conducting research, teaching future engineers, managing industrial risks, or shaping policy, this volume offers not only methods and models but also a guiding philosophy: safety and reliability are evolving practices that demand ongoing attention, creativity, and collaboration.

Liverpool, UK
Sydney, Australia
Xi'an, China
Chengdu, China
September 2025

He Li
Mohammad Yazdi
Ke Feng
Hong-Zhong Huang

Contents

Reliability Analysis of Aircraft Grooved Joints Based on Response Surface Methodology 1
Baixu Chen, Changqing Su, and Xin Xu

Comprehensive Evaluation Techniques of Basic Reliability Based on Data Fusion of Similar Products 17
Yi Jiao

Mission Reliability Comprehensive Evaluation Technique Based on Multi-stage Information 27
Haijing Song, Ming Su, and Zimeng Xu

Preliminary Exploration of Exemption Methods for Single-Fault Criteria in Nuclear Power Plants 37
Jin An, Lin Yan, and Jian Yang

Accuracy Lifetime Prediction Methodology for RV Reducer 47
Liangbo Yao, Cheng Qiu, Yikun Yang, Yu Gong, and Jinfeng Li

Derating Design of Control Circuits Based on Harpness Simulation 61
Xin Wang, Jun-Wen Shi, and Xiang-Yang Yu

AK-SYSC: A Kriging-Based Approach for Multi-failure-Mode Structural Systems Reliability Analysis 71
Liangli He, Xinfa Chen, and Xinyao Li

Risk-Informed Intelligent Design Optimization for Nuclear Power Plants Using Probabilistic Safety Analysis 83
Ming Wang, Jiaoshen Xu, and Bing Zhang

Research on Fault Data-Driven Quality Risk Assessment Method for Aviation Equipment 97
Jiake Li and Yiyi Zhao

A Preliminary Summary of Failure Mode and Effects Analysis of Digital Control Systems 111
Bohao Tian, Jun Yang, Chenyu Jiang, and Zhanyu He

Research on the Reliability Assurance System in the Product Development Process 125
Yinping Ren, Tongyu Zhang, and Lili Chang

An Architectural Design for the Reliability and Safety of Battery Management System 147
Xinpeng Wang, Yiqun Du, and Bo Meng

Research on Reliability Classification Technology of High-Temperature Gas-Cooled Reactor 155
Yiming Wang, Yongsheng Wen, Dingqiang Li, Zhijun Li, and Changliang Liu

Maintenance Optimisation for K-out-of-N: G System Using Deep Reinforcement Learning 169
Ning Zhang, Yifan Zhou, Ze Li, and Ziqiang Huang

Investigation on Quantitative Evaluation Technology for System Integrated Support 189
Yuhan Liu, Xiaoming Gao, Baidong Jin, and Pengwei Hu

Goal-Oriented Reliability Program Framework for Space Science and Application Payloads 199
Kuanyu Gao, Peng Li, Man Fang, and Hongen Zhong

Study on the Effectiveness Evaluation of Cross-Flooding Hatches in Damage Stability Design of Large Passenger Vessels 213
Yi-ming Huang, Gang Chen, Yong Guo, Zhang-jian Wang, Xin-yi Fan, Jia-ning Li, Gao-shuai Wang, Ai-hua Gao, and Fu-tong Li

Calculation and Analysis of Fatigue Life Damage of Diesel Engine Crankshaft 229
Shufei Xue, Hualiang Zhang, and Gongye Yu

Exploratory Research on Optimizing Equipment Support Supply Chain Based on Boundaryless Organization Theory 247
Tianqiao Yan, Yabin Wang, Jinguo Wang, and Cenyu Hu

Research on High-Efficiency Milling Process and Machining Parameter Optimization of Aircraft Wing Rib Components 261
Bo Pang, Jie Chen, and Jinglei Wang

ESPAC-Based Linear Array Microtremor Tomography Method and Its Application in Goaf Detection 277
Shasha Liang, Xinyue Wang, and Ziqi Wang

A Study on Fault Prediction of Simulated Die Press Machine Based on Machine Learning ... 301
Jun Feng Lai, Hui Yu, Yue Zhao, and Jia Liu

Digital Twin-Driven Condition Monitoring for Servo Valves ... 317
Chaoyi Zhang, Tongrong Zhang, Zijun Gao, Ying Zhong, Xiaobang Wang, and Wenzheng Xu

YOLO-MVCP: A Lightweight Fault Diagnosis Method of Rolling Bearing Based on STFT Time–frequency Graph and MobileViT Network Pruning ... 333
Yu Wang, Yue Li, and Shufeng Zhang

Design of System-Level Self-health Energy Management of Manned Spacecraft ... 349
Qian Liu, Fangyuan Wen, Zhen Lv, Minjun Mao, and Rong Fan

A PINN-CNN Hybrid Architecture for Vibration-Based Damage Detection in Airship Envelopes ... 361
Fangyuan Zhao, Yiwei Cheng, Keqiang Xie, and Yuanhang Wang

Coupling Simulation Analysis of High Temperature Heater Under Multiple Physical Fields ... 375
Junbin Zhang, Shubin Jia, Ye Huang, Jiufen Chen, and Yingdong Chen

Research on Instantiation Method of Object-Oriented Bayesian Network for Fault Diagnosis Based on Dynamic Weight Parameter Transfer ... 391
Shigang Zhang, Mengqiao Chen, Yongxuan Fang, Xudong Suo, and Xu Luo

From Phenomena to Mechanisms: A Time–Space–Matter Multidimensional Analysis of Aerospace Electronic Failures ... 403
Zongsheng Xie, Dawei Zhu, Mengjiang Liu, and Danye Li

Research on the Health Monitoring Technology of Preload for Complex Aerospace Equipment Fastening Devices ... 417
Wu Kai, Luo Zhijun, Gu Lixiang, Zuo Wenjia, and Liu Yuchang

Probabilistic Life Estimation Framework for Mechanical Component ... 431
Xuekang Li, Mingtao Wu, Yongqiang Wang, Zuwang Gan, and Haikun Zhao

Uncertainty Propagation in Bridge Construction Carbon Emission Quantification ... 439
Hafiz Muhammad Hazib, Imran Shabbir, Weizong Lai, Yue Pan, and Jianjun Qin

Gas Turbine Rotor Systematic Upgrading Technology Research and Engineering Application 453
Huayun Zhu, Jiangsheng Zheng, Yuming Zhang, Hongchen Zhao, Jiaming Wei, Linfeng Xue, Jibing Lan, and Yongfeng Sui

A Storage Life Assessment Method for Missile Based on Critical Component Inspection Date 469
Guiming Tao, Qirui Wang, and Hui Li

Research on Internal Stress Damage Detection Strategies for Optical Components 477
Chendong Lei, Peng Yang, Xinyu Luo, Han Yu, and Guanjun Liu

Study of Properties of High Zn Content Aluminium Alloys for Thin Wall Deep Cavity Casting 489
Guiming Tao, Yi Wang, and Anhang Liu

Nanoindentation-Based Finite Element Characterization Method for 25Cr2Ni2MoV Rotor Steel Welded Joints 507
Guo Sujuan, Pan Zijian, Zhao Jian, and Zhu Mingliang

Quantification of Risk and Control of Bidding for a Nuclear Power Plant Project's Marine Water Intake Project 519
Mingfang Chen, Tao Wang, Xiaodong Guo, and Xinhua YU

Research on Accelerated Test Method for Reliability of a Certain Refrigeration Device 531
Dayong Rui, Cheng Sun, and Yisheng Zhu

Research on Degradation Model of Marine Seawater Centrifugal Pump Based on Main Mechanism of Cavitation Erosion 541
Xiaoli Yang, Yong Wang, Yutong Luo, and Hongwei Hu

Generative AI in Safety-Critical Systems: Risks of Sycophancy, Hallucination, and Trust Degradation 557
Mohammad Yazdi, Esmaeil Zarei, Sidum Adumene, and Amin Beheshti

Human-Centred Intelligence in Manufacturing Safety: From Observation to Smart Conversations 583
Mohammad Yazdi

Sloshing Dynamics and Its Impact on Ship Stability and Safety 609
Kombo Theophilus-Johnson, Sidum Adumene, and Azubuike John Chuku

A Computational-Based Comparative Fatigue Analysis of Composite and Steel Risers 625
Morrison Inegiyemiema, Sidum Adumene, and Mohammad Yazdi

Implementation of Process Safety Management System in the Power Plant Industry Using the Monte Carlo Probabilistic Method 655
Mahdi Mozaffari, Behrouz Behnam, Adel Gholami, and Adel Mohammad Pour

Prioritizing HSE Management Risks in Bridge Construction Projects Using Monte Carlo Modeling 679
Alireza Amanzadeh, Adel Gholami, and Mahdi Mozaffari

Fault Prediction Methods for Locomotive Turbochargers 701
Yu Xing, Weida Wei, He Li, and Junyu Guo

About the Editors

Dr. He Li obtained his Ph.D. degrees from the University of Electronic Science and Technology of China, China (2021), and the University of Lisbon, Portugal (2025) and has been Researcher at Liverpool John Moores University, UK (2024–). He is Fellow of the International Society of Engineering Asset Management (ISEAM Fellow), Technical Committee Member of the European Safety and Reliability Association (Marine Engineering, ESRA Fellow), and Editor/Associate Editor/Guest Editor/Editorial Board Member for more than 10 Journals. He has been selected as Marie Curie Fellow and recognized as World's Top 2% Scientist (2024, 2025). Dr. Li has a selection of publications, e.g., monographs and journal/conference papers with several highly cited/hot papers and best paper awards. His research focuses on the reliability and maintainability of marine energy systems. e-mail: h.li@ljmu.ac.uk

Prof. Ke Feng is Full Professor at Xi'an Jiaotong University, China. He is Marie Curie Fellow, a World's Top 2% Scientist, and received a Ph.D. degree from the University of New South Wales, Australia. He worked at the University of British Columbia and the National University of Singapore in 2022 and 2023, respectively. His main research interests include digital twins, vibration analysis, structural health monitoring, dynamics, tribology, signal processing, and machine learning. He is recognized as Emerging Leader (2023) by the *Measurement Science and Technology Journal.* He has been Associate Editor and Guest Editor of several journals, including IEEE Transactions on Industrial Informatics, Information Fusion, Mechanical Systems and Signal Processing, IEEE Transactions on Industrial Cyber-Physical Systems, etc. e-mail: kefeng@xjtu.edu.cn

Dr. Mohammad Yazdi is a Lecturer and Researcher at Macquarie University, Australia. He earned a dual Ph.D. degree from Memorial University of Newfoundland, Canada, and Macquarie University, Australia. His research interests and professional background converge at the nexus of system safety, risk assessment, resilience, process integrity, and asset management, especially concerning renewable and non-renewable energy infrastructure. With an impressive track record of leading large-scale energy projects and technology-rich initiatives, Mohammad offers invaluable support to asset operators, developers, and maintainers, and has been recognized as a World's Top 2% Scientist for many years. e-mail: mohammad.yazdi@mq.edu.au

Prof. Hong-Zhong Huang is Full Professor and Director of the Center for System Reliability and Safety, at the University of Electronic Science and Technology of China. He has held visiting appointments at several universities in the USA, Canada, and Asia. He received a Ph.D. degree in Reliability Engineering from Shanghai Jiao-tong University, China. He has published more than 200 journal papers and five books in the fields of reliability engineering, optimization design, fuzzy sets theory, and product development. His main research interests include reliability design, optimization design, condition monitoring, fault diagnosis, and life prediction. e-mail: hzhuang@uestc.edu.cn

Reliability Analysis of Aircraft Grooved Joints Based on Response Surface Methodology

Baixu Chen, Changqing Su, and Xin Xu

Abstract Grooved joints are commonly utilized as tensile joints in aircraft structures. While fulfilling the stiffness and strength requirements, the reliability assessment of these grooved tensile joints is critical and should not be overlooked. Response surface methodology (RSM), a statistical approach for experimental design and analysis, is employed to establish mathematical models that relate input and output variables. The optimization of grooved joint connection structures is analyzed using RSM, with the Box-Behnken matrix design method applied for the experimental setup. Response surfaces and second-order regression equations are derived through fitting based on the generated design points. Additionally, response surface functions are formulated using the least squares regression method. From a reliability standpoint, limit state equations are developed, and the reliability, along with the sensitivity of relevant random influence parameters, is calculated. The findings of this study offer valuable insights for subsequent dimensional optimization design.

Keywords Response surface methodology · Grooved joints · Reliability analysis · Reliability and sensitivity calculations

1 Introduction

Aircraft connecting joints are essential for linking structural components and enabling the transfer and distribution of loads. Nevertheless, fasteners within these joints are prone to stress concentrations, which can result in structural deterioration and fatigue damage. Consequently, the strength of connecting joints is a critical subject in investigations pertaining to aircraft safety and reliability. Slotted tension joints are frequently employed in aircraft connections, being secured to load-bearing

B. Chen (✉) · C. Su · X. Xu
Shenyang Aerospace University, Shenyang, China
e-mail: chenbaixu_2001@163.com

Safety Engineering of Shenyang Aerospace University, Shenyang, China

H. Li et al. (eds.), *Data-Driven Methods for Reliability and Safety Engineering: Applications in Industrial Systems*, Springer Series in Reliability Engineering,
https://doi.org/10.1007/978-3-032-22873-4_1

elements through multiple fasteners [2]. The determination of limit state equations and the execution of reliability analyses for grooved joints using response surface methodology (RSM) represent an effective analytical approach.

Anjum et al. [1] utilized the response surface method to formulate the limit state equation for mid-span deflection of truss bridges and evaluated their reliability under normal service limit states. Yang et al. [2] proposed a first-order reliability method that computes the reliability of individual probabilistic constraints under extreme conditions and optimizes design accordingly. Fang et al. [3] developed the limit state function for horizontal bearing reliability analysis of slope foundation piles based on the stochastic response surface method, calculating reliability indices and failure probabilities using the first and second-order moment methods. Lan et al. [4] introduced a calculation method for collapse zone extent founded on reliability indices obtained through RSM, while Cai et al. [5] outlined a comprehensive reliability analysis procedure grounded in RSM.

This research concentrates on the reliability assessment of aircraft grooved tensile joints, applying response surface methodology principles to design experiments on dimensional parameters that influence joint performance. Design points are generated, and response surface software is employed to construct response surfaces for each dimensional parameter alongside the corresponding maximum stress values. Interactions among dimensional parameters are examined to assess their significance, and the effects of parameter variations on maximum stress are investigated. Quadratic regression analysis is conducted on the collected data to derive fitting equations, from which limit state equations are formulated for reliability evaluation. The results provide valuable guidance for future dimensional optimization in the design of aircraft joints.

2 Response Surface Methodology

2.1 Response Surface Theory

Response surface methodology (RSM) [6] is a statistical technique used to model the relationships between input and output parameters, enabling the prediction of key indicators such as response, performance, and reliability under varying operational conditions. This approach involves constructing response surface models to approximate system behavior. During sensitivity analyses, RSM fits a response surface that approximates the limit state, providing an approximate expression for the system function. To perform response surface analysis, a set of design variables must first be identified [7]. These variables may encompass material properties, geometric characteristics, load parameters, and others. The response variables, which represent the structural performance metrics to be predicted, are then determined. Subsequently, the response surface model is developed to analyze the relationship between the design variables and the response variables. Response surface functions can take the form of linear functions, polynomial functions, neural networks, or other function forms. A key component of the response surface method is test design [8], Due to the complexity of the design space, numerous factors must be considered when selecting test points within it. Since test point selection affects accuracy to a certain extent, the classical response surface method employs the star design approach, which selects test points around the sampling center at a specific distance along both positive and negative directions of the random variable coordinate axis. Various experimental design methods exist, with commonly used ones including full factorial design (two-level or multi-level), partial factorial design, and central composite design [9]. The steps of the response surface method design experiment are shown in Fig. 1.

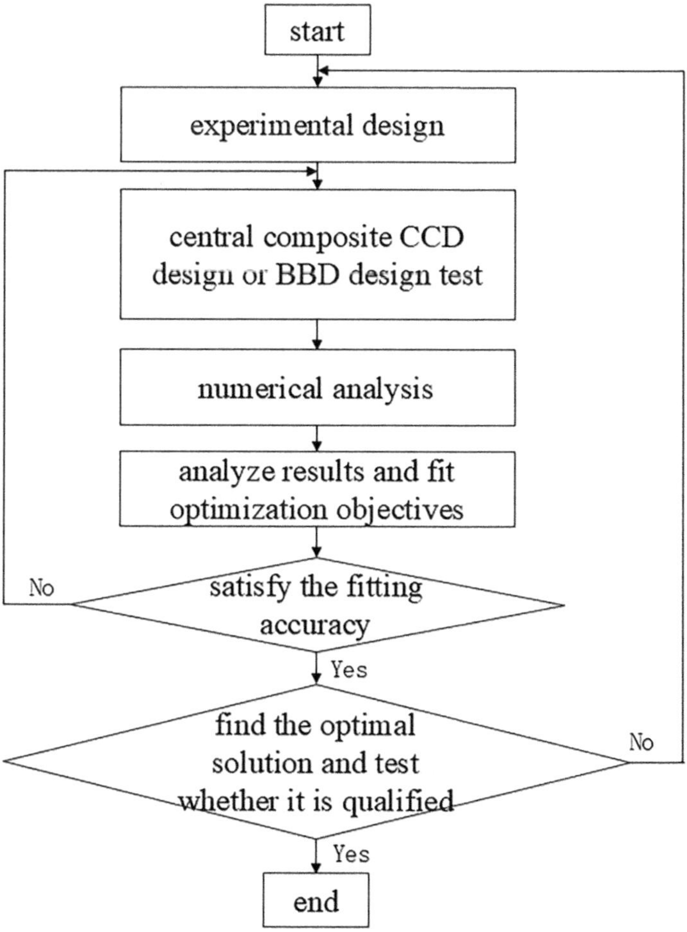

Fig. 1 Step-by-step diagram of response surface methodology design of experiments

2.2 Reliability Calculation Based on Response Surface Methodology

The stress-strength interference theory model is utilized to calculate the probability that strength exceeds stress. It is assumed that, in the response surface function S is the stress intensity factor of the connected structure and R is the permissible stress intensity of the material, which are equivalent to the stress x_m and strength x_n in the stress-strength interference model, respectively, and Z is known as the limit state function, which can represent the three states of the structure; $g(X)$ is greater than 0 for the safe stage; $g(X)$ is less than 0 for the failure state; $g(X)$ is equal to 0 for the limit state. The strength reliability limit state equation [6] can be defined from the strength theory failure criterion.

$$Z = g(X) = R(X) - S(X)$$

The reliability of the structure can be measured directly by using the reliability index. If the basic random parameter vector X obeys the normal distribution.

From the above equation, it can be seen that β reflects the degree of reliability of the structure, which is called the structural reliability index, the larger β is, the smaller the failure probability P_f is, and the higher the structural reliability is.

3 Aircraft Grooved Joint Analysis Results

3.1 Finite Element Analysis

The structure of the aircraft's grooved tensile joint is relatively intricate, consisting of 8 fastener holes in the base plate and 5 bolt holes on each side of the two sidewall plates. In the structural dimension design, the bolt hole diameter is specified as 12 mm. The outer diameter of the steel washer is 25 mm, with an inner diameter of 12 mm. The combined thickness of the bolts and washers is 1 mm. For the wall plates, the radius of the fastener holes near the bolt ends on both sides is set to 4 mm, while the distance from the bolt ends to the fastener holes on the opposite side is set to 7 mm, which is equivalent to the fastener hole diameter. The dimensional design model is illustrated in Fig. 2.

Geometric modeling of the slotted tension joint was performed in the FEA software ANSYS using a bottom-up approach. The slotted joint consists of an end section with a central bolt, two side wall plates with fastener holes, and a base plate. The slotted joint features 8 fastener holes in its base plate and 5 fastener holes on each side wall plate. Its structure was shown in Fig. 3. The model uses the SOLID187 solid element. The base material is aluminum alloy, while the bolt material is high-toughness alloy steel.

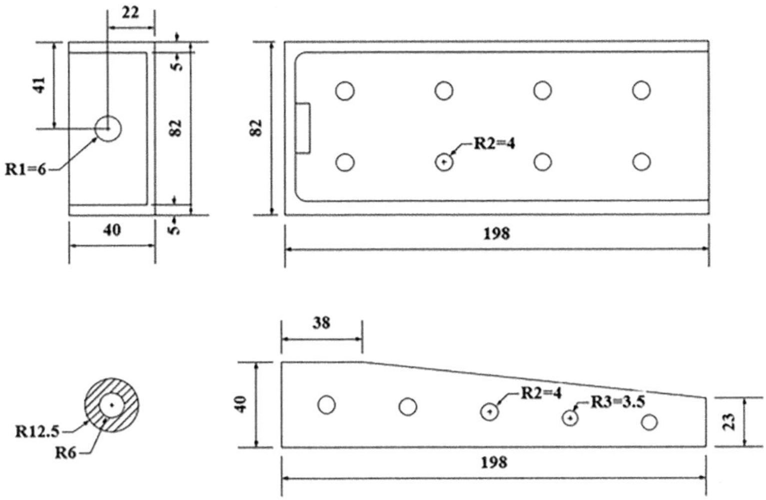

Fig. 2 Dimensioning of grooved joint

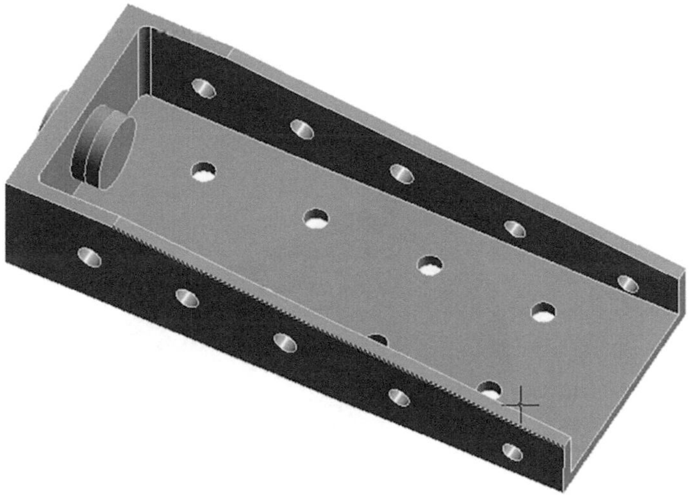

Fig. 3 Schematic diagram of groove joint structure

Full constraints were applied to the fastener holes of the slotted tension joints during simulation. The simulation condition involved applying a horizontal tensile force of 15,000 N to the bottom surface of the bolt. This force was transmitted to the slotted joint through contact between the bolt and fastener holes, with load constraints applied as presented in Fig. 4.

After establishing the finite element model of the grooved tension joint connection structure, simulation calculations were performed to generate the stress cloud

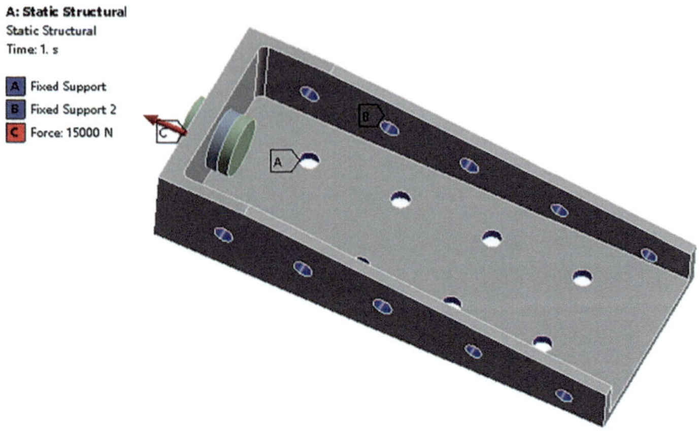

Fig. 4 Slotted joint load constraint schematic diagram

diagram. Calculations were conducted in the solver, and the post-solution equivalent stress results were used to assess whether strength requirements are met. The generated stress cloud diagram is presented in Fig. 5.

Analysis of the stress cloud diagram in Fig. 5 reveals a maximum stress of 307.96 MPa, which is lower than the material's yield strength of 445 MPa, thus satisfying the design requirements.

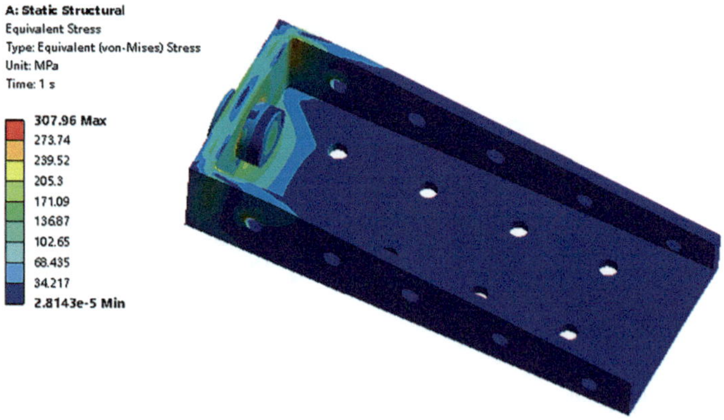

Fig. 5 Stress map of grooved joint

3.2 *Response Surface Analysis of Grooved Joints*

The Box-Behnken Design principle was applied to design the experimental program, providing complete steps and data processing results. To determine the optimal maximum stress under different factor conditions, the input parameters selected for factors influencing grooved joints include: diameter of end bolt holes (D1), diameter of fastening holes (D2, D3), chamfering radius (RD), and modulus of elasticity (EX1). The selected output parameter is the maximum stress value. The Box-Behnken test design is shown in Table 1.

The selected parameters and response values were input into the response surface module of the finite element software in the corresponding order to generate test points, yielding a total of 41 sets of test results as presented in Table 2.

The maximum stress model was developed and analyzed with the joint parameters replaced by letters, respectively.

The data related to testing the significance of the model and other relevant data, such as testing the significance of the model, were obtained by performing a significance test and ANOVA on the model [10], which were shown in Table 3.

Tables 3 and 4 showed that the model was highly significant with a coefficient of determination of R^2 of 0.9828, which indicated that the model was well fitted and could fit the experimental results more intuitively. A non-significant lack-of-fit term indicates small model error, while the adjusted coefficient of determination ($RAdj^2 = 0.9656$) confirms strong model correlation, sufficient for theoretical analyses and predictions.

Analysis of each factor's influence on maximum stress reveals that both primary and secondary terms exert a highly significant effect. Among the five influencing factors, their impact on maximum stress follows the order: D > E > A > B > C, corresponding to chamfer radius > elastic modulus > end bolt hole diameter > large fastener hole diameter > small fastener hole diameter. This indicates that these five factors influence maximum stress in a non-linear manner.

Table 1 Experimental design table for 5-factor 3-level response surface analysis

Factor	level		
	− 1	0	1
A: D1—End bolt hole diameter (mm)	10.8	12	13.2
B: D2—Fastener hole large diameter (mm)	7.2	8	8.8
C: D3—Fastener hole small diameter (mm)	6.3	7	7.7
D: RD—Radius of chamfer (mm)	2.7	3	3.3
E: EX1—Modulus of elasticity (MPa)	63,900	71,000	78,100

Table 2 Response surface to ultimate maximum stress design and results

Number	D1	D2	D3	RD	EX1	Maximum stress
1	12	8	7	3	71,000	307.96
2	12	7.2	6.3	3	71,000	341.74
3	12	8.8	6.3	3	71,000	342.37
4	12	7.2	7.7	3	71,000	349.32
5	12	8.8	7.7	3	71,000	345.23
6	12	8	7	2.7	63,900	279.48
7	12	8	7	3.3	63,900	285.7
8	12	8	7	2.7	78,100	272.82
9	12	8	7	3.3	78,100	309.67
10	12	8	6.3	3	71,000	296.66
11	10.8	8	7.7	3	71,000	301
12	10.8	8	6.3	3	71,000	294.67
13	13.2	8	7.7	3	71,000	272.03
14	13.2	7.2	7	2.7	71,000	299.72
15	12	8.8	7	2.7	71,000	316. 11
16	12	7.2	7	3.3	71,000	272.92
17	12	8.8	7	3.3	71,000	299.84
18	12	8	7	3	63,900	282.84
19	10.8	8	7	3	78,100	268.67
20	10.8	8	7	3	63,700	300.93
21	13.2	8	7	3	71,000	290.46
22	13.2	8	6.3	2.7	71,000	297.29
23	12	8	7.7	2.7	71,000	282.52
24	12	8	6.3	3.3	71,000	306.07
25	12	8	7.7	3.3	63,900	289.42
26	12	7.2	7	3	63,900	289.1
27	12	8.8	7	3	78,100	262.72
28	12	7.2	7	3	78,100	305.43
29	12	8.8	7	3	71,000	309.93
30	10.8	8	7	2.7	71,000	347.96
31	10.8	8	7	3.3	71,000	309.29
32	13.2	8	7	2.7	71,000	335.55
33	13.2	8	7	3.3	71,000	332.35
34	10.8	7.2	7	3	71,000	331.06
35	10.8	8.8	7	3	71,000	331.25
36	13.2	7.2	7	3	71,000	280.23
37	13.2	8.8	7	3	71,000	365.83

(continued)

Table 2 (continued)

Number	D1	D2	D3	RD	EX1	Maximum stress
38	12	8	6.3	3	63,900	331.06
39	12	8	7.7	3	63,900	318.49
40	12	8	6.3	3	78,100	369.46
41	12	8	7.7	3	78,100	302.61

Table 3 Results of regression analysis for the maximum stress model and regression coefficients

Std. dev.	4.58	R-squared	0.9828
Mean	300.44	Adj R-squared	0.9656
C.V. %	1.52	Pred R-squared	0.9120
PRESS	149.38	Adeq precision	24.578

3.3 *Interaction of Factors in Grooved Joints*

Within the response surface software, response surfaces were plotted based on the generated design points. These plots can intuitively reflect how factor interactions influence maximum stress [11], facilitating the identification of optimal process parameters and parameter interactions [12].

Results from the AB interaction diagram analysis are presented in Fig. 6. Within the AB interaction surface, the relationship between joint stress and the factors is relatively complex, featuring obvious curvature that indicates a strong nonlinear relationship. As the end bolt hole diameter increases, maximum stress tends to first increase and then decrease. Considering only the interaction between these two factors, maximum stress is greater when the large fastener hole diameter is approximately 7.6–8.4 mm and the end bolt hole diameter is around 12–12.6 mm.

The AD interaction diagram is presented in Fig. 7. Within the AD interaction surface, maximum stress first increases and then decreases as the end bolt hole diameter and chamfer radius increase. This indicates a significant interaction between the end bolt hole diameter and chamfer radius, with maximum stress being greater when the chamfer radius is approximately 3–3.24 mm and the end bolt hole diameter is around 11.4–12.6 mm.

The AE interaction diagram is presented in Fig. 8. Within the AE interaction surface, maximum stress first increases and then decreases as the elastic modulus increases, indicating an obvious interaction between these two parameters. Maximum stress is higher when the elastic modulus is approximately 67,450–71,000 MPa and the end bolt hole diameter is around 12–12.6 mm.

As observed in the BD interaction surface (Fig. 9), maximum stress tends to first increase and then decrease as the fastener hole diameter and chamfer radius increase. This indicates a significant interaction between the fastener hole diameter and chamfer radius. Considering only this interaction, maximum stress is higher

Table 4 Maximum stress modelling and ANOVA results

Variance	Sum of Squares	Degrees of freedom	Mean Square	F value	P-value (Prob > F)	Significance
Model	16,757.24	20	1196.96	57.18	< 0.0001	Highly significant
A-D1	422.50	1	422.50	21.14	0.0004	Highly significant
B-D2	350. 14	1	350. 14	16.73	0.0011	Highly significant
C-D3	140.78	1	340.78	21.88	0.0021	Highly significant
D-RD	1873.00	1	1873.00	89.48	< 0.0001	Highly significant
E-EX1	926.64	1	926.64	44.27	< 0.0001	Highly significant
AB	3.42	1	3.42	0.16	0.6921	
AC	27.84	1	27.84	0.37	0.5692	
AD	234.55	1	234.55	11.21	0.0048	Highly significant
AE	0.88	1	0.88	0.042	0.8402	
BC	3.42	1	3.42	0.046	0.8394	
BD	238.39	1	238.39	11.39	0.0045	Highly significant
BE	181.98	1	181.98	8.69	0.0106	significant
CD	4.27	1	4.27	0.057	0.8209	
CE	78.77	1	78.77	0.094	0.7720	
DE	27.72	1	27.72	1.32	0.2691	
A2	6038.64	1	6038.64	288.49	< 0.0001	Highly significant
B2	5622.05	1	5622.05	258.59	< 0.0001	Highly significant
C2	5243.20	1	5243.20	210.45	< 0.0001	Highly significant
D2	4322.80	1	4322.80	206.52	< 0.0001	Highly significant
E2	3346.57	1	3346.57	159.88	< 0.0001	Highly significant
Residual	293.05	14	20.93			
Misfit term	248.55	10	24.85	0.26	0.2278	Not significant
Error term	44.50	4	11.12			
Sum	17,050.28	28				

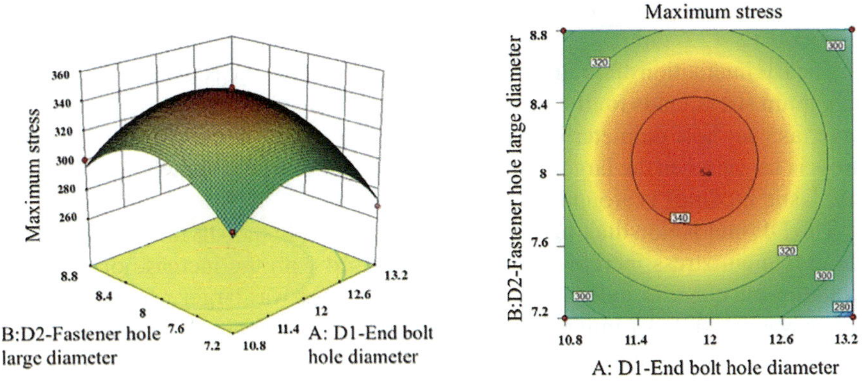

Fig. 6 Effect of AB interaction on maximum stresses

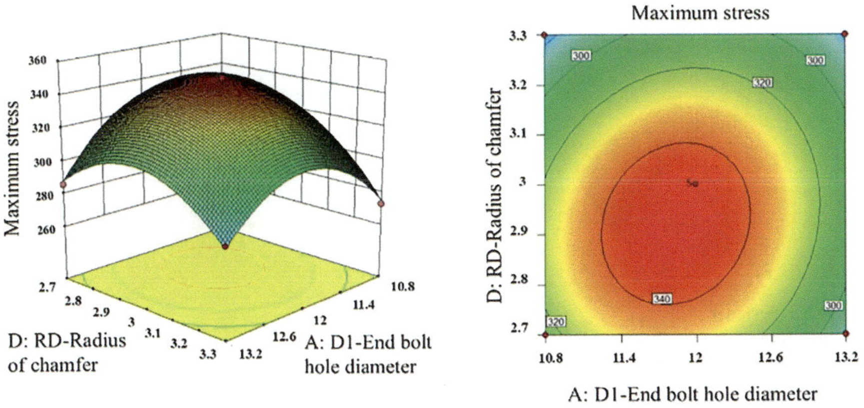

Fig. 7 Effect of AD interaction on maximum stresses

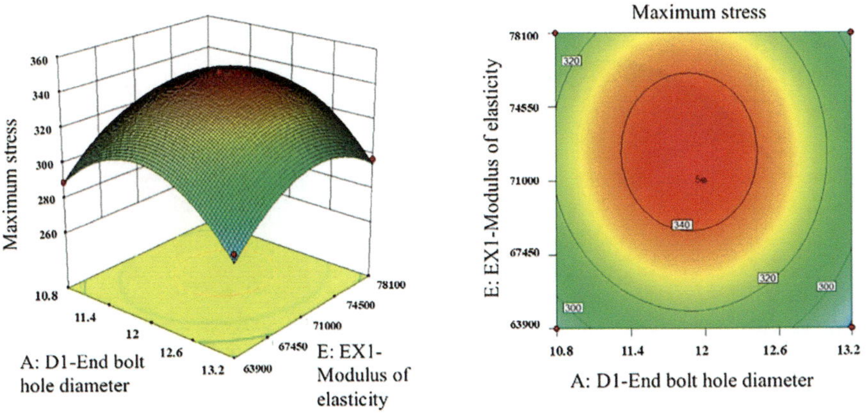

Fig. 8 Effect of AE interaction on maximum stresses

when the fastener hole diameter is approximately 8–8.4 mm and the chamfer radius is around 12–12.6 mm.

The BE interaction diagram is plotted in Fig. 10. Within the BE interaction surface, maximum stress tends to first increase and then decrease as the elastic modulus and large fastener hole diameter increase. Considering only their interaction, maximum stress is larger when the large fastener hole diameter is approximately 8–8.4 mm and the elastic modulus is around 67,450–71,000 MPa.

The DE interaction diagram was shown in Fig. 11. In the DE interaction surface, the maximum stress increased and then decreased with the increase of the elastic modulus and chamfer radius, and the maximum stress was larger when the chamfer radius was about 3–3.1 mm and the elastic modulus was about 71,000–74,550 MPa, considering only the interaction of the two.

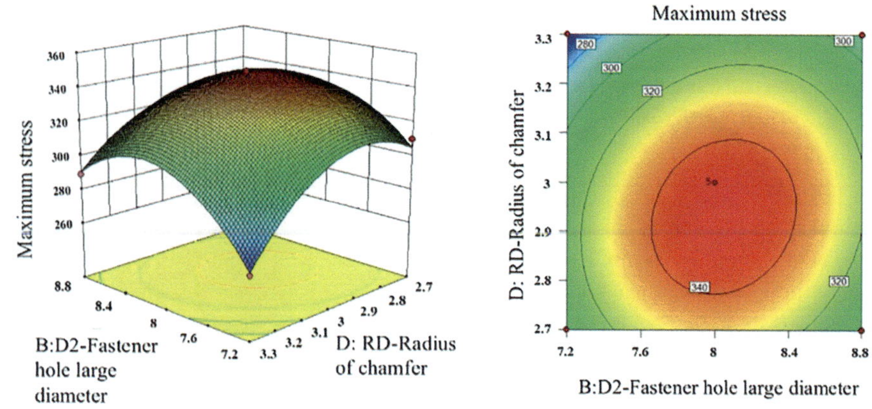

Fig. 9 Effect of BD interaction on maximum stresses

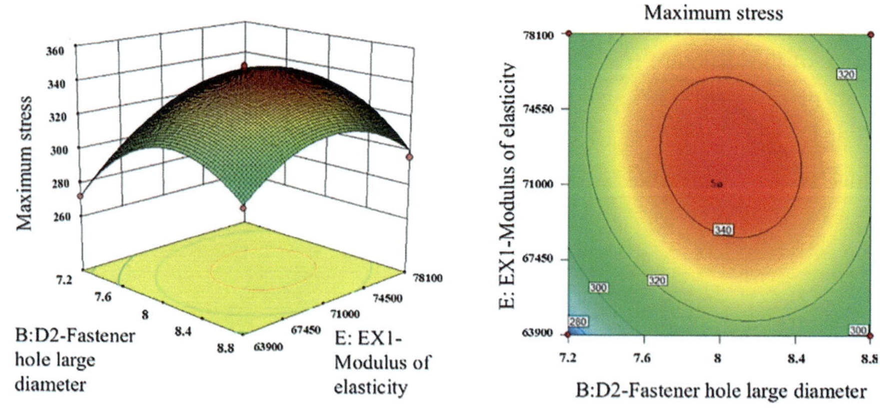

Fig. 10 Effect of BE interaction on maximum stresses

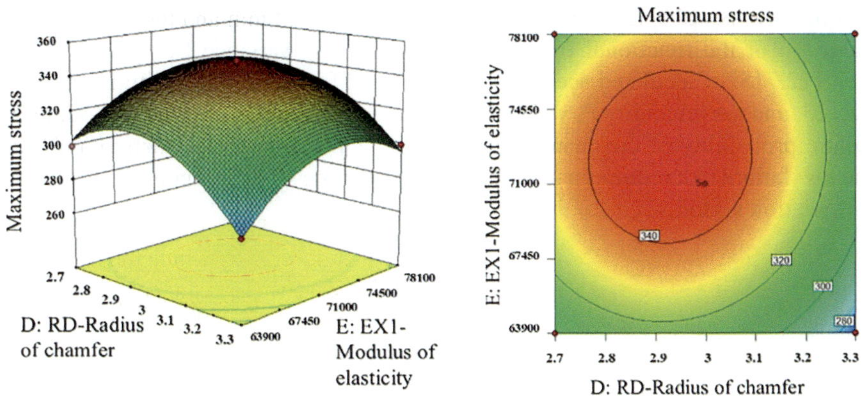

Fig. 11 Effect of DE interaction on maximum stresses

3.4 *Reliability Calculation Based on Response Surface Methodology*

In this study, the end bolt hole diameter (D1) is selected as a random variable for analysis, along with the large fastener hole diameter (D2), small fastener hole diameter (D3), chamfer radius (RD), grooved joint material properties (elastic modulus EX1), and stress intensity R. The distribution types of these random variables are presented in Table 5.

Grooved joint failure under external forces occurs when the maximum stress exceeds the yield limit. Using the design point data generated in Sect. 3.2, a total of 41 sample data sets for grooved joints were obtained via BBD test point design to derive the second-order response surface regression equation [13].

A method for solving the reliability index and structural reliability of grooved joint structures based on the primary second-order method of moments. The function $Z = g(X)$ was expanded at a certain point by Taylor series to solve the reliability index of the structure. According to this method, the mean value μ_g and variance D_g of each random parameter in the connection structure of the grooved joint can

Table 5 Distribution of random variables for slotted joint

Parameter symbolic	Dispersion pattern	Mean value	Standard deviation
D1	Normality	12	0.4
D2	Normality	8	0.4
D3	Normality	7	0.35
RD	Normality	3	0.15
EX1	Normality	71,000	3550
R	Normality	455	10

be obtained. By carrying out the above calculations for the grooved joint, the mean value μ_g of the functional function was 249.37, the variance D_g of the function was 112.7152. Calculated according to Eq. 1 and Eq. 2, the reliability index β was 2.2123, and the reliability P_s was 0.98653. Based on the reliability index β value, the failure probability of the channel-type joint is approximately 1.34%, which meets the low failure risk requirements for critical aerospace structural components within the aviation sector.

$$\beta = \mu_g/\sigma_g \tag{1}$$

$$R = \Phi(\beta) \tag{2}$$

The partial derivatives μ_g and D_g of μ and D with respect to the mean vector $\partial R/\partial \mu^T$ and the variance vector $\partial R/\partial D^T$ were calculated according to the formulae to obtain the following mean and variance sensitivities for each random parameter [14–16].

The reliability sensitivity results for each random parameter of the grooved tension joints are represented in bar charts as shown in Fig. 12.

Analysis of the sensitivity bar chart reveals that among these influencing parameters, the end bolt hole diameter exhibits the largest absolute sensitivity value. This indicates that the mean value of this parameter exerts the greatest impact on the reliability of the joint connection structure, and its positive influence should be fully considered during design, manufacturing, and usage processes. From the perspective of the negative impact of random parameters on the joint, the mean sensitivity of

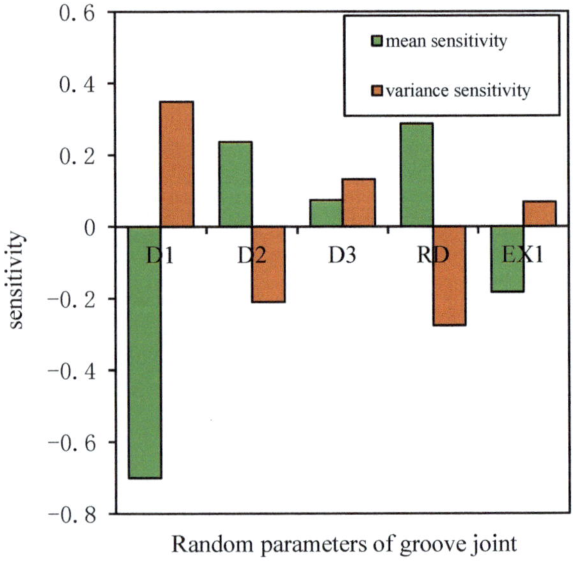

Fig. 12 Reliability sensitivity of various random parameters of grooved joints

chamfering radius and the variance sensitivity of diameter of end bolt holes are positive and maximal. Therefore, when designing, manufacturing, and applying groove joints, the negative impact of the mean of chamfering radius and the variance of diameter of end bolt holes on reliability should be controlled first.

Calculations in this study are based on ideal service conditions and do not account for extreme environments such as high-altitude high-temperature flight or salt spray corrosion effects. The sensitivity analysis addresses the independent influence of a single parameter only, without considering the combined effects of multiple factors.

Future research directions could include reliability testing of joints under extreme conditions to enhance the applicability of the metrics in complex operating environments. Establishing a sensitivity analysis model incorporating multiple factors and optimising design parameter combinations could further improve the reliability redundancy of the joints.

4 Conclusion

In this study, response surface methodology (RSM) principles are employed to design experiments on the dimensional parameters influencing the reliability of grooved joints. Design points are generated and analyzed using response surface software, which produces response surfaces for each dimensional parameter and the corresponding maximum stress values. This analysis evaluates the significance of interactions between dimensional parameters and examines how changes in these parameters affect the maximum stress.

(1) A multiple regression fitting is applied to the obtained data from the response surface analysis of the grooved joints, resulting in quadratic regression polynomial fitting equations.
(2) A second-order response surface equation for the grooved joint is derived using RSM to establish the limit state equation. The reliability of the grooved joint is then computed, along with the reliability sensitivity of statistical characteristics for each random parameter.
(3) Reliability sensitivity values are ranked based on their magnitude, allowing the identification of sensitive parameters that exert a significant impact on reliability, and less sensitive parameters that have a minimal influence.

References

1. Anjum F, Naz MY, Ghaffar A (2021) Moisture and temperature response of structural and lithology based thermophysical and energy saving traits of limestone using experimental and least-square fitting methods. J Therm Sci 30:551–561

2. Guo J, Wang Z, Li H, Yang Y, Huang CG, Yazdi M, Kang HS (2024) A hybrid prognosis scheme for rolling bearings based on a novel health indicator and nonlinear Wiener process. Reliab Eng Syst Saf 245:110014
3. Fang T, Huang X, Yu K (2024) Multi-objective optimization of intake parameters based on response surface methodology. In: 2024 international conference on aerospace engineering and aviation technology, Beijing, China, p 012139
4. Yazdi M, Mohammadpour J, Li H, Huang HZ, Zarei E, Pirbalouti RG, Adumene S (2023) Fault tree analysis improvements: a bibliometric analysis and literature review. Qual Reliab Eng Int 39(5):1639–1659
5. Cai ZL, Shou SL, Zhou JW (2023) System reliability analysis of landslides involving sliding surfaces using intelligent multiple response surfaces. Front Earth Sci 10:1053634
6. Wei YX, Bai GC, Song LK (2019) The estimation of reliability probability of structures based on improved iterative response surface methods. KSCE J Civ Eng 23(9):4063–4074
7. Yang S, Yang H, Bai S (2021) Reliability analysis deflector of underwater wall detection device based on finite element and response surface methodology. In: 2021 international conference on mechanical engineering and applied composite materials, Shanghai, China, October 2021, p 052047
8. Li H, Yazdi M (2022) Advanced decision-making methods and applications in system safety and reliability problems. Studies in systems, decision and control, vol 211
9. Li DQ, Zheng D, Cao ZJ (2016) Response surface methods for slope reliability analysis: review and comparison. Eng Geol 203:3–14
10. Zhang W, Xiao Y (2019) An adaptive order response surface method for structural reliability analysis. Eng Comput 36(5):1626–1655
11. Wan Z, Gao Y, Tao W (2024) Structural reliability analysis using generalized distribution reconstruction method with novel improvements. ASCE-ASME J Risk Uncertainty Eng Syst A Civil Eng 10(2):04024028
12. Zhang Z, Wang CX, Zhang Q (2024) Analysis and optimization of the dynamic response characteristics of aircraft cargo rack. Vibroengineering Procedia 56:15–21
13. Li DX, Li L, Xu L (2022) Stability analysis of upstream and downstream dam slopes with water level drawdown using response surface function. Geotech Geol Eng 40(6):3107–3123
14. Yang Z, Yan Y, Pak U (2023) Thermal reliability assessment and sensitivity analysis of 18,650 cylindrical lithium-ion battery. J Energy Storage 59:106504
15. Li H, Yazdi M (2022) Reliability analysis of correlated failure modes by transforming fault tree model to Bayesian network: a case study of the MDS of a CNC machine tool. Advanced decision-making methods and applications in system safety and reliability problems: approaches, case studies, multi-criteria decision-making, multi-objective decision-making, fuzzy risk-based models. Springer International Publishing, Cham, pp 15–28
16. Yazdi M (2024) Mathematical models for industrial system reliability. In: Advances in computational mathematics for industrial system reliability and maintainability. Springer Nature Switzerland, Cham, pp 17–42

Comprehensive Evaluation Techniques of Basic Reliability Based on Data Fusion of Similar Products

Yi Jiao

Abstract Basic reliability, as a crucial indicator of equipment quality, significantly influences aspects such as maintainability, testability, and supportability. However, the growing levels of equipment reliability, alongside shorter test cycles and cost constraints, have led to a challenge of insufficient basic reliability test samples, complicating the assessment process. This paper addresses two primary factors affecting basic reliability: technical state and environmental conditions. By leveraging data generated from similar products in comparable environments, the necessary test data for basic reliability assessment is transformed. Additionally, reliability test results under different conditions are integrated, resulting in the development of a comprehensive evaluation methodology. Through field-based basic reliability assessments of equipment, the technique is applied in technical engineering contexts, with results demonstrating an increase in the effective sample size for basic reliability.

Keywords Reliability · Data credibility · Data conversion · Comprehensive evaluation

1 Introduction

Basic reliability refers to the duration or probability that a product will operate without failure under specified conditions, typically quantified by Mean Time Between Failures (MTBF). As an intrinsic design characteristic, it not only forms a critical component of the equipment's design attributes but also serves as a key determinant limiting the operational performance of aviation equipment. Basic reliability testing, verification, and assessment are integral to aviation airborne equipment reliability engineering. This process primarily involves gathering basic reliability test data, including effective operating time and failures/accountable failures

Y. Jiao (✉)
Reliability Research Centre, Chinese Flight Test Establishment, Xi'an, China
e-mail: 283098133@qq.com

H. Li et al. (eds.), *Data-Driven Methods for Reliability and Safety Engineering: Applications in Industrial Systems*, Springer Series in Reliability Engineering,
https://doi.org/10.1007/978-3-032-22873-4_2

[1], throughout the equipment's life cycle. Due to the exponential distribution characteristics and cost control constraints, a one-sided confidence lower limit model is typically employed to evaluate the basic reliability indicators of equipment.

With the advancement of high-tech solutions and the improvement in aviation equipment reliability, basic reliability analysis and assessment of high-performance airborne equipment during performance verification testing face numerous new challenges. Presently, many types of equipment encounter reduced test cycles and compressed development budgets, resulting in a widespread issue where high-reliability products lack sufficient test samples for reliability assessment. Traditional statistical analysis methods struggle to provide scientifically sound conclusions under these constraints.

The problem of small sample sizes has long been a challenge in reliability assessment. In theory, the larger the data volume, the closer the assessment results are to the true reliability level [2]. Consequently, the approach to addressing the "small sample" problem in this paper is not a breakthrough in terms of theoretical methodology [3], but rather through the effective use of samples from similar products in comparable environments to mitigate the sample size issue. This study focuses on basic reliability by investigating methods for converting data from different sources within the same phase (such as historical field test data from similar products and initial operational data from user units [4]) and methods for fusing reliability data from the same product at different test phases (such as reliability verification tests [5–7]). These strategies collectively form a comprehensive model for basic reliability assessment. Additionally, by applying this model in the engineering context of basic reliability assessment for airborne equipment across various aircraft models, this approach successfully increases the effective sample size for basic reliability.

2 Research Approach

The primary research content of this paper is organized as follows:

Step 1: Convert data from different sources within the same stage, specifically utilizing historical reliability field test data and prior data from similar products within the same aircraft series.

Step 2: Integrate data from different stages of the same product, particularly the reliability acceptance data obtained during the certification phase from other tests of the same product.

Step 3: Develop a comprehensive small-sample basic reliability evaluation model based on conjugate (same distribution family) theory.

The research approach is illustrated in Fig. 1.

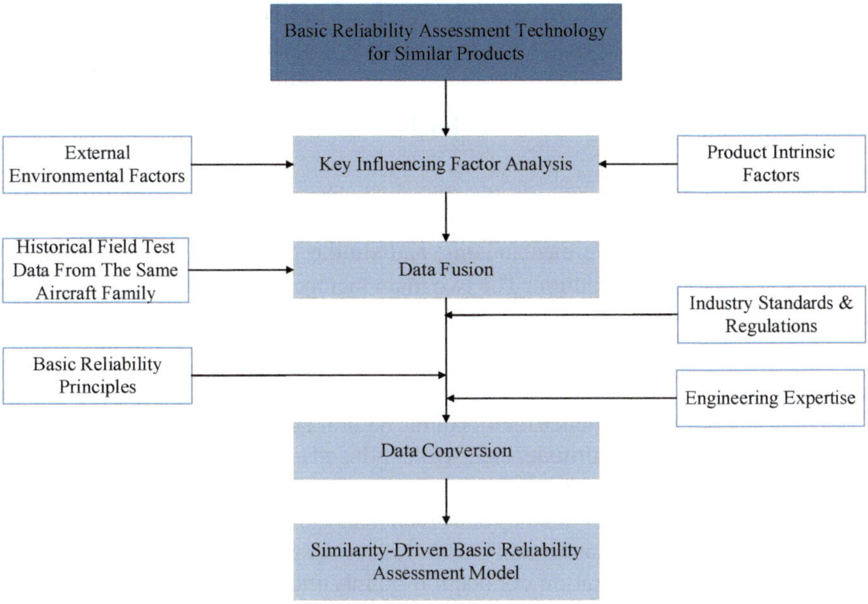

Fig. 1 Research approach framework diagram

3 Research Content

The data objects for small-sample basic reliability evaluation consist of data from different sources at the same stage and data from distinct stages of the same product. The primary focus is on integrating data from various sources within the same stage, such as field test data, including historical information from similar products across a series of aircraft, and initial operational tests conducted by units. This is achieved through similarity-based data conversion algorithms. Subsequently, data from different stages of the same product, such as laboratory reliability qualification test data, are integrated using data fusion algorithms based on cross-entropy theory. Together, these methodologies establish a comprehensive evaluation model for small-sample basic reliability.

3.1 *Reliability Data Conversion Method for Similar Products in Aircraft Series*

3.1.1 Analysis of Influence Factors for Basic Reliability

For data from different sources at the same stage, such as historical prior reliability data of similar products in aircraft series, which primarily exhibit the same distribution type, consistent failure mechanisms, and similar carrier environments, can be used as sources for data adoption. The two main factors affecting the basic reliability level of equipment include the platform environment and product entity. Therefore, key influence factors are determined based on these two factors, as follows:

(a) The platform environment is divided into the natural environment and induced environment. The influence factors of the natural environment include: temperature, humidity, flight altitude, etc., while induced environment influence factors mainly include: vibration, shock, etc.
(b) The product entity is divided into the technical state and the installation connection. The technical state mainly includes: component composition and quantity, etc., while installation connection mainly includes fixing and connection methods, etc.

3.1.2 Principles for Adopting Reliability Data

Based on the key influence factors of basic reliability identified in the analysis, after analyzing their effects on basic reliability, principles for adopting data from similar products are established. Besides ensuring that products are developed and manufactured by the same unit, other requirements are as follows:

(a) Platform environment: The operational envelope and test envelope have not undergone significant changes resulting in more severe environmental conditions, and single indicator changes do not exceed 10%;
(b) Product entity: Product connection and fixing methods have not undergone significant changes resulting in more severe environmental conditions, and the proportion of newly improved parts does not exceed 10% of the original prototype.

3.1.3 Reliability Data Conversion Method

Environmental factors are a critical parameter in reliability analysis, as they quantify how the failure rate of the same airborne equipment varies across environments of differing severity, thereby reflecting environmental stress levels. To address this, a conversion method for natural environmental conditions has been developed, in which k_1 represents the platform environment conversion coefficient. When reliability data are derived from prototype components, data from improved or modified parts are excluded, and only information from fully identical components is used to ensure consistency. Given that the baseline reliability model is a series system, operating time can be converted by weighting the failure rates of prototype and improved components, while also accounting for differences in product quality systems through the introduction of a quality coefficient. Accordingly, a corresponding conversion model is proposed, where the product entity conversion coefficient is defined as a function of the failure rate of unchanged components, the total failure rate of the new product, and a quality coefficient. This quality coefficient is primarily determined with reference to GJB 450B-2021 General Requirements for Equipment Reliability Work and GJB/Z 299C-2006 Handbook for Reliability Prediction of Electronic Equipment.

For equipment reliability assessment, the evaluation of basic reliability Mean Time Between Failures (MTBF) requires two types of test data: operating time and responsible failures. During the assessment, a one-sided confidence lower limit evaluation model for basic reliability is used. According to the established conversion method for basic reliability assessment test data, the operating time is converted. Effective cumulative operating time is calculated by adding the reliability test data of the adoptable parts and applying the developed conversion coefficient calculation model.

Effective cumulative responsible failures: Failures belonging to improved parts or not related to the new airborne equipment are not counted, while failures of retained parts are all included without conversion. Through data normalization, merging, and determination of cumulative responsible failures, using the required test data for basic reliability assessment of equipment and adoptable test data, a reliability assessment model based on similar products in the aircraft series is formed by integrating the effective cumulative time conversion coefficient calculation model and the reliability assessment model.

3.1.4 Fusion Method for Laboratory and Field Reliability Data

For data from different stages of the same product, a data fusion algorithm based on cross-entropy theory is adopted, such as for laboratory reliability qualification test data. The cross-entropy method ensures that field test data serves as the primary structure and won't be "consumed" by an excessive amount of laboratory data. For the same product, reliability qualification test data and field test data have consistent user/confidence levels, similar failure distribution density curves, and both types of data can pass distribution fitting tests. Therefore, a data fusion algorithm based on cross-entropy theory can be adopted, with field test data as the primary structure, to integrate reliability qualification test data, thereby increasing the sample size for reliability assessment. The main steps are as follows:

(a) Using cross-entropy theory, construct distance functions for the two types of test data.

(b) The larger the symmetric cross-entropy between two distribution functions, the greater the difference, i.e., the smaller the similarity. Accordingly, a similarity function is constructed.
(c) Based on similarity, determine the weight coefficients for each data set.
(d) Using a linear weighting approach, provide the failure density function after fusion.
(e) Finally, after integrating the density function, the Mean Time Between Failures (MTBF) of the product after fusion can be calculated as:

4 Case Application

4.1 Assessment Requirements

This paper uses the basic reliability indicator Mean Time Between Failures (MTBF) of Equipment A installed on Aircraft Type A as an example. The requirement is N hours; the field test results for Equipment A are cumulative operation time of 1.05N hours with 1 failure.

4.2 Basic Reliability Assessment

First is the reliability data conversion and assessment based on similar products in the aircraft series. The main process is as follows:

Step 1: Determine reliable data sources. According to the data adoption principles, the data sources are determined to be Equipment B installed on Aircraft Type B and Equipment C installed on Aircraft Type C.
Step 2: Determine data conversion coefficients. For Equipment B installed on Aircraft Type B, the operation time is 4.39N hours with 2 responsible failures, yielding calculated values of $k_1 = 0.896$ and $k_2 = 0.917$. For Equipment C installed on Aircraft Type C, the operation time is 0.76N hours with no responsible failures, yielding calculated values of $k_1 = 0.923$ and $k_2 = 0.941$.
Step 3: Mean Time Between Failures (MTBF) assessment. The MTBF assessment results for Equipment A installed on Aircraft Type A are shown in Table 1.

Second is the reliability data fusion and assessment oriented to laboratory tests. The main process is as follows.

Step 1: Determine fusible data sources. The fusible data source is the reliability qualification test of Equipment A installed on Aircraft Type A, with test data as follows: 6.75N hours of operation with 3 responsible failures.
Step 2: Determine the weights of laboratory and field data.

Table 1 Comprehensive assessment results of MTBF for Equipment A installed on Aircraft Type

No.	Data source	Effective time	Responsible faults	Assessment result
1	Type A Aircraft	1.05N	1	0.35N
2	Type A & B Aircraft	5.16N	3	0.94N
3	Type A & C Aircraft	1.79N	1	0.60N
4	Type A & B & C Aircraft	5.88N	3	1.07N

Table 2 Comprehensive assessment results of MTBF for Equipment A installed on Aircraft Type A

No.	Data source	Effective time	Responsible faults	Assessment result
1	Type A Aircraft	1.05N	1	0.35N
2	Type A & B Aircraft	5.16N	3	0.94N
3	Type A & C Aircraft	1.79N	1	0.60N
4	Type A & B & C Aircraft	5.88N	3	1.07N
5	Type A & B, & C Aircraft + reliability qualification testing	–	–	1.13N

Step 3: Comprehensive assessment of Mean Time Between Failures (MTBF). The comprehensive assessment results of MTBF for Equipment A installed on Aircraft Type A are shown in Table 2.

4.3 Analysis of Assessment Results

After verification, the assessment results derived from the reliability data conversion method for similar products and the laboratory-field reliability data fusion method are proven reliable. By adopting data from similar products in the aircraft series and fusing reliability qualification test data, the effective time in effective test samples required for basic reliability assessment increased by nearly five times, while failure samples also increased by two times. It can be observed from Table 2 that the MTBF assessment result has increased from 0.35N to 1.13N, indicating that the larger the volume of data used for evaluation, the higher the confidence level of the assessment result, and it is closer to the post-delivery continuous MTBF assessment result of 1.28N for this aircraft type. In practical engineering applications, the assessment results obtained by using the comprehensive evaluation method for basic reliability based on data fusion of similar products proposed in this paper are more accurate.

5 Conclusion and Prospects

This paper introduces a similarity conversion method for basic reliability indicators based on data from different sources within the same stage, as well as a reliability fusion method for integrating data from various stages of the same product. These approaches lay the theoretical groundwork for both basic reliability and mission reliability assessments in engineering contexts. The comprehensive technology developed for basic reliability using small samples from similar products and environmental conditions offers methodological support and serves as a reference for data validation and fusion across other field test domains. When applied to the basic reliability assessment of equipment within model engineering applications, the results demonstrate an increase in the effective sample size, thereby enhancing the comprehensiveness, objectivity, and authenticity of the assessment outcomes. This methodology holds significant value and prospects for engineering applications, refer to for more information [8, 9].

This paper mainly achieves the conversion of reliability data based on environmental factors and quality coefficients. The determination of some coefficients in the model has certain subjectivity, and further consideration can be given to more objective methods for converting reliability data in the future. In addition, this paper uses the cross-entropy method for reliability data fusion, which does not take into account the uncertainty of reliability information. Future research can explore methods incorporating uncertainty to fuse reliability data, thereby improving the rationality of the fusion method.

References

1. Blond KE (2023) An enterprise system engineering analysis of KC-46A maintenance program decision-making. Ph.D. thesis, Colorado State University
2. International Air Transport Association (2010) Aircraft health monitoring and reliability program guidance material, Manual, 2010, 1st edn. IATA, Montreal, Canada
3. Liu Q, Cheng J, Tan Y (2018) Darlington tube fault prediction based on KPCA and Mahalanobis distance. J Air Force Eng Univ (Nat Sci Ed) 19(5):71–77
4. Song D, Lv C, Qi L (2018) Health assessment of hydraulic system under variable working conditions based on health baseline and Mahalanobis distance. Syst Simul Technol 13(3):202–208
5. Luo W, Lan H, Ma Z (2018) Multi-model switching control of nonlinear systems based on fractal measure. Control Eng 18(4):539–542
6. Guo J, Wang Z, Li H, Yang Y, Huang CG, Yazdi M, Kang HS (2024) A hybrid prognosis scheme for rolling bearings based on a novel health indicator and nonlinear Wiener process. Reliab Eng Syst Saf 245:110014
7. Huang P, Gu Y, Li H, Yazdi M, Qiu G (2023) An optimal tolerance design approach of robot manipulators for positioning accuracy reliability. Reliab Eng Syst Saf 237:109347
8. Li H, Yazdi M (2022) Advanced decision-making methods and applications in system safety and reliability problems. Studies in systems, decision and control, vol 211

9. Yazdi M, Mohammadpour J, Li H, Huang HZ, Zarei E, Pirbalouti RG, Adumene S (2023) Fault tree analysis improvements: a bibliometric analysis and literature review. Qual Reliab Eng Int 39(5):1639–1659

Mission Reliability Comprehensive Evaluation Technique Based on Multi-stage Information

Haijing Song, Ming Su, and Zimeng Xu

Abstract The traditional methods of mission reliability evaluation require too many test samples, as they use flight time as mission time. To solve this, a small-sample MR assessment technique is proposed in this study. Statistics and test design are combined, and Bayes' theorem is applied to merge information from previous similar tests (prior distributions) with new test data (posterior distributions). A new method is also proposed to determine the minimum test time (T) and maximum allowed failures for fixed-duration tests. By applying the technique to real equipment, it illustrates that the MR evaluation accuracy and mission success rates have been improved, and the need for excessive testing has been reduced.

Keywords Reliability evaluation · Minimum sample size · Parameter estimation · Bayesian estimation · Time censored test

1 Introduction

Mission reliability refers to the ability of equipment to perform specified functions within a prescribed set of mission profiles, typically measured by Mission Reliability (MR). As a critical indicator of equipment, it significantly impacts combat and operational support effectiveness. Current mission reliability assessments adapt basic reliability evaluation methods by directly using flight time as mission execution time. By collecting mission reliability test data and applying one-sided confidence limit models, equipment mission reliability indicators are evaluated. However, this approach faces several issues: exponential distribution models do not match the binomial distribution characteristics of mission reliability data; mission planning and test profiles fail to meet sample selection requirements specified in development

H. Song (✉) · M. Su
Reliability Research Centre, Chinese Flight Test Establishment, Xi'an, China
e-mail: 2452534073@qq.com

Z. Xu
College of Civil Aviation, Nanjing University of Aeronautics and Astronautics, Nanjing, China

H. Li et al. (eds.), *Data-Driven Methods for Reliability and Safety Engineering: Applications in Industrial Systems*, Springer Series in Reliability Engineering,
https://doi.org/10.1007/978-3-032-22873-4_3

requirements; and specialised mission reliability test planning requires numerous test samples.

Currently, many researchers have achieved certain accomplishments in related fields. Brataas et al. conducted an empirical case study on the Norwegian tax report system, examining how independent organisations coordinate software performance testing in systems of systems. The study found that applying high-reliability organisation theory helps these organisations bridge incompatible worldviews, improving collective management and preparedness for workload peaks [1]. Cao et al. proposed a Dynamic Bayesian Network-based method to analyse the reliability of ship standby systems, explicitly considering time redundancy. Their results showed that this approach captures the dynamic effects of redundant backup and time redundancy, offering practical guidance for ship reliability engineering [2]. Chen et al. developed a dynamic two-stage abort policy for safety–critical systems performing performance-based missions, integrating real-time monitoring and mission progress to guide abort decisions. Their analysis, supported by unmanned aerial vehicle case studies, demonstrated that this policy improves mission safety, reliability, and cost-effectiveness across diverse operational scenarios [3]. Wang et al. investigated reliability analysis for dynamic k-out-of-n phase-AND mission systems, introducing a phase combination theory to merge phases and streamline analysis. The authors developed improved decision diagram-based methods, demonstrating enhanced efficiency and effectiveness through case studies [4]. Chowdury et al. introduced a time-efficient solution for multi-task reliability redundancy allocation using an online transfer parameter estimation-based multifactorial evolutionary algorithm (MFEA-II). Their results showed that MFEA-II outperformed basic MFEA, genetic algorithms, and particle swarm optimisation in both reliability and computation time across multiple test scenarios [5]. Fang et al. developed a comprehensive reliability modelling and analysis method for reconfigurable ship electronic information systems, integrating AADL-based modelling, SMA-DE hybrid optimisation, and a GO method with a co-causal failure correction algorithm. Their approach enabled effective mission reliability allocation and quantitative analysis, validated through application to a typical ship system task [6]. Yılmaz and Gültekin analysed the reliability of repairable multi-state phased mission systems using a Markov approach based on states, considering exponential repair and failure rates. The authors calculated and compared reliability for two-level and three-level systems, highlighting the value of intermediate states in complex system analysis [7]. Guan et al. introduced a multi-objective optimisation and reliability assessment method for multi-layer radiation shielding in deep space missions, using a genetic algorithm to balance dose reduction, mass, and thickness. Their case study showed that the optimised shielding design significantly reduced failure rates and improved electronic component reliability, supporting lightweight, reliable space mission development [8]. Jing et al. developed a reliability analysis and emergency maintenance optimisation method for multi-state airborne power supply systems, utilising a dependent competing importance measure and semi-Markov process. Their approach, validated through a case study, improved reliability assessment accuracy and provided practical guidance for emergency maintenance under budget constraints [9]. Jingyu et al. proposed an operation trial programming method for UAV swarms based on mission reliability, constructing reliability models for various mission types and optimising trial programmes using a multi-objective quantum particle swarm algorithm. Their simulation case confirmed the method's effectiveness in balancing reliability and cost for UAV swarm operation trials [10]. Lee developed a reliability-based design optimisation framework for high-energy laser weapons, employing an artificial neural network ensemble to efficiently predict penetration time distributions under uncertainty. The framework determines optimal laser power for target reliability, with sensitivity analysis revealing absorptivity uncertainty as the most influential factor [11]. Li et al. explored dynamic prediction of bridge reliability indices and load effects using monitoring data, integrating Bayesian updated density evolution filtering and the first-order second-moment method. The team aimed to enhance prediction accuracy for existing bridges by fusing dynamic linear models, probability density evolution, and particle filtering [12]. Li et al. developed a reliability modelling approach for multi-state phased mission systems with random phase durations and dynamic combined phases, using a multi-valued decision diagram framework. The group sought to improve reliability analysis by introducing a path probability evaluation method based on the Markov regenerative function [13]. Liu et al. assessed the reliability of the self-matching task for measuring the self-prioritization effect by applying a multiverse analysis across diverse datasets and indicators. The authors aimed to clarify the reliability paradox in cognitive tasks, revealing moderate group-level effects but limited individual-level reliability [14]. Ma et al. explored reliability assessment with small samples by integrating subjective information and basic test data of input parameters. The authors proposed methods to transform incomplete prior information into conjugate priors using maximum entropy and combined these with test data through Bayesian synthesis [15]. Ming et al. analysed the reliability and safety of road vehicles and trains meeting on a single-level rail-cum-road bridge under crosswind, using a simulation platform and an improved fractional moment-based maximum entropy method. The group aimed to address uncertainties in system parameters and highlight key factors affecting operational safety [16]. Mou et al. designed a compact miniature vapor compression refrigerator for aircraft, targeting reliable thermal management under extreme temperature shifts. The team evaluated system performance and control strategies, confirming effective temperature regulation for sensitive onboard equipment in harsh environments [17]. Porthin et al. introduced a task reliability index to assess operator performance and failure probability in control room simulators, integrating plant-centred and human-centred perspectives. The authors formalised the use of simulator evidence, applying Bayesian updating to enhance human error probability estimation [18]. Rongxing et al. proposed a reliability analysis framework for phased mission systems under mixed uncertainty and common cause failure, using a hybrid model of dynamic fault trees and generalized stochastic Petri nets. The team incorporated interval theory and combined genetic algorithm with Monte Carlo simulation to quantify uncertainty and simplify reliability boundary computation [19]. Sezgin proposed a scenario-driven evaluation model for UAV mission reliability, integrating large language models and retrieval-augmented generation to enhance decision-making in the Internet of Drones. The study demonstrated improved adaptability and operational effectiveness by analysing drone telemetry and mission data in real time [20]. Shi et al. developed a calculation method for sortie mission reliability of shipborne unmanned vehicle groups, constructing a network model and introducing a novel approach based on equal-length minimal paths and disjoint products. The authors validated the method's accuracy and efficiency through comparative analysis with Monte Carlo simulations [21].

This paper addresses mission reliability by integrating statistics and experimental design theories to develop mission reliability test techniques. Based on the similarity between prior test profiles and typical equipment profiles, we apply the Bayesian theorem to conduct a comprehensive evaluation of prior distributions and samples, proposing a small-sample mission reliability assessment model that fuses prior and posterior density functions. We also conducted engineering applications of the small-sample mission reliability evaluation technique for a specific equipment type, solving problems of unsuitable reliability assessment models and high specialised test sample requirements, thereby achieving increased probability of successful mission execution and comprehensive improvement of mission reliability.

2 Minimum Sample Size Determination for Time Censored Test

When product lifetime follows an exponential distribution, the key considerations in test design are the selection of the total test time T (the cumulative exposure time of all specimens under test stress) and the acceptance criterion r (the maximum allowable number of cumulative relevant failures across all specimens). Accordingly, this section introduces a generalised concept of "minimum sample size," defined as a systematic approach for determining the required total test time T and the corresponding acceptance number r for product approval. GJB 899A-2009 specifies two categories of time-censored test plans: the standard plan and the short-term high-risk plan. The standard plan adopts producer's risk alpha and consumer's risk beta in the range of 10–20%, whereas the short-term high-risk plan sets both alpha and beta at 30%. Assuming a true mean time between failures (MTBF), the probability of product acceptance can be described using the Poisson distribution, and this probability must satisfy predefined risk constraints. Time-censored testing allows the test duration to be fixed in advance, enabling a trade-off among test time T, producer's risk alpha, consumer's risk beta, the upper test limit theta_0, and the lower test limit theta_1. Given alpha, beta, and the discrimination ratio d (or equivalently theta_0 and theta_1), the corresponding T and r can be determined. For time-censored tests with replacement, T and r are obtained from the specified relationships, where the test time is governed by the risk equations and the minimum acceptance number r is selected to satisfy the acceptance probability requirements. In practice, directly solving the acceptance probability equation is computationally intensive, as it requires iterative

Table 1 Simplified standard time-censored test plans

Plan No.	Decision risk (%)				d	T	r
	Nominal		Actual				
	α	β	α'	β'			
9	10	10	12.0	9.9	1.5	45.0	36
10	10	20	10.9	21.4	1.5	29.9	25
11	20	20	19.7	19.6	1.5	21.5	17
12	10	10	9.6	10.6	2.0	18.8	13
13	10	20	9.8	20.9	2.0	12.4	9
14	20	20	19.9	21.0	2.0	7.8	5
15	10	10	9.4	9.9	3.0	9.3	5
16	10	20	10.9	21.3	3.0	5.4	3
17	20	20	17.5	19.7	3.0	4.3	2

Table 2 Simplified short-term high-risk time-censored test plans

Plan No.	Decision Risk (%)				d	T	r
	Nominal		Actual				
	α	β	α'	β'			
19	30	30	29.8	30.1	1.5	8.1	6
20	30	30	28.3	28.5	2.5	3.7	2
21	30	30	30.7	33.3	3.0	1.1	0

evaluation of T for different values of r. To address this, GJB 899A-2009 provides tabulated standard test plans. Typically, practitioners select an appropriate plan based on alpha, beta, and d to directly obtain T and r. Standard plans (alpha and beta at 10%–20%) and short-term high-risk plans (alpha = beta = 30%) are provided in the standard, and simplified versions are often summarised in tables. Once the cumulative test time T (expressed in unit-hours) reaches the specified value, the product is rejected if the observed number of relevant failures exceeds r, and accepted if it is less than or equal to r (Tables 1 and 2).

3 Comprehensive Reliability Evaluation Based on Multiple-Stage Information

For products with exponentially distributed lifetimes, the lifetime at stage k is assumed to follow an exponential distribution characterised by a failure rate at that stage. Under this assumption, the number of failures observed during testing at each stage follows a Poisson process. A reliability growth hypothesis is adopted in which the product has a defined reliability level at stage k, and this reliability increases from stage k to stage k + 1 by a growth increment reflecting improvements made between stages. The failure rate at each stage is treated as a random variable, and its prior distribution is modelled using a Gamma distribution due to its conjugacy with the Poisson likelihood. At the initial stage, where little system information is available, a non-informative prior is used to avoid biasing the analysis. As test data are collected, Bayesian updating is applied, resulting in a posterior Gamma distribution for the failure rate at each stage, from which the posterior mean and variance can be obtained. Information fusion across stages follows a sequential updating principle: the posterior distribution at stage k becomes the prior distribution at stage k + 1. The prior mean at stage k + 1 is defined as a function of the posterior mean at stage k and a reliability growth factor, while the prior variance at stage k + 1 is set equal to the posterior variance at stage k. This growth factor represents the effectiveness of corrective actions implemented between stages, with larger values indicating more effective reliability improvements. Empirical studies have shown that this factor typically falls within a moderate range, consistent with international standards. Accurate estimation of the growth factor is therefore central to combining reliability information across stages and can be achieved using classical estimation approaches based on stage-wise failure rate estimates or alternative quantile-based methods for time-censored exponential tests at a specified confidence level.

3.1 ML-II Method

Assume the test information collected at stage k is available. Under the second form of maximum likelihood estimation, this information is treated as a sub-sample drawn from the corresponding marginal density, which is obtained by integrating out the failure-rate parameter under the Gamma–exponential (or Gamma–Poisson) framework. Using this marginal density, the likelihood for the stage-k data can be written in closed form; applying the standard integral identity of the Gamma function then allows the expression to be simplified substantially. Next, by substituting the cross-stage parameter linkage introduced earlier (i.e., the relationship that connects the prior at stage k + 1 to the posterior at stage k via the growth factor), the resulting marginal likelihood depends on only one unknown parameter, which can be denoted by a single scalar to streamline optimisation. For convenience, this scalar can be defined explicitly in terms of the prior/posterior hyperparameters and the stage-k test exposure and failures, with intermediate quantities introduced to keep the algebra readable. Because this estimator uses information from both the previous and current stages, it typically yields a more accurate and stable assessment than single-stage estimation; however, the main limitation is that the optimisation has no simple closed-form solution and therefore requires numerical procedures. If the objective is to estimate only the prior parameters at stage k (without cross-stage fusion), the formulation reduces to a single-stage case by fixing the linkage terms accordingly, after which the remaining parameters can be obtained via numerical iteration.

4 Case Study

Assume an exponential lifetime model and set up the MTBF hypothesis test with a lower (reject) limit theta_1 and an upper (accept) limit theta_0. Given the specified producer's risk alpha and consumer's risk beta, the objective is to determine the acceptance number r (the maximum allowable relevant failures) and the minimum total test time T required for the time-censored test. Using the standard relationship for exponential, time-censored demonstration testing, the required r is first obtained from the risk settings. Next, as r becomes sufficiently large, the test-time expression can be approximated using a chi-square distribution quantile. From the chi-square table, the required quantile value is 66.0757, which leads directly to the calculated minimum test time T. The resulting plan can therefore be stated as "accept if failures are less than or equal to r within total test time T; otherwise reject." Based on Table 1, the computed plan aligns closely with a standard time-censored plan corresponding to r = 36 failures and the associated total test time T shown in the standard plan table.

4.1 Case Study for Determining Minimum Sample Size

A certain aerospace product has undergone three stages of development, with test data as shown in Table 3. The reliability analysis using the method in this section is as shown in Table 3.

The mean and variance of the product's lifetime at each stage are shown in Table 4.

Table 3 Test data for each stage of an aerospace product

Stage	Test time	Number of failures	Failure times (h)
1	150	6	10, 30, 60, 101, 128, 150
2	540	9	202, 230, 270, 301, 330, 380, 430, 502, 540
3	920	7	610, 660, 740, 801, 830, 870, 920

Table 4 Mean and variance of product lifetime at each stage

Stage	Mean $E(X_k)$	Variance $Var(X_k)$
1	79.83	3089
2	353.89	13,984
3	775.86	12,570

Table 5 Estimated values of the growth factor for each stage

Method	Etha 1	Etha 2
Classical estimation	0.7744	0.5439
F-quantile estimation landa = 0.005	0.9488	0.8620
F-quantile estimation landa = 0.01	0.9430	0.8486
F-quantile estimation landa = 0.1	0.8925	0.7393

Using standard properties of the Gamma distribution, the mean and variance can be expressed directly in terms of its shape and scale parameters. Accordingly, once the posterior mean and variance of the failure rate are obtained at each stage, these moment relationships can be inverted to estimate the corresponding Gamma parameters. Applying this procedure to the three testing stages yields the estimated Gamma distribution parameters reported for each stage, which summarise the updated belief about the stage-wise failure rate after incorporating the observed test data.

According to the classical estimation, F-quantile estimation, and ML-II estimation method, the estimated values of the growth factor are shown in Table 5.

The results show that the F-quantile estimation method is significantly affected by the confidence level; the ML-II method, which comprehensively considers sample and prior information, yields more reliable results; the classical method is inherently simple to compute.

Once the reliability change factor is determined, it can be used to convert the posterior information from the previous stage into the prior information for the current stage, and then combined with the current stage's test data for data fusion analysis.

5 Conclusion

(1) A novel mission reliability (MR) evaluation technique is specifically designed for situations with limited test data in the study. By integrating Bayesian statistics, the method intelligently combines valuable information from prior, similar tests (forming "prior distributions") with newly collected test results ("posterior distributions"). This fusion overcomes the limitations of traditional methods that require large sample sizes and often use mismatched statistical models (like exponential distribution for inherently pass/fail mission data).

(2) The proposed techniques directly address critical practical issues faced in equipment development and testing. The method for determining the minimum test duration (T) and acceptable failure count (r) for fixed-time tests provides clear, standardized planning guidance, reducing uncertainty and potential over-testing. More importantly, the comprehensive Bayesian-based multi-stage evaluation method solves the fundamental problems of model mismatch and excessively high sample demands for specialized MR tests, refer to [22, 23] for more information.

(3) The application of the method to a real aerospace product case study validated its practicality and effectiveness. Results show the approach successfully integrated test data across multiple development stages using the reliability growth factor concept. The successful application confirms the method's ability to provide reliable MR estimates even with constrained test resources, offering a significant advantage over conventional approaches.

References

1. Brataas G, Braskerud P, Tondel IA, Kjærnsrod S (2025) Organizational factors of software performance testing for systems of systems: a case study using high-reliability organization theory to understand an outage. J Syst Softw 229(Nov), Art no. 112444
2. Cao EZ, Long ZD, Ma Y, Huang YX, Liang XF (2025) A DBN-based reliability analysis method for ship standby system considering time redundancy. Ocean Eng 333(Jul), Art no. 121528
3. Chen J, Gao H, Fang C (2025) Optimal two-stage abort policy considering performance-based missions. Reliab Eng Syst Saf 257(PA):110803–110803
4. Guo J, Wang Z, Li H, Yang Y, Huang CG, Yazdi M, Kang HS (2024) A hybrid prognosis scheme for rolling bearings based on a novel health indicator and nonlinear Wiener process. Reliab Eng Syst Saf 245:110014
5. Chowdury MAM, Nath R, Rauniyar A, Shukla AK, Muhuri PK (2025) A time-efficient solution approach for multi/many-task reliability redundancy allocation problems using the online transfer parameter estimation based multifactorial evolutionary algorithm. Reliab Eng Syst Saf 264(PA):111175–111175
6. Fang X, Wang X, Feng L, Zuo J, Liu S (2024) Mission reliability modeling and analysis methods for reconfigurable ship electronic information systems. Appl Sci 14(21):9626–9626
7. Yılmaz S, Gültekin ÖE (2023) Reliability analysis of repairable multistate phased mission systems with Markov approach based on states. Eng Comput 40(5):1041–1062
8. Guan S, Fu G, Wan B, Wang X, Chen Z (2025) Multi-objective optimization and reliability assessment of multi-layer radiation shielding for deep space missions. Aerospace 12(4):337–337
9. Jing YF, Jiao J, Yuan Y, Chen ZW, Lu C, Dui H (2025) Reliability analysis and emergency maintenance strategy optimization of the multi-state airborne power supply system based on dependent competitive importance measure. Reliab Eng Syst Saf 263(Nov), Art no. 111274
10. Jingyu W, Ping J, Dingxing X, Jianjun Q (2025) Programming UAV swarm operational test based on mission reliability. J Northwestern Polytech Univ 43(2):368–380
11. Lee U (2025) Reliability-based design optimization (RBDO) framework for high-energy laser weapons (HELWs) using artificial neural network ensemble. Expert Syst Appl 287(Aug), Art no. 128202
12. Huang P, Gu Y, Li H, Yazdi M, Qiu G (2023) An optimal tolerance design approach of robot manipulators for positioning accuracy reliability. Reliab Eng Syst Saf 237:109347
13. Li XY, Li H, Xiong X, Li M, Yazdi M, Zarei E (2025) Reliability modeling of multi-state phased mission systems with random phase durations and dynamic combined phases. Reliab Eng Syst Saf 253:110524–110524
14. Liu Z, Hu M, Zheng Y, Sui J, Peng HC (2025) A multiverse assessment of the reliability of the self-matching task as a measurement of the self-prioritization effect. Behav Res Methods 57(1):37–37
15. Ma Z, Zhu J, Xu N (2003) Reliability assessment using multiple information sources. Chin J Comput Phys 20(5):391–398

16. Ming L, Jin Z, Yijun Z, Mengxue W, Yongle L (2025) Reliability analysis of road vehicle and train-running safety when meeting on a single-level rail-cum-road bridge under crosswind conditions. J Bridge Eng 30(8)
17. Mou XZ, Li YX, Zhong H, Xu S, Zhang JY (2025) Design and performance analysis of an airborne miniature vapor compression refrigerator for extreme temperature environments. Appl Therm Eng 274(Sep), Art no. 126793
18. Porthin M, Podofillini L, Dang VN (2024) Task reliability index for operator performance and failure probability assessment in control room simulators. Reliab Eng Syst Saf 251:110390–110390
19. Li H, Peng W, Adumene S, Yazdi M (2023) Cutting edge research topics on system safety, reliability, maintainability, and resilience of energy-critical infrastructures. In: Intelligent reliability and maintainability of energy infrastructure assets. Springer Nature Switzerland, Cham, pp. 25–38
20. Sezgin A (2025) Scenario-driven evaluation of autonomous agents: integrating large language model for UAV mission reliability. Drones 9(3):213–213
21. Shi H, Wang N, Liu Q (2024) Calculation method for sortie mission reliability of shipborne unmanned vehicle group. J Mar Sci Eng 12(8):1309–1309
22. Li H, Yazdi M (2022) Advanced decision-making methods and applications in system safety and reliability problems. Studies in systems, decision and control, vol 211
23. Yazdi M, Mohammadpour J, Li H, Huang HZ, Zarei E, Pirbalouti RG, Adumene S (2023) Fault tree analysis improvements: a bibliometric analysis and literature review. Qual Reliab Eng Int 39(5):1639–1659

Preliminary Exploration of Exemption Methods for Single-Fault Criteria in Nuclear Power Plants

Jin An, Lin Yan, and Jian Yang

Abstract The Chinese regulation titled "Design Safety Regulations for Nuclear Power Plants" (HAF102) mandates the application of the single fault criterion to all safety system combinations within nuclear power plant designs. This requirement aims to ensure that the designed systems maintain high reliability when subjected to design basis accidents. However, the universal implementation of the single fault criterion in design basis accident analyses may introduce overly restrictive design limitations. To address this issue, the present study examines the design of a passive residual heat removal system on the secondary side of a specific reactor type and proposes a methodology for exemption from the single-fault criterion. This approach is founded on a comprehensive assessment incorporating reliability analysis, risk evaluation, operating experience, and the interaction of safety systems. The proposed method guarantees that the safety system's reliability is commensurate with the overall plant risk, thereby allowing simplification of the plant design in certain cases without compromising safety.

Keywords Single fault criterion · Exemption · Risk assessment

1 Introduction

The Single Fault Criterion (SFC) defines a failure condition in which the malfunction of a single component, including any resultant failures, inhibits the execution of a safety function. According to the Chinese regulation HAF102, the application of the single fault criterion is mandated for every safety combination within the design of nuclear power plants. A safety combination is defined as an assembly of devices intended to perform the necessary actions in response to a specific hypothetical initiating event, aiming to prevent the consequences of anticipated operational occurrences and design basis accidents from exceeding prescribed limits [1–6].

J. An (✉) · L. Yan · J. Yang
China Nuclear Power Engineering CO., Ltd., Beijing, China
e-mail: angle.anjin@163.com

H. Li et al. (eds.), *Data-Driven Methods for Reliability and Safety Engineering: Applications in Industrial Systems*, Springer Series in Reliability Engineering,
https://doi.org/10.1007/978-3-032-22873-4_4

The primary objective of the SFC is to ensure high reliability of safety-critical functions and to guarantee their continued performance despite the occurrence of a single failure. Nonetheless, the application of the SFC has not invariably resulted in safety system designs whose reliability appropriately corresponds to the frequency of safety challenges encountered by the plant. Risk assessment studies have demonstrated that strict adherence to the SFC can lead to redundant system components that, while providing an acceptable safety margin, contribute minimally to actual risk reduction. Although maintaining sufficient safety margins remains a fundamental objective, the assumption of the worst-case single failure for all design basis accidents may impose undue constraints on licensees. In recognition of this, the U.S. Nuclear Regulatory Commission introduced risk-informed alternatives to the SFC in 2005, though widespread implementation is still pending.

This paper examines the design of the main steam system and the passive residual heat removal system on the secondary side of a particular reactor type, investigating practical methodologies for exempting the single-fault criterion in nuclear power plant designs that do not fully comply with this requirement.

2 Brief Introduction of System Design

In the development of a certain reactor type, PRS is installed to cope with design basis accidents (DBA). The PRS comprises three independent trains connected to the reactor coolant system's steam generators. The cooling capacity of every train is 0.5% of full power, i.e., 16.2 MW, so that the whole system is adequate to ensure that the reactor can be taken to a safe shutdown state in 72 h. During normal operation and DBA condition, the PRS system is standing by. The deployment of the PRS system requires the successful closure of the main steam isolation valve (MSIV) and other valves to form a closed loop. The simplified flowchart of the main steam system design for this reactor type is shown in Fig. 1. When the DBA is a secondary side break, the break loop is unavailable. If it is assumed that one MSIV in the intact loop fails to close, relying solely on one column of PRS is not enough to mitigate the accident. Therefore, this design does not meet the single fault criterion.

This issue can be solved by adding an isolation valve to each SG pipeline or increasing the design capacity of PRS. However, considering simplification of design, improvement of economy, reduction of layout difficulty, and reduction of design uncertainty, it is proposed to first demonstrate the possibility of excluding the specific fault of the main steam isolation valve refusal to close from a single fault analysis.

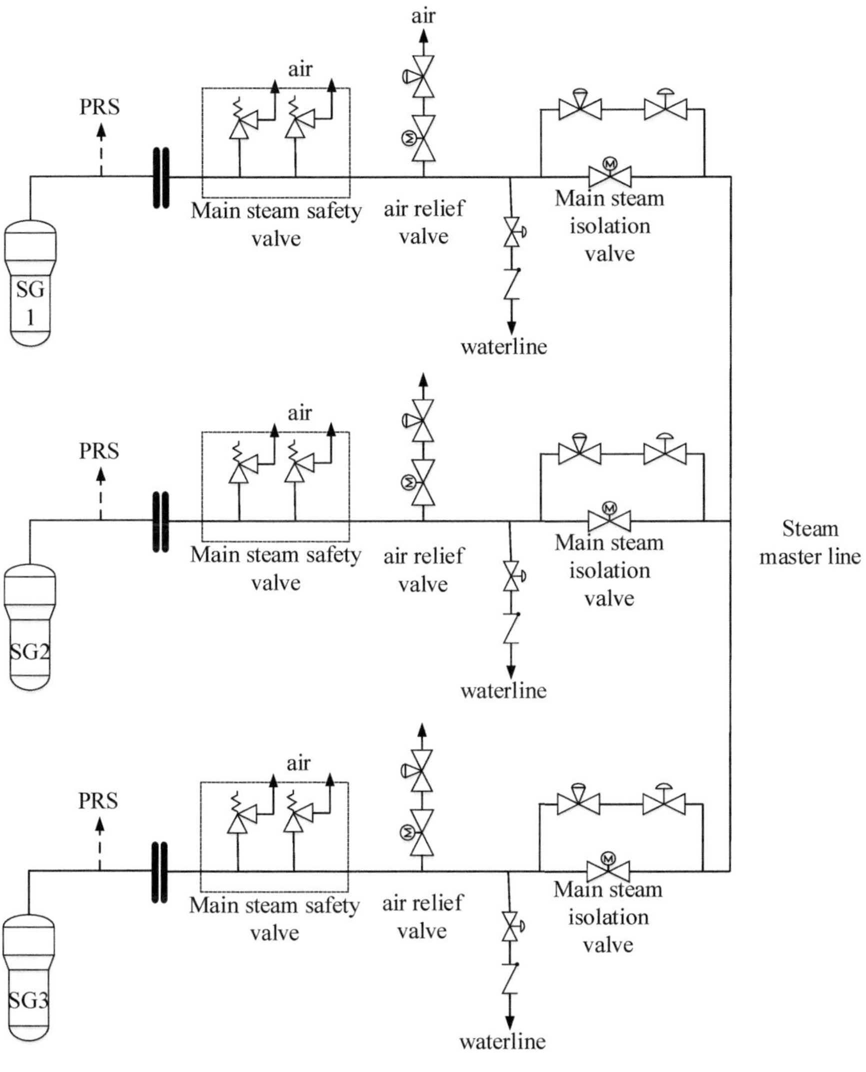

Fig. 1 The simplified flowchart of MSS

3 Exemption Method for Single Fault Criterion

Chinese regulations and standards provide provisions and specific guidelines for exempting the single fault criterion under certain circumstances. Clause 5.4.5.3 of HAF102 states that situations failing to meet the single fault criterion must be exceptionally rare and thoroughly justified through safety analysis. This paper aims to demonstrate, based on the reliability of the MSIV closure function, operational experience, risk assessment, and their integrated evaluation, the justification for excluding a particular fault from the single-fault analysis.

3.1 Reliability of MSIV Closure Function

The closure of MSIV is a necessary condition for the input of the PRS system. In the design of this reactor type, a gas–liquid linkage main steam isolation valve is used. Although the structure is relatively complex, the closing of this valve is driven by high pressure difference and does not require external power. The control mechanism is designed redundantly, and the control power is also a safety level DC power with a redundant design. The gas–liquid linkage MSIV has accumulated rich operating experience in China. Operational data indicate that gas–liquid MSIVs in Chinese plants have completed 13,042(n) actuations over 200 reactor-years. Only 1(x) closure failure was recorded, involving a delayed response time beyond technical specifications. Based on the data from NUREG/CR-6928 as prior information and combined with domestic operational experience, the Bayesian analysis method was used to calculate the MSIV rejection failure probability. The calculation process is as follows:

Prior information is specified as a failure probability of $q0 = 7.61E{-}04$ with a Beta prior distribution, $q0 \sim Beta(0.5, 656.5)$. Because the Beta family is conjugate to binomial data, observing evidence in the form of x failures over n demands (assuming the binomial model conditions hold) preserves the distributional form: a Beta prior updates to a Beta posterior. Accordingly, the posterior parameters are obtained by adding the observed failures to the prior "success" shape parameter and adding the observed non-failures to the prior "failure-free" shape parameter, which directly yields the posterior failure probability q1. Using this update rule, the posterior failure probability is $q1 = 1.1E{-}04$.

In addition, by conducting failure mode and effects analysis (FMEA) of MSIVs, the subcomponents that affect the quick closing function were identified. Based on this, a sub-component level fault tree model was constructed, and the main causes of MSIV refusal to close were identified as valve stem transmission failure and valve body failure.

3.2 Risk Analysis of DBA Relevant Sequences

The design basis accidents that challenge the single-fault criterion include: large steam line break (LSLB), small steam line break (SSLB), steam generator tube rupture (SGTR), and feed water line break (FLB).

Taking the accident LSLB as an example, after the accident occurred, the flow rate on the secondary side of SG rapidly increased, resulting in an increase in reactor heat removal, triggering an emergency shutdown and closing the main steam isolation valve. According to the accident process, the startup feedwater system (TFS) is first triggered to supply water to the steam generator in an active manner. Under the compensation injected through the full-pressure water tank (HPMT) in the primary side, the unit is in a safe and stable state, and the accident is mitigated. At this point, assuming the MSIV of an intact SG loop fails to close, the design of the TFS system can still ensure that the core residual heat can be removed while supplying water to a steam generator. If the combination of the TFS system and/or atmospheric relief valve fails, resulting in unsuccessful thermal conduction on the active secondary side, the residual heat in the core can still be removed through feed-and-bleed. The development of the event sequence is shown in Fig. 2, with the highlighted yellow part indicating the sequence development of MSIV isolation failure.

Table 1 shows the quantitative results of the four types of operating conditions that challenge SFC. It can be seen from the table that the results of each sequence are relatively low.

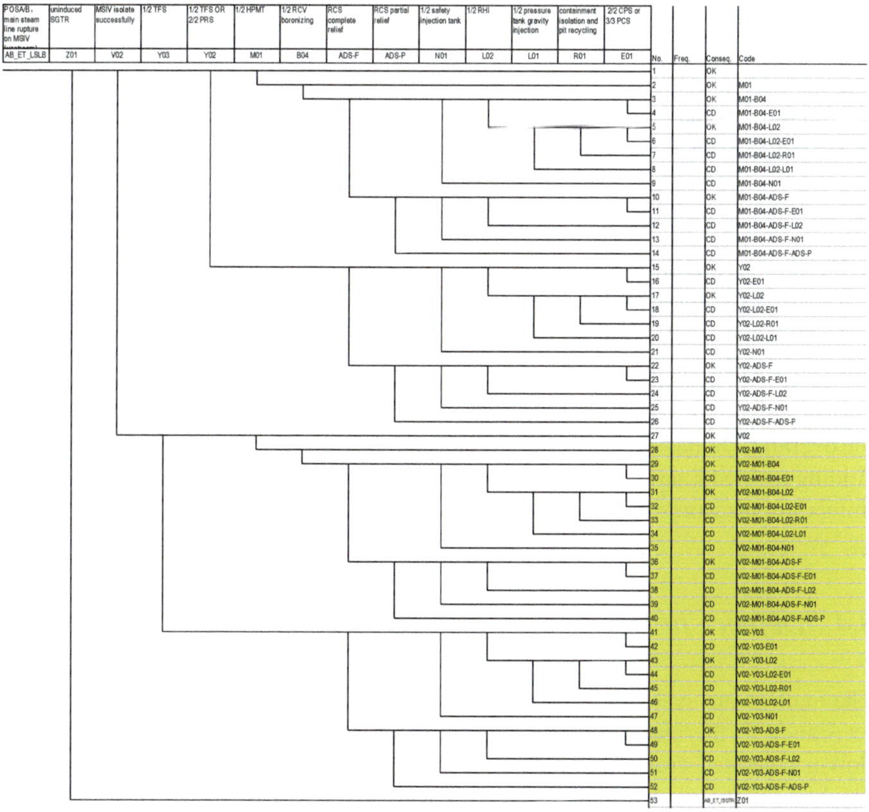

Fig. 2 The event tree of LSLB

Table 1 The quantitative results

Condition	Sequence frequency (/ry)	TFS (/ry)	Feed-and-bleed (/ry)
LSLB with one MSIV failure to close	2.3E−07	7.0E−10	1.9E−12
SSLB with one MSIV failure to close	2.3E−05	7.0E−08	1.9E−10
SGTR with one MSIV failure to close	2.3E−06	7.0E−09	1.9E−11
FLB with one MSIV failure to close	8.5E−07	2.6E−09	6.9E−12

Table 2 The results of sensitive analysis

Case	Base	Case 1	Case 2
CDF (/ry)	1.01E−07	2.41E−07	1.00E−07
CD-LSLBU (/ry)	1.27E−09	1.36E−09	1.27E−09
CD-SSLB (/ry)	1.19E−09	1.44E−08	1.16E−09
CD-SGTR (/ry)	3.58E−09	1.15E−07	3.30E−09
CD-FLB (/ry)	6.21E−10	6.79E−10	6.21E−10

3.3 Sensitive Analysis

This section intends to conduct a sensitivity analysis for two cases:

Case 1: Assuming that a main steam isolation valve is unavailable;
Case 2: Assuming that a main steam isolation valve is added to each SG pipeline to meet the single-fault criterion.

The sensitive analysis results are shown in Table 2.

The sensitivity analysis results indicate that assuming one MSIV is unavailable, the overall CDF has not yet exceeded 1E−06/reactor year, and its impact is minimal. Adding an MSIV has a very limited effect on the overall CDF reduction because the mitigation measures for the accident are sufficient.

3.4 Defence in Depth

The main steam system pipeline is classified as a high-energy pipeline and exhibits resistance to thermal fatigue, stress corrosion cracking, localized corrosion, and flow-accelerated cracking, resulting in a low rupture frequency. In the event of a pipeline break accident, the activation of the active residual heat removal system on the secondary side effectively prevents the initiation of the engineered safeguard system. Should this active measure fail, reliance is placed on the passive residual heat removal system on the secondary side alongside passive safety injection on the primary side. The effectiveness of the former depends on the operation of the passive residual heat removal system and the isolation of primary equipment on the secondary side, whereas the latter depends on associated equipment on the primary side. These two systems operate independently, and both constitute safety-level passive systems. The failure to meet the single fault criterion pertains solely to the passive residual heat removal system on the secondary side. In extreme scenarios where all aforementioned measures fail, an active safety injection system is designed to mitigate the accident and prevent core damage. This system functions independently and remains unaffected by the failure of the main steam isolation valve. In summary, the defense-in-depth design implemented to address relevant accidents is adequate. The defense-in-depth levels are detailed in Table 3, with the sections highlighted in yellow denoting cases where the single fault criterion is not satisfied.

Table 3 Defence in depth levels

Condition	Level 1	Level 2	Level 3a	Level 3b
SLB/FLB	HEP LBB	TFS + ARV	ISO	
			ISO + PRS	
			HPMT + ADS4 + ACC + PSI + PCS	
				ADS2/3 + HPMT + RHI + PCS
SGTR	Pipe material improvement In-service inspection	RCV + TFS + ARV + ADS1	ISO	
			ISO + PRS	
			HPMT + ADS4 + ACC + PSI + PCS	
				ADS2/3 + HPMT + RHI + PCS

Note
ARV Atmospheric relief valve; *ISO* Break loop isolation; *HPMT* Full pressure makeup water tank; *ADS1/2/3/4* 1/2/3/4 automatic depressurization system; *ACC* Safety injection tank; *PSI* Passive core cooling system; *PCS* Passive containment heat removal system; *RHI* Residual heat removal and active injection system; *LBB* Leak before break technology; *HEP* High energy pipeline

4 Conclusion

Risk analysis indicates that for all four scenarios challenging the single fault criterion, namely large steam line break, small steam line break, steam generator tube rupture, and feedwater line break with main steam isolation valve failure, accident mitigation remains feasible through either the turbine feedwater system or feed-and-bleed operation when the passive residual heat removal system is unavailable. The frequency of failure sequences related to these scenarios is low. Assuming the unavailability of one main steam isolation valve, the total core damage frequency is calculated as 2.41×10^{-7} per reactor year, which remains below the threshold of 1×10^{-6} per reactor year. The addition of an extra main steam isolation valve results in a minimal reduction in overall core damage frequency.

Quantitative reliability analysis demonstrates the high dependability of the main steam isolation valve. Further measures to enhance valve reliability during operation include the installation of a fatigue life monitoring system on the valve stem and optimization of inspection procedures and intervals for the valve.

In summary, although the design does not fully comply with the single-fault criterion, it maintains acceptable risk levels with infrequent failure sequences. The high reliability of the main steam isolation valve ensures that the safety system's reliability is commensurate with the plant's risk profile. Considering factors such as design simplification, economic efficiency, and reduced layout complexity, exemption from the single fault criterion may be justified.

5 Discussion and Outlook

The purpose of the single-fault criterion is to enhance the reliability of safety functions that are critical to safety. Therefore, reliability is a decisive threshold when considering exemptions from the single-fault criterion. This paper proposes a strict justification logic for specific high-reliability equipment in particular low-risk scenarios. In practical engineering applications, if the reliability of equipment is questionable or only at a general level, conservative engineering practices should be prioritized, that is, strictly adhering to the single-fault criterion by adding redundancy and employing diversification methods to improve the overall safety margin of the system, rather than seeking exemptions. Refer to [7, 8] for more information.

Current efforts focus on refining the application scope of the single fault criterion in selected areas by supplying risk-informed data within the regulatory framework, without altering the existing regulatory structure for single fault criterion implementation. Further research will explore the use of risk-informed approaches to select single-fault event sequences in design basis accident analyses.

Acknowledgements This research was supported by grant KY25010. The financial assistance provided was instrumental in facilitating the completion of this study.

References

1. Yazdi M (2024) Reliability-centered design and system resilience. In: Advances in computational mathematics for industrial system reliability and maintainability. Springer Nature Switzerland, Cham, pp 79–103
2. GB/T 13626-2008 (2008) Application of the single failure criterion to safety systems in nuclear power plant
3. CNSA (2016) Design safety regulations for nuclear power plants
4. Zhao XY, Wan LK, Hu LS et al (2020) A kind of analysis method of single failure criterion based on reliability
5. Tang B (2024) Evolving nuclear power concepts and methods in practice: an examination from a historical perspective. Nucl Saf 23(3):1–20
6. SECY-05-0138 (2005) Risk-informed and performance-based alternatives to the single failure criterion
7. Li H, Yazdi M (2022) Advanced decision-making methods and applications in system safety and reliability problems. Studies in systems, decision and control, vol 211
8. Yazdi M, Mohammadpour J, Li H, Huang HZ, Zarei E, Pirbalouti RG, Adumene S (2023) Fault tree analysis improvements: a bibliometric analysis and literature review. Qual Reliab Eng Int 39(5):1639–1659

Accuracy Lifetime Prediction Methodology for RV Reducer

Liangbo Yao, Cheng Qiu, Yikun Yang, Yu Gong, and Jinfeng Li

Abstract To address the critical challenges associated with limited predictive accuracy and narrow applicability in existing precision degradation and lifetime prediction models for high-precision RV reducers, this study introduces a novel methodology through three primary contributions. First, a three-stage Gamma process-based accuracy degradation model is developed by systematically capturing the characteristic deterioration patterns of RV reducers, effectively enhancing the stochastic representation of change points beyond the capabilities of conventional approaches. Second, a Monte Carlo bootstrap-based data augmentation method is implemented to improve the reliability of degradation trajectory estimation under small-sample conditions. Third, a lifetime prediction framework, driven by accelerated testing for accuracy degradation and incorporating the equivalent lifetime calculation method, is developed to overcome the applicability limitations of traditional models. Case study validation confirms that the proposed three-stage model reduces relative error in remaining useful life estimation by 75%. This methodology offers precise support for proactive health management of in-service RV reducers in industrial robotics and high-precision transmission applications.

Keywords RV reducer · Gamma process · Prediction of accuracy lifetime · Monte Carlo bootstrap method · Change points

1 Introduction

For RV reducers, accuracy serves as the sole criterion for defining effective service life. Any deviation or loss in precision signifies the reducer's failure to fulfill high-accuracy transmission requirements, rendering it unsuitable for precision applications. Unexpected failure of such precision reducers can lead to substantial economic losses and, in some cases, catastrophic incidents. Therefore, it is essential to predict

L. Yao (✉) · C. Qiu · Y. Yang · Y. Gong · J. Li
Yanqi Lake Institute of Basic Manufacturing Technology Co., Ltd, Beijing, China
e-mail: yaoliangbo@pcmi.com.cn

H. Li et al. (eds.), *Data-Driven Methods for Reliability and Safety Engineering: Applications in Industrial Systems*, Springer Series in Reliability Engineering,
https://doi.org/10.1007/978-3-032-22873-4_5

the degradation of transmission accuracy and implement timely maintenance strategies before precision loss occurs. This approach aims to extend service life and ensure safe and reliable operation throughout the intended functional period.

However, the accuracy degradation process in RV reducers is highly complex [1], influenced by multiple interrelated factors such as material properties, environmental conditions, and fluctuating operational loads. These factors contribute to the difficulty in accurately capturing degradation behavior through purely physics-based models. With the advancement of computational methodologies, data-driven degradation models have shown considerable promise in predicting the remaining useful life (RUL) of components with poorly defined degradation mechanisms and complex operational environments.

A number of studies have explored stochastic process-based approaches in this context. For instance, Zhou Kun [2] performed high-stress accelerated degradation testing and analyzed the time-evolution characteristics of accuracy degradation in RV reducers, modeling the process using a Wiener framework. Li et al. [3, 4] utilized variation data in transmission accuracy and backlash during degradation to construct several reliability assessment models for RV reducers, employing Gamma processes and Weibull distributions. Xu et al. [5] considered transmission error as the primary indicator of accuracy degradation and proposed a single-stage Gamma process-based model. Additionally, by integrating vibration data collected during the degradation process, a Gaussian process regression model optimized via genetic algorithms was developed. This approach enabled real-time updating of the posterior parameters of the accuracy degradation model and facilitated prediction of the transmission accuracy of RV reducers at arbitrary time points based on signal characteristics.

To address the phase-dependent characteristics observed in equipment degradation processes, Zhang et al. [6] proposed a RUL prediction method based on cumulative and test change point analysis and drift Brownian motion. They created a multi-stage drift Brownian motion model and made more reasonable predictions of equipment remaining life. Ma [7] developed a multi-change point RUL prediction method based on Wiener process, which well reflected the degradation process of rolling bearings. At the same time, the multivariate parameter optimization analysis model was studied to improve the applicability of the model. Liao [8] established a degradation model for two-stage and multi-stage degradation products based on the nonlinear drift-driven Wiener process and considered the uncertainty of the degradation state of the change point, and achieved a high accuracy of RUL prediction. Huang [9] introduced a dynamic feature-based multi-phase degradation model to estimate health indicators and subsequently predict RUL. Wang et al. [10] modelled the rotational error degradation of harmonic reducers using a Gamma-based multi-phase degradation process and incorporated Gaussian regression to achieve real-time vibration feature-driven RUL prediction. This approach significantly improved prediction accuracy compared to single-phase models. Wang et al. [11] proposed a hierarchical Bayesian change-point Gamma process model where the shape parameter was modelled as a function and estimated accordingly, demonstrating its effectiveness through simulation and degradation data analysis of N-channel MOSFETs. Gao et al. [12] developed a two-phase Wiener model incorporating measurement error and individual variability at change-points to better characterize the transmission accuracy degradation of RV reducers, thus improving reliability assessments and RUL prediction accuracy. Ren et al. [13] proposed a RUL prediction method for RV reducers that integrates DBN (Deep Belief Network), SVDD (Support Vector Data Description), and TCN (Temporal Convolutional Network). DBN extracts degradation features, SVDD accurately locates the starting point of degradation, and TCN improves prediction accuracy. Experiments verify that this method can effectively extract the degradation features of reducers and bearings, and the remaining service life prediction accuracy is significantly better than traditional methods.

In summary, existing degradation and prediction models for RV reducer precision typically assume fixed, single-stage model parameters. However, experimental studies have shown that the accuracy degradation process of RV reducers conforms to the typical "bathtub curve" of the wear process, with three stages: running-in period, normal wear period, and rapid wear period, showing the characteristics of a multi-stage accuracy degradation process [14]. These multi-phase characteristics arise from different wear mechanisms and coupled factors at each stage. Therefore, for research on the accuracy life prediction of RV reducers, it is necessary to fully consider the stage characteristics of their accuracy degradation processes. By analysing the failure mechanism and evolution patterns of different degradation stages, a segmented accuracy degradation model is established to improve the accuracy and reliability of the accuracy life prediction.

A multi-stage degradation model, extending conventional single-phase approaches through the application of change-point statistical theory, offers a comprehensive statistical framework for characterizing degradation processes exhibiting distinct phase-specific behaviors. Owing to their non-negative, independent increment nature and monotonically increasing sample paths, Gamma processes are particularly suitable for representing cumulative damage or degradation phenomena, such as corrosion and wear. Building upon the properties of the Gamma process, this study develops a multi-stage accuracy degradation model and an associated accuracy life prediction method for RV reducers.

2 Gamma Process-Based Accuracy Degradation Model

2.1 Parameter Estimation Method for the Gamma Stochastic Process Model

Based on the Gamma process, let X(t) denote the degradation amount of equipment at time t, and x(t) denote the corresponding incremental performance loss. The degradation amount X(t) and loss increment x(t) have the following characteristics:

A Gamma process can be defined by three assumptions: the degradation (or cumulative loss) starts at zero, the loss increments over any time interval are independent, and each increment follows a Gamma distribution with a time-dependent shape parameter and a constant scale parameter. Under these assumptions, the process has a closed-form probability density function determined by the shape function alpha(t), the constant scale beta, and the Gamma function. When the Gamma process is used to model product degradation leading to failure, failure is assumed to occur once the cumulative degradation crosses a specified threshold, and reliability at time t is therefore the probability that degradation has not yet reached that threshold; this can be expressed using the incomplete Gamma function. To capture realistic, time-varying degradation, the shape parameter alpha(t) is often modelled as a power function of time with constants a and b; when b equals one, alpha(t) increases linearly with time and the process is termed a stationary Gamma process, which yields a simplified density expression. Based on observed degradation increments, a likelihood function for the failure probability density can be constructed, and taking the natural logarithm produces a log-likelihood that is more convenient for estimation. Parameter estimation is then performed by maximizing this log-likelihood with respect to the unknown parameters (typically a and beta), which is achieved by differentiating the log-likelihood, setting the resulting score equations to zero, and solving them numerically; the digamma function appears naturally in these derivatives due to differentiation of terms involving the Gamma function.

Due to the presence of change points in the accuracy degradation process of RV reducers, the degradation process can be divided into three stages (see Fig. 1), each characterized by a different degradation rate. As a result, a single Gamma process is insufficient to accurately describe the entire degradation behavior. Therefore, a multi-stage degradation model based on the Gamma process is proposed.

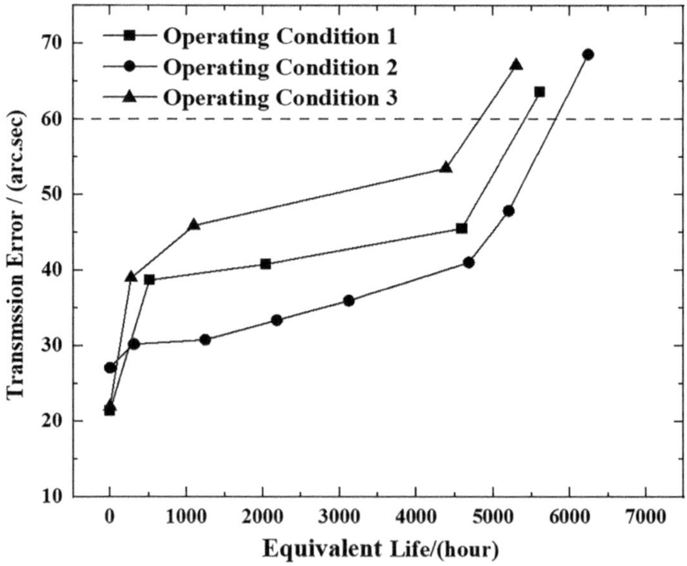

Fig. 1 Degradation process trajectory of transmission accuracy of RV reducers

Using standard properties of the Gamma distribution, if the degradation increment over a unit interval is modelled as x(t) following a Gamma distribution with time-varying shape parameter alpha(t) and constant scale parameter beta, then its moments (in particular, the expectation) can be written directly in terms of alpha(t) and beta. Based on this result, the expected accuracy degradation of the RV reducer per unit time can be expressed as a function of the Gamma-process parameters, providing an interpretable measure of the average degradation rate. To capture non-stationary behaviour observed in practice, the accuracy degradation is then modelled as a three-stage process with two change points that partition the timeline into three regimes. Each stage is characterised by its own Gamma-process parameters, namely a stage-specific shape parameter (governing how degradation accumulates) and a scale parameter (governing dispersion), so the full model is defined by the two change points and the parameter set for each stage. Consistent with this piecewise formulation, the probability density function for the RV reducer's degradation is also expressed in a three-stage (piecewise) form, with the density in each interval determined by the corresponding stage parameters.

2.2 Small-Sample Expansion Method for Precision Degradation Curves Based on Monte Carlo Bootstrap

As illustrated in Fig. 1, the available accuracy degradation data samples for the RV reducer are limited in quantity, rendering them inadequate for precise modeling of the degradation process. To address this limitation, a sample augmentation method based on the Monte Carlo bootstrap technique is developed to expand the accuracy degradation dataset. And the workflow is shown in Fig. 2. In order to ensure that the augmented samples accurately represent the degradation characteristics of RV reducers and to enhance the reliability of the expanded data, two boundary constraints are integrated into the sample generation algorithm:

(1) All generated degradation trajectories must fall within the 95% confidence interval of the original data;
(2) All values on the generated degradation trajectories must be non-negative, ensuring that no degradation value falls below zero.

In accordance with the Monte Carlo bootstrap principles and the specified boundary constraints, a MATLAB-based implementation is developed to execute the required computations. The corresponding results are presented in Figs. 3 and 4.

Figure 3 shows a portion of the samples expanded using the Monte Carlo bootstrap method. It can be observed that the confidence interval bandwidth gradually increases over time, with a particularly significant divergence beyond 5000 h. This trend aligns with the "bathtub curve" wear-out failure theory in mechanical systems, where the increase in uncertainty during the later stages reflects the cumulative effects of individual variability and random failure factors.

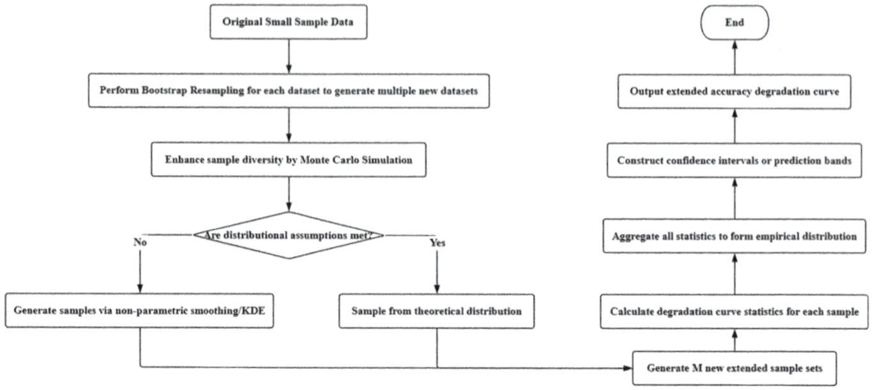

Fig. 2 Workflow of small sample expansion process based on Monte Carlo bootstrap method

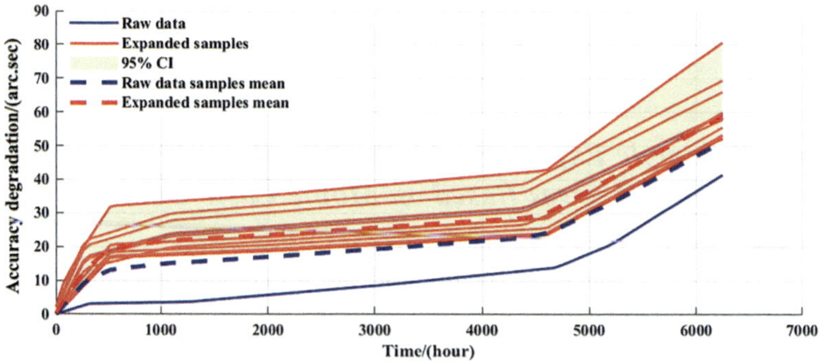

Fig. 3 Small sample expansion results of transmission accuracy degradation curve based on Monte Carlo self-help method

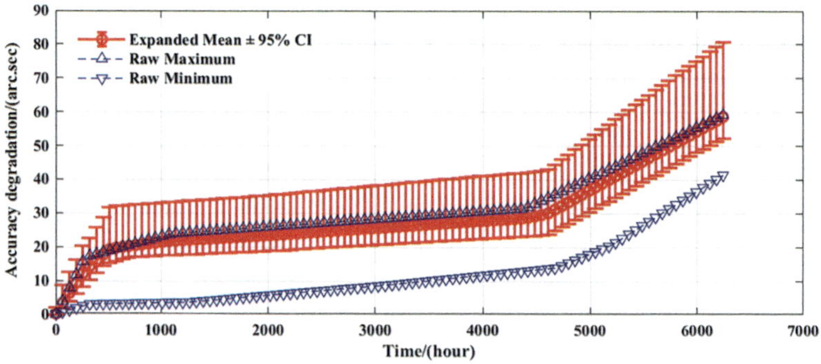

Fig. 4 Comparison of statistical characteristics of transmission accuracy degradation curves with a small sample

Figure 4 presents the statistical characteristics of the transmission precision degradation curves after small-sample expansion. The gap between the original maximum curve and the 95% CI upper bound of the expanded samples indicates that the Monte Carlo bootstrap method captures potential extreme degradation scenarios not included in the original data. At 6000 h, the expanded upper bound is approximately 15% higher than the original maximum value. The near-perfect overlap between the original mean curve and the expanded mean curve (error < 3%) demonstrates that the expansion method preserves the unbiased nature of the original data and introduces no systematic error during the small-sample expansion process.

2.3 Change Point Partitioning and Identification

As illustrated in Fig. 1, although the three accuracy degradation curves are partitioned into three stages by two change points, the positions of these change points differ across the curves. Similarly, the change point locations vary among the expanded sample curves of accuracy degradation shown in Fig. 3. This variability arises from inherent differences between product batches, along with variations in test conditions and environmental factors. To facilitate research and analysis, while accounting for this variability, the present study posits that the positions of the change points conform to a normal distribution, as illustrated in Fig. 5.

This study employs a sliding window technique to compute the second derivatives of each degradation curve and identify curvature change points. Subsequently, a normal distribution is fitted to the inflection points from all curves, outputting the mean (μ) and standard deviation (σ) for the inflection points of each stage. The corresponding calculation results are presented in Fig. 6.

The calculation results presented in Fig. 6 show that the estimated times for the first and second change points are 641.2 h and 4534 h, respectively. The standard deviations for the positions of these two change points are 32.4 and 140, indicating significant variability in their locations. Furthermore, the variability is notably greater for the second change point. This increased dispersion is attributed to accelerated wear in the precision reducer with extended operating time, resulting in greater uncertainty in failure time.

3 Model Parameter Estimation Results

The accuracy degradation data presented in Fig. 3 is segmented into three distinct stages based on the two estimated change points identified in Fig. 6. Subsequently, maximum likelihood estimation (MLE) is applied to determine the location and shape parameters of the Gamma process corresponding to each stage. The estimated parameters for the Gamma process across the three stages are summarized in Table 1.

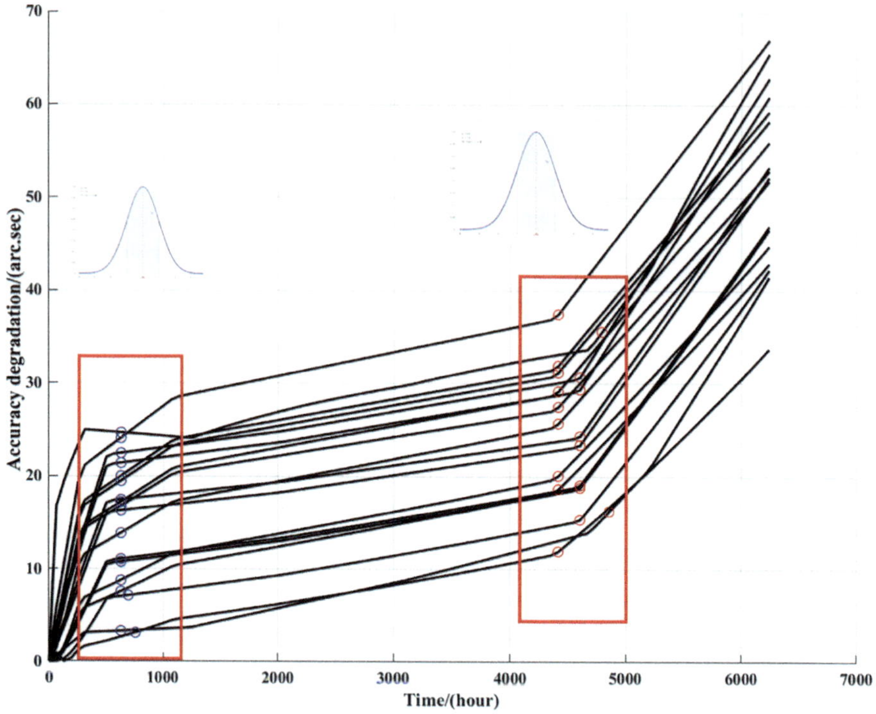

Fig. 5 Schematic diagram of transmission accuracy degradation curve, stage division, and change point identification

The observed degradation behaviour across the defined time zones shows clear stage-wise variation. In the initial time zone, the degradation level is relatively low, with a value of 0.016 corresponding to an associated measure of 0.7158, indicating a slow and stable early-stage degradation process. In the intermediate time zone, degradation increases sharply, with the time-zone indicator rising to 0.0772 and the corresponding value reaching 29.6724, reflecting an accelerated degradation phase. In the final time zone, the indicator further increases to 0.2928, while the associated value decreases to 18.5397, suggesting that although degradation continues, its rate moderates compared with the second stage. This pattern supports the suitability of a multi-stage degradation model, as the degradation dynamics differ substantially across the three time zones.

Assuming the specified failure threshold for the transmission accuracy deviation of this series of precision planetary cycloidal pinwheel reducers is $D_f = 42$ arc.sec, the accuracy life reliability curve based on transmission accuracy can be derived as shown in Fig. 7.

To evaluate the effectiveness of the three-stage accuracy degradation model, a conventional single-stage accuracy degradation model based on the Gamma process was established. The fitted accuracy degradation curve and reliability curve are

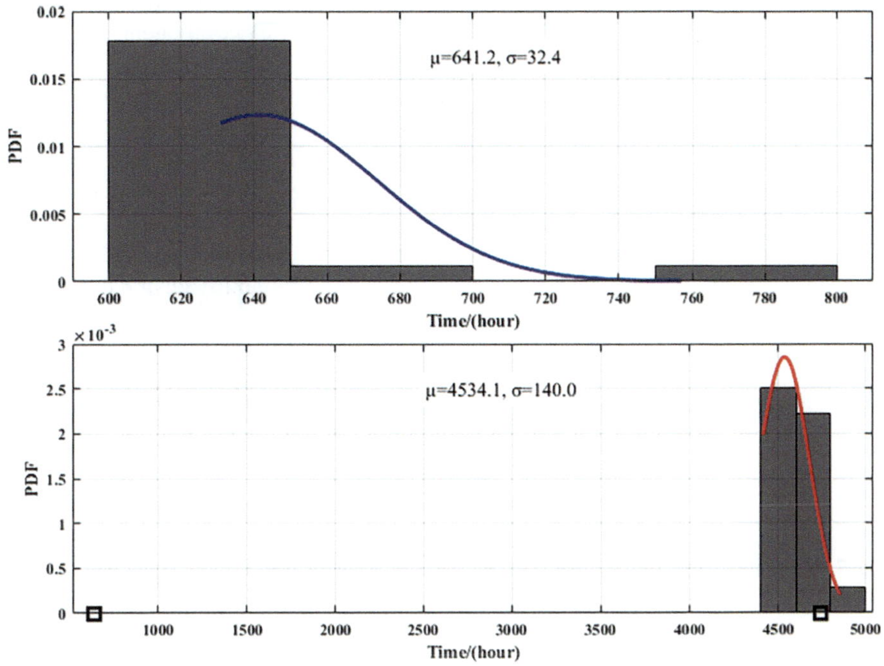

Fig. 6 Schematic diagram of the transmission accuracy degradation curve change point distribution identification result

Table 1 Parameter estimation results of the three-stage Gamma process with accuracy

No	Stages	Alpha	Betha
1	Stage 1	0.016	0.7158
2	Stage 2	0.0772	29.6724
3	Stage 3	0.2928	18.5397

presented in Fig. 8, with the calculated shape parameter and scale parameter of the accuracy degradation model being 0.0162 and 1.9738, respectively.

4 Experimental Validation

Using the accuracy degradation data under Operating Condition 1 from Fig. 1, the probability density functions of the predicted remaining accuracy life at different sampling times, obtained from two Gamma process-based degradation models, are presented in Fig. 9. A comparative analysis between the predicted mean values and the actual remaining accuracy life is shown in Fig. 10.

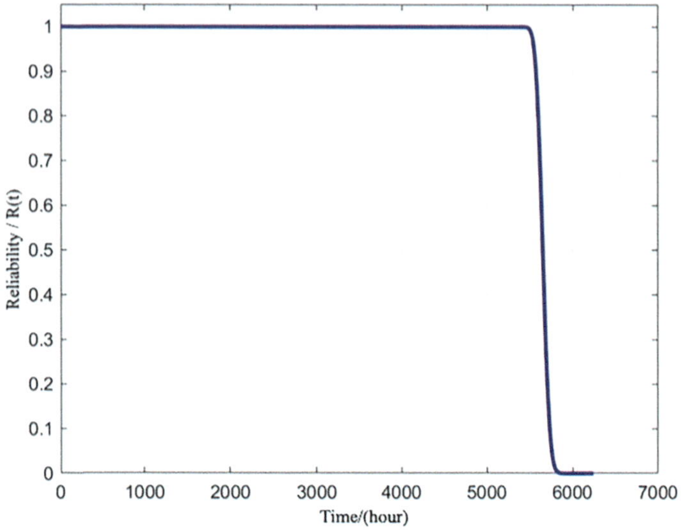

Fig. 7 Reliability curve of the three-stage RV reducer accuracy degradation model based on the Gamma process

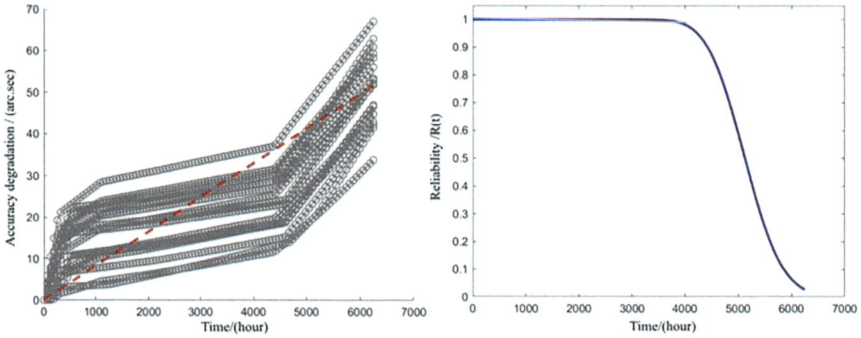

Fig. 8 Reliability curve of the single-stage precision reducer accuracy degradation model based on the Gamma process

Table 2 summarizes the actual RUL versus predicted RUL at all sampling points throughout the lifecycle. The analysis reveals that the three-stage Gamma process-based accuracy degradation model achieves a mean absolute error (MAE) of 94 h for remaining accuracy life predictions. This represents a 75% reduction in MAE compared to the single-stage model (379 h). Furthermore, the standard deviations of prediction bias for the two models are 31.3 h and 173.7 h, respectively. These results demonstrate that the three-stage model more accurately captures the actual degradation behaviour of RV reducers, delivering predictions significantly closer to ground-truth values. Notably, larger prediction deviations occur at sampling points

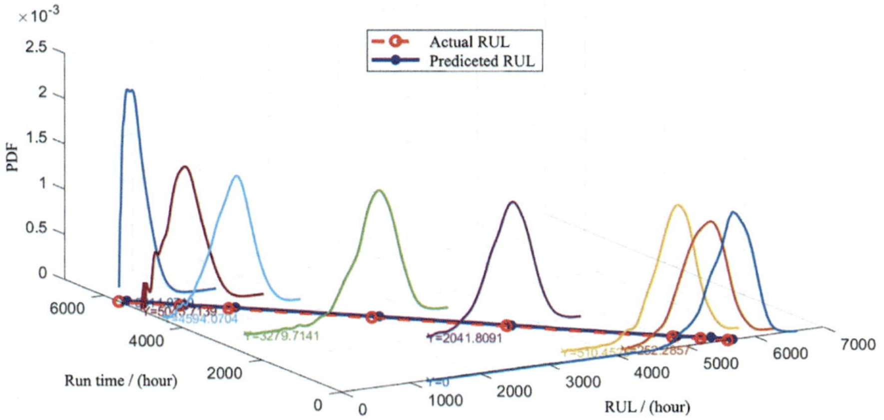

Fig. 9 PDF of the TUL of the three-stage accuracy degradation model based on the Gamma process

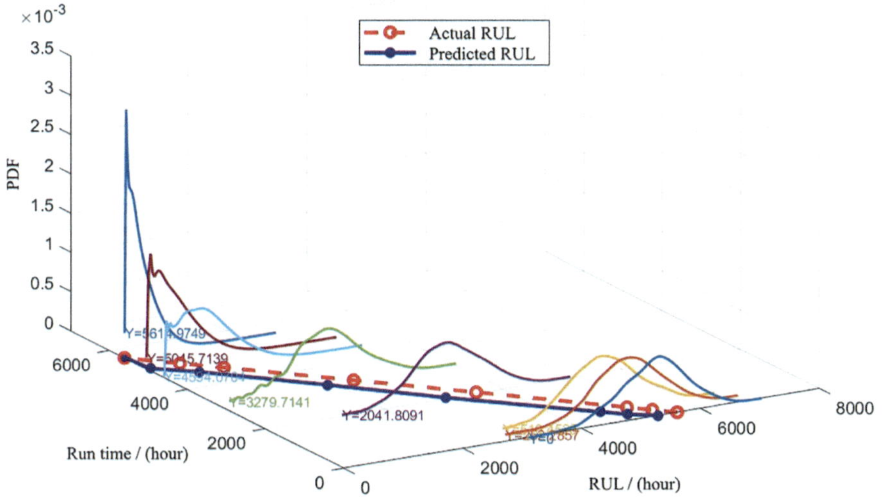

Fig. 10 PDF of the RUL of the single-stage accuracy degradation model based on the Gamma process

2, 4, 5, and 8 for the three-stage model: (1) Sampling points 2, 4, and 5(proximal to change points) exhibit elevated errors primarily attributed to uncertainty in change point estimation, reflecting inherent variability across products and operating conditions. (2) Sampling point 8 shows increased deviation due to amplified failure uncertainty during terminal degradation stages, aligning with empirical failure progression patterns. Critically, Figs. 9 and 10 demonstrate that the PDF of predicted RUL for the three-stage model exhibits higher probability concentration at every sampling

Table 2 Statistics of actual remaining life and predicted life at different sampling points

Sample points	Actual RUL (h)	Predicted RUL based on 3-stages modeling (h)	Predicted RUL based on 1-stage modeling (h)
1	5615	− 64	327
2	5363	− 151	423
3	5105	− 63	447
4	3573	− 73	507
5	2335	− 107	441
6	1021	− 105	410
7	569	− 73	502
8	0	− 118	28
Mean	/	− 94	379
Standard deviation of bias	/	31.3	173.7

interval. This indicates greater prediction confidence (reduced dispersion), enhanced prognostic precision, and statistically robust life estimation.

The three-stage Gamma process framework significantly enhances the effectiveness of proactive health maintenance for in-service precision RV reducers by: (1) Quantifying degradation-phase transitions via change points. (2) Reducing life prediction MAE by 75% over conventional methods. (3) Providing uncertainty-quantified failure prognostics ($\sigma = 31.3$ h).

5 Conclusion

To address the limitations of existing RV reducer accuracy degradation models, particularly their limited precision and restricted applicability in accuracy life prediction, a methodology based on accelerated degradation data is proposed in this study. A prognostic model grounded in the Gamma process is established, enabling accurate life prediction across varying operating conditions while substantially improving the prediction accuracy of the remaining accuracy life for precision RV reducers. The primary conclusions are outlined as follows:

(1) The limited accuracy degradation data from small samples are expanded using the Monte Carlo bootstrap method. This technique comprehensively incorporates stochastic variability resulting from differences in product batches and operational environments, thereby increasing both the quantity and diversity of the dataset to overcome the constraints of limited data.

(2) A multi-stage degradation model based on the Gamma process is constructed by integrating the inherent progression characteristics of RV reducers and accounting for the random occurrence of change points. This modeling approach

enhances consistency with the actual physical degradation process and improves the robustness and reliability of the model, for more info refer to [15, 16].

(3) Validation using real-time operational data confirms that the multi-stage Gamma process-based degradation model and the proposed remaining accuracy life prediction framework offer improved prognostic accuracy (75%) and higher confidence levels, which could significantly reduce the risk of unplanned failures in industrial robots.

Acknowledgements This study was supported by the High-Quality Development Special Project of the Ministry of Industry and Information Technology, China (2023ZY01050) and technology research and development project of the Taihang National Laboratory, China (CXPT-2022-054).

References

1. Liu C (2002) Research on prediction and degradation law of transmission accuracy of RV reducer. Ph.D. thesis, Chongqing University
2. Zhou K, Ye N, Wu Jh (2022) High stress accelerated degradation test and reliability analysis of RV reducer. J Harbin Inst Technol 54(7):37–44
3. Li J, Yang Y, Wang X (2023) Reliability evaluation of RV reducers based on multi degradation data. J Mech Transmission 47(5):82–87
4. Li J, Yang Y, Yao L (2022) Reliability evaluation of RV reducers based on transmission performance degradation data. Mech Eng Autom 5:7–9
5. Hang X, Nie YX, Wen DJ et al (2025) Degradation and reliability assessment of accuracy life of RV reducers. J Mech Transmission, 1–9
6. Zhang Z, Hu C, Gao Y (2017) A residual useful life prediction approach for equipments with multi-state stochastic degradation. J Syst Eng 32(1):1–7
7. Ma CY (2021) Research on prediction method of rolling bearing RUL on multi-change point. Master thesis, Chongqing University
8. Liao GB (2022) Research on remaining useful lifetime prediction method of equipment based on multi-phase Wiener process. Master thesis, Chongqing University
9. Huang ZY (2023) Prediction method of remaining useful life for rolling bearing based on multi-degradation stage assessment. Master thesis, Central South University
10. Guo J, Wang Z, Li H, Yang Y, Huang CG, Yazdi M, Kang HS (2024) A hybrid prognosis scheme for rolling bearings based on a novel health indicator and nonlinear Wiener process. Reliab Eng Syst Saf 245:110014
11. Li H, Peng W, Adumene S, Yazdi M (2023) Cutting edge research topics on system safety, reliability, maintainability, and resilience of energy-critical infrastructures. In: Intelligent reliability and maintainability of energy infrastructure assets. Springer Nature Switzerland, Cham, pp 25–38
12. Gao X, Ye N, Zhou K et al (2023) Two stage wiener modeling and reliability analysis considering measurement errors. China Mech Eng 34(09):1067
13. Yazdi M (2024) Reliability-centered design and system resilience. In: Advances in computational mathematics for industrial system reliability and maintainability. Springer Nature Switzerland, Cham, pp 79–103
14. Yao LB, Qiu C, Gong Y (2024) Study on accuracy degradation experimental of RV reducer. In: Proceedings of 2024 6th international conference on robotics, intelligent control and artificial intelligence (RICAI), Nanjing, Dec 2024, pp 504–508

15. Li H, Yazdi M (2022) Advanced decision-making methods and applications in system safety and reliability problems. Studies in systems, decision and control, vol 211
16. Yazdi M, Mohammadpour J, Li H, Huang HZ, Zarei E, Pirbalouti RG, Adumene S (2023) Fault tree analysis improvements: a bibliometric analysis and literature review. Qual Reliab Eng Int 39(5):1639–1659

Derating Design of Control Circuits Based on Harpness Simulation

Xin Wang, Jun-Wen Shi, and Xiang-Yang Yu

Abstract Derating design, as a classical and extensively applied method in circuit reliability engineering, plays a pivotal role in maintaining the stable performance of electronic circuits. Nevertheless, traditional derating methodologies are limited by several challenges, including a high degree of subjectivity, the absence of standardized quantitative criteria, computational complexity, and significant difficulty in accurately assessing transient electrical stresses during component evaluation. To overcome these limitations, this study incorporates the simulation analysis module of Harpness software to enhance the derating design process. By utilizing simulation-based calculations of electrical stress under both standard operating conditions and extreme scenarios, the proposed approach facilitates precise assessment of transient electrical stresses and component junction temperatures. The simulation platform outputs data tables that provide detailed stress analysis and derating evaluations for all components, thereby substantially improving the accuracy and efficiency of the derating design process. Particular attention is given to the simulation methodology for FPGA device derating, which applies equivalent circuit 1 odelling and equivalent load techniques to enable accurate simulation analysis.

Keywords Derating design · Components · Simulation · Reliability

---

X. Wang
The Military Representative Bureau of the Navy Equipment Department of the People's Liberation Army Navy stationed in Shenyang Area, Shenyang, China
e-mail: wangxin@zcsca.com

J.-W. Shi (✉)
Beijing Zhongcheng Renhe Technology Development Co., LTD., Beijing, China
e-mail: shijunwen@zcsca.com

X.-Y. Yu
Qingdao Campus of Naval Aviation University of the People's Liberation Army of China, Qingdao, China
e-mail: yuxiangyang@zcsca.com

H. Li et al. (eds.), *Data-Driven Methods for Reliability and Safety Engineering: Applications in Industrial Systems*, Springer Series in Reliability Engineering,
https://doi.org/10.1007/978-3-032-22873-4_6

1 Introduction

Derating design, as a classical and extensively applied methodology in circuit reliability engineering, is fundamental for ensuring stable circuit performance. However, traditional derating methods present notable limitations in accurately assessing electrical stresses, including voltage, current, and power experienced by components. These evaluations often depend heavily on the subjective judgment of designers, leading to significant variability and a lack of standardized and precise quantitative criteria [1]. Moreover, conventional approaches are frequently computationally demanding and time-consuming, typically providing only approximate steady-state estimates of electrical stress. Accurately quantifying transient electrical stresses remains particularly challenging, thereby introducing potential risks into circuit reliability design.

In the context of rapid advancements in science, technology, and the industry, the demand for enhanced product reliability has become increasingly stringent. In highly competitive domestic and international markets, achieving technological sophistication and superior reliability is critical for ensuring product excellence. These factors are essential for optimizing product performance and substantially influence lifecycle costs, thereby directly affecting market competitiveness and long-term viability.

Within this framework, the advancement of research and practical applications in reliability engineering has emerged as an urgent and significant endeavor. This pursuit is closely tied to the sustained development of the national economy, serving as a fundamental force in driving industrial transformation and economic growth. It also supports continued technological innovation, contributing to national advancement. Additionally, it is indispensable for the stable development of national defense, playing a vital role in safeguarding national security and enhancing the operational performance of equipment. Reliability is formally defined as "the ability of a product to perform its required functions under specified conditions and within a designated time period".

Years of global experience in reliability engineering have demonstrated that reliability design significantly impacts product reliability. According to Japanese electronics industry statistics, design flaws account for 80% of product unreliability issues, followed by component selection (15%) and manufacturing processes (5%). Meanwhile, U.S. Naval Electronics Laboratory data indicates that design contributes to 40% of reliability failures, with components (30%), operation/maintenance (20%), and manufacturing (10%) as secondary contributors. It has been frequently observed in practice that many component failures stem from flawed design methodologies. These statistics underscore that enhancing product reliability fundamentally hinges on robust reliability design practices. Key reliability design methodologies encompass derating design, redundancy design, dynamic design, and thermal design. Among these, derating design is particularly effective in reducing component failure rates, as the majority of electronic components demonstrate significantly lower failure rates when operating stresses are maintained below their rated thresholds.

This study focuses on the application of reliability-oriented derating design in electronic test equipment, intending to systematically enhance product reliability through the implementation of optimized stress control strategies [4].

2 Introduction to the Derating Design Grades of Components

Reliability-oriented derating design for electronic test equipment primarily focuses on ensuring that both electrical and thermal operating stresses experienced by individual components remain significantly below their respective rated limits. This approach aims to prolong service life, decrease failure rates, and improve overall operational reliability. Electronic equipment typically comprises a combination of electronic components and mechanical elements, with electronic components including integrated circuits, resistors, capacitors, relays, and various other passive and active devices.

Each category of electronic component generally operates most reliably within an optimal stress range, where changes in stress levels can have a substantial impact on failure rates and are more readily manageable during the design process. However, excessive derating can yield unintended negative consequences, such as deviations in component performance characteristics, an increase in the total number of components, or challenges in sourcing suitable parts. These issues can, in turn, impair normal system operation and compromise overall reliability.

Derating design practices are commonly classified into three levels, depending on the criticality of the product and specific application requirements:

(a) Grade I Derating

Grade I represent the maximum derating level, yielding the most substantial improvement in component operational reliability. Beyond this threshold, further derating provides diminishing reliability returns while complicating equipment design implementation.

This grade applies to critical systems where component failure could jeopardize safety, cause mission failure, or result in significant economic losses.

(b) Grade II Derating

Grade II constitutes moderate derating, offering significant reliability improvement with greater design feasibility compared to Grade I.

This level is suitable for systems where component failure might damage equipment and associated support infrastructure.

(c) Grade III Derating

Grade III represents minimal derating and offers the highest relative efficiency in enhancing reliability per unit of derating. However, it results in lower absolute reliability improvements when compared to Grades I and II.

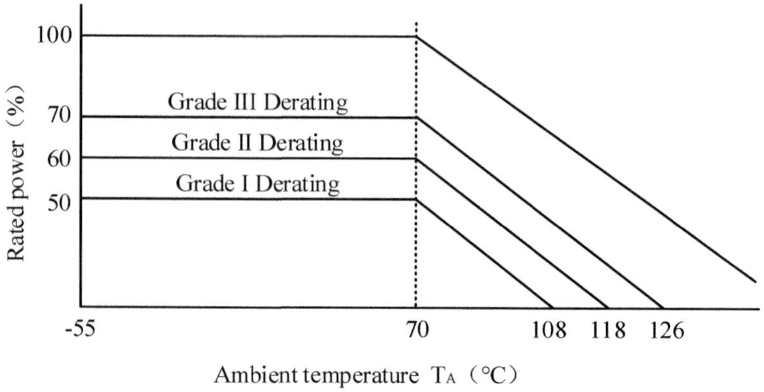

Fig. 1 Example of a derating curve of a thin-film resistor

Table 1 Example of derating criteria for integrated circuit amplifiers

Derating parameter	Downgrading level		
	Grade I Derating	Grade II Derating	Grade III Derating
Power supply voltage	0.70	0.80	0.80
Input voltage	0.60	0.70	0.70
Output current	0.70	0.80	0.80
Power	0.70	0.75	0.80
Maximum junction temperature	80	95	105

This category applies to non-critical systems where component failure would minimally impact mission success, permit cost-effective repairs, and not endanger personnel safety [2].

Figure 1 is an example of the derating curve of a thin-film resistor, and Table 1 is an example of the derating criterion for an integrated circuit amplifier.

3 Derating Design Methodology Utilizing Harpness Software

Method Description

Harpness software is a self-developed domestic EDA software. The software provides an integrated model for electrical system design, simulation verification, and reliability design.

The Harpness-based simulation methodology provides an advanced approach for calculating electrical component stresses. It performs these calculations under both normal operating conditions and extreme scenarios, such as power supply voltage fluctuations and current overloads, using computational modeling.

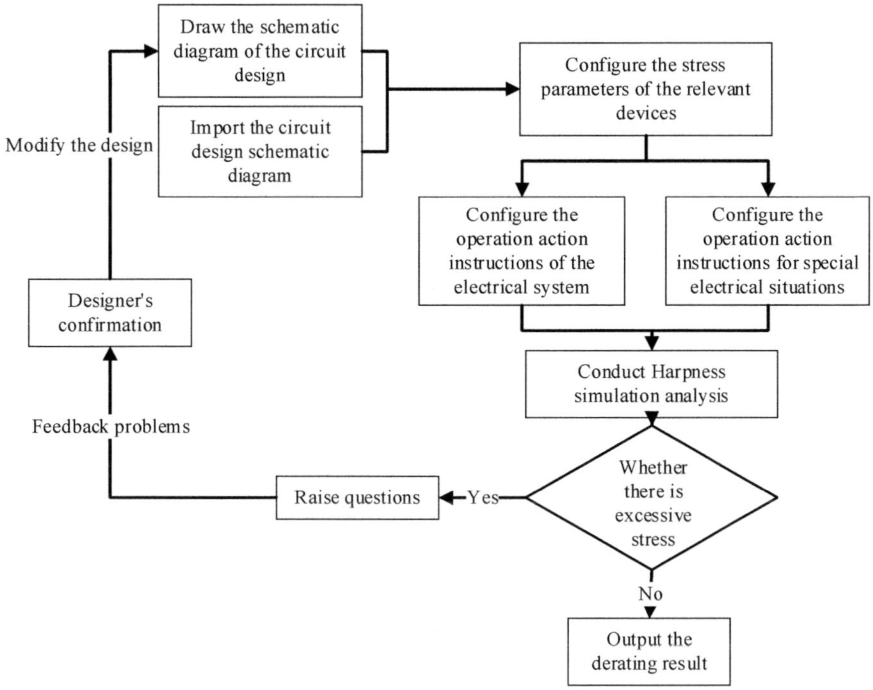

Fig. 2 The basic flowchart of Harpness analysisobtain the corresponding analysis data

This methodology enables precise analysis of transient electrical stresses and junction temperatures for each component in the circuit, while facilitating automated data extraction via simulation software interfaces. The platform produces detailed stress analysis reports and derating compliance tables for all components, leading to substantial improvements in derating design accuracy and efficiency. The operational procedure is illustrated in Fig. 2.

As illustrated in Fig. 2, the circuit design schematic must first be created within the Harpness software. This can be achieved either through manual drafting or by importing the design. Once the schematic is completed, derating parameters for various components should be configured. These include the components' maximum operating condition ratings, such as voltage and current, the calculation formulas for derating parameters, and component-specific environmental parameters such as ambient temperature.

After configuration is complete, the circuit system's command actions must also be configured based on the normal operating switch sequence. After adding the completed instruction action, the Spice simulation is run through the analysis method of circuit transient simulation (the Spice simulator is integrated internally into the Harpness software) for simulation analysis.

The actual parameters are compared with the rated values of the input component parameters, then the results are output, and the qualified derating grade is determined.

Simulation analysis can both calculate whether component stress parameters meet design requirements and check whether circuit functional characteristics meet design requirements. If a component exceeds the safe working condition range, it will be displayed in the output result, and the brand, name, model, and other information of the component will be marked. The design can be refined by modifying the circuit schematic, adjusting the component models and specifications, and subsequently re-executing the Harpness simulation tool. This iterative process continues until all stress-related issues within the circuit components are fully resolved.

The Processing of Derating Design Simulation for FPGA Chip Classes

Derating design simulations for FPGA devices requires specific processing. Given that the simulation model of such chips cannot be fully reconstructed, simulations are typically conducted using an "equivalent" approach. The I/O interface circuitry of the chip is equivalently modeled to replicate the functionality of FPGA-like components. Based on the circuit design requirements, the input and output methods of the FPGA chip's I/O interface are determined, and the corresponding interface circuits are constructed.

The interface circuit is set up as shown in the following Figs. 3 and 4.

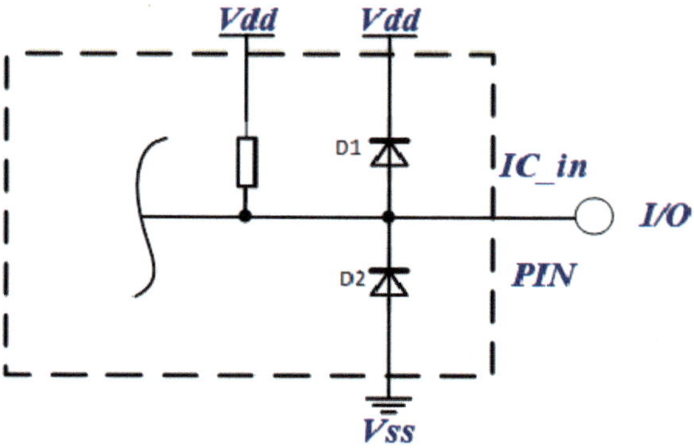

Fig. 3 Pull-up resistor input interface

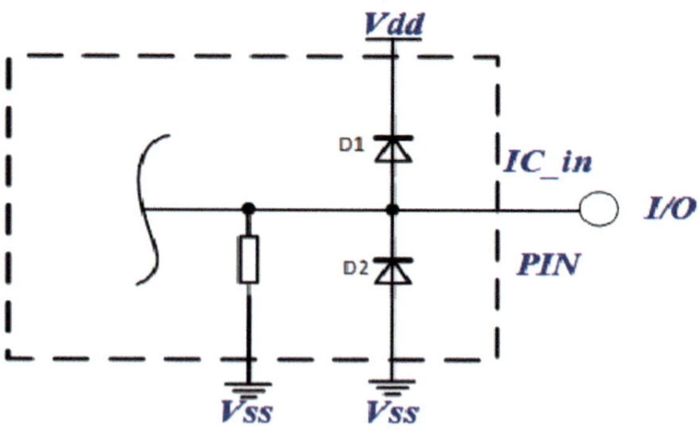

Fig. 4 Pull-down resistor input interface

Upon completion of the I/O interface circuit design, it is also essential to calculate the actual power consumption of the FPGA chip. Using an equivalent load approach, the power consumption is incorporated into the FPGA chip itself, following which the derating design simulation is performed to obtain the corresponding analysis data.

Engineering Example

Based on an engineering example, the following introduces the derating design process in engineering through the Harpness software.

First, import the circuit schematic diagram (as shown in Fig. 5) into the Harpness software to generate the circuit effect diagram as shown in Figs. 6, 7, and 8.

Then, based on the actual working conditions of the chip, build the component model, configure the stress parameters of the component simulation model, add the corresponding power supply and zero potential points on the circuit schematic diagram, conduct simulation analysis, and obtain the simulation analysis data. The exported simulation data is output at its maximum value. If you want to view the detailed data information of the device, you can measure it separately through an oscilloscope in the software.

The method of derating design through Harpness software only considers the influence of the power of FPGA chip-like devices on the circuit and whether the current and voltage values in the circuit exceed the rated values of each pin of the FPGA chip-like devices during circuit operation, without considering the actual control

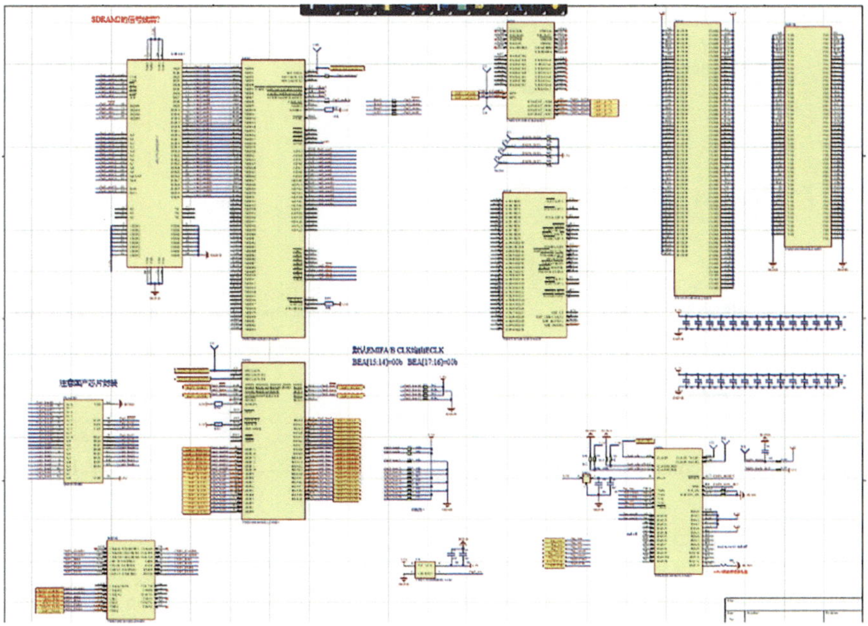

Fig. 5 Circuit schematic diagram of Altium Designer

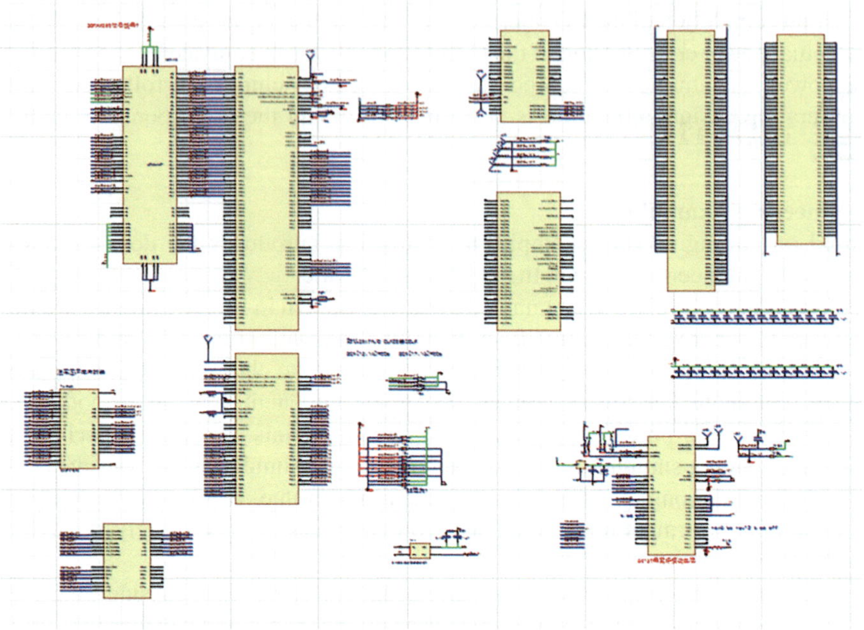

Fig. 6 Import the circuit effect diagram of the Harpness software

```
* LF347 operational amplifier "macromodel" subcircuit
* created using Parts release 4.01 on 07/05/89 at 13:19
* (REV N/A)
* connections:   non-inverting input
*                | inverting input
*                | | positive power supply
*                | | | negative power supply
*                | | | | output
*                | | | | |
.subckt LF147    1 2 3 4 5
*
  c1   11 12 3.498E-12
  c2    6  7 15.00E-12
  dc    5 53 dx
  de   54  5 dx
  dlp  90 91 dx
  dln  92 90 dx
  dp    4  3 dx
  egnd 99  0 poly(2) (3,0) (4,0) 0 .5 .5
  fb    7 99 poly(5) vb vc ve vlp vln 0 14.15E6 -10E6 10E6 10E6 -10E6
```

Fig. 7 Example of component code setup

Serial number	Component information		Stress parameter	Use stress		Strength		Stress ratio(%)	Downgrading level	Remarks
	Name	Position number		Stress value	Source	Rated value	Source			
1	CAP	C2	Dc working voltage	3.3	Electrical simulation	50	Data manual	6.6	I	
			Maximum rated ambient temperature	85	Simulation calculation	125	Data manual	/	I	
2	CAP	C1	Dc working voltage	3.3	Electrical simulation	50	Data manual	6.6	I	
			Maximum rated ambient temperature	85	Simulation calculation	125	Data manual	/	I	
3	Res2	R22	Voltage	5	Electrical simulation	50	Data manual	10	I	
			Power	0.02	Electrical simulation	0.05	Data manual	40	I	
			Ambient temperature	50	Simulation calculation	150	Data manual	/	I	

Fig. 8 Example of exported data

function of the FPGA chip-like devices. When conducting simulation calculations for junction temperature, refer to the actual operating temperature of the circuit.

This project adopted both the derating design method based on software simulation and the traditional manual calculation method. One engineer completed the entire derating design work in three days with the help of Harpness software. When a three-person team used the traditional derating design method to carry out the calculation work, it took them ten days to complete the derating design. The efficiency of completing the derating design was improved by 90%.

4 Conclusion

Derating design, a well-established strategy in circuit reliability engineering, effectively minimizes device failure rates and enhances product dependability by reducing operational stress levels [3, 5–7]. This study introduces an innovative application of the Harpness software simulation module to advance the derating design process, thereby addressing limitations associated with conventional approaches, such as procedural complexity and reliance on empirical estimations. When components do not satisfy derating requirements, circuit modifications or component substitutions can be implemented in a timely manner, significantly improving both the accuracy and efficiency of the derating procedure. This methodology establishes a new technical framework for circuit reliability design while fully complying with the provisions outlined in the "GJB/Z35 Component Derating Criteria".

References

1. Bo X, Jing M (2007) Derating design of sld control circuit based on pspice simulation. Foreign Electron Meas Technol 26(1):22–26
2. GJB/Z35 (1993) Criteria for Derating Components (In Chinese)
3. Guo J, Wang Z, Li H, Yang Y, Huang CG, Yazdi M, Kang HS (2024) A hybrid prognosis scheme for rolling bearings based on a novel health indicator and nonlinear wiener process. Reliab Eng Syst Saf 245:110014
4. Jie DZ (2004) Reliability derating design of military electronic measuring instruments. Electron Qual 12:49–51+75
5. Li H, Peng W, Adumene S, Yazdi M (2023) Advances in failure prediction of subsea components considering complex dependencies. In: Intelligent reliability and maintainability of energy infrastructure assets. Cham, Springer Nature Switzerland, pp 93–105
6. Li H, Peng W, Adumene S, Yazdi M (2023) Using global average pooling convolutional siamese networks for fault diagnosis of planetary gearboxes. In: Intelligent reliability and maintainability of energy infrastructure assets. Cham, Springer Nature Switzerland, pp 73–91
7. Li H, Peng W, Adumene S, Yazdi M (2023) Advances in intelligent reliability and maintainability of energy infrastructure assets. Intell Reliab Maint Energy Infrastruct Assets:1–23

AK-SYSC: A Kriging-Based Approach for Multi-failure-Mode Structural Systems Reliability Analysis

Liangli He, Xinfa Chen, and Xinyao Li

Abstract In engineering structures, many systems are subject to multiple failure modes, making the accurate and efficient analysis of system reliability a significant challenge. Among the existing estimation methods, AK-SYS and AK-SYSi approaches based on Kriging surrogate model have shown competitive performance. However, the high false identification rate associated with AK-SYS method and the omission of minimum or maximum mode information in AK-SYSi method limit both accuracy and efficiency. To enhance estimation accuracy and computational efficiency, this study proposes a novel method, referred to as AK-SYSC method, which incorporates a refined learning function U_c. AK-SYSC method addresses the limitations of AK-SYS approach by mitigating the risk of false identification of extreme modes caused by non-convergent Kriging models, which can otherwise lead to reduced estimation efficiency and inaccurate predictions of system failure probability. Furthermore, by fully utilizing the minimum or maximum value information neglected in AK-SYSi method, the proposed approach enhances the identification speed of the system limit state surface, thereby further improving computational efficiency. Three numerical examples serve to validate the performance and exactness of the proposed AK-SYSC method.

Keywords System reliability assessment · Multiple failure modes · Adaptive kriging model · Refined learning function

1 Introduction

Structural system reliability analysis involving multiple failure modes is inherently more challenging and computationally demanding than the assessment of a single failure mode. In such systems, each failure mode is represented by a performance function that defines its corresponding failure domain, governed by a set of random

L. He · X. Chen (✉) · X. Li
Institute of Systems Engineering, China Academy of Engineering Physics, Mianyang, Sichuan, People's Republic of China

H. Li et al. (eds.), *Data-Driven Methods for Reliability and Safety Engineering: Applications in Industrial Systems*, Springer Series in Reliability Engineering,
https://doi.org/10.1007/978-3-032-22873-4_7

input variables with a joint probability distribution. The overall system failure probability is obtained by integrating over the system failure domain, with its exact definition depending on the system configuration. In series systems, failure occurs when any single failure mode is activated, whereas in parallel systems, failure is realised only when all modes fail simultaneously.

For many real-world engineering applications, complex systems can be decomposed into combinations of series and parallel subsystems. Accordingly, this study concentrates on these two fundamental system configurations as the building blocks of more intricate reliability structures.

Conventional reliability analysis methods developed for single failure modes are often insufficient for systems with multiple interacting modes. As a result, recent research has increasingly focused on multi-mode system reliability problems. Existing approaches can be broadly categorised into three groups: sampling-based techniques, moment-based methods, and surrogate-model-based strategies, particularly those relying on Kriging models. Although Monte Carlo simulation provides a straightforward and conceptually robust solution, its high computational cost severely limits its practicality for engineering-scale problems. To improve efficiency, advanced sampling techniques such as importance sampling [5] and subset simulation [7, 9] have been extended from single-mode to system reliability analysis. Nevertheless, these methods still face difficulties in achieving an optimal balance between computational effort and estimation accuracy.

Moment-based methods estimate system failure probability by approximating probability moments of the performance functions. Two-point and three-point estimation techniques [12, 13] can significantly increase the number of required performance function evaluations as the dimensionality and number of failure modes grow. Similarly, approaches such as the Zhou–Nowak method [19] and the Zhao–One methods [16, 17] lead to a substantial computational burden. As observed in single-mode applications, moment methods remain sensitive to high-dimensional input spaces and strong nonlinearities, which limit their effectiveness for complex multi-mode structural systems.

The emergence of Kriging modeling has attracted significant attention due to its favorable balance between computational efficiency and accuracy [2, 4, 6, 18]. An early implementation of Kriging for system reliability, the Efficient Global Reliability Analysis (EGRA) method [1], established a framework where individual Kriging models are developed for each failure mode, with only sufficiently accurate models contributing to the final failure probability estimation. Building upon EGRA, the AK-SYS method [3] introduced three distinct surrogate modeling strategies. The first strategy extends the EGRA methodology, while the second constructs a single Kriging model for the system performance function based on extreme values across failure modes, though the irregular nature of such functions often complicates modeling. The third strategy implements a composite criterion to develop surrogate models for each failure mode according to their impact on the system failure domain. While this approach demonstrates superior efficiency among the three strategies, it experiences elevated misclassification rates that undermine both accuracy and computational performance [15]. To mitigate this issue, AK-SYSi method was proposed in [15] using a refined learning function. However, AK-SYSi method neglects critical minimum and maximum value information, which plays an essential role in series and parallel systems, thereby affecting estimation efficiency.

To reduce the false identification rate of the AK-SYS method (specifically the third strategy) and to fully exploit extreme value information, namely minimum values for parallel systems and maximum values for series systems—a dedicated learning function is introduced. The Kriging-based system reliability framework built upon this learning function is referred to as the AK-SYSC method. The proposed learning function systematically evaluates four possible scenarios arising from surrogate model predictions and their associated estimation accuracy, with separate treatment for series and parallel system configurations. Importantly, this function can correctly determine the state of each failure mode even when the Kriging model has not yet fully converged, while explicitly preserving critical extreme value information. For failure modes with uncertain or ambiguous classification, the method adopts a conservative strategy by selecting the minimum value across all modes to steer the adaptive learning process. This targeted use of extreme information effectively guides sample enrichment and leads to a substantial reduction in failure mode misclassification rates.

The remainder of this paper is structured as follows. A brief overview of AK-SYS and AK-SYSi methods for structural system reliability is provided in Sect. 2. Section 3 introduces a new learning function within the proposed AK-SYSC framework, which systematically investigates and compares four distinct scenarios in the estimation process for both series and parallel systems, and outlines the computational procedure for evaluating system failure probability using AK-SYSC. Section 4 presents a numerical test and two engineering applications to illustrate the improved efficiency and accuracy of the proposed method. Concluding remarks are given in Sect. 5.

2 An Overview of AK-SYS and AK-SYSi for System Reliability

AK-SYS

In order to efficiently estimate $P_f^{(s)}$ based on an adaptive Kriging, three strategies are proposed by AK-SYS method, i.e., the Component strategy, the Composite model strategy, and the Composite criterion strategy. As one of the most competitive approaches, the Composite criterion approach integrates the ideas of the two previous methods. Following the construction of initial Kriging models for all failure modes, the adaptive learning process is selectively directed toward those modes that contribute significantly to the system failure domain.

Among the three approaches, the Composite Criterion method has been demonstrated to be the most efficient, as validated by the case studies presented in Ref. [3].

However, several limitations have been identified, as discussed in Ref. [15]. AK-SYS is prone to misclassification during system reliability estimation, potentially compromising both computational efficiency and result accuracy. This problem often stems from determining the system state using an unconverged Kriging model, leading to inaccuracies in pinpointing the governing failure mode. To mitigate this drawback, AK-SYSi was introduced in Ref. [15] to enhance state recognition reliability.

AK-SYSi

From the viewpoint of the most readily identifiable failure mode, AK-SYSi based on an enhanced learning function was introduced.

Therefore, a novel learning function that reduces identification errors in AK-SYS method and incorporates system extreme value information should be developed to overcome deficiencies in both methods.

3 A Novel Learning Function in AK-SYSC

To address identification inaccuracies in AK-SYS and incorporate the extreme value information overlooked by AK-SYSi, a new learning function is put forward.

For the series system

Consider a system with multiple potential failure modes, where each performance function is approximated by a corresponding Kriging surrogate model. For a series system, failure occurs if any single mode fails, while the system is considered safe only when all modes remain within their safe operating limits. As a result, confirming system failure requires the correct identification of at least one failed mode, whereas confirming system safety requires accurate verification that all modes are safe. Based on the Kriging surrogate representation, the system failure domain indicator for a series system is defined accordingly.

Estimation of the system failure probability using Kriging models involves analysing four distinct surrogate-state scenarios, which reflect different combinations of predicted system states and their associated identification accuracy.

Case I.

When one or more Kriging models predict a failed state at a given sample point and this prediction is correct, the system can be conclusively classified as failed at that point. The learning function in this case directly targets the failed mode with the most critical contribution for refinement.

Case II.

When all Kriging models predict failure, but this prediction is incorrect, determining the true system state becomes challenging. In this situation, the minimum system response at the sample point is used to characterise the system limit state. The learning function is constructed based on this minimum value, and the mode associated with it is selected for refinement.

Case III.

When all Kriging models predict a safe state and this prediction is correct, the system can be confidently identified as safe at the sample point. The learning function in this case reflects the safe margin, and the mode most relevant to maintaining this margin is selected for updating.

Case IV.

When one or more Kriging models predict a safe state, but this identification is incorrect, it becomes difficult to determine whether the system is truly safe or has failed. Using the same learning strategy as in Case II may reproduce the misclassification issues observed in the original AK-SYS method. To avoid this, a dedicated learning function is introduced that explicitly accounts for extreme value behaviour and prevents false identification. The corresponding mode selected for refinement is the one most critical to resolving the ambiguity.

By integrating all four cases, a unified learning function for series systems is obtained, together with a consistent rule for selecting the failure mode to be updated. This strategy ensures accurate system state identification while effectively guiding adaptive refinement in multi-mode system reliability analysis.

For the parallel system

For a parallel system, structural safety is maintained provided that at least one mode remains safe. Conversely, system failure occurs only upon the failure of all constituent modes. Therefore, verifying system safety only requires the correct identification of at least one mode remaining in a safe state. Conversely, to assess whether the system has failed, it is necessary to accurately confirm that all modes are in a failed state. Accordingly, the failure domain indicator function of a parallel system is expressed via Kriging model.

Based on the novel learning function, the AK-SYSC method estimates system reliability through the following procedure.

Step 1.

Generate an initial candidate pool by sampling a predefined number of points from the joint probability density function of the input random variables.

Step 2.

Initiate a Design of Experiments (DoE) stage. For each performance function, select an initial subset of samples from the candidate pool. These samples are evaluated using the exact performance models to obtain the corresponding responses. The resulting datasets are used as the initial training sets for constructing the Kriging surrogate models.

Step 3.

Construct an initial Kriging surrogate model for each failure mode based on its corresponding training dataset.

Step 4.

Evaluate the learning function for all samples in the candidate pool using the current Kriging models. Based on the system type (series or parallel), compute the learning function values by appropriately incorporating extreme value information across all failure modes.

Step 5.

Assess the convergence of the current Kriging models with respect to the system failure probability. If convergence has not been achieved, proceed to the next step; otherwise, move directly to the final reliability estimation step.

Step 6.

Identify a new training point from the candidate pool by selecting the sample associated with the most critical learning function value.

Step 7. Determine the specific failure mode that requires refinement at the selected sample point.

Step 8. Update the training dataset of the identified failure mode by adding the new sample point and its exact model response. Reconstruct the corresponding Kriging model and return to the learning function evaluation step.

Step 9. Once convergence is achieved, estimate the system failure probability using the final Kriging models and the system failure indicator function.

4 Examples

The effectiveness of the proposed AK-SYSC method is demonstrated through three numerical case studies, with results benchmarked against direct Monte Carlo simulation (MCS) as the reference solution, as well as the AK-SYS and AK-SYSi methods. To ensure an impartial and consistent comparison, all methods are initialized using identical training samples and the same candidate pool. The candidate pool is generated via Latin hypercube sampling (LHS) [10, 11], which is employed for its low-discrepancy properties and its ability to enhance the robustness and statistical reliability of the comparative results. The first case study considers a roof truss structure subjected to external loading (Fig. 1). The outer load is treated as a random variable, while the internal loads are derived functions of this external load. System reliability analysis of the truss is conducted by considering two distinct failure modes, each associated with a different structural response mechanism. The random variables governing loads and material or geometric properties follow prescribed probability distributions, with their parameters summarised in Table 1. Given the series system configuration of the truss, system failure is defined as the occurrence of failure in any single component. Accordingly, the overall system failure probability is evaluated based on this series-system assumption, providing a representative testbed for assessing the accuracy and efficiency of the proposed AK-SYSC method in comparison with existing approaches.

The results estimated by four different methods are presented in Table 2. With 32 initial samples and 20 repetitions, the table indicates that the proposed method achieved the lowest average Ncall, demonstrating its superior efficiency among all four approaches.

Figure 2 shows the fault identification rate of the minimum mode throughout the updating processes. Figure 2 shows that AK-SYSC method has a lower false identification rate compared with AK-SYS method. Meanwhile, Figs. 3 and 4 demonstrate convergence plot of failure probability evaluated by AK-SYS and proposed AK-SYSC method, using 8 and 32 initial sample points, respectively. As shown in Fig. 3,

Table 1 Distribution information of the inputs in example 2

Inputs	Mean	Standard deviation	Distribution
Uniform load	20,000	1400	Normal
Length	12	0.12	Normal
Cross-sectional area	9.82×10^{-4}	5.982×10^{-5}	Normal
Cross-sectional area	0.04	0.0048	Normal
Tensile strength	3.35×10^{8}	4.02×10^{7}	Normal
Compressive strength	1.34×10^{7}	1.608×10^{6}	Normal

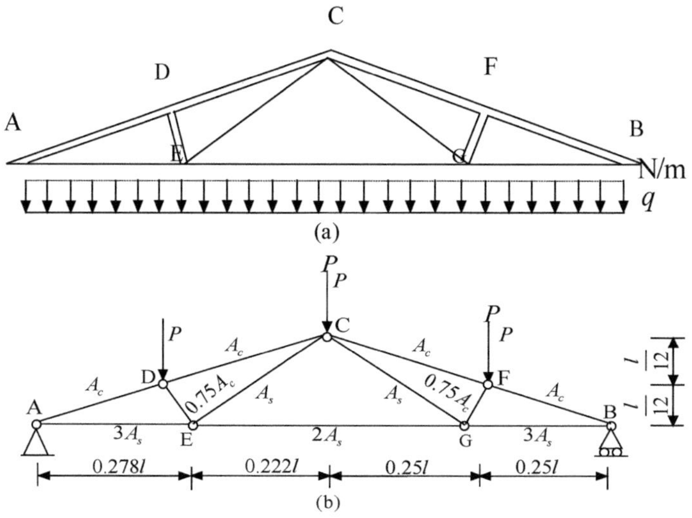

Fig. 1 A roof truss structure

Table 2 Results of example 2

	MCS	AK-SYS	AK-SYSi	AK-SYSC
Initial samples	–	32	32	32
Accurate replicates	20	20	20	20
g1	10^6	130	135.18	100.35
G2	10^6	89	76.22	64.20
Average Ncall	2×10^6	219	211.4	164.55
P	0.001623	0.001556	0.001602	0.001623
Std	4.7e–5	3.5e–4	2.2e–4	3.3e–4

when starting with only 8 initial points, the AK-SYS method fails to converge to the reference value even after incorporating over 200 additional samples. In contrast, the AK-SYSC method attains an accurate result with just 120 updates under the same initial conditions. Similarly, with 32 initial samples (Fig. 4), the proposed method converges after only 62 added points, whereas the AK-SYS method requires more than 80 updates to achieve comparable accuracy. Thus, Figs. 2, 3 and 4 illustrate that the AK-SYS method may cause false identifications in the estimating process, which may influence the estimating efficiency even the final accuracy of analysis result, and the main reason is that the identification for the system state is predicated by the non-convergent Kriging model, which causes the identification error of the extreme failure mode. While the proposed AK-SYSC method avoids the irrationality of the AK-SYS method that the non-convergent Kriging model may cause the false identifications of the extreme mode, and further leads to results in the lower estimating efficiency even the false prediction of the system failure probability.

Overall, the results demonstrate that proposed AK-SYSC method offers superior accuracy and computational efficiency for system failure probability estimation compared to other approaches.

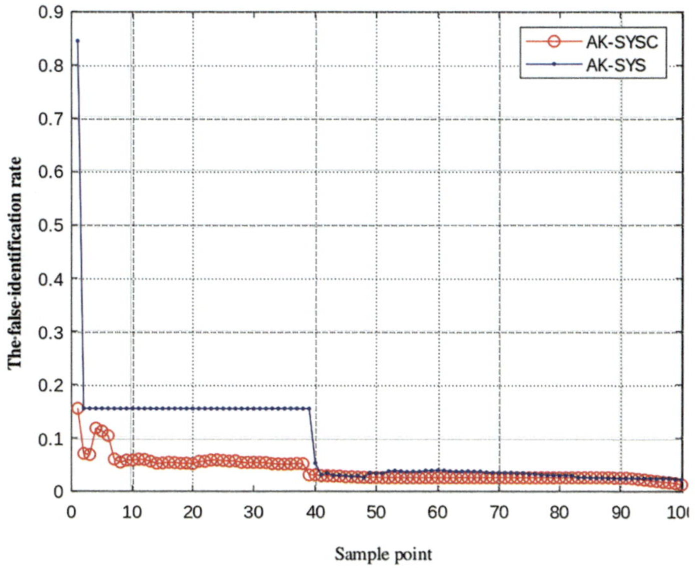

Fig. 2 The false identification rate with the updating process in example 2

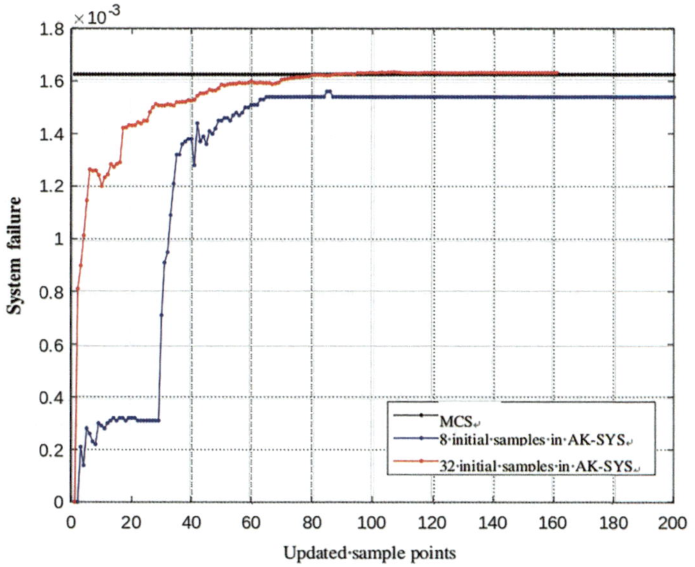

Fig. 3 The system failure probability obtained by AK-SYS with different initial samples

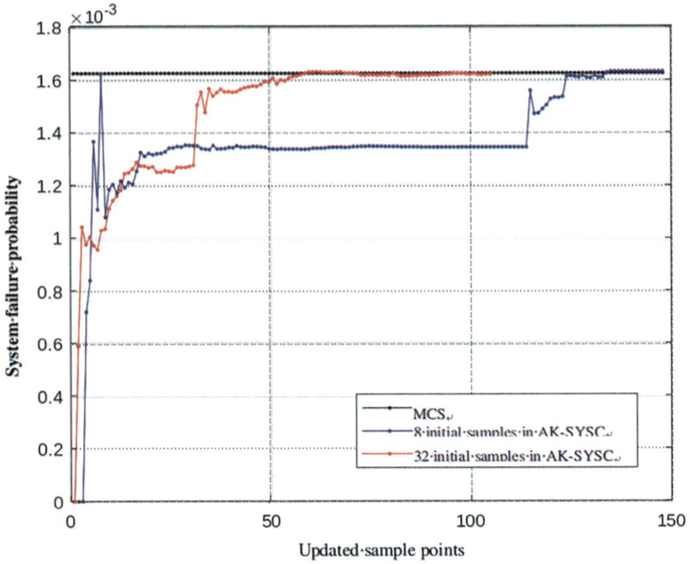

Fig. 4 The system failure probability obtained by AK-SYSC with different initial samples

5 Conclusion

To overcome the limitations of the AK-SYS and AK-SYSi methods, this study introduces the AK-SYSC approach for system reliability analysis, built upon a novel learning function, denoted as Uc. The AK-SYSC method is specifically designed to reduce the false identification rate observed in AK-SYS, while simultaneously making full use of the minimum or maximum value information that is largely neglected in AK-SYSi. In developing the proposed learning function Uc, four distinct scenarios associated with surrogate model predictions and their estimation accuracy are systematically analysed for both series and parallel system configurations. These scenarios are then integrated into a unified formulation of Uc. Importantly, the learning function remains effective irrespective of the convergence status of the current Kriging models, enabling reliable discrimination between failure and safe states. By explicitly exploiting extreme value information, Uc enhances both the accuracy and computational efficiency of system failure probability estimation.

AK-SYSC method mitigates the issue of inaccurate mode identification caused by non-convergent Kriging models, thereby preventing erroneous predictions of system failure probability and avoiding reductions in estimation efficiency. Moreover, by incorporating minimum or maximum value information, the identification speed of the system limit state surface is increased, which further improves the computational efficiency of system failure probability estimation.

An engineering application case is presented to validate the efficacy of proposed approach. Some main findings are outlined in the following:

First, a high false identification rate during the estimation process can adversely affect the accuracy and decrease estimation efficiency. The proposed AK-SYSC method significantly improves both accuracy and efficiency by reducing these misclassifications, refer to [8, 14].

Second, the use of minimum or maximum value information contributes to faster identification of critical regions, thereby enhancing the efficiency in evaluating failure probability. As a result, AK-SYSC method demonstrates superior performance compared to AK-SYSi method, which omits such information.

Finally, the proposed method operates independently of the current Kriging model's convergence state, thereby offering enhanced robustness and stability in system reliability analysis.

References

1. Bichon B, Mcfarlan J, Mahadevan S (2011) Efficient surrogate models for reliability analysis of systems with multiple failure modes. Reliab Eng Syst Saf 96:1386–1395
2. Echard B, Gayton N, Lemaire M (2011) AK-MCS: an active learning reliability method combining kriging and Monte Carlo simulation. Struct Saf 33(2):145–154
3. Fauriat W, Gayton N (2014) AK-SYS: an adaptation of AK-MCS method for system reliability. Reliab Eng Syst Saf 123:137–144
4. Feng KX, Lv ZZ, Wang L (2020) A novel dual-stage adaptive kriging method for profust reliability analysis. J Comput Phys 419:109701
5. Gong C, Zhou W (2018) Importance sampling-based system reliability analysis of corroding pipelines considering multiple failure modes. Reliab Eng Syst Saf 169:199–208
6. Guo J, Wang Z, Li H, Yang Y, Huang CG, Yazdi M, Kang HS (2024) A hybrid prognosis scheme for rolling bearings based on a novel health indicator and nonlinear wiener process. Reliab Eng Syst Saf 245:110014
7. Li DQ, Yang ZY, Cao ZJ, Au SK, Phoon KK (2017) System reliability analysis of slope stability using generalized subset simulation. Appl Math Model 46:650–664
8. Li H, Yazdi M (2022) Advanced decision-making methods and applications in system safety and reliability problems, Studies in systems, Decision and control, vol 211
9. Li H, Peng W, Adumene S, Yazdi M (2023) Advances in intelligent reliability and maintainability of energy infrastructure assets. In: Intelligent reliability and maintainability of energy infrastructure assets, pp 1–23
10. Olsson A, Sandberg G, Dahlblom O (2003) On Latin hypercube sampling for structural reliability analysis. Struct Saf 25(1):47–68
11. Owen AB (1994) Controlling correlations in Latin hypercube samples. J Am Stat Assoc 89(428):1517–1522
12. Rosenblueth E (1981) Two-point estimates in probability. Appl Math Model 5(5):329–335
13. Seo HS, Kwak BM (2002) Efficient statistical tolerance analysis for general distributions using three-point information. Int J Prod Res 40(4):931–944
14. Yazdi M, Mohammadpour J, Li H, Huang HZ, Zarei E, Pirbalouti RG, Adumene S (2023) Fault tree analysis improvements: a bibliometric analysis and literature review. Qual Reliab Eng Int 39(5):1639–1659
15. Yun WY, Lu ZZ, Zhou YY (2019) AK-SYSi: an improved adaptive kriging model for system reliability analysis with multiple failure modes by a refined U learning function. Struct Multidiscip Optim 59:263–278
16. Zhao YG, Ono T (2001) Moment method for structural reliability. Struct Saf 23:47–75
17. Zhao YG, Ono T (2004) On the problems of the fourth moment method. Struct Saf 26:343–347

18. Zheng PJ, Wang CM, Zong ZH (2017) A new learning method based on the learning function U of AK-MCS reliability analysis method. Eng Struct 148:185–194
19. Zhou JH, Nowak AS (1988) Integration formulas to evaluate functions of random variables. Struct Saf 5(4):267–284

Risk-Informed Intelligent Design Optimization for Nuclear Power Plants Using Probabilistic Safety Analysis

Ming Wang, Jiaoshen Xu, and Bing Zhang

Abstract This study presents a systematic framework for enhancing nuclear power plant safety through intelligent design optimization, guided by Probabilistic Safety Assessment (PSA) importance analysis results. Taking the Internal Event Level 1 PSA model of the Hua-long Pressurized Reactor (HPR1000) nuclear power plant as a reference, the methodology identifies high-priority components for intelligent system upgrades. A case analysis demonstrates the potential for significant risk reduction through the implementation of advanced systems, including prognostic health management for steam isolation valves in the Secondary Passive Heat Removal System (ASP), as well as the incorporation of smart substation technologies to address Loss of Off-site Power (LOOP) risks. These targeted upgrades correspond to critical components identified through PSA, resulting in quantifiable improvements in plant safety metrics. The proposed framework establishes a risk-informed decision matrix that supports the prioritization of intelligent design investments, offering nuclear operators quantitative guidance for maximizing safety benefits. The approach is validated through system modeling of the HPR1000, confirming its applicability for next-generation nuclear reactor safety enhancement.

Keywords Probabilistic Safety Assessment (PSA) · Intelligent design optimization nuclear power plant safety · Prognostic Health Management (PHM) · Risk-informed · Decision making · Loss of Off-site Power (LOOP) · Secondary Passive Heat Removal System (ASP)

M. Wang (✉) · J. Xu · B. Zhang
State Key Laboratory of Nuclear Power Safety Technology and Equipment, China Nuclear Power Engineering Co., Ltd, Shenzhen, Longgang, China
e-mail: wangming25436@163.com

J. Xu
e-mail: xjsh01@163.com

B. Zhang
e-mail: xiaohan1123@163.com

H. Li et al. (eds.), *Data-Driven Methods for Reliability and Safety Engineering: Applications in Industrial Systems*, Springer Series in Reliability Engineering,
https://doi.org/10.1007/978-3-032-22873-4_8

1 Introduction

With the application of new generation information technologies such as artificial intelligence (AI), the Internet of Things (IoT), and big data, intelligent design has become a core direction in the evolution of nuclear power technology. It is of crucial importance for enhancing the safety, economic efficiency, and operational effectiveness of nuclear power plants (NPPs). At present, the whole world is gradually promoting the intelligent transformation of NPPs, which can also help promote the safety and economic performance of NPPs. As an example.

Currently, the global community is actively advancing the intelligentization process of NPPs, which also serves as a key pathway to boost the safety and economic performance of NPPs. For example, the the US Nuclear Regulatory Commission formulated the Artificial Intelligence Strategic Plan for fiscal years 2023–2027 [2], In China, intelligent nuclear power has been incorporated into the "14th Five - Year Plan" for the energy sector [7]. Major domestic nuclear power construction enterprises attach great importance to the development of intelligent nuclear power and have established a diverse product system. Meanwhile, a number of intelligent systems have been gradually implemented in NPPs, such as predictive maintenance platforms integrating fault diagnosis and health management (PHM), intelligent operation procedure controls. Moreover, the Smart Nuclear Power (SNP) led by entities such as China General Nuclear Power Corporation (CGN) has significantly elevated the intelligent level of nuclear power plants through extensive digital and intelligent system integration.

Despite these advancements, the adoption of intelligent design in nuclear facilities remains constrained by several challenges. Newly developed digital technologies must meet the rigorous standards imposed by the nuclear industry, with some applications requiring tailored development for site-specific requirements. In contrast to traditional industrial sectors, the nuclear industry operates within a more limited market scale, compounded by high research and development costs, thereby necessitating careful cost-benefit analyses prior to implementation. Additionally, extensive intelligent retrofits of structures, systems, and components (SSCs) may result in considerable cost escalation.

Probabilistic Safety Analysis (PSA) is the basic way of assessing the safety of a nuclear power plant, thus it can also help generate some risk-informed design optimization approaches. By finding the major contributors of the risk of a nuclear power plant via PSA, the intelligent design solutions for enhancing overall plant safety can then be devised accordingly. This paper offers a systematic way to generate intelligent design proposals using PSA, with the Internal Events Level 1 PSA model of the HPR1000 reactor as the analytical reference case. The resulting recommendations offer practical guidance for advancing intelligent design within nuclear power plant systems.

2 Design Concept Framework

As an important fundamental technological element of PSA, importance analysis plays a vital role in multiple key aspects of safety management. Specifically, it supports the detection of safety vulnerabilities, the identification of influential factors behind severe accident scenarios, and the development of plant-specific risk reduction plans, all of which are derived from the results of PSA importance analysis. Furthermore, based on the analytical outcomes of PSA importance analysis, a dedicated framework is designed to generate intelligent design recommendations, with the specific goal of lowering the risk values associated with the engineering aspects of a particular plant. The detailed implementation steps are illustrated in Fig. 1.

Selection of the Baseline PSA Model

Selecting an appropriate baseline PSA model exerts a critical influence on the reliability of importance analysis results. Given the direct impact of model selection on analysis validity, two criteria govern baseline PSA model choice:

The baseline risk model should feature high accuracy, preferably secure regulatory endorsement (e.g., from national nuclear safety authorities) or undergo rigorous peer review validation. Typical examples encompass licensing—compliant PSA models deployed in nuclear power plant certification processes, or operational PSA models utilized for Configuration Risk Management (CRM, i.e., managing risks associated with plant configuration changes).

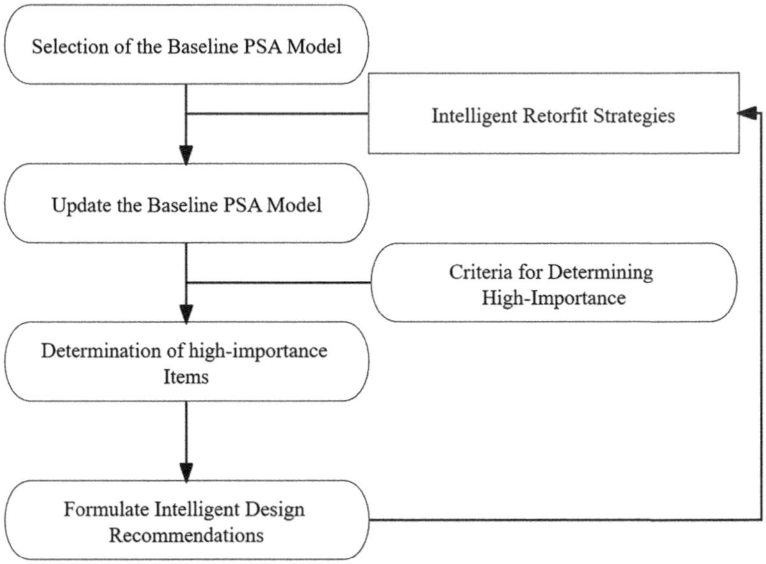

Fig. 1 Risk-informed intelligent design optimization framework

The design characteristics of the PSA model must align with the technical characteristics of nuclear power plants implementing intelligent designs, particularly regarding nuclear power generation technical specifications (e.g., reactor core design, thermal—hydraulic parameter thresholds).

Update the Baseline PSA Model

For intelligent design features explicitly considered in nuclear power plants, their potential impacts on PSA model results should be incorporated into the baseline risk model when analysis indicates safety implications. Taking the "Automation of Accident Operation Procedures" as an example, this intelligent design enables automatic system diagnostics for three accident types: loss of support function accidents, fire accidents, and thermo-hydraulic parameter abnormality accidents. The automated diagnostic function significantly reduces operator workload during accident conditions, necessitating dedicated analysis of its impact on Human Reliability Analysis (HRA) parameters in the baseline risk model.

Criteria for Determining High-Importance

In risk-informed design processes, the identification of high-importance items and elements typically relies on two metrics: Fussell-Vesely (FV) importance and Risk Achievement Worth (RAW) [10, 11]. Their specific definitions are as follows:

FV importance: Represents the proportion of the total frequency of all minimum cut sets (MCS) containing a specific basic event to the total quantitative risk value of the nuclear power plant (e.g., Core Damage Frequency, CDF). A higher value indicates a more significant impact of random failures of the basic event on safety.

where, MCS_i, Minimum cut sets containing basic event i; Q, Total quantitative risk value of the nuclear power plant; Q_b, Benchmark quantitative risk value of the nuclear power plant.

Risk Achievement Worth (RAW): Measures the multiple by which the total quantitative risk value increases when a basic event is confirmed to occur (i.e., its unavailability is set to 1),A higher RAW value signifies a stronger impact of random failures of the basic event on safety.where, $P_i = 1$, Unavailability of basic event i is set to 1.

Referring to the screening criteria commonly used in risk-informed technical applications, an item shall be classified as a high safety importance level if the FV/RAW factors from the importance analysis in the PSA model meet any of the following conditions:

When the FV importance of a single item is greater than 0.005, the sum of the FV importance of all its basic events (including common cause failures (CCFs)) shall be taken;

When the RAW importance of a single item is greater than 2.0, the maximum RAW of the basic events shall be taken.

The above screening criteria (FV > 0.005 and RAW > 2.0) are derived from the American industry technical standard ASME/ANS RA-Sa-2013 [1], which explicitly designates these two criteria as the core basis for determining risk significance to support the evaluation of Probabilistic Safety Assessment (PSA) model results. In addition, the Nuclear Energy Institute (NEI) Document 00–04 [9] also adopts these criteria to identify components with high safety significance, thereby guiding the safety classification of Structures, Systems and Components (SSC), which stands as a typical industry practice for the application of such criteria.

It should be noted that the above screening criteria are not applicable to the screening of high importance initiating events. The frequency of initiating events may be greater than 1, in which case the RAW of the initiating event is not accurate. Therefore, as a supplement to the above screening criteria, based on engineering experience (in the industrial practice of PSA, it is generally considered that if the failure risk of a specific accident scenario accounts for less than 1% of the total risk quantification result, the scenario is deemed to be at a relatively low risk level), in this study, the FV importance of initiating events greater than 0.01 is used as the screening criterion for high importance initiating events. At the same time, if different initiating events represent the same equipment, their FV values need to be merged before applying the criteria.

Determination of high-importance Items

Quantitative calculations are performed on the selected model to derive the importance analysis results (FV and RAW) for initiating events, equipment failure events, and human failure events (HFEs). A rationality analysis of these results is required to identify uncertainties arising from modeling assumptions and other factors, thereby determining the items/events that truly make higher contributions to risk. Based on these analysis results and the criteria specified above, a list of high-importance initiating events, equipment, and human error events can be identified.

Formulate Intelligent Design Recommendations

Among the various intelligent design schemes for nuclear power plants, their design objectives are characterized by diversity: some schemes take enhancing safety as the core requirement, while others are developed from the perspectives of improving operational and maintenance convenience or optimizing the economic efficiency of nuclear power plants. Therefore, after identifying high-importance items/events, it is necessary to systematically screen the intelligent design schemes and evaluate their potential for safety gain from the following dimensions:

Whether they can reduce the frequency of key initiating events or eliminate their root causes;

Whether they can lower the failure probability of important equipment or establish failure prevention mechanisms;

Whether human actions engineering optimization measures can be applied to reduce the likelihood of significant HFEs.

By establishing a mapping relationship between "high-importance factors and Intelligent Retrofit Strategies", targeted suggestions can be formulated: priority should be given to intelligent design schemes that are directly related to risk reduction of high-importance items/events, and among such schemes, key implementation objects should be screened in the order of importance.

3 Case Analysis

Case Overview

In this study, the importance analysis results derived from the Internal Events Level 1 PSA model for a typical HPR1000 nuclear power plant under CGN are selected as the analysis case [8]. This PSA model is developed in accordance with the requirements of industry PSA technical standards, covering event scopes such as power operation, low-power operation, and shutdown conditions. The equipment reliability data involved in the model mainly refers to the China Equipment Reliability Data Report [6], while the data on CCFs of equipment is referenced from INL/EXT-21-62940 [3]. Furthermore, the applicability of relevant data has undergone rigorous review by domestic nuclear regulatory authorities, which guarantees the authority and reliability of the data sources.

Outcomes of Importance Analysis

Given the substantial number of high-importance items identified using the criteria outlined above, the case study focuses on the top 5 component failure events, top 5 HFEs (by importance contribution), and initiating events with an FV exceeding 0.1 for streamlined analysis. Key findings are summarized as shown in Tables 1, 2 and 3.

Suggestions for Intelligent Design

This study was conducted at the initial stage of the CGN SNP Project. Its core objective was to provide decision support for the project team in key aspects, such as the selection of intelligent design schemes and resource allocation, from the perspective of safety improvement benefits. Based on this, combined with the high-importance items identified previously, this study puts forward the following recommendations to offer directional technical guidance for the project team. These recommendations are mainly categorized into two types:

1. High-importance Items with Matching Mature Intelligent Design Schemes

The previously identified high-importance items should be prioritized during the selection of application objects for the schemes. Typical application scenarios are detailed as follows:

For high-importance equipment, including the ASP3110VV/ASP1110VV/ASP2110VV steam isolation valves and the LHA1101TB switchboard, was incorporated into the scope of PHM related schemes. Real-time operational data (e.g., pressure, temperature, and valve position feedback) were collected to evaluate performance degradation trends; machine learning algorithms were leveraged to predict potential faults, and proactive maintenance strategies were formulated in advance through early warning mechanisms. Implementation of these measures can effectively enhance the operational reliability of the target equipment, reduce the probability of transient initiating events triggered by nuclear power unit trip or shutdown due to equipment failures, prevent the recurrence of similar operational incidents,

Table 1 High-importance component failure events analyzed by FV and RAW metrics

Ranking	Event code	Description	Risk implication
FV #1, RAW #4	RGL_4_STICK	Stuck 4 or more control rods	In various accident scenarios, the reactor shutdown function is a core safety barrier. When 4 or more control rods get stuck, this critical shutdown function is completely disabled. This direct failure can trigger two major hazardous outcomes: Core damage (CD) or anticipated transient without scram (ATWS). Even in the case of ATWS, there remains a definite probability of eventual CD, making this event the top risk factor affecting the safety of the nuclear power unit.
FV #2	KDS_SIG_FAIL	Failure of signal acquisition and distribution cabinet (KDS)	The KDS is responsible for integrating and transmitting key operational signals of the nuclear plant, covering critical parameters such as steam generator (SG) pressure, SG water level, and pressurizer pressure. Once the KDS malfunctions, the normal collection and transmission of these signals will be interrupted, which directly leads to the failure of the associated safety injection system to activate automatically. This significantly impairs the plant's ability to respond to accidents in a timely and effective manner, posing a severe threat to the stable and safe operation of the entire unit.
FV #3	EDG_CCF_OP	Common cause operation failure of emergency diesel generators (EDGs)	After a loss of offsite power (LOOP) event, EDGs and station blackout (SBO) diesel generators are the core backup power sources for accident mitigation systems. If these generators suffer from common cause operation failure, all mitigation systems relying on their power supply will be unable to start. Since the PSA model does not include mobile diesel engines as alternative power sources, conservative safety assumptions stipulate that this failure will directly result in CD, significantly escalating the plant's accident risk.

(continued)

Table 1 (continued)

Ranking	Event code	Description	Risk implication
FV #4	RPC_4_CCF	CCF of 4 reactor protection cabinets (RPC)	The CPU units inside the RPC are mainly responsible for issuing reactor shutdown protection signals and activating engineered safety features (ESFs). When more than 3 CPU units experience CCF, the RPC loses the ability to send normal shutdown signals, and ESFs also fail to start as required. This directly hinders the initiation of post-accident mitigation functions, thereby increasing the overall safety risk of the nuclear power unit.
FV #5	PGR_FAIL	Failure of power grid restoration	In case of LOOP failure, bringing back the external power grid is crucial in supplying power to both repair devices and equipment as well as the accident mitigation systems and equipment. If there is no timely recovery of the grid, then these systems that depend on the grid would be kept out of service for a long time, which extends the duration of the accident response window and increases the difficulty of controlling the accident. Indirectly speaking, this also leads to a greater risk of a core damage or other major accident consequences.
RAW #1	ASP3110VV_ EL	Steam inlet motor isolation valve ASP3110VV external leakage	The inlet motor isolation valve is a key part of the secondary passive heat removal system (ASP). Its failure through external leaks could therefore cause the lack of usable cooling capacity of the ASP. In this case, when the secondary circuit cooling system fails, the ASP would be ineffective in cooling the primary circuit, leading to an increased chance that the core overheats, resulting in a hypothetical severe event such as CD.
RAW #2	ASP1110VV_ EL	Steam inlet motor isolation valve ASP1110VV external leakage	
RAW #3	ASP2110VV_ EL	Steam inlet motor isolation valve ASP2110VV external leakage	
RAW #5	LHA1101TB_ OPFAIL	Failure of distribution panel LHA1101TB operation	Following an LOOP event, the LHA1101TB distribution panel will deliver electrical support to key accident mitigation devices during accidents; and if it does not function properly, then all the accident mitigation devices that run on this panel cannot be used. According to the assumptions of the PSA model, a failure leading to CD directly results from this one and constitutes a risk element affecting the plant's post-accident recovery.

Table 2 High-importance HFEs analyzed by FV and RAW metrics

Ranking	Event code	Description	Risk implication
FV #1, RAW #1	OP_LHSI_HC1_FAIL	Failure of simultaneous injection operation in cold leg and hot leg of low pressure safety injection system	During the long-term phase of large/medium loss of coolant accidents (LOCA), operators must perform simultaneous injection of the cold leg and hot leg using the low-pressure safety injection system. This operation is critical to prevent boron crystallization on fuel elements, which would otherwise cause local high temperatures and damage the fuel. This HFE is incorporated into multiple LOCA event trees, and since LOCA initiating events have a high frequency, this error significantly increases the risk of fuel damage and core overheating.
FV #2, RAW #5	OP_ASG1_LINK_FAIL	Failed to manually connect auxiliary Feedwater system (ASG) tanks in other columns (columns 2 and 3)	The ASG system is responsible for replenishing water to the steam generator (SG) to maintain its normal heat exchange function. When ASG column 1 fails, the inability of operators to manually connect ASG tanks in columns 2 and 3 directly reduces the water supply reliability of the ASG system. This failure is included in the fault trees of multiple functional events, and nearly 30 event trees specifically model the water replenishment function of the ASG. As a result, this human error severely impairs the plant's ability to maintain SG water level, increasing the risk of SG dry-out and subsequent core damage.
FV #3	OP_ASG2_LINK_FAIL	Failed to manually connect auxiliary ASG tanks in other columns (columns 1 and 3)	Similar to OP_ASG1_LINK_FAIL, once the ASG column 2 malfunctions, then manual connections between ASG tanks in columns 1 and 3 are prevented, which leads to reduction in the ASG water supply capacity. This, subsequently, leads to a less redundant and reliable ASG water replenishment function. During an accident where SG water replenishment is required, this error could lead to an insufficient SG water level, thus lowering core cooling capability and increasing the accident severity.

(continued)

Table 2 (continued)

Ranking	Event code	Description	Risk implication
FV #4	OP_SBO_DG_START_FAIL	Failed to manually start up station blackout (SBO) diesel generators	After a LOOP event, if emergency diesel generators (EDGs) also fail, SBO diesel generators become the last resort power source for accident mitigation systems. Operators' failure to manually start SBO diesel generators means all mitigation systems relying on diesel power cannot be activated. Since the PSA model does not include mobile diesel engines as backups, this error directly leads to CD, making it a critical risk factor for triggering severe accidents.
FV #5, RAW #3	OP_RHR_S_RESTART_FAIL	Failed to manually re-start up residual heat removal (RHR) pump	During the low-power operation phase, if a small LOCA occurs and the residual heat removal mode needs to be restored but fails, manually restarting the RHR pump is essential to remove residual heat from the core. Operators' failure to complete this restart operation results in the inability to effectively dissipate core residual heat, leading to a continuous rise in core temperature and a high probability of CD.
RAW #2	OP_ISO_DIL_FAIL	Failed to manually isolate dilution source or implement corrective action in case of successfully carry out extra borating system (RBS)	During boron dilution events, the automatic isolation function of the false dilution source could fail; manually isolating the dilution source was deemed to be an essential fallback measure to avoid continuous cooling of the primary circuit; if operators did not isolate the false source or take any remedial actions for an isolated dilution source, then the primary circuit would continue to be diluted, pushing the reactor into supercritical operation. Such an event poses serious risk of the core losing control of reactivity
RAW #4	OP_IRWST_S_EHR_FAIL	Failed to manually switch to containment heat removal system (EHR)	The in-containment refueling water storage (IRWST) is utilizing different kinds of cooling methods, and when necessary, it is cooled by the EHR system as a backup. If any of the methods of cooling the IRWST fail, the operators have to switch manually to the EHR system to keep the cooling capacity of the IRWST; if such switching is not completed on time, then the cooling capacity of the IRWST is lost, which may result in the IRWST heating up or ceasing to function, which can lead to failure of containment integrity and also cause over-pressurization of containment.

Table 3 Initial events with FV > 0.1 and their associated risks

Ranking	Event code	Description	Key equipment involved	Risk implication
1	LOCA	Large, medium and small breaks in primary circuit	Pipeline within the pressure boundary of primary circuit	LOCA has two prominent risk characteristics: First, the frequency of LOCA initiating events is relatively high compared to other severe accident initiators. Second, the probability of accident mitigation measures failing during LOCA is also high. Large and medium breaks in the primary circuit can quickly lead to massive coolant loss, causing rapid core overheating. Although small breaks result in slower coolant loss, they may still lead to core damage if mitigation systems (such as safety injection systems) fail to operate effectively. Additionally, LOCA involves complex thermal-hydraulic processes, which increase the uncertainty of safety system operation and further exacerbate accident risks.
2	LOOP	Loss of offsite power	Power grid, main transformer voltage transformer, auxiliary transformer, overhead lines, etc.	LOOP is a high-risk initiating event with obvious chain reaction characteristics. Once off-site power is lost, most safety-related systems in the nuclear plant that depend on external power—Including core cooling systems, containment ventilation systems, and monitoring systems—Are immediately affected. Although the plant is equipped with backup power sources like EDGs, if these backup systems also fail (e.g., due to CCF), it will lead to a complete loss of power (SBO). This makes it impossible to activate accident mitigation systems, significantly reducing the plant's ability to control accidents. Any subsequent secondary accidents will be more difficult to mitigate.

and mitigate the potential risk increment arising from the performance degradation of safety mitigation systems. From the perspective of contributions to reducing transient event frequency, operational experience data of all in-service nuclear power plants owned by CGN during 2016–2023 were analyzed in this study (with a total of 180.4 reactor-years). This dataset served as the statistical basis for calculating the frequency of transient initiating events in the baseline PSA model, where the frequency was derived by dividing the number of historically recorded transient events by the total statistical reactor-years. The results indicated that the aforementioned intelligent design schemes could reduce at least 7 relevant incidents within the statistical scope. If such incidents were eliminated, the frequency of transient initiating events would decrease by approximately 3.88×10^{-2} per reactor-year, leading to a reduction in the CDF by about 1.37×10^{-9} per reactor-year, corresponding to a reduction rate of roughly 1.75%.

For high-importance HFEs, should be be prioritized in the he design of automation schemes for accident operating procedures. Group control technology and automatic diagnostic logic were adopted. This scheme is expected to significantly reduce the diagnostic workload of operators under accident conditions, shorten the time required for diagnosis and operation, improve the execution reliability of operator actions, and thereby exert a positive effect on the safety level of nuclear power units. A preliminary estimation was conducted using OP_LHSI_HC1_FAIL (Failure of simultaneous injection operation in the cold leg and hot leg of the low-pressure safety injection system) as an example: this scheme is projected to reduce the event diagnosis time by approximately 20%. However, given the long time window of this HFE, its impact on Human Error Probability (HEP) is minimal, and the resultant influence on the CDF is negligible. It should be noted that such automation schemes are not designed for a single HFE; expanding their application scope will yield a more significant effect on CDF reduction.

2. High-importance Risk Items without Matching Mature Intelligent Design Schemes

For high-importance risk items lacking mature intelligent solutions at present (e.g., the CCF of EDGs), control rod sticking, etc.), continuous tracking and research are recommended. Once feasible intelligent design schemes become available, the project team may reconsider these high-importance items in the context of technical decision-making.

4 Conclusion

This paper studies and applies a kind of methodology based on risk informed design optimization, which can provide helpful suggestions for the intelligent design of the nuclear power plant. The method provides accurate identification of risk significant factors, supports designers to formulate strategies in the most suitable way and define reasonable analytical limits; in the intelligent design objectives selection stage, safety contribution analysis can be adopted as a complementary index of early screening; but meanwhile the practical engineering usage need consideration of the other factors such as technical feasibilities, economics, engineer constrains and so forth together with failures causes and historic operating conditions.

This study exemplifies the practical application of risk-informed methodologies in the design optimization of nuclear power plants. The approach developed enables accurate identification of risk-significant factors, thereby supporting intelligent design teams in formulating targeted design strategies and defining appropriate analytical boundaries. For the selection of intelligent design objectives, safety contribution analysis serves as a supplementary criterion for preliminary screening. However, in practical engineering applications, it is essential to incorporate multidimensional information, including component failure mechanisms and historical operational data, while also considering technical feasibility, economic implications, and engineering constraints to develop feasible and implementable design solutions.

Although the proposed method may identify a considerable number of high-importance components, limitations in available resources and technical capacity necessitate prioritization. It is advisable to perform cost-benefit and technical priority analyses to select key components for further investigation, employing strict screening thresholds to optimize resource allocation. It is important to note that the outcomes of importance analysis are highly sensitive to model completeness and data uncertainty. Conventional PSA models often lack the capability to simulate the dynamic behavior of intelligent systems or to account for emerging failure modes such as software failures and cyberattacks. To address this, sensitivity analyses should be incorporated into the evaluation process, and alternative techniques such as Monte Carlo simulation may be utilized to validate the robustness and applicability of critical assumptions and parameters [3-5].

The high-importance screening thresholds proposed herein are applicable exclusively to internal event PSA models. Given the distinct characteristics of external event risk models, FV/RAW metrics may have limited relevance in such contexts. For instance, the impact of additional electrical equipment introduced through intelligent design may influence fire risk via changes in ignition frequencies at the compartment level, rather than through direct FV/RAW analysis. Consequently, it is recommended to develop context-specific screening criteria by integrating intelligent design modifications with the technical parameters of external event PSA.

As directed by PSA, the intelligent design process for nuclear power plants is naturally iterative. Following the identification and application of intelligent design improvements, they should be incorporated back into the PSA model. After this update, further quantifications can be carried out and provide new emerging risk-significant items that would help guide additional design improvements.

References

1. American Society of Mechanical Engineers & American Nuclear Society (2013) Addenda to ASME/ANS RA-S-2008: standard for level 1/large early release frequency probabilistic risk assessment for nuclear power plant applications, ASME/ANS RA-Sa-2013. American Society of Mechanical Engineers, New York, NY, USA
2. Dennis M (2022) Artificial intelligence strategic plan, fiscal years 2023–2027: draft report for comment. United States Nuclear Regulatory Commission, Washington, DC, USA
3. Guo J, Wang Z, Li H, Yang Y, Huang CG, Yazdi M, Kang HS (2024) A hybrid prognosis scheme for rolling bearings based on a novel health indicator and nonlinear wiener process. Reliab Eng Syst Saf 245:110014
4. Li H, Yazdi M, Huang HZ, Huang CG, Peng W, Nedjati A, Adesina KA (2023) A fuzzy rough copula Bayesian network model for solving complex hospital service quality assessment. Complex Intell Syst 9(5):5527–5553
5. Li H, Peng W, Adumene S, Yazdi M (2023) An intelligent cost-based consequence model for offshore Systems in Harsh Environments. In: Intelligent reliability and maintainability of energy infrastructure assets. Springer Nature Switzerland, Cham, pp 107–117
6. Ma Z, Kvarfordt KJ (2021) 'CCF parameter estimations, 2020 update', INL/EXT-21-62940. Idaho National Laboratory, Idaho Falls, ID, USA
7. National Development and Reform Commission, National Energy Administration (2022) The 14th Five-Year Plan for a Modern Energy System [S]. China Market Press, Beijing, China
8. National Nuclear Safety Administration (2023) Report on equipment reliability of Chinese nuclear power plants. General Office of Ministry of Ecology and Environment, Beijing, China
9. Nuclear energy institute (2005) 10 CFR 50.69 SSC categorization guideline, NEI 00–04, rev. 0. Nuclear Energy Institute, Washington, DC, USA
10. Wang M, Zhang B, Xiao L et al (2024) PSA-based intelligent design suggestions for nuclear power plant. In: Proceedings of 17th international conference probabilistic safety assessment and management & asian symposium on risk assessment and management (PSAM17 & ASRAM 2024), Sendai, Japan, pp 7–11
11. Zhang J, Lu Y (2023) Application of intelligent innovation in accident procedure of nuclear power plant. In: book application of intelligent innovation in accident procedure of nuclear power plant, 1st edn. Springer, Cham, Switzerland, pp 253–262

Research on Fault Data-Driven Quality Risk Assessment Method for Aviation Equipment

Jiake Li and Yiyi Zhao

Abstract As a fundamental component of national security and military capability, the operational quality of aviation equipment exerts a direct influence on flight safety, mission success, and the service life of the equipment. With ongoing advancements in aviation technology and increasingly rigorous user demands regarding operational performance, the accurate evaluation of aviation product quality has become a critical aspect of comprehensive quality management. In response to these evolving requirements, this study analyzes failure and application data generated during the operation of aviation equipment to quantitatively assess current operational quality and associated risks. An optimized Risk Priority Number (RPN) algorithm is employed to establish a systematic quality risk assessment methodology. This approach facilitates both manufacturers and end-users in evaluating operational quality by providing objective performance insights. It enhances manufacturers' initiatives for quality improvement while offering users a data-driven evaluation framework. In doing so, the methodology reduces the subjectivity inherent in conventional assessment techniques and improves the effectiveness of operational data utilization for advancing aviation product quality.

Keywords Fault data-driven · Aviation equipment operational quality · RPN optimization · Three-dimensional risk assessment · Mission-cycle characterization

1 Current Status of Operational Quality Risk Assessment

Traditional approaches to operational quality risk assessment for aviation equipment predominantly rely on probabilistic statistics and expert judgment, exhibiting a strong emphasis on qualitative analysis. Representative methods include Fault Tree Analysis (FTA), which constructs fault trees to investigate system failure causality and compute associated probabilities; Failure Modes and Effects Analysis (FMEA) [1],

J. Li · Y. Zhao (✉)
AVIC China Aero-Polytechnology Establishment (CAPE), Beijing, China
e-mail: 715836150@qq.com

H. Li et al. (eds.), *Data-Driven Methods for Reliability and Safety Engineering: Applications in Industrial Systems*, Springer Series in Reliability Engineering,
https://doi.org/10.1007/978-3-032-22873-4_9

which identifies potential failure modes and evaluates their risk based on severity, occurrence probability, and detectability; and the Analytic Hierarchy Process (AHP) [2], which applies hierarchical modeling and expert scoring to assign weights to risk factors. Although these techniques are widely utilized, they demonstrate limited capacity to leverage real-time operational data and lack adaptability to dynamic and evolving environments.

With the advancement of artificial intelligence, big data, and the Internet of Things (IoT), newer methodologies have emerged. These include machine learning algorithms, support vector machines, deep neural networks, and Bayesian networks, which offer the capability to extract patterns from unstructured datasets. These data-driven methods are more applicable for evaluating individual products under complex operational conditions. Drawing on studies related to quality Key Performance Indicators (KPIs), delivery KPIs, and planning or management KPIs proposed by the International Aerospace Quality Group (IAQG) [3], this section focuses on the interpretation and significance of operational quality metrics for aviation products. However, due to the inherent complexity of aviation systems, such methods often struggle to capture comprehensive features necessary for holistic and integrated quality assessments.

Consequently, this study integrates operational data with systemic equipment evaluation, incorporating in-service failures, design-manufacturing requirements, and post-failure improvements. By refining the RPN algorithm [7], a unified operational-end data-driven risk assessment methodology is established that fully leverages real-world data, dynamically adapts to evolving requirements, and enables quantitative risk analysis for complex aviation systems and their components.

2 Research on Characterization Methods for Aviation Equipment Operational Quality

2.1 Definition and Connotation of Operational-End Quality Data

Current classifications of data exhibit variability; this study adopts a categorization approach based on the direction of data circulation across seven distinct lifecycle phases: design, manufacturing, testing, delivery, operation, maintenance, and retirement. Within this framework, two primary data sources are defined. The first is the development-end, which encompasses the design, manufacturing, testing, and delivery stages. The second is the operational end, which includes data generated during the operation, maintenance, and retirement phases. The various types of data associated with these sources are illustrated in Fig. 1.

Based on current equipment quality definitions and evolving quality requirements, development-end conformance quality no longer fully meets contemporary operational quality standards. Instead, operational quality must be evaluated using

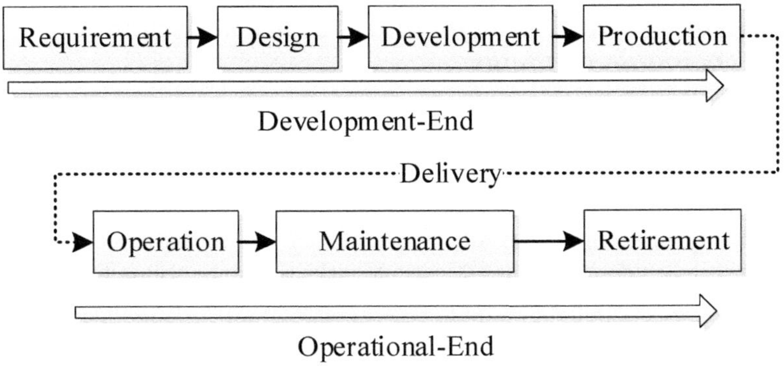

Fig. 1 Data circulation direction

operational-end data (including in-service performance metrics and failure records), with findings fed back to development-end stakeholders to enhance manufacturers' quality assurance capabilities. Thus, operational-end data forms, as depicted in Fig. 2, establish a closed-loop feedback mechanism to drive continuous quality improvement.

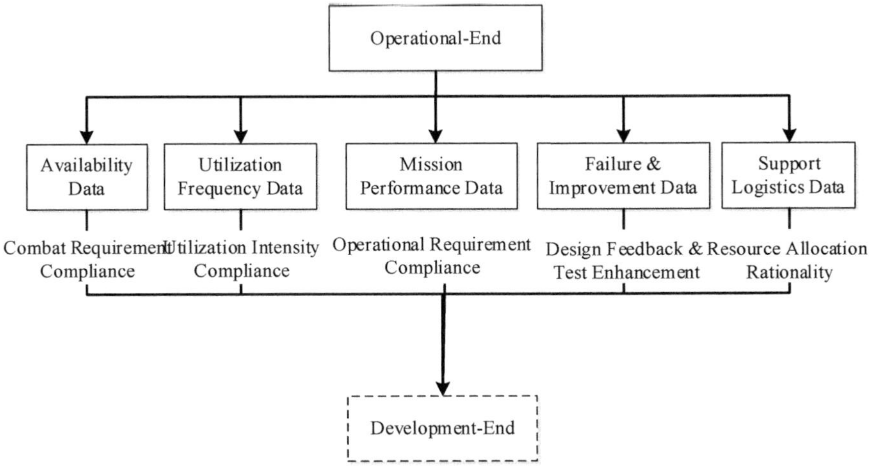

Fig. 2 From operational-end to development-end data transmission

2.2 Mission-Cycle-Based Quantitative Characterization of Operational Quality

Taking combat aircraft mission cycles as an exemplar, an equipment operational task cycle model [4] is established. Beginning with ground inspection, fixed-wing aircraft proceed to taxi and takeoff roll phases. Upon completion of engine warm-up, the wheels lift off to initiate takeoff. In accordance with mission requirements, the aircraft must reach combat zones to execute tasks before returning to base, thereby completing a single-cycle sortie. If mission demands recur, it enters sortie turnaround preparation for repeated takeoff, task execution, and landing (Refer to Fig. 3).

The integration of the mission-cycle model with operational-end data achieves comprehensive coverage across all mission phases, forming a task-based operational data generation framework that delineates specific data correspondences: ground inspection maps to availability data, failure/improvement data, and support logistics data; taxi and takeoff roll associate with availability data and failure/improvement data; in-flight task execution corresponds to mission performance data and failure/improvement data; landing preparation aligns solely with failure/improvement data;

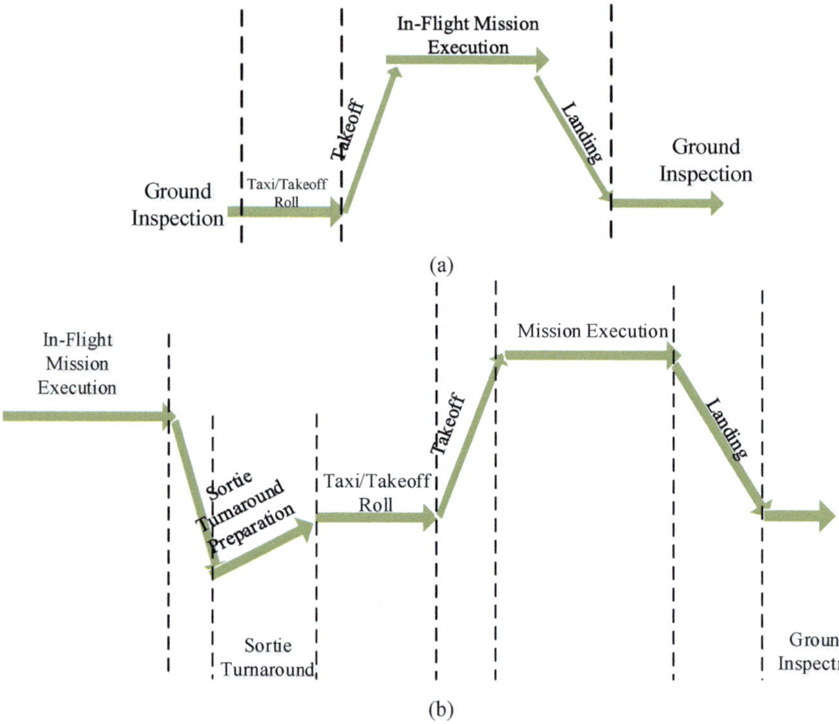

Fig. 3 Aircraft mission execution process **a** single-flight mission execution process, **b** re-sortie mission execution process

while sortie turnaround preparation incorporates utilization frequency data, failure/ improvement data, and support logistics data.

Table 1 reveals that fault and improvement data constitute the most critical elements for assessing and characterizing aviation equipment operational quality. With fault data as the primary driver and the other four data categories as supplementary inputs, metrics are employed to characterize mission-cycle states across all phases. These metrics, integrated with their computational dimensions, are fed back to the development end, establishing a fault-data-driven quantitative characterization system for operational quality. This system leverages auxiliary operational data to achieve full mission-cycle coverage, with operational-end data flows illustrated in Fig. 4.

Quantitative characterization dimensions achieve full mission-cycle coverage by correlating with operational data across all phases, enabling assessment of task executability, completability, continuous execution, and effective fulfillment based on defined metric requirements, thereby validating the completeness and rationality of assessment dimensions and indicators. Within this framework, improvement timeliness is quantified through recurrent failure occurrences and mean improvement cycles, culminating in the characterization indicator system depicted in the Fig. 5.

The Aviation Equipment Operational Quality Index synthesizes multiple coupled factors, constructing a five-dimensional characterization framework comprising eight metrics through the integration of operational quality dimensions and mission-cycle parameters. Computational methods and value ranges for each metric are specified in the Table 2.

Table 1 Data output utilization by flight phase

	Ground inspection	Taxi/takeoff roll	In-flight mission execution	Landing preparation	Sortie turnaround preparation
Mission performance data			●		
Availability data	●	●			
Utilization frequency data					●
Failure and improvement data	●	●	●	●	●
Support logistics data	●				●

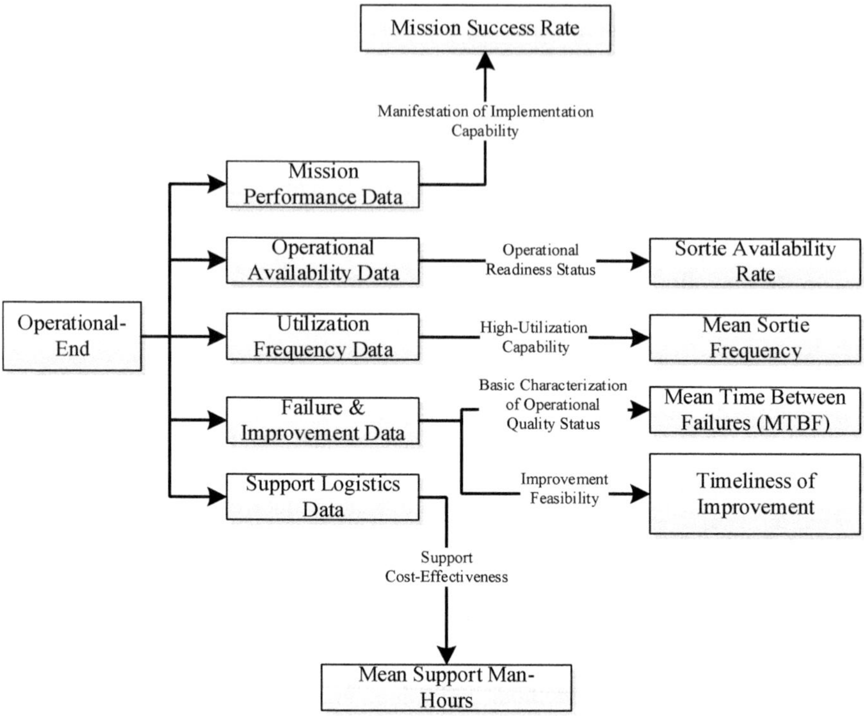

Fig. 4 Quantitative characterization dimensions of operational quality

3 Operational Quality Risk Assessment Model

3.1 Optimization of Risk Assessment Methodology

Risk assessment serves as a fundamental technique in safety engineering for evaluating production risks [5]. It quantifies potential risks to enable visualized comprehension of the risk magnitudes associated with engineering risk sources. The methodology operates on the principle of decomposing risk levels into two impact factors: the probability of occurrence (P) and the accident consequences (L) should the risk materialize. The risk level (R) is expressed as:

$$R = P * L \tag{1}$$

R denotes the risk level of a specific risk source,
P represents the occurrence probability associated with that risk source,
L signifies the consequential loss if the accident materializes.

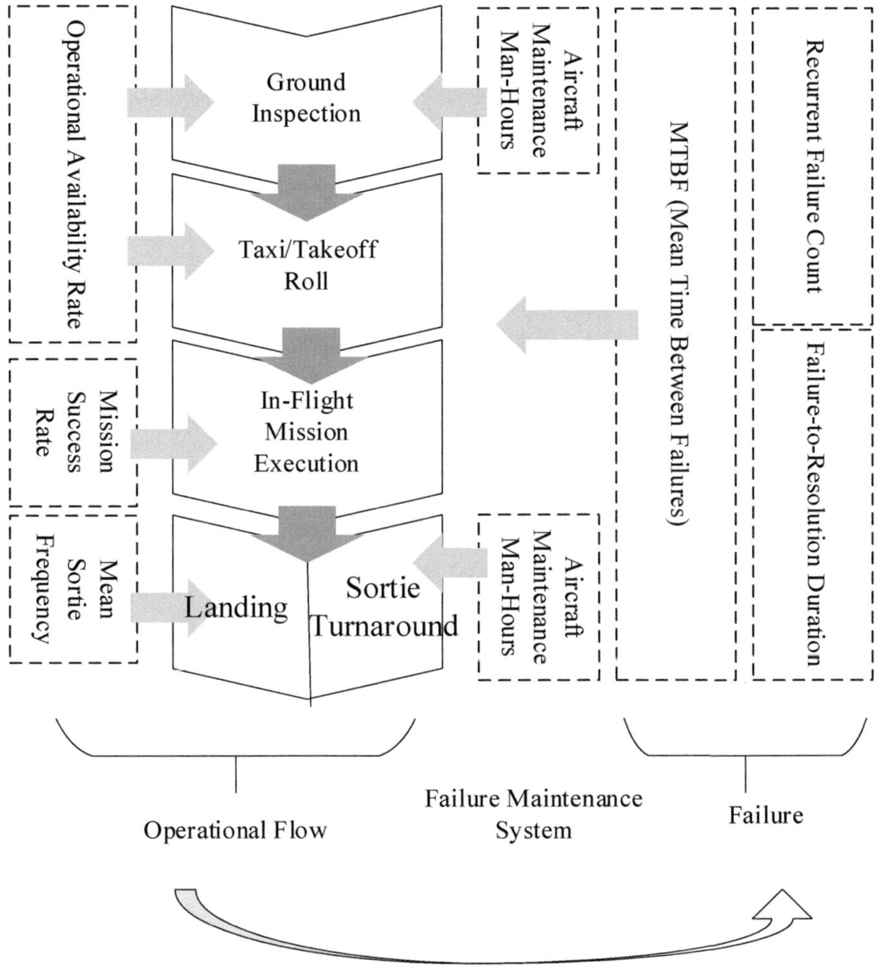

Fig. 5 Framework of characterization indicators for operational quality

Thus, when a risk source exhibits both high occurrence probability (*P*) and severe consequential loss (*L*), the risk level (*R*) is determined to be critically high. Conversely:

High P with low L yields moderate R,

High L with low P similarly results in moderate R.

This establishes a two-dimensional risk characterization model (Fig. 6), where extreme risk emerges only when both probability and consequence simultaneously reach critical thresholds.

The two-dimensional risk characterization model exhibits a critical gap: a risk source with both high occurrence probability and severe consequences may present lower actual risk if readily correctable, compared to risks that are less amendable to

Table 2 Operational quality characterization for aircraft

Metric	Required data	Calculation method	Value range
Operational availability rate	• Operation-ready aircraft times (a_1) • Non-operation-ready aircraft times (a_2) • Operation-ready days (a_3) • Non-operation-ready days (a_4)	$(a_1*a_3)/(a_1*a_3+a_2*a_4)$	[0,1]
Mission success rate	• Successful sorties count (b_1) • Required sorties count (b_2) • Total missions (b_3)	(b_2*b_1) $/(b_2*b_3)$	[0,1]
Mean maintenance man-hours	•Failure-to-recovery duration (c_1) • Total failures (c_2)	$(\sum c_1)/c_2$	$(0, +\infty)$
Mean sortie frequency	• Aircraft deployed per mission (d_1) • Total sorties per mission (d_2)	$(\sum d_1)/d_2$	$(0, +\infty)$
MTBF (mean time between failures) [8]	• Total operational hours (e_1) • Total failures (e_2)	e_1/e_2	$(0, +\infty)$
Recurrent failure ratio [6]	• Repeated failures under same mechanism (m_1) • Total failures of the product (m_2)	m_1/m_2	[0,1]
Mean improvement time	• Failure-to-resolution duration (r_1, hours) • Total failures (r_2)	$(\sum r_1)/r_2$	$(0, +\infty)$

correction. However, this model cannot differentiate between such scenarios, thereby impeding comparative risk assessment across equipment operational statuses. To address this limitation, this study introduces correctability as a third dimension, extending the framework into a three-dimensional risk characterization model. This enhancement enables more precise characterization of risks and their influencing factors within operational quality status, as illustrated in Fig. 6.

In this study, the Correctability of identified risks from the three-dimensional risk magnitude model is employed. After normalization via current standardization and weighting methods, this parameter is multiplied by the values of P and L. Meanwhile, the calibration constant μ is introduced to ensure that the three dimensions exhibit distinct impact weights across different types of evaluated objects. The core purpose of normalization is to enable comparative analysis of parameters from the three dimensions on a unified scale. Given the extensive existing research on normalization and weighting methods, and considering that the focus of this paper lies in proposing

Fig. 6 Two-dimensional risk characterization model

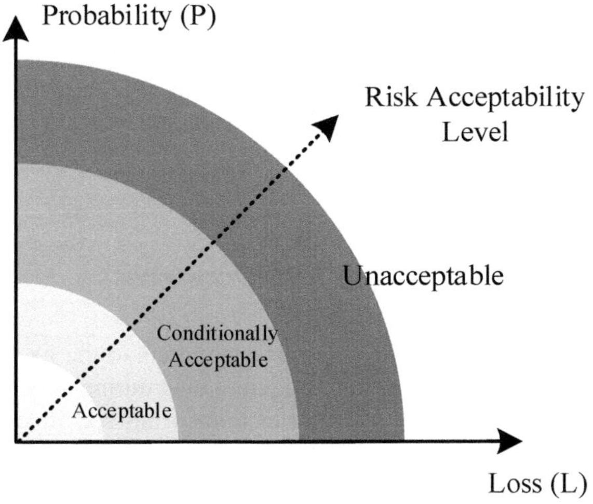

risk measurement dimensions by integrating data and indicators, detailed elaboration on the normalization and weighting methods between dimensions is omitted herein. For engineering applications, it is recommended to adopt relatively simple data processing methods, such as the Delphi method and the Analytic Hierarchy Process (AHP), refer to Fig. 7.

1. *Three-Dimensional Risk Assessment Model*

Risk level quantifies risk magnitude, where higher values indicate greater risk exposure and lower values denote reduced risk. Based on fault data analysis

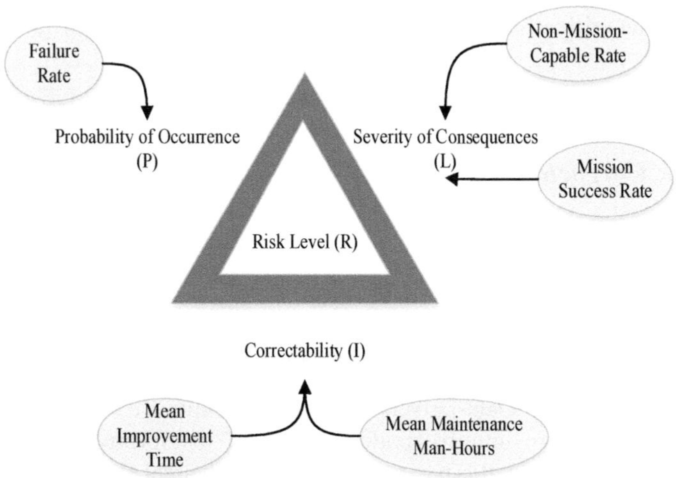

Fig. 7 Three-dimensional risk assessment model

and established operational quality characterization dimensions, this study defines computational methods for three core risk parameters:

P: Probability of mission-impacting risk occurrence.

L: Severity of consequences from materialized risks.

I: Correctability of identified risks.

3.2 Probability of Risk Occurrence

The probability of risk occurrence refers to the likelihood that aviation equipment fails to meet operational requirements during service, quantified through mission failures and equipment malfunctions within specified timeframes. Specifically, faults detected during scheduled inspections or maintenance compromise operational readiness, while in-mission failures directly impact task execution. Consequently, Mean Time Between Failures (MTBF) serves as the primary input for probability calculation (P), effectively substituting failure rate as a measure of risk occurrence probability.

3.3 Severity of Risk Consequences

This dimension assesses potential damage magnitude upon risk materialization, where aviation equipment failures compromise success rates across all mission phases. For in-flight operations, mission success rate quantifies failure impacts, while ground phases (inspection/taxi) utilize non-mission-capable aircraft days (NMCAD) to characterize mission disruption probability from faults/maintenance, thereby integrating mission-critical and operational availability metrics for consequence severity (L).

3.4 Risk Correctability

Correctability (I) encompasses both the difficulty of root cause rectification and the complexity of maintenance following a failure. Derived from established operational quality metrics, Mean Improvement Time captures the efficiency of systemic resolution, while Mean Maintenance Man-hours reflects the complexity of onsite repairs. Together, these two metrics jointly quantify the ease of risk mitigation. Implementation of Fault Data-Driven Operational Quality Risk Assessment.

To facilitate a more intuitive understanding of the formula logic, given the constraints of limited data, this paper presents a hypothetical case to verify the rationality of the newly added risk dimension. Taking two types of products and their failure modes with significantly different correctability in aero-engines as examples—seal leakage of seals and wear of bearings—the results calculated by the traditional two-dimensional risk model indicate that the risk of seal leakage is far higher than that of bearing wear. However, seal leakage can usually be addressed by rapidly replacing faulty components or adopting other expedited repair methods to mitigate its impacts. On the other hand, after applying the three-dimensional risk model, the poor correctability of bearing wear becomes more prominent, which explains why the actual risk of bearings was significantly underestimated in traditional assessments. Therefore, resources can be allocated to bearings for preventive maintenance or technological improvement based on practical needs, so as to support the early detection of potential problems.

Empirical evaluation enables manufacturers to qualitatively identify weaknesses, prioritize improvements, and reveal risks while verifying the methodology's engineering implementability and technical validity. This establishes benchmark metrics for longitudinal comparison and sustainable assessment data accumulation. To achieve these objectives, a structured implementation workflow was executed as shown in Fig. 8.

Upon finalizing the evaluation targets, an assessment work plan is formulated to specify timelines, evaluation methodologies, operational requirements, organizational structure, and reporting dissemination protocols. This plan is jointly developed by the evaluation technical support agency and the chief designer department through consultation, with equipment users retaining authority to specify additional requirements. Subsequently, an evaluation working group is established, comprising representatives from the product chief designer department, contracted R&D entities, and professional evaluation agencies, while military users and industrial regulatory authorities may participate as appropriate, refer to Table 3.

Implementation of Fault Data-Driven Operational Quality Risk Assessment. The kick-off meeting establishes evaluation objectives, scope, and significance. Data collection is completed through field inspections, document reviews, questionnaire surveys, and baseline data recording. The working group consolidates and processes collected data using the established methodology to derive risk level classifications and vulnerability analyses. Based on these outcomes, a preliminary report is generated covering risk levels and critical risk rankings, with the final report finalized after stakeholder consultations and revisions. Report structure includes: background, objectives, methodology overview, conclusions, analysis, and raw data appendices. The closing meeting involves report presentation, advisory discussions, and action planning. Evaluation conclusions are promptly communicated to assessed units to clarify performance status and implement improvement measures for high-risk areas per working group recommendations [9–12].

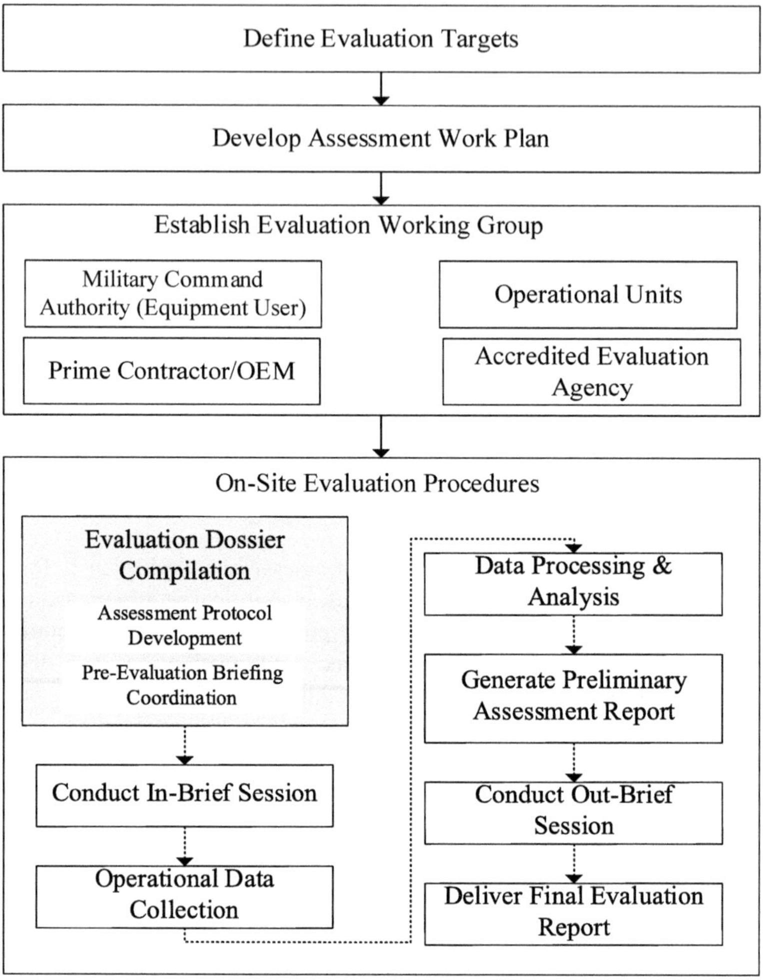

Fig. 8 Risk assessment implementation workflow

4 Conclusion

In response to the "high-standard, high-demand" expectations associated with aviation product quality, this study proposes an operational quality risk assessment methodology for aviation equipment by refining the RPN algorithm. The approach incorporates Mean Time Between Failures (MTBF) as a core metric to assist both manufacturers and end-users in evaluating operational quality, quantitatively

Table 3 Operational data outputs by phase

Organizations participating in the rating process	Responsibilities
Military command authority (equipment user)	Defines evaluation requirements; oversees organization, guidance, and deployment of assessment activities when necessary
Operational units	Provides operational data; validates risk assessment accuracy; monitors corrective actions for vulnerabilities
Prime contractor/OEM	Coordinates subordinate contractors; participates in data collection and reviews; submits self-audit reports; reports progress on corrective actions
Accredited evaluation agency	Executes data collection, field inspections, and report drafting; facilitates stakeholder communication; tracks implementation of corrective measures

assessing risk levels, identifying critical vulnerabilities, and strengthening data governance capabilities. This method enhances overall quality standards, provides analytical support for user-side quality assessment, reduces subjectivity inherent in traditional evaluation approaches, and optimizes the utility of operational data to support continued improvements in aviation quality.

Acknowledgements I would like to express my sincere gratitude to the institution for providing essential academic resources and infrastructural support throughout this research. My profound appreciation also extends to my colleagues for their invaluable assistance and collaborative insights. Furthermore, I acknowledge the scholars whose citations enriched this work, as well as the experts who generously contributed their time and expertise to review this manuscript.

References

1. Sae J (2009) Potential failure mode and effects analysis in design and manufacturing. SAE Int 1739
2. Saaty TL (1980) The analytic hierarchy process. McGraw-Hill
3. AS9100 (2009) Quality management systems—requirements for aviation, space, and defense organizations, IAQG
4. Smith J et al (2010) Mission-cycle characterization for aircraft operations. J Aviat Technol 5(3):45–60
5. Kaplan S, Garrick BJ (1981) On the quantitative definition of risk. Risk Anal 1(1):11–27
6. Jones M, Brown K (2012) Analysis of recurrent failures in aviation systems. Proc Int Conf Aerospace Reliability, pp 112–125
7. Wang Y et al (2010) Optimization of risk priority number in FMEA. Qual Reliab Eng Int 26(4):305–318
8. IEC (2006) Equipment Reliability Testing—Part 4: Statistical Procedures for Exponential Distribution, IEC, pp 60605–4

9. Guo J, Wang Z, Li H, Yang Y, Huang CG, Yazdi M, Kang HS (2024) A hybrid prognosis scheme for rolling bearings based on a novel health indicator and nonlinear Wiener process. Reliab Eng Syst Saf 245:110014
10. Li H, Yazdi M, Huang HZ, Huang CG, Peng W, Nedjati A, Adesina KA (2023) A fuzzy rough copula Bayesian network model for solving complex hospital service quality assessment. Complex Intell Syst 9(5):5527–5553
11. Li H, Peng W, Adumene S, Yazdi M (2023) An intelligent cost-based consequence model for offshore systems in harsh environments. In: Intelligent reliability and maintainability of energy infrastructure assets, pp 107–117. Springer Nature Switzerland, Cham
12. Yazdi M (2024) Reliability-centered design and system resilience. In: Advances in computational mathematics for industrial system reliability and maintainability, pp 79–103. Springer Nature Switzerland, Cham

A Preliminary Summary of Failure Mode and Effects Analysis of Digital Control Systems

Bohao Tian, Jun Yang, Chenyu Jiang, and Zhanyu He

Abstract With the modernization of Digital Instrumentation and Control (DI&C) systems in nuclear power plants, nuclear industry has entered a transformative era. However, this digitization transition also introduces critical challenges related to the reliability, security, and safety integrity of software and hardware components. Failure Mode and Effects Analysis (FMEA), as a foundational risk assessment technique mandated by IEC 61508/61513 standards, is critical for ensuring the functional safety, cybersecurity resilience, and operational robustness of digital instrumentation and control systems in nuclear power plants. In the paper, a preliminary summary of the failure mode and effects analysis of digital control systems in both hardware and software dimensions is conducted. A taxonomy of failure modes of digital system components is summarized with the literature review on the leading FMEA reports issued in recent years. The key challenges facing by FMEA for probabilistic risk analysis have also been discussed.

Keywords Failure mode and effects analysis · Digital control system · Safety and reliability · Nuclear power plants

B. Tian · J. Yang (✉)
School of Electric Power Engineering, South China University of Technology, Guangzhou, China
e-mail: youngjun51@hotmail.com

C. Jiang · Z. He
China Electronic Product Reliability and Environmental Testing Research Institute (China CEPREI Laboratory), Ministry of Industry and Information Technology Key Laboratory of Quality and Reliability Engineering Technology of Civil Aircraft and Aero-Engine Guangzhou, Guangzhou, China

H. Li et al. (eds.), *Data-Driven Methods for Reliability and Safety Engineering: Applications in Industrial Systems*, Springer Series in Reliability Engineering,
https://doi.org/10.1007/978-3-032-22873-4_10

1 Introduction

Nowadays, the nuclear industry is exploring innovative digital technologies in advanced nuclear reactors to replace analog Instrumentation and Control (I&C) systems. These digital systems offer significant advantages, including enhanced functionality, greater flexibility for upgrades, and potentially higher reliability. However, the inherent complexity characterized by the intricate hardware configurations, software interactions and novel failure modes distinct from analog predecessors also introduces big challenges for ensuring the robustness and safety of digital systems in nuclear power plants. Consequently, rigorous analytical methodologies like Failure Mode and Effects Analysis (FMEA) become indispensable.

The FMEA of digital control systems has been a hot topic in the past two decades. For instance, Karol Kowal et al. [1] performed a systematic assessment of failure modes in High Temperature Engineering Test Reactor (HTTR) facilities using the FMEA method for system design improvements. Antti Pakonen et al. [2] combined process FMEA with model-checking techniques to pinpoint failure modes and design defects that could lead to false activation. Weitao Yang et al. [3] used FMEA to assess failure modes and key risks in complex systems. Li et al. [4] proposed a dynamic FMEA to analyze the reliability of digital control systems in nuclear power plants. The simulation technique is augmented in the dynamic FMEA to model the interactive behaviors between the hardware, software, and firmware in digital control systems. Han Bao et al. [5] proposed a redundancy-guided STPA method by integrating fault tree analysis and System Theoretic Process Analysis (STPA) to effectively identify the Common Cause Failures (CCFs) in software and leverage risk assessments for digital safety system. A generic failure modes and effects analysis approach is proposed by Tsong-Lun Chu et al. [6] for reliability modeling of digital I&C systems.

The FMEA can provide a systematic, proactive framework for exhaustively identifying potential failure mechanisms within digital I&C components and subsystems, evaluating their effects on overall system function and plant safety. However, there is no consensus reached in the industry due to the complex nature of digital I&C systems in Nuclear Power Plants (NPPs). It is therefore that a preliminary summary of the failure mode and effects analysis is carried out for digital control systems in NPPs in the paper.

The rest of the paper is organized as follows. Section 2 introduce the basic concepts involved in failure mode and effects analysis. Section 3 summarizes the FMEA of digital instrumentation and control systems in NPPs. The FMEA of software is presented in Sect. 4. Finally, discussions and conclusions are drawn in Sect. 5.

2 Failure Mode and Effects Analysis Method

Failure Mode and Effects Analysis (FMEA) was first introduced in the 1950s in the U.S. aerospace manufacturing sector. Initially, it was used to address quality and reliability issues in military products. With the continuous development of technology and increasing demand for product reliability, FMEA has been widely applied across various industries such as aerospace, nuclear, petrochemical, oil and gas and other industrial manufacturers [7]. Generally, the FMEA can be applied at each stage of production life cycle including design, manufacturing, installation, operation and decommissioning. According to stages and objectives of the analysis, FMEA can be categorized into several types including system FMEA (SFMEA), design FMEA (DFMEA), and process FMEA (PFEMA) [8].

The design FMEA adopts a bottom-up approach, progressively identifying and assessing the failure modes of systems, components, or equipment, and their consequences. In contrast, the system FMEA (also known as functional FMEA) uses a top-down analysis approach, focusing on the impact of system function failures during the early conceptual stage of system. Process FMEA is utilized to analyze potential failures within manufacturing and assembly processes. It is aimed at identifying potential failure modes in each step of the process for improving process efficiency and reducing the likelihood of failures during system operation. The characteristics of the various FMEA are summarized in Table 1.

In digital control systems, both design FMEA and functional FEMA are leveraged to ensure hardware integrity and software interactions for integrated system-logic validation. The integration of design FMEA (DFMEA) and system FMEA in the design and operation of digital instrumentation and control systems in nuclear power plants is a critical and increasingly sophisticated practice. By combining Functional FMEA (FFMEA) and Design FMEA (DFMEA), it is possible to systematically identify and assess the potential failure modes of equipment, systems, or components, as well as their impact on system performance. These methods not only help identify key risks but also provide support for improvements during the design phase, thereby enhancing system reliability and safety. The effective implementation of FMEA provides a reliable framework for risk management of complex digital systems,

Table 1 Comparison between SFMEA, DFMEA and PFMEA

Type of FMEA	Characteristics	Phase of application
SFMEA	Identify and mitigate system-related deficiencies, such as safety or subsystem integration	Early conceptual/architectural design stage
DFMEA	Identify potential failure modes caused by design deficiencies in subsystems	Detailed design and prototyping stage
PFMEA	Analyze potential failures within manufacturing and assembly processes	Process planning and validation stage

especially when combined with other analytical methods, such as Fault Tree Analysis (FTA), which further improves the efficiency and accuracy of failure analysis.

In the context of Probabilistic Risk Assessment (PRA) for digital I&C applications, SFMEA, DFMEA, and PFMEA should be implemented as an integrated methodology to combine complementary perspectives in risk identification. For example, we can use SFMEA to establish the high-level system failure definitions needed to structure Event Tree sequences. The incorporation of high-level failures via SFMEA can ensure an accurate traceability of digital system failure consequences to plant safety impacts. DFMEA provides granular component failure data, such as 'output stuck' or 'spurious actuation', which forms the basic events for Fault Tree quantification. Finally, PFMEA serves to validate the as-built reliability assumptions inherent in the risk models by addressing manufacturing-induced defects that could undermine the assumed integrity of the digital hardware.

3 Literature Review of FMEA of Digital Control Systems

Since the advent of digital systems in nuclear power plants in the late 1980s and early 1990s, there have been numerous research efforts dedicated to the FMEA of digital control systems. This marks the period when digital systems began significantly replacing analog systems in NPPs, driving regulatory concern and focused research by agencies like NRC, ORNL, EPRI, and OECD/NEA, etc. A series of reports and publications are issued during the past few decades. Here some leading FMEA reports are summarized and highlighted.

EPRI-3002000509 (EPRI, 2013 [9, 14]*)*

In the report, six methods including functional FMEA, design FMEA, a top-down method using fault tree analysis (FTA), Hazard and Operability (HAZOP) analysis, Systems Theoretic Process Analysis (STPA), and Purpose Graph Analysis (PGA) are reviewed by the project team from Electric Power Research Institute (EPRI) to improve current practices in failure mode and effects analysis of digital instrumentation and control systems. For each of the methods studied, the report provides a comprehensive examination of their strengths and weaknesses, encompassing a thorough methodological description, a detailed step-by-step procedure and illustrative case examples. It is concluded that more work is required to apply the improved hazard analysis methods to resolving regulatory uncertainty regarding understanding and managing digital system and software failure modes.

NKS-330 (Nordic Nuclear Safety Research, 2015 [10])

In the report, a failure modes taxonomy specifically tailored for digital instrumentation and control systems has been developed jointly with the OECD/NEA Working Group on Risk Assessment. Within the failure modes taxonomy, the failure modes are divided into five categories, including basis components, module, I&C unit, division and system. The applicability of the newly developed taxonomy for PSA objectives is validated on a four-channel redundant reactor protection system. The research findings reveal that modeling at the module level is suggested as the preferred granularity for PSA analysis.

NKS-341 (Nordic Nuclear Safety Research, 2015 [11]*)*

In the report, the failure modes and data collection for software reliability anlysis is outlined. By classifying failure modes and integrating operating data on software failures, the report proposed a method for quantification of software reliability for the purpose of probabilistic safety assessment in nuclear power plants. The major contribution of the report lies in proposing a clear classification of software failure modes and incorporating practical operational experience into the reliability assessment process.

NKS-361 (Nordic Nuclear Safety Research, 2016 [12]*)*

The NKS-project MODIG aims to develop a consensus approach for reliability analysis of digital I&C systems in nuclear power plants. A survey of the defense-in-depth (DiD) framework has also made to integrate probabilistic and deterministic methods into digital I&C licensing. The major contributions of the study also include failure data collection for software failure probability quantification and a practical application of probabilistic safety assessment.

NEA/CSNI/R (2014) 16 (Organization for Economic Cooperation and Development, Nuclear Energy Agency, Committee on the Safety of Nuclear Installation, 2015 [13]*)*

The report conducted an in-depth investigation of FMEA of digital instrumentation and control systems in nuclear power plants. A taxonomy of failure modes of digital components for the purposes of probabilistic risk assessment is also developed. The taxonomy can be served as the foundation for reliability modelling and quantification of digital I&C systems. It also assists in defining a structure for data collection and probabilistic risk assessment studies.

ORNL/TM-2010/32 (Oak Ridge National Laboratory, 2010 [14]*)*

The intent of the report is to create a unified framework for classifying generic failure modes and mechanisms for digital instrumentation and control systems and to gather operational experience and data to support reliability analysis in the context of probabilistic safety assessment. Particular attention is put on the module-level failure modes by drawing on databases from both nuclear and non-nuclear industries.

NUREG/CR-6942 (Nuclear Regulatory Commission/The Ohio State University, 2007 [15])

In the report [15], the FMEA of a digital feedwater control system is presented for benchmark analysis between Markov/CCMT and DFM. Although the failure states and data for the example system components are well documented, obtaining detailed information about potential failures remains essential for the broader digital control system analysis.

NUREG/CR-6962 (*Brookhaven National Laboratory*, 2008 [16])

The report [16] mandated that the impact of component failure modes on digital control systems and even entire nuclear power plants must be explicitly represented in probabilistic models. To achieve this, three levels of detail for the FMEA are defined for the case study of digital feedwater control system (DFWCS), which are system level, module level, and major-component-of-module level. The system-level FMEA entails a comprehensive analysis covering the DFWCS as a top-level system. The module-level FMEA focuses on the key modules within the DFWCS such as the Main CPU, Backup CPU, various controllers for valves/pumps, PI Controller, etc. The component-level FMEA describes the lowest-level FMEA targeting internal components of DFWCS modules.

NUREGIIA-0254 (IRSN and U. S. NRC, 2011 [17])

The French Institute for Radiological Protection and Nuclear Safety (IRSN) and the U.S. Nuclear Regulatory Commission (NRC) have jointly conducted a study to evaluate the applicability of FMEA in identifying faults attributable to complex logic in digital I&C systems for safety functions. The study concluded that FMEA is of limited use for complex logic, particularly for software faults identification. It is also highlighted that optimal combination of FMEA and fault tree analysis is required for further investigations.

INL/EXT-21-65055 (Idaho National Laboratory, 2021 [18])

The report [18] documented the current quantitative results of industry-average performance for components and Initiating Events (IEs) at U.S. commercial nuclear power plants. Consistent with NUREG/CR-6928 and its prior updates, this report addresses four categories of events: component unreliability, unavailability of components or trains due to test or maintenance outages, special event probabilities encompassing operational issues, and initiating event frequencies.

A summary of taxonomies of digital system failure modes (Tsong-Lun Chu [6])

In the paper, a thorough comparison of twelve failure mode taxonomies from nine organizations is conducted. The failure modes are systematically defined and classified across a hierarchical framework that spans five progressively detailed levels: the overall system, its constituent divisions, individual I&C units, functional modules within those units, and ultimately the discrete components themselves. The software failure modes are also in detail discussed.

Based on the above reports analysis and review, a comprehensive multi-layered failure modes taxonomy is summarized as shown in Table 2.

Table 2 Failure mode and effects analysis of digital I&C systems in NPPs

Level	Items		Failure mode	Failure effect
System	RPS System		Output of 1 instead of 0 Output of 0 instead of 1 No output in the required time frame	
	Entire RPS/ESFAS		No activation signal	
	Entire digital feedwater control system		Loss of automatic control	
	Sensors + entire RPS system + trip circuit breakers + Manual actuation by human operators		Failure to actuate on demand	
Division	A single channel of RPS		Failure to actuate on demand	
I&C unit	Acquisition process unit (APU)		Loss of all functions (outputs) of the I&C unit Loss of a specific function Spurious output (one function) Spurious output (all functions)	
	Voting unit (VU)			
	Trip logic unit((TLU))			
	Safety logic unit(SLU)			
	Load driver			
	Input/output devices (PI/O)			
Module	I&C modules with digital outputs	Digital input modules Digital output modules	Hang, Crash (no output)	Fatal failure
			Output fails to 1 Output fails to 0 Output stuck to current value Output fails to the opposite state Delayed output Random output	Non-fatal failure
		Processing module	Hang, crash (no output)	Fatal failure
			Wrong output Delayed output Random output	Non-fatal failure
	I&C modules with analog outputs	Analog input modules, analog output modules	Hang, crash (no output)	Fatal failure
			Output fails to MAX	Non-fatal failure
			Output fails to MIN/0	

(continued)

Table 2 (continued)

Level	Items		Failure mode	Failure effect
			Output fails to an erroneous value (out of range) Delayed output Random output (output fluctuates, in range, between minimal and maximal value)	
			Drifted output (output is x% more than actual value)	Non-significant or non-functional effect, with plausible or implausible behaviour
Basic component	CPU		CPU Half CPU Logic Error CPU Data Corruption CPU Crash Permanent CPU Damage	
	RAM		All data is lost Some data is corrupted or lost	
	ROM		Some data corrupted Some data lost	
	A/D,D/A		Lack of output signal Inaccurate fluctuation of output signal Output signal high Output signal low Inaccurate output signal Noisy output analogue Signal	

4 FMEA Taxonomy of Software

Software, as a critical component embedded in digital systems, is inherently different from hardware in terms of failure characteristics and behavior. Unlike hardware failures which are usually known and can be analyzed based on physical characteristics, software failures typically arise from design flaws, logic errors, or unexpected interactions between components. Given the unique characteristics and complexities of software, the FMEA and dependability analysis of digital software is typically conducted in a separate and distinct manner. Software FMEA is usually performed using the functional approach, by which the functional failure characteristics of software such as computational errors, logic issues, data handling problems, and interface failures are focused. For instance, Phillip McNelles et al. [19] conducted a comprehensive FMEA analysis for FPGA systems, encompassing both the chip hardware and embedded software during the design phase. By integrating FMEA with fault tolerance techniques, Seung Jun Lee et al. [20] proposed a new method for safety assessment of digital I&C systems in nuclear power plants. By combining with software reliability testing, Jeongil Seo et al. [21] proposed a method for optimizing test cases based on the importance of failures. A rigorous software FMEA process was developed by Stadler et al. [22] to drive the identification of risks to reliability and safety. The various safety assessment methods and tools for FMEA of software-based systems are developed and evaluated in report [23]. Hecht et al. [24] discussed the failure modes and effects analysis for a software-intensive satellite control system. A detailed-level software FMEA analysis approach is proposed by G. Y. Park et al. [25], in which a template is developed for failure modes analysis for software architecture. The failure modes of software for dependability assessment of safety instrumentation and control systems in nuclear power plants was in detailed discussed in the IAEA nuclear energy series report [26]. The failure modes of software embedded in digital control systems are summarized in Table 3.

Figure 1 illustrates a fundamental divergence in failure effects between hardware and software within digital instrumentation and control systems. The failure mode taxonomy for digital I&C systems is divided into five levels, which are system, division, I&C unit, module, and basic component. The failure mode classification framework is also consistent with the contents as summarized in Table 2 and Table 3.

System Level. The system level is at the topmost tier, which encompasses the entire digital I&C system as a single integrated entity.

Division Level. Division refers to the sub-systems within the overall digital I&C system. It is segmented by functional scope (e.g., reactor protection division, process control division).

I&C Unit Level. The I&C unit represents a discrete functional unit within a division, which can be further divided into interconnected modules (e.g., a safety logic unit, a communication gateway unit).

Module Level. A module refers to a specialized hardware/software component within a I&C unit that is responsible for a specific task (e.g., analog input module, logic processing module).

Component Level. The component level is at the lowest tier, referring to individual physical or logical elements (e.g., microchips, capacitors, software subroutines) that compose a module.

Table 3 Failure mode and effect of software

Level	Items	Failure modes
System/Division (the failure modes of the collection of all software of a digital system)	RPS	Output of 1 instead of 0 Output of 0 instead of 1 No output in the required time frame
	System with redundant and/or diverse	Full failures partial failures
	System with multi-functions	Single function failures Correlated multiple function failures Complex multiple function failures
	Failure to generate signal in time (omission failure) Spurious signal Adverse effects on other functions	
Microprocessor (the failure modes of a software running on a microprocessor of a digital system)	Software stalls Software runs as usually but with wrong outputs Software runs with misleading commands to the user, incomplete or incorrect display of information	
	Hang Abort Missing operation Extra operation Erroneous operation	
	Data Acquisition	Incorrect value Incorrect validity No value No validity
	Logic Processing	Failure to actuate spurious failure
	Voting Logic	Incorrect voting No vote
	Priority Actuation Logic	Incorrect priority No priority
Sub-level (the level of software elements, which include input, output, communication, resource allocation, and processing elements.)	Internal failures	Function failures Attribute failures Function set failures
	Interaction failures	Input/output failure modes Multiple interaction failure modes Support failure modes
		Timing/order failure Interrupt induced failure Omission of a function or an attribute Unintended function or attribute Incorrect implementation of a function or an attribute Data error

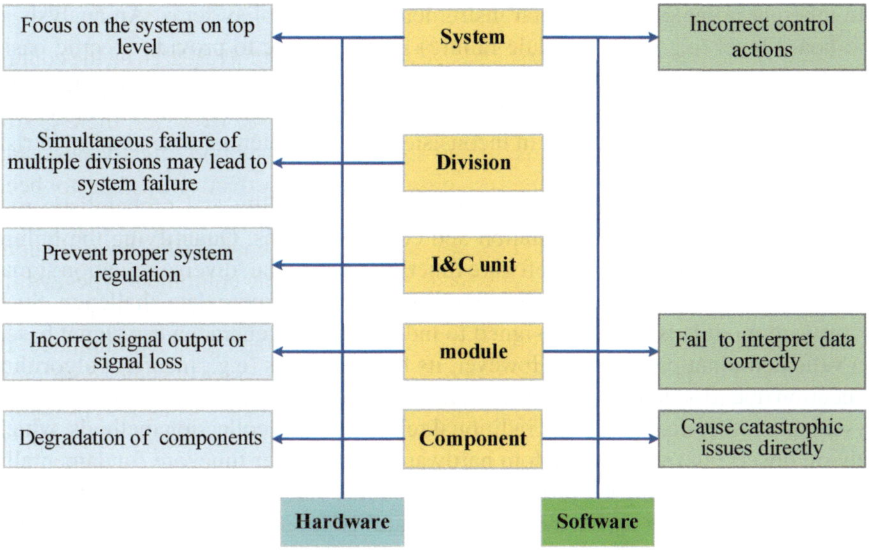

Fig. 1 Failure effects of hardware and software with digital I&C systems

This multi-level taxonomy ensures that the failure manifestations can be captured at an appropriate granularity while at the same time maintaining traceability from component-level faults to system-level consequences. It is also aligned with the established practices in probabilistic safety assessment where failure modes at lower levels serve as failure causes or effects at higher levels.

5 Discussions and Conclusions

In the paper, a preliminary summary of the failure mode and effects of digital control systems is conducted. The pilot FMEA reports endorsed by OECD/NEA, EPRI, ORNL are reviewed and summarized. A taxonomy of failure modes of digital components and software is presented based on the literature review of existing research on failure mode and effects analysis of digital control system in nuclear industry. Although there are significant advancements and improvements in FMEA of digital instrumentation and control systems in nuclear power plants, the following issues remain prominent and require a further resolution.

The absence of standardized guideline to determine the appropriate level of abstraction for FMEA. Concurrently, the OECD/NEA CSNI Working Group on Risk Assessment (WGRisk) and the U.S. Nuclear Regulatory Commission (NRC) has recommended a generic taxonomy of failure modes for digital instrumentation and control systems. However, the failure modes and mechanisms of some key digital components, particularly at the module level, are still missing and unclear due to the inherent complexity of digital instrumentation control systems. An analysis at the board level (e.g., CPU module failure) is too generic to provide specific basic events (latch-up in a specific register) for fault trees. Conversely, analyzing at the gate-level generates an unmanageable number of failure modes with no supporting reliability data. This gap results in inconsistencies and potential inaccuracies in risk assessments.

Data collection for quantification of software reliability and probabilistic risk assessment of digital instrumentation and control systems. Quantifying the failure probability of a safety–critical software function, such as the 'diverse actuation signal generation' in a digital reactor trip system, highlights the core data challenge. Such a software module is usually designed to independently generate a trip signal based on validated plant parameters. However, its failure modes (e.g., incorrect algorithm execution due to a latent logic flaw or data corruption in a shared memory buffer) are systematic rather than random. Traditional reliability data collection methods, which rely on observing frequent, random hardware failures over time, are fundamentally inadequate for capturing the probability of such rare, design-induced software faults. Consequently, how to derive a credible probability for a software failure mode that may have never been observed in operation but could have severe consequences remains an open and significant challenge for risk analysis of safety–critical digital systems.

Integration of FMEA in probabilistic risk assessments of digital I&C systems. A practical integration gap arises when transferring FMEA results for digital I&C systems into the PSA model. The FMEA lists multiple detailed component-failure modes (e.g., sensor drift, microprocessor lock-up), but the PSA model typically requires a single probabilistic basic event that represents the failure of a defined system function. The key challenge is the absence of a standardized methodology to logically aggregate these diverse failure modes into unified functional failure events, and synthesize their individual likelihoods while accounting for interdependencies into a credible probability suitable for PSA quantification. The lack of a consistent method to aggregate or map the detailed FMEA modes into a PSA-ready probability leads to inconsistent integration.

To address the aforementioned issues, a more consensus-based taxonomy and approach that supports PRA practice and captures the critical dependencies and design features is necessary. Besides, a tiered-mapping approach can be adopted. The integration process begins with a system-level FMEA to identify high-level functional failures, such as incorrect safety-actuation output. These high-level failures are then decomposed using a regulator-endorsed, generic taxonomy, for instance, the module-level failure modes (e.g., processing error, communication timeout) recommended by the U.S. NRC to derive a focused and manageable set of module-level failure causes. Industry collaboration can subsequently establish standardized reliability benchmarks for these generic modes. A hybrid evidence-based framework can also be employed. Limited operational data (e.g., reboot logs) are complemented by process-based evidence (e.g., test-coverage metrics, static-code-analysis results) and architecture-based reliability models (e.g., accounting for process-watchdog and recovery mechanisms). These qualitative and quantitative sources are then fused via a Bayesian belief network to produce an informed probability range (rather than a point estimate) for use in PSA.

Acknowledgements This work is supported by the fundings from National Foreign Experts Program under Grant No. H20250123, 2025 Annual Special Innovation Projects for Institution of Higher Education in Guangdong Province under Grant No. 2025KTSCX007, and Guangzhou Municipal Science and Technology Project under Grant No. 2024A04J3613.

References

1. Kowal K, Torabi M (2021) Failure mode and reliability study for electrical facility of the high temperature engineering test reactor. Reliab Eng Syst Saf 210:107529
2. Pakonen A, Buzhinsky I, Björkman K (2021) Model checking reveals design issues leading to spurious actuation of nuclear instrumentation and control systems. Reliab Eng Syst Saf 205:107237
3. Yang WT, Li Y, He C (2022) Fault injection and failure analysis on Xilinx 16 nm FinFET Ultrascale + MPSoC. Nucl Eng Technol 54(6):2031–2036
4. Li YK, Wang X, Lin M (2020) Application of the dynamic FMEA in the reliability analysis of DCS. In: Proceedings of the E3S web of conferences, vol 194, pp 01018
5. Bao H, Shorthill T, Zhang H (2020) Hazard analysis for identifying common cause failures of digital safety systems using a redundancy-guided systems-theoretic approach. Ann Nucl Energy 148:107686
6. Yue M, Postma W (2012) A summary of taxonomies of digital system failure modes. DigRel Task Group
7. Mutlu NG, Altuntas S (2019) Risk analysis for occupational safety and health in the textile industry: integration of FMEA, FTA, and BIFPET methods. Int J Ind Ergon 72:222–240
8. Kania A, Cesarz-Andraczke K, Odrobiński J (2018) Application of FMEA method for an analysis of selected production process. J Achiev Mater Manuf Eng 91(1):34–40
9. Alto P (2013) EPRI Hazard Analysis Methods for Digital Instrumentation and Control Systems
10. Authén S, Holmberg JE, Zamani L et al (2015) NKS-330, Guidelines for reliability analysis of digital systems in PSA context-Final report
11. Bäckström O, Holmberg JE, Jockenhövel-Barttfeld M et al (2015) NKS-341, Software reliability analysis for PSA: failure mode and data analysis
12. Authén S, Bäckström O, Holmberg JE et al. (2016) NKS-361, Modeling of Digital, MODIG-Interim report
13. NEA/CSNI/R (2014) 16, Faliure Modes Taxonomy for Reliability Assessment of Digital Instrumentation and Control Systems for Probabilistic Risk Analysis (2015)
14. Korsah K, Cetiner SM, Muhlheim MD, Poore III WP (2010) ORNL/TM-2010/32, An Investigation of Digital Instrumentation and Control System Failure Modes
15. AldemirI T, Stovsky MP, Kirschenbaum J et al (2007) NUREG/CR-6942. Dynamic Reliability Modeling of Digital Instrumentation and Control Systems for Nuclear Reactor Probabilistic Risk Assessment
16. Chu TL, Martinez-Guridi G, Yue M, Lehner J et al (2008) NUREG/CR-6962. Traditional Probabilistic Risk Assessment Methods for Digital Systems
17. Gassino J, Regnier P, Birla S et al (2011) NUREGI/IA-0254. Suitability of Fault Modes and Effects Analysis for Regulatory Assurance of Complex Logic in Digital Instrumentation and Control Systems
18. Ma ZG, Wierman TE, Kvarfordt KJ et al (2021) INL/EXT–21-65055. Industry-Average Performance for Components and Initiating Events at U.S. Commercial Nuclear Power Plants: 2020 Update

19. McNelles P, Zeng ZC, Renganathan G (2017) Failure mode taxonomy for assessing the reliability of Field Programmable Gate Array based Instrumentation and Control systems. Ann Nucl Energy 108:198–228
20. Lee SJ, Jung W, Yang JE (2016) PSA model with consideration of the effect of fault-tolerant techniques in digital I&C systems. Ann Nucl Energy 87:375–384
21. Seo J, Kang HG, Lee EC (2020) Experimental approach to evaluate software reliability in hardware-software integrated environment. Nucl Eng Technol 52(7):1462–1470
22. Stadler JJ, Seidl NJ (2013) Software failure modes and effects analysis. In: Proceedings annual reliability and maintainability symposium (RAMS). IEEE, pp 1–5
23. Haapanen P, Helminen A (2002) Failure mode and effects analysis of software-based automation systems
24. Hecht M, Shokri E, Meyers S, Nguyen E, Greenwell W, Lam A (2008) Failure modes and effects analysis for a software-intensive satellite control system. The Aerospace Corporation, El Segundo, California, USA
25. Park GY, Kim DH, Lee DY (2014) Software FMEA analysis for safety-related application software. Ann Nucl Energy 70:96–102
26. International Atomic Energy Agency (2018) Dependability Assessment of Software for Safety Instrumentation and Control Systems at Nuclear Power Plants, IAEA Nuclear Energy Series No. NP-T-3.27, IAEA, Vienna

Research on the Reliability Assurance System in the Product Development Process

Yinping Ren, Tongyu Zhang, and Lili Chang

Abstract With the progression of reliability engineering, high reliability has increasingly become a fundamental requirement for product development across all stakeholders. Reliability is systematically integrated into design, production, and management processes. The reliability objectives established during the design phase must be realized through appropriate manufacturing practices. This study seeks to elucidate the concept and connotation of reliability assurance throughout the product development lifecycle and constructs a reliability assurance system for product development. The proposed system delineates specific work items along with their implementation guidelines, offering both management decision-making support and technical engineering foundations to effectively ensure the fulfillment of product reliability objectives.

Keywords Product development process · Reliability assurance system · Work items · Implementation guidelines

1 Introduction

Awareness of the importance of ensuring reliability throughout product development has been steadily increasing, making the assurance of product reliability a critical concern. Reliability engineering is not an isolated activity but is deeply embedded within the design, manufacturing, and managerial frameworks. The process begins with the rigorous definition of reliability indicators during the conceptual design phase, followed by the execution of targeted tasks to verify that these benchmarks are achieved. This study first clarifies the definition and connotation of reliability assurance across the entire product development lifecycle. Subsequently, by drawing on risk management principles and considering factors that influence the implementation of reliability requirements, a reliability assurance system is established.

Y. Ren (✉) · T. Zhang · L. Chang
Beijing Great Wall Electronic Equipment Co., LTD, Beijing, China
e-mail: renyinping2514@126.com

H. Li et al. (eds.), *Data-Driven Methods for Reliability and Safety Engineering: Applications in Industrial Systems*, Springer Series in Reliability Engineering,
https://doi.org/10.1007/978-3-032-22873-4_11

This system, based on the reliability tasks specified in GJB450B-2021, includes detailed work items and implementation criteria. The proposed framework provides a scientific basis for conducting reliability assurance activities throughout product development and offers both management decision-making support and technical engineering guidance to effectively ensure product reliability.

2 Definition and Connotation of Reliability Assurance in Product Development Process

2.1 Definition of Reliability Assurance in Product Development Process

The definition of reliability assurance throughout product development refers to the capability to ensure that a developed product meets specified reliability requirements under the constraints of various factors within the product development reliability assurance system.

2.2 The Connotation of Reliability Assurance in the Product Development Process

The connotation of this system rests on three pillars: the temporal dimension, which covers the entire lifecycle from design to implementation; the structural dimension, comprising specific work items and their corresponding execution guidelines; and the influencing factors, such as infrastructure and management controls. These elements collectively form the foundation for identifying risks and enforcing reliability standards throughout the development process.

These three basic connotations provide a clearer definition of reliability assurance throughout product development, guiding both research and practical implementation of reliability assurance during development.

3 Reliability Assurance System for Product Development Process

Drawing on the concept and connotation of reliability assurance throughout product development, this study analyzes reliability assurance activities in the development process and constructs a reliability assurance system applicable to current product development. Research is conducted across three dimensions: infrastructure, reliability-related tasks, and management-control activities, with the aim of defining the content of reliability assurance work during product development.

3.1 Infrastructure

Infrastructure [1] refers to the material facilities essential to the product development process. These facilities serve as the material foundation for ensuring product reliability and meeting design requirements, encompassing organization, personnel allocation, equipment and facilities, measurement and testing, and standardization.

Organization: The establishment of specialized or independent reliability-focused organizations or roles within product development units is essential to ensure product reliability. Personnel allocation: During product development, the influence of personnel and institutional arrangements is significant [2]. Factors such as personnel assignment, work competence, physiological conditions, and psychological factors may lead to inadequate management or process errors, thereby affecting product reliability. Equipment and facilities: Equipment constitutes the fundamental hardware for transforming raw materials into products and executing production processes, serving as a critical mechanism to ensure manufacturing process reliability and guarantee product quality. Measurement and testing: Measurement activities permeate all stages of new product development, including raw material testing, production process control, product quality inspection, and environmental monitoring. These activities play a crucial role in verifying that components, processes, equipment, and products conform to design and process specifications [3]. Standardization: The establishment of management systems and operating procedures facilitates the identification and mitigation of hidden risks, monitoring of major hazard sources, implementation of preventive mechanisms, and regulation of production activities. This ensures that production processes comply with relevant laws, regulations, and standards, maintains personnel, equipment, materials, and the environment in optimal condition [4], and supports the smooth execution of production processes while achieving reliability objectives.

3.2 Reliability Work

Reliability refers to a product's ability to perform its specified functions under defined conditions and within a designated time frame. According to GJB450B-2021 General Requirements for Equipment Reliability Work, reliability activities in the product development process are defined to include four key components: reliability requirements, reliability design and analysis, reliability testing and evaluation, and design input evaluation and feedback.

Reliability requirements: Define qualitative and quantitative reliability requirements to serve as inputs for reliability design and a basis for assessment and verification. Reliability design and analysis: Ensure the effective implementation of product design reliability requirements through reliability design and analysis methods.

Reliability testing and evaluation: Conduct reliability testing and evaluation in accordance with reliability design and implementation requirements. This identifies defects in product design, materials, processes, and manufacturing, while verifying compliance with quantitative reliability requirements. Evaluation and feedback of design input: Manufacturers verify design schemes in advance of manufacturing in response to product reliability design requirements provided by designers. This includes evaluating material performance and the feasibility of implementing design requirements during manufacturing.

3.3 Management and Control Work

During product development, development units integrate various reliability activities through planned management practices and effective control methods to meet the reliability indicators specified in the design [5]. These efforts coordinate and ensure that products satisfy reliability requirements, with three key components: planning and scheduling, process control, and information management.

Planning and scheduling: Systematically plan, schedule, and manage the product development process to ensure product reliability.

Process control: Monitor and supervise reliability activities throughout the entire product development process to ensure product reliability, while facilitating feedback and adjustments for design improvements.

Information management: Establish, monitor, and maintain reliability information systems for the product development process to achieve reliability design indicators.

4 Reliability Assurance System for Product Development Process

This section integrates factors influencing the implementation of reliability requirements during product development with the reliability tasks specified in GJB450B-2021 [6]. It constructs specific tasks applicable to reliability assurance in China's product development process, aiming for comprehensive coverage without omissions. To clarify the structural hierarchy prior to detailed tabulation, the system is organized such that the three primary dimensions—infrastructure, reliability work, and management—decompose into specific work items, which are further operationalized by executable implementation guidelines. This cascading framework ensures that high-level reliability objectives are systematically translated into actionable engineering protocols and verifiable criteria across the development lifecycle. Development and production units may make deletions based on actual circumstances, as outlined in Table 1.

5 Reliability Assurance Work Project for the Product Development Process

Following the hierarchical framework established in the previous section, Table 2 operationalizes the "Work Items" listed in Table 1 into specific "Implementation Guidelines." While Table 1 identifies the essential tasks required at each developmental stage, Table 2 defines the executable criteria for verifying these tasks. This distinction ensures that the system provides not only a managerial checklist but also a granular engineering standard for execution and assessment, serving as the basis for subsequent reliability evaluations.

The product development process, particularly for equipment in China, encompasses the demonstration stage, scheme design, engineering development, finalization stage, and production stage. This process involves the determination, implementation, verification, and refinement of equipment tactical and technical indicators as well as general quality characteristic indicators, while establishing a comprehensive closed-loop control system throughout all stages. The reliability assurance system for product development proposed in this study is examined from a 5M1E perspective, encompassing nearly all factors that influence the achievement of reliability objectives and providing a comprehensive synthesis and integration. In practical operations, organizational departments allocate responsibilities according to the operation of the quality management system. The technical department emphasizes reliability assurance during product design, the manufacturing department focuses on reliability assurance in production processes, and functional departments oversee reliability management throughout product development. This approach aligns with the principle that reliability is designed, produced, and managed. Effective implementation of reliability assurance throughout product development can be enhanced by addressing two primary aspects: resource investment and reliability assurance activities [11–13]. Resource investment must be secured and coordinated, and reliability assurance tasks should be selected appropriately based on the constraints of the development stage and available resources to prevent resource waste and avoid delays in the development cycle.

Table 1 Reliability assurance work project for product development process

Reliability assurance system	Reliability assurance work	Reliability assurance work project
Infrastructure	Organization	Establish independent reliability work and organizational management structures or positions
		Establish a dedicated reliability testing management organization or position
		Establish a dedicated reliability information management organization or position
		Establish a dedicated management organization or position for ensuring the reliability of electronic components
		Establish a dedicated after-sales reliability improvement and growth management organization or position
	Personnel configuration	Set up a dedicated leader responsible for the reliability work of the unit
		Equipped with a comprehensive reliability team
		Equip reliability testing personnel
		Set up reliability information management personnel
		Establish a comprehensive reliability training system
		Establish a comprehensive system for evaluating the work of reliability professionals
	Equipment and facilities	Configure supporting production and manufacturing equipment
		Ensure the number of devices
		Ensure the progressiveness of production and processing equipment [7]
		Ensure the reliability of the equipment [6, 8]

(continued)

Table 1 (continued)

Reliability assurance system	Reliability assurance work	Reliability assurance work project
		Ensure the supply of basic resources for reliability work
		Establish a long-term mechanism for ensuring reliable work resources
	Measurement and testing	Adopting correct and standardized measurement methods and tools
		Introduce and update new measurement and testing tools
		Management and maintenance of measuring and testing equipment [9]
	Standardization [4]	Standardization of basic management
		Standardization of infrastructure security conditions
		Standardization of the work environment and occupational health
Reliability work	Reliability requirements	Determine the qualitative and quantitative requirements for reliability
		Determine the requirements for reliability work projects
		Reliability allocation
	Reliability Design and Analysis	Reliability modeling
		Reliability prediction
		Failure modes, effect and criticality analysis (FMECA)
		Fault tree analysis (FTA)
		Sneak analysis
		Circuit tolerance analysis
		Development of reliability design criteria and compliance checks
		Selection and control of components, standard parts, and raw materials
		Determine reliability critical products

(continued)

Table 1 (continued)

Reliability assurance system	Reliability assurance work	Reliability assurance work project
		Determine the impact of functionality, packaging, storage, loading and unloading, transportation, and maintenance on product reliability
		Vibration simulation analysis
		Temperature simulation analysis
		Electrical stress simulation analysis
		Durability analysis
		Software reliability requirement analysis and design
		Process analysis and control of reliability critical products
	Reliability testing and evaluation	Environmental stress screening. (ESS)
		Reliability development test
		Reliability qualification test
		Reliability acceptance test
		Reliability analysis and evaluation
		Life test
		Software reliability testing
	Evaluation and feedback of design inputs	Evaluation and feedback of material properties
		Verification and feedback of reliability design requirements
		Verification and feedback of processing procedures and processes
Management and control work	Planning and scheduling	System planning and management reliability work
		Develop a work plan that meets product reliability requirements and characteristics
		Establish applicable reliability assurance standards
	Process control	Develop requirements for reliability work management

(continued)

Table 1 (continued)

Reliability assurance system	Reliability assurance work	Reliability assurance work project
		Implement reliability growth management
		Conduct reliability design verification
		Establish a FRACAS based on the entire product production process
		Establish a reasonable and effective reliability workflow
		Organize milestone reviews
		Capable of controlling technical status
		Establish a fault review organization
		Adaptability of design improvement
		Management of the production site
		Perform reliability management on the subcontractor and supplier
	Information management	Establish and operate a comprehensive reliability information system
		Establish a reliability information database
		Establish standardized operational requirements for information systems
		Effective operation of reliability information system

6 Application

The engineering applicability of this system is validated using a shipborne electronic product development unit as the research subject. Weight allocation for reliability assurance tasks during development is determined through the Analytic Hierarchy Process (AHP). The overall reliability assurance level throughout the product development lifecycle is quantified using the radar chart minimum area method in conjunction with the implementation rules for reliability assurance tasks. A difference analysis method is employed to evaluate the impact of individual reliability

Table 2 Guidelines for the implementation of reliability assurance work projects in the product development process

Reliability assurance work project	Implementation guidelines
Establish independent reliability work and organizational management structures or positions	Establish a specialized reliability organization or position with independent functions, and be responsible for organizing the reliability work of the unit
	Develop job responsibilities with clear tasks and collaborate with other departments to carry out reliability work
Establish a dedicated reliability testing management organization or position	Establish specialized institutions or positions responsible for the reliability testing work of the unit
	Carry out reliability testing work in accordance with system requirements, overall planning, and organization
Establish a dedicated reliability information management organization or position	Establish a specialized organization or position responsible for reliability assurance information management
	Develop clear job responsibilities, coordinate and organize reliability information work
Establish a dedicated management organization or position for ensuring the reliability of electronic components	Establish a specialized organization or position responsible for managing component reliability information
	Develop clear job responsibilities, coordinate, and organize the reliability work of electronic components
Establish a dedicated after-sales reliability improvement and growth management organization or position	Establish specialized agencies or positions responsible for improving and increasing product after-sales reliability
	Establishing clear job responsibilities can coordinate and organize after-sales reliability work
Set up a dedicated leader responsible for the reliability work of the unit	There are dedicated leaders responsible for the reliability work of this unit
Equipped with a comprehensive reliability team	The three-level technical personnel team of reliability chief engineer, supervisor, and engineer is fully equipped and has clear responsibilities
	Third-level technical personnel have complete abilities and can meet the needs of carrying out reliability assurance work
Equip reliability testing personnel	Experimental technology and management personnel can comprehensively organize the operation of the laboratory

(continued)

Table 2 (continued)

Reliability assurance work project	Implementation guidelines
	The number and professional skills of experimental technicians meet the requirements for the operation of the experiment
Set up reliability information management personnel	Information management personnel can comprehensively organize the operation of reliable information institutions
	Trained personnel are responsible for the collection, processing, storage, and transmission of information
Establish a comprehensive reliability training system	Develop reliability training management systems at all levels and types, as well as a certification system for employment
	Original records, relevant textbooks, on-site pictures, and other materials of organized training
Establish a comprehensive system for evaluating the work of reliability professionals	Establish a reliability professional competence evaluation method and conduct regular performance evaluations
	The level setting and job benefits of reliability professionals are comparable to those of other departments
Configure supporting production and manufacturing equipment	Configure appropriate process equipment based on product characteristics
	Develop equipment functions with clear tasks and coordinate the completion of specified tasks
Ensure the number of devices	The unit has corresponding purchasing power for equipment
	Ensure the smooth progress of the product development process with sufficient quantity
Ensure the progressiveness of production and processing equipment	The unit has the ability to timely introduce and update new equipment
	The new equipment meets the requirements of product reliability design and can improve the accuracy of product processing
Ensure the reliability of the equipment	The unit has corresponding technical conditions for regular maintenance and management of equipment
	The unit is equipped with corresponding technical personnel to maintain equipment performance and ensure stable working condition of the equipment
Ensure the supply of basic resources for reliability work	There is a clear demand for resource support, such as cost, site, and policy for reliability work

(continued)

Table 2 (continued)

Reliability assurance work project	Implementation guidelines
	Ensure the implementation of reliability work requires resources to meet the needs of carrying out reliability assurance work
Establish a long-term mechanism for ensuring reliable work resources	Relevant plans and schedules for ensuring reliable work resources
	Reliable resource guarantee projects have been arranged in previous conditional construction and technological upgrades
Adopting correct and standardized measurement methods and tools	Equipped with corresponding testing tools
	Having corresponding detection technology and personnel
	The testing tool meets the product design requirements
Introduce and update new measurement and testing tools	Ability to introduce and update advanced detection technologies and tools
	Having corresponding technical personnel to carry out and implement testing work
	The inspection and testing technology meets the product design requirements
Management and maintenance of measuring and testing equipment	Closed-loop management and maintenance of testing equipment
	Having corresponding management personnel
Standardization of basic management	Develop clear, open, and document-based management systems and operating procedures
	The management system and operating procedures can play a role and are feasible
	The system and regulations are approved by the authorized person and undergo regular review and modification
Standardization of infrastructure security conditions	The safety conditions of the infrastructure comply with relevant regulations
	The relevant infrastructure can be securely placed without affecting usage and operation
	Infrastructure placement conditions meet safety standards
Standardization of the work environment and occupational health	Reasonable layout of homework environment
	Detection, management, and monitoring of hazardous factors related to occupational health in the workplace.[10]
	Health surveillance
Determine the qualitative and quantitative requirements for reliability	Reliability requirement research

(continued)

Table 2 (continued)

Reliability assurance work project	Implementation guidelines
	Reliability requirement demonstration
	Reliability requirement review
Determine the requirements for reliability work projects	Identify and preliminarily determine the requirements for reliability work projects
	Tailoring work projects based on product type and characteristics
	Reliability work project review
Reliability allocation	Select appropriate reliability allocation techniques and auxiliary software tools based on product type and characteristics
	Analyze the rationality of reliability allocation and gradually carry out allocation optimization work based on the deepening of design work
Reliability modeling	Select appropriate reliability modeling techniques and auxiliary software tools based on product type and characteristics
	Establish basic reliability model and task reliability model
Reliability prediction	Select appropriate reliability prediction techniques and auxiliary software tools based on product type and characteristics
	Analyze the rationality of reliability prediction, and gradually carry out prediction optimization work based on the deepening of design work
Failure modes, effect and criticality analysis (FMECA)	Select appropriate FMECA technology methods and auxiliary software tools based on product type and characteristics
	Ensure the full lifecycle of FMECA work and generate FMECA reports
	FMECA data is accurate and results are effective, conduct FMECA review
	FMECA results management closed loop, proposing improvement measures and providing guidance
Fault tree analysis (FTA)	Select appropriate fault tree techniques and auxiliary software tools based on product type and characteristics
	Conduct FTA for key faults that affect safety and tasks, ensure accurate fault tree data, effective results, and generate FTA reports
	FTA result management closed loop, proposing improvement measures and providing guidance

(continued)

Table 2 (continued)

Reliability assurance work project	Implementation guidelines
Sneak analysis	According to the design phase, initiate SA as early as possible, gradually update and form SA report
	Ensure the accuracy and effectiveness of SA data, propose improvement measures, and provide guidance
Circuit tolerance analysis	According to the design phase, conduct circuit tolerance analysis and generate SA report
	Ensure the accuracy and effectiveness of circuit tolerance design results, propose improvement measures, and provide guidance
Development of reliability design criteria and compliance checks	Determine reliability design criteria based on product type, characteristics, and engineering experience
	Implement the established reliability design criteria and form a reliability design criteria compliance report
	Combined with design review, review the compliance of reliability design criteria
Selection and control of components, standard parts, and raw materials	Develop and confirm the selection and control requirements for components, standard parts, and raw materials based on product types, characteristics, and engineering experience
	Develop and revise a catalog of preferred components, standard parts, and raw materials
	Conduct reliability review based on the design phase
Determine reliability critical products	Form a list of reliability critical products, clarify corresponding control methods, and testing requirements
	Conducting reliability reviews and making reasonable adjustments can serve as input for conducting analytical work
Determine the impact of functionality, packaging, storage, loading and unloading, transportation, and maintenance on product reliability	Analyze and evaluate the impact of functional testing on reliability
	Ensure the practicality of analysis and evaluation results, and guide product design improvement
Vibration simulation analysis	Select appropriate vibration simulation techniques and auxiliary software tools based on product type and characteristics

(continued)

Table 2 (continued)

Reliability assurance work project	Implementation guidelines
	Analyze the vibration simulation results, identify design weaknesses, propose improvement measures, and generate an analysis report
	Ensure the accuracy, effectiveness, and practicality of the analysis results, and provide guidance for product design improvement
Temperature simulation analysis	Select appropriate temperature simulation techniques and auxiliary software tools based on product type and characteristics
	Analyze temperature simulation results, identify design weaknesses, propose improvement measures, and generate an analysis report
	Ensure the accuracy, effectiveness, and practicality of the analysis results, and provide guidance for product design improvement
Electrical stress simulation analysis	Select appropriate electrical stress simulation techniques and auxiliary software tools based on product type and characteristics
	Analyze the results of electrical stress simulation, identify weak links in the design, propose improvement measures, and form an analysis report
	Ensure the accuracy, effectiveness, and practicality of the analysis results, and provide guidance for product design improvement
Durability analysis	Select appropriate analysis methods and auxiliary software tools based on product type and characteristics to conduct durability analysis
	Based on the in-depth design work, iterate durability analysis and generate a durability analysis report
	Ensure the accuracy, effectiveness, and practicality of the analysis results, and provide guidance for product design
Software reliability requirement analysis and design	Carry out the determination and allocation of software reliability requirements
	Determine software reliability design criteria based on software product type and characteristics

(continued)

Table 2 (continued)

Reliability assurance work project	Implementation guidelines
	Carry out software reliability design and analysis work, closely integrate with functional design, and identify weak links
	Conduct reliability assessments on purchased and developed software to effectively control reliability risks within an acceptable range
Process analysis and control of reliability critical products	Conduct process analysis based on the characteristics of reliability-critical products, develop a general process plan, and process quality control requirements
	Develop process control documents and conduct reviews
	Integrate with the actual manufacturing process, analyze and update relevant documents, take corrective measures for deviations from reliability levels, and ensure reliability levels
Environmental stress screening (ESS)	Determine the screening timing and product hierarchy, and conduct tests according to the environmental stress screening test plan, outline, or relevant standards
	Clarify the departments and responsibilities for implementing and supervising the experiment
	Ensuring the practicality of environmental stress screening test results can eliminate early product failures and provide guidance for product design and improvement
Reliability development test	According to the characteristics and types of the product, conduct tests in accordance with the reliability development test plan, outline, or relevant standards
	Ensure the accuracy and validity of the test results, analyze the test results and data, and implement necessary improvements
	Record the faults that occurred during the experiment and execute FRACAS
Reliability qualification test	Conduct tests according to the reliability identification test plan, outline, or relevant standards based on product characteristics and types
	Combined with product FMECA, record faults that occur during testing and execute FRACAS
	Review the experimental outline, experimental arrangement, and experimental results

(continued)

Table 2 (continued)

Reliability assurance work project	Implementation guidelines
Reliability acceptance test	Conduct tests according to the reliability acceptance test plan, outline, or relevant standards based on product characteristics and types
	Combined with product FMECA, record faults that occur during testing and execute FRACAS
	Review the experimental outline, experimental arrangement, and experimental results
Reliability analysis and evaluation	Determine the analysis and evaluation object and timing
	Develop an analysis and evaluation plan based on the specific situation of the product, and carry out and implement reliability analysis and evaluation as early as possible
	Review the reliability analysis and evaluation results
Life test	Determine the test subjects and timing, and develop a test plan based on the specific situation of the product
	Review the experimental outline and results
	Record faults that occur during testing and execute FRACAS
	Ensuring the accuracy, effectiveness, and practicality of the test results provides guidance for product design and improvement
Software reliability testing	Determine the software reliability testing object and timing
	Specify the testing plan based on the specific situation of the software, and determine the testing type and method
	Review the test outline and test results
	Ensuring the accuracy, effectiveness, and practicality of test results provides guidance for software design and improvement
Evaluation and feedback of material properties	Master the technical methods of material performance evaluation and equip evaluation software tools
	Conduct comprehensive evaluation of material properties using product development, testing, and usage information
	Feedback the evaluation results to the design team, support reliability design analysis, design improvement, and growth management

(continued)

Table 2 (continued)

Reliability assurance work project	Implementation guidelines
Verification and feedback of reliability design requirements	Master the technical methods for evaluating reliability design requirements and equip evaluation software tools
	Comprehensive evaluation of reliability design requirements using product development, testing, and usage information
	Feedback the evaluation results to the design team, support reliability design analysis, design improvement, and growth management
Verification and feedback of processing procedures and processes	Master the technical methods for evaluating process and process requirements, and equip evaluation software tools
	Conduct comprehensive evaluation of process and technological requirements using product development, testing, and usage information
	Provide feedback on the evaluation results to the design team, support reliability design analysis, design improvement, and growth management
System planning and management reliability work	The top-level documents and requirements related to the overall planning of the reliability assurance work of our unit
	Develop corresponding management systems to ensure the implementation of reliability planning in practical work
	Comprehensive planning of product reliability assurance work projects
Develop a work plan that meets product reliability requirements and characteristics	The reliability assurance work plan reflects the characteristics of the product, with clear input and output, and complete project content
	The time nodes of the reliability assurance work plan meet the relevant requirements and are coordinated with the model design schedule
	The reliability assurance work plan is continuously revised and improved with the development progress and military requirements
Establish applicable reliability assurance standards	Develop applicable work standards based on the reliability assurance system
	The work standards are in line with the characteristics of our unit, and have operability and pertinence

(continued)

Table 2 (continued)

Reliability assurance work project	Implementation guidelines
Develop requirements for reliability work management	Develop management requirements and control plans for reliability testing verification, component control, and other related work
	Capable of carrying out management and control work in accordance with management requirements and control plan systems
Implement reliability growth management	Develop a reliability growth plan as early as possible based on the development progress
	Determine reliability growth targets and select reliability growth models based on product types and characteristics
Conduct reliability design verification	Determine the reliability design verification object, timing, and requirements
	Form a reliability design verification report and conduct risk assessment
Establish a FRACAS based on the entire product production process	Establish a sound FRACAS
	Establish a dedicated fault reporting and analysis committee and effectively carry out its work
Establish a reasonable and effective reliability workflow	Establish a reliable workflow that is in line with the characteristics of the unit and can be matched with the model development program
	Establish a system for workflow and strictly implement it, iterate reliability design and product design, and promote design improvement
Organize milestone reviews	Clearly define the reliability status and requirements of products in each milestone
	Conduct reliability thematic review during milestone review, and the review results have a closed-loop control mechanism
Capable of controlling technical status	Master the technical status and changes in the product development process
	Master the technical status and changes after product finalization and delivery
	The information covers the entire lifecycle process of model development, production, testing, and use
Establish a fault review organization	Establish a sound fault review organization
	Capable of conducting regular fault review work
	The fault review results are correct and effective

(continued)

Table 2 (continued)

Reliability assurance work project	Implementation guidelines
Adaptability of design improvement	Maintain a smooth information exchange channel between the manufacturer and the designer
	Timely adjust production progress to meet design requirements
	Timely obtain design improvement information and adjust the manufacturing process to ensure product reliability meets design requirements
Management of the production site	Manage the entire production process on site
	Manage production site equipment and personnel
	Regularly inspect safety hazards on the production site
Perform reliability management on the subcontractor and supplier	Clearly state the reliability requirements of the model product to the supplier and include them in the contract
	Develop a reliability control plan for the subcontractor and conduct effective supervision and management
Establish and operate a comprehensive reliability information system	Establish a reliability information system based on product development and production characteristics
	Can provide effective information to support management consulting
	The process of summarizing, analyzing, reporting, and publishing reliability information is clear, with complete steps and a closed-loop operation
Establish a reliability information database	Establish a reliability information database and accumulate historical product information
	The database information covers the main reliability assurance work requirements and can support the implementation of reliability assurance work
Establish standardized operational requirements for information systems	Establish a reliability information application mechanism with clear requirements for collection, processing, analysis, and publication
	The information covers the entire lifecycle process of model development, production, testing, and use

(continued)

Table 2 (continued)

Reliability assurance work project	Implementation guidelines
Effective operation of reliability information system	The process of summarizing, analyzing, reporting, and publishing reliability information is clear, with complete steps and a closed-loop operation
	Information systems can provide effective information to support subsequent experimental verification and management consulting

assurance tasks, identify factors with significant influence on reliability throughout development, and prioritize their execution. Analysis and calculations indicate that FMEA, ESS, and reliability acceptance testing exert the most substantial influence on fulfilling reliability design requirements and represent the critical tasks for practical implementation. The prominence of these specific tasks stems from their critical roles at distinct lifecycle stages: FMEA proactively eliminates design-inherent failure modes, ESS exposes latent manufacturing defects through stress screening, and reliability acceptance testing serves as the definitive validation gate to ensure compliance with quantitative specifications before delivery. Assigning a dedicated leader to oversee the unit's reliability assurance activities and establishing a structured reliability assurance team have relatively lower impact on the development process; however, these elements remain essential and should not be neglected in practical operations.

7 Conclusion

In response to the growing emphasis on achieving high reliability requirements throughout product development, this study defines and elucidates the concept and connotation of reliability assurance within the development process. By integrating risk management principles and factors influencing the implementation of reliability requirements with the guidelines specified in GJB450B-2021, a reliability assurance system is established, comprising specific work items and implementation guidelines. Practical examples are employed to validate the system's applicability, and the influence of various factors on reliability assurance is systematically analyzed. This framework guides product development units to effectively implement reliability assurance tasks and enhance overall reliability assurance levels. Expert scoring and fuzzy evaluation methods are frequently applied in the verification of the practical examples. Future research will focus on refining and improving the defined tasks within the system, with particular emphasis on quantitative assessment of reliability assurance levels throughout product development, aiming to provide robust references for the scientific execution of reliability assurance activities during product development.

References

1. Fang P, Jian Z, Xiao FC (2009) A brief discussion on corrosion and protection of mechanical equipment and infrastructure. Inn Mongla Petrochem (12):12–13
2. Guo QX, Bao ZC (2001) Research on the mechanism and reliability of human error. China Saf Sci J 11(1):22–26
3. Fu CZ (1999) Handbook of measurement and testing technology for mechanical manufacturing. Machinery Industry Press
4. Shan YY, Wen C (2012) Theory and practice of safety production standardization in the mechanical manufacturing industry. J Dalian Jiaotong Univ 33(4):105–108
5. Hua S, Rui L, Guo LX (2012) Analysis of reliability management in equipment development process. J Acad Equip 23(5):109–112
6. GJB450B-2021 (2021) General requirements for equipment reliability work, National Military Standard of the People's Republic of China
7. Li XQ (1991) Several experiences on introducing mechanical equipment, Municipal Technology, pp 121
8. He ZJ, Cao HR, Zi YY (2014) Development and reflection on reliability assessment of mechanical equipment operation. J Mech Eng 50(2):171–186
9. Li YS (2007) How to manage measuring and testing equipment in a foundry. China Foundry Equip Technol 1:10–14
10. Zhang ZB, Sun QY, Guo JY (2008) Research on the influencing factors and quantification of occupational hazard supervision in workplaces. J Saf Sci Technol 4(2):35–39
11. Guo J, Wang Z, Li H, Yang Y, Huang CG, Yazdi M, Kang HS (2024) A hybrid prognosis scheme for rolling bearings based on a novel health indicator and nonlinear Wiener process. Reliab Eng Syst Saf 245:110014
12. Li H, Yazdi M, Huang HZ, Huang CG, Peng W, Nedjati A, Adesina KA (2023) A fuzzy rough copula Bayesian network model for solving complex hospital service quality assessment. Complex Intell Syst 9(5):5527–5553
13. Li H, Peng W, Adumene S, Yazdi M (2023) An intelligent cost-based consequence model for offshore systems in harsh environments. In: Intelligent reliability and maintainability of energy infrastructure assets, pp 107–117. Springer Nature Switzerland, Cham

An Architectural Design for the Reliability and Safety of Battery Management System

Xinpeng Wang, Yiqun Du, and Bo Meng

Abstract To improve the reliability and safety of power batteries, an enhanced architectural design for a battery management system is proposed. This design incorporates advanced methodologies and algorithms, including machine learning and a dynamic safety response mechanism, to establish the system architecture and functional capabilities. The proposed design facilitates effective fault diagnosis and prevention, thereby enhancing the operational safety and overall efficiency of complex battery-powered products.

Keywords Architectural design · Reliability · Safety · Battery management system

1 Introduction

The deployment of new power batteries in various complex products is expanding rapidly, driven by global energy transitions and heightened environmental awareness. Battery management systems (BMS) are essential for ensuring safe battery operation by monitoring critical parameters such as voltage, current, and temperature in real time, thereby preventing overcharging, over-discharging, and overheating [1]. In complex applications, high-voltage battery systems pose additional challenges to reliability and safety, which in turn affect the operational integrity of the entire product. Consequently, the development of advanced fault diagnosis methods integrated within BMS has become a focal point in research on battery reliability and safety in complex systems [2].

X. Wang (✉) · Y. Du
No. 92942 Unit of PLA, Beijing, China
e-mail: 1204297054@qq.com

B. Meng
Logistic Department of PLA Northern Combat Area Navy, Qingdao, Shandong, China

H. Li et al. (eds.), *Data-Driven Methods for Reliability and Safety Engineering: Applications in Industrial Systems*, Springer Series in Reliability Engineering,
https://doi.org/10.1007/978-3-032-22873-4_12

Li et al. implemented real-time monitoring and accurate estimation of battery status via a BMS to effectively prevent safety issues, including overcharging, over-discharging, and thermal runaway [3]. Additionally, Li et al. proposed a machine learning-based fault prediction model, demonstrating significant improvements in fault detection accuracy and timeliness [4]. These approaches enhanced battery operational efficiency and overall system safety. Nevertheless, existing BMS techniques exhibit limitations under complex failure scenarios and extreme operating conditions. For example, Geng et al. reported that traditional BMS fault diagnosis algorithms may fail under elevated temperature or voltage conditions, leading to misjudgments or omissions [1]. Furthermore, the advent of advanced battery technologies, such as solid-state and lithium-sulfur batteries, introduces novel failure modes and operational characteristics that differ markedly from conventional lithium-ion batteries. These developments necessitate adaptations in existing BMS techniques to meet the specific requirements of emerging battery technologies.

This study presents a systematic BMS architectural design incorporating machine learning-based fault diagnosis and safety management strategies. The proposed design aims to provide effective fault detection and prevention for power batteries, thereby enhancing the operational safety and overall efficiency of complex products and improving the user experience.

2 System Architecture

An architectural design for a battery management system (BMS) is proposed, incorporating advanced fault diagnosis and safety management functionalities. The BMS architecture consists of three primary components: a hardware platform, a software framework, and a communication protocol. The hardware platform comprises various sensors, including voltage, current, and temperature sensors, a microcontroller unit (MCU), a battery unit, and an actuator. These elements function collaboratively to monitor critical battery parameters in real time and execute control commands to protect the batteries from potential damage. Data collected by the sensors is transmitted to the MCU for real-time processing and analysis using preconfigured algorithms to assess battery status.

The sensors are responsible for continuously monitoring voltage, current, and temperature to ensure the safe and efficient operation of the batteries. The MCU serves as the central processing unit, receiving sensor data, processing and analyzing the information through the embedded algorithms, and generating real-time judgments regarding battery status. Based on these assessments, the MCU issues appropriate control commands. The battery unit provides the power supply for the system, while the actuator executes commands from the MCU, such as adjusting the charge current or initiating the cooling system, to maintain safe and reliable battery operation.

Software framework: It consists of data processing algorithms, a fault diagnosis module, safety decision logic, and a user interface. Data processing algorithms extract useful information from sensor data and evaluate battery health status. The fault diagnosis module adopts the machine learning technique to predict and identify possible faults based on historical and real-time data. According to the diagnosis results, the safety decision logic will take corresponding safety measures such as adjusting the charge strategy, starting the cooling system, or disconnecting batteries in severe cases. The user interface provides a visual platform for operators to monitor the status of batteries, set system parameters, and receive fault alarms.

1. Data processing algorithms: They are responsible for extracting useful information from the data collected by sensors and evaluating the health status of batteries. The common algorithms include filtering, data fusion, and status estimation.
2. Fault diagnosis module: The machine learning technique is adopted to predict and identify possible faults based on historical and real-time data. For instance, pattern recognition can be used to detect any anomaly in batteries beforehand and take preventive measures.
3. Safety decision logic: It takes corresponding safety measures based on the results from the fault diagnosis module. For instance, if batteries are detected to be overheating, the system will adjust the charge strategy or start the cooling system. In severe cases, the system may even disconnect batteries to prevent an accident.
4. User interface: It offers a visual platform for operators to monitor the status of batteries, set system parameters, and receive fault alarms. It is normally graphical so that operators can have an immediate glimpse of the operational status of batteries, and then implement necessary operations.

Communication protocol: It coordinates the communication between hardware components and software applications. The system employs the Controller Area Network (CAN) protocol, conforming to industrial standards. The CAN protocol can not only guarantee reliable and real-time data transmission, but also support the communication of multiple MCUs and sensors when the central host is not involved. This will enhance the redundancy and anti-interference of a system, making it particularly significant in automotive applications.

To explicitly define the operational boundaries, the hardware platform functions as the physical execution layer responsible for precise data acquisition and actuator enforcement, while the software framework operates as the logical control layer executing complex state estimation. This structural separation is integrated via a deterministic data-flow mechanism that synchronizes physical sensing with algorithmic processing, ensuring that logical fault decisions utilize temporally aligned battery status data.

The system design focuses on modularization and scalability for better applicability to different types and scales of battery systems. This integrated architecture enables the BMS to effectively manage normal battery operation, provide critical protection, and significantly enhance battery safety and reliability (Fig. 1).

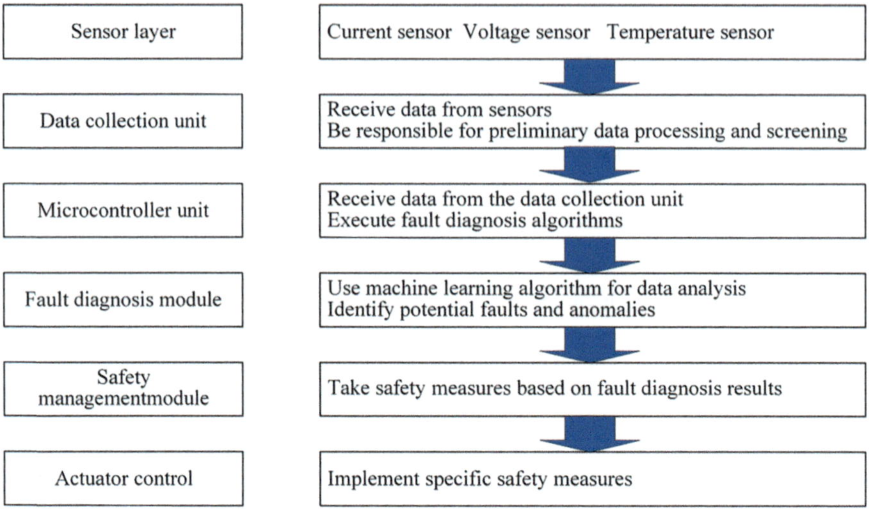

Fig. 1 Overall framework of a battery management system

3 Functionality Implementation

In this study, the functional implementation of the battery management system (BMS) emphasizes improving the accuracy and response speed of fault diagnosis while enhancing overall battery safety through a dynamic safety management mechanism. The implementation is centered on three primary modules: the fault diagnosis module, the safety management module, and the user interaction interface.

The fault diagnosis module utilizes machine learning algorithms to identify and predict potential battery faults. This module continuously collects real-time data on battery voltage, current, and temperature, which are then employed to train the predictive models. These models are capable of recognizing abnormal battery conditions, and the system triggers an alert upon detection of any potential fault signals. Additionally, data analysis is performed to identify probable causes and recommend solutions for the detected faults. The fault diagnosis module encompasses three core functionalities:

1. Data collection and preprocessing: This module collects the data of voltage, current, and temperature of the battery unit in a real-time manner. After being collected from sensors, the data are preprocessed, e.g., noise filtering and data normalization, so as to ensure the quality of input data.
2. Algorithm training and application: The preprocessed data are used to train the machine learning models. These models are used to predict potential faults by identifying the abnormal operational status of batteries.
3. Fault detection and response: If any potential fault signal is detected, the system will immediately trigger an alarm. In the meanwhile, the system may give the possible causes and suggested solutions for the fault after data analysis.

Safety management module: It executes corresponding safety measures according to the fault diagnosis results, including dynamic adjustment of charge and discharge strategies. For instance, it may limit charge current or adjust the depth of discharge to prevent excessive stress on batteries. While detecting several faults, e.g., short circuit or overheating of batteries, the module can automatically disconnect batteries from the system to avoid worse safety incidents. Safety management has three core functionalities:

1. Dynamic adjustment strategy: The charge and discharge strategies are dynamically adjusted according to the diagnosis results. For instance, limiting charge current or adjusting depth of discharge can prevent excessive stress on or loss of batteries.
2. Emergency protection measure: While detecting any severe fault (e.g., short circuit or overheating of batteries), the module can automatically disconnect batteries from the system to prevent any safety accident. This mechanism effectively avoids some dangers, including fire or explosion caused by battery overheating.
3. System integration: The Safety management module is integrated with other systems of a vehicle (e.g., vehicle-mounted cooling system), and collaborates with them to effectively manage battery temperature. For instance, if the battery temperature is too high, the system will start the cooling system to lower the temperature.

User interaction interface: and receives timely fault diagnosis and safety warning. The information displayed on the interface includes real-time battery status (voltage, current, and temperature), fault diagnosis results, and safety warnings. Furthermore, users can adjust BMS parameter settings via the interface according to operational conditions and personal preferences. The user interaction interface has three core functionalities:

1. Real-time status monitoring: The interface displays the real-time status of batteries, including voltage, current, and temperature. The data are displayed graphically to give the user a straightforward glimpse of battery operation.
2. Fault and safety warning: When the system detects any fault or potential danger, a warning will appear in the interface together with detailed results of fault diagnosis and suggestions on safety. For instance, if batteries are overheating, the interface will display a reminder to the user and recommend reducing the load or checking the cooling system.
3. Parameter adjustment: Users may adjust the settings of BMS parameters through the interface to different operation conditions and personal preferences. For instance, the user may set different charge strategies or temperature thresholds for the optimal performance and life of batteries.

Through the implementation of the above functionalities, the BMS in this study will not only achieve effective fault diagnosis and prevention of batteries, but also enhance the operational safety and efficiency of batteries. By integrating these functionalities, the BMS can improve user interaction experience and system adaptability while realizing advanced fault handling.

4 Discussion

The battery management system (BMS) proposed in this study demonstrates significant efficiency in fault diagnosis and safety management for power batteries. By integrating advanced machine learning algorithms with a dynamic safety response mechanism, the system can accurately predict and identify potential battery faults and implement necessary safety measures in real time, thereby substantially enhancing operational safety. Despite these advantages, several technical and application challenges have emerged during system implementation [11–14].

Firstly, while machine learning improves fault diagnosis accuracy, model training and optimization heavily depend on large volumes of high-quality data. In practical operation, batteries generate complex and variable data, posing challenges for data processing and model training. Moreover, ensuring the generalization capability of the model to maintain accuracy across different and novel battery types remains a significant concern. To address these challenges, the system incorporates synthetic minority over-sampling techniques to mitigate the impact of data imbalance inherent in rare fault events. Furthermore, the robustness of the diagnostic model is substantiated through rigorous k-fold cross-validation and noise-injection testing, ensuring that the algorithm maintains high detection fidelity even when encountering unseen failure modes or environmental interference in safety–critical scenarios. Secondly, although the dynamic safety management strategy facilitates effective fault response, its performance under extreme operating conditions requires further enhancement. Additionally, the design of the user interface is critical. The interface must convey accurate information regarding battery status and faults while providing clear visualization and ease of operation. Optimizing the interface can further enhance the usability and operational efficiency of the BMS.

While the proposed BMS shows promising results in power battery fault diagnosis and safety management, future research will focus on improving data processing capabilities, enhancing model generalization, developing robust measures for extreme conditions, refining user interface design, and advancing overall system performance. Continued technological innovation and practical application will contribute to further improvements in system reliability, operational safety, and user experience.

5 Conclusion

This study proposes an enhanced architectural design for the reliability and safety of the battery management system, encompassing the hardware platform, software framework, and communication protocol. The hardware platform enables real-time monitoring of battery status through integrated sensors and a microcontroller unit. The software framework incorporates machine learning and related techniques, combined with advanced fault diagnosis and safety management functionalities. The communication protocol employs the controller area network protocol to ensure reliable and real-time data transmission, significantly improving battery safety and operational reliability. Future research will focus on enhancing data processing efficiency, optimizing safety measures under extreme operating conditions, and improving user interface customization and functionality to achieve broader adaptability across diverse battery types and applications.

References

1. Xuena G, Zebin L (2023) Research on the distributed fault diagnosis algorithm. Sci technol Inf 21(22):9–12
2. Zhao Z et al (2023) Power electronics-based safety enhancement technologies for lithium-ion batteries: an overview from battery management perspective. IEEE Trans Power Electron (2023)
3. Xianfeng L, Chenggang H, Liming B et al (2024) A joint real-time SOC estimation method for energy storage batteries. Zhejiang Electric Power 43(05)
4. Ning L, Wei Z, Zelin G et al (2022) Fault prediction of intelligent electricity meter based on multi-classification machine learning model. Electr Energy Manage Technol 02:6–11
5. See KW et al (2022) Critical review and functional safety of a battery management system for large-scale lithium-ion battery pack technologies. Int J Coal Sci Technol 9(1):36
6. Olabi AG et al (2022) Battery thermal management systems: recent progress and challenges. Int J Thermofluids 15:100171
7. Hanhua L (2023) Research and design of electric vehicle power battery safety management system. Int Combust Eng Parts 01:33–35
8. Yuhua C (2020) Analysis on safety management strategy of power battery of electric vehicle. J Henan Inst Educ (Nat Sci Edn) 29(03):36–38
9. Liu FM, Liu T, Fu YZ (2015) An improved SOC estimation algorithm based on artificial neural network. In: 2015 8th international symposium on computational intelligence and design (ISCID), December 12–13, 2015. Hangzhou, China: IEEE, pp 152–155
10. Yuan DL, Zhang C, Tang SF et al (2021) Ferric ionascorbic acid complex catalyzed calcium peroxide for organic wastewater treatment: optimized by response surface method. Chinese Chem Lett 32(11):3387–3392
11. Guo J, Wang Z, Li H, Yang Y, Huang CG, Yazdi M, Kang HS (2024) A hybrid prognosis scheme for rolling bearings based on a novel health indicator and nonlinear wiener process. Reliab Eng Syst Saf 245:110014
12. Li H, Yazdi M, Huang HZ, Huang CG, Peng W, Nedjati A, Adesina KA (2023) A fuzzy rough copula Bayesian network model for solving complex hospital service quality assessment. Compl Intell Syst 9(5):5527–5553
13. Li H, Peng W, Adumene S, Yazdi M (2023) An intelligent cost-based consequence model for offshore systems in harsh environments. In: Intelligent reliability and maintainability of energy infrastructure assets. Springer Nature Switzerland, Cham, pp 107–117

14. Li H, Yazdi M (2022) What are the critical well-drilling blowouts barriers? A progressive DEMATEL-game theory. Advanced decision-making methods and applications in system safety and reliability problems: approaches, case studies, multi-criteria decision-making, multi-objective decision-making, fuzzy risk-based models. Springer International Publishing, Cham, pp 29–46

Research on Reliability Classification Technology of High-Temperature Gas-Cooled Reactor

Yiming Wang, Yongsheng Wen, Dingqiang Li, Zhijun Li, and Changliang Liu

Abstract To enable the rational allocation of maintenance and management resources, enhance the reliability of systems and components, and ensure the safe, stable, and economical operation of nuclear power plants, hierarchical equipment management is essential. With the advancement of equipment reliability management and maintenance technologies, methods for equipment reliability classification have gradually evolved. This paper examines the methodology and process of equipment reliability classification and presents a pilot study on the cabin negative pressure ventilation system of a high-temperature gas-cooled reactor. The analysis demonstrates that the high-temperature gas-cooled reactor provides a viable basis for implementing equipment reliability classification. Integrating classifications based on equipment operating frequency and environmental conditions into the reliability classification process facilitates the optimization of preventive maintenance programs, thereby enhancing both safety and economic performance.

Keywords Equipment management · Reliability classification · Preventive maintenance

1 Introduction

The safety of nuclear power plants fundamentally relies on the reliability of structures, systems, and components that fulfill the safety objectives of the facility. According to HAD 102/03 [3], structures, systems, and components must be safety-classified during the design and construction phases from the perspective of achieving the reactor's safety functions. Traditionally, safety classification categorizes structures, systems, and equipment into "safety-related" and "non-safety-related" based on their functional contributions to safety. The primary purpose of such classification is to establish graded design standards and specify differentiated requirements for SSCs according to their safety significance [4, 8, 9].

Y. Wang (✉) · Y. Wen · D. Li · Z. Li · C. Liu
Chinergy Co., Ltd., Beijing, China

H. Li et al. (eds.), *Data-Driven Methods for Reliability and Safety Engineering: Applications in Industrial Systems*, Springer Series in Reliability Engineering,
https://doi.org/10.1007/978-3-032-22873-4_13

For nuclear power plants, operational economic efficiency must be maximized while maintaining safety. To allocate maintenance and management resources effectively, enhance system and component reliability, and ensure safe, stable, and economical operation, graded management of equipment is essential. Traditional reliability-centered maintenance (RCM) methods begin with system function analysis and assess equipment criticality through failure mode and effects analysis (FMEA) [1, 2, 6, 7]. This approach, however, may inadequately identify critical equipment from a system function perspective, potentially excluding important assets from analysis. With the advancement of equipment reliability management and maintenance technologies, equipment reliability classification methods have been developed [9, 10]. These methods use the severity of consequences from individual equipment failures as the benchmark for equipment criticality classification, with the core focus on determining equipment criticality. In evaluating consequences, it is assumed that hot standby, cold standby, redundant systems, and installed spare parts are capable of performing their required functions.

Given the significant influence of equipment reliability classification on both safety and economic performance, this study investigates methods to enhance the safety and efficiency of high-temperature gas-cooled reactors and improve their market competitiveness. The research focuses on the equipment reliability classification methodology and analysis process, using the cabin negative pressure ventilation system as a case study to derive practical insights and establish a foundation for subsequent work.

2 Analysis Method

In the process of equipment reliability classification, equipment is divided into key equipment, important equipment, and general equipment according to the identification conditions corresponding to the consequences of equipment failure. The equipment reliability classification process is shown in Fig. 1.

Collecting Information

To carry out equipment reliability classification, the following design and operation data need to be collected in advance:

Design documents, including system design manuals, technical specifications, etc.;

Design drawings, including system flow chart, electrical wiring diagram, instrumentation and control logic diagram, etc.;

Operation procedures, including system operation procedures, emergency operating procedures, etc.;

Safety classification report;

Maintenance experience and operation experience of systems and component.

System Boundary

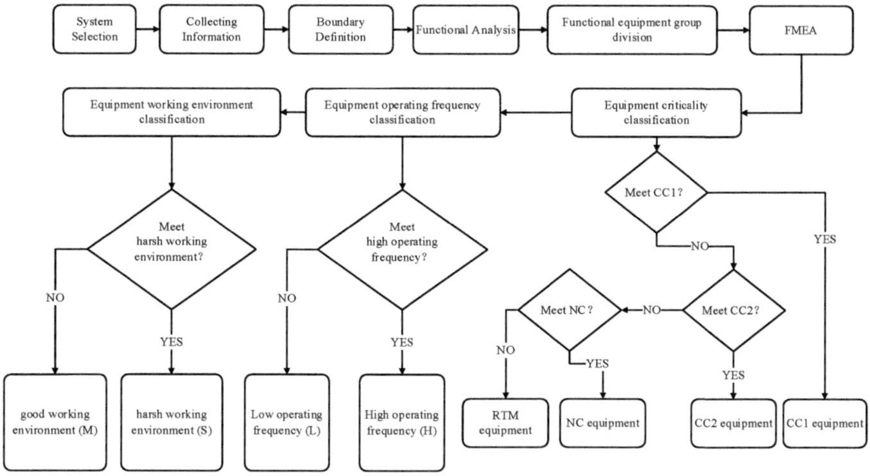

Fig. 1 Flow Diagram of equipment reliability Classification

In principle, the system boundary is consistent with the system KKS code of the nuclear power plant. The following method is usually used to define the system boundary:

1. Divide the boundaries by machinery: All mechanical equipment in the system is included in the analysis scope;
2. According to the electrical division boundary: the circuit breaker supplying power is included in the boundary, but the busbar is not included;
3. According to the instrumentation boundary: the entire instrument loop is contained within the boundary, even if its output may be sent outside the analysis boundary.

Functional Analysis

The functions of the system should be determined based on the system design manual, equipment technical specifications, and system flow charts. These system functions should be described in as much detail as possible, with quantitative data used where applicable. If necessary, the functions should be combined and classified according to the consequences of their failure and the judgment conditions.

Functional Equipment Group Division

Based on the main equipment under the system function, functional equipment groups (hereinafter referred to as "FEG") are divided, and the supported system functions are allocated to FEGs.

The FEGs coding rule is: P-XXX-YY-FEG-ZZZ. Among them, the first field "P" indicates the unit to which it belongs; the second field "XXX" is the system code; the third field "YY" is the serial number; the fourth field "FEG" is the abbreviation of the functional equipment group; the fifth field "ZZZ" indicates the serial code of the functional equipment group.

Failure Mode and Effect Analysis

Based on the operating characteristics of the equipment in the system and combined with the equipment's failure mode, determine the possible failure mode of the equipment in this system.

For equipment in the same functional group, the reliability of the main equipment is usually analysis first, and the reliability of other equipment can be analyzed according to the impact on the main equipment. The consequences of a single failure of the same equipment under different working conditions may be different. In order to ensure the accuracy of the equipment reliability classification, all normal operating conditions involved in the system must be considered. The consequences of failure should be analyzed based on the design documents such as the system design manual, and the specific functions, functional failures, failure modes, and the impact on the functional group, the system, and the nuclear power plant after the failure of each equipment in the FEGs should be analyzed, and the analysis results should be recorded.

Equipment Criticality Classification

Equipment criticality classification is based on the severity of the consequences caused by a single equipment failure. In the process of equipment criticality classification, according to the identification conditions corresponding to the consequences of equipment failure, it is divided into critical equipment (CC, including CC1 and CC2), important equipment (NC) and run-to-failure equipment (RTM). Among them, CC1 equipment, also known as critical sensitive equipment, refers to critical equipment whose single failure causes the nuclear power plant to shut down or shut down or stop heating or reduce power or power fluctuation exceeds the preset fluctuation range; CC2 equipment refers to equipment that plays a key role in the nuclear safety and unit power generation of the nuclear power plant. Failure of such equipment may lead to the loss of important functions that support the safety of the nuclear power plant or its degradation; NC equipment refers to equipment that plays an important role in the nuclear safety and unit power generation of the nuclear power plant, or equipment that can avoid major losses and reduce costs through maintenance; RTM equipment refers to other equipment except critical and important equipment.

Equipment Operating Frequency Classification

In order to optimize the preventive maintenance program based on the equipment reliability classification, the equipment operating frequency classification is added to the equipment reliability classification. The equipment operating frequency is divided into high operating frequency (H) and low operating frequency (L).

If the operating frequency of the equipment meets any of the following criteria, the operating frequency of the equipment is defined as high operating frequency (H):

1. Continuous operation;

2. Frequent switching;
3. Frequent manual throttling state;
4. In a state of continuous or frequent power supply.

Equipment Working Environment Classification

To optimize the preventive maintenance program based on the equipment reliability classification, the equipment working environment classification is added to the equipment reliability classification. The equipment working environment is divided into a harsh working environment (S) and a good working environment (M).

If the working environment of the equipment meets any of the following criteria, the working environment of the equipment is considered to be a harsh working environment (S):

High temperature environment;
High pressure environment;
High vibration environment;
High humidity environment;
High radiation environment;
Corrosive environment;
Spray, steam, and water scouring, erosion, dusty environment.

3 High Temperature Gas-Cooled Reactor Pilot Study

The high-temperature gas-cooled reactor exhibits high inherent safety, strong environmental adaptability, elevated core outlet temperatures, modular construction, and considerable potential for commercial applications in cogeneration and high-temperature process heat. It is regarded as one of the most promising fourth-generation reactors with significant development prospects. To maintain the cabin in a negative pressure state during primary circuit pipeline breakage accidents, a cabin negative pressure ventilation system (NPV) is implemented to purify radioactive helium within the cabin. This system effectively controls the accumulation of radioactive materials in the control area and prevents their release to the external environment, making it a critical component of the high-temperature gas-cooled reactor. Compared with traditional pressurized water reactors, the cabin negative pressure ventilation system represents a unique design feature of the high-temperature gas-cooled reactor. Consequently, this study selects the cabin negative pressure ventilation system for pilot analysis in equipment reliability classification.

System Boundary

In principle, the NPV system boundary and the KKS code of the NPV system are consistent. The following method is used to define the analysis boundary of the NPV system:

1. Mechanical boundary: all mechanical equipment such as pumps, valves, etc., with relevant numbers in the NPV system flow chart;
2. Electrical boundary: all motors and electrical switches with relevant numbers of the NPV system on the NPV system flow chart;
3. Instrumentation and control boundary: all local indicating instruments with relevant numbers of the NPV system on the NPV system flow chart, various analog measuring instruments for sending to DCS, various switching measuring instruments for sending to DCS, etc.

Functional Analysis

According to the design specification of the NPV system, the function of the cabin negative pressure ventilation system is to maintain the cabin in a negative pressure state, so that the accumulation of radioactive substances in the control area and the release to the external environment are effectively controlled. Its main functions are as follows:

Maintain the cabin in a negative pressure state;

After a primary-circuit pipeline breakage accident occurs in the cabin, the radioactive helium inside the cabin is purified to maintain the cabin in a negative pressure state based on the ALARA principle;

When personnel enter the cabin to carry out maintenance operations, ventilation should be carried out.

Functional Equipment Group Division

According to the design documents of the NPV system, FEGs are divided, and each FEG is coded. According to the impact of functional failure, the criticality level of each functional equipment group is analyzed. The functions and failure impacts of some FEGs are shown in Table 1.

Failure Mode and Effect Analysis

According to the NPV system design manual and combined with the equipment failure status, the specific functions, functional failures, failure modes, impacts on the functional groups after failure, impacts on the NPV system, and impacts on the high-temperature gas-cooled reactor of each device in the FEGs are analyzed, and the analysis results are recorded.

Table 1 Function and failure effect of partial FEGs

FEGs code	Functions of FEGs	Impact of function failure
3-NPV-01-FEG-001	Maintain the cabin in a negative pressure state	Failure to maintain the cabin in a negative pressure state
3-NPV-01-FEG-002	Purify the radioactive helium inside the cabin after a small primary pipeline breaks in the cabin	If a small primary pipeline breaks in the cabin, the radioactive helium inside the cabin may be discharged into the environment

In the analysis, combined with the design characteristics of the high-temperature gas-cooled reactor, the state of the equipment after failure was studied and summarized as follows:

1. A manual valve, manual switch or knife gate is in its original position after a fault;
2. The air valve remains in place after failure;
3. The electric valve does not consider the failure modes of accidental opening and accidental closing;
4. The electric valve remains in the de-energized position after power is lost.

In the analysis, the above four principles of equipment failure status are adopted (except for specific designs).

Equipment Criticality Classification

Taking FEGs as the unit, equipment criticality classification starting from the equipment list. Based on the consequences caused by a single equipment failure, the criticality level of equipment is divided into CC1, CC2, NC, and RTM according to the identification conditions corresponding to the consequences of equipment failure. The criticality classification process and results of some equipment in the NPV system are shown in Table 2.

Equipment Operating Frequency Classification

According to the working frequency of each device and the corresponding judgment criteria of the working frequency, the working frequency of the equipment is analyzed. The working frequency classification process and results of some equipment in the NPV system are shown in Table 2.

Equipment Working Environment Classification

According to the working environment of each device and the corresponding judgment criteria of the working environment, the working environment of the device is analyzed. The working environment classification process and results of some devices in the NPV system are shown in Table 2.

4 Analysis Results

Through the above analysis, the reliability classification results of the high temperature gas cooled reactor NPV system are obtained. These results indicate that reliability-significant equipment is concentrated in a limited number of functional equipment groups that directly support the negative-pressure function and radioactive helium purification. Within these FEGs, assurance tasks with strong diagnostic or screening capability—most notably FMEA and environmental stress screening (ESS)—naturally dominate the overall impact on system reliability. FMEA provides a systematic mapping from functional failures to equipment-level failure modes and

Table 2 Equipment reliability classification

Equipment code	Equipment name	FEGs code	Criticality code	Criticality determination basis	Working frequency	Working environment	Critical-ity level
NPV10AH001	Air purification unit	3-NPV-01-FEG-001	NC-g	Equipment failure leads to unacceptable increase in repair, replacement or operating costs	H(continuous operation)	M(good working environment)	NC
NPV10AH002	Air purification unit	3-NPV-01-FEG-001	NC-g	Equipment failure leads to unacceptable increase in repair, replacement or operating costs	H(continuous operation)	M(good working environment)	NC
NPV10AN001	Variable frequency centrifugal exhaust fan	3-NPV-01-FEG-001	NC-g	Equipment failure leads to unacceptable increase in repair, replacement or operating costs	H(continuous operation)	S(high vibration environment)	NC
NPV10AN002	Variable frequency centrifugal exhaust fan	3-NPV-01-FEG-001	NC-g	Equipment failure leads to unacceptable increase in repair, replacement or operating costs	H(continuous operation)	S(high vibration environment)	NC
NPV10AA001	Control valve	3-NPV-01-FEG-001	NC-g	Equipment failure leads to unacceptable increase in repair, replacement or operating costs	L(low operating frequency)	M(good working environment)	NC

(continued)

Table 2 (continued)

Equipment code	Equipment name	FEGs code	Criticality code	Criticality determination basis	Working frequency	Working environment	Critical-ity level
NPV10AA002	Control valve	3-NPV-01-FEG-001	NC-g	Equipment failure leads to unacceptable increase in repair, replacement or operating costs	L(low operating frequency)	M(good working environment)	NC
NPV20AH001	Air purification unit	3-NPV-01-FEG-002	CC2-h	Equipment failure leads to the inability to carry out the filtration and venting of leaked helium gas in the event of a minor pipe breakage accident	L(low operating frequency)	M(good working environment)	CC2

(continued)

Table 2 (continued)

Equipment code	Equipment name	FEGs code	Criticality code	Criticality determination basis	Working frequency	Working environment	Critical-ity level
NPV20AA101	Electric duct sealing valve	3-NPV-01-FEG-002	CC2-i	May cause the rupture disk in the compartment to burst, resulting in radioactive helium gas entering the outdoor environment without purification treatment	L(low operating frequency)	M(good working environment)	CC2
NPV20AA501	Fire valve	3-NPV-01-FEG-002	CC2-i	May cause the rupture disk in the compartment to burst, resulting in radioactive helium gas entering the outdoor environment without purification treatment	L(low operating frequency)	M(good working environment)	CC2

consequences, and therefore directly drives the identification of CC2 equipment in key ventilation and purification paths. ESS is primarily assigned to equipment operating in harsh environments or under high operating frequency, where latent defects are more likely to be revealed and where early removal of defective items has a disproportionate effect on preventing safety-related functional losses.

The NPV system has a total of 250 devices, which are divided into 15 FEGs. Among them, there are 0 CC1 level devices, 22 CC2 level devices, 228 NC level devices, and 0 RTM level devices.

In the pilot study, the relative importance of different assurance tasks was assessed under the implicit assumption that safety-related consequences carry higher weight than purely economic losses, and that equipment with high operating frequency or harsh working environments should be prioritised for proactive screening. Under this weighting scheme, tasks that are directly linked to the prevention of safety function degradation (such as FMEA on CC2 equipment) and to the early detection of environment-induced failures (such as ESS on equipment in S-class environments) receive higher priority. Expert scoring is mainly used to distinguish borderline cases, for example between NC and CC2 classifications or between assigning ESS versus routine inspection. Although such expert-based assessments are inevitably subject to uncertainty arising from differences in experience and risk perception, the qualitative prioritisation remains robust as long as the scoring preserves the dominance of safety-driven criteria over purely economic ones. A more detailed quantitative uncertainty analysis of the scoring process, such as sensitivity analysis on task weights or probabilistic modelling of expert judgements, is left as an extension for future work.

Further description of equipment reliability classification is as follows:

1. Equipment reliability classification includes all equipment in principle, but equipment such as pipelines and pipeline equipments, factory doors, etc., which have extremely low failure rates and no failure experience feedback in the industry, do not need to be classified;
2. The use of functional equipment group method improves the analysis efficiency and can directly determine the criticality level of key equipment functions within a group. Equipment in the same functional group can be scheduled for preventive maintenance at the same time.
3. In order to facilitate the formulation and optimization of the PM outline after equipment reliability classification, the classification process adds equipment operating frequency classification and working environment classification. Adding equipment operating frequency classification and working environment classification to the equipment reliability classification process will help to optimize the preventive maintenance outline based on the equipment reliability classification in the future.

Based on the equipment reliability classification results, the four levels of equipment—CC1, CC2, NC, and RTM—can be managed differently. CC1 and CC2 level equipment affect the safety and power generation of the high temperature gas-cooled reactor, so they need to be managed with priority. NC level equipment neither affects safety nor power generation, so maintenance investment can be minimized. RTM level equipment is generally not included in the maintenance plan. For the optimization of the preventive maintenance program, CC1 and CC2 level equipment are recommended to use a combination of preventive and predictive maintenance methods; NC level equipment is recommended to use predictive maintenance; and RTM level equipment is subject to inspection management, with corrective maintenance performed after failure.

5 Conclusions

This paper studies the equipment reliability classification method and conducts a case analysis using the cabin negative pressure ventilation system of a high temperature gas-cooled reactor compartment as an example, more here [2, 5–7]. The summary is as follows.

1. The high temperature gas-cooled reactor has the foundation for carrying out equipment reliability classification work. The equipment reliability classification method can be applied to the high temperature gas-cooled reactor to optimize the equipment maintenance strategy and equipment management. In particular, the three-layer architecture that links physical infrastructure, reliability analysis tasks, and plant management/control activities provides a clear pathway for embedding the classification results into existing equipment management processes.
2. According to the analysis results, the NPV system is divided into 15 FEGs, 0 CC1-level devices, 22 CC2-level devices, 228 NC-level devices, and 0 RTM-level devices.
3. Equipment reliability classification includes all equipment in principle. However, equipment with extremely low failure rates and no failure experience feedback in the industry, such as passive equipment such as pipelines and pipeline equipments and factory doors, does not need to be classified for equipment reliability.
4. In order to optimize the preventive maintenance program based on the equipment reliability classification, the equipment working frequency classification and equipment working environment classification were added to the equipment reliability classification.

References

1. Guan G (2013) Research on the characteristics and application of the reliability classification method of AP1000 nuclear power plant. Nucl Power Eng 34(6):88–91
2. Guo J, Wang Z, Li H, Yang Y, Huang CG, Yazdi M, Kang HS (2024) A hybrid prognosis scheme for rolling bearings based on a novel health indicator and nonlinear wiener process. Reliab Eng Syst Saf 245:110014

3. HAD 102/03 (1986) Classification of safety functions and components for boiling water reactors, pressurized water reactors and pressure tube reactors
4. INPO (2003) AP-913: equipment reliability process description
5. Li H, Yazdi M (2022) What are the critical well-drilling blowouts barriers? A progressive DEMATEL-game theory. In: Advanced decision-making methods and applications in system safety and reliability problems: approaches, case studies, multi-criteria decision-making, multi-objective decision-making, fuzzy risk-based models. Springer International Publishing, Cham, pp 29–46
6. Li H, Yazdi M, Huang HZ, Huang CG, Peng W, Nedjati A, Adesina KA (2023) A fuzzy rough copula Bayesian network model for solving complex hospital service quality assessment. Complex Intell Syst 9(5):5527–5553
7. Li H, Peng W, Adumene S, Yazdi M (2023) An intelligent cost-based consequence model for offshore systems in harsh environments. In: Intelligent reliability and maintainability of energy infrastructure assets. Springer Nature Switzerland, Cham, pp 107–117
8. Moubray J (1997) RCM II, reliability centered maintenance, 1st edn. Industrial Press Inc., South Norwalk
9. Smith AM (1993) Reliability-centered maintenance, 1st edn. McGraw-Hill

Maintenance Optimisation for K-out-of-N: G System Using Deep Reinforcement Learning

Ning Zhang, Yifan Zhou, Ze Li, and Ziqiang Huang

Abstract In high-dimensional stochastic degradation environments, condition-based maintenance optimization for k-out-of-n systems is extremely difficult, making conventional dynamic programming techniques computationally unfeasible. This study uses a Gamma process degradation framework with scheduled inspections to formulate the maintenance problem as a Markov decision process. We propose a continuous-action Soft Actor-Critic (SAC) approach in which maintenance actions reflect the number of components that need to be replaced. We compare our SAC methodology with three discrete-action deep reinforcement learning methods: a Double Deep Q-Network (DDQN) for component quantity selection, a DDQN01 variation that provides binary maintenance decisions for individual components, and Proximal Policy Optimization (PPO). A threshold-based Monte Carlo method is also used for baseline comparison. When applied to a 3-out-of-5 feedwater pump configuration, the SAC technique outperforms threshold-based policies by about 3% on long-term average costs. Discrete-action methods produce undesirable intervention options, albeit having faster convergence rates. According to our sensitivity analysis, the benefits of deep reinforcement learning are greatest in situations with little redundancy or large-scale deployments, whereas traditional threshold rules are adequate for smaller, redundancy-rich systems. These results support the effectiveness and scalability of both continuous-action SAC and discrete-action deep reinforcement learning variants for optimizing multi-component maintenance, and they point to potential future uses in multi-objective cost structures and dependent failure scenarios.

N. Zhang · Y. Zhou (✉) · Z. Li · Z. Huang
School of Mechanical Engineering, Southeast University, Nanjing, China
e-mail: yifan.zhou@seu.edu.cn

N. Zhang
e-mail: 230228064@seu.edu.cn

Z. Li
e-mail: 220230400@seu.edu.cn

Z. Huang
e-mail: 220240407@seu.edu.cn

H. Li et al. (eds.), *Data-Driven Methods for Reliability and Safety Engineering: Applications in Industrial Systems*, Springer Series in Reliability Engineering,
https://doi.org/10.1007/978-3-032-22873-4_14

Keywords Maintenance · K-out-of-N: G System · Deep reinforcement learning · Soft actor-critic

1 Introduction

In reliability engineering, the issue of optimizing maintenance strategies for k-out-of-n systems remains a key one. These systems, which are the foundation of many industrial implementations such as electronic circuit configurations, satellite communication networks, petroleum pipeline systems, and aircraft navigation technologies, need at least k operational components out of a total of n components to remain functional [8]. Although inherent redundancy increases system reliability, it also introduces complicated scenarios for maintenance decisions that necessitate specifically arranging interventions to guarantee operational availability while keeping cost-effectiveness.

For k-out-of-n systems, traditional maintenance optimization techniques usually rely on fixed threshold approaches, in which component replacement or repair happens when predefined deterioration thresholds are reached [1, 4, 10–12]. The stochastic nature of component deterioration or environmental changes is not taken into account by these traditional approaches, which presume predictable degradation patterns. Usually, this oversimplification leads to suboptimal decisions, including premature maintenance interventions or delayed responses that compromise system reliability. In order to handle these limitations, dynamic condition-based maintenance (CBM) frameworks for multi-component systems that make use of Markov decision processes (MDPs) have been developed. These frameworks improve dependability performance and cost-effectiveness by facilitating adaptive maintenance plans that react to real-time system conditions and effectively capture the probabilistic aspects of degradation processes.

MDPs offer a suitable structure for modelling sequential decision-making processes in uncertain environments, capturing the probabilistic transitions between system states and the extended impacts of maintenance decisions. A number of research efforts have effectively utilized MDPs for the maintenance of k-out-of-n systems through the application of dynamic programming methods. Hao et al. [3] proposed a CBM policy integrating component reallocation and system replacement, optimized through value iteration to reduce costs and extend system lifetime. Ning et al. [9] investigated joint optimization of maintenance and triggering mechanisms for a protective device system using MDPs and value iteration to minimize expected costs and improve reliability. Zheng et al. [22] applied value iteration for the joint optimization of CBM and spare provisioning, while Zhang et al. [20] addressed CBM under failure dependence using value iteration to resolve inspection and maintenance challenges. Wang and Zhu [16] discretized degradation states and employed MDPs to jointly optimize CBM and inventory control, thereby reducing computational complexity. Additionally, Xu et al. [17], Sun et al. [13], and Wang et al. [15] have applied dynamic programming to solve MDPs in this setting.

MDP models are useful for determining optimal maintenance thresholds in low-dimensional CBM, but their huge state and action spaces make them difficult to use in high-dimensional scenarios. Conventional techniques such as value iteration and policy iteration require an exhaustive examination for each state-action pair, which results in significant computational complexity and infeasibility in practice.

As a practical solution for large-scale maintenance optimization problems based on the MDP model, researchers have increasingly turned to Reinforcement Learning (RL), especially Deep Reinforcement Learning (DRL), to overcome the aforementioned concerns. Unlike traditional value iteration or policy iteration methods, RL methods learn optimal policies directly from interaction data, making them more adaptable in complex environments. Yang et al. [18] developed an improved RL method with re-learning and pre-learning to optimize CBM strategies for redundant systems. Zhang et al. [19] introduced a Kernel Reinforcement Learning (KRL) method for multi-state system maintenance, integrating uncertainty modeling and high-value data screening.

Effective policy optimization in high-dimensional state spaces is made possible by DRL techniques, which would be computationally expensive for conventional approaches. For k-out-of-n systems, the combination of MDPs and DRL offers significant benefits, especially when the state space grows exponentially with the number of components. SAC is distinguished from other DRL algorithms by its high sample efficiency, stability, and effectiveness in continuous control situations. SAC is an off-policy reinforcement learning technique that uses entropy regularization to improve exploration while increasing expected rewards. It can effectively balance the trade-off between exploration and exploitation due to this characteristic, which makes it especially useful for tasks involving high-dimensional and continuous action spaces.

In summary, this study presents a novel DRL-based maintenance optimization technique for k-out-of-n systems. The stochastic component degradation processes and the effects of condition-based preventive maintenance (PM) and corrective maintenance (CM) are depicted using an MDP framework. A SAC-based algorithm with a reduced action space is created to handle the huge continuous state space. Actions in this algorithm are defined as the number of components that must be maintained at each inspection. Especially in high-dimensional systems, this method significantly reduces the dimensionality of the action space, facilitating more effective learning and accelerating convergence. Comprehensive sensitivity analyses are carried out to evaluate the effect of important parameters, and the obtained policy is extensively analysed to reveal its performance across various system settings.

2 Problem Description

This study examines a multi-component system with independent and identical components grouped in a k-out-of-n structure, with the system considered functional if at least k components are operational. To simplify the modeling of this system, the following assumptions are made:

The system consists of n independent, identical, and non-repairable components. When at least $n - k + 1$ components are in a failed state, the system is deemed to have failed.

Component degradation is modelled as a gamma process.

Inspections are used to detect system and component failures. Each inspection is perfect and instantaneous.

All maintenance actions are completed instantly. The component is returned to initial state after PM or CM.

In this study, these modeling assumptions are adopted primarily to isolate the effect of the maintenance policy and to keep the state–action space tractable. The k-out-of-n structure with independent, identical components is widely used in the maintenance literature and provides a neutral baseline in which performance differences can be attributed to the decision policy rather than to structural heterogeneity. Assuming independent Gamma degradation processes neglects common-cause shocks and load-sharing effects, and may therefore underestimate the benefit of proactive replacements in highly coupled systems; however, it enables a transparent evaluation of the proposed DRL framework. Perfect and instantaneous inspections avoid confounding the learning process with detection delays or misclassification errors and can be interpreted as reasonably accurate condition monitoring with negligible inspection time relative to the inspection interval. Similarly, instantaneous maintenance abstracts away detailed shop-floor logistics and reflects situations where component downtime is short compared with the planning horizon. These assumptions tend to yield optimistic cost estimates because real systems may exhibit delayed repairs, imperfect detection, or component interactions, but they provide a consistent benchmark for comparing DRL-based policies with threshold-based rules and can be relaxed in future work by incorporating inspection errors, maintenance durations, or dependent degradation into the MDP formulation.

The degradation process of each component is time-dependent and is described by a Gamma stochastic process. A Gamma distribution determines the degree of degradation at time t. The degradation processes are continuous and monotonic, indicating the effects of aging or wear. $x(t)$ represents the cumulative degradation of a component at time t. When the degradation level is above a predefined threshold L_f, failure occurs. The following expression describes how deterioration changes over time:

$$x(t + \Delta t) - x(t) \sim Ga(\lambda \Delta t, \beta) \tag{1}$$

where λ and β represent the shape parameter and scale parameter of the Gamma distribution, respectively.

This paper considers two maintenance actions. Component deterioration levels can only be observed during inspection. The periodic inspection interval is T. At each inspection, a decision must be made on whether to replace certain components based on the degree of degradation. Component failures can be detected instantly and necessitate CM. PM means replacing a component whose deterioration is less than the failure threshold L_f. CM occurs when a component has already failed. If the component fails within the detection interval, the system failure time is calculated using the Gamma bridge sampling method mentioned in reference [14]. This method determines the current value by using the past and future values of the Gamma distribution. The costs associated with maintenance actions include:

PM cost per component c_p;

CM cost per failed component c_c;

System failure cost c_f, incurred whenever the number of failed components equals or exceeds $n - k + 1$.

The goal of the optimization is to identify a maintenance strategy that minimizes the long-term average cost per unit time, considering all maintenance actions and failure-related penalties.

$$Minimize\ C = \lim_{T \to \infty} \sum\nolimits_{t=1}^{T} (C_{pm,t} + C_{cm,t} + C_{f,t})/T \tag{2}$$

3 MDP-Based Maintenance Optimization Via the SAC Method

3.1 MDP Framework

To cast the maintenance optimisation problem within an RL framework, it is modelled as an MDP defined by the tuple (S, A, P, R, γ), as described below.

State space (**S**): The state is defined by the degradation levels of all system components at the current inspection time, $\mathbf{S} = [x_1, x_2, \ldots, x_n]$.

Action space ($\mathbf{A}_s$): For a given state $s \in \mathbf{S}$, the action specifies the number of components to be maintained at a decision epoch, $\mathbf{A}_s = [0, 1, \ldots, n]$. At each inspection time t, a maintenance action $a \in \mathbf{A}_s$ is selected based on the observed system state. Components are replaced sequentially, starting with those exhibiting the highest degradation levels and proceeding toward those with lower degradation [21]. When multiple components share the same degradation level and only a subset requires maintenance, the components to be replaced are selected randomly from this group.

State transition probability (**P**): This represents the probability of transitioning from the current system state $s \in \mathbf{S}$ to the next state $s' \in \mathbf{S}$ after applying a chosen maintenance action.

Reward function (R): The reward r_t is defined in terms of the total cost incurred during each inspection cycle, capturing the economic impact of maintenance decisions.

$$r_t = -(c_p\, n_{pm} + c_c\, n_{cm} + I_{\text{system failure}}(c_f + (n - k + 1)c_c)) \tag{3}$$

where n_{pm} and n_{cm} denote the quantities of components replaced through preventive and corrective actions, respectively, and $I_{\text{system failure}}$ signifies a system failure occurring within the inspection interval.

Discount factor (γ): The discount factor $\gamma \in (0,1)$ balances the weight between immediate and future costs.

This MDP formulation enables systematic learning of optimal maintenance policies by balancing system degradation, maintenance actions, and associated costs over time.

3.2 SAC Reinforcement Learning Algorithm

The previous analysis demonstrates that the developed decision-making agent functions within a continuous action space, and that the state space $\boldsymbol{S}$ of the maintenance system is both continuous and high-dimensional, capturing the dynamic degradation of equipment as well as system-level attributes. Given these properties, the SAC algorithm, a policy-based DRL method well-suited for continuous control problems, is adopted.

SAC is an actor-critic algorithm that directly models a stochastic policy $\pi(a|s)$, in contrast to value-based approaches like DQN or DDQN and policy-based methods like PPO. The objective is to maximize both the expected cumulative reward and the entropy of the policy, which promotes exploration and prevents premature convergence to suboptimal deterministic policies. This is particularly useful in complex maintenance environments with uncertainty and non-deterministic transitions.

The SAC algorithm simultaneously updates three key components: a soft Q-function that estimates the expected return, a value-related mechanism that stabilizes the learning process, and a policy network that generates actions by sampling from a Gaussian distribution. A central feature of SAC is the introduction of a temperature parameter α, which explicitly controls the trade-off between reward maximization and entropy maximization. Through this mechanism, SAC encourages sufficient exploration while still converging toward high-reward policies. The objective is

$$\pi^* = \arg\max_\pi \mathbb{E}_{(s_t,a_t)} \sim \rho_\pi \left[\sum_{t=0}^{\infty} \gamma^t (r(s_t, a_t) + \alpha \mathcal{H}(\pi(\cdot \mid s_t))) \right] \tag{4}$$

where $\mathcal{H}$ is the entropy of the policy, and α is the temperature parameter. To improve generalization capability, neural networks are employed to approximate both the Q-value function and the policy function. The soft Q-network is trained by minimizing a temporal-difference-based loss $J_Q(\theta)$, which incorporates a target soft Q-network to enhance training stability. The parameters of the target network are updated in a controlled manner to prevent rapid oscillations during learning.

$$J_{Q}(\theta)=\mathbb{E}_{(s_t,a_t,r_t,s_{t+1})\sim D}\left[\frac{1}{2}\Big(\mathcal{Q}_{\theta}(s_t,a_t)-\widehat{Q}(s_t,a_t)\Big)^{2}\right] \tag{5}$$

Here, $\widehat{Q}(s_t, a_t)$ denotes the soft Bellman target defined in Eq. (6)

$$\begin{aligned}\widehat{Q}(s_t,a_t)&=r(s_t,a_t)+\gamma\,\mathbb{E}_{s_{t+1}\sim p}[V(s_{t+1})]\\ V(s_{t+1})&=\mathbb{E}_{a_{t+1}\sim\pi_\phi(\cdot|s_{t+1})}[\mathcal{Q}_{\overline{\theta}}(s_{t+1},a_{t+1})-\alpha\log\pi_\phi(a_{t+1}\mid s_{t+1})]\end{aligned} \tag{6}$$

where Q with target-network parameters $\overline{\theta}$ denotes the target soft Q-network, and θ and $\overline{\theta}$ are the parameters of the Q-value network and the target soft Q-network, respectively. In environments with continuous action spaces, the SAC algorithm adopts a reparameterization strategy for policy outputs. This approach enables efficient and low-variance gradient estimation when updating the policy network. The policy loss is formulated to encourage actions that both yield high expected returns and maintain high entropy, thereby reinforcing the balance between exploitation and exploration. The policy loss function is defined as follows:

$$J_{\pi}(\phi)=\mathbb{E}_{s_t\sim D,a_t\sim\pi_\phi}(\cdot|s_t)[\alpha\log\pi_\phi(a_t\mid s_t)-\mathcal{Q}_\theta(s_t,a_t)] \tag{7}$$

where Φ is the parameter of the policy network. In addition, the temperature parameter α is not fixed but is automatically adjusted through minimization of $J(\alpha)$ during training. This is achieved by optimizing it with respect to a target entropy level, allowing the algorithm to adaptively regulate exploration as learning progresses.

$$J(\alpha)=\mathbb{E}_{a_t\sim\pi_\phi}[-\alpha\log\pi_\phi(a_t\mid s_t)-\alpha\mathcal{H}_0] \tag{8}$$

where $\mathcal{H}_0$ is the target entropy. The main steps of the SAC algorithm can be summarized as follows:

Step 1: The policy network, Q-function networks, target Q-networks, and replay buffer are initialized.

Step 2: At each decision step, the agent selects an action according to the current policy.

Step 3: The agent then interacts with the environment by executing the action, receiving a reward, and transitioning to the next state, which in this context follows the degradation dynamics of the system.

Step 4: The resulting state–action–reward–next-state transitions are stored in a replay buffer and later sampled for off-policy training, enabling stable and data-efficient learning.

Step 5: Adjust the Q-function parameters and policy weights, modify the temperature setting, and update the parameters of the target network. Mini-batches are sampled to update the actor and critics using stochastic gradient descent. Target critic networks are updated using an exponential moving average.

Step 6: Output parameters.

4 Numerical Examples

This section utilizes a feedwater pump system [21] to demonstrate the effectiveness of the proposed optimal maintenance strategy. The system consists of five identical pumps, where at least three must be functioning properly for the system to operate normally. Each pump experiences degradation following a Gamma process, and a pump is deemed to have failed when its degradation level surpasses the failure threshold L_f. The parameters governing the Gamma degradation process, maintenance costs, routine inspection intervals, and the failure threshold are provided in Table 1. Section 4.1 assesses the long-term average cost performance across different optimization algorithms to confirm the effectiveness of the proposed approach. Section 4.2 analyzes the policy structures under different DRL algorithms, while Sect. 4.3 examines the robustness of the proposed algorithm across different parameter settings.

All DRL methods are applied within a simulation environment to optimize maintenance policies. To evaluate the effectiveness of the SAC method, three alternative algorithms, two DDQN variants and the PPO method, were implemented under identical simulation conditions, with experimental parameters detailed in Table 2. One DDQN variant employs a one-dimensional action output specifying the number of components to repair, whereas the second utilizes an action vector equal in length to the number of components, indicating for each whether repair is required; this variant is denoted as the DDQN01 algorithm. The PPO algorithm similarly defines an action space corresponding to the simultaneous repair of multiple components. All three alternative methods operate within discrete action spaces, while SAC employs a continuous action representation. In addition, a threshold-based PM policy is included, with the threshold optimized using Monte Carlo simulation (MC).

4.1 Comparison of Different Policy Optimization Algorithms

Table 3 presents a comparative summary of the performance of the maintenance optimization algorithms. To ensure reproducibility, each method is tested across five separate trials, and the derived policies are then simulated over 100,000 steps to calculate the long-term average cost rate. The Monte Carlo baseline, employing a threshold-based preventive maintenance policy with an optimized preventive maintenance threshold of $L_p = 8.0$, serves as the benchmark. The results presented in Table 3 indicate that all DRL algorithms achieve lower long-run average costs compared to

Table 1 presents the parameters related to pump degradation, inspection, and maintenance

Parameters	λ	β	T	L_f	c_p	c_c	c_f
	1.5	2.0	1	10	1	1.2	5

Table 2 Hyperparameter settings for different DRL methods

	SAC	DDQN	DDQN01	PPO
TargetUpdateFrequency		200	200	
DiscountFactor	0.95	0.90	0.90	0.90
MiniBatchSize	128	128	64	128
LearnRate	1e-4	5e-4	2e-3	1e-2
NodeNum	128	64	128	128
ExperienceBuffer	1e6	1e5	1e6	
TargetSmoothFactor	5e-4			
ExperienceHorizon				256
ClipFactor				0.2
NumEpoch				10

Table 3 Comparison of long-run average cost rates for different algorithms

	Mean	Min	95% CI
MC	1.662	1.661	[1.661, 1.662]
SAC	1.614	1.612	[1.612, 1.617]
PPO	1.653	1.620	[1.618, 1.687]
DDQN	1.640	1.619	[1.611, 1.669]
DDQN01	1.655	1.623	[1.605, 1.705]

the Monte Carlo baseline. SAC achieves the lowest long-run average cost rate, with its entire confidence interval below the MC baseline. DDQN comes next, while PPO and DDQN01 yield smaller, less consistent gains. SAC's continuous action space thus delivers the most reliable cost reduction.

4.2 *Analysis of strategy structure*

In Fig. 1, the maintenance policies learned by SAC, PPO, DDQN, and DDQN01 are compared with respect to the degradation levels of components 2 and 5, while components 1, 3, and 4 are held at fixed degradation levels of 2.5, 5.0, and 7.5, respectively. For the SAC algorithm, when three components have degraded to a moderate level and one component is in a slightly worse state, a single-component repair is scheduled. Once three components are in a significantly degraded state, the policy schedules proactive replacement of two components. This ensures low failure risk under the 3-out-of-5 rule, where the system remains operational unless three or more components fail, even if some components are still degraded. Premature intervention drives up maintenance costs, whereas overly delayed repairs risk system breakdown. The PPO algorithm adopts a more conservative maintenance strategy, initiating component replacements earlier under generally healthy system conditions. It proactively replaces two components when three components enter the inferior state. In contrast, DDQN defers dual-component interventions until either two components are degraded and one has failed, or three components are degraded, thereby minimizing the risk of system failure. Finally, DDQN01 is the most restrained policy, scheduling simultaneous replacement of two components only when both are on the verge of failure.

4.3 Sensitivity Analysis

A sensitivity analysis is conducted, in which each key factor, the minimum required number of functioning components k, the total number of components n, and failure cost c_f is varied independently across predetermined levels while holding the remaining parameters unchanged. For every parameter setting, SAC, PPO, DDQN, and DDQN01 are trained and evaluated in five independent replications, with each learned policy assessed over 10^6 simulation steps. The resulting long-run average cost rates are aggregated into boxplots. Due to the high consistency of MC simulation outcomes across multiple trials, visual representation in boxplots alongside other DRL algorithms proved infeasible. Consequently, MC results are tabulated separately in Table 4. The results are identical to those for $k = 3$ and $n = 5$, therefore, the table only shows the results for $c_f = 5$.

Table 4 and Fig. 2 indicate that the SAC algorithm achieves the lowest long-run average cost in most scenarios and attains the second-best performance under certain parameter variations. Across all three levels of failure cost, SAC consistently yields the smallest average cost rate. As the failure cost decreases, the performance gap between the threshold-based policy and SAC narrows. In contrast, higher failure costs strengthen component interdependence, under which threshold-based rules, by ignoring cross-component state interactions—incur substantially higher costs. The remaining three discrete-action deep reinforcement learning methods also outperform the threshold-based strategy in most cases. An increase in the parameter k governing reliability requirements further amplifies the cost advantage of SAC. When n is varied, SAC delivers the best or second-best long-run average cost rate in nearly all settings. The threshold-based strategy performs best only under particularly stringent reliability requirements, where a restricted action set allows maintenance efforts to focus exclusively on the most severely degraded components, thereby reducing cost. Overall, SAC demonstrates stable superiority and strong robustness across all examined parameter domains. Under high failure-cost conditions, deep reinforcement learning, especially SAC, achieves significant cost savings, while even under lower penalty regimes, DRL methods retain a non-negligible advantage, highlighting their adaptability to varying cost sensitivities. By contrast, the Monte Carlo–based technique attains optimal or near-optimal performance only under extremely

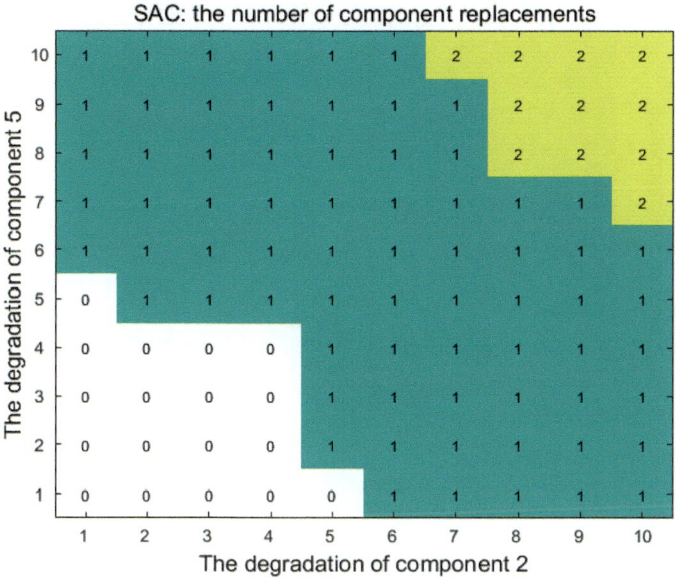

(*a*)

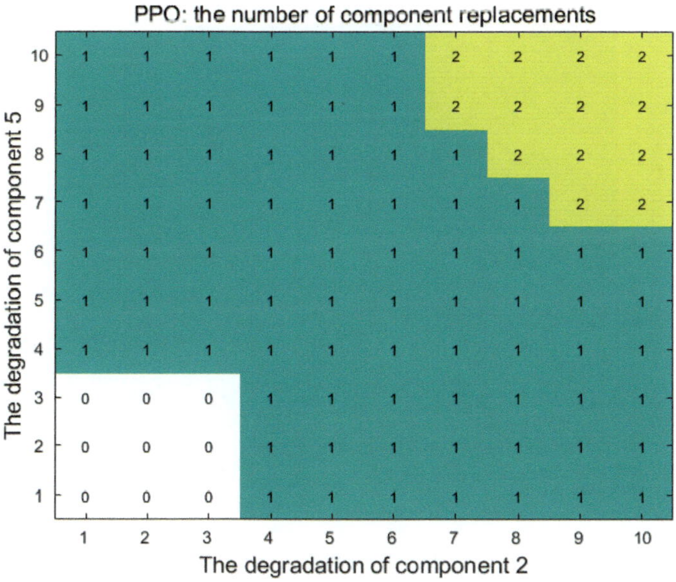

(*b*)

Fig. 1 **a** The maintenance strategy of SAC, **b** The maintenance strategy of PPO, **c** The maintenance strategy of DDQN, **d** The maintenance strategy of DDQN01

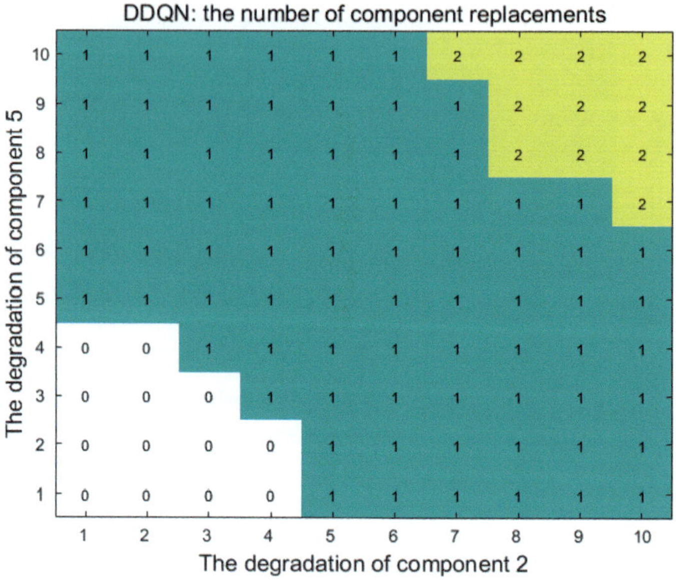

(*c*)

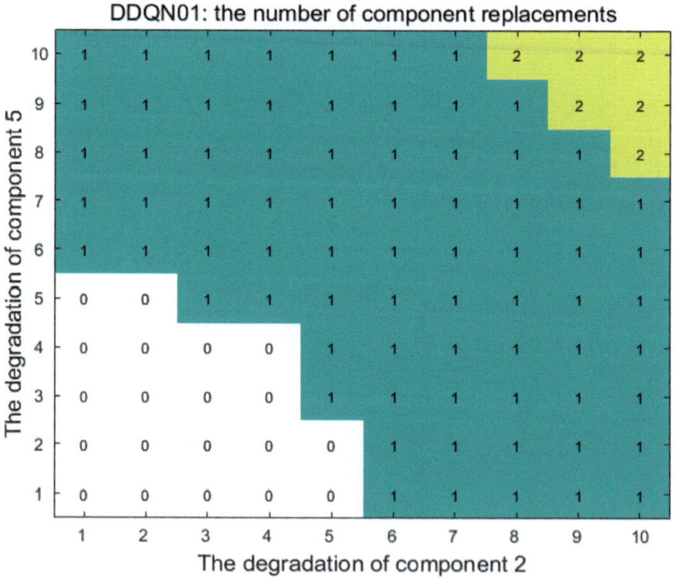

(*d*)

Fig. 1 (continued)

Table 4 Long-run average cost rates for the MC algorithm

	Mean	Min	Std
$c_f = 5$	1.662	1.661	0.0007
$c_f = 4.5$	1.651	1.650	0.0007
$c_f = 4$	1.640	1.638	0.0010
$k = 2$	1.501	1.500	0.0004
$k = 4$	2.052	2.051	0.0005
$n = 4$	1.591	1.590	0.0006
$n = 6$	1.844	1.844	0.0002

strict reliability requirements (e.g., $n = 4$ or $k = 4$) or when component coupling is weak ($c_f = 4$).

As shown in Fig. 2, SAC consistently achieves the lowest median cost and the narrowest interquartile ranges, indicating superior robustness to variations in failure costs, minimum operational component requirements, and total component numbers. PPO attains the greatest cost savings under highly constrained, low-redundancy conditions, whereas DDQN provides stable improvements with moderate variance across all scenarios [2, 5–7]. In contrast, DDQN01 exhibits substantial instability due to its significantly larger action space, resulting in higher variance compared with the other methods and occasional extreme outlier cost values.

5 Discussion on Uncertainty and Policy Robustness

When deployed to real systems, the proposed framework faces several additional sources of uncertainty. The parameters of the Gamma degradation model and the cost structure are typically estimated from finite historical data and expert judgement, so the transition dynamics and reward function used during training are only approximations of the true environment. Even under fixed model parameters, stochastic variance in the Gamma process and variability in operating conditions can cause realized degradation paths and failure events to deviate from those observed in simulation. These factors may lead to differences between the expected long-run average cost under the learned policy and its actual performance after implementation.

The sensitivity analysis partially addresses robustness by varying key structural parameters such as the k-out-of-n configuration, the total number of components, and the failure cost. The results indicate that SAC retains a clear cost advantage over the threshold-based policy and competitive DRL baselines across a wide range of settings, suggesting that the learned policy is not overly tuned to a single parameter configuration. A more systematic treatment of uncertainty would require explicit incorporation of parameter and model uncertainty into the optimization problem, for example through Bayesian updating of degradation parameters, distributionally

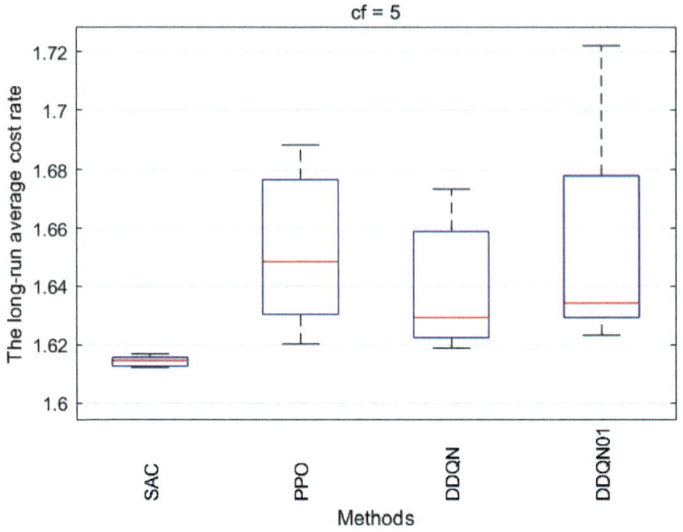

(a)

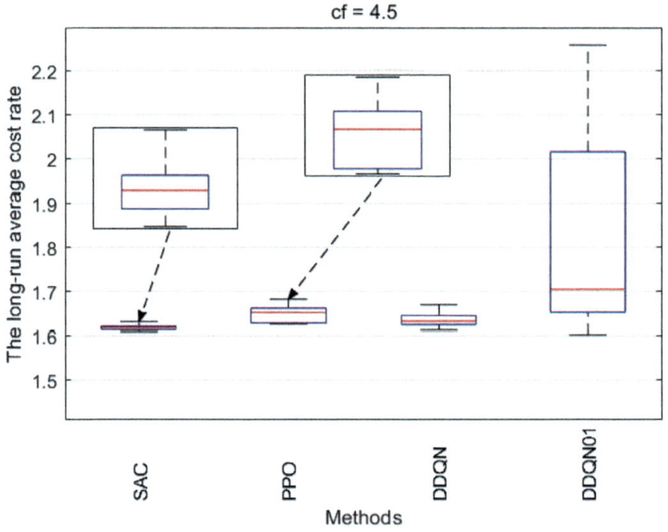

(b)

Fig. 2 Long-run average cost rates of SAC, PPO, DDQN, and DDQN01 under different parameter settings: (a) $c_f = 5$, (b) $c_f = 4.5$, (c) $c_f = 4$, (d) $k = 2$, (e) $k = 4$, (f) $n = 4$, and (g) $n = 6$

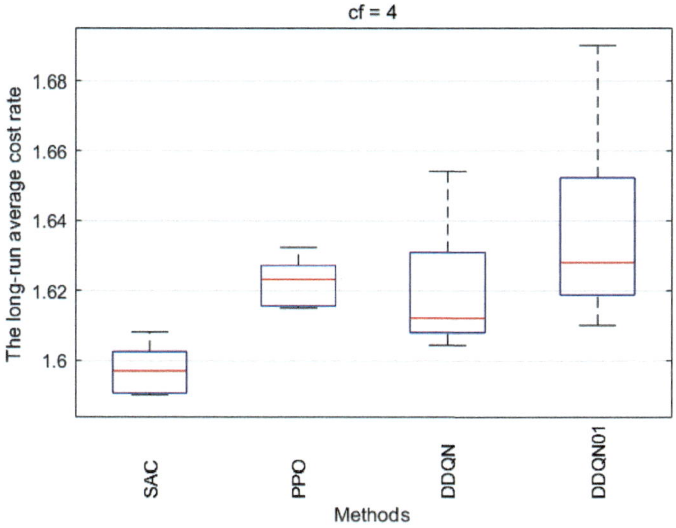

(*c*)

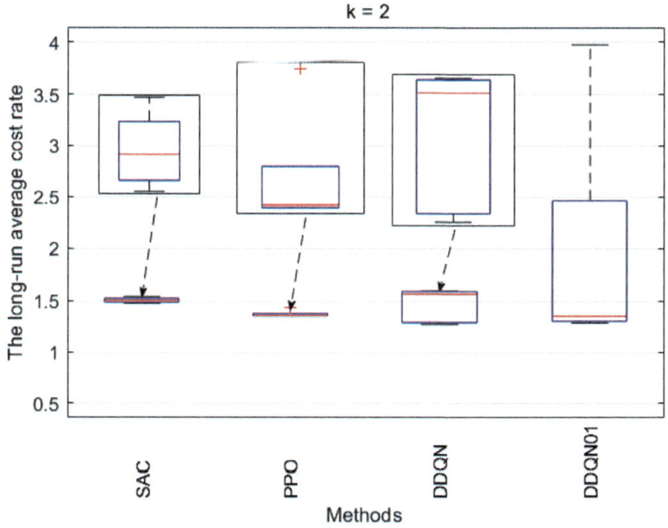

(*d*)

Fig. 2 (continued)

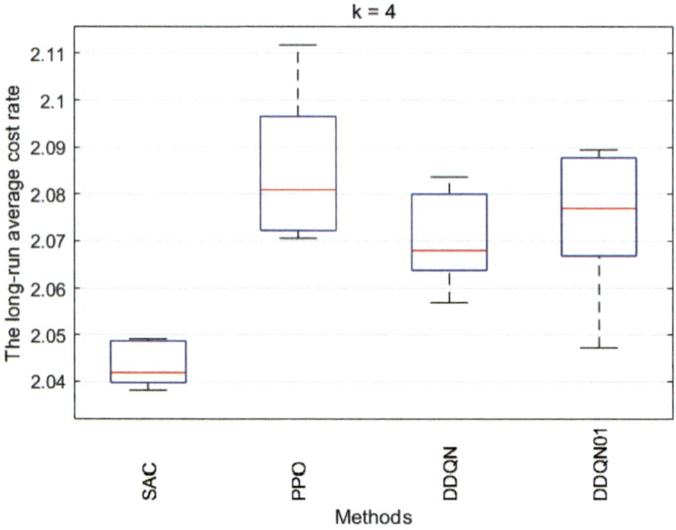

(*e*)

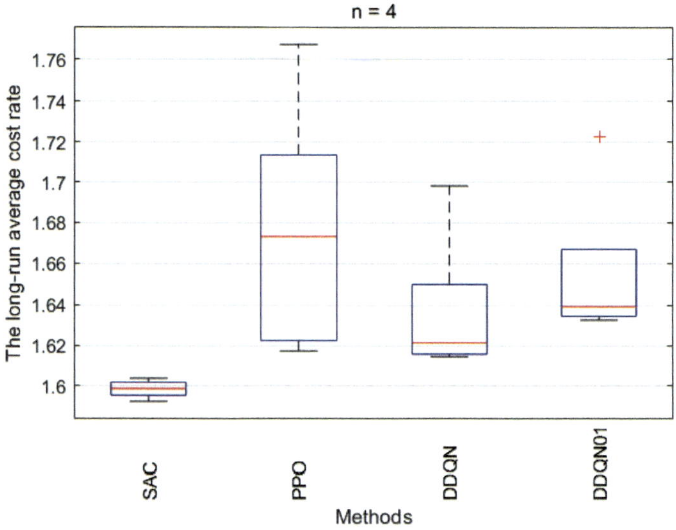

(*f*)

Fig. 2 (continued)

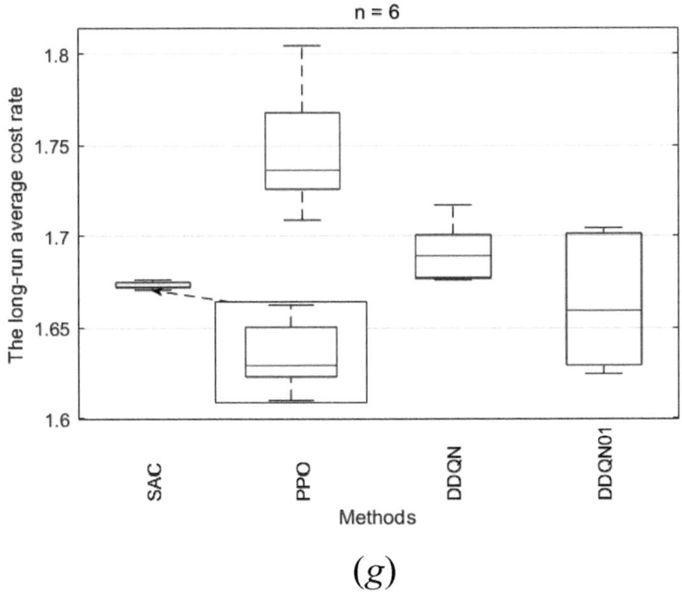

Fig. 2 (continued)

robust or risk-sensitive reinforcement learning formulations, or domain randomization during training. These extensions would allow the learned policy to hedge against model misspecification and estimation error, and constitute an important direction for future research, especially in applications with scarce failure data or rapidly changing operating conditions.

6 Conclusion

A SAC framework for condition-based maintenance of *k*-out-of-*n* systems is developed to address the dimensionality limitations of traditional dynamic programming. By representing actions as the total number of component replacements, SAC efficiently learns in a continuous state-action space. Applied to a 3-out-of-5 feedwater pump system, SAC achieves a reduction in long-run average costs relative to the threshold policy and outperforms three discrete-action deep reinforcement learning baselines, namely PPO, DDQN, and DDQN01. Analysis of the policy structure indicates that SAC's continuous control enables the determination of optimal intervention thresholds, effectively balancing early failure prevention with replacement costs more efficiently than fixed-threshold or discrete-action approaches. Sensitivity analyses reveal that deep reinforcement learning methods perform best in low-redundancy or large-scale systems, where threshold-based maintenance policies are less effective, whereas simple threshold policies remain adequate for small systems with ample redundancy. In addition, the current formulation assumes known degradation and cost parameters, perfect inspections, and instantaneous maintenance actions. These simplifications can be relaxed using existing methods. For instance, parameter uncertainty in the degradation model can be handled by Bayesian updating of the Gamma process parameters or by training DRL agents in a domain-randomized environment so that the learned policy is robust to model misspecification. In practice, cost parameters are usually estimated offline from historical maintenance and production data (and, if needed, expert judgement) and then treated as fixed inputs to the optimization model. Imperfect inspections and noisy condition monitoring can be modeled within a POMDP framework by introducing explicit observation models and belief-state updates, enabling the agent to optimize decisions under partial observability. Non-negligible maintenance durations can be incorporated via semi-Markov decision processes or by augmenting the state space with repair progress and system downtime, allowing the policy to balance intervention timing and availability loss. Dependent failures and load-sharing effects can be captured through multivariate degradation models or common-cause shock structures, and solved using DRL on factored or multi-agent MDP formulations. Exploring these extensions within the DRL framework will further enhance the practical applicability and robustness of the proposed maintenance optimization approach.

Acknowledgements The work is supported by the National Natural Science Foundation of China (Grant No. 72071044) and the China Scholarship Council.

References

1. Atashgar K, Abdollahzadeh H (2017) A joint reliability and imperfect opportunistic maintenance optimization for a multi-state weighted k-out-of-n system considering economic dependence and periodic inspection. Qual Reliab Eng Int 33(8):1685–1707
2. Guo J, Wang Z, Li H, Yang Y, Huang CG, Yazdi M, Kang HS (2024) A hybrid prognosis scheme for rolling bearings based on a novel health indicator and nonlinear wiener process. Reliab Eng Syst Saf 245:110014
3. Hao Y, Zhu X, Kuo W (2024) Optimization of condition-based maintenance with multiple times of component reallocation using Markov decision process. IEEE Trans Reliab 73(1):131–141
4. Jiang A, Huang Z, Xu J et al (2021) Condition-based opportunistic maintenance policy for a series–parallel hybrid system with economic dependence. J Qual Maint Eng 28(3):584–605
5. Li H, Yazdi M (2022) What are the critical well-drilling blowouts barriers? A progressive DEMATEL-game theory. In: Advanced decision-making methods and applications in system safety and reliability problems: approaches, case studies, multi-criteria decision-making, multi-objective decision-making, fuzzy risk-based models. Springer International Publishing, Cham, pp 29–46
6. Li H, Yazdi M, Huang HZ, Huang CG, Peng W, Nedjati A, Adesina KA (2023) A fuzzy rough copula Bayesian network model for solving complex hospital service quality assessment. Complex Intell Syst 9(5):5527–5553

7. Li H, Peng W, Adumene S, Yazdi M (2023) An intelligent cost-based consequence model for offshore systems in harsh environments. In: Intelligent reliability and maintainability of energy infrastructure assets. Springer Nature Switzerland, Cham, pp 107–117
8. Liu Y, Wang J, Liu P (2024) A condition-based maintenance policy with non-periodic inspection for k-out-of-n: G systems. Reliab Eng Syst Saf 241:109640
9. Ning R, Wang X, Zhao X et al (2024) Joint optimization of preventive maintenance and triggering mechanism for k-out-of-n:F systems with protective devices based on periodic inspection. Reliab Eng Syst Saf 251:110396
10. Rezaei E, Jafary B, Fiondella L (2022) Optimal maintenance policies for linear consecutive k-out-of-n:F systems susceptible to dependent failures. Comput Ind Eng 173:108657–108657
11. Sangroudi HA (2018) Wind power performance optimization considering redundancy and opportunistic maintenance. Mendel Soft Comput J 1–8
12. de Smidt-Destombes KS, van der Heijden MC, van Harten A (2009) Joint optimisation of spare part inventory, maintenance frequency and repair capacity for k-out-of-N systems. Int J Prod Econ 118(1):260–268
13. Sun Q, Ye Z-S, Chen N (2018) Optimal inspection and replacement policies for multi-unit systems subject to degradation. IEEE Trans Reliab 67(1):401–413
14. Van Noortwijk JM (2009) A survey of the application of gamma processes in maintenance. Reliab Eng Syst Saf 94(1):2–21
15. Wang C, Ning G, Deng Q et al (2025) Joint optimization of cooperative maintenance and inventory control for multiple k-out-of-n:F systems considering component interchange and shared spare parts. Reliab Eng Syst Saf 262:111181
16. Wang J, Zhu X (2021) Joint optimization of condition-based maintenance and inventory control for a k-out-of-n:F system of multi-state degrading components. Eur J Oper Res 290(2):514–529
17. Xu J, Liang Z, Li Y-F et al (2021) Generalized condition-based maintenance optimization for multi-component systems considering stochastic dependency and imperfect maintenance. Reliab Eng Syst Saf 211:107592
18. Yang A, Qiu Q, Zhu M et al (2022) Condition-based maintenance strategy for redundant systems with arbitrary structures using improved reinforcement learning. Reliab Eng Syst Saf 225:108643
19. Zhang D, Zhang Y, Li P et al (2025) Kernel reinforcement learning for sampling-efficient risk management of large-scale engineering systems. Reliab Eng Syst Saf 260:111022
20. Zhang N, Tian S, Cai K et al (2022) Condition-based maintenance assessment for a deteriorating system considering stochastic failure dependence. IISE Trans 55(7):687–697
21. Zhao S, Wei Y, Cheng Y et al (2025) A state-specific joint size, maintenance, and inventory policy for a K-out-of-N load-sharing system subject to self-announcing failures. Reliab Eng Syst Saf 257:110855
22. Zheng MM, Lin J, Xia TB et al (2023) Joint condition-based maintenance and spare provisioning policy for a K-out-of-N system with failures during inspection intervals. Eur J Oper Res 308:1220–1232

Investigation on Quantitative Evaluation Technology for System Integrated Support

Yuhan Liu, Xiaoming Gao, Baidong Jin, and Pengwei Hu

Abstract Frequent post-delivery support challenges impose significant demands on both users and manufacturers, particularly regarding personnel, technology, cost, and project schedules. A major contributing factor is that most existing supportability evaluations are qualitative rather than quantitative, and systematic reviews throughout the development process are often insufficient. In response to these issues, this paper examines quantitative evaluation methodologies for integrated support domestically and internationally, identifies representative techniques such as continuous support maturity assessment and support effectiveness evaluation, and provides a reference framework for future research on quantitative approaches to integrated support.

Keywords Quantitative evaluation of integrated support · Continuous support maturity · Support effectiveness evaluation

1 Introduction

Analysis of the use and maintenance support issues of several systems indicates that, during the model development process, there is a tendency to prioritize functional performance while neglecting supportability design. Development efforts often emphasize the "usefulness" of system performance parameters, while insufficient attention is given to supportability considerations and the development of supporting resources. Attributes such as "usability," "durability," and "practicality" have not been accorded equivalent importance. The lack of systematic analysis, evaluation, and inspection of elements contributing to support effectiveness during development has resulted in significant post-delivery support risks and adversely affected overall system performance. This paper examines representative quantitative evaluation

Y. Liu (✉) · B. Jin · P. Hu
China Aerospace Standardization Institute, Beijing, China

X. Gao
China Aerospace Science and Technology Corporation, Beijing, China

H. Li et al. (eds.), *Data-Driven Methods for Reliability and Safety Engineering: Applications in Industrial Systems*, Springer Series in Reliability Engineering,
https://doi.org/10.1007/978-3-032-22873-4_15

technologies for integrated support, including continuous support maturity assessment and support effectiveness evaluation, aiming to provide novel perspectives for future research in integrated support.

These two perspectives are selected because they jointly capture the process–result duality that is essential for integrated support. Continuous support maturity focuses on the life-cycle evolution of product support strategies and processes, providing a structured view of how support capabilities are planned, implemented, and improved over time. Support effectiveness, in contrast, focuses on the quantitative outcomes of support activities in terms of availability, reliability, cost, and responsiveness. Considering both dimensions helps to align process design with outcome-oriented evaluation, and offers a more transparent analytical logic for understanding how support decisions influence system-level performance.

2 Concept of Continuous Support Maturity Level

To facilitate effective continuous support across the entire system life cycle and to assess the level of support at each stage, the U.S. Department of Defense (DoD) published the Product Support Manager Guidebook in April 2011. This guide formally introduced a twelve-level conceptual model of Sustainment Maturity Levels (SML), which standardizes and evaluates support processes throughout the system life cycle, thereby mitigating implementation risks. The Sustainment Maturity Levels reflect the system's compliance with support requirements and embody the principle of continuous improvement. In April 2016, the DoD updated the Product Support Manager Guidebook; however, the conceptual framework, definition, and twelve maturity levels of SML remained unchanged. The overall structure of the Product Support Manager Guidebook is illustrated in Fig. 1.

The U.S. military's Sustainment Maturity Level (SML) model consists of 12 levels, corresponding to 5 stages of the life cycle. From the perspective of the Sustainment Maturity Model proposed in the Product Support Manager Guidebook, its direct evaluation objects are the system and its key subsystems. The evaluation content focuses on the state of the system's implementation of the product support strategy (including supportability design and the obtained product support package) to achieve sustainment indicators. The main assessment points defined in the model include the adequacy of early-stage supportability design, the implementation of the product support package in the later stages, and the formation of sustainment capabilities. This model was proposed by the U.S. Department of Defense to better implement the requirement of "improving life-cycle sustainment" in the National Defense Authorization Act and Defense Acquisition Reform Act. Its role is to assist product support managers in conducting more scientific and effective life-cycle sustainment management by providing a universal, consistent, and repeatable way to express the maturity of product support packages.

The Sustainment Maturity Level (SML) is used to record the status of a system when implementing a product support strategy, including the design scheme and

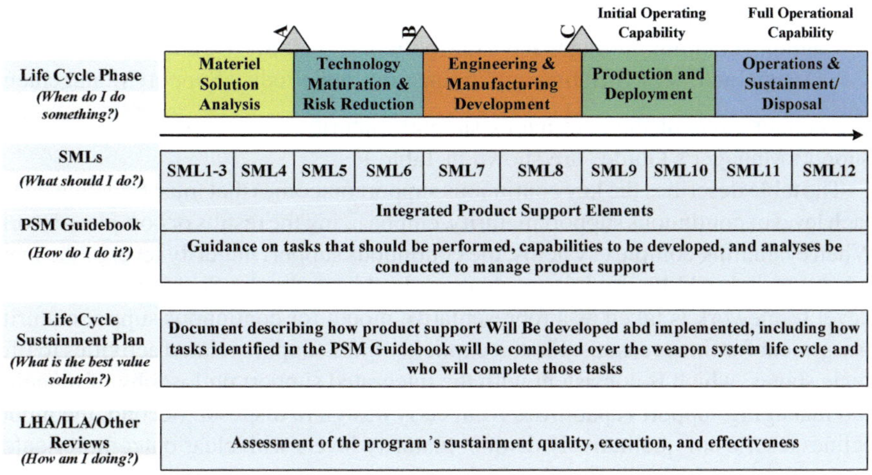

Fig. 1 Graphic overview of the PSM guidebook's structure

the corresponding Product Support Package (PSP, which refers to the collection of support elements, sustainment process contracts, and agreements to achieve and maintain equipment readiness) to realize sustainment outcome indicators.

The Sustainment Maturity Level serves as a standard for assessing the maturity of system support strategies. It is used to define the scope and scheduling of product support activities, determine the maturity level of support plans at key lifecycle milestones, and evaluate alignment with user mission requirements.

It provides product support managers with a structured framework and rigorous methodology, enhancing their ability to evaluate project performance based on the implementation status of product support.

The Sustainment Maturity Level has the following characteristics:

(1) It applies to all possible support schemes, which can be independently undertaken by traditional research units or commercial departments, or jointly completed by both, and is not limited to a specific support scheme;
(2) It is system-wide and can also be applied to major subsystems, providing a universal, consistent, and repeatable method for establishing and understanding the maturity of the Product Support Package;
(3) It is life-cycle oriented and complements the continuous optimization of system design strategies.

The Sustainment Maturity Level has the following functions:

(1) Optimizing product support schemes and seeking the best support solutions;
(2) Evaluating and reducing risks in the implementation process of the Product Support Package. Meanwhile, the Product Support Manager Guide also proposes 12 sustainment elements, namely Maintenance Planning and Management, Supply Support, Support Equipment, Facilities and Infrastructure,

Manpower and Personnel, Training and Training Support, Technical Data, Packaging, Handling, Storage, and Transportation (PHS&T), Computer Resources, Design Interfaces, Sustainment Engineering, and Product Support Management.

The connotations of the continuous support maturity levels proposed in *Product Support Manager's Guide* are shown in Table 1.

The table describes the key continuous support outcomes that must be achieved for each level of continuous support maturity, emphasizing the results or completed work. When evaluating complex systems, the continuous support maturity level of the main subsystems should be used for evaluation. In this study, the Sustainment Maturity Level framework is taken as a representative model for continuous suppot maturity assessment for three reasons. First, it explicitly links support-related activities to life-cycle stages, which is consistent with the integrated support philosophy of planning and managing support capabilities from early design to disposal. Second, the model defines a series of qualitatively distinct maturity levels with clear outcome-oriented descriptions, which facilitates practical assessment and communication among stakeholders. Third, the SML concept has been applied and refined in multiple defence programs, providing a relatively mature reference for analysing how supportability

Table 1 The connotations of continuous support maturity levels

Level	Connotation
1	Identify supportability characteristics and optional solutions for continuous support.
2	Identify conceptual product support and maintenance solutions.
3	Elaborate on conceptual product support, continuous support, and supportability requirements in detail to underpin conceptual product support and maintenance solutions.
4	Determine supportability objectives and requirements for key performance parameters, key system characteristic parameters, etc.; identify new or better technical requirements for equipment systems or chains.
5	Incorporate the supportability design characteristics for achieving key performance parameters and key system characteristic parameters into the design requirement documents.
6	Complete the maintenance solutions and continuous support strategies; the life-cycle continuous support plan is approved.
7	Embed supportability characteristics into the design; complete system supportability and subsystem maintenance task analysis.
8	Demonstrate and verify product support capabilities, and validate and confirm chain management methods.
9	Demonstrate and verify the product support package in the operational environment.
10	Deploy the initial product support package at the site of use; measure relevant performance indicators against availability, reliability, and cost metrics.
11	Measure the continuous support performance against the usage requirements; enhance the product support level through continuous process improvement.
12	Fully apply the product support package, including base maintenance capabilities.

design, product support packages, and sustainment outcomes are connected in a quantitative evaluation context.

3 Evaluation of Support Effectiveness

In Logistics Testing and Evaluation in Flight Testing, the NATO Research and Technology Organization identified ten key support elements: maintenance planning, manpower and personnel, support equipment, supply support, technical data, training and training support, computer resource support, facilities, packaging, handling, storage and transportation, and design interfaces. The report also established objectives for evaluating these support elements, including associated measurement parameters, metrics, analytical techniques, and evaluation tools. However, this framework addresses only individual support elements, which represent portions of the overall support system and do not encompass the entire system.

The report Design and Evaluation of Supportability for DoD Weapon Systems: A Guide to Improving Reliability and Reducing Logistics Footprint, published by the U.S. Department of Defense (DoD), introduced a comprehensive framework for enhancing system reliability, maintainability, and supportability while minimizing the support footprint and associated life-cycle costs. The central objective is to achieve an optimal trade-off among system effectiveness, support footprint, and cost. The U.S. Army Headquarters Command, in AR700–127 Integrated Logistics Support, identifies five top-level performance-based support parameters: operational availability, mission reliability, cost, logistics response time, and logistics footprint. In July 2006, the Joint Requirements Oversight Council (JROC) established mandatory Key Performance Parameters (KPPs) for combatant equipment readiness and sustainment. Subsequently, in a March 2007 memorandum, the DoD required that four objectives related to equipment readiness be determined at the earliest stages of program decision-making, refined throughout design and development, and maintained as baseline objectives until system retirement. Currently, five evaluation attributes have been established: cost, schedule, performance, funding, and Life-Cycle Sustainment (LCS) capability. Metrics for LCS capability include equipment availability, equipment reliability, ownership cost, and average downtime. Tony Park notes in Introduction to Logistics Testing and Evaluation that the overarching goal of logistics support evaluation is to maximize operational availability within life-cycle cost constraints.

The U.S. Air Force, in AFI 10–602 Determination of Mission Capability and Support Capability Requirements, specifies parameters for assessing mission success and support capability, including Reliability, Availability, Maintainability (RAM), deployability, and interoperability. The directive also defines mission success and support capability parameters for nine categories of systems, such as spacecraft, missiles, and automated information systems, with parameter definitions tailored to each object category:

(1) Availability and sustainability metrics: Operational Availability (Ao), Mission Capable Rate (MCR), Utilization Rate (UR), Equipment System Repair Time per Flight Hour (ESRT/FH), Uptime Ratio (UTR);
(2) Mission reliability metrics: Weapon System Reliability (WSR), Breakdown Rate (BR), Combat Rate (CR), Mean Time Between Catastrophic Failures (MTBCF);
(3) Logistics reliability metrics: Mean Time Between Maintenance (MTBM);
(4) Maintainability metrics: Mean Down Time (MDT), Fix Rate (FR), Mean Repair Time (MRT);
(5) Manpower metrics: Maintenance Man-Hours per Life Unit (MMH/LU), Maintenance Personnel per Usage Unit (MP/U);
(6) Deployability considerations: Deployment scale, logistics, and sustainment support.

The evaluation of support capability in the above documents is essentially an evaluation of support resources, which separates support capability from combat readiness and mission success, does not consider the impact of support on the equipment system level, and has not formed a unified consensus on the connotation of support capability or support effectiveness.

Compared with advanced foreign levels, the main gaps in support effectiveness in China currently are:

(1) Both domestic and foreign studies have conducted certain research on the connotation and parameter analysis models of integrated support core capabilities for various equipment or equipment support effectiveness. However, there is still no unified consensus in China on the connotation of support effectiveness, and a universal parameter system for equipment support effectiveness has not been proposed.
(2) International simulation models for equipment supportability have demonstrated the ability to model systems involving multiple equipment types, while also emphasizing interoperability and integration with acquisition frameworks. These mature tools are widely used across the R&D and production phases. In contrast, domestic modeling technologies remain at an early stage, with most tools still in prototype form. Current systems mainly focus on single-type equipment and lack the technical capability to simulate support scenarios for complex, multi-type systems.

Support effectiveness refers to the dynamic performance level of an equipment support system. From the reviewed literature, although the U.S. military began analysing and evaluating support in the twentieth century, there is relatively little interpretation of the connotation of support effectiveness. The definitions of equipment support effectiveness have not been unified domestically and internationally.

Song Tailiang et al. proposed that to make capabilities more directly and conveniently measurable, the concept of effectiveness is used to evaluate support capabilities. Effectiveness reflects the comprehensive ability of equipment to complete all tasks and is represented by a single value between 0 and 1.0. Therefore, effectiveness can also be used for performance evaluation, and the support effectiveness of equipment can be assessed based on support performance. The ability of equipment to complete a specific task is called mission effectiveness; from a support perspective, it can be referred to as the support effectiveness for executing that task.

Zhou Yang et al. defined support effectiveness as the ability of a support system to economically and effectively ensure equipment readiness and mission sustainability under expected usage conditions. Their definition emphasizes the coordination between equipment design and the support system.

Zhang Kun et al. described equipment support effectiveness as a quantitative assessment of support processes, support system modes, task execution quality, and support response speed, all within defined conditions and timeframes based on relevant standards. Cui Shuang et al. defined that support effectiveness refers to the actual effect exerted by the equipment support system in completing support tasks. Meanwhile, they explored an approach and steps for evaluating equipment support effectiveness based on "wargame drills".

Kong Fang et al. developed an evaluation index system for equipment support effectiveness based on the OODA loop. A corresponding evaluation model was established to reflect the realities of modern warfare, enabling a quantitative assessment of the overall effectiveness of support systems [3, 7]. Regarding the parameter system, some standards have the following relevant provisions:

The General Requirements for Integrated Equipment Support classifies supportability parameters into three categories: supportability parameters for equipment systems, supportability design characteristic parameters for equipment, and supportability parameters for support systems and their resources. For example, parameters for support systems mainly include Mean Logistics Delay Time (MLDT) and Mean Administrative Delay Time (MADT); support resource parameters mainly include spare parts satisfaction rate, spare parts utilization rate, support equipment satisfaction rate, support equipment utilization rate, etc. [5, 8].

Equipment Supportability Analysis stipulates that support system parameters mainly include Mean Logistics Delay Time (MLDT) and Mean Administrative Delay Time (MADT), which directly impact combat readiness and are important measurement parameters for evaluating support system effectiveness; support resource parameters mainly include spare parts utilization rate, spare parts satisfaction rate, support equipment utilization rate, support equipment satisfaction rate, etc., which are the result of trade-offs between economy and combat readiness, directly affecting the planning of support resources and the design of support systems.

The Demonstration of Reliability, Maintainability, and Supportability Requirements for Equipment specifies that RMS parameters are divided into comprehensive parameters, reliability parameters, maintainability parameters, testability parameters, and support system and resource parameters. Reasonable RMS parameters should be determined based on three factors: the type of equipment and its operational characteristics, as well as the practices of similar equipment across services and arms; the complexity of the equipment and its product hierarchy; and the characteristics of the product.

The General Requirements for Equipment Reliability Work specifies various parameters for measuring equipment maintainability, such as Mean Time To Repair (MTTR), Maximum Time To Repair (MaxTTR) [6, 9, 11], Mean Preventive Maintenance Time (MPMT), etc. The equipment operating hours or service life used to calculate certain parameters can be converted into other time units, work cycles, number of firings, mileage, etc., according to the equipment type.

Based on the above research, the support effectiveness parameter system studied in this paper is shown in Fig. 2. The construction of this parameter system follows several guiding principles. First, the selected indicators must be traceable to mission needs and system-level performance, ensuring that support effectiveness is not evaluated in isolation from combat readiness and operational objectives. Second, the indicators should be measurable or at least estimable based on data that can be realistically obtained in typical equipment programs, such as failure records, maintenance logs, and resource consumption statistics. Third, the system seeks to balance completeness and practicality by covering system performance, support processes, and support resources, while avoiding excessive redundancy and overlap among indicators. Finally, the parameter structure is organised in a hierarchical manner, so that higher-level effectiveness measures (such as overall availability or life-cycle sustainment capability) can be decomposed into lower-level parameters that are more directly controllable through design and support decisions.

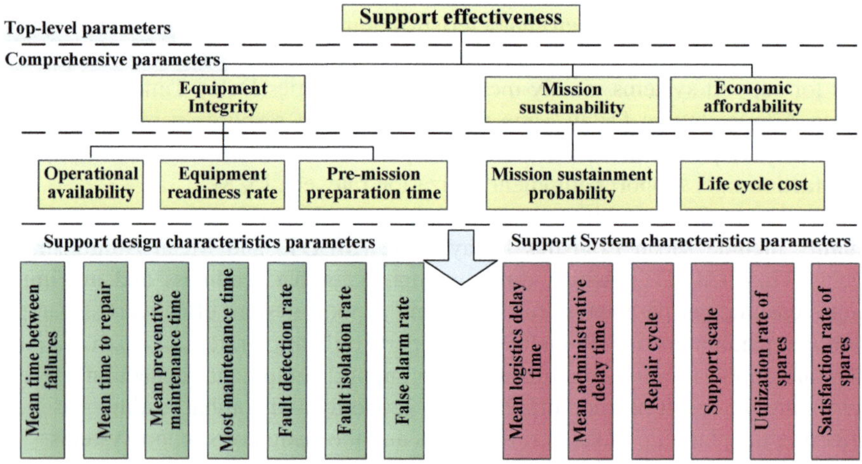

Fig. 2 General parameter system of equipment support effectiveness

4 Conclusion

This paper examines quantitative evaluation technologies for integrated support in both domestic and international contexts. It reviews the current progress of representative methods, including the Sustainment Maturity Level (SML) framework and support effectiveness evaluation. Furthermore, it systematically analyzes the structure of the SML model, identifies key sustainment support elements, defines the conceptual basis of support effectiveness, and outlines associated parameter systems. The study provides a valuable reference for future research on quantitative evaluation methodologies in integrated support, aiming to advance the transition from post-event assessments toward full life-cycle, process-oriented, and high-precision evaluation frameworks. At the same time, the paper recognises that the current parameter system and modelling assumptions are still subject to uncertainties arising from data quality, contextual differences, and simplified treatment of indicator interactions. Subsequent work should therefore focus on embedding sensitivity analysis, expert-driven weighting schemes, and probabilistic evaluation techniques into the quantitative assessment process, so as to obtain more robust and decision-oriented measures of integrated support effectiveness.

References

1. Cui S, Li BG, Dong DP (2023) Research on evaluation method of equipment support effectiveness based on wargame deduction system. Ship Electron Eng 43(8):10–14
2. DoD (2011) Product support manager guidebook
3. Guo J, Wang Z, Li H, Yang Y, Huang CG, Yazdi M, Kang HS (2024) A hybrid prognosis scheme for rolling bearings based on a novel health indicator and nonlinear wiener process. Reliab Eng Syst Saf 245:110014
4. Kong F, Zhu LJ, Hun JJ (2024) Evaluation of equipment support effectiveness for combined arms brigade based on ooda loop. Ship Electron Eng 44(3):124–129
5. Li H, Yazdi M (2022) What are the critical well-drilling blowouts barriers? A progressive DEMATEL-game theory. In: Advanced decision-making methods and applications in system safety and reliability problems: approaches, case studies, multi-criteria decision-making, multi-objective decision-making, fuzzy risk-based models. Springer International Publishing, Cham, pp 29–46
6. Li H, Yazdi M (2022) Reliability analysis of correlated failure modes by transforming fault tree model to Bayesian network: a case study of the MDS of a CNC machine tool. In: Advanced decision-making methods and applications in system safety and reliability problems: approaches, case studies, multi-criteria decision-making, multi-objective decision-making, fuzzy risk-based models. Springer International Publishing, Cham, pp 15–28
7. Li H, Yazdi M, Huang HZ, Huang CG, Peng W, Nedjati A, Adesina KA (2023) A fuzzy rough copula Bayesian network model for solving complex hospital service quality assessment. Complex Intell Syst 9(5):5527–5553
8. Li H, Peng W, Adumene S, Yazdi M (2023) An intelligent cost-based consequence model for offshore systems in harsh environments. In: Intelligent reliability and maintainability of energy infrastructure assets. Springer Nature Switzerland, Cham, pp 107–117

9. Li H, Peng W, Adumene S, Yazdi M (2023) Cutting edge research topics on system safety, reliability, maintainability, and resilience of energy-critical infrastructures. In: Intelligent reliability and maintainability of energy infrastructure assets. Springer Nature Switzerland, Cham, pp 25–38
10. Song TL, Wang YL, Fang Y (2014) General theory of equipment great support view. Beijing, National Defense Industry Press
11. Yazdi M (2024) Reliability-centered design and system resilience. In: Advances in computational mathematics for industrial system reliability and maintainability. Springer Nature Switzerland, Cham, pp 79–103
12. Zhang K, Jia YX, Jiao Y (2020) Evaluation method of typical parameters for aviation equipment support effectiveness. Fire Control Command Control 45(7):110–114
13. Zhou Y, Zeng ZY, Zhou Y et al (2014) Research on simulation-based support effectiveness evaluation of military aircraft. Aviat Maint Eng 3:63–65

Goal-Oriented Reliability Program Framework for Space Science and Application Payloads

Kuanyu Gao, Peng Li, Man Fang, and Hongen Zhong

Abstract Space science and application payloads encounter significant reliability and safety challenges due to technological complexity, cost constraints, and limited development timelines. Traditional reliability programs define requirements and work items across mission phases; however, several critical issues persist, including disconnected functional and reliability design processes, ambiguous definitions of reliability missions and requirements, and the application of inappropriate reliability metrics. This study proposes a goal-oriented reliability program framework for space payloads, transitioning from conventional process-oriented approaches to goal-driven reliability assurance. Initially, widely adopted reliability program standards, including GJB 450B, GJB 1909, GJB 9421, and GB/T 35439, are reviewed. Multiphase processes and static verification and validation procedures characterize these standards. Additionally, contemporary NASA reliability and maintainability (R&M) documents, as well as SpaceX and Huawei reliability workflows, are analyzed. Based on this analysis, the limitations of existing standards are identified in the context of space science and application mission requirements. Subsequently, a novel reliability program framework is proposed to enhance scientific and engineering value

K. Gao · P. Li · M. Fang · H. Zhong
Technology and Engineering Center for Space Utilization, Chinese Academy of Sciences, Beijing, China
e-mail: gaokuanyu24@mails.ucas.ac.cn

M. Fang
e-mail: fangman@csu.ac.cn

H. Zhong
e-mail: zhonghongen@csu.ac.cn

K. Gao · M. Fang · H. Zhong
University of Chinese Academy of Sciences, Beijing, China

P. Li (✉)
Department of Mechanical Engineering, Division of Mechatronic System Dynamics, KU Leuven, Ghent, Belgium
e-mail: mscalipeng@163.com; peng.li@kuleuven.be

H. Li et al. (eds.), *Data-Driven Methods for Reliability and Safety Engineering: Applications in Industrial Systems*, Springer Series in Reliability Engineering,
https://doi.org/10.1007/978-3-032-22873-4_16

while minimizing safety and reliability risks. The goal-oriented workflow encompasses reliability requirement analysis, constraint identification, hierarchical decomposition, and program planning. A central feature is the iterative decomposition framework guided by the application value–planning principle. This study provides methodological references and insights for developing general reliability program plans using a requirements-driven and goal-decomposition approach.

Keywords Reliability · Space payloads · Standards · Systems engineering

1 Introduction

Space science and application payloads, as integral components of all space missions, are essential for both the success of scientific exploration and the accuracy of collected data, as demonstrated by missions such as the James Webb Space Telescope [1] and the Zhu-Rong Mars rover [2]. The extreme and unforgiving space environment imposes stringent reliability requirements on these payloads. Key challenges include operation in high vacuum, exposure to significant temperature variations, electromagnetic radiation, and high-energy particle impacts. Additionally, launch-induced vibration and shock, microgravity, and atomic oxygen erosion in orbit further compromise payload reliability [3]. The inherent irreparability of space systems necessitates that mission-critical subsystems maintain exceptionally high reliability throughout their operational lifetimes, thereby increasing verification complexity. Historical failures underscore these risks. For instance, the 2018 Soyuz-FG rocket accident in Russia was attributed to a deformed sensor pin [4], while eight failures among 263 global launches in 2024 were primarily caused by insufficient verification in both new and established systems [5]. These cases indicate that comprehensive testing of both novel and mature products is essential to identify and mitigate risks, enhance reliability, and ensure mission success.

To address these challenges, aerospace agencies and enterprises internationally have established systematic reliability management frameworks. NASA, the U.S. Department of Defense, and the European Space Agency have developed numerous standards, including MIL-STD-810H Environmental Engineering Considerations and Laboratory Tests, and ECSS-Q-ST-30C Dependability [6]. These standards encompass full-lifecycle reliability control methods from design and testing to on-orbit monitoring, ensuring stable operation of spacecraft and payloads under extreme environmental conditions. China has also accumulated substantial experience in space reliability, establishing top-level standards such as GJB 450B General Requirements for Materiel Reliability Program, GJB 150A Laboratory Environmental Test Methods for Military Equipment, and GJB 1027 Test Requirements for Launch Vehicles, Upper Stages, and Spacecraft. Furthermore, leading global technology enterprises have developed mature reliability engineering practices in electronics and communication sectors. Apple Inc. employs a full-lifecycle closed-loop methodology to construct a comprehensive reliability framework, forming an integrated reliability ecosystem from design to decommissioning. Foxconn implements end-to-end reliability management through triple inspection protocols, automated production line enhancements, supplier-tiered oversight, Six Sigma management, and extreme environment testing. The reliability management practices of these advanced enterprises offer valuable insights for the engineering and management of spacecraft payload reliability.

Due to the characteristics of space science and application payloads, which often involve high-precision measurements, adaptability to extreme environments, and multidisciplinary design, their reliability requirements are more mission-specific [7]. In some primary space science missions in Europe and the United States, special attention is given to scientific requirements (the demands of science on engineering). The scientific requirements document serves as a central element throughout the entire development cycle of both the spacecraft and the ground-space system, and is subject to continuous refinement and updates [8]. This reflects the concept of ensuring the quality of scientific output. In China's aerospace engineering, great importance is attached to status management, with technical status strictly controlled as a baseline to ensure quality. Traditional reliability standards for aerospace products are process-oriented, emphasizing margin design, qualification testing, and acceptance testing. Given the "small-sample" characteristics of aerospace systems, these standards prescribe phase-specific reliability requirements and tasks for stages such as conceptual design, prototyping, and final product development [9]. Scholars in the academic field have carried out in-depth research on the space reliability program. Yang et al. [10] examined the standard framework of aerospace reliability engineering and proposed a "five-in-one" reliability concept encompassing design, manufacturing, testing, maintenance, and management. They advocated for model-based design, enhanced system robustness, and the development of foundational digital capabilities to address the needs of complex aerospace systems. Han et al. [11] conducted a comparative study of domestic and international aerospace quality and reliability standards, proposed a four-tier framework encompassing basic terminology, quality management systems, process control, and methods and tools, and highlighted the critical role of digitalization and intelligence in standard development. Although the existing reliability standards and engineering practices are relatively mature, there are still certain shortcomings in specialized reliability planning for space science and application payloads. Zhu et al. [12] revealed the limitations of the traditional "qualification test + acceptance test" model and suggested establishing a full-process testing model of "development testing to expose defects—qualification testing to verify capability—acceptance testing to screen defects". Existing reliability programs suffer from poor adaptability to complex space science and application payloads, as well as from incomplete full-process verification systems. Furthermore, they often fail to reconcile dynamic scientific requirements with fixed engineering reliability designs.

This paper analyzes typical domestic and international reliability standards and advanced reliability practices, reveals the limitations of current aerospace reliability standards in engineering applications, explores a goal-oriented new paradigm of reliability program from the perspective of application value, and proposes a reliability planning scheme applicable to space science and application payloads, intending to provide a scientific and reasonable planning basis for subsequent missions.

2 Reliability Standards Analysis

Reliability is a fundamental attribute in engineering, ensuring that a product or system operates without failure under specified conditions and within a designated time frame. In the 1990s, the concept of Reliability System Engineering (RSE) was introduced, emphasizing the prevention and control of faults. RSE integrates reliability, maintainability, and other quality attributes through lifecycle management, employing the V-model and digital twin technologies to establish a closed-loop system [13]. More recently, Belief Reliability Theory has further enhanced the scientific foundation of RSE by introducing three core principles: margin reliability, degradation inevitability, and the uncertainty principle [14].

In reliability engineering, distinct standards are developed to address diverse applications and ensure system and device reliability and maintainability. In China, aerospace standards are primarily derived from national, military, and enterprise-level regulations. The formulation of national aerospace standards is overseen by three principal Technical Committees (TCs) under the Standardization Administration of China: TC425, TC570, and TC312. TC425 is responsible for space technology standards, TC570 oversees manned spaceflight standards, and TC312 manages space science standards. Internationally, ISO TC20, NASA, and the European Cooperation for Space Standardization (ECSS) serve as primary authorities in the development of aerospace standards.

2.1 Equipment Reliability Standards

In equipment development and aerospace engineering, reliability standard systems form the core safeguard for ensuring product performance and safety throughout their lifecycle. GJB 450B-2021, as the core guiding document for military equipment reliability, establishes a structured full-lifecycle process, specifying general requirements and 37 reliability tasks categorized into five series (requirements definition, management, design/analysis, testing/evaluation, and operational improvement). It clarifies responsibilities between purchasers and contractors but focuses on hardware systems, requiring supplementary software reliability standards.

The GJB 1909 series, with GJB 1909A-2009 as its framework, creates a dynamic parameter conversion system for nine equipment types (e.g., missiles and satellites). It translates operational targets/thresholds into contractual specifications/minimums through phase-based indicator freezing and hierarchical model management, yet excludes emerging fields like Non-Terrestrial Network communication payloads.

For specialized applications, GJB 9421 addresses risk control in manned space experiments. Covering 34 tasks across four phases (project initiation to data analysis), it manages device reliability and biological/radiation safety through environmental testing and compatibility verification. Integrating reliability management with scientific objectives requires supplemental standards for software reliability. GB/T 35439, the first comprehensive Safety, Reliability, and Maintainability (SRM) standard in China for space station payloads, introduces a risk matrix that quantifies risk using severity and probability metrics and categorizes payloads into four control tiers.. It integrates redundancy design and cabin maintenance protocols, but needs coordination with GJB 5000 to address software reliability gaps. These four standards establish a hybrid technical framework combining general and specialized specifications, supporting reliability design across diverse scenarios through parallel theoretical and practical approaches.

2.2 New Reliability Standards of NASA

NASA structures its Reliability and Maintainability (R&M) framework across three document tiers: policies, requirements, and technical standards. Guided by strategic objectives, this system integrates R&M goals into lifecycle project management through tiered specifications and operational guidelines. Policies establish strategic directions for R&M implementation. Requirements define actionable frameworks and compliance criteria. Technical Standards provide validated methodologies and process tools. These tiers translate strategic goals into operational actions through phased implementation, forming a top-down closed-loop management system. Table 1 summarizes the core requirements and implementation priorities of NASA's key R&M documents.

NPD 8720.1C positions R&M as a core element of systems engineering, mandating full-cycle integration of design verification and achieving availability objectives through cross-domain integration (safety, risk management). NPR 7120.5F embeds R&M requirements into project management workflows. Phase division and milestone reviews align R&M activities with the Safety and Mission Assurance Plan and lifecycle cost control systems, forming a structured governance mechanism. NASA-STD-8729.1A innovatively establishes a "goal-driven" framework, dynamically linking 14 sub-goals with 49 implementation strategies (e.g., fault tree analysis, reliability modeling) based on mission risk levels. Its core mechanism ensures R&M is embedded in system design from the conceptual phase through a "goal-strategy-verification evidence" hierarchy. It allows flexible compliance validation via design evidence (test reports/models), unifying regulatory compliance with engineering effectiveness.

Table 1 NASA R&M documents

Number	NPD 8720.1C	NPR 7120.5F	NASA-STD-8729.1A
Title	NASA reliability and maintainability (r&m) program policy	NASA space flight program and project management requirements	NASA reliability and maintainability (R&M) standard for spaceflight and support systems
Core content	R&M policy: Establish and implement reliability and maintainability requirements	Program management: standardize lifecycle management of spaceflight projects	R&M standard: build an "objective–strategy" framework with outcome orientation
Key implementation points	Integrate into systems engineering; define maintenance needs; use standardized documentation	Include in safety and mission assurance plan; define project phases and review points	Decompose into 14 objectives and 49 strategies; validate through analysis and testing

2.3 *SpaceX Reliability Standards*

The commercial satellite reliability standard system serves as a critical foundation for supporting large-scale, low-cost development. In contrast to traditional satellites that prioritize ultra-high reliability at the single-unit level, commercial satellites adopt a novel framework by moderately relaxing individual reliability requirements. This is achieved through constellation-level redundancy, process consistency control, and on-orbit verification technologies, thereby balancing cost, quality, and schedule.

Take SpaceX's Starlink constellation as an example: it breaks the traditional "finalize design before launch" approach, adopting a rapid iterative strategy integrating R&D, manufacturing, launch, and application. This creates a dynamic closed-loop reliability verification system. Initial satellites validate core functions with simplified testing, using early on-orbit data to drive iterative upgrades in subsequent versions, reducing single-satellite verification cycles by over 50%. Constellation redundancy compensates for individual satellite reliability limitations. For component reliability assurance, SpaceX innovates with a "simplified component screening and board-level joint verification" model, maintaining reliability while cutting costs, and flexibly adopting aerospace standards or automotive standards as qualification criteria.

SpaceX's practices shift from "absolute reliability" to "system robustness," achieving coordinated optimization of reliability, cost, and efficiency through technological iteration, data-driven methods, and architectural innovation. This provides vital insights for developing reliability standards suited to the mega-constellation era.

Table 2 Huawei reliability workflow node tasks

Node	Name	Objectives	Output
TR1	Reliability requirement analysis and decomposition	Define and analyze reliability goals; derive and allocate requirements from standards and issues	1. Reliability requirement analysis specification 2. Requirements list 3. Requirement decomposition and allocation
TR2	Reliability requirement design	Design implementation based on requirements; define interfaces; plan for interdependencies	Reliability design specification
TR3	Development and realization of reliability requirements	Execute design tasks (e.g., part selection, schematic, PCB, software)	Detailed design documentation
TR4 TR5	Acceptance of reliability requirements	Verify reliability goals through testing and evaluation	Test and reliability prediction report
TR6	Initiation of reliability certification	Certify product to meet standards and enable market access	Certification report

2.4 Huawei Reliability Standards

Huawei's reliability engineering adopts a closed-loop process guided by Technical Review (TR) milestones and is deeply integrated with the Integrated Product Development (IPD) system. Phase-specific tasks are summarized in Table 2. The framework employs a matrix-based organizational structure and a layered process system, coordinated by Global Process Owners to align core processes. This structure facilitates cross-domain collaboration across supply chains and IT systems. A "Blue Army" mechanism validates process resilience, while a "sunset policy" removes redundant steps, supported by FFR metrics for continuous optimization. This preventive design-oriented paradigm embeds reliability at the design stage, significantly enhancing the global competitiveness of core products (e.g., 5G base stations, Kirin chips).

2.5 Summary

A systematic analysis of China's domestic reliability standards, NASA's updated reliability frameworks, and innovative reliability methodologies adopted by commercial aerospace leaders like SpaceX and Huawei reveals three critical challenges facing traditional reliability standards in engineering practice, demanding urgent methodological innovation.

(1) Separated Functional and Reliability Designs

Reliability programs are typically initiated after the completion of product design, which limits the effectiveness and completeness of reliability parameter verification. Space science and application tasks have high demands for components' performance, specifications, and quality levels. However, due to cost or other factors, high-grade components are limited in space use. As a result, space science and application payloads often use commercial off-the-shelf products(COTS) and life-sensitive components. This makes the verification and evaluation tasks for the payloads heavy.

(2) Ambiguous Reliability Mission and Requirement Definitions

Disparate disciplinary requirements contribute to ambiguity in defining reliability parameters related to microgravity levels, radiation exposure thresholds, and space station positioning accuracy. The development of space science and application payloads must balance standardized technical platforms with the specific needs of individual experiments, employing customized design methodologies that support the integration of advanced technologies. Ambiguity in reliability specifications often results in suboptimal designs that do not adequately address the complexities of on-orbit operational conditions. Furthermore, insufficient consideration of interrelated environmental factors such as prolonged microgravity, extreme thermal cycling, and high-energy particle radiation significantly increases the risk of failure during both orbital verification and operational deployment phases.

(3) Inappropriate Reliability Metrics

The implementation paradigm of space science and application missions has evolved beyond traditional single-mission constraints, establishing an ecosystem characterized by dynamic iteration, multi-domain collaboration, and open innovation. Through coordinated operations among space stations, scientific satellites, and deep-space probes, these missions now achieve comprehensive exploration capabilities spanning near-Earth orbits to the solar system's frontiers. However, establishing effective reliability metrics remains challenging due to the inherent complexity of space science and application payloads. Current reliability methodologies struggle to address multi-physics interactions, cumulative environmental effects (e.g., radiation, thermal extremes, microgravity), and compatibility with iterative hardware/software upgrades. A critical need exists for establishing differentiated, quantifiable reliability metrics that accommodate operational diversity while ensuring verifiable performance validation across evolving mission profiles.

Collectively, these observations reveal a structural gap between existing reliability standards and the mission logic of space science and application payloads. Traditional standards and workflows are predominantly process-oriented: they organise reliability work as phase-based task lists and compliance checkpoints, emphasising "which activities are performed in which phase". However, they provide limited guidance on how these activities trace back to explicit mission value, scientific objectives, or payload-level contributions to system-of-systems performance. In contrast, space science payloads are ultimately judged by goal-driven criteria such as data quality, availability during critical observation windows, and safety margins for on-orbit operations. Bridging this gap requires a reliability program paradigm that starts from mission value and decomposes it systematically into requirements, constraints, and targeted reliability strategies, rather than aggregating pre-defined task lists aligned only with lifecycle phases.

3 Reliability Program Framework

This chapter advances a paradigm shift in reliability engineering for space science missions, transitioning from process-oriented compliance to goal-driven, value-based practice. Compared with conventional reliability programs, which directly map standards to phase-specific activities, the proposed framework reverses the logical sequence. Rather than starting with "which tasks must be performed in each phase", it first clarifies "why" reliability is required in terms of mission value, payload role, and application scenarios, and only then defines "what" reliability measures and evidence are necessary. This reordering from a "phase–task–partial objective" logic to a "goal–constraint–strategy–task" logic is intended to address the gap identified in the standards analysis. It provides an explicit argument pathway from scientific and application objectives to engineering-level reliability implementation, thereby aligning process execution with goal-driven reliability assurance. The framework defines customized and granular reliability requirements by systematically decomposing scientific goals into quantifiable engineering parameters. Implementation strategies align subsystem reliability targets with the mission's operational architecture, ensuring traceability from component-level specifications to overarching mission requirements. This approach specifies what tailored reliability measures should be designed, validated, and monitored to ensure mission success while optimizing resource allocation across the mission lifecycle.

3.1 Principles

The overarching aims of space science and application payloads in mission-specific scenarios are considered fulfilled when contractual requirements are met through comprehensive implementation of reliability strategies across all system levels, supported by verifiable evidence. Analysis and validation shall adhere to the following principles:

(1) Application Value Maximization Principle

Focused on optimizing payload effectiveness, this principle improves output stability and longevity by identifying critical vulnerabilities, refining redundancy design, and implementing lifecycle management. It ensures baseline mission reliability while maximizing scientific returns, resource efficiency, and technological synergies.

(2) Goal-Driven Design Principle

Transitioning from process-based to goal-oriented methodologies, this approach directly addresses mission-level technical requirements by focusing on deliverable outcomes rather than prescribing standardized reliability tasks. The system implements a top-down hierarchical architecture of "application value–planning principle" to conduct hierarchical iterative decomposition of overarching requirements. This methodology emphasizes application value-driven requirements decomposition, with verification and validation activities focusing on implementation strategies for multi-level requirement fulfillment.

(3) Risks Classification and Payloads Categorization Principle.

Risk management, task prioritization, and impact domain assessment guide the determination of tailored reliability workflows. Mission-specific characteristics and payloads risk levels dictate customized adjustments to reliability tasks and processes, ensuring alignment with classified application values and strategic priorities.

3.2 *Workflow*

The workflow for executing space science and application payloads reliability engineering shall follow four defined phases, as shown in Fig. 1.

(1) Reliability Requirements Analysis

Reliability requirement analysis clarifies scientific mission success criteria through task analysis and functional modeling, supporting goal and strategy development in the hierarchical "application value-planning principle" framework. Using overall mission requirements as the baseline, it systematically organizes technical requirements by linking application values with key design features, establishing reliability goals and implementation strategies to ensure space science and application payloads achieve objectives during mission cycles. Inputs include overall indicators, engineering requirements, and multi-dimensional design constraints from decomposition. The process includes collaboration with stakeholders, risk-based management, use of historical data, and the definition of environmental constraints.

Reliability task analysis supports environmental adaptability objective decomposition through mission/ lifetime/ environmental profile analysis while providing inputs for functional modeling. Functional modeling focuses on design compliance and fault tolerance capability, guiding reliability improvements by identifying design vulnerabilities through a collaborative "designer-led reliability engineering" approach. Mission success requires aligning scientific, engineering, and reliability

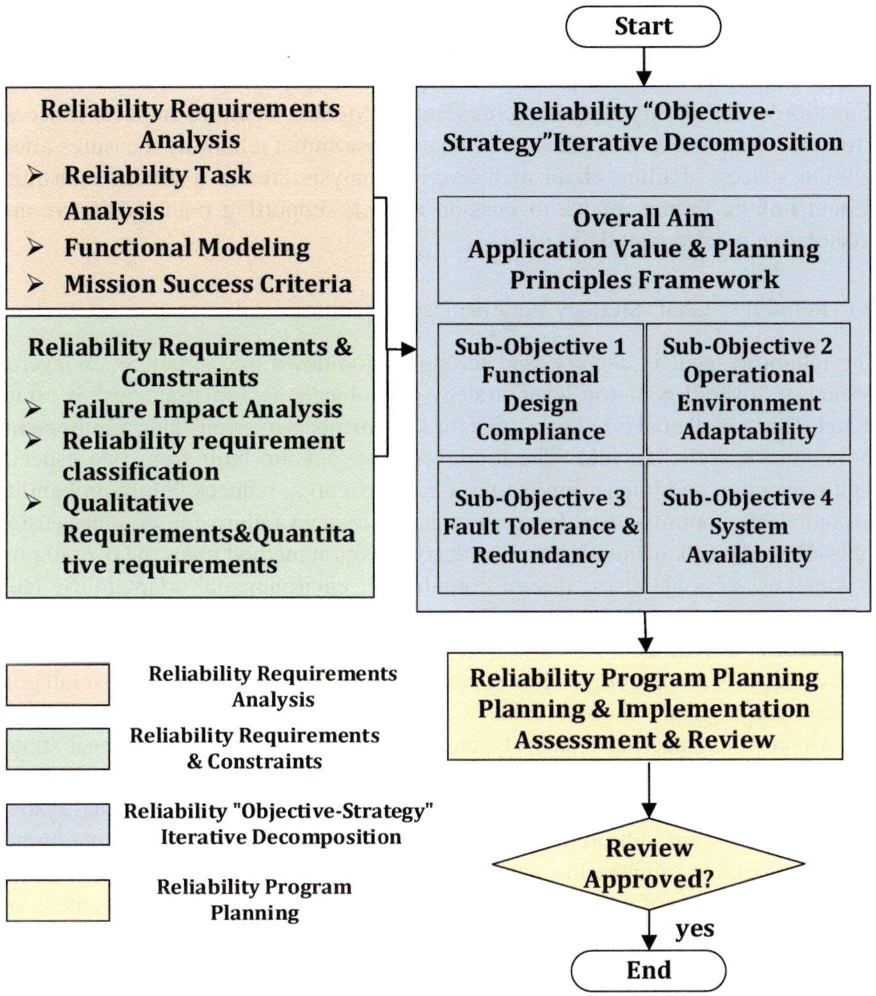

Fig. 1 Reliability Program Organization and Implementation Process

goals, with a verification system built on metrics like performance, stability, and data validity under technical and resource constraints.

(2) Reliability Requirements Constraints

Reliability requirements constraints systematically identify potential risks and establish a failure correlation grading system to inform goal and strategy development within the "application value-planning principle" hierarchical framework. This constraint system is based on resource limitations, risk profiles, and operational scenarios specific to space science and application payloads. Inputs include advanced engineering metrics for cross-domain payloads, COTS component risks, schedule/

budget limits, and maintenance strategies. This ensures that reliability goals align with mission resources and complex system design.

The execution follows a progressive logic flow: Objective (Mission)—Strategy (Function)—Reliability Requirements (Failure Modes)—Failure Impacts (Success Criteria)—Requirement Grading. It evaluates how unmet reliability measures affect mission success. Failure chain and severity analysis creates a graded constraint system linking failure modes to mission impact, supporting both qualitative and quantitative requirement definitions.

(3) Reliability Goal -Strategy Iterative Decomposition

The reliability goal is decomposed through a top-down tree structure. A layered system of "overall goal–top-level strategy–sub-objectives–sub-strategies" is established. The overall goal is to ensure that the system meets mission value requirements throughout its full lifecycle. The top-level strategies are built from two aspects: fault prevention and impact mitigation. Fault prevention reduces failure probability through design optimization. Impact mitigation reduces failure consequences using fault-tolerant mechanisms. A heuristic decomposition method maps the overall goal to four key sub-objectives: design compliance, environmental adaptability, fault tolerance, and system availability. These are further broken down into executable bottom-level strategies.

Each layer of this decomposition serves a distinct conceptual role. The overall goal captures the payload's mission value in an integrated form that combines scientific performance, engineering feasibility, and safety expectations. The top-level strategies organise the dominant reliability logics—preventing faults versus limiting their impact—into a structure that is understandable to both mission planners and system engineers. The four sub-objectives act as a bridge between these abstract strategies and concrete engineering actions, ensuring that design rules, environmental mitigation measures, fault-tolerant architectures, and availability management are coherently aligned. Bottom-level strategies then translate these sub-objectives into specific, actionable activities, such as derating guidelines, redundancy allocation, environmental screening and margin design, diagnostic coverage targets, and on-orbit monitoring plans.

The process uses dynamic trade-offs between mission risk tolerance and constraints. Goals and strategies are refined iteratively, which can refer to the appendix of NASA-STD-8729.1A. Each strategy is based on a higher-level goal and guides the next level's sub-objectives, forming a measurable reliability baseline. This closed-loop logic supports engineering under the "application value–planning principle" framework.

(4) Reliability Program Planning

Reliability program planning is structured through a hierarchical framework of objectives and strategies. It includes a comprehensive set of technical activities such as design, analysis, testing, and verification. The planning process follows systems engineering methodologies. It coordinates multiple technical interfaces by applying the lowest acceptance-level strategy to ensure the effective implementation of the overall reliability objective. To meet the diverse needs of space science missions, application payloads, and commercial space programs, whether aiming at mission success, result maximization, or cost-effective high reliability, a payload classification and control model is established in accordance with GB/T 35,439. This model supports the balance among risk, criticality, and cost. All reliability design, testing, and management activities are carried out under the framework defined by GJB 450B.

Program performance is assessed using milestone reviews and standardized evidence chain verification. Technical verification is completed under the supervision of the chief engineer of the system. A three-dimensional review process is used to evaluate the completeness of execution, compliance with standards, and acceptability of residual risk. This review confirms whether the residual risk remains within thresholds and whether the approach aligns with the principle of application value and planning. Based on the review outcome, a two-way decision process is applied. If the outcome is positive, the program proceeds to implementation and acceptance [15]. If the outcome is negative, it initiates iterative refinement of the objectives and strategies and updates the reliability baseline.

4 Conclusion

Current reliability standards encounter three principal challenges in the engineering implementation of space science and application missions. These include the disconnect between product performance design and reliability programming, ambiguities in the definition of reliability tasks and requirements, and limitations in the demonstration and applicability of reliability metrics. This study reviews major domestic reliability standards such as GJB 450B, GJB 1909, GJB 9421, and GB/T 35439, as well as leading international standards including NASA's NPD 8720.1C, NPR 7120.5F, and NASA-STD-8729.1A. Commercial satellite reliability standards and reliability workflows from advanced technology enterprises such as Huawei are also examined. Based on this review, critical limitations inherent in conventional reliability standards are identified and summarized.

In response, this paper introduces a dual closed-loop control mechanism grounded in the "application value planning principle" and adopts a structured verification methodology to ensure alignment with mission objectives. A hierarchical and iterative decomposition architecture is proposed, also based on the "application value planning principle," to guide comprehensive reliability planning. This approach facilitates dynamic adjustment and configuration of the reliability program, thereby establishing a tailored reliability framework for space science and application payloads.

The proposed framework addresses the inflexibility commonly found in traditional reliability engineering practices. It ensures the scientific coherence of overall project objectives while affording technical flexibility at the subsystem level. This design philosophy aims to avoid redundant resource allocation and simultaneously safeguard essential reliability parameters. In doing so, it fosters a standardized and collaborative model between system-level and subsystem-level design, embedding reliability considerations deeply within the product development lifecycle. Consequently, the framework offers a practical and scalable engineering solution that improves the probability of mission success for space science and application payloads during on-orbit operations.

Acknowledgements The work described in this paper was supported by National Key Research and Development Program—NQI (2022YFF0610100), China Electric Institute State Key Laboratory of Environmental Adaptability for Industrial Products (2025EASKJ-008), and HORIZON-MSCA-2024-PF-01-01 under the Marie Skłodowska-Curie grant agreement (101206130).

References

1. Sharma AK (2022) James webb space telescope. Res-J Sci Educ 27(8):1355–1369
2. Wang L, Zhu Y, Zhang B et al (2024) Design and implementation of payload system of the Zhurong Mars rover. Scientia Sinica (Technologica) 54(1):149–164
3. Gao M, Zhao G, Gu Y (2022) Recent progress in space science and applications of China's space station in 2020–2022. Chinese J Space Sci 42(4):503–510
4. Zhang Z, Li P, Zhang W et al (2022) Communication fault analysis of an aerospace payload and reliability improvement measures. Environ Technol 40(5):63–67
5. Liu J, Chu H, Dong J (2025) Global space launch activities in 2024. Aerospace China 1:19–27
6. Sun X (2012) Review of military reliability assurance standards. Electron Prod Reliab Environ Test 30(1):61–65
7. Guan Q, Hang C, Li S et al (2015) Review on reliability of electronic devices in deep space environment. Syst Eng Electron 1–15
8. Gu Y (2022) Thoughts on space science development. Bulletin of Chinese Acad Sci 37(8):1031–1049
9. Li J, Jia Y (2025) Enlightenment of China's deep space exploration mission planning and implementation. J Eng Stud 1–9
10. Yang S, Yang Z, Cheng H et al (2024) Innovation and practice of aerospace reliability engineering. Aerospace China 4:34–39
11. Han X, Du B, Zhao P (2023) The construction plan for the system of aerospace quality and reliability standards. In: Annual collection of outstanding papers in China's standardization, pp 246–253
12. Zhu F, Chen Q, Yang Y et al (2016) A discussion on the development test for spacecraft products. Spacecraft Environ Eng 33(5):470–478
13. Wang Z (2021) Current status and prospects of reliability systems engineering in China. Front Eng Manage 8(4):492–502
14. Kang R, Wang Z (2022) Reliability systems engineering: a research review and prospect. Acta Aeronautica et Astronautica Sinica 43(10):180–190
15. Li P, Gao K, Xiao X, Chen J, Zhang L, Zhong H (2026) Spaceborne health management standards for autonomous operations: a review and outlook. Acta Astronaut. 245:64–82

Study on the Effectiveness Evaluation of Cross-Flooding Hatches in Damage Stability Design of Large Passenger Vessels

Yi-ming Huang, Gang Chen, Yong Guo, Zhang-jian Wang, Xin-yi Fan, Jia-ning Li, Gao-shuai Wang, Ai-hua Gao, and Fu-tong Li

Abstract With the implementation of the new SOLAS 2020 damage stability requirements, multiple strategies are being actively investigated to enhance the damage stability performance of large passenger vessels. Among these, cross-flooding devices such as pipes and ducts are commonly employed [1, 2]. This study focuses on evaluating the effectiveness of cross-flooding hatches and proposes a methodology for performance assessment, addressing the current lack of explicit regulatory guidance concerning this specific type of device. Conventional pipe-based formulations from IMO are utilized to estimate the cross-flooding time associated with hatches installed on a 150,000 GT cruise vessel. In parallel, a time-domain

Y. Huang · G. Chen (✉) · Y. Guo · Z. Wang · X. Fan · J.-n. Li · G. Wang · A. Gao
Shanghai Waigaoqiao Shipbuilding Co., Ltd, Shanghai, P.R. China
e-mail: chengang@chinasws.com

Y. Huang
e-mail: huangyiming@chinasws.com

Y. Guo
e-mail: guoyong@chinasws.com

Z. Wang
e-mail: wangzhangjian@chinasws.com

X. Fan
e-mail: fanxinyi@chinasws.com

J.-n. Li
e-mail: lijianing@chinasws.com

G. Wang
e-mail: wanggaoshuai@chinasws.com

A. Gao
e-mail: gaoaihua@chinasws.com

Y. Huang · F. Li
School of Naval Architecture Ocean and Energy Power Engineering, Wuhan University of Technology, Wuhan, P.R. China
e-mail: 15971501830@163.com

H. Li et al. (eds.), *Data-Driven Methods for Reliability and Safety Engineering: Applications in Industrial Systems*, Springer Series in Reliability Engineering,
https://doi.org/10.1007/978-3-032-22873-4_17

simulation using computational fluid dynamics is conducted to model the flooding process from initiation to equilibrium. The results demonstrate that while theoretical pipe equations offer conservative estimates useful for preliminary design, time-domain simulations capture the actual dynamic response more accurately, predicting a reduced time to reach equilibrium. These findings highlight the need for further investigation and methodological refinement regarding cross-flooding hatch calculations, particularly in light of the absence of detailed regulatory frameworks.

Keywords Cross-flooding hatch · Damage stability · Large passenger vessel

1 Introduction

The increasing trend toward larger cruise ship designs, coupled with growing passenger and crew capacities, has intensified the emphasis on stability safety. With the enforcement of the SOLAS 2020 stability regulations, the design of large cruise ships is subject to more rigorous criteria, presenting significant challenges for compliance. In order to meet these updated requirements using conventional approaches, strategies such as scaling the principal dimensions from reference vessels and increasing the density of watertight compartments below the bulkhead deck are frequently employed. However, these methods often result in considerable cost escalation. As a result, the deliberate integration of targeted cross-flooding devices within the general arrangement layout has emerged as an effective alternative for improving damaged stability performance in large cruise ship designs [3–7].

Cross-flooding devices function by hydraulically linking flooded compartments to adjacent balancing tanks through structural elements such as cross-flooding pipes or ducts. This configuration serves to counteract asymmetric flooding resulting from hull breaches, enabling the vessel to correct excessive heel angles and restore hydrostatic equilibrium within the specified time limits. By mitigating dangerous heel conditions and reducing their duration to transient states excluded from survival probability index (Si) calculations, these systems contribute to improved Si values and consequently enhance the attained subdivision index (A-index).

Resolution MSC.362(92) specifies that the current regulations for cross-flooding focus on evaluating the time period from the activation of the flooding system after damage occurs until the water levels in the flooded compartment and the equalizing compartment reach equilibrium. If this process is completed within 60 s, it is considered instantaneous flooding, and dynamic effects may be disregarded. If the duration exceeds 60 s, the most critical angle of heel during the gradual flooding of the equalizing compartment must be assessed. For passenger ships, an additional requirement applies: if the equalization time is within 10 min, only the smaller value between the final heel state and the most dangerous heel state during the process needs to be evaluated.

Fig.1 Example of cross-flooding hatch on cruise ship

The common applications of cross-flooding device are cross-flooding pipe and cross-flooding duct, while IMO guidelines of MSC.362(92) provide details of calculation parameters and methods correspondingly. However, these two device types do not encompass all solutions. A 'cross-flooding hatch' is now widely used. This hatch maintains fire integrity under normal conditions but is designed to dislodge when submerged during flooding, thereby enabling cross-flooding. Figure 1 shows an example of cross-flooding hatch applied on cruise vessel.

Current regulations lack guidance for calculation of cross-flooding hatch. In practice, designers and naval architect program process the cross-flooding hatches as zero-length cross-flooding pipes by directly adopting parameters from pipe-related calculations. This empirical approach lacks theoretical validation, and the mismatch between applied parameters and actual conditions leads to computational deviations that may compromise design outcomes and construction costs.

To analysis this difference, this article employs numerical simulation methods to investigate a large cruise ship equipped with cross-flooding hatches. Through time-domain simulation of the complete cross-flooding process, the actual equilibrium restoration time is obtained. Comparative analysis is also conducted between these simulation results and formula-based calculations derived from the zero-length pipe assumption methodology.

2 Numerical Methods

2.1 *Volume of Fluid Multiphase Method*

Multiphase fluids, in which multiple fluids simultaneously flow within a given domain, are of critical importance in various industrial applications. By associating fluid phases with gases, liquids, or solids, the range of problems addressable through numerical fluid simulations is significantly expanded. Upon enabling the Volume of Fluid (VOF) model in STAR-CCM+, the software automatically activates the separated flow model. In this framework, all phases are considered smaller than the characteristic solution scale and are permitted to coexist throughout the computational domain. The model assumes that the solution reflects a time-averaged behavior of the fluid system, a concept referred to as "interpenetrating continua." Each phase is delineated by an interface, the position and shape of which evolve dynamically with phase movement. The VOF approach further presumes that the computational mesh possesses adequate resolution to accurately capture the interface between the different phases.

In the simulation of cross-flooding, both water and air are involved as two distinct phases. The water phase is defined as phase i, while the air phase is distributed as 1-i. As the cross-flooding, the relative motion between the moving water and the escaping air generates inter-phase slip ($\boldsymbol{v}_{d,i}$). Furthermore, by incorporating a user-defined custom source term of phase i (S_{α_i}) and, the relationship between the surface area vector ($\boldsymbol{a}$) and the velocity (v) can be accurately described.

2.2 *Dynamic Fluid Body Interaction (DFBI)*

STAR-CCM+ provides the capability to model the motion of rigid bodies using continua. The coupling between the continuum and the fluid boundary is referred to as Dynamic Fluid Body Interaction (DFBI). Within a rigid body, the relative positions of internal points remain unchanged, and the motion of the rigid body can be calculated based on its center of mass. It allows for the predefinition of forces such as gravity and fluid forces, as well as the customization of forces and moments. During computation, the motion of the rigid body in the fluid is considered through methods such as deforming the fluid boundary (DFBI deformation), rigidly moving the entire fluid mesh (DFBI rotation and translation, as well as DFBI embedded rotation), or superimposing mesh motion via a defined body.

In the simulation of the cross-flooding process, the roll motion requires activating the center of mass. The solution is performed in the body-fixed coordinate system centered at the origin of the center of mass. The relationship between the resultant moment (**n**) acting on the cross-flooding ship model and the ship's moment of inertia (**M**) and angular acceleration ($\boldsymbol{\omega}$).

The moment of inertia tensor is expanded as:

The matrix of the moment of inertia tensor is symmetric along the diagonal. Therefore, this matrix can be defined by one set of principal components (Mxx, Myy, Mzz) specifying the diagonal directions and another set of non-diagonal components (Mxy, Mxz, Myz).

The motion options provided by DFBI can apply physical constraints to the body's motion. These constraints allow certain degrees of freedom while restricting others.

2.3 Computational Grid and Numerical Solution

In this study, the calculation model of the cross-flooding device in the large passenger vessel engine room is a simplified version of a certain large passenger vessel engine room. Considering that the shape of the superstructure has a relatively small impact on the transverse flooding device, details above the waterline are ignored during the modelling process. In addition, to further simplify the calculation, the internal components of the engine room are neglected, and only important structures such as internal openings, bulkheads, and pillars are modelled.

2.4 Cruise Ship Model

A certain large passenger vessel has an overall length of 321 m, a width of 36.4 m, and a designed loading draft of 8.46 m. The main design parameters are shown in Table 1, and the actual model is shown in Fig. 2.

Table 1 Cruise ship main dimensions

Length overall Loa/m	Length between perpendiculars Lpp/m	Beam B/m	Scantling draft Td/m
321.0	285.2	36.4	8.46

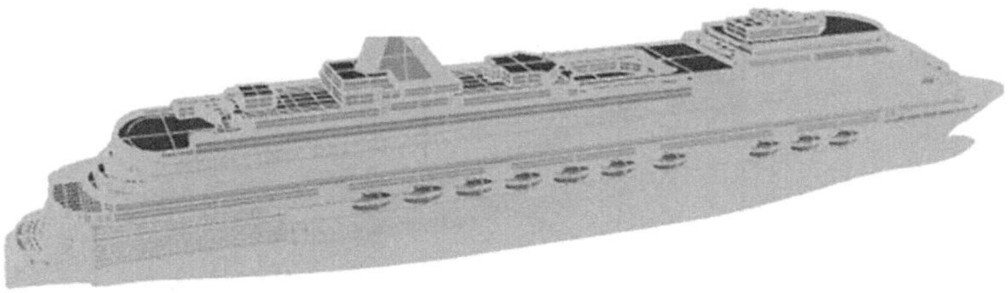

Fig. 2 The object cruise ship in research

2.5 Grid Model

The simulation model of the cross-flooding device in the engine room of a large cruise ship consists of a background domain that can simulate seawater changes, an overlapping domain that can follow the hull movement, and a cabin domain that is connected to the background domain and equipped with transverse flooding holes, as shown in Fig. 3.

The grid division is shown in the XZ-plane cross-section of the overall grid in Fig. 4 and the grid on the cruise-ship surface in Fig. 5. The basic grid size of the background domain is 20 m. Volume refinement of the free-surface in the Z-direction and volume refinement of the interpolation area with a custom size of 10% are carried out, resulting in a total of 1.13 million grids. The overset domain has a base size of 4 m. Refinement includes volume refinement at the bow/stern and free surface (Z-direction), plus surface refinement on the hull (custom size 10%), totaling approximately 0.85 million cells. The grid resolution was selected to balance numerical accuracy and computational cost while preserving the dominant physics of progressive flooding and free-surface evolution. A relatively coarse base size is used in the background domain because it mainly serves as an outer reservoir and to accommodate far-field free-surface adjustment, whereas the overset domain is refined to reliably capture hull–fluid coupling and local free-surface deformation induced by roll motion. The cabin domain employs a substantially finer uniform base size to resolve internal openings, bulkheads, and the cross-flooding process, which are the primary drivers of equalization time. With this multi-domain strategy, the mesh density is concentrated where transient flooding process and vessel dynamic response of the damaged cruise ship are expected to dominate, reducing unnecessary cell counts in regions with limited influence on the equalization response.

The basic grid size of the cabin domain is uniformly set to 0.2 m, and the engine-room model has 1.65 million grids. The grid division is shown in Fig. 6.

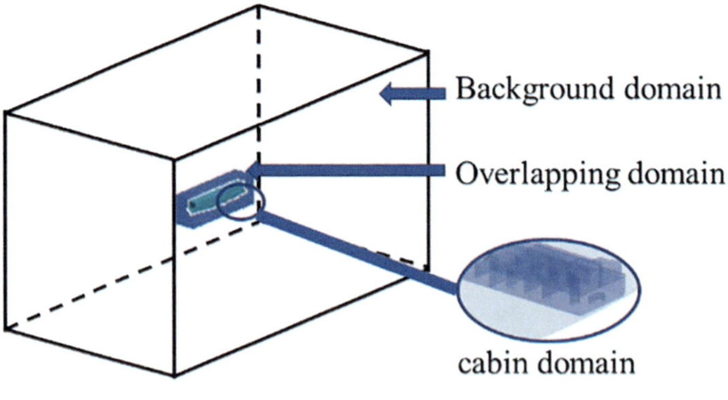

Fig. 3 Computational model of cross-flooding in the engine room of a large cruise ship

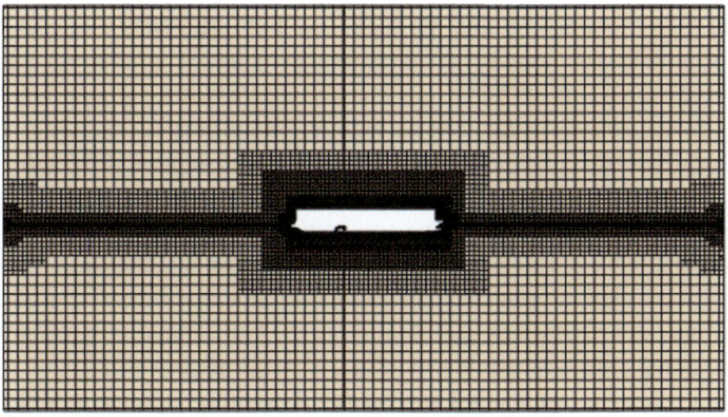

Fig. 4 XZ cross-section

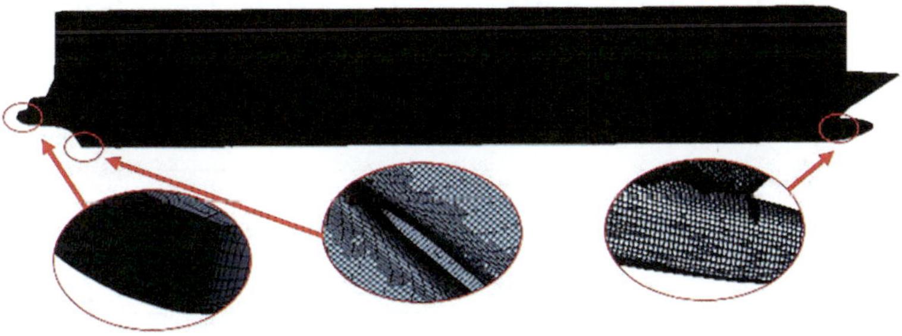

Fig. 5 Surface mesh

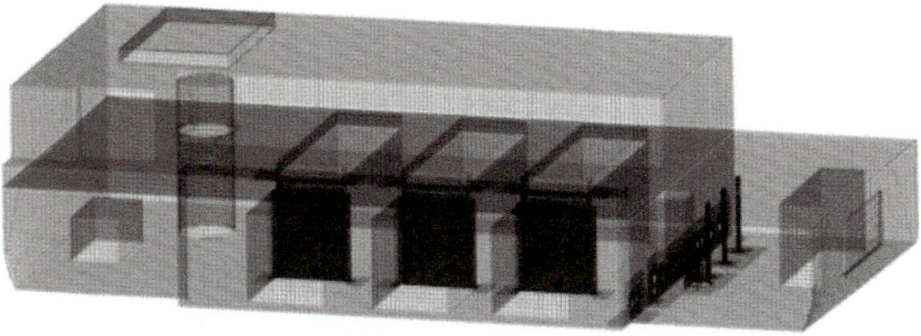

Fig. 6 Cabin model mesh

2.6 Initial Conditions and Boundary Conditions

The initial conditions and boundary condition setting parameters for the computational model are illustrated in Fig. 7. The numerical method combining the VOF wave and overset grid is employed, and the realizable k-ε turbulence model is utilized to establish the model of the transverse flooding system in the engine room compartment area of a large cruise ship. When using the VOF wave method to conduct numerical simulation of the cross-flooding device, only two Eulerian phases, namely water and air, are established. After the solution is completed, the volume distribution of water is derived.

VOF Wave: Configure a flat wave to simulate still water, with the Z-direction height of the flat wave surface set at $-$ 8.55 m (a certain loading draft). This continuous setting is applied to both the background domain and the overlapping domain, with the volume distribution as depicted in Fig. 8. The cabin domain is not assigned a continuum value and is separately configured to be filled with air.

DFBI: Set the roll moment of inertia for the computational model to 2.01×10^{10} kg m^2. The roll moment of inertia for a large cruise ship can be obtained using empirical formulas 5 and 6. Additionally, configure the model's centre of mass based on the principal dimensions. Activate the Z-axis movement and X-axis rotation for the computational model. The DFBI constraints were defined to retain the dominant degrees of freedom affected by asymmetric flooding, namely heave and roll, while suppressing secondary motions that have limited influence on the cross-flooding

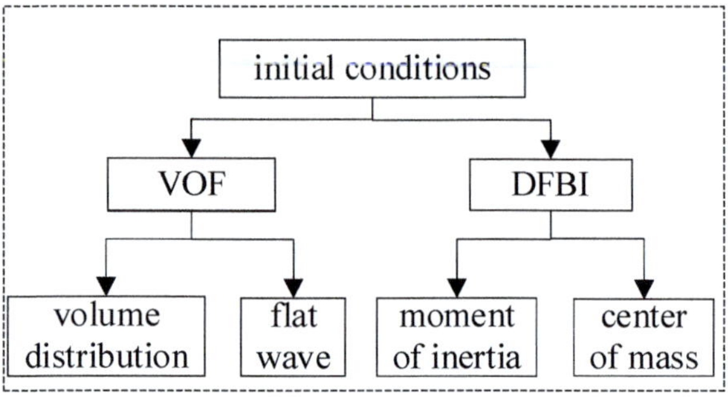

Fig. 7 Initial conditions

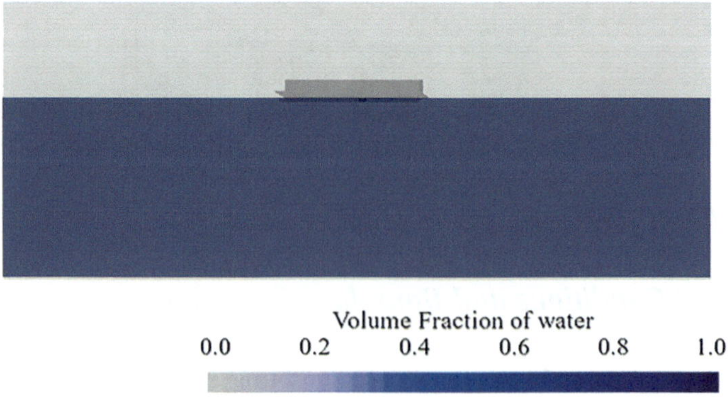

Fig. 8 Initialization solution

equalization time in the present still-water damage scenario. This constrained-motion setup improves numerical robustness and interpretability by isolating the coupling between transverse flooding and vessel dynamic response. The roll inertia is prescribed based on empirical estimates consistent with the principal dimensions, ensuring that the simulated roll dynamics remain physically representative for the studied loading condition.

The background domain is configured with the upper surface as a pressure outlet, the side surfaces parallel to the XZ plane as a velocity inlet and a pressure outlet respectively, the front and rear surfaces parallel to the YZ plane as symmetry planes, and the bottom surface as a velocity inlet. The overlapping domain assigns the outer surfaces as overset grids, nested within the background domain, with the cruise ship's surface set as walls, and interfaces generated at the ship's starboard side and the damaged surface of the cabin. Additionally, the upper surface of the cabin domain is set as a pressure outlet with an area corresponding to the operational condition, while the remaining surfaces are designated as walls. The specific boundary conditions for the computational model are illustrated in Fig. 9. These boundary conditions are chosen to represent an open-water environment while minimizing artificial reflections and undesired confinement effects. The pressure outlet on the upper boundary provides a stable reference pressure and allows air to leave the domain during transient free-surface motion, which is essential for VOF-based flooding simulations. Velocity inlet boundaries on the lateral and bottom faces are used to maintain a numerically stable open boundary for the still-water initialization and to accommodate inflow induced by the flooding-driven volume redistribution. Symmetry planes at the fore and aft faces reduce computational cost.

To improve transparency of the numerical setup, a screening sensitivity check is recommended for key numerical parameters that may affect the predicted equalization time, including (i) cabin-domain mesh size near internal openings and cross-flooding paths, (ii) global time step through a Courant-number-based control in the

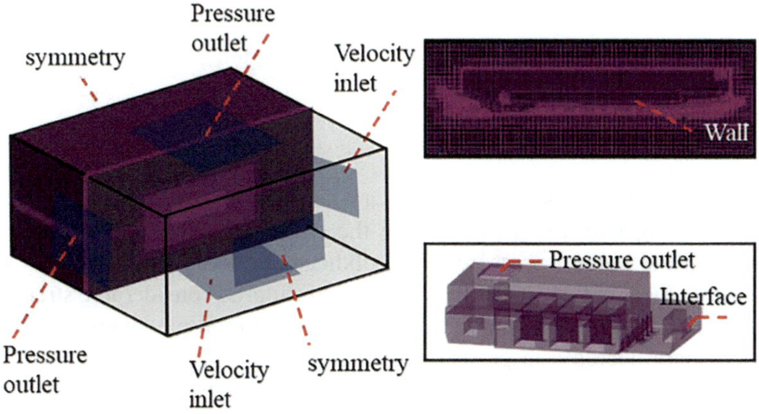

Fig. 9 Boundary conditions

free-surface region, and (iii) DFBI constraint settings. In practice, the sensitivity can be evaluated by repeating the simulation with a modest refinement of the cabin-domain base size (e.g., reducing the base size by 20–30%) and a reduced time step, and then comparing integral outputs such as the time to equilibrium, the peak heel angle, and the roll decay characteristics. If these outputs change marginally (typically within a few percent), the baseline setup can be regarded as sufficiently resolved for engineering decision-making; otherwise, additional refinement should be applied to the regions that control hydraulic losses and phase-interface dynamics.

3 Results and Discussion

3.1 *Flooding Process*

Following the forced opening of the valve in the auxiliary machinery room due to incoming water pressure, flooding was initiated. The internal mechanical equipment, positioned directly in front of the seawater entry point, experienced substantial turbulence as seawater surged into the space and propagated toward the engine room via the manholes. Intense oscillations occurred between the bulkhead and the first tier of machinery. Given that the diversion capacity of the manholes was significantly lower than the seawater inflow rate, a considerable volume of water accumulated within the auxiliary machinery room. Only a limited amount of seawater was able to pass into the balancing compartment through the designated flooding openings. As a result, the auxiliary machinery room was rapidly inundated, causing a subsequent decline in the flow rate at the breach. The vessel's heel angle reached its maximum during this phase. As the water level in the balancing compartment gradually rose, the overall intensity of water movement diminished. The vessel began to regain an upright posture, with the amplitude of rolling motions progressively decreasing until a stable equilibrium was achieved refer to Figs. 10 and 11.

3.2 *Pressure*

The influx of seawater resulted in significant seawater pressure within the auxiliary machinery room, with the area opposite the entrance, the side of the machinery, experiencing the highest pressure. The bulkheads between compartments and the pillars within the auxiliary machinery room also endured considerable stress. As the seawater flowed into the engine room, it encountered multiple layers of machinery, causing a sharp decrease in pressure, with the maximum pressure diminishing by half. Significant pressure was only observed at the bulkhead and the first layer of machinery facing the water flow. Moreover, a slight negative pressure emerged after the seawater was channelled into the engine room, as depicted in Fig. 12.

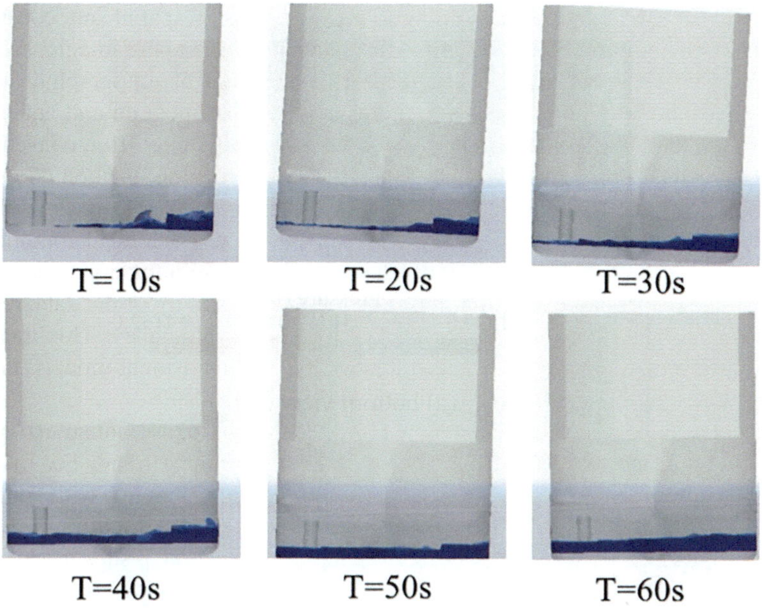

Fig. 10 Crossing-flooding

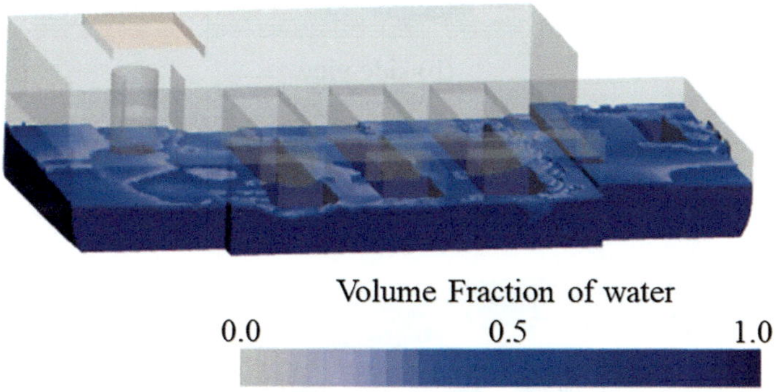

Fig. 11 Completed state of crossing-flooding (60 s)

Upon initiating the counterbalance to the flooding in the auxiliary machinery room, the seawater flow rate was rapid. As it passed through the flooding holes, the channel narrowed abruptly, accelerating the seawater flow, which was then obstructed by the first layer of machinery in the engine room. This resulted in turbulent water movement within the space and significant wall shear stress, as depicted in Fig. 13. When designing the crossing-flooding holes, considerations such as surface roughness should be considered to optimize the efficiency of the crossing-flooding holes in balancing the seawater, further similar studies can be found out from [8–12].

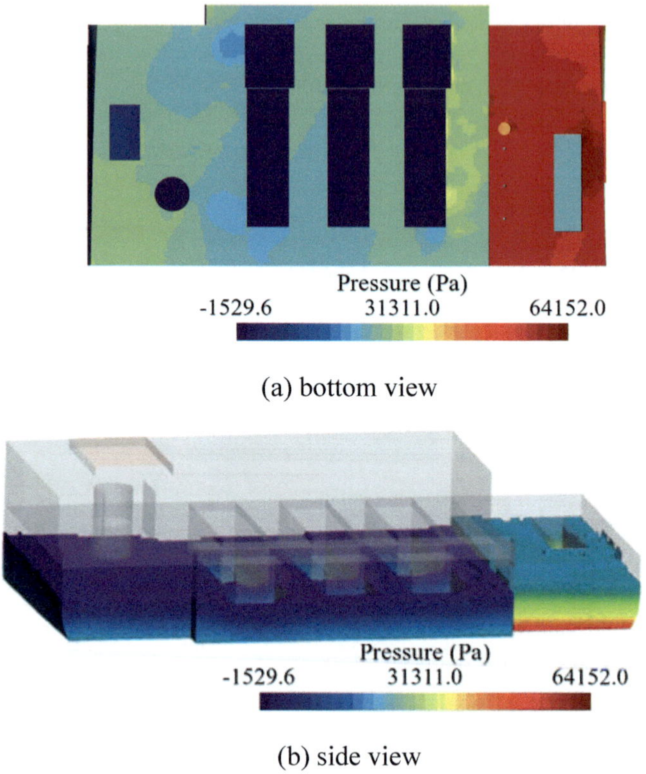

Fig. 12 Pressure distribution

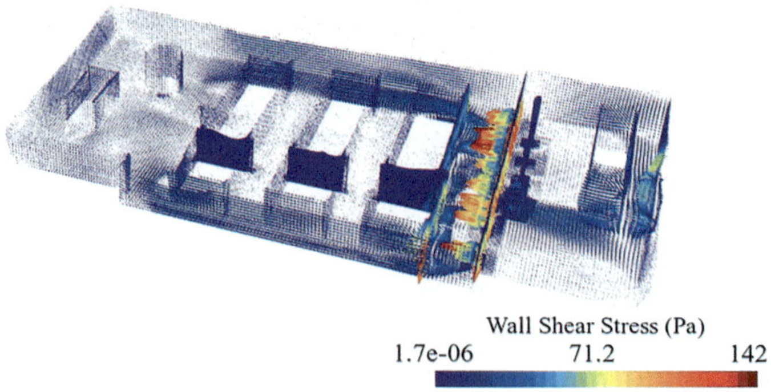

Fig. 13 Wall shear stress

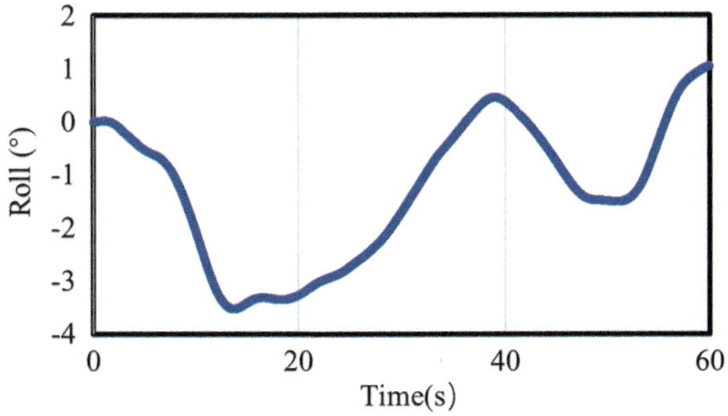

Fig. 14 Roll time history curve

3.3 *Roll*

The roll motion curve of the engine room model during hull breach flooding is illustrated in Fig. 14. After the auxiliary machinery room was flooded, the water flow impacted the machinery directly opposite the seawater inlet, causing the cruise ship to exhibit a slight positive trim angle. Subsequently, the water rapidly rushed in, causing the ship to list swiftly to the starboard side. The water entered the engine room through the manholes in the bulkhead, and the curve displayed a gentle corner. However, as the mass of seawater in the damaged compartment increased, the angle of heel quickly escalated to the maximum heel angle. Thereafter, as the engine room counterbalanced the flooding in the auxiliary machinery room, the amplitude of the cruise ship's rolling motion gradually diminished. Ultimately, at 13.7 s, the cruise ship began to rectify the large heel angle caused by the flooding in the auxiliary machinery room. The manholes located on the bulkhead between the auxiliary machinery room and the engine room served as effective crossing-flooding orifices, balancing the distribution of seawater flooding into the compartments, and fulfilling the stipulated response time for the cruise ship's crossing-flooding system.

3.4 *Rule Calculation Comparison*

The same case is calculated accordingly via NAPA program. By using the rule formula indicated in MSC.362(92), parameters are gained as per real arrangement on ship, and the equivalent parameters are listed in Table 2.

And the calculation results gained by NAPA is shown in Fig. 15, and the cross-flooding time is 208 s, which exceed 60 s but within 10 min.

Table 2 Equivalent parameters used in calculation acc. to MSC.362(92)

Width	Height	D-equi.	S-equi.	t	t/D
0.4	0.6	0.497	0.194	0.03	0.06

PROBABILISTIC DAMAGE STABILITY

CASE	STAGE	RHASE	SIDE	T m	TR m	HEEL deg	SFACSOL	CROSSTM s
DS/SDSP2.1.0	CROSS	EQ	P	8.970	-0.838	0.0	1.0000	208
DS/SDSP2.1.0	CROSS	2	P	8.881	-0.676	-1.5	1.0000	208
DS/SDSP2.1.0	CROSS	1	P	8.775	-0.479	-3.2	1.0000	208

Fig. 15 Equilibrium time read by quasi-static simulation

Fig. 16 Equilibrium time read by time-domain simulation results

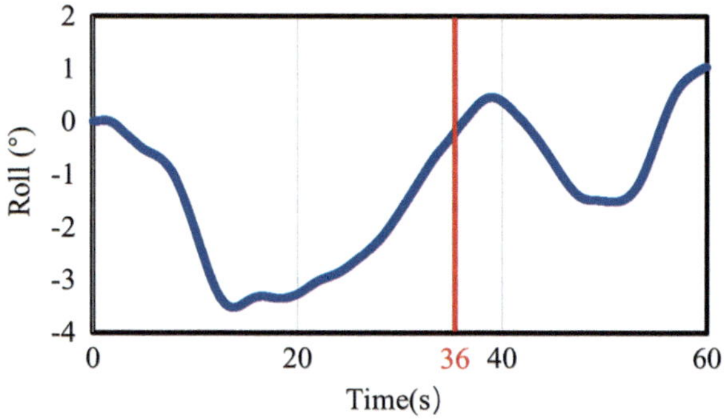

While looking at the time domain simulation results, at time of 36 s, equilibrium state is gained which is within 60 s and is more less than 208 s gained by NAPA, and this results in higher $s_{\text{intermediate}}$, refer to Fig. 16.

4 Conclusions

This This study examines a large cruise vessel featuring cross-flooding hatches installed on its longitudinal bulkheads. A comparison between time-domain computational fluid dynamics simulations and conventional rule-based calculations indicates that the most recent IMO circular, MSC.362(92), does not provide explicit guidance on calculating the time to reach equilibrium when cross-flooding hatches are employed on bulkheads.

While modeling the hatch as a zero-length pipe and applying conventional pipe-based formulations to determine the time constant k produces usable results, the rationale behind the selection of parameters and the application of the formula remains insufficiently defined.

The CFD time-domain simulation validates the effectiveness of the design and placement of the cross-flooding hatch within the studied vessel. However, it also demonstrates that the time required to reach equilibrium is notably shorter than the duration predicted using the conventional pipe-based method. For future design decisions, this result suggests that conservative pipe-based formulations should be applied with caution when cross-flooding hatches are used as functional equivalents of pipes/ducts. In particular, designers should explicitly document the assumed hydraulic representation (e.g., equivalent diameter/area and loss assumptions) and demonstrate that the selected method does not artificially penalize the equalization-time category defined in MSC.362(92). Where CFD is used to support compliance arguments, it is recommended to report a verifiable safety margin by accounting for key modelling uncertainties (mesh/time-step dependence, turbulence closure, ventilation assumptions, and hatch opening behaviour) and by presenting bounded estimates of equalization time rather than a single point value. These practices would improve auditability for classification and approval, and provide a clearer basis for developing future regulatory guidance specific to cross-flooding hatches.

These findings highlight the importance of further investigation into the calculation of the time constant k for cross-flooding hatches, as well as the development of alternative methodologies to evaluate the performance and effectiveness of such devices.

References

1. Li FT, Gan J, Huang YM, Jing JP, Wu WG (2023) Experimental study on transverse flooding device model considering the influence of ventilation conditions. China Shipbuild 64(4):16–27
2. International Maritime Organization [IMO], “MSC 85/26”, 1974/2020
3. Haussalo J (2018) Implementation of new damage stability regulations in the design of small polar cruise ship. Aalto University, Helsinki
4. Ruponen P, Kurvinen P, Saisto I (2013) Air compression in a flooded tank of a damaged ship. Ocean Eng 57:64–71
5. Palazzi L, De Kat J (2004) Model experiments and simulations of a damaged ship with air flow taken into account. Mar Technol SNAME News 41(1):38–44
6. Gap Z, Gao Q, Vassalos D (2013) Numerical study of damaged ship flooding in beam seas. Ocean Eng 61:77–87
7. Ruponen P (2007) Progressive flooding of a damaged passenger ship, vol 94. Mechanical Engineering Konotekniikan Osasto, p 128
8. Guo J, Wang Z, Li H, Yang Y, Huang CG, Yazdi M, Kang HS (2024) A hybrid prognosis scheme for rolling bearings based on a novel health indicator and nonlinear Wiener process. Reliab Eng Syst Saf 245:110014
9. Li H, Yazdi M, Huang HZ, Huang CG, Peng W, Nedjati A, Adesina KA (2023) A fuzzy rough copula Bayesian network model for solving complex hospital service quality assessment. Complex Intell Syst 9(5):5527–5553
10. Li H, Peng W, Adumene S, Yazdi M (2023) An intelligent cost-based consequence model for offshore systems in harsh environments. In: Intelligent reliability and maintainability of energy infrastructure assets. Springer Nature Switzerland, Cham, pp 107–117
11. Li H, Yazdi M (2022) What are the critical well-drilling blowouts barriers? A progressive DEMATEL-game theory. In: Advanced decision-making methods and applications in system safety and reliability problems: approaches, case studies, multi-criteria decision-making, multi-objective decision-making, fuzzy risk-based models. Springer International Publishing, Cham, pp 29–46

12. Li H, Peng W, Adumene S, Yazdi M (2023) Cutting edge research topics on system safety, reliability, maintainability, and resilience of energy-critical infrastructures. In: Intelligent reliability and maintainability of energy infrastructure assets. Springer Nature Switzerland, Cham, pp 25–38

Calculation and Analysis of Fatigue Life Damage of Diesel Engine Crankshaft

Shufei Xue, Hualiang Zhang, and Gongye Yu

Abstract A fatigue life damage assessment is conducted on diesel engine crankshafts. Finite element analysis is carried out using ABAQUS software to investigate the stress distribution under various operating conditions. Based on the simulation results, Fe-safe software is employed to evaluate the fatigue life, identifying that stress concentrations are predominantly located at the transition fillet between the crankpin and the main journal. Design optimization strategies are proposed to enhance fatigue resistance in these critical regions. This study provides valuable insights into improving diesel engine performance and operational safety through a detailed examination of crankshaft stress behavior and fatigue life characteristics.

Keywords Diesel engine crankshaft · Fatigue life · Damage calculation

1 Introduction

Fatigue life assessment of diesel engine crankshafts is conducted using ABAQUS finite element analysis to examine stress distribution patterns under varying operational conditions. Subsequent evaluation through Fe-safe software identifies critical stress concentration zones, particularly at the crankpin–main journal transition fillet. Design optimization measures are formulated to enhance fatigue resistance in these key regions. This systematic investigation advances diesel engine reliability and safety by offering a comprehensive characterization of crankshaft stress behavior and fatigue performance. As fundamental power units in both transportation and industrial sectors, diesel engines exert considerable influence on overall system efficiency. The crankshaft, as a principal component, converts piston-induced reciprocating motion into rotational energy while subjected to complex bending and torsional loads. ABAQUS-based finite element analysis determines the stress distribution characteristics under actual working conditions. Fe-safe software [1] enables accurate fatigue

S. Xue (✉) · H. Zhang · G. Yu
China Institute of Marine Technology and Economy, Beijing, China
e-mail: bjutxsf@163.com

H. Li et al. (eds.), *Data-Driven Methods for Reliability and Safety Engineering: Applications in Industrial Systems*, Springer Series in Reliability Engineering,
https://doi.org/10.1007/978-3-032-22873-4_18

life prediction through detailed analysis. Quantitative damage evaluation facilitates performance optimization and mitigates the risk of mechanical failure. Key stress concentration areas are systematically identified, providing a foundation for targeted structural improvements [2].

2 Establishment of Simulation Model of Diesel Engine Crankshaft

Based on the structural characteristics of diesel engine crankshafts, the assembly can be effectively approximated as multiple crank segments, as illustrated in Fig. 1. This simplification enables the evaluation of overall crankshaft strength through analysis of individual crank components. Static analysis is conducted using the integrated ABAQUS and FE-safe software platforms. A detailed finite element model of a single crank is developed with boundary conditions corresponding to actual operational parameters. The model facilitates comprehensive comparison and analysis of stress distribution patterns, deformation characteristics, and fatigue life performance.

Traditional finite element analysis of crankshafts typically employs three model types: full crankshaft models, single-throw models, and quarter-turn models. While full crankshaft models provide the most comprehensive simulation of actual working conditions, they demand substantial computational resources. The quarter-turn model offers improved computational efficiency through reduced complexity. This study adopts the single-throw crankshaft model, balancing computational efficiency with analytical accuracy. The model dimensions include: 209 mm from crank center to main journal center (L1), 70 mm from crank arm center to main journal center (G), 418 mm between main journal centers (L), and 188 mm crank pin width. The SolidWorks-generated single-throw model is converted and imported into ABAQUS simulation software, as depicted in Figs. 2 and 3.

The single-throw crankshaft model is configured with solid material properties based on the 34CrNi3MoA steel characteristics. Key material parameters include density, elastic modulus, Poisson's ratio, yield strength, and tensile strength, with detailed values provided in Table 1.

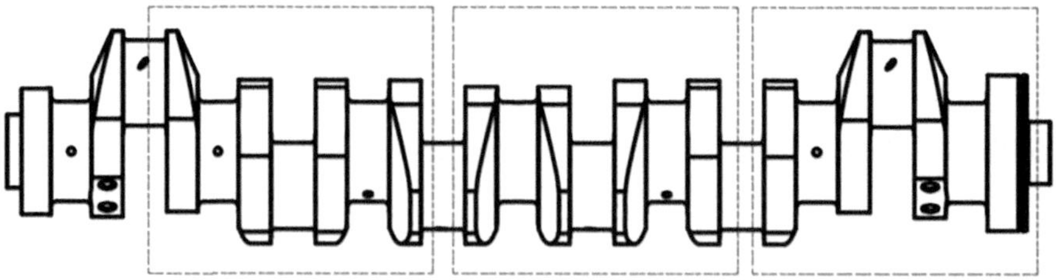

Fig. 1 Crankshaft geometry model

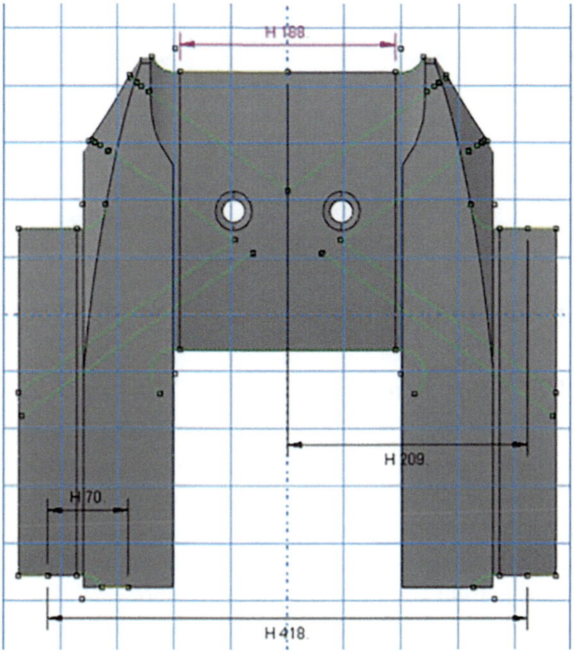

Fig. 2 Single-throw crankshaft size model

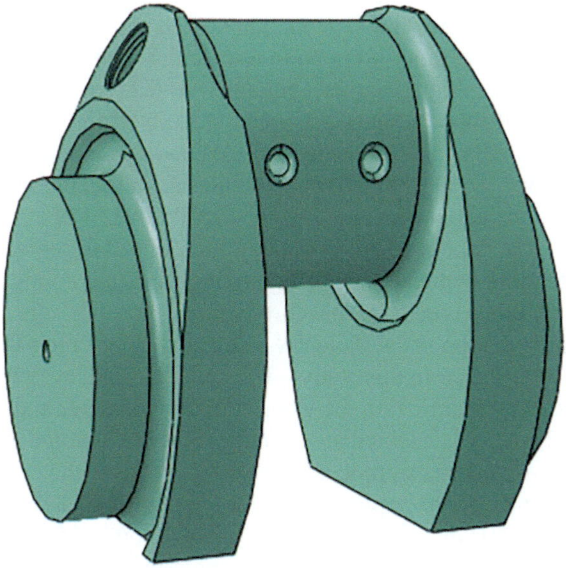

Fig. 3 Single-throw crankshaft geometry model

Table 1 Performance parameters of 34CrNi3MoA material

Parameter type	Numeric
Elastic modulus E (GPa)	210
Poisson's ratio (μ)	0.3
Material density ρ (kg/m^3)	7850
Yield strength (MPa)	950
Tensile strength (MPa)	1089

3 Simulation Analysis of Stress in Diesel Engine Crankshaft

3.1 Simulation Analysis of Crankshaft Bending Stress

In order to obtain the nominal bending moment of crankshaft bending fatigue.

Where: M_{-1} is the nominal working bending moment (N m); D^2 is the cylinder diameter (m); L_1 is the distance from the center of the crank to the center of the main journal (m); G is the distance from the center of the handle arm to the center of the main journal (m); p is the burst pressure (MPa); K is the support coefficient (full support K = 0.75, non-full support K = 0.85, rolling bearing K = 0.90); L is the distance between the centers of two adjacent main journals (m).

Through the measurement of the single-turn crankshaft model, it can be obtained: L1 = 0.209 m, G = 0.07 m, L2 = 0.418 m. According to experience, in diesel engine design, the ratio of crankpin to cylinder diameter is between 0.35 and 0.45. Taking 0.4, the cylinder diameter calculated is 0.47 m, that is, D = 0.47 m. At the same time, the full support method is used to obtain K = 0.75, and the gas explosion pressure p = 25 MPa. From this, the nominal bending moment of the crankshaft is calculated to be M_{-1} = 113,855.73 N m. At the same time, the width of the crankpin is 0.188 m, and the equivalent applied force is 605,615.61 N.

The bending loading condition is defined to represent the nominal firing-induced bending demand under a full-support bearing configuration. The peak cylinder gas pressure p is taken as the representative burst pressure at the rated operating condition, and the support coefficient K is selected according to the adopted bearing support type (full support in this study). The cylinder diameter D is derived from the commonly used geometric ratio between crankpin diameter and cylinder diameter (0.35–0.45), where 0.4 is adopted as a nominal design value for preliminary loading estimation.

A worked example is provided to illustrate the conversion from nominal bending moment to the applied force used in the finite element model. Using $L_1 = 0.209$ m, $G = 0.07$ m, $L_2 = 0.418$ m, $D = 0.47$ m, $p = 25$ MPa, and $K = 0.75$, Eq. (1) yields $M_{-1} = 113,855.73$ N m. In the single-throw model, this nominal bending moment is applied as an equivalent force pair across the crankpin width $b = 0.188$ m. Therefore, the resultant force magnitude is $F = M_{-1}/b = 113,855.73/0.188 \approx 605,615.61$ N, which is the load level imposed in ABAQUS for the bending case.

The bending load solution task was submitted using a multi-core processing configuration. Upon completion, the software's general post-processor generated the

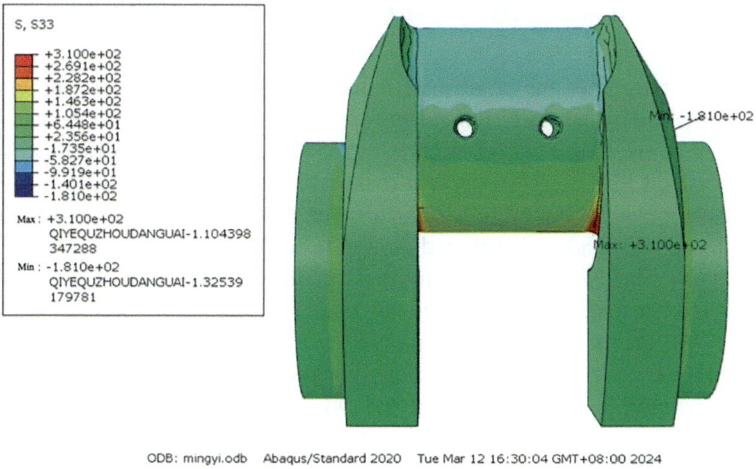

Fig. 4 Single crankshaft bending stress distribution cloud diagram

stress distribution contour for the single-throw crankshaft. Analysis revealed significant stress concentration at the journal transition fillet, with peak stress occurring specifically at the lower crank pin transition fillet, as illustrated in Fig. 4.

The analysis confirms that maximum bending stress consistently occurs at the transition fillet, aligning with established empirical data. This critical region, located between the main journal and connecting rod journal, experiences significant stress concentration due to geometric discontinuity. The abrupt diameter change disrupts stress streamline distribution, creating localized stress elevation exceeding nominal levels. This phenomenon is particularly pronounced at fillet regions where material must withstand combined bending and torsional loads from crankshaft rotation and alternating forces. Consequently, transition fillets represent both the highest stress concentration zones and the most probable initiation sites for fatigue cracks in crankshaft design. To improve the robustness of the numerical setup, a brief sensitivity check is recommended for the inputs that most strongly govern the nominal loads and local stress concentration: (i) the burst pressure p and support coefficient K in Eq. (1), (ii) the geometric ratio used to infer D, and (iii) the mesh density at the transition fillet where peak stresses occur. Practically, this can be performed by varying p and K within a small engineering range and refining the fillet-region mesh, then comparing peak von Mises stress and predicted life. If the ranking of critical locations and the life order of magnitude remain stable, the loading definition is considered sufficiently representative for design comparison.

3.2 Simulation Analysis of Crankshaft Torsional Stress

The torsional fatigue of the crankshaft is mainly caused by the periodic torsional stress experienced by the crankshaft during operation. In order to make the calculation results close to the actual working conditions, according to the traditional limited width journal oil film pressure stress distribution law, while ignoring the influence of the pressure peak at the oil hole and chamfer, the force boundary conditions are assumed to be: the load is distributed along the main journal axis according to the quadratic parabola law q_x; along the journal circumference within an angle of 120°, it is distributed according to the cosine law $q_{x,\theta}$.

where: Q_c is the load on the crankpin (N); x is the distance from the axial center of the crankpin $-L \sim L$ (mm), L is half the length of the crankpin; θ is the circumferential angle of the crankpin $-60° \sim 60°$; R is the radius of the crankpin (mm).

In ABAQUS software, complex loads can be applied by expression field. According to the formula, the local coordinate system x1y1z1 needs to be established first, which is obtained by translating the initial coordinate system 165 mm along the x-axis. Since the formula is a piecewise function and there are angle variables, the formula needs to be deformed and integrated first.

By measuring the single-throw crankshaft model, we can obtain: L = 120 mm, R = 94 mm, Qc = 1.2e6 N. In ABAQUS, an expression field is established to set the load and distribute the pressure to the crank pin. The set model is submitted for analysis to obtain a torsional stress distribution cloud diagram, as shown in Fig. 5. The maximum stress is generated at the transition fillet of the main journal, which is 138.5 MPa.

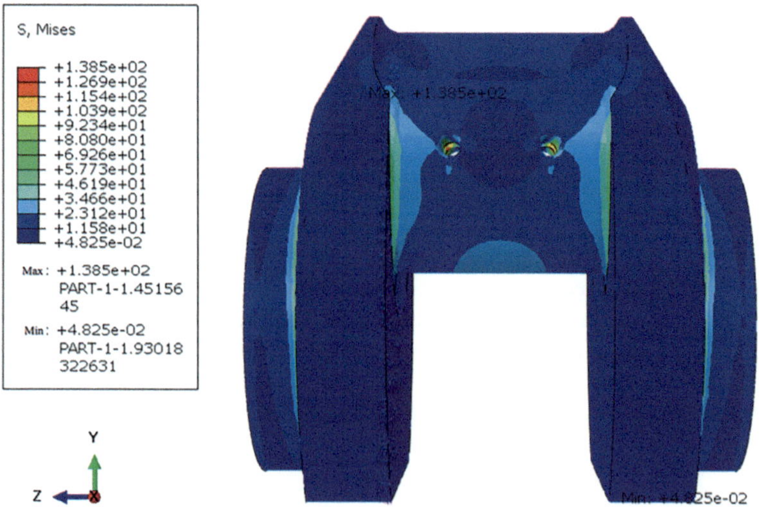

Fig. 5 Cloud chart of torsional stress distribution of single-crankshaft

4 Numerical Simulation of Fatigue Life of Diesel Engine Crankshaft

The stress simulation results obtained from ABAQUS are imported into FE-SAFE for fatigue life assessment. During this data transfer process, stress units are systematically converted from Pascals (Pa) to Megapascals (MPa) to maintain consistency with the finite element analysis framework. This unit standardization ensures dimensional homogeneity throughout the fatigue evaluation procedure. For consider the comprehensive view, refer to Figs. 6, 7, 8, 9, 10, 11, 12, 13, 14 and 15 as the following.

The fatigue analysis is performed on the complete single-crank model since it constitutes a homogeneous component with uniform material properties (34CrNi3MoA steel), eliminating the need for partial analysis. The material is created in the FE-SAFE material library by inputting all relevant mechanical parameters previously specified for 34CrNi3MoA steel.

The fatigue life assessment employs the conventional stress-life (S-N) approach, a well-established engineering methodology for fatigue prediction. This technique utilizes experimentally derived S-N curves, where the abscissa represents stress amplitude (S) and the ordinate denotes the corresponding fatigue life (N) in cycles. The fundamental assumption underlying this method postulates a deterministic relationship between applied stress levels and material fatigue endurance. The analysis procedure involves interpolating component life expectancy from these characteristic curves under specified loading conditions.

34CrNi3MoA parameters: mainly including Von Mises, R ratio, S-N fatigue algorithm, elastic modulus (20,600 MPa), Poisson's ratio 0.3, tensile strength (1089 MPa), E-N curve parameters ($b = -0.069$, $S'f = 1258.92$ MPa), S-N curve parameters (10000,1E7, 690.128378.381).

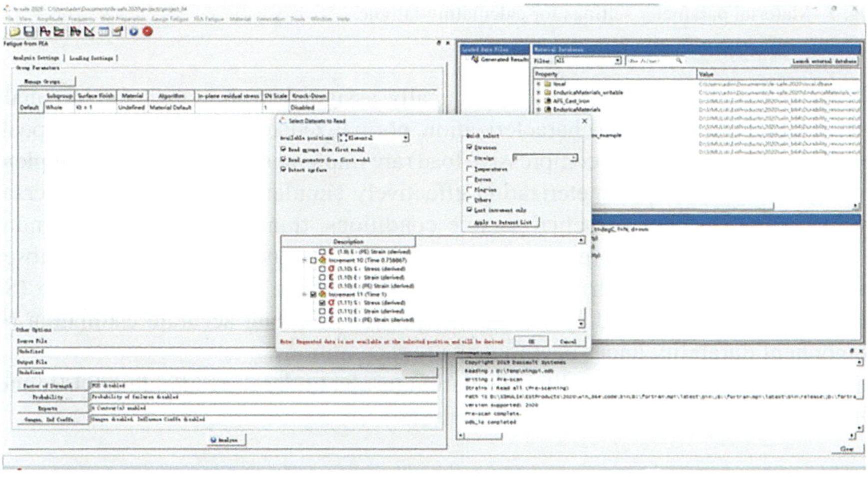

Fig. 6 Import of finite element calculation model

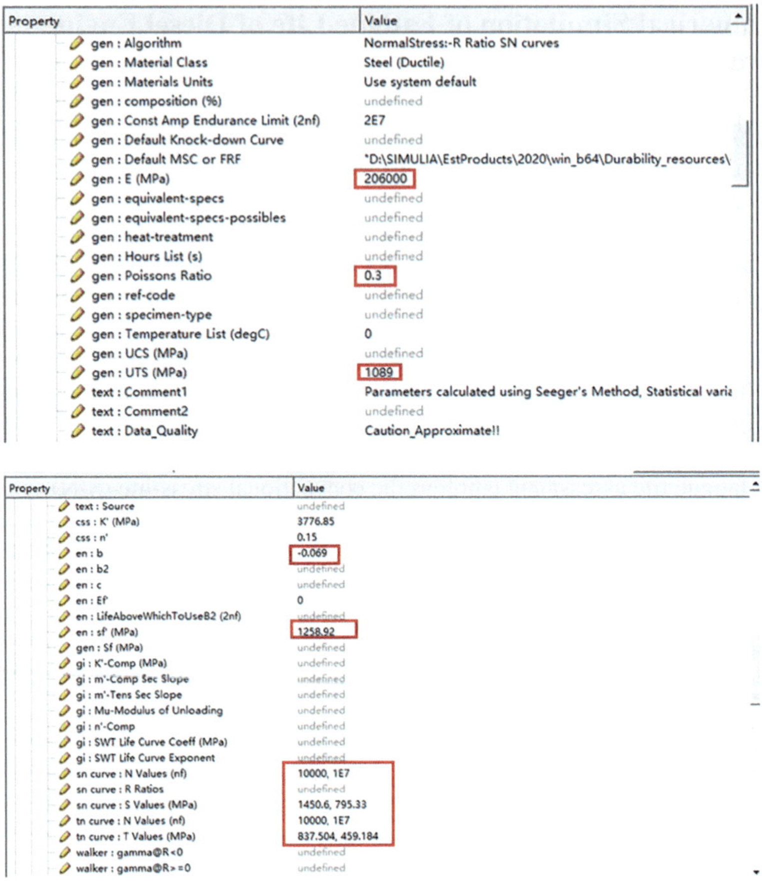

Property	Value
gen : Algorithm	NormalStress:-R Ratio SN curves
gen : Material Class	Steel (Ductile)
gen : Materials Units	Use system default
gen : composition (%)	undefined
gen : Const Amp Endurance Limit (2nf)	2E7
gen : Default Knock-down Curve	undefined
gen : Default MSC or FRF	*D:\SIMULIA\EstProducts\2020\win_b64\Durability_resources\
gen : E (MPa)	206000
gen : equivalent-specs	undefined
gen : equivalent-specs-possibles	undefined
gen : heat-treatment	undefined
gen : Hours List (s)	undefined
gen : Poissons Ratio	0.3
gen : ref-code	undefined
gen : specimen-type	undefined
gen : Temperature List (degC)	0
gen : UCS (MPa)	undefined
gen : UTS (MPa)	1089
text : Comment1	Parameters calculated using Seeger's Method, Statistical vari
text : Comment2	undefined
text : Data_Quality	Caution_Approximate!!

Property	Value
text : Source	undefined
css : K' (MPa)	3776.85
css : n'	0.15
en : b	-0.069
en : b2	undefined
en : c	undefined
en : Ef'	0
en : LifeAboveWhichToUseB2 (2nf)	undefined
en : sf' (MPa)	1258.92
gen : Sf (MPa)	undefined
gi : K'-Comp (MPa)	undefined
gi : m'-Comp Sec Slope	undefined
gi : m'-Tens Sec Slope	undefined
gi : Mu-Modulus of Unloading	undefined
gi : n'-Comp	undefined
gi : SWT Life Curve Coeff (MPa)	undefined
gi : SWT Life Curve Exponent	undefined
sn curve : N Values (nf)	10000, 1E7
sn curve : R Ratios	undefined
sn curve : S Values (MPa)	1450.6, 795.33
tn curve : N Values (nf)	10000, 1E7
tn curve : T Values (MPa)	837.504, 459.184
walker : gamma@R<0	undefined
walker : gamma@R>=0	undefined

Fig. 7 Material parameter settings for calculating fatigue

The material properties are systematically assigned to the target fatigue analysis model. For cyclic load characterization, normalized amplitude values of 1 (peak tensile load) and − 1 (peak compressive load) are implemented to define the complete load spectrum. This parameterization effectively simulates the full stress reversal cycle experienced during actual service conditions, transitioning from maximum tensile to maximum compressive stress states. The fatigue life prediction is subsequently computed through numerical integration of the prescribed stress-life (S-N) curve with the defined cyclic loading profile, enabling accurate estimation of component durability under operational stress conditions.

Finally, output the fatigue life calculation results to the corresponding folder and click on Analysis to start fatigue life analysis.

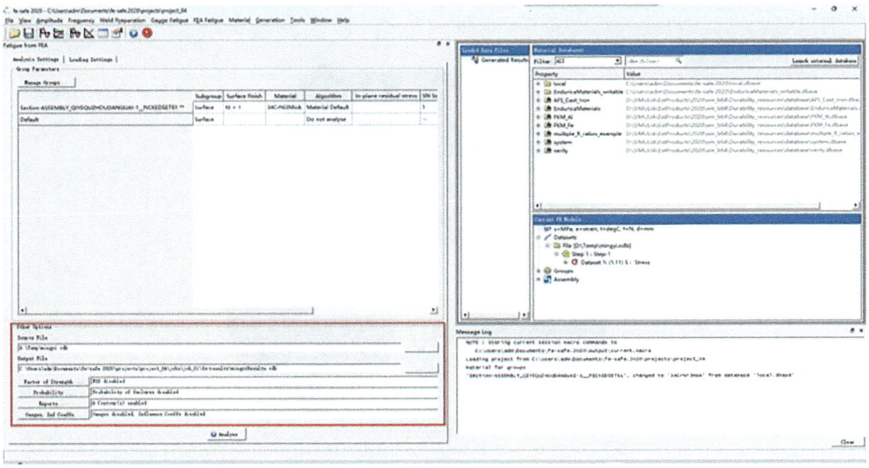

Fig. 8 Calculation of fatigue life and output setting of results

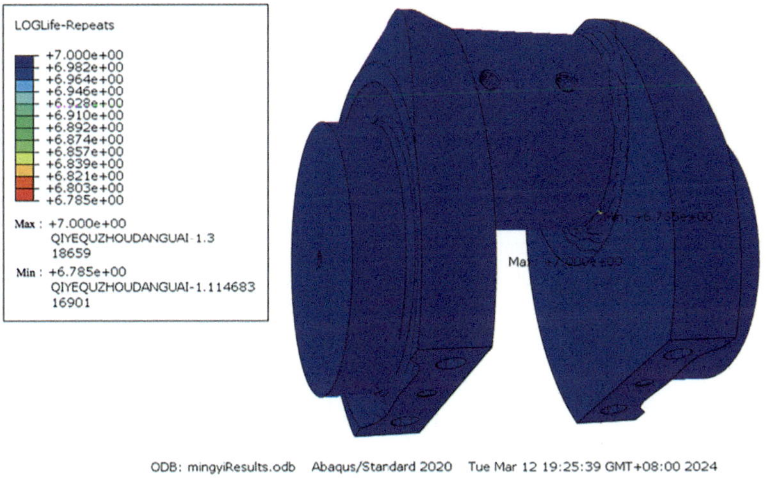

Fig. 9 Fatigue life under bending stress

4.1 Fatigue Life of Crankshaft Under Bending Stress

The fatigue analysis results for the experimental model (shown in the accompanying figure) reveal critical damage initiation at the transition fillet beneath the crank pin, with a computed fatigue life of 6,095,368 cycles. From a design perspective, this result indicates that bending-driven high-cycle fatigue is governed by the stress concentration at the crankpin–journal transition fillet. Geometric discontinuities in this region amplify the local stress through a high stress concentration factor, and even

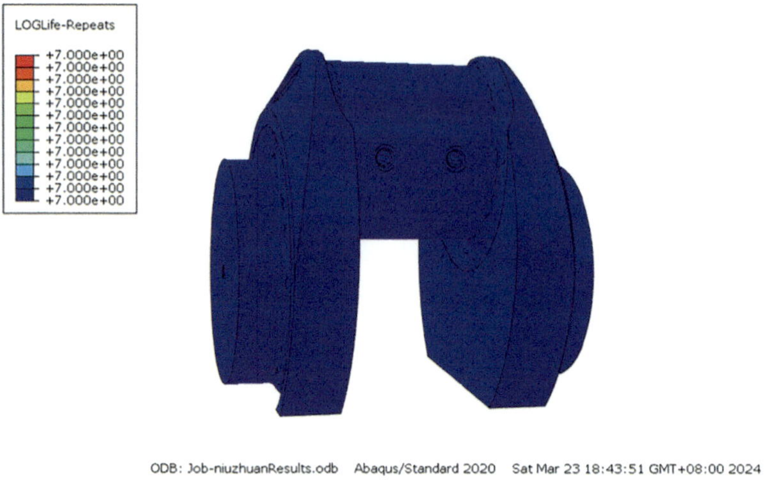

Fig. 10 Fatigue life under torsional stress

Fig. 11 Schematic diagram of parameter selection

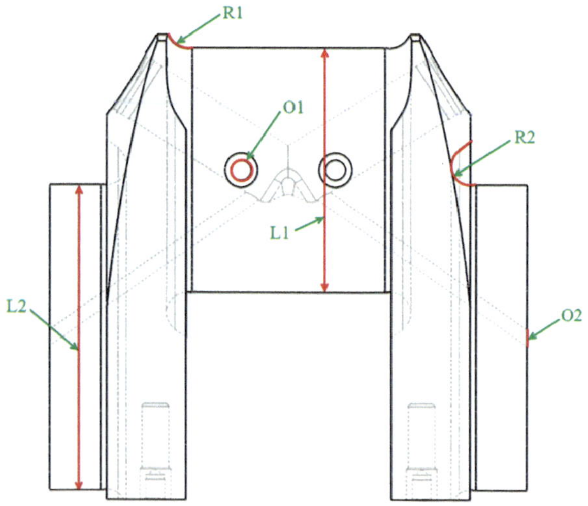

moderate reductions in the peak stress can translate into a disproportionate increase in life under an S-N framework. Therefore, geometric optimization should prioritize (i) increasing the fillet radius at the critical crankpin-side transition, and (ii) reducing stress raisers associated with oil-hole features near the fillet (e.g., by limiting oil-hole diameter, optimizing hole location away from the peak stress path, and applying appropriate chamfering or surface finishing). These measures are expected to lower local alternating stress and delay crack initiation, which is the dominant contributor

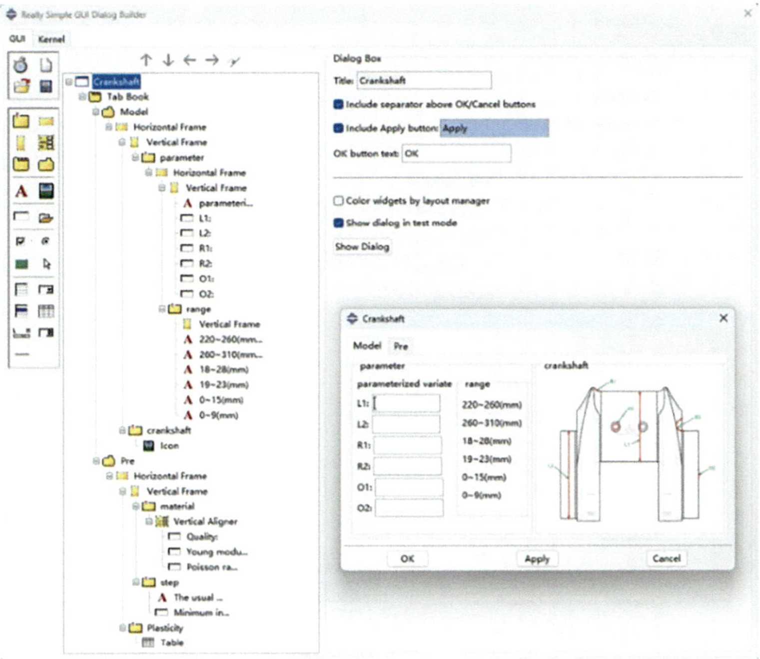

Fig. 12 GUI tab creation

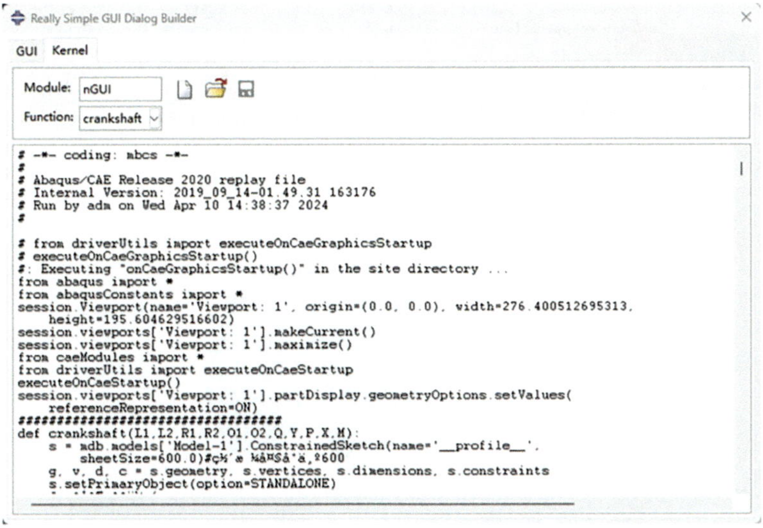

Fig. 13 The association between dialog boxes and kernel scripts

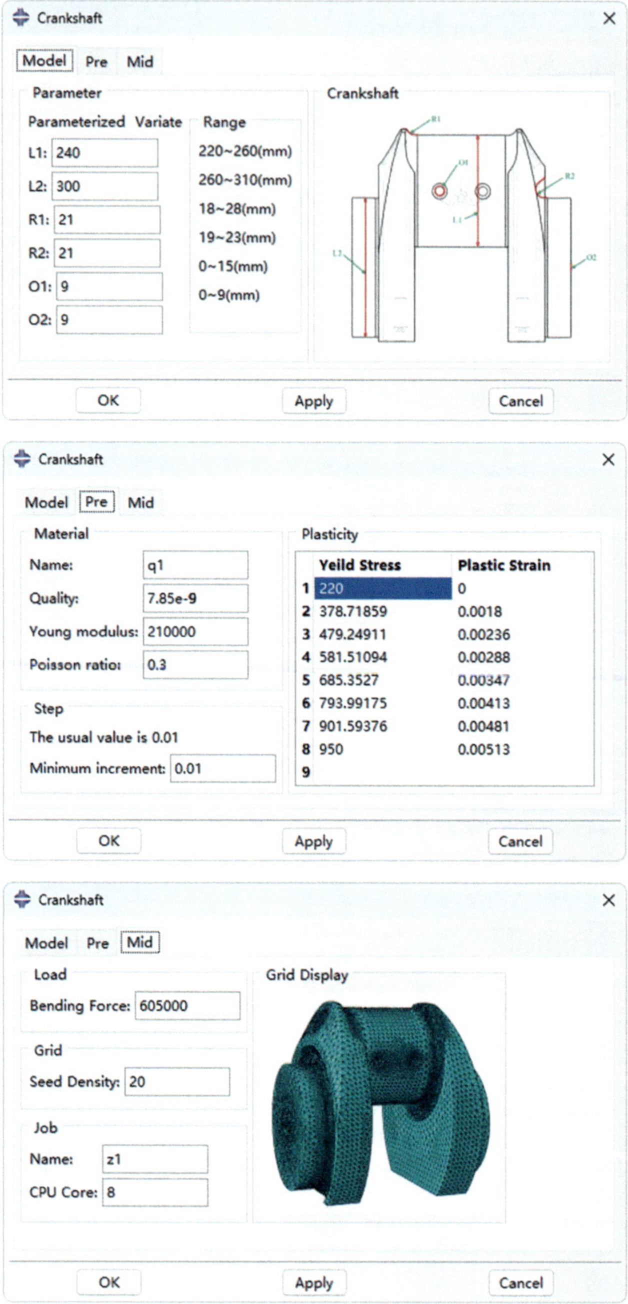

Fig. 14 GUI interface display

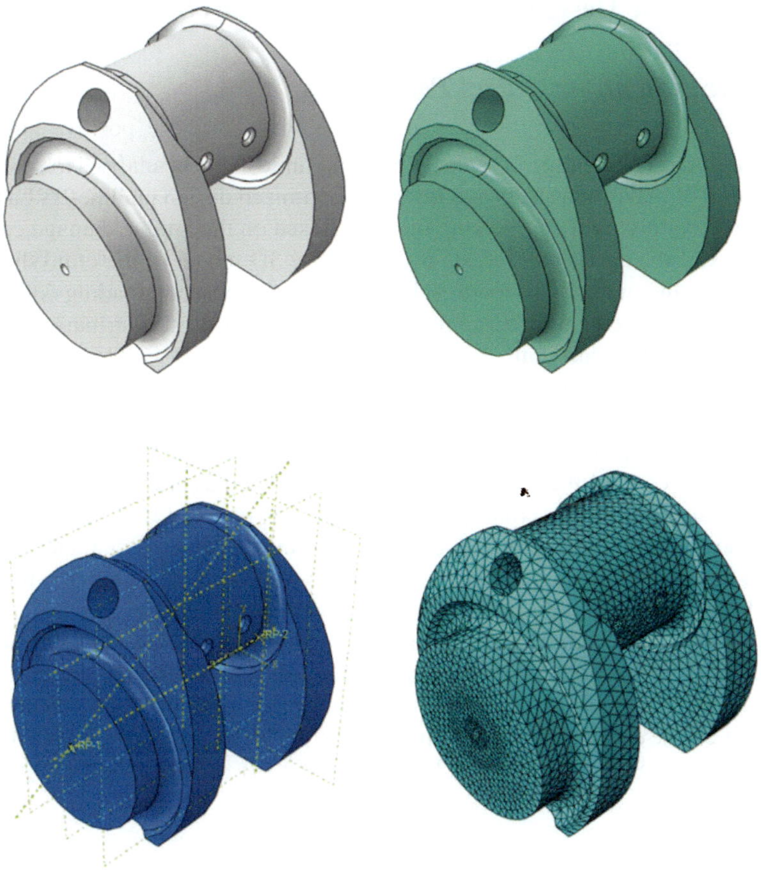

Fig. 15 Example of parameterized crankshaft model

to the predicted life in the present analysis. This localized failure mechanism corresponds precisely with the previously identified maximum stress concentration zone, validating the computational predictions through empirical observations.

4.2 Fatigue Life of Crankshaft Under Torsional Stress

Following the established methodology for bending stress analysis, the computational framework was adapted to evaluate torsional stress effects while maintaining identical procedural parameters. The resultant analysis (illustrated in the accompanying figure) demonstrates no observable fatigue damage under the prescribed torsional loading conditions. This outcome suggests superior torsional fatigue resistance compared to bending stress performance in the examined configuration.

5 Parametric Design Scheme for Crankshaft Structure

5.1 Selection of Structural Parameters

In the parametric modeling of crankshafts, using finite element analysis to determine the location of maximum stress and selecting optimized design parameters based on this is a commonly used engineering method. Based on the fatigue damage calculation results of the crankshaft, create a preliminary 3D model of the crankshaft and set corresponding material properties, boundary conditions, and loading conditions. After using finite element software for simulation, obtain stress distribution maps of various parts of the crankshaft.

After obtaining the stress and fatigue life distribution of the crankshaft, it can be observed that the stress concentration areas are mainly located at the transition corners of the crankshaft pin, the main journal, and the oil hole. Therefore, the variable parameters of the crankshaft can be determined as crank pin diameter L1, main journal diameter L2, crank pin fillet radius R1, main journal fillet radius R2, crank pin oil passage diameter O1, and main journal oil passage diameter O2. The selected parameters are marked on the schematic diagram as shown in the figure.

At the same time, it can be observed that each variable parameter has a corresponding range, and exceeding the corresponding range will result in model creation failure, which in turn will affect subsequent analysis. Based on the relevant data of the model and considering the operability of subsequent analysis, the range of each parameter can be determined as: 220 mm $\leq$ L1 $\leq$ 260 mm, 260 mm $\leq$ L2 $\leq$ 310 mm, 18 mm $\leq$ R1 $\leq$ 28 mm, 19 mm $\leq$ R2 $\leq$ 23 mm, 0 $\leq$ O1 $\leq$ 15 mm, 0 $\leq$ O2 $\leq$ 9 mm.

5.2 Establishment of Parameterized Modeling Interface for Crankshaft

ABAQUS is a widely used finite element analysis software in engineering applications, featuring multiple functions including the Remote Simulation Gateway (RSG). The RSG function facilitates simulation computing on remote resources, allowing efficient management and execution of simulation tasks. Users can configure and pre-process simulation tasks locally before submitting them to remote servers for computation. This functionality supports the development of a GUI interface for parametric crankshaft modeling, as depicted in the figure. The interface is accessed by clicking the "ABAQUS" button in the "Plug-ins" menu to open the RSG dialog box, which contains three dedicated tabs for different modeling functions.

The first tab implements two vertically aligned group boxes with customized dimensions and labels. The left group box functions as the variable parameter input panel, while the right group box visually presents these parameters on the schematic diagram. Six dedicated text box controls are integrated into the left panel, each corresponding to specific geometric parameters. A supplementary group box in the right panel displays the allowable range for each variable parameter. Particular attention is given to maintaining keyword consistency between interface controls and kernel script variables. The right panel also incorporates a parameter selection schematic diagram, with its dimensions optimized for clear visualization within the interface space [3–7].

The second tab features a tripartite layout comprising material elasticity parameters, material plasticity parameters, and minimum analysis increment step inputs. Each section combines descriptive labels with dedicated parameter entry fields. For material plasticity definition, a tabular input format with nine data rows facilitates comprehensive material property specification, ensuring complete characterization of the material's plastic behavior under various loading conditions.

The third tab integrates text box controls for configuring variable parameters across three functional modules: load application, mesh generation, and job submission. Crankshaft bending analysis requires specific parameterization, including the application of bending stress at reference point RP1 while maintaining complete fixation at the opposing end, establishing the RP1 stress magnitude as a primary variable parameter. The mesh generation process incorporates global seed size as a critical variable for grid density control, while job configuration includes CPU core allocation as an adjustable computational parameter. These input controls are systematically arranged in the left panel of the interface group box, complemented by a meshing schematic diagram in the right panel to enhance operational clarity and interface functionality.

Finally, click on the Kernel tab to associate the dialog box created in the GUI tab with the functions in the kernel script, and map the keywords of each control in the dialog box to the variables in the kernel script, as shown in the figure.

After everything is set up, save the GUI interface to "Plug ins", click on the input parameters to draw the corresponding crankshaft and complete the finite element analysis.

5.3 *Output of Simulation Results*

The parametric crankshaft modeling GUI interface is established following the described methodology. The modeling process is initiated by selecting the "Crankshaft" button within the "Plug-ins" menu, enabling users to input structural parameters within the specified design ranges along with corresponding material properties and minimum analysis step parameters. This streamlined interface facilitates rapid model generation, with a representative example illustrated in the accompanying figure.

6 Conclusion

This study presents a fatigue life assessment of a diesel engine crankshaft, focusing on its complex load-bearing characteristics during operation and analyzing the influence of bending and torsional stresses on structural integrity and durability. The principal conclusions are summarized as follows:

The crankshaft exhibits a maximum bending stress concentration of 310 MPa at the transition fillet between the main journal and connecting rod journal. This critical location experiences elevated stress levels due to geometric discontinuity, where a sudden reduction in cross-sectional diameter disrupts the uniform flow of stress. Such discontinuity intensifies localized stress, particularly under the combined effects of bending and torsional loads during engine operation. This transition fillet constitutes the most likely initiation site for fatigue cracks within the crankshaft assembly.

The highest torsional stress, measured at 138.5 MPa, also occurs at the main journal transition fillet. This stress concentration is attributed to complex dynamic loading conditions, including cylinder combustion pressure and inertial forces from the connecting rod. These operational loads collectively exacerbate stress localization in the geometrically sensitive region adjoining the main journal and adjacent crankshaft components.

Fatigue life simulations performed using Fe-safe software under bending and torsional loading conditions reveal that the crankshaft endures a bending fatigue life of 6,095,368 cycles. This value falls below the 10^7 cycle threshold commonly regarded as the criterion for infinite life in high-cycle fatigue scenarios, indicating the necessity for structural enhancement. Conversely, the torsional fatigue life meets the 10^7 cycle benchmark, suggesting negligible fatigue failure risk under torsional loading conditions. These results establish bending stress as the primary factor limiting the crankshaft's fatigue performance.

Based on the identified hot spot, design optimisation should primarily target the crankpin-side transition fillet and nearby oil-hole features. Within the feasible manufacturing ranges, increasing the fillet radius and improving surface integrity (e.g., controlled finishing at fillet and hole edges) are recommended as first-order measures, while oil-hole diameter and placement should be treated as second-order levers constrained by lubrication requirements.

The fatigue predictions are also subject to uncertainty stemming from load definition (e.g., the assumed burst pressure p, support coefficient K, and the equivalence used to convert bending moment to applied forces), modelling idealizations in the torsional pressure field, and material S–N parameter variability. Therefore, for design verification, it is recommended to report a bounded life estimate by conducting a small sensitivity study on p/K and on key geometric variables (R_1, O_1), and by applying an engineering safety factor when translating numerical life to allowable service life.

A parametric crankshaft modeling interface has been developed to streamline model generation through simplified input of structural parameters, material data, and basic analysis step configurations. This instruction-based finite element modeling approach allows for efficient bending fatigue life evaluation of crankshafts with varying design specifications. The methodology operates through automated kernel calls to ABAQUS, minimizing computational resource usage during file generation. This implementation significantly expedites the modeling and mechanical property prediction processes, while reducing manual effort and the likelihood of human error, thereby improving research efficiency and accuracy.

References

1. Yi T, Wu J, Diao A et al (2008) Fatigue life of diesel engine exhaust pipe under vibration based on finite element and FE-SAFE. Intern Combust Engine Eng (03):76–80
2. Huang Y, Gao H, Liu P (2012) Analysis of fatigue strength and life of diesel crankshaft based on multi-body dynamics. J Nav Univ Eng (5):54–57, 81
3. Guo J, Wang Z, Li H, Yang Y, Huang CG, Yazdi M, Kang HS (2024) A hybrid prognosis scheme for rolling bearings based on a novel health indicator and nonlinear Wiener process. Reliab Eng Syst Saf 245:110014
4. Li H, Yazdi M, Huang HZ, Huang CG, Peng W, Nedjati A, Adesina KA (2023) A fuzzy rough copula Bayesian network model for solving complex hospital service quality assessment. Complex Intell Syst 9(5):5527–5553
5. Li H, Peng W, Adumene S, Yazdi M (2023) An intelligent cost-based consequence model for offshore systems in harsh environments. In: Intelligent reliability and maintainability of energy infrastructure assets. Springer Nature Switzerland, Cham, pp 107–117
6. Li H, Yazdi M (2022) What are the critical well-drilling blowouts barriers? A progressive DEMATEL-game theory. In: Advanced decision-making methods and applications in system safety and reliability problems: approaches, case studies, multi-criteria decision-making, multi-objective decision-making, fuzzy risk-based models. Springer International Publishing, Cham, pp 29–46
7. Li H, Peng W, Adumene S, Yazdi M (2023) Cutting edge research topics on system safety, reliability, maintainability, and resilience of energy-critical infrastructures. In: Intelligent reliability and maintainability of energy infrastructure assets. Springer Nature Switzerland, Cham, pp 25–38

Exploratory Research on Optimizing Equipment Support Supply Chain Based on Boundaryless Organization Theory

Tianqiao Yan, Yabin Wang, Jinguo Wang, and Cenyu Hu

Abstract Within the framework of the traditional bureaucratic model, the equipment support supply chain often exhibits inefficiencies related to inadequate information flow, limited interdepartmental collaboration, and suboptimal resource allocation. These challenges frequently result in the formation of "information islands" and the emergence of the "bullwhip effect" throughout the supply process. To mitigate these deficiencies, boundaryless organization theory is introduced as a strategic approach for optimizing the equipment support supply chain. A comprehensive office is established within the organizational structure to facilitate horizontal resource integration, while an information platform is employed to ensure effective vertical information flow. These measures collectively address the horizontal fragmentation and vertical disconnection inherent in conventional supply chain models, thereby providing novel insights and actionable strategies for enhancing supply chain coordination and operational efficiency.

Keywords Boundaryless organization theory · Equipment support · Supply chain

1 Introduction

In the contemporary context of increasingly complex and dynamic military equipment support, the efficient functioning of the equipment support supply chain is critical to maintaining equipment reliability and ensuring combat readiness. With the continuous advancement of modern warfare and the escalating sophistication of military systems, traditional supply chains encounter significant obstacles. These include inefficient information dissemination, diminished coordination efficiency resulting from organizational boundary constraints, and suboptimal resource distribution. Collectively, these limitations undermine the timeliness and overall effectiveness of equipment support.

T. Yan · Y. Wang (✉) · J. Wang · C. Hu
Army Engineering University of PLA, Shijiazhuang, China
e-mail: wangyabin123@163.com

H. Li et al. (eds.), *Data-Driven Methods for Reliability and Safety Engineering: Applications in Industrial Systems*, Springer Series in Reliability Engineering,
https://doi.org/10.1007/978-3-032-22873-4_19

The root of these challenges can be attributed to the inherent conflict between the bureaucratic model's control-centric organizational logic and the collaborative operational nature of supply chains. Boundaryless organization theory offers a conceptual solution to this structural incompatibility. It advocates for the dissolution of restrictive boundaries within and across organizations, including vertical boundaries (hierarchical levels), horizontal boundaries (interdepartmental divisions), external boundaries (institutional barriers), and geographical boundaries (spatial constraints). This approach facilitates the unrestricted flow of information, resources, and competencies. Its successful implementation in domains such as media, education, and corporate management underscores its potential to enhance the adaptability and efficiency of supply chains operating within bureaucratic structures.

2 Boundaryless Organization Theory

2.1 Concept of Boundaryless Organization Theory

The boundaryless organization theory, introduced by management scholar Ron Ashkena and colleagues in the 1990s [1], aims to enhance organizational flexibility, collaborative efficiency, and innovation capacity by eliminating the rigid constraints of traditional organizational hierarchies. The theory addresses structural and operational barriers, both internal and external to the organization, across four primary dimensions. First, vertical boundaries are minimized by reducing hierarchical stratification and facilitating the decentralization of decision-making authority. Second, horizontal boundaries are mitigated by encouraging interdepartmental collaboration and dissolving functional divisions. Third, external boundaries are addressed by fostering strong cooperative relationships with external stakeholders, including suppliers and customers. Fourth, geographical boundaries are transcended through the strategic utilization of global resources to achieve efficient allocation and coordination.

The core of the boundaryless organization theory is to break down rigid barriers within and outside the organization, which is highly consistent with the essence of "cross subject collaboration and high timeliness requirements" in the supply chain. The internal and external boundaries in the supply chain, such as departmental segmentation and subject barriers, are the core causes of information distortion and coordination delays, such as data loss caused by multi-level transmission and decision-making lag caused by rigid barriers, which are more prominent in high uncertainty scenarios. The borderless organization advocates "eliminating boundaries" by building information sharing platforms, flat collaboration mechanisms, and strategic trust alliances to eliminate internal boundaries, reduce information transmission links, break external boundaries, promote real-time supply–demand docking, alleviate distortion and delay problems from the root, and achieve "information direct access" and "decision-making linkage" at supply chain nodes. This connection not only provides a practical landing point for the theory of boundaryless organizations in supply chain scenarios, but also provides organizational theoretical support for optimizing supply chain behavior, deepening the inherent correlation between "organizational form supply chain efficiency" and expanding the depth of conceptual intersection between the two.

2.2 Application Practice of Boundaryless Organization Theory

The literature [2] examined the 21st Century Business Herald as a case study, revealing the transformation path of a hierarchical media organization. The traditional three-tier vertical structure of "Editor-in-Chief—Department Director—Reporter" resulted in a decision-making cycle of up to seven days for major topics, while the rate of resource sharing between departments (such as the Financial News Department and the Integrated Media Center) was below 30%. By implementing a flattened hierarchy (compressed to two layers of "Editor-in-Chief—Project Team") and establishing cross-functional teams (with the Editorial Center coordinating planning, the Vertical Center deepening expertise, and the Regional Center executing implementation), the organization increased the response speed for topic selection to 48 h. Furthermore, through the joint construction of a data platform with financial institutions (external boundary penetration), the output of in-depth reports increased by 50%. This case demonstrated that breaking vertical and horizontal boundaries can significantly enhance the agility of hierarchical organizations.

The literature [3] examined the field of vocational education, highlighting that vertical hierarchical differences (such as the decision-making gap between school management and enterprise executives) and horizontal professional barriers (such as goal conflicts between theoretical teachers and practical mentors) extended the curriculum development cycle to as long as twelve months, with the proportion of practical content remaining below 40%. By implementing flat management (granting frontline teachers autonomy in curriculum design) and establishing a cross-boundary communication platform combining online seminars with offline workshops, a certain industry college achieved knowledge complementarity between theoretical and practical teachers, shortened the curriculum development cycle to six months, and increased the skill proficiency rate of graduates by 35%. This case demonstrated that cross-boundary collaboration mechanisms could effectively activate the collaborative potential of different entities within bureaucratic systems.

The literature [4] used Smith Company as a case study to reveal the innovation bottleneck in bureaucratic enterprises. Under the traditional "leader decision-making—employee execution" model, the tacit knowledge of frontline employees (such as equipment operation experience) was difficult to transform into organizational capabilities, resulting in a product iteration cycle of up to eighteen months. By implementing dynamic power allocation (shifting from centralized management to autonomous decision-making by members) and building a virtual collaboration platform, the company's CI team achieved an average annual growth of 50% in innovation projects, and the iteration cycle was shortened to twelve months. The study demonstrated that when power shifted from "positional authority" to "capability authority," organizations could unleash a higher level of collaborative innovation capability.

In response to the problems of traditional sports resource management, which was characterized by sole government supply and closed school operations, reference [5] proposed introducing social forces (such as sports clubs and folk sports inheritors) by penetrating external boundaries to establish a diversified supply system of "government + school + society." For example, in a certain region, cross-school booking and utilization of sports venues among primary schools, secondary schools, and universities was achieved by establishing a sports resource sharing platform, increasing the utilization rate of venues from 30 to 70%. In addition, social coaches were introduced to offer specialized courses such as rock climbing and hip-hop dance, thereby meeting the diverse needs of students. This case demonstrated that breaking external boundaries could enhance the resource acquisition capabilities and service innovation potential of bureaucratic organizations.

Based on the aforementioned literature, it can be concluded that the boundaryless organization theory, through the combined strategies of de-hierarchization, cross-boundary collaboration, and resource sharing, can effectively improve the response speed, resource utilization, and innovation vitality of bureaucratic organizations, thereby demonstrating its applicability to organizational structure reform. Applying the boundaryless organization theory to the optimization of equipment support supply chains is expected to open new avenues for refining equipment support processes, enhancing support efficiency, and strengthening the coordination and responsiveness of equipment support.

3 Problems Existing in the Current Equipment Support Supply Chain

The equipment support supply chain comprises multiple stages, including demand reporting, material procurement, factory production, and material supply. This study focuses on demand reporting and material supply as examples to examine existing problems and explore optimization directions. At present, military forces generally adopt a hierarchical structure (see Fig. 1) for the demand reporting and material supply tasks of equipment support. Under this model, strict hierarchical division, standardized process design, and clear allocation of responsibilities ensure the stability, controllability, and compliance of resource allocation.

However, in the context of increasingly stringent requirements for practical training, the efficiency of equipment support must be further accelerated and

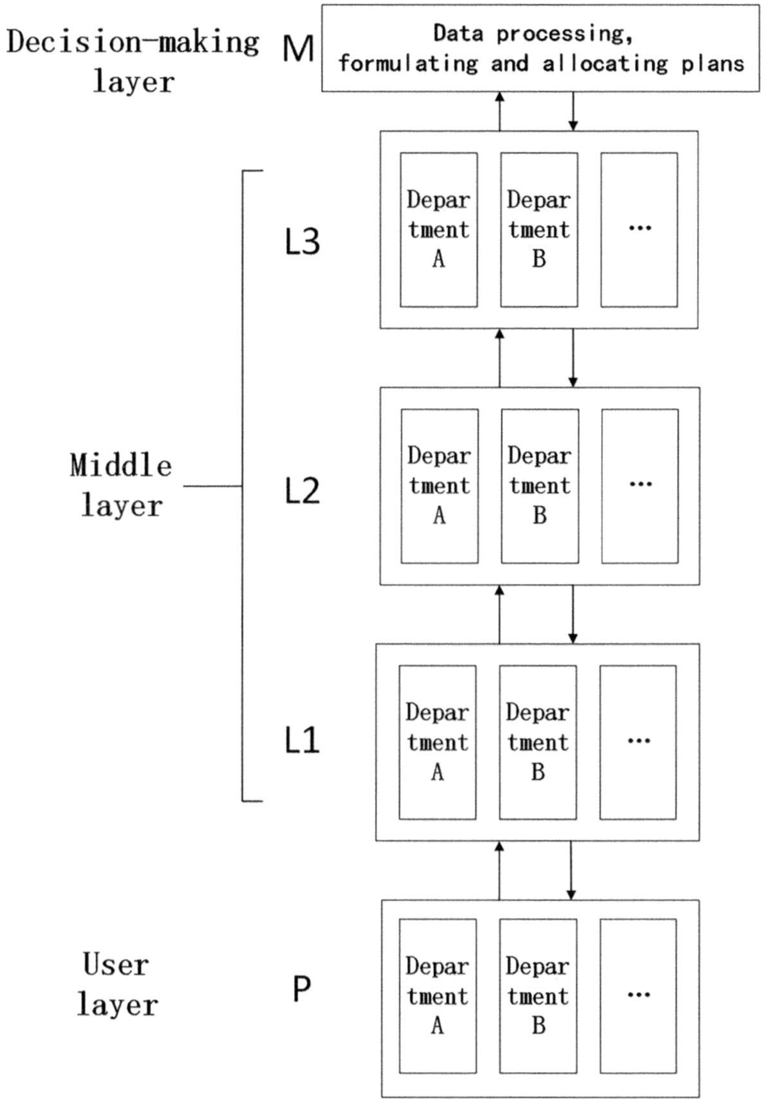

Fig. 1 Traditional equipment support supply chain model

upgraded. The existing equipment support model no longer meets current demands, and the main issues are as follows.

3.1 Multiple Approval Levels and Long Supply Cycles

Under the hierarchical management model, the process of equipment supply strictly follows the principle of vertical hierarchical progression. Demand reporting, instruction transmission, and material allocation must pass sequentially through multiple levels, including grassroots units, middle-level management departments, and high-level decision-making bodies. At each level, procedural operations such as content review and hierarchical countersignature must be completed. Moreover, differences in policy interpretation and implementation standards among different levels often lead to repeated returns and revisions of requirement documents due to non-standard expressions, incomplete attachments, and other issues, further prolonging response times. In addition, the bidirectional transmission of information is constrained by hierarchical barriers, making it difficult for grassroots units to obtain timely feedback on the progress of higher-level approvals, and for higher-level units to rapidly capture changes in actual frontline demand. This mismatch of supply and demand information creates an inefficient cycle of "demand submission—multi-level approval—slow feedback," significantly reducing the timeliness of equipment supply.

3.2 Involving Multiple Departments, Communication and Coordination Are Challenging

Equipment supply support is a systematic process encompassing multiple stages, including demand justification, planning, procurement execution, warehouse management, transportation and distribution, and handover and acceptance. Each stage requires the collaborative participation of different functional departments. In theory, this multi-department collaborative model could improve efficiency through division of labor; however, within a bureaucratic framework, functional boundaries between departments have become obstacles. Business departments often prioritize completing tasks within their own domains as the core objective, showing limited initiative in cross-departmental collaboration. For instance, demand departments may overestimate requirements to secure their own usage, while equipment departments focus primarily on meeting technical specifications and supply timeliness. Furthermore, the structural contradiction between the vertical command chain and the horizontal collaboration network exacerbates coordination challenges: grassroots units must follow vertical guidance from higher-level business departments while also obtaining resources from contract-level support departments, which can easily lead to a "dual leadership" dilemma in decision-making. Although various coordination meetings and joint mechanisms are frequently employed, their lengthy organizational processes and unclear division of responsibilities often result in inefficient problem-solving, ultimately creating a vicious cycle of "more departments, more difficult coordination, and lower efficiency."

3.3 *Data Is Not Interoperable, Leading to Significant Planning Errors*

In the hierarchical management system, equipment data is dispersed and stored in independent systems at each level along the vertical chain of demand planning, procurement, warehousing, and transportation, forming natural "information islands." Due to the dynamic changes in equipment usage scenarios, statistical biases often occur in the loss data of grassroots units. For example, hidden losses during daily maintenance may not be recorded in a timely manner, and excessive consumption during emergency tasks may be difficult to report accurately. These deviations are further amplified into a significant "bullwhip effect" during upward transmission because of inconsistent data format requirements and statistical standards across different levels. Such distorted data directly affects higher-level judgments of equipment demand, creating a serious disconnect between planned and actual needs: overestimation can lead to inventory backlogs and resource waste, while underestimation may cause material shortages that affect task execution. Moreover, the lack of information exchange reinforces the phenomenon of "information silos." Demand units find it difficult to obtain real-time data on equipment models, inventory quantities, and storage locations from higher levels, forcing them to submit requests based on historical experience or vague recollection, which often results in redundant applications. Meanwhile, supply departments are unable to track real-time consumption dynamics at the front line, making it difficult to adjust procurement and distribution plans in a timely manner. These information barriers hinder cross-level and cross-departmental data sharing and collaborative decision-making, ultimately undermining the scientific basis and effectiveness of equipment supply plans.

3.4 *Hierarchical Support Leads to Low Supply Efficiency*

Within the hierarchical system, equipment supply typically follows a "push" support model in which demand units passively await authorization and resource allocation from higher levels, rather than actively acquiring information or coordinating resources based on operational needs. Under this model, demand units often lack visibility into critical information, including approval status, equipment types and quantities, and logistics updates. The approval process may span several weeks or even months, during which grassroots units are unable to engage with suppliers to coordinate delivery schedules or to initiate equipment maintenance and task adjustments in parallel. Consequently, preparations are frequently rushed upon receipt of final notifications, resulting in significant delays in operational readiness.

In addition, the hierarchical resource allocation mechanism lacks adaptability. Equipment and materials are distributed rigidly based on staffing quotas assigned to each level, and dynamic redistribution across units or departments requires elaborate approval processes. This rigidity hinders responsiveness to the fluid demands of diverse operational scenarios. The static allocation system often results in a paradoxical coexistence of surplus and deficiency: some units accumulate excess equipment due to changes in mission requirements, while others encounter material shortages stemming from temporary task assignments. Such imbalances contribute to reduced overall supply chain efficiency.

More importantly, limited information transparency diminishes the engagement and accountability of demand units, constraining their capacity to supervise and provide feedback throughout the supply process. This opacity perpetuates a passive support culture characterized by dependency and minimal initiative, ultimately reinforcing a detrimental cycle of inefficiency, dissatisfaction, and further performance decline.

4 Optimization Model of Equipment Support Supply Chain Based on Boundaryless Organization Theory

4.1 Modeling Approach

The application of the boundaryless organization theory in military supply chain research by the United States military began relatively early and achieved significant results. During the promotion of supply chain management reform after 2010, a cross-service Joint Requirements Oversight Committee (JROC) was established to break down vertical barriers between military branches such as the Army, Navy, and Air Force, and to report equipment requirements directly to the Department of Defense. The JROC integrated equipment demand data from various branches, avoided duplicate procurement, and adjusted supply chain priorities in real time. This reform reduced the hierarchical approval process and shortened demand response time by more than 30%.

During the development of the F-35 Lightning II fighter jet, a joint team of approximately 200 members—comprising Air Force, Navy, and Marine Corps combat personnel, engineers, and cost analysts—was led by the U.S. Department of Defense Acquisition Office to participate in requirement definition, design validation, and test evaluation throughout the entire process. The team developed a unified digital R&D platform through which all suppliers, including first-, second-, and third-tier contractors, shared design data, process standards, and test results in real time. Joint meetings between military branches and enterprises were held regularly to synchronize information on requirement changes. While meeting the core requirements of the three major military branches (such as stealth performance and avionics compatibility), secondary requirements (such as aircraft carrier adaptability for the Navy) were differentiated through modular design. The F-35 project systematically broke the horizontal boundaries of "service segmentation" and "enterprise barriers" in the traditional military supply chain through a cross-service joint requirements team and a cross-enterprise technology sharing platform, becoming a benchmark case for horizontal collaboration in the military equipment supply chain.

By analyzing the problems in equipment supply and support under the current model, it can be concluded that horizontal isolation and vertical inconsistency are significant factors affecting the efficiency of the equipment support supply chain. The boundaryless organization theory provides a potential solution to this problem. By blurring horizontal and vertical boundaries, it promotes information integration and sharing, eliminates the limitations of "information islands," mitigates the impact of the "bullwhip effect," and accelerates information transmission while maintaining its accuracy. The above effects can be explained by a simple coordination-and-information mechanism. In a hierarchical supply chain, demand information typically passes through multiple organizational boundaries, and each handoff introduces delays, re-interpretation, and local buffering decisions (e.g., safety allowances and workload-driven adjustments). This repeated "filtering and amplification" increases information distortion and lengthens feedback loops, which in turn elevates forecast error and triggers further over-ordering, forming the bullwhip effect. Boundaryless organization principles reduce these distortions by (i) decreasing the number of boundary crossings (fewer approvals and fewer re-aggregation steps), (ii) enabling cross-functional joint decision-making that discourages silo-based local optimization, and (iii) establishing a shared, near-real-time information baseline via integrated digital platforms. As a result, both coordination delays and uncertainty-driven buffering are reduced, leading to more stable procurement and allocation behaviour across the equipment support supply chain.

4.2 Model Construction

Under the guidance of the comprehensive department, interdepartmental barriers are removed to achieve horizontal information integration. An information platform is introduced as a channel for information transmission to facilitate vertical information flow. These measures enhance the efficiency of equipment support in both horizontal and vertical dimensions. The optimized model of the equipment support supply chain is illustrated in Fig. 2.

4.3 Comparative Analysis

In terms of horizontal comparison, under the traditional model, each level required the collaboration of at least three departments to complete equipment support tasks, and four levels necessitated the cooperation of at least twelve departments. Such a wide involvement of personnel often created additional work bottlenecks. In the optimized model, with the establishment of comprehensive departments, equipment support tasks are primarily undertaken by four departments, resulting in more centralized

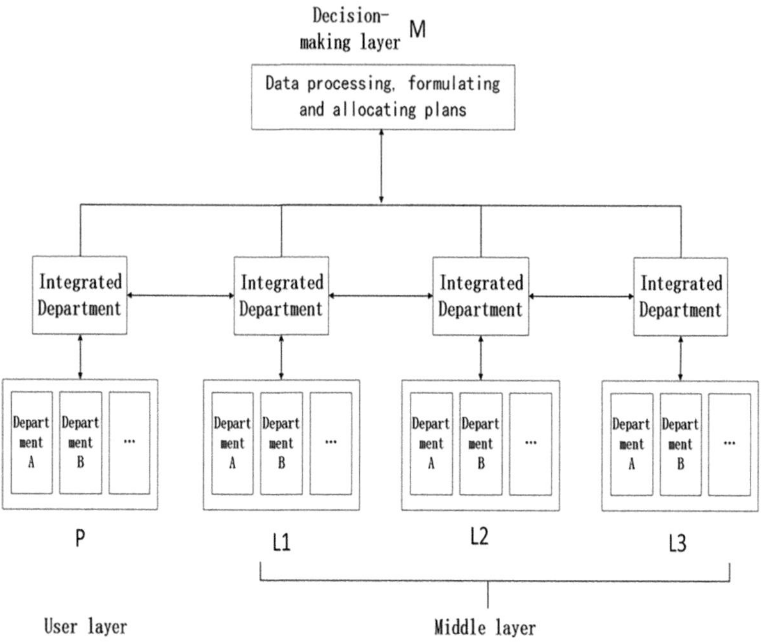

Fig. 2 Optimization model of equipment support supply chain

functions and smoother communication and coordination both between and within levels.

It can be observed that the reported data is significantly distorted, which may lead to serious misjudgments by the decision-making layer in the subsequent allocation process. In the optimized model, there is only one level of data reporting from the user layer to the decision-making layer, and the data received at the decision-making layer is closer to the actual needs, providing a more accurate reference for subsequent allocation.

Overall, compared with traditional models, the optimization coefficients for the two main parameters, the number of collaborative departments and the data amplification factor, in the optimized model are 33.3% and 36.4%, respectively. This indicates that applying the boundaryless organization theory to optimize the operating model of the equipment support supply chain has significantly improved horizontal business workload efficiency and enhanced vertical data accuracy, refer to Table 1.

Table 1 Comparison of main parameters between two models

Parameters	Number of departmental collaborations	Data amplification factor
Traditional equipment support supply chain model	12	3.8416
Optimization model of equipment support supply chain	4	1.4
Optimization coefficient	33.3%	36.4%

4.4 Effect Analysis

Based on the boundaryless organization theory, the optimization model of the equipment support supply chain, developed through the integration of process reengineering and information platforms, has resulted in several key improvements.

First, the model enables effective integration and transmission of information. From a horizontal perspective, the establishment of a general office, designation of responsible personnel within departments, coordination of inter-office functions, and assignment of dedicated staff for organizing departmental information collectively address previous shortcomings such as fragmented operations and limited information flow. These measures enhance the efficiency of information aggregation and dissemination, thereby improving operational effectiveness at the execution level. From a vertical perspective, information technology platforms facilitate the direct synchronization of user-level data with intermediate and decision-making tiers, enabling accurate assessment of demand by higher-level authorities. This mechanism partially resolves the issue of demand uncertainty and provides a data-driven basis for proactive equipment deployment planning [9–11].

Second, the model supports real-time dynamic interaction of information. During demand submission, intermediate and decision-making levels are able to monitor real-time data from subordinate units, conduct analytical evaluations, and incorporate redundancy into their planning to ensure data accuracy. Throughout the supply process, both user and intermediate levels gain real-time visibility into allocation plans across all organizational tiers, allowing early calculation of equipment shortages and ensuring adequate lead time for emergency market procurement or direct sourcing from manufacturers[12, 13].

Third, the model facilitates process supervision of information transmission. Under the traditional model, information flow across different levels remained opaque, giving rise to two primary issues: (1) stepwise transmission of data often resulted in amplified distortions, contributing to the "bullwhip effect"; and (2) inefficiencies at any single node within the supply chain could significantly impede overall performance, with such bottlenecks proving difficult to identify. The optimized model addresses these issues by leveraging the transparency of the information platform, which renders the activities of each level visible to both superiors and subordinates. This transparency ensures accuracy in data transmission and enables collaborative supervision of work progress, ultimately enhancing operational efficiency across the supply chain. Implementation considerations and challenges.

Despite the expected benefits, implementing a boundaryless equipment support supply chain may face several practical challenges. First, cultural resistance can arise when departments shift from task completion within functional silos to cross-boundary joint accountability, especially if performance evaluation and incentives remain department-centric. Second, authority redistribution is often non-trivial: the proposed comprehensive office and cross-functional coordination mechanisms require clear rules on decision rights, escalation paths, and responsibility boundaries to prevent ambiguity or "dual leadership" conflicts. Third, digital integration requirements are stringent in practice. Achieving real-time data synchronization requires unified data standards, interoperable interfaces among legacy systems, and robust data governance (including access control, auditability, and confidentiality management), which may introduce non-negligible implementation cost and schedule risks.

For real-world adoption, a phased deployment strategy is recommended. Pilot the model in a limited set of high-frequency supply scenarios, define a clear responsibility matrix for cross-department collaboration, and align performance indicators with end-to-end supply outcomes (timeliness, accuracy, and responsiveness). In parallel, establish a minimum viable digital platform that supports a single source of truth for demand, inventory, and allocation status, and then expand integration scope iteratively based on operational feedback and risk assessment.

5 Conclusion

This study addresses the challenges of horizontal fragmentation and vertical disarticulation within the equipment support supply chain. By incorporating the boundaryless organization theory, an optimized supply chain model is proposed that facilitates integrated information flow, real-time dynamic interaction, and process-level monitoring of data transmission. The developed model offers both theoretical foundation and methodological guidance for enhancing the operational efficiency of the equipment support supply chain.

References

1. Ashkenas R, Ulrich D, Jick T, Kerr S (1995) The boundaryless organization: breaking the chains of organizational structure. Jossey-Bass
2. Yan X (2024) Exploration and research on cross-departmental collaboration mechanism under the boundaryless organizational structure of financial media. Knowl Econ 698(34):173–175
3. Tong YY, Han X (2024) Research on the construction of diversified faculty in industry colleges based on the boundaryless organization theory. Educ Occup 1066(18):89–95

4. Ping QW, Chen X, Li J (2017) Research on the co-evolution of team knowledge integration and team power allocation in the context of boundaryless organizations. Acta Manag Sin 14(11):1616–1623
5. Liu NW, Yao L, Bai CL (2016) Reform of school sports resource management from the perspective of boundaryless organization theory. J Phys Educ 23(6):20–25
6. Xu C, Jiang HM (2014) Optimization and integration of equipment maintenance material supply chain. Logist Technol 3:455–457
7. Mofokeng TM, Chinomona R (2019) Supply chain partnership, supply chain collaboration and supply chain integration as the antecedents of supply chain performance. S Afr J Bus Manag 50(1):1–10
8. Archibong U, Harvey B, Baxter C et al (2015) Career progression of equality and diversity professionals in UK national health service organizations: a knowledge review. Psychol Issues Organ Cult 5(4):8–29
9. Li H, Peng W, Adumene S, Yazdi M (2023) An intelligent cost-based consequence model for offshore systems in harsh environments. In: Intelligent reliability and maintainability of energy infrastructure assets. Springer Nature Switzerland, Cham, pp 107–117
10. Li H, Yazdi M (2022) What are the critical well-drilling blowouts barriers? A progressive DEMATEL-game theory. In: Advanced decision-making methods and applications in system safety and reliability problems: approaches, case studies, multi-criteria decision-making, multi-objective decision-making, fuzzy risk-based models. Springer International Publishing, Cham, pp 29–46
11. Li H, Peng W, Adumene S, Yazdi M (2023) Cutting edge research topics on system safety, reliability, maintainability, and resilience of energy-critical infrastructures. In: Intelligent reliability and maintainability of energy infrastructure assets. Springer Nature Switzerland, Cham, pp 25–38
12. Guo J, Wang Z, Li H, Yang Y, Huang CG, Yazdi M, Kang HS (2024) A hybrid prognosis scheme for rolling bearings based on a novel health indicator and nonlinear Wiener process. Reliab Eng Syst Saf 245:110014
13. Li H, Yazdi M, Huang HZ, Huang CG, Peng W, Nedjati A, Adesina KA (2023) A fuzzy rough copula Bayesian network model for solving complex hospital service quality assessment. Complex Intell Syst 9(5):5527–5553

Research on High-Efficiency Milling Process and Machining Parameter Optimization of Aircraft Wing Rib Components

Bo Pang, Jie Chen, and Jinglei Wang

Abstract Aircraft rib components, which serve as critical elements within wing structures, present significant challenges in efficient machining due to their intricate geometries, thin-walled low-rigidity characteristics, and high material removal demands. These factors have a direct influence on aircraft development timelines. This study addresses the issues of process instability, limited machining efficiency, and scalability constraints in CNC manufacturing. A systematic framework is proposed that integrates optimized process planning, milling stability analysis, and machining parameter optimization. A vacuum-embedded clamping system secured with hexagonal screws is developed to improve structural stability during operations. A standardized two-stage, four-step machining strategy is established to streamline tooling configurations, toolpath planning, and cutting parameter selection, thereby supporting batch manufacturing. Milling stability is assessed through calibration of cutting force coefficients and single-factor experimentation, while hammer-impact testing is employed to extract frequency response functions for the tool and workpiece. Stable machining regions are identified using a regenerative chatter model and semi-discrete method, with results represented through stability lobe diagrams. A parameter optimization model is formulated to maximize productivity by adjusting spindle speeds, axial and radial depths of cut, and feed rates within system-defined constraints. CNC simulation is employed to refine toolpaths, and experimental validation indicates that rough machining material removal rates increase by over 50%. The primary contributions of this study include: (1) implementation of a unified clamping mechanism and standardized machining protocol to improve repeatability; (2) deformation mitigation through residual stress redistribution and optimized toolpath planning; and (3) stability-based parameter optimization that effectively balances productivity and machining precision. This integrated approach offers a scalable and high-efficiency solution for the manufacturing of rib components, significantly decreasing production cycles while maintaining high standards of quality in aerospace applications.

B. Pang · J. Chen (✉) · J. Wang
Inner Mongolia University of Technology, Hohhot, People's Republic of China
e-mail: 287766156@qq.com

H. Li et al. (eds.), *Data-Driven Methods for Reliability and Safety Engineering: Applications in Industrial Systems*, Springer Series in Reliability Engineering,
https://doi.org/10.1007/978-3-032-22873-4_20

Keywords Aircraft structural components · High-efficiency machining · Milling parameter optimization · Process optimization

1 Introduction

Milling operations play a fundamental role in contemporary manufacturing. To address challenges related to machining quality and operational efficiency, ongoing research has increasingly focused on the development of predictive models and simulations for milling processes, with particular emphasis on stability analysis. In the context of CNC machining for typical aircraft structural components, ribbed parts function as essential structural elements within the wing box assembly. The integration of the front beam, middle beam, rear beam, and inter-beam ribs forms the core structural framework of the wing box. More than 70 such components are incorporated throughout a single aircraft, highlighting the importance of advancing efficient machining strategies for ribbed components.

The wingbody flange, a critical element of the wing assembly in large aircraft, serves as a vital structural interface responsible for bearing significant stress loads. This component facilitates the integration of key load-bearing elements, including wing panels, ribs, beams, and trusses. Its geometric complexity and the precision of its spatial positioning directly influence the assembly accuracy of these major structural elements. As a result, the wingbody flange significantly affects the overall manufacturing timeline of the aircraft. Owing to its irregular and complex cross-sectional geometry, this component is particularly susceptible to deformation defects during mechanical processing. This recurring technical bottleneck represents a widely encountered obstacle that continues to impede the advancement of multiple aircraft research and development initiatives.

1.1 Dynamic Modeling and Stability Prediction in Milling Processes

Extensive research has been conducted globally on the dynamic modeling of milling systems and the prediction of machining stability, yielding significant advancements. The dynamic modeling of milling processes extends beyond static analysis by incorporating the dynamic characteristics of both the tool and workpiece systems. This approach facilitates accurate prediction of key parameters such as cutting forces and deformations, which are critical for enhancing machining performance and minimizing the risk of instability.

Chatter vibrations during milling operations pose a substantial challenge, as they negatively affect surface quality, reduce productivity, and accelerate the wear of both tools and machines. Therefore, preventive measures are essential in industrial settings. Predictive analysis of milling stability plays a fundamental role in mitigating chatter-related issues. Existing analytical frameworks for milling dynamics can be broadly classified into analytical and time-domain numerical approaches. Analytical methods address delay differential equations through approximation techniques, effectively balancing computational efficiency with acceptable accuracy in determining stability boundaries.

Qin [1] reported that the bottom of storage boxes is fabricated using thin-walled melon flap welding techniques. Given the large and thin geometry of melon petals, a stability analysis method was developed to address chatter issues in mirror milling of thin-walled melon flaps. Yu [2] explored the dynamic modeling and stability prediction of multi-delay five-axis milling in thin-walled blades under time-varying parameters. The study focused on four principal areas: calibration of force coefficients in five-axis milling of thin-walled blades, development of dynamic models for multi-delay systems, stability analysis under time-varying influences, and corresponding experimental validation. Nie [3] examined chatter mechanisms in the milling of thin-walled parts using variable-pitch end mills. By accounting for excitation force delays induced by inter-tooth angle variations and the time-varying nature of dynamic parameters in thin-walled structures, a delay-time-varying dynamic model was established to predict process stability and guide optimization of tool structures and process parameters. Zhang [4] concentrated on corner radius end mills, establishing a milling force model through mechanical principles. Cutting force coefficients were obtained experimentally, and the dynamic characteristics of thin-walled workpieces were analyzed using experimental modal analysis. This yielded the transfer function and modal parameters of the tool–workpiece vibration system.

1.2 Investigation into Optimization of Milling Process Parameters

This study concentrates on improving milling efficiency for aircraft wing rib components via a systematic methodology that integrates dynamic modeling and stability analysis. A semi-discretization approach is applied to evaluate machining stability under operational conditions, and hammer-impact tests alongside cutting force calibration experiments are conducted to characterize the milling system's dynamic parameters. Chatter stability boundaries were mapped under practical machining constraints. Subsequently, a multi-objective optimization framework is established, which prioritizes maximum material removal rate (MRR) through coordinated adjustments of axial/radial cut depths, spindle speed, and feed per tooth. The proposed model, validated through experimental trials, demonstrates significant improvements in process robustness and productivity.

Yi [5] applied a BP neural network to construct an energy consumption prediction model for CNC machine tools. The structure of the BP neural network was optimized using the dung-beetle-roach optimization (DBO) algorithm, resulting in a DBO-BP neural network-based CNC machine tool energy consumption prediction model. Deng [6] proposed a method to predict the milling stability and determine optimal machining parameters based on a bootstrap aggregating (bagging) procedure and the non-dominated sorting genetic algorithm-II(NSGA-II). Li [7] proposed a GA-BP multi-objective optimization method based on NSGA-II. An optimization model targeting tool life and carbon emission was established, with the constructed GA-BP neural network model employed as the objective function to optimize and solve the Pareto optimal solution set.

Lv [8] proposed a cutting parameter optimization model based on the processing stability of the process system. Various influencing factors in processing were considered, and multiple constraints, constraint ranges, objective functions, and design variables for cutting parameter optimization were determined. A maximum production efficiency objective function was established, with rotational speed, feed per tooth, cut depth, and cut width as variables. Finally, milling experiments confirmed that the surface machining accuracy of the workpiece improved after optimizing stable milling parameters.

Yang [9] analyzed problems including poor process stability, low efficiency, high manufacturing risk, and difficulty adapting to mass production in the manufacturing of aircraft wing-body flange strip structural components. Research was conducted across process design, clamping strategy, deformation control, cutting technology, and simulation processing. Based on cutting experiment analyses of aircraft wing-body flange strip structural components, a finite element prediction model for cutting these components was established.

2 Methodology

2.1 Structural Characteristics and Process Evaluation of Aircraft Rib Components

Rib components function as primary load-bearing elements in aircraft wing assemblies, forming the core skeletal framework of wing boxes. Using a specific rib component (Fig. 1) as an example, the wing box structure integrates front, middle, and rear spars. These spars are interconnected by inter-spar ribs to define aerodynamic contours. These ribs require stringent dimensional accuracy and flatness at mounting interfaces, with inter-spar ribs being the most geometrically complex variants in outsourced aerospace manufacturing.

The component features intricate geometries, tight tolerances, and significant manufacturing challenges. Fabricated from 7050 aluminum alloy extruded profiles (Fig. 2), the raw material has an "H"-shaped cross-section, with a maximum length of 2.6 m and a 50 mm-thick web. A 90% material removal rate (MRR) during machining intensifies technical challenges, especially in balancing efficiency and structural integrity.

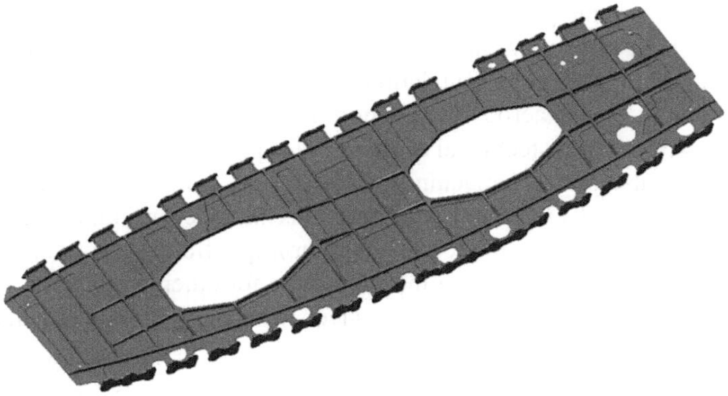

Fig. 1 Aircraft wing inter-spar ribs

Fig. 2 Wing profile blank cross-sectional diagram

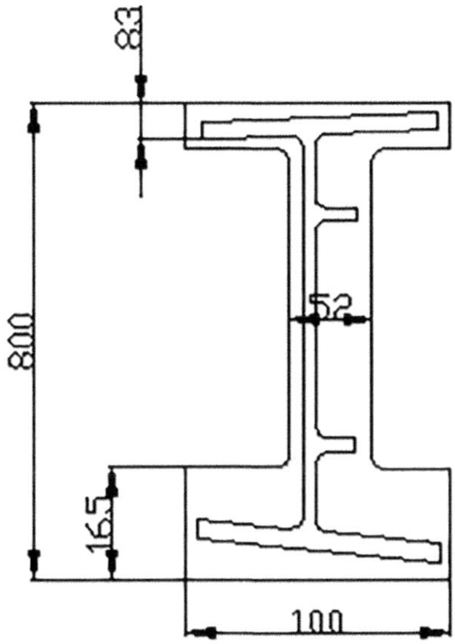

The rib component exhibits significant machinability challenges in CNC operations due to its inherent structural characteristics. With an overall dimension exceeding 2 m and a thin-walled configuration (typical of aerospace load-bearing structures), the part's dual role as a primary wing load-bearing member and its large-scale geometry amplifies susceptibility to deformation and vibration during machining. These combined factors—scale effects, reduced rigidity from thin-walled features, and strict functional requirements, necessitate advanced process controls to mitigate manufacturing risks.

High-efficiency and high-precision machining of aircraft wing rib parts poses a significant challenge for both cutting processes and equipment. Appropriate cutting schemes and clamping positioning strategies must be selected to achieve high-precision and high-efficiency cutting. A reliable and user-friendly clamping system forms an important hardware foundation for achieving high-precision machining of wing rib parts. Due to aerodynamic optimization design, its irregular arc transition shape introduces new technical challenges for positioning and clamping during machining. It is also prone to machining deformation, complicating surface quality control. During actual clamping, considerations include providing reliable positioning and support for structural component turning, offering auxiliary support for thin-walled parts, and enabling continuous cutting for outer contour machining. To meet these requirements, attention must be given to positioning adjustment, clamping adjustment, and structural adjustment.

2.2 *Challenges in CNC Machining*

Traditional machining practices have long relied on empirically determined cutting parameters, which prioritize operational safety over efficiency. Although these conservative settings reduce risks, they inherently restrict productivity improvements. To address this bottleneck, systematic exploration of stability boundaries in cutting processes is crucial for identifying optimal parameters that balance safety and material removal rates.

Regenerative Chatter Mechanism: During metal cutting, intense tool-workpiece interactions induce vibration patterns that imprint waviness on machined surfaces. Subsequent tool passes over these textured surfaces intensify dynamic instabilities—a phenomenon termed regenerative chatter, which represents a dominant self-excited vibration mode in machining systems.

Dynamic Modeling for Stability Enhancement: Tackling chatter requires foundational analysis of its regenerative principles. This involves constructing dynamic models of milling systems and developing computational methods for stability lobe prediction. These models facilitate quantitative evaluation of parameter-dependent stability thresholds.

Critical Dynamics Parameters: Accurate determination of stability limits requires characterization of system dynamics, including: natural frequency spectrum of the tool-workpiece assembly; modal stiffness (k) and damping ratios; cutting force coefficients specific to aluminum alloys.

2.3 *Milling Stability of Aircraft Rib Components*

Intense dynamic interactions between cutting tools and workpieces during metal removal operations generate surface undulations. These periodic surface irregularities subsequently engage in tool-workpiece recontact phenomena during successive cutting passes. The regenerative effect arises from phase differentials between vibration-induced surface patterns across consecutive machining cycles, forming a predominant self-excitation mechanism in machining systems. This section constructs a dynamic model of regenerative chatter and applies the semi-discretization methodology for milling stability evaluation.

To accurately analyze the stability of the milling system, this project will employ impact hammer tests and cutting force coefficient calibration experiments to determine the dynamic parameters of the milling system, including the system's natural frequency ω_n, modal stiffness k, modal damping ς, and the cutting force coefficients for 7050 aluminum alloy, comprising the tangential cutting force coefficient K_{tc}, radial cutting force coefficient K_{rc}, tangential ploughing force coefficient K_{te}, and radial ploughing force coefficient K_{re}. Through the integration of theoretical analysis and experimental methods, stability limit diagrams for high-speed milling of 7050 aluminum alloy will be generated.

2.4 Dynamic Modeling of Milling Systems

Practical machining operations necessitate explicit consideration of dynamic interactions between cutting tools and workpieces. Based on Altintas' foundational theory on machining dynamics, the milling system is abstracted into a two-degree-of-freedom (2DOF) vibration model along mutually perpendicular X- and Y-axis directions, as schematically shown in Fig. 3.

Fig. 3 Dynamic model of multi-axis milling system

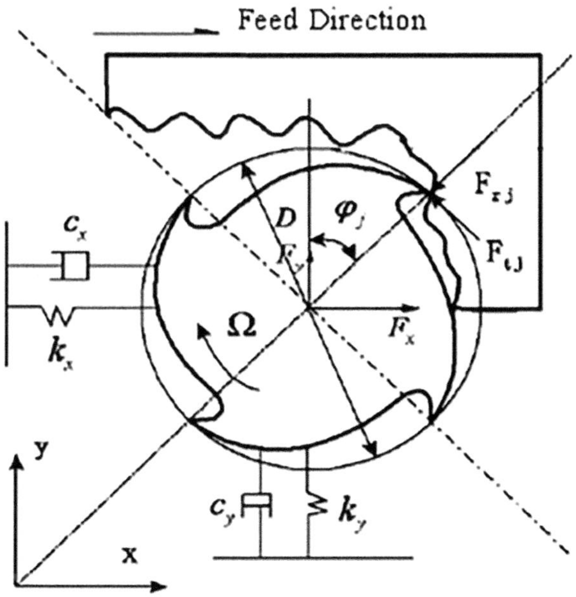

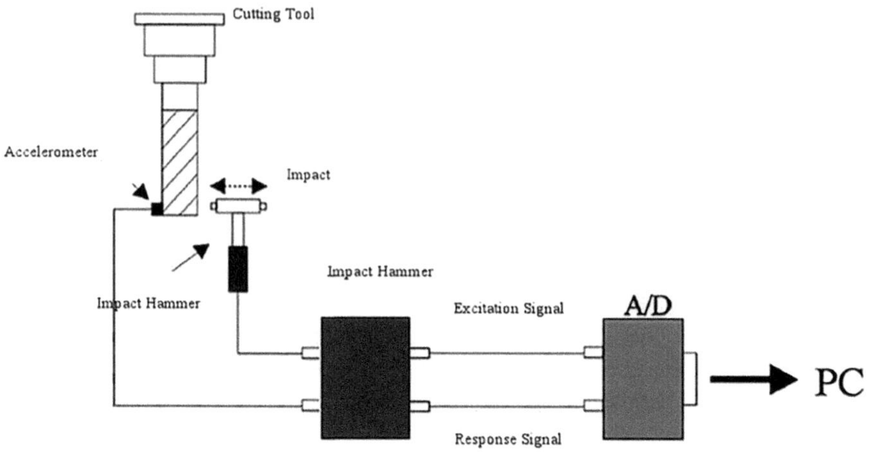

Fig. 4 Impact hammer testing for tool modal parameter identification

2.5 Experimental Identification of Tool System Modal Parameters

The impact hammer testing was conducted on a five-axis machining center of specific model. Figure 4 shows the schematic diagram of the impact hammer testing setup for tool modal parameter measurement. The testing involves applying excitation to the tooltip point using an impact hammer, generating force and acceleration signals. These signals are captured by an accelerometer, amplified through a charge amplifier, and then recorded by a data acquisition unit for computer processing. The on-site configuration of the impact hammer testing is illustrated in Fig. 5.

2.6 Stability Analysis *Via* Semi-discretization Method

In general, physical objects possess an infinite number of modes. However, most vibrational energy is concentrated in lower-order modes. Higher-order modes contain negligible energy compared to the first few modes. Because external excitation frequencies rarely exceed the range needed to activate higher-order modes beyond the second, this analysis focuses solely on the first-order vibrational mode of the cutting tool. The data recorded by the multi-channel data recorder is imported into analytical software to derive the Frequency Response Function (FRF) of the tool system.

The 2-degree-of-freedom dynamic model of the milling system lays a theoretical foundation for describing tool-workpiece vibration interactions, the calibration of milling force coefficients provides necessary quantitative inputs for stability calculations, and the stability lobe diagram generated via the semi-discretization method

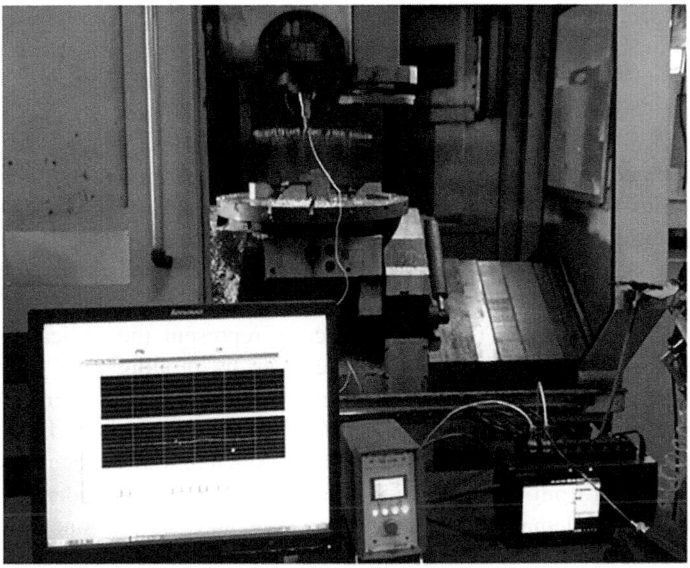

Fig. 5 Frequency response testing of tool-workpiece system

converts these dynamic characteristics into visual stable/unstable boundaries. These steps together form a sequential basis for the subsequent parameter-optimization framework, ensuring that the optimized parameters are rooted in both theoretical stability and experimental system characterization.

2.7 Calibration Experiments for Milling Force Coefficients

The calibration of cutting force coefficients primarily involves two approaches: one method utilizes an established orthogonal cutting database to retrieve coefficient values, while the other employs experimental calibration. The former method is highly cumbersome due to the extensive time required for database construction. Consequently, rapid experimental calibration is generally preferred. The most widely adopted methodology, proposed by Altintas, involves conducting milling tests with varying feed rates while maintaining other parameters constant. The average milling force per cycle is measured and expressed as a linear function of the feed rate f. Cutting force coefficients are subsequently determined via linear regression.

This experiment was conducted on a five-axis machining center (specific model). A Kistler triaxial dynamometer measured milling forces at a sampling frequency of 4000 Hz. Measurement ranges were 1500 N for the X/Y axes and 3000 N for the Z-axis. A Kistler charge amplifier and a high-speed isolated 8-channel data acquisition system were used for signal conditioning and recording. The tool was a 20 mm diameter solid carbide end mill operating in dry climb milling mode.

To minimize interference from variables such as cutting parameters, machine dynamics, and tool geometry—all known to influence cutting force coefficients—five test groups were conducted under identical external conditions with only feed rates varied. The experimental setup is shown in Fig. 6.

By performing linear regression on the experimental results, the cutting force regression outcomes under various cutting parameters are obtained, enabling the determination of the final milling force coefficients.

With a cutting width of 4 mm, the experimentally derived dynamic parameters are input into a MATLAB program based on the semi-discretization method. This program generates the system stability lobe diagram (SLD), as shown in Fig. 7.

Stability lobe diagrams (SLDs) graphically represent the critical boundaries between stable and unstable machining regimes by plotting axial depth of cut (ADOC) against spindle speed. As shown in Fig. 7, these diagrams employ spindle speed *n* (rpm) as the abscissa and tool's ADOC ap (mm) as the ordinate. For any given spindle speed, the corresponding critical ADOC a_{plim}) is determined by the vertical projection from the speed axis intersecting the SLD boundary curve. This characteristic curve demarcates stable (below curve) and chatter-prone (above curve) machining zones.

The SLD reveals spindle speed-dependent variations in permissible ADOC, demonstrating that elevated rotational velocities enable increased material removal rates while maintaining process stability. Strategic spindle speed selection based on SLD analysis provides theoretical foundations for maximizing cutting efficiency without inducing chatter vibrations. Higher speeds (e.g., 15,000–25,000 rpm) often permit greater ADOC values than lower-speed operations.

Fig. 6 Machining test setup

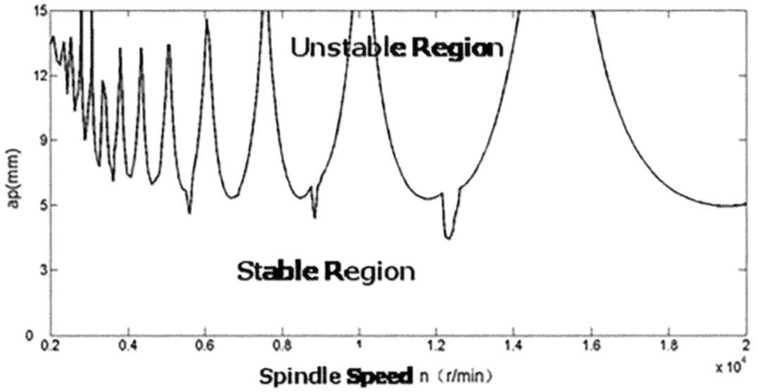

Fig. 7 Stability lobe diagram (SLD) of the system

Calculate a_{plim} through SLD intersection analysis at target spindle speeds. Maintain operational parameters 15–20% below SLD-predicted instability thresholds. Adjust feed rates and tool paths according to SLD-derived stability constraints.

The methodological framework for parameter optimization is constructed through a logically sequential integration of dynamic modeling, experimental calibration, and stability boundary mapping. First, the establishment of a 2-DOF dynamic model based on regenerative chatter theory provides the fundamental analytical structure for describing the tool-workpiece interaction. Subsequently, the experimental identification of system modal parameters (via hammer-impact tests) and the calibration of cutting force coefficients furnish the model with accurate, case-specific dynamic properties and force input data. Finally, these elements are synthesized through the semi-discretization method to generate the Stability Lobe Diagram (SLD), which visually demarcates the stable and unstable machining regimes as a function of spindle speed and axial depth of cut. This SLD serves as the direct and critical constraint boundary for the subsequent optimization stage. Therefore, the dynamic model defines the theoretical basis, the calibrated parameters ensure empirical accuracy, and the generated SLD provides the operational constraints. Together, they form a robust, three-pillar foundation for the parameter-optimization framework, ensuring that the quest for maximal material removal rate is rigorously bounded by process stability limits.

3 Results

3.1 Optimization Method

The material removal rate (MRR) and tool life (T_{tool}) during the processing are selected as the core goals for optimizing the cutting parameters. On the basis of ensuring that the surface quality of the workpiece meets the standards, efforts are made to maximize the processing efficiency and extend the service life of the tool.

The service life of cutting tools T_{tool} can be characterized by a Taylor series expansion that includes all design parameters, and the correlation coefficients and constant terms can be obtained by referring to the corresponding data tables.

The core optimization objective focuses on the coordinated improvement of processing efficiency, product quality, and production cost. Among these, the material removal rate (MRR) serves as an indicator for measuring processing efficiency, tool life reflects cost control, and stability constraints ensure product quality meets standards. A multi-objective linear weighted synthesis method is employed to construct the optimization model, achieving balanced optimization of these three aspects.

3.2 Parameter Optimization for Pocket Milling Operations

The procedural framework for optimizing rough cavity milling parameters comprises the following systematic phases:

Establish equipment specifications including CNC machine tools, fixtures, and milling cutters in accordance with component manufacturing requirements. Configure radial engagement within 30–70% of cutter diameter through empirical evaluation. Modal analysis of the machine-tool system is conducted to acquire frequency response functions (FRFs) in orthogonal X/Y directions during roughing operations. Stability lobe diagrams (SLDs) are generated through chatter stability simulations under specified radial engagement parameters, as illustrated in Fig. 8. Optimal spindle speed (15,000 rpm) and axial depth-of-cut (4 mm) are selected within SLD-defined stable zones, ensuring compliance with rotational speed and cutting velocity constraints. Required machining layers (6 layers in this application) are calculated based on total stock removal and predetermined axial depth-of-cut, with uniform depth distribution implemented across layers where feasible [10–13]. Dynamic simulations are employed to determine chip load parameters while maintaining operational limits for machine power, torque constraints, feed rate boundaries, dimensional accuracy, and surface integrity requirements.

The pre-optimization configuration yielded a material removal rate (MRR) of 52.5 cm^3/min, while the optimized parameters elevated MRR to 99.7 cm3/min—representing a 1.9-fold productivity improvement. This advancement enables significant reductions in production lead times and associated manufacturing costs [14, 15].

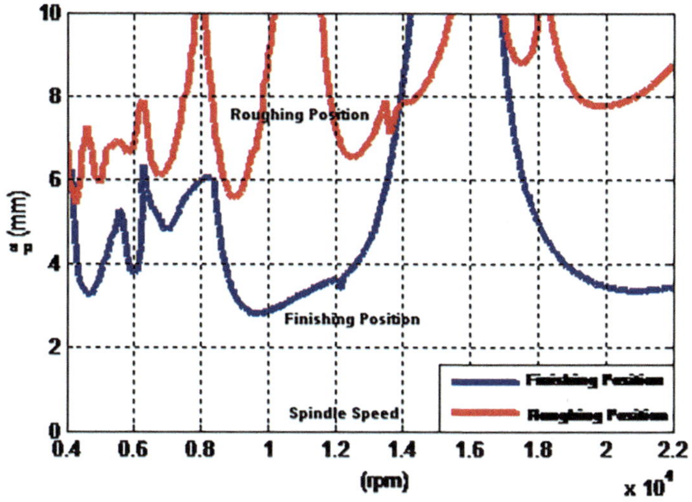

Fig. 8 Simulation results of chatter stability lobes with optimized cavity cutting parameters

The technical implementation involves three critical phases: Systematic characterization of cavity geometries in aerospace structural components informs process design decisions. Developing representative test specimens enables empirical derivation of high-performance machining protocols and parameter specifications. Implementing optimized finishing strategies ensures final components meet stringent aerospace quality standards for dimensional accuracy and surface integrity.

3.3 *Interpretation of Optimization Results and Discussion of Limitations*

The significant improvement in material removal rate (MRR), from 52.5 cm^3/min to 99.7 cm^3/min, can be attributed to the systematic exploitation of the system's dynamic stability characteristics. The core mechanism lies in the strategic selection of a spindle speed (15,000 rpm) that corresponds to a high-stability lobe (see Fig. 8). At this speed, the phase relationship between the current tool vibration and the waviness left on the workpiece surface from the previous tooth passage becomes destructive to chatter growth, thereby increasing the critical axial depth of cut (ADOC). The optimization framework leveraged this phenomenon, allowing the ADOC to be safely increased to 4 mm—a parameter previously deemed risky under empirical settings. This synergistic increase in both spindle speed and ADOC, governed by the SLD boundary, is the direct physical reason for the near-doubling of MRR while simultaneously suppressing regenerative chatter. The stability is maintained because the operating point (spindle speed, ADOC) is deliberately positioned within the SLD's stable region, well below the chatter threshold.

While the proposed framework demonstrates substantial efficacy for the specified rib component in 7050 aluminum alloy, its practical application warrants consideration of several limitations. First, the cutting force coefficients and stability boundaries are material-specific. The optimization results are directly valid for 7050-T7451 aluminum alloy; application to other aerospace alloys (e.g., titanium or nickel-based superalloys) would require re-calibration of force coefficients and re-generation of SLDs due to differing material machinability and dynamic interactions. Second, the stability lobes are intrinsically tied to the Frequency Response Functions (FRFs) of the specific machine-tool-workpiece system used in this study. Variability in machine tool dynamics, tool holder interface stiffness, or even tool wear state across different machining centers would alter the FRFs and thus the SLD. Therefore, the optimal parameters are not universally portable; the methodology, however, is. Third, the geometric scalability of the approach needs validation. This study focused on a rib with a characteristic thin web (50 mm thick). Components with significantly thicker webs or more irregular, asymmetric spar connections may exhibit different dynamic behaviors and mode shapes, potentially requiring an expanded multi-modal analysis beyond the first-order mode considered here. Addressing these limitations forms the basis for future work to generalize the framework for a broader class of aerospace structural components.

Based on milling stability theory, this study developed an efficient machining parameter optimization method for aviation rib parts. Through experimental testing and theoretical analysis, the structural characteristics of large aviation rib parts and the dynamic performance of processing equipment have been thoroughly characterized. A systematic study on the stable area of the milling process was carried out for CNC machine tools for aviation rib parts. Based on stable domain analysis results, the optimized configuration of cutting parameters is realized. Optimized processing parameters were applied to the CNC program of the parts. While ensuring no flutter in the processing process, it significantly improves production efficiency and processing quality. This process optimization solution provides reliable technical support for efficient and precise machining of aviation structural parts.

The optimized spindle speed aligns with the high-stability zone in the SLD, enabling a larger axial depth of cut to be achieved without chatter. This synergistic adjustment directly enhances MRR while maintaining process robustness, which explains why the optimized parameters improve MRR and stability. The optimization results of this study are specific to 7050 aluminum alloy, and validation is required for other aerospace materials. Additionally, results dependent on FRFs may vary across different machining centers, and the framework needs adaptation for rib components with extremely thick webs or irregular spar connections—these are the limitations of this study.

4 Conclusion

Considering the structural characteristics of large aircraft rib components and the operational capabilities of machining equipment, this study proposes a milling parameter optimization methodology grounded in stability analysis. The research emphasizes the identification of milling stability regions specific to CNC machining centers utilized in aircraft rib manufacturing, followed by the systematic optimization of process parameters within these stability domains. This methodology enables significant enhancements in machining efficiency while ensuring cutting operations remain free from chatter.

The optimized high-efficiency milling strategy was applied to the production of representative rib components. Implementation results demonstrated a 50% increase in material removal rate during rough machining relative to conventional processing techniques.

Acknowledgements The authors would like to thank Inner Mongolia University of Technology for providing the funds to carry out this work.

References

1. Qin X, Liu X, Zhang S (2024) Stability analysis for mirror milling of thin-walled melon flap on box bottom. J Tianjin Univ Sci Technol 57(11):1161–1169
2. Yu C (2024) Dynamic modeling and stability prediction of multi-time delay five-axis milling of thin-walled blades under the action of time-varying modal parameters. Xi'an Technological University
3. Nie W (2022) Research on delay-time varying dynamics behavior and stability of variable pitch end mills for milling thin-walled parts. Harbin University of Science and Technology
4. Zhang J, Liu C (2018) Forced vibration response during the milling of thin-walled workpieces J Tsinghua Univ (Sci Technol) 58(11):961–965
5. Yi W, Yin R, Tian Y (2024) Research on energy consumption prediction and multi-objective optimization of cutting parameters in CNC milling. J Chongqing Univ Technol (Natural Sci) 38(3):240–249
6. Deng C, You Q, Zhao Y (2024) Research on the stability prediction and optimization of CNC milling based on bagging-NSGAII under the influence of multiple factors. Adv Eng Sci 56,(4):238–249
7. Li C, Yin R (2024) Research on optimization of CNC milling parameters based on GA-BP and NSGA-II. Tool Eng 58(3):86–92
8. Lv Y (2019) Research on high-efficiency milling process and optimization of processing parameters for TC18 titanium alloy frame parts. Zhejiang University, Zhejiang
9. Yang T (2018) Research on processing deformation of aircraft wing-body edge strips. Harbin Institute of Technology, Heilongjiang
10. Guo J, Wang Z, Li H, Yang Y, Huang CG, Yazdi M, Kang HS (2024) A hybrid prognosis scheme for rolling bearings based on a novel health indicator and nonlinear Wiener process. Reliab Eng Syst Saf 245:110014
11. Li H, Yazdi M, Huang HZ, Huang CG, Peng W, Nedjati A, Adesina KA (2023) A fuzzy rough copula Bayesian network model for solving complex hospital service quality assessment. Complex Intell Syst 9(5):5527–5553

12. Li H, Peng W, Adumene S, Yazdi M (2023) An intelligent cost-based consequence model for offshore systems in harsh environments. In: Intelligent reliability and maintainability of energy infrastructure assets. Springer Nature Switzerland, Cham, pp 107–117
13. Li H, Yazdi M (2022) What are the critical well-drilling blowouts barriers? A progressive DEMATEL-game theory. Advanced decision-making methods and applications in system safety and reliability problems: approaches, case studies, multi-criteria decision-making, multi-objective decision-making, fuzzy risk-based models. Springer International Publishing, Cham, pp 29–46
14. Li H, Peng W, Adumene S, Yazdi M (2023) Cutting edge research topics on system safety, reliability, maintainability, and resilience of energy-critical infrastructures. In: Intelligent reliability and maintainability of energy infrastructure assets. Springer Nature Switzerland, Cham, pp 25–38
15. Li H, Peng W, Adumene S, Yazdi M (2023) Advances in failure prediction of subsea components considering complex dependencies. In: Intelligent reliability and maintainability of energy infrastructure assets. Springer Nature Switzerland, Cham, pp 93–105

ESPAC-Based Linear Array Microtremor Tomography Method and Its Application in Goaf Detection

Shasha Liang, Xinyue Wang, and Ziqi Wang

Abstract Collapse columns and goafs present significant risks to coal mine safety. Accurate detection of their geometrical configurations is essential for ensuring safe operations and guiding mining design. Traditional Spatial Autocorrelation (SPAC) detection techniques, however, face notable limitations, including narrow applicability, vulnerability to noise interference, and reduced resolution caused by one-dimensional model splicing. Using the Talahao Coal Mine as a case study, a linear array configuration is implemented built upon background noise imaging and the Extended Spatial Autocorrelation (ESPAC) method to collect microtremor data. A specialized noise suppression algorithm is developed to eliminate various sources of ambient and instrumental noise, thereby enhancing the signal-to-noise ratio (SNR) of the microtremor recordings. By means of human–computer interaction, dispersion curves between station pairs are extracted, fitted, and assessed for quality. To support quality control, the theoretical waveform of the spatial autocorrelation function is incorporated. A special point constraint method is introduced to guarantee the dependability of the dispersion data and enable the successful extraction of low-frequency surface waves. Based on this processed data, tomographic inversion is conducted to derive a two-dimensional structural model beneath the survey line. The imaging process accounts for topographic variations, resulting in seamless model stitching and improved resolution. Potential goaf locations are identified through the spatial distribution of shear wave low-velocity zones. This information serves as a foundation for delineating hazardous regions and developing mitigation strategies, thereby contributing to the reduction of structural instability risks during mining operations.

Keywords Ambient noise · Spatial autocorrelation · Passive velocity tomography · ESPAC · Goaf detection

S. Liang
Inner Mongolia Autonomous Region Seismological Bureau, Hohhot, China

S. Liang · X. Wang (✉) · Z. Wang
School of Mathematical Sciences, Inner Mongolia University, Hohhot, China
e-mail: selene.wang@qq.com

H. Li et al. (eds.), *Data-Driven Methods for Reliability and Safety Engineering: Applications in Industrial Systems*, Springer Series in Reliability Engineering,
https://doi.org/10.1007/978-3-032-22873-4_21

1 Introduction

The Earth's surface and subsurface are continuously affected by weak vibrations known as microtremors, which are not generated by seismic sources. Within the domain of natural seismology, such vibrations are typically referred to as seismic background noise or ground pulsations [1]. Surface wave tomography techniques that utilize background seismic noise to extract Rayleigh-wave Green functions through cross-correlation of long-duration ambient seismic noise recordings. These functions contain valuable information regarding the structure of the shallow to middle crust [2, 3].

Microtremor imaging based on background noise primarily encompasses two methodologies. The first is the seismic interferometry approach, which retrieves approximate Green functions between station pairs through cross-correlation functions (NCF). This method is predominantly employed for imaging large-scale subsurface velocity structures. The second and more widely used approach is the Spatial Autocorrelation (SPAC) method, initially introduced by Aki [4], commonly referred to as microtremor exploration. This technique is particularly effective in imaging localized shallow subsurface velocity structures [5]. Comparative studies indicate that the signal derived from time-domain cross-correlation between two sensor nodes represents the spatial autocorrelation signal, which is phase-corrected in accordance with the Rayleigh wave dispersion relationship between the stations [6]. This finding demonstrates that spatial autocorrelation corresponds to time-domain cross-correlation at zero time delay, the two are equivalent in effect [5].

In terms of array deployment, the most commonly employed configurations for the SPAC method include equilateral triangular and nested triangular arrays [1]. To enhance the method's adaptability under varying working conditions, in 2003 Okada presented the Extended Spatial Autocorrelation (ESPAC) method [7], which compensates for the limitations of conventional regional monitoring systems by calculating the phase velocities at different frequencies [8]. Additionally, Chávez-García et al. introduced a SPAC technique by linear arrays, demonstrating that with sufficiently long recordings, adequate wave sampling from all directions can be achieved [8, 9].

A landmark contribution to this field was made by Shapiro et al., whose published a paper in 2004 on extracting Rayleigh waves from the correlation of background noise [10] laid the foundation for modern seismic background noise research. In engineering applications and shallow geophysical exploration, surface wave imaging based on ambient noise has matured and is now widely applied for large-scale subsurface structural imaging. For example, the SPAC method combined with the layer-stripping inversion technique is employed to determine S-wave velocity structures at various depths [11]. This methodology has also been successfully implemented in geothermal exploration, concealed fault detection [12], and urban subsurface space mapping and planning [13, 14].

In the context of mine goaf detection, imaging studies have utilized the ESPAC method alongside other geophysical techniques. For instance, Yunfeng Chen et al.

developed an asynchronous station background noise imaging method, effectively overcoming the limitations of traditional synchronous multi-station recordings and revealing a strong correlation between mineral resource distribution and seismic velocity characteristics [15]. The Institute of Geology and Geophysics, Chinese Academy of Sciences, was the first to apply dense-array passive-source noise imaging in the Kalatongke mining region of Xinjiang, obtaining a shallow three-dimensional shear wave velocity structure within the 0–1.3 km depth range, thereby validating the method's capability to resolve fine-scale subsurface structures in mining environments [16].

Additionally, a research team from Peking University constructed a three-dimensional velocity model (0–500 m depth) of the Habute Gai mining area using a combination of active-source (artificial hammering) and passive-source (ambient noise) observations. Their results demonstrated the feasibility of integrated active–passive source imaging for identifying goafs and concealed structures in geologically complex settings [17]. Other studies have implemented passive seismic imaging in coal mines using ambient noise seismic interferometry in conjunction with dense array deployments [18]. Practical application cases, such as linear array ambient noise imaging for mining environments [8], passive tomography for coal mine disaster assessment [19], and goaf detection in the Harwusu Open-Pit Mine [20, 21], provide significant reference value for optimizing integrated geophysical exploration strategies for goaf detection in mining operations.

Microtremor imaging of background noise has significant advantages and is very feasible and applicable in reflecting underground spatial structures, exploring geological disasters, etc. [22]. As a non-intrusive method, it uses short-period dispersion diagrams from natural vibrations to reveal crustal shear wave velocity distribution, working even without earthquakes and avoiding environmental interference and safety risks of traditional artificial seismic sources [10]. Yet, it faces challenges: weak, complex ambient noise signals are hard to extract and rely on advanced technologies and equipment; the spatiotemporal uncertainty of noise sources may cause imaging biases.

Collapse columns and goafs are common geological phenomena in coal mining, prone to surface collapse and other situations, posing serious hazards to coal mine production and safety. Therefore, accurate detection of collapse columns and identification of the spatial morphology of goafs are crucial for safe production and mining design in mining areas. However, traditional cross-correlation calculations are processed in the time domain, and the results still require time–frequency analysis for surface wave extraction, leading to large errors. Additionally, the conventional SPAC method can only obtain a one-dimensional average structure beneath the array. It is necessary to deploy regular microtremor arrays one by one and splice one-dimensional models into two-dimensional profiles to obtain detection results, which have low resolution and are not conducive to distinguishing lateral anomalies [23]. This study selects the Talahao Coal Mine in Inner Mongolia, China, as a research case for passive velocity tomography. A linear array configuration based on ambient noise imaging and the ESPAC method is employed to record microtremor data, enhancing information acquisition capabilities. During data processing, key steps such as dispersion curve extraction, discrete dispersion point fitting, and quality evaluation between any two stations are achieved through human–computer interaction, fully exploiting data information to obtain the lowest possible frequency surface waves. The theoretical spatial autocorrelation function waveform is introduced as an auxiliary tool, combined with real-time generated dispersion curves for quality control, improving the accuracy and efficiency of dispersion curve picking. Under the constraints of quality control and prior information, bad data points and those outside the detection frequency range are eliminated through special point constraint methods to ensure high-quality dispersion data. The apparent velocity structure varying with depth is calculated from the phase velocity dispersion curves to complete data tomographic inversion for linear and approximately linear microtremor arrays, directly obtaining a two-dimensional structural model beneath the survey line. Terrain undulations are considered in the imaging results to achieve seamless stitching, scientific and rational resolution assurance, while reducing environmental impact, field workload, and improving exploration efficiency. According to the distribution characteristics of shear wave low velocity zones, potential goaf locations are identified, and preliminary inferences about the underground structure of the mining area are made. Potential collapse hazard zones can be delineated in advance, providing a basis for eliminating disaster risks and selecting scientific governance methods, thereby reducing the risk of structural instability in mining operations.

2 Principles and Methods of Microtremor Tomography

2.1 Comparative Analysis of Linear Arrays and Areal Arrays

The resolving power of the Vs profile generated by the linear array may be affected by various factors [24]. To ensure the dependability of dispersion curves derived from linear arrays via the SPAC method, In 2022, Mi et al. computed the dispersion spectrum using multiple approaches to verify the accuracy of the extracted phase velocities [25]. Kangdong Wang et al. conducted a comparison between the dispersion curves of linear arrays and regional arrays during the detection of mine goafs (Fig. 1d) [8]. When the central points of the linear array and regional array are identical (Fig. 1c). Although the dispersion spectrograms differ slightly, their dispersion curves are similar. This result demonstrates the applicability of linear arrays in mining area exploration, taking into account factors such as terrain and the flexibility of node deployment. Therefore, we use linear arrays to record ambient noise for imaging the subsurface of mining areas.

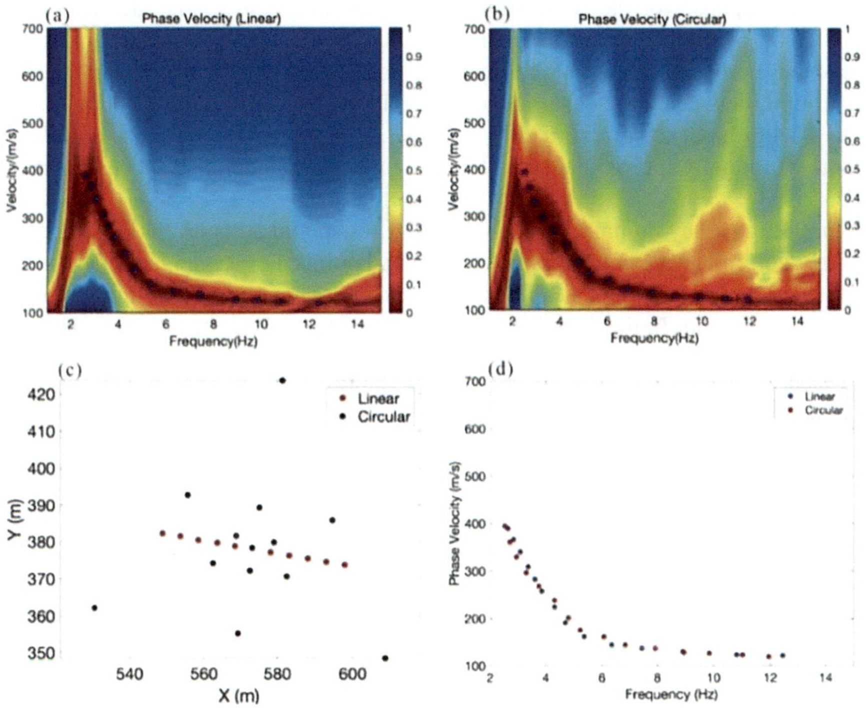

Fig. 1 **a** The linear array. **b** The regional monitoring array. **c** The layout of stations in linear and regional monitoring array. **d** Comparison of dispersion curves from the linear array and regional monitoring array [8]

2.2 Theory and Methods for Extracting Surface Wave Dispersion Using Spatial Autocorrelation

For irregular or linear arrays, the spatial autocorrelation coefficient is calculated by fitting the Bessel function to determine phase velocity, establishing a correlation between spatial autocorrelation coefficient and distance [5]. The ESPAC method utilizes recorded data from multiple seismic stations, computing the cross-correlation function between station pairs. By analyzing the signal coherence between these stations, it is possible to extract wave velocity information of the subsurface medium, particularly the S-wave velocity structure. Although termed "spatial autocorrelation," this method processes data recorded at two different stations. Temporal and spatial stationarity imply that the signal recorded at one station is a time-delayed (or advanced) [6] version of the signal at another station. Since these sensors lie much closer to one another than to earthquakes [10], and the faster propagation of S-waves, recordings within a certain range of stations can be regarded as the same signal propagating from one station to another. Therefore, the spatial variation in the signal's phase can be obtained through cross-correlation, and this spatial variation in phase for the same signal is referred to as the spatial autocorrelation coefficient [6]. When noise sources are spatially homogeneous, the dominant energy in the noise cross-correlation function corresponds to a time delay equal to the wave travel time between two sensors. Cross-correlating and stacking long-term continuous background noise recorded by the two sensors makes the cross-correlation function gradually approximate their Green's function, achieving its "reconstruction" [6].

In the frequency domain, Green's function for spatial autocorrelation represents the correlation of wavefields between two spatial points. For Rayleigh surface waves, Bessel and Hankel functions can be used to express the Green's function within the frequency domain. In the frequency domain, the G(r,ω) describes the correlation of wavefields at a distance r between two spatial points for an angular frequency ω. For Rayleigh surface waves in the homogeneous medium, the Green's function follows the form of Eq. (1).

$$G(r, \omega) = A \times H_0^{(1)}(kr) + B \times H_0^{(2)}(kr) \tag{1}$$

where r denotes the distance between two sensor nodes, ω stands for angular frequency, and k is the wavenumber of the Rayleigh surface wave (k = ω/c, c is the phase velocity of the Rayleigh wave). $H_0^{(1)}$ and $H_0^{(2)}$ correspond to the first and second kind Hankel functions respectively, A and B are constants associated with source and medium characteristics.

The Hankel functions are related to the Bessel function of the first kind $J_0(x)$ and the unction $Y_0(x)$ as shown in Eqs. (2) and (3):

$$H_0^{(1)}(x) = J_0(x) + iY_0(x). \tag{2}$$

$$H_0^{(2)}(x) = J_0(x) - iY_0(x). \tag{3}$$

Therefore, the Green's function can be further expressed as Eq. (4):

$$G(r, \omega) = A(J_0(kr) + iY_0(kr)) + B(J_0(kr) - iY_0(kr)) \tag{4}$$

In practical applications, only the outgoing wave is typically considered. Therefore, the Green's function can be simplified as Eq. (5):

$$G(r, \omega) \propto H_0^{(1)}(kr) = J_0(kr) + iY_0(kr). \tag{5}$$

In ambient noise imaging, If the ambient noise sources are homogeneously dispersed, the spatial autocorrelation function of the noise between two sensor nodes can be approximately represented as the real component of Green's function, as Eq. (6). This condition can be satisfied through spatial and temporal averaging of observations from stations at different azimuths in a microtremor array.

$$G(r,\omega) \approx Re(G(r,\omega)) = J_0(kr). \tag{6}$$

In summary, based on the assumption that microtremor signals are stability and randomly distributed in both time and space, the spatial autocorrelation of continuous waveforms between two microtremor arrays separated by a distance r is calculated [26]. The real component of the resulting waveform is equivalent to the zeroth-order Bessel function. After obtaining the autocorrelation function between stations, the phase velocity dispersion information of Rayleigh waves at different frequencies can be derived using the following Eq. (7):

$$\rho(r,\omega) = J_0\left(\frac{2\pi f}{C(f)} r\right). \tag{7}$$

In the equation, r represents the distance between the two sensors, f denotes the frequency, $C(f)$ represents the surface wave phase speed, ρ is the azimuthally averaged autocorrelation coefficient, and J_0 is Bessel function of the first kind of order zero. This formula is applicable when the fundamental mode of the surface wave dominates, and the vertical component of the recordings is used to extract the surface wave dispersion via spatial autocorrelation1.

2.3 *Inversion Principle of Dispersion Curves*

Surface waves' phase velocity $C(f)$ depends on frequency f, expressed as $C(f) = \frac{2\pi f}{k}$, where $\omega = 2\pi f$ denotes angular frequency and k refers to wavenumber. The phase velocity $C(f)$ is obtained by solving the dispersion equation of surface waves, which is typically represented as Eq. (8):

$$D(k, \omega, V_s(z), \rho(z)) = 0. \tag{8}$$

where D is the dispersion function, $V_s(z)$ is the transverse wave velocity of the subsurface medium, and $\rho(z)$ is the density.

The surface wave travel time t is the propagation time of the surface wave from the source to the station and is related to the surface wave propagation path L and the frequency-dependent phase velocity $C(f)$.

$t_{obs,i}$ is the ith observed travel time,$t_{cal,i}$ is the ith theoretical travel time, $R(m)$ is the model regularization term (e.g., smoothing constraints), and λ is a regularization parameter used to balance data fitting and model complexity. The Damping Parameter (DP) is used to balance data fitting and model smoothness in background noise inversion imaging. The damping parameter can avoid unreasonable velocity fluctuations in the inversion results. Different damping parameters may be required for different geological conditions and data quality. In practice, the inversion results under different parameters can be compared through several tests, and the λ value that best meets the geological cognition and data characteristics can be selected to guarantee the precision and dependability of inversion outcomes.

After extracting the dispersion curve or travel time data of the surface wave, the initial transverse wave velocity model $V_s(z)$ and density model $\rho(z)$ are set according to the geological data, and the theoretical dispersion curve or travel time data are calculated using the initial model, and the velocity model is adjusted through iterative optimization to match the theoretical data with the observed data, and to generate the 2D or 3D transverse wave velocity structure of subsurface medium.

3 Data Processing and Analysis

The background noise spatial autocorrelation micromotion imaging process mainly covers three steps: data pre-processing, dispersion curve extraction, and dispersion curve inversion imaging. In the single-station data pre-processing process, to improve the accuracy of the dispersion curve extraction, it is often necessary to screen the waveform, de-instrument the response, de-mean, de-linearize the trend, and normalize and whiten the spectrum of the original micromotion recordings [27]. After pre-processing, to reduce the amount of computation, the method of calculating the spatial autocorrelation coefficient in segments is adopted, and then the complete autocorrelation coefficient curve is obtained by superposition. On this basis, the dispersion curve is extracted through fitting of the Bessel function. The processing and calculation process is shown in Fig. 2.

3.1 Data Acquisition and Data Pre-processing

Data were collected at a coal mine situated in Dongsheng District, Ordos City, within the Inner Mongolia Autonomous Region. 90 three-component node instruments were used and arranged according to a station spacing of 7 m. The survey line is roughly 630 m long,; and Fig. 3a shows the distribution of the equipment. The data collection lasted for 1020 min. The starting time of data timing was 16:00 on September 7, 2024, and the ending time of data timing was 9:00 am on September 8, 2024. A total of 15,300,000 data points were collected, with a sampling rate of 250 Hz and a sampling interval of 0.004 s. After the remote viewing and data reception were completed, the data were segmented, and subsequent processing and analysis were carried out using data imaging software.

First, the original three-component SAC files recorded were organized to obtain multiple discrete files for each station, as shown in Fig. 3b. From these, the vertical component (Z-component) of the three components was selected and integrated into a complete background noise file. Subsequently, an initial pre-screening of this file was performed to check whether the waveforms were normal and whether the data discontinuities were severe, etc., a process known as initial pre-screening.

Fig. 2 Processing flow of background noise extracted surface wave dispersion data

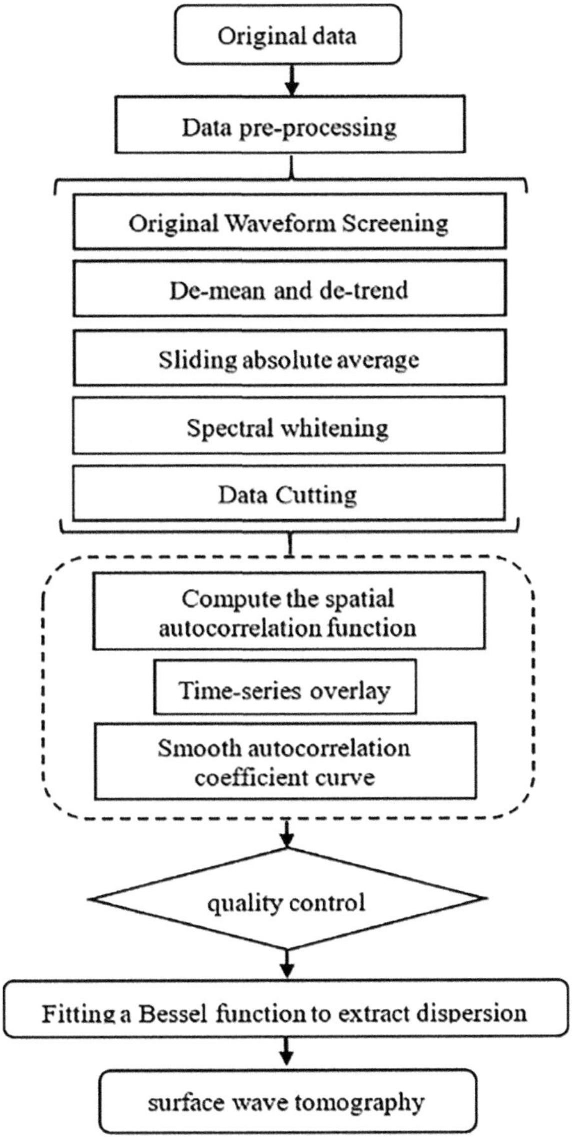

After pre-screening, eligible stations are selected to participate in the pre-processing calculation.

The signals recorded by seismometers contain interferences such as instrumental response, DC offsets, and linear drifts, which can affect the accuracy of seismic analysis. The data selected in this paper is free of instrumental response. Taking the original waveform from a certain station as an example, the processing process is demonstrated as follows: DC offsets and linear drifts are eliminated through demeaning

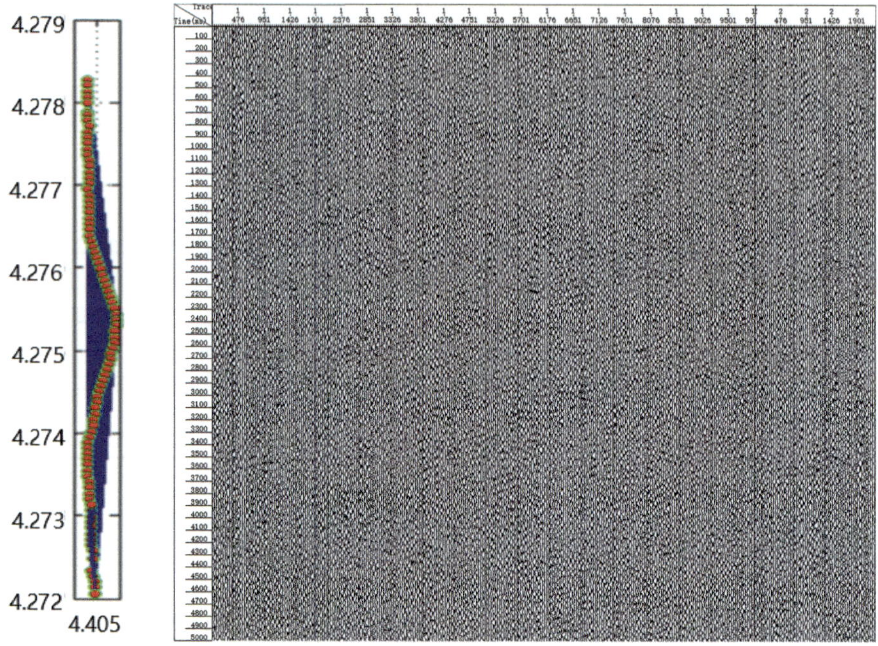

Fig. 3 **a** Spatial distribution map of node positions; **b** Micro motion observation waveforms

($X_{De-mean} = x - \mu$) and detrending ($X_{Detrent} = x - (a \cdot t + b)$), where $a \cdot t + b$ is the fitted linear trend, and t represents time); the moving absolute average method is used to smooth the data, retaining effective signals, reducing noise, and improving the SNR; spectral whitening is applied to balance the energy of each frequency component, enhance weak high-frequency signals, and improve data resolution. The processing results of each step are shown in Fig. 4a–f, respectively.

3.2 Calculation of Spatial Autocorrelation Coefficient

When calculating noise files, it is necessary to ensure that the time series lengths of the two stations are consistent. Direct calculation of long-duration complete files occupies substantial memory and reduces efficiency. Consequently, segmented processing is implemented: the station data are divided equally by time windows, 2 min noise data are taken each time for sliding absolute value averaging, the spatial autocorrelation coefficients between stations are calculated, and then the long-term autocorrelation coefficients are superimposed. This method can reduce memory usage and improve efficiency.

For the calculation of autocorrelation coefficients between stations, first, a 2-min window is used for segmented calculation, and then the correlation coefficients of

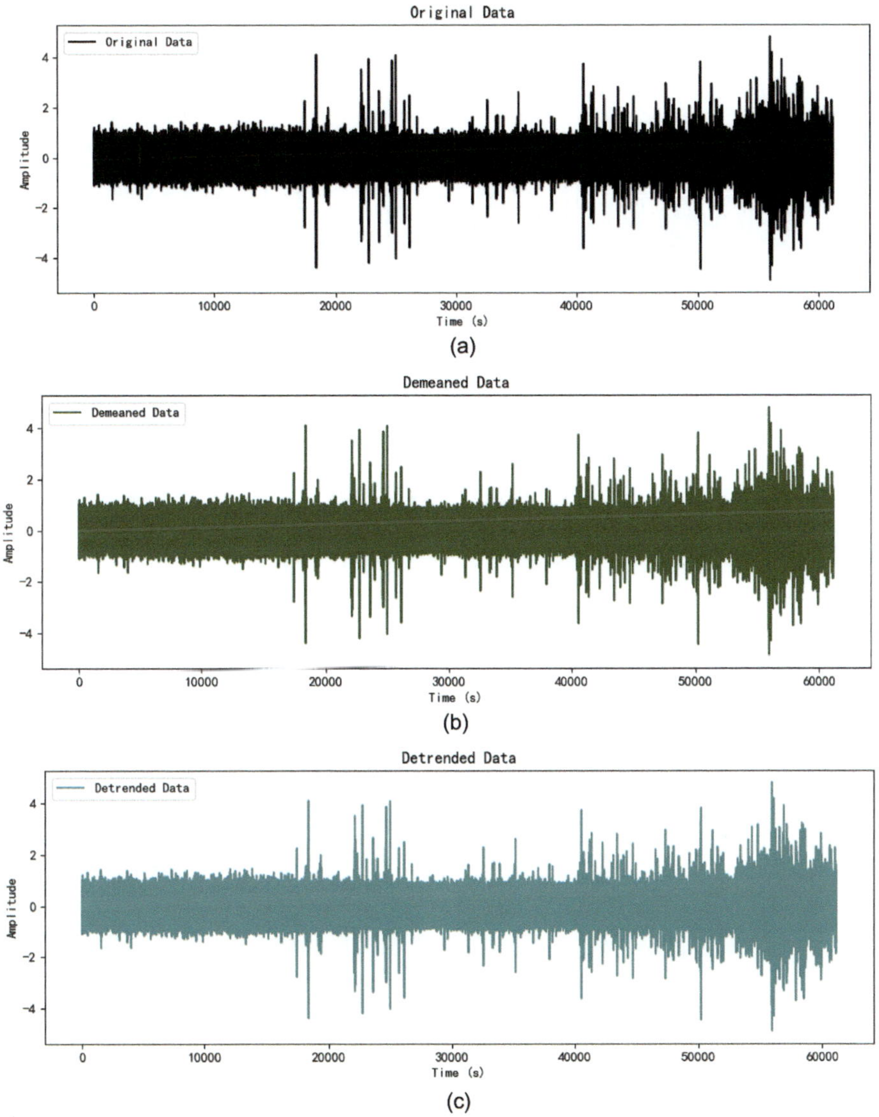

Fig. 4 **a** Original data waveform; **b** Waveform of De-meaned data; **c** De-trending data waveform; **d** Sliding absolute average data waveform; **e** Sliding absolute average versus raw data waveforms; **f** Spectrum whitening data waveform

all windows are superimposed to obtain the final result, forming a spatial autocorrelation coefficient curve (Fig. 5). To facilitate subsequent extraction work, the curve is smoothed as shown in Fig. 6. The same method can be applied to process curves of other stations with different spacings, providing high-quality data for subsequent research.

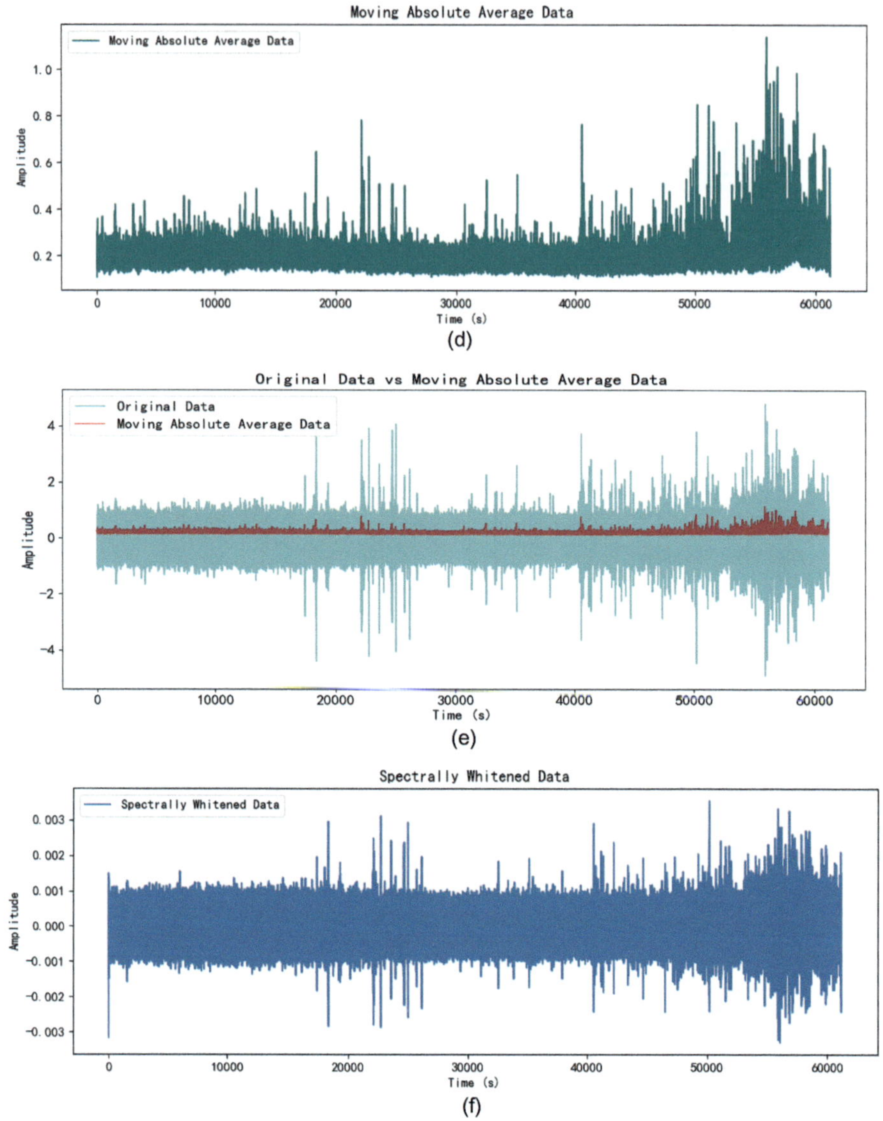

Fig. 4 (continued)

3.3 *Dispersion Curve Extraction*

After obtaining the autocorrelation coefficient, the dispersion curve can be extracted according to Bessel function of the first kind of order zero. As can be seen from Eq. (7), when r is a fixed value, ρ(r,f) is a function of frequency, and at this time, the first to judge the range of resolvable frequency band; the selection of this range is mainly

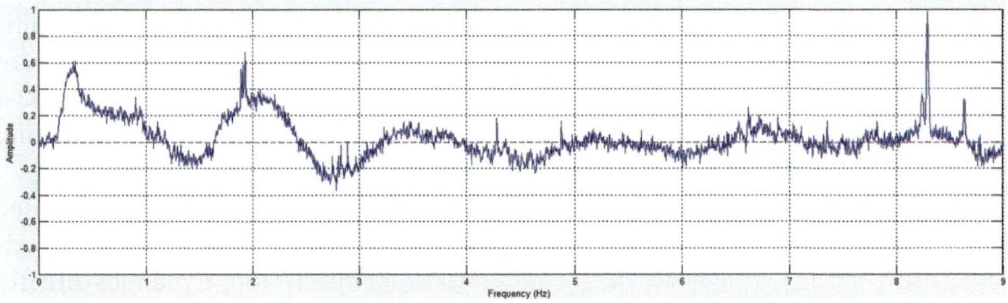

Fig. 5 Plot of spatial autocorrelation coefficient between two stations

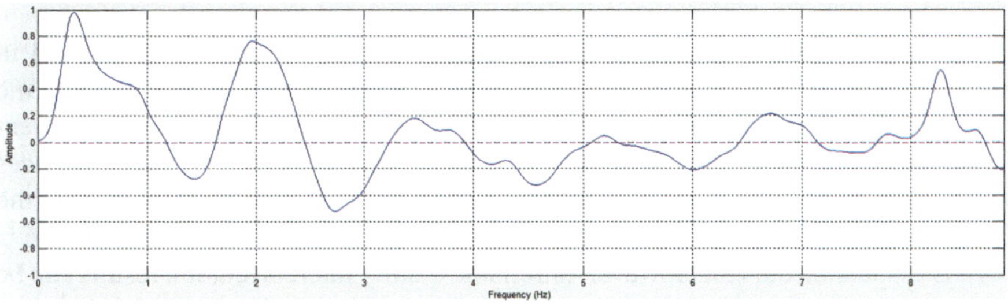

Fig. 6 Plot of spatial autocorrelation coefficient between two stations after smoothing

based on the spatial autocorrelation coefficient curve and the Bessel curve similar to the portion of the judgment, and due to the attenuation of the high-frequency signals resulting in the extraction of the effect of the deterioration of the method according to the method of the current test can be resolved in a range of frequency bands is about 1–9 Hz, as shown in Fig. 7, by fitting the spatial autocorrelation coefficient in the effective frequency band to the corresponding section of the standard Bessel curve, the dispersion curve in this frequency band (Fig. 8) can be obtained.

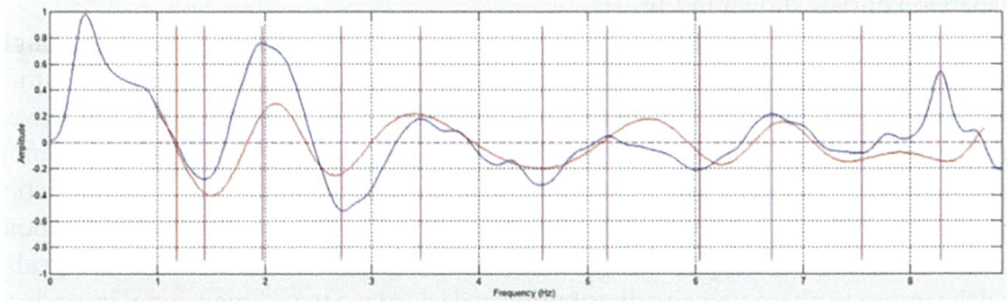

Fig. 7 Fitting of the spatial autocorrelation coefficient to the first-kind Bessel function; the blue curve represents the calculated spatial autocorrelation curve, and the red curve denotes the theoretical spatial autocorrelation curve (first-kind Bessel function)

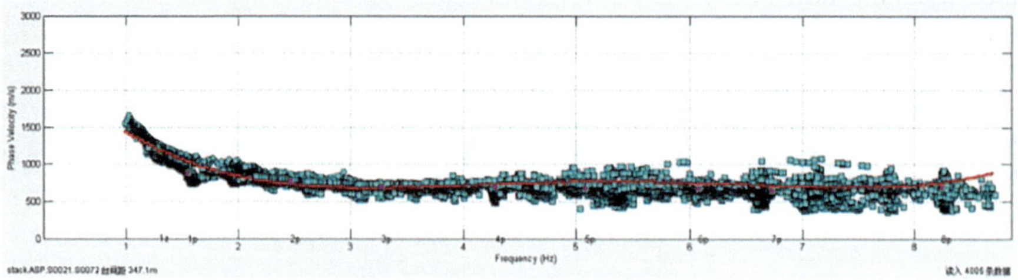

Fig. 8 The observed dispersion points and their polynomial fitting curve, the blue scatter points are all extracted surface wave dispersion curve points, while the red discontinuous line is the average value of all dispersion curves (2853 bars)

The paper utilize a similar dispersion curve extraction method to that used in Aki's article, using special points. The special points used here are not just the extreme value points, zeros, and minima of one spatial autocorrelation coefficient curve, but points on the spatial autocorrelation coefficient curves at many distances between different stations are used, so that more points can be found to constrain the dispersion curves. Theoretically, for 90 stations, it should be possible to extract $C_{90}^{2} = 4005$ dispersion curves. However, unreliable quality intercorrelation results can be presented on all the paths. To guarantee the trustworthiness of stratigraphic imaging results, we screened these dispersion curves based on the SNR test and the station spacing and extracted 1400 dispersion curves and 2800 dispersion curves for the comparison of the effects, respectively. The calculation steps are as follows: ① Find the values of the independent variables at the zero, maximum and minimum points of the zeroth-order Bessel function, which are denoted as x; ② Find the station pair spacing r and frequency f corresponding to the zeros and maximum and minimum points of the spatial autocorrelation coefficients averaged azimuthally for different station pairs of spacings. At different frequency points can be obtained by substituting x, r, and f into the above equation. Connecting the frequency-point phase velocities extracted from the same autocorrelation curve, the dispersion curve graph shown in Fig. 9 can be obtained. Then, polynomial interpolation is used to obtain continuous dispersion curves shown in Fig. 10.

Rayleigh waves of different frequencies propagate at varying speeds through different strata, causing a pulse wave to gradually transform into a wave train over time, a phenomenon known as dispersion [28]. Low-frequency surface waves penetrate deeper and reflect the structure of deeper layers, while high-frequency surface waves can better reflect shallow structures. Leveraging this characteristic, the extracted dispersion curves can be used to analyze the subsurface geological structures at different depths [28]. Firstly, the inversion area is divided into grids, and the group or phase velocity distribution is derived via two-dimensional inversion; subsequent inversion of their dispersion curves [29], yields the apparent surface-wave velocity structure model of the subsurface medium beneath the array.

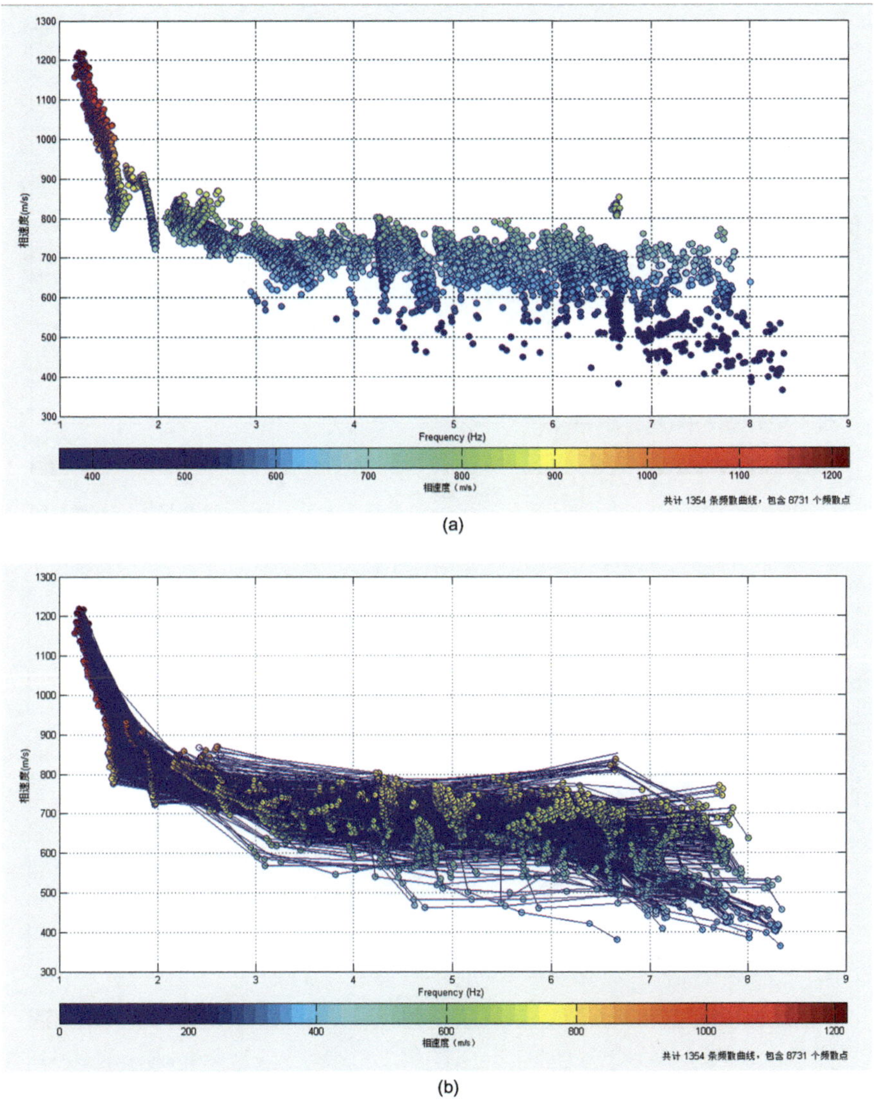

Fig. 9 Comparison between original surface wave dispersion points and dispersion curves

The ESPAC-based signal processing pipeline proceeds as follows: raw ambient noise data undergo preprocessing including demeaning, detrending, and spectral whitening; subsequent segmented windowing is applied to calculate spatial autocorrelation coefficients between station pairs, with long-term coefficients obtained via superposition; the resulting autocorrelation curves are smoothed to reduce noise interference; dispersion curves are then extracted by fitting smoothed curves to the zeroth-order Bessel function within the resolvable frequency band; quality control involving

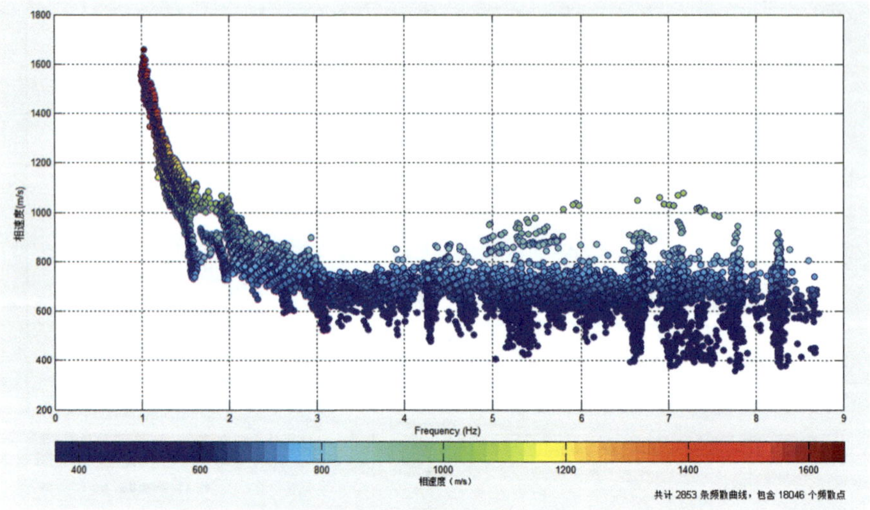

(c)

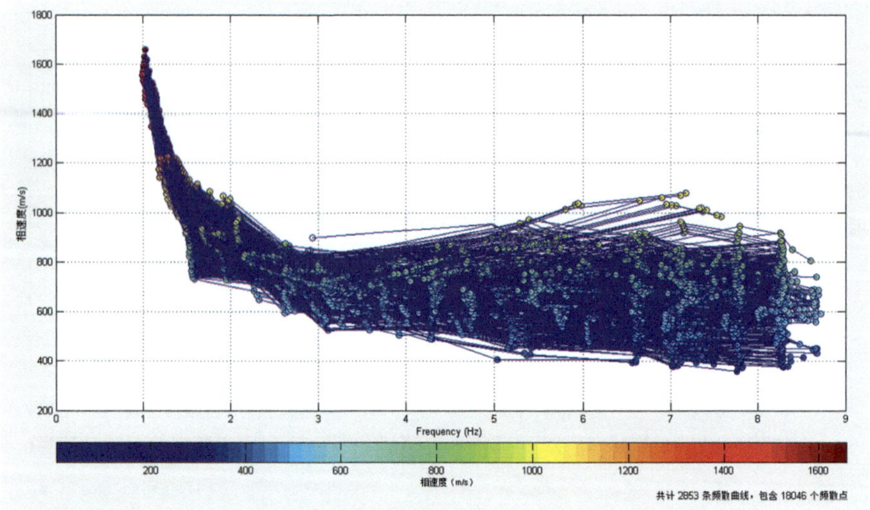

(d)

Fig. 9 (continued)

SNR checks and validation against theoretical Bessel curves ensures only reliable dispersion picks are retained, where windowing maintains computational efficiency without losing temporal resolution, smoothing enhances curve fitting accuracy, and quality control minimizes errors in final dispersion data.

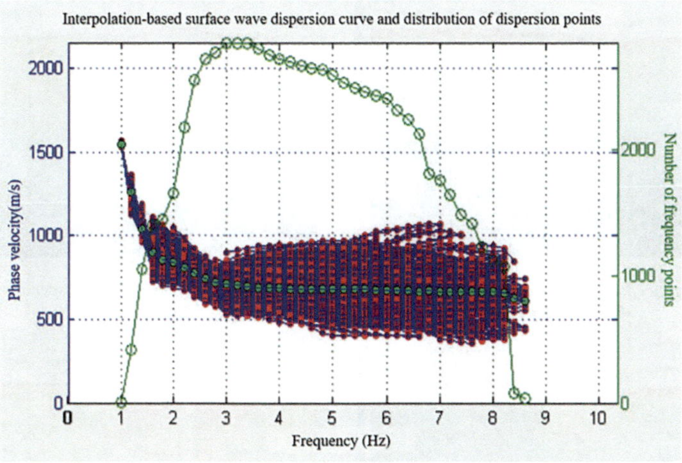

Fig. 10 The continuous dispersion curve image is obtained using the polynomial interpolation method

4 Results and Discussion

In the actual imaging operation, the number of dispersion curves extracted and the different damping parameter settings will have an impact on the imaging results. In the paper, we extracted 1354 dispersion curves and 2853 dispersion curves, respectively, and adopted four different damping parameters of 0.20, 0.15, 0.10, and 0.07 for the comparison of the imaging results and the underground transverse wave velocity structure imaging comparison diagrams shown in Fig. 11 were obtained.

As can be seen from the above figure, the shallow part of the surface is mainly characterized by a low-speed layer, which gradually transitions to a medium- and high-speed layer, with local low-speed anomalies (areas marked by red circles). In the range of 50 m to 150 m below the survey line, there are several obvious low-speed anomalies, some of which show a wide funnel-like pattern from the shallow layer to the deep layer. It is assumed that these low-velocity anomalies are caused by the underground mining triggered stratum collapse and rock fragmentation, forming the trap column, and then led to the subsidence of each velocity layer, and the velocity value of local strata was significantly reduced, which resulted in the comprehensive low-velocity effect. In addition, there is a clearly defined anomalous low-velocity circle near a depth of 200 m, which is inferred to be the underground mining area formed by the crushed rocks and cavities retained after mining. Comparison shows that increasing dispersion curves can significantly improve the resolution of the inversion model, precisely characterize changes in subsurface structures, and enhance the reliability and stability of results, but it will increase computational load. Larger damping parameters enhance model stability at the expense of resolution and fitting accuracy. Conversely, smaller parameters increase resolution and accuracy but heighten susceptibility to noise interference. In practical applications, it is necessary

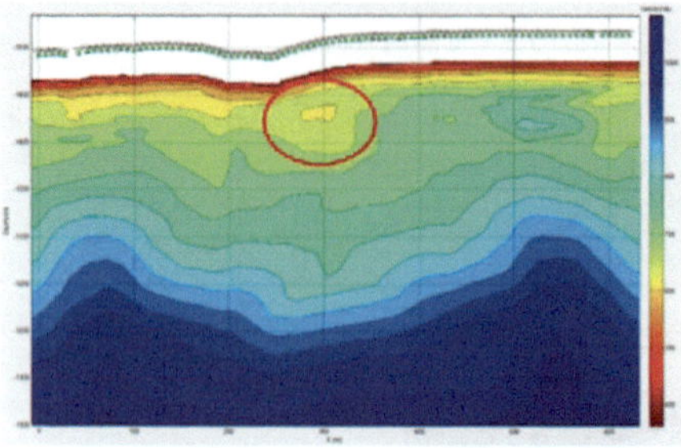

(a) The imaging effect of 1,354 dispersion curves at a damping of 0.20

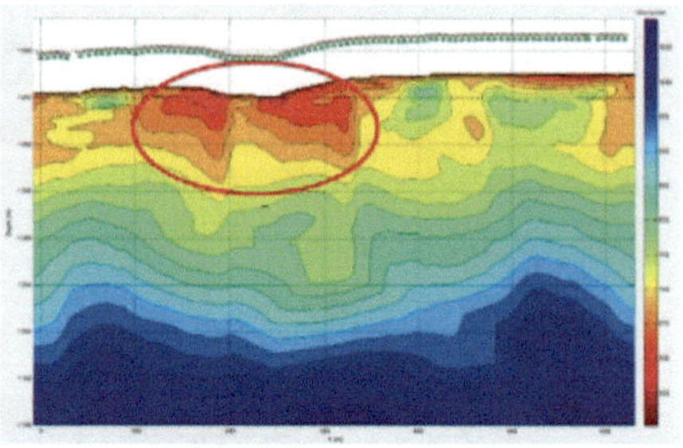

(b) The imaging effect of 2,853 dispersion curves at a damping of 0.20

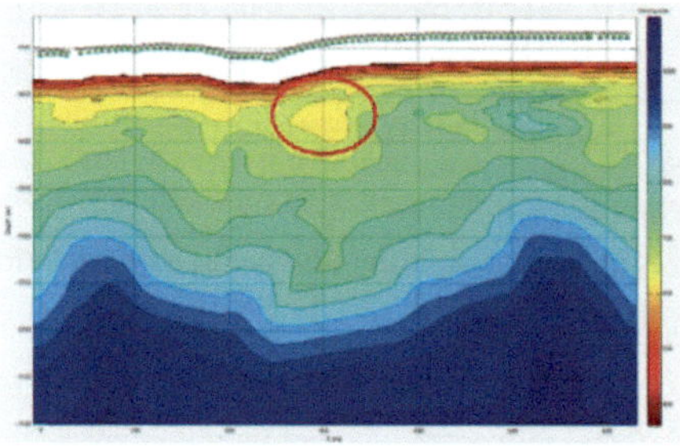

(a) The imaging effect of 1,354 dispersion curves at a damping of 0.15

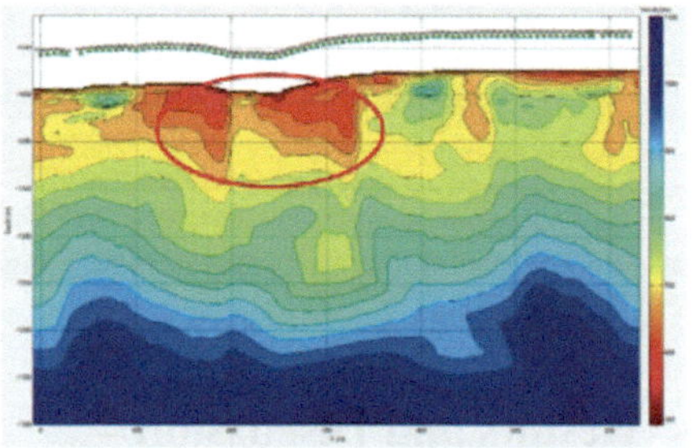

(b) The imaging effect of 2,853 dispersion curves at a damping of 0.15

Fig. 11 Imaging comparison of 1354 and 2853 dispersion curves under varying damping parameters

to balance data volume, computational efficiency, and result stability according to specific needs [30–35].

This study presents a 2D microtremor tomography detection method based on the ESPAC technique. Through a human–computer interaction model, data processing procedures such as surface wave dispersion curve extraction, discrete dispersion point fitting, and quality evaluation of microtremor data between any two stations are completed to fully exploit data information and obtain the lowest possible frequency surface waves. During data processing, the theoretical spatial autocorrelation function waveform is referenced to assist in picking dispersion curves, improving the quality and efficiency of dispersion curve extraction. Real-time generation of dispersion curves and constraints from prior information enables timely data quality control to eliminate bad data points and those outside the detection frequency range, retaining high-quality dispersion data. The obtained surface wave dispersion curve data are interpolated using polynomials to obtain continuous dispersion curves, laying the foundation for inversion imaging of subsurface Vs structures. Grounded on the Rayleigh surface wave dispersion curves obtained at each microtremor measurement point, the apparent velocity structure varying with depth is calculated from the phase velocity dispersion curves to complete tomographic inversion of data from linear and approximately linear microtremor arrays. This directly yields a 2D structural model

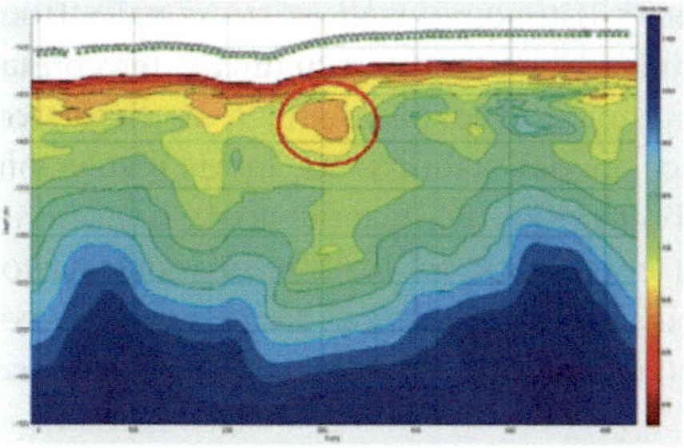

(a) The imaging effect of 1,354 dispersion curves at a damping of 0.10

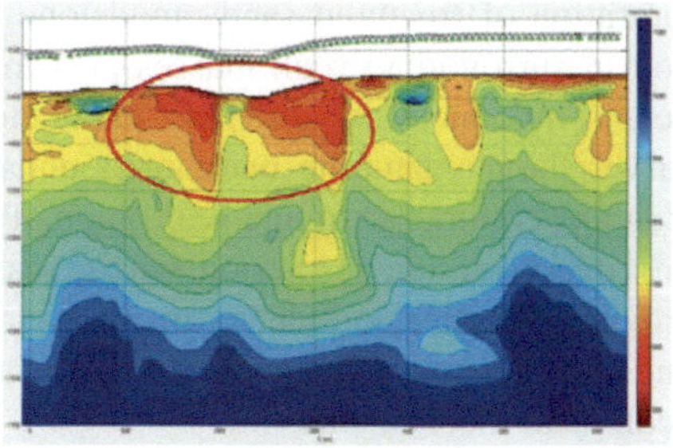

(b) The imaging effect of 2,853dispersion curves at a damping of 0.10

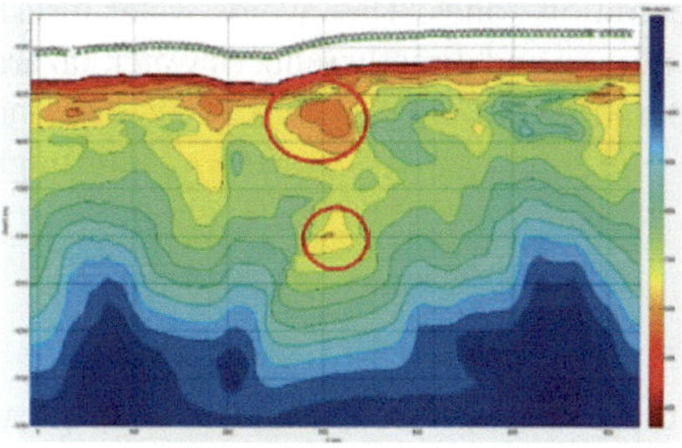

(a) The imaging effect of 1,354 dispersion curves at a damping of 0.07

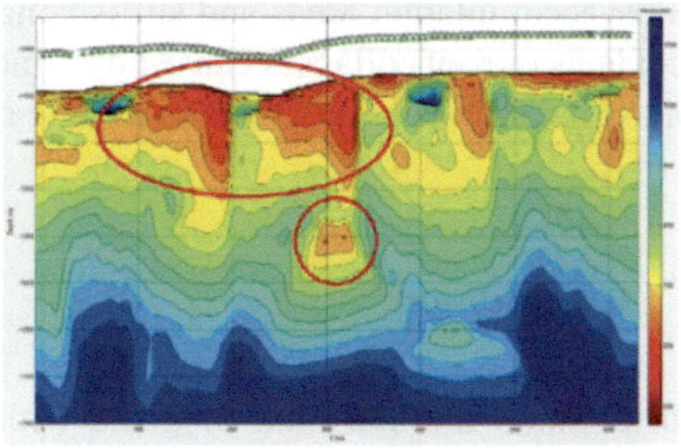

(b) The imaging effect of 2,853 dispersion curves at a damping of 0.07

Fig. 11 (continued)

below the survey line, with topographic undulations considered in the imaging results to achieve seamless, scientifically reasonable interpretation that ensures resolution.

In the Talahao Coal Mine, the roof and floor rocks of coal seams are predominantly sandy mudstone and fine-grained sandstone, followed by siltstone. These rocks have extremely low compressive strength, with even lower shear and tensile strengths. The compressive strength of sandy mudstone decreases significantly when saturated, and most rocks soften, deform, or even disintegrate upon contact with water. By analyzing surface wave signals in ambient noise, the spatial location, morphological scale, and burial depth of goafs can be non-invasively inverted. Compared with traditional exploration methods, since the apparent shear wave velocity is typically sensitive to changes in rock mass density, the profile can naturally and reliably reflect variations in rock properties. This method can more finely reveal goaf boundaries, roof thickness, and surrounding rock integrity. By comparing imaging results under different numbers of dispersion curves and different damping parameters, microtremor velocity anomalies in the goaf roof strata are identified.

Observed low-velocity anomalies are linked to specific geological mechanisms: reduced velocities in goaf-related zones stem from rock fracturing and fragmentation induced by mining, while further velocity reductions may result from water saturation of fractured sandy mudstone and siltstone, as well as stress redistribution in surrounding rock masses post-mining. Uncertainty in anomaly interpretation is influenced by damping parameter choices and the number of dispersion curves: larger damping parameters enhance model stability but lower resolution, reducing confidence in fine-scale anomaly boundaries, while more dispersion curves improve inversion resolution but increase computational load, with more curves providing higher confidence in low-velocity zone delineation compared to fewer ones; such considerations support the reliability of inferred goaf locations and geological mechanisms.

Potential goaf locations are identified on the basis of the distribution features of low Vs zones. This determination provides data support for assessing potential goaf water accumulation areas and stress concentration zones. This avoids water inrush accidents or roof collapses caused by blind mining and directly reduces water hazard and structural instability risks in mining operations.

5 Conclusion

To facilitate the detection of goafs within the Talahao Coal Mine, a near-linear array was deployed above the mining site. By applying spatial autocorrelation analysis of background noise, true two-dimensional tomographic imaging beneath the survey line was successfully realized. The inclusion of topographic undulations in the modeling process enables more intuitive identification and analysis of subsurface anomalous zones. This approach differs significantly from conventional microtremor measurement techniques. Specifically, the ESPAC method adopted in this study exhibits resistance to electromagnetic interference, greater flexibility in array deployment, and broader applicability across various geological conditions.

Furthermore, the implementation of a human–computer interaction approach for surface wave dispersion extraction improves the overall quality of the data, reduces on-site operational workload, and enhances processing efficiency. This approach eliminates errors associated with the stitching of one-dimensional models, thereby preserving resolution and improving the reliability of the results.

With continued advancements in exploration technologies, integrated multi-method geophysical exploration has emerged as a prominent trend. For instance, the combination of ambient noise spatial autocorrelation imaging with reflection seismology allows for the complementary strengths of both techniques to be fully utilized. Reflection seismology offers high-resolution detection of subsurface formation interfaces, while ambient noise spatial autocorrelation imaging provides distinct advantages in identifying subsurface Vs structures. The synergistic integration of these methods enables a more comprehensive and accurate characterization of subsurface geological formations. Such integration delivers reliable geological information to support applications in geological engineering, resource exploration, and mining safety assessment.

We express our gratitude to the teachers of Inner Mongolia Regional Seismological Bureau and Inner Mongolia University for their patient guidance.

Funding: This study received financial support from the Natural Science Foundation of Inner Mongolia Autonomous Region, China (Grant No. 2024MS04028), with the Article Processing Charge (APC) also funded by the same foundation.

References

1. Liu QH, Li LY, He ZQ, Hu G, Wang KM, Gong Y (2016) Inversion of S-wave velocity structure near the surface by spatial autocorrelation technique of microtremors. Acta Seismol Sin 38(1):86–95
2. Lin F-C, Moschetti MP, Ritzwoller MH (2008) Surface wave tomography of the western United States from ambient seismic noise: Rayleigh and Love wave phase velocity maps. Geophys J Int 173(1):281–298
3. Li LL, Wang WT, Zhu LB, Chen HP, Wang QD, Miao P, Wang J (2014) Processing seismic ambient noise data to obtain reliable surface wave dispersion measurements. Earthq 34(3):108–116
4. Aki K (1957) Space and time spectra of stationary stochastic waves, with special reference to microtremors. Bull Earthq Res Inst 35:415–456
5. Gao J, Zhang HJ, Zha HS et al (2023) The effect of different arrays and noise source distribution on microtremor imaging and its application in solute mine cavity detection. Chinese J Geophys 66(6):2498–2506
6. Lu LY (2021) Revisiting the cross-correlation and spatial autocorrelation (SPAC) of the seismic ambient noise based on the plane wave model. Rev Geophys Planetary Phys 52(2):123–163
7. Okada H (2003) The microtremor survey method; Society of exploration geophysicists, geophysical monograph series n. 12; Society of Exploration Geophysicists, Tulsa, OK, USA, p 135
8. Wang KD, Qian JW, Zhang HJ, Gao J, Bi DC, Gu N (2022) Seismic imaging of mine tunnels by ambient noise along linear arrays. J Appl Geophys 203:104718
9. Chávez-García FJ, Rodríguez M, Stephenson WR (2005) An alternative approach to the SPAC analysis of microtremors: exploiting stationarity of noise. Bull Seismol Soc Am 95(1):277–293
10. Shapiro NM et al (2005) High-resolution surface-wave tomography from ambient seismic noise. Science 307:1615–1618
11. He HB, Li QH, Fan XP (2021) Research on S-wave velocity structure of shallow stratum with stripping method from array records of microtremors. J Disaster Prevention Mitigation Eng 30(1):103–108
12. Xu PF, Li SH, Du JG et al (2013) Microtremor survey method: a new geophysical method for dividing strata and detecting buried fault structures. Acta Petrol Sin 29(5):1841–1845
13. Xu PF, Shi W, Ling SQ et al (2012) Mapping spherically weathered "Boulders" using 2D microtremor profiling method: a case study along subway line 7 in Shenzhen. Chinese J Geophys (in Chinese) 55(6):2120–2128
14. Xu P, Ling S, Long G, Qiao G, Shen Q, Yao J, Zhang H (2021) ESPAC-based 2D mini-array microtremor method and its application in urban rail transit construction planning. Tunnelling Underground Space Technol 115:104070. https://doi.org/10.1016/j.tust.2021.104070
15. Chen Y, Saygin E, Kennett B et al (2023) Next-generation seismic model of the Australian crust from synchronous and asynchronous ambient noise imaging. Nat Commun 14:1192
16. Du P, Wu J, Li Y, Wang J, Han C, Lindsay MD, Yuan H, Zhao L, Xiao W (2020) Imaging Karatungk Cu-Ni Mine in Xinjiang, Western China with a passive seismic array. Minerals 10:601. https://doi.org/10.3390/min10070602

17. Guo W (2025) Collaborative efforts among government, universities, and enterprises: Peking University's School of Earth and space sciences aids in solving mineral resource exploration challenges. Guangming Online. Available at: https://edu.gmw.cn/2025-04/23/content_37984506.htm. Accessed on 18 May 2025
18. Gu N, Chamarczuk M, Gao J, Malinowski M, Zhang H (2021) Passive seismic structure imaging of a coal mine by ambient noise seismic interferometry on a dense array. Acta Geol Sin-Engl Ed 95:37–39. https://doi.org/10.1111/1755-6724.14826
19. Cao AY, Dou LM, Cai W, Gong SY, Liu S, Jing GC (2015) Case study of seismic hazard assessment in underground coal mining using passive tomography. Int J Rock Mech Min Sci 78:1–9
20. Yan J, Zhang LS, Hong HT, Huang QC, Lei YH, Liao XL, Gui LB, Yang X (2023) Application of ambient noise and dense seismic array imaging techniques in goaf detection beneath coal mines at Haerwusu. CT Theory Appl Res 32(04):461–470. https://doi.org/10.15953/j.ctta.2023.023
21. Li KP, Shao J (2018) The application of comprehensive exploration methods in the exploration of gypsum mine goafs. Resource Inf Eng 33(3):47–48
22. Zhou X, Wang WJ, Li Y et al (2021) Research and application of microtremor in the geological disaster. Periodical Ocean Univ China 51(8):58–64
23. Zhou Q (2021) Research on microseismic exploration technology based on seismic interferometry. Master's Thesis, Wuhan University, Wuhan, China
24. Mi B, Xia J, Shen C, Wang L, Hu Y, Cheng F (2017) Horizontal resolution of multichannel analysis of surface waves. Geophysics 82:EN51-EN66. https://doi.org/10.1190/geo2016-0202.1
25. Mi B, Xia J, Tian G, Shi Z, Xing H, Chang X, Xi C, Liu Y, Ning L, Dai T et al (2022) Near-surface imaging from traffic-induced surface waves with dense linear arrays: an application in the urban area of Hangzhou, China. Geophysics 87:B145–B158. https://doi.org/10.1190/geo2021-0184.1
26. Li HL (2020) Two-dimensional shear wave velocity structure of the shallow crust studied by the microseismic survey. Comput Techniques Geophys Geochem Exploration 42(5):563–568
27. Li N (2017) Study on the crustal and upper mantle velocity structure beneath the Northern Section of the South-North seismic zone using surface wave imaging method based on dense array. Doctoral Dissertation, Institute of Geophysics, China Earthquake Administration, Beijing, China
28. Li QL, Lei XD, Zhu L, Li C, Chen BL (2022) Study on the abnormal response features of underground space based on microtremor survey. Geotechn Investigation Surveying 1:72–78
29. Li JP, Zhong WX, Li HY, Li XF et al (2015) Application research on velocity structure of surface medium and its changes in the Chinese mainland using ambient noise. China Earthq Eng J 37(1):235–241
30. Guo J, Wang Z, Li H, Yang Y, Huang CG, Yazdi M, Kang HS (2024) A hybrid prognosis scheme for rolling bearings based on a novel health indicator and nonlinear Wiener process. Reliab Eng Syst Saf 245:110014
31. Li H, Yazdi M, Huang HZ, Huang CG, Peng W, Nedjati A, Adesina KA (2023) A fuzzy rough copula Bayesian network model for solving complex hospital service quality assessment. Complex Intell Syst 9(5):5527–5553
32. Li H, Peng W, Adumene S, Yazdi M (2023) An intelligent cost-based consequence model for offshore systems in harsh environments. In: Intelligent reliability and maintainability of energy infrastructure assets. Springer Nature Switzerland, Cham, pp 107–117
33. Li H, Yazdi M (2022) What are the critical well-drilling blowouts barriers? A progressive DEMATEL-game theory. Advanced decision-making methods and applications in system safety and reliability problems: approaches, case studies, multi-criteria decision-making, multi-objective decision-making, fuzzy risk-based models. Springer International Publishing, Cham, pp 29–46

34. Li H, Peng W, Adumene S, Yazdi M (2023) Cutting edge research topics on system safety, reliability, maintainability, and resilience of energy-critical infrastructures. In: Intelligent reliability and maintainability of energy infrastructure assets. Springer Nature Switzerland, Cham, pp 25–38
35. Li H, Peng W, Adumene S, Yazdi M (2023) Advances in failure prediction of subsea components considering complex dependencies. In: Intelligent reliability and maintainability of energy infrastructure assets. Springer Nature Switzerland, Cham, pp 93–105

A Study on Fault Prediction of Simulated Die Press Machine Based on Machine Learning

Jun Feng Lai, Hui Yu, Yue Zhao, and Jia Liu

Abstract Reliability and fault prediction are essential for enhancing the operational performance of simulated tablet presses. In this study, multiple machine learning algorithms are applied to fault prediction to improve both prediction accuracy and computational efficiency. The algorithms employed include K-nearest neighbor classifiers, support vector machine classifiers, Naive Bayes classifiers, Gaussian process classifiers, a parameter-tuned Random Forest model, and a SMOTE-BO-XGBoost ensemble model. A comprehensive evaluation is conducted using multi-dimensional performance metrics, including accuracy, precision, recall, F1 score, and first error rate, to compare the effectiveness of the models. The results indicate that the parameter-tuned Random Forest model achieves superior performance in terms of both accuracy and F1 score, demonstrating that parameter optimization plays a significant role in enhancing model performance. In addition, the SMOTE-BO-XGBoost ensemble model exhibits excellent performance in minimizing the first error rate, underscoring its robustness and predictive accuracy. These findings confirm the feasibility and reliability of applying machine learning techniques to fault prediction for simulated tablet presses, and they offer valuable guidance for the selection and optimization of predictive algorithms in future applications.

Keywords Machine fault · Reliability · Integrated model · Predictive model

J. F. Lai · H. Yu
School of Science, Inner Mongolia University of Technology, Hohhot, China

Y. Zhao
School of Forestry, Inner Mongolia Agricultural University, Hohhot, China

J. Liu (✉)
School of Aeronautics, Inner Mongolia University of Technology, Hohhot, China
e-mail: liujiajing1979@163.com

H. Li et al. (eds.), *Data-Driven Methods for Reliability and Safety Engineering: Applications in Industrial Systems*, Springer Series in Reliability Engineering,
https://doi.org/10.1007/978-3-032-22873-4_22

1 Introduction

Reliability has become a critical indicator in modern industrial production because of its direct influence on equipment performance and product quality. With the rapid advancement of industrial automation, the increasing complexity of mechanical systems and the variety of operating environments have imposed more rigorous requirements on equipment reliability. Tablet presses, which are essential industrial machines widely used in high-precision manufacturing sectors such as automotive, aerospace, electronics, and medical industries, play a vital role in ensuring product consistency and dimensional accuracy. Even minor faults in these systems can lead to serious consequences, including production delays, deterioration in product quality, higher maintenance costs, and disruptions in the supply chain. Therefore, investigating fault prediction methods for simulated tablet presses is of great importance for improving equipment reliability and production efficiency, and it serves as a representative case study in the broader field of mechanical fault prediction.

Due to the nonlinear and time-varying nature of tablet press faults, issues such as mechanical wear and control system irregularities often occur in complex combinations. Traditional approaches for analyzing multisource sensor data, including vibration, temperature, and pressure signals, generally result in low predictive accuracy and delayed fault detection. These challenges highlight the necessity of developing more advanced fault prediction techniques for tablet press applications.

In recent years, machine learning has achieved notable progress in the area of fault prediction. Data-driven methods such as artificial neural networks, support vector machines, and deep learning have shown effectiveness in predictive maintenance and early fault detection by analyzing operational data to identify fault patterns and anticipate potential failures. These approaches have been widely adopted for estimating the remaining service life of complex mechanical systems and for diagnosing faults. Considering the characteristics of tablet press operational data and the limitations of traditional methods, machine learning offers a promising solution for fault prediction in this context. Its application not only bridges existing technical gaps but also lays the groundwork for innovative strategies aimed at improving the reliability of similar industrial equipment.

1.1 The Applications of Machine Learning Technologies in the Fields of Fault Prediction and Reliability Analysis

Wang [1] demonstrated the effectiveness of machine learning models in improving prediction accuracy and reducing both equipment downtime and maintenance costs through case studies involving various data-driven approaches. These findings highlighted the potential of data-driven methodologies for enhancing the reliability analysis of future power equipment. Wang et al. [2] applied ten machine learning classification algorithms to identify rock types based on composite directional drilling parameters and evaluated the performance of each model. The results indicated that Random Forest, AdaBoost, CatBoost, K-nearest neighbor, support vector machine, and ExtraTree classifiers performed effectively, with AdaBoost achieving the highest overall performance. These outcomes confirmed the feasibility and reliability of machine learning techniques in rock type identification. Ma et al. [3] used exhaustive feature selection on annealing parameter datasets and combined cross-validation with grid search for hyperparameter optimization. They compared several machine learning models and found that the support vector regression model performed best in predicting pull-off force, while the random forest model excelled in predicting elongation force. These results laid a foundation for enhancing the reliability of bonding wires. Yu et al. [4] addressed the class imbalance and accuracy issues in software defect prediction by proposing a random forest model combined with SMOTE and the edited nearest neighbor algorithm. When compared with other popular algorithms, this model demonstrated high efficiency in handling class imbalance while avoiding overfitting. Liu et al. [5] developed a machine learning-based fault prediction model for building machines. They collected operational parameters and information to build a multidimensional and multi-layer dataset. By using algorithms like random forests, support vector machines, and recurrent neural networks for in-depth data analysis, they identified feature patterns related to faults. Cross-validation and hyperparameter tuning ensured the model's generalization and prediction accuracy. Experiments showed that the model could predict fault types and locations in advance, providing early warning for maintenance personnel and effectively reducing fault frequency. Wang et al. [6] constructed a model for predicting the strength of gently dipping ore pillars using support vector machines (SVM), extreme learning machines (ELM), and LightGBM. They optimized the model with genetic programming (GP) and an improved quantum particle swarm algorithm (IQPSO), establishing a nonlinear mapping between the strength and its influencing factors. The results showed that the model's predictive performance improved after optimization with GP and IQPSO. Comparisons with classical ore pillar strength theories confirmed the model's accuracy, offering a new approach for predicting the strength of gently dipping ore pillars. Wang et al. [7] developed a power system operational reliability prediction model using a continuous collaborative machine learning algorithm. They first established a mechanism for precise prediction of power system operational reliability. Then, they created a mechanism for the precise prediction of core elements in power system operational reliability under chronological order. Finally, they utilized a power system operational reliability prediction function to output optimal prediction results. Verification results indicate that the model meets the intelligent transformation requirements for power system operational reliability prediction and significantly optimizes the intelligent and controllable perception mechanism for power system operational reliability prediction. Chen et al. [8] introduced a BP-Adaboost ensemble learning algorithm and proposed an accurate, efficient, and simple method for analyzing the overall reliability of transmission tower-line structures. They first established a finite element analysis model of transmission tower-line structures, analyzed the failure criteria, and derived the corresponding limit state function. Then, they combined the BP-Adaboost ensemble learning algorithm with the Monte Carlo simulation method to build a high-precision prediction model to replace the complex finite element model. This led to a new method for analyzing the overall reliability of transmission tower-line structures based on the BP-Adaboost model. Finally, through numerical examples, they verified and applied the proposed method, which demonstrated higher accuracy and efficiency. These studies highlight the broad applications and significant advantages of machine learning technologies in enhancing industrial equipment reliability and reducing fault-related downtime and maintenance costs.

1.2 Classification and Comparison of Machine Learning Algorithms

Machine learning algorithms can be categorized into supervised learning, unsupervised learning, semi-supervised learning, and reinforcement learning. Supervised learning, which includes models like random forests and support vector machines, is widely used for fault classification and regression analysis. Unsupervised learning is suitable for discovering fault patterns in unlabeled data. Semi-supervised learning optimizes the learning process by using limited labeled data. Reinforcement learning demonstrates good adaptability and decision-making capabilities in dynamic environments. Zhang et al. [9] proposed an intelligent monitoring system based on reinforcement learning to address the insufficient real-time and low efficiency of traditional power distribution network electrical equipment monitoring methods. The system, through multi-source heterogeneous data perception, feature extraction, deep reinforcement learning decision-making, and intelligent control, enables online equipment condition assessment, fault early warning, and operation and maintenance optimization. Experimental results indicate that this system surpasses traditional methods in equipment fault prediction accuracy, abnormal diagnosis timeliness, and operation and maintenance strategy reliability, thus strongly supporting the safe, reliable, and economical operation of power distribution networks.

This study aims to combine machine learning technology to explore fault prediction methods for simulated die press machines. Through the construction of machine learning-based fault prediction models, real-time monitoring of equipment operation and early fault warnings can be realized, thus reducing maintenance costs and enhancing equipment reliability and production efficiency.

2 Methodology

The selection of the six machine learning algorithms (K-nearest neighbors, support vector machine, Naive Bayes, Gaussian process classifier, parameter-tuned Random Forest, and SMOTE-BO-XGBoost) is tailored to the characteristics of the simulated tablet press dataset, which includes multi-source sensor features (e.g., pressure, vibration, temperature) with nonlinear relationships and potential class imbalance. K-nearest neighbors is chosen for its non-parametric nature, requiring no prior assumption about data distribution and enabling effective capture of local feature relationships in operational data. Support vector machine is selected for its ability to handle nonlinear data via kernel functions, which aligns with the complex interactions between tablet press operational parameters. Naive Bayes, based on the assumption of feature independence, is included for its computational efficiency, suitable for rapid baseline predictions. Gaussian process classifier is adopted for its capacity to provide probabilistic predictions, supporting preliminary uncertainty assessment. The parameter-tuned Random Forest is selected due to its robustness to overfitting, ability to model feature interactions, and suitability for handling the multimodal features (e.g., pressure, speed) observed in the dataset. The SMOTE-BO-XGBoost ensemble model is chosen to address class imbalance—a key challenge in fault prediction—via SMOTE, while Bayesian optimization optimizes hyperparameters to enhance performance on high-dimensional sensor data, matching the dataset's multi-feature structure.

2.1 K-Nearest Neighbors

The core idea of this method lies in determining the categories to which the K nearest neighbor samples of a sample belong. If most of the K nearest neighbor samples of a given sample belong to a specific category, the sample also belongs to that category. The specific theory is as follows:

The detailed procedure of this experiment is shown in Table 1.

Table 1 Summary table of K-Nearest Neighbors algorithm procedure

Procedure:
Input: Training dataset, test sample, number of neighbors K
[5] **Calculate Distances:** Compute the Euclidean distance d_i between the test sample and each training sample [6] **Select Nearest** K **Neighbors:** Sort the training samples based on and select the K samples with the smallest distances as neighbors [7] **Voting:** Determine the prediction by majority voting among the labels of the K neighbors. The label with the most votes becomes the predicted result **Output:** Return the predicted label

Table 2 Summary table of Support Vector Machine (SVM) algorithm procedure

Procedure:
Input: Training labels **Formulate Optimization Problem** **Introduce Kernel Function:** Gaussian kernel: **Solve Dual Problem:** **Determine Support Vectors** **Calculate Bias Term *b*** **Generate Final Classification Decision Function:**
Output: Classification decision function

2.2 *Support Vector Machine*

The method is based on the VC dimension theory and the structural risk minimization principle, achieving the goal of data classification through a supervised learning approach. In essence, it aims to find an optimal classification boundary. When data cannot be linearly separated in the original space, the method unfolds based on kernel functions to find a trade-off that minimizes both risk and the confidence interval. It then maps the data into a high-dimensional space through nonlinear mapping, enabling the application of linear learning machines to solve nonlinear problems. The specific theory is as follows:

Let be the predicted classification value for the i-th sample, n be the total number of samples, is the Lagrange multiplier obtained from Support Vector Machine (SVM) training, representing the voting weight of the i-th training sample. The larger this value is, the greater the influence of the sample on the classification boundary, is the true class label of the i-th sample., is the kernel function, which measures the similarity between the sample x and the training sample in the high-dimensional space., is the feature indicators, b be the threshold, b is the threshold, which adjusts the position of the classification boundary, and σ is the parameter of the kernel function, which affects the smoothness of the space after sample mapping. The smaller this value is, the more dispersed the distribution of samples in the high-dimensional space, and the more complex the classification boundary will be. The predicted class of a new sample x is finally expressed as:

The essence of is a weighted combination of sample points and support vectors in the high-dimensional space.

The parameters and b are obtained by solving the following equation:

If a Gaussian radial basis function (RBF) is selected as the kernel function for the support vector machine (Table 2).

2.3 *Naive Bayes*

The experiment uses Gaussian Naive Bayes, which combines Bayes' theorem with Gaussian distribution assumptions and feature independence. It predicts whether

Table 3 Summary table of Naive Bayes algorithm procedure

Procedure:
Input: Training dataset **Calculate Prior Probability**: For each class y, compute $P(y)$ as the proportion of samples in class y **Calculate Conditional Probability** **Calculate Posterior Probability:** **Predict Class**
Output: Predicted label

tablet press samples are faulty based on maximum posterior probability. The core process involves estimating parameters from training data and making predictions via maximum posterior probability.

Assuming that feature x_i follows a Gaussian distribution given class y, and features are independent (the "naive" assumption).

The detailed procedure of this experiment is shown in Table 3.

2.4 *Gaussian Process Classifier*

In this experiment, a Gaussian process classifier is used to map the outputs of a Gaussian process to class labels using a probabilistic approach. It is assumed that there exists an underlying latent function that follows a Gaussian process prior. The classifier employs a radial basis function (RBF) kernel, where the length-scale parameter controls the smoothness of the function. Model training is performed by maximising the log marginal likelihood, which represents the joint probability of the observed labels given the inputs, integrating over the latent function. This likelihood consists of two components: the data likelihood and the Gaussian process prior. The posterior distribution is obtained by combining the likelihood with the prior, resulting in an updated belief about the latent function after observing the data. For binary classification, the relationship between the latent function and the observed class labels is modelled using a sigmoid link function, which converts latent values into class probabilities. For a new input sample, the predictive probability is computed by marginalising over the posterior distribution of the latent function, yielding the probability that the input belongs to a particular class.

The detailed procedure of this experiment is shown in Table 4.

2.5 *Random Forest*

The method is composed of multiple decision trees. First, sample subsets are randomly drawn with replacement from the training set for training each decision

Table 4 Summary table of Gaussian process classification algorithm procedure

Procedure:
Input: Training dataset **Define Gaussian Process:** **Log Marginal Likelihood:** **Posterior Probability::** **Sigmoid Function:** **Predict Class Probability:**
Output: Predicted class probability for the test sample

Table 5 Summary table of Random Forest algorithm procedure

Procedure:
Input: Training dataset, where are labels, parameter grid param_grid **Split Dataset:** Divide into training set (80%) and test set (20%) **Initialize Random Forest Model:** Set up the random forest model with initial parameters **Tune Parameters:** Define the parameter grid param_grid. Use grid search to traverse parameter combinations, evaluating each with fivefold cross-validation to obtain the best parameter set best_params
Output: Best parameter combination best_params and the trained random forest model

tree, where each tree is constructed by recursively partitioning the feature space. Second, nodes of the tree need to randomly select optimal features and thresholds, and determine whether to split the tree by measuring the contribution of features to classification.

H(D) represents the entropy of the dataset D, which is a quantitative measure of the degree of chaos in the samples, and is the subset after splitting by feature ratio to D represents the proportion of sample size in the sub dataset, which is used for weighted calculation of the entropy after splitting. Each decision tree independently predicts the class of a sample.

Here, T is the number of trees, and is the indicator function. Model parameters are optimized via grid search and cross-validation. The detailed procedure of this experiment is shown in Table 5.

2.6 XGBoost Ensemble Model

In XGBoost, the objective function comprises a loss term, measuring the discrepancy between predicted and actual values, and a regularization term, which prevents overfitting. The model optimizes the objective function via second-order Taylor expansion, using information gain to select split points that maximize gain. Leaf node weights are calculated from gradients and second derivatives, while the regularization term optimizes model performance by controlling tree complexity and weight magnitude. XGBoost excels in classification and regression, enhancing prediction

Table 6 Summary figure of XGBoost ensemble model algorithm procedure

Procedure:
Input: Training dataset **SMOTE for Imbalanced Samples:** For each sample in the minority class, find its k-nearest neighbors using Euclidean distance Generate a new balanced dataset with equal positive and negative samples **Bayesian Optimization for Hyperparameter Tuning:** Initialize a Gaussian process Define the acquisition function: Iterate T times: Find using the acquisition function Update the dataset: Update the Gaussian process Obtain the optimal parameters **Build XGBoost Model:** In each boosting round, add a new subtree model Compute first and second-order gradients Optimize the objective function: Generate a new subtree using gradient statistics and greedy algorithms to minimize the loss Assign weights to the new subtree and accumulate results:
Output: Predictions

accuracy and model generalization. Before prediction, data imbalance is addressed using SMOTE, followed by parameter tuning with Bayesian optimization, and finally, prediction using the XGBoost model. The detailed procedure of this experiment is shown in Table 6.

3 Results

3.1 Data Source

The dataset utilized in this experiment is the simulated tableting machine dataset from the Kaggle platform, with its properties detailed in Table 7.

3.1.1 Experimental Process and Results

First, histograms and kernel density estimation plots were used to observe the feature distributions (Figs. 1 and 2). Most features showed unimodal distributions, indicating data concentration, while Pressure and Speed exhibited multimodal distributions, suggesting possible multiple modes or categories. Correlation matrix analysis revealed weak linear relationships between features.

Table 7 Data property information

Feature Attribute	Meaning
Pressure	Pressure level during operation
Vibration	Vibration level detected during operation
Temperature	Temperature changes during operation
Speed	Equipment rotational speed
Vibration	Vibration level detected during operation
Humidity	Humidity level in the equipment's operating environment
Maintenance Cycles	Number of maintenance cycles completed by the machine

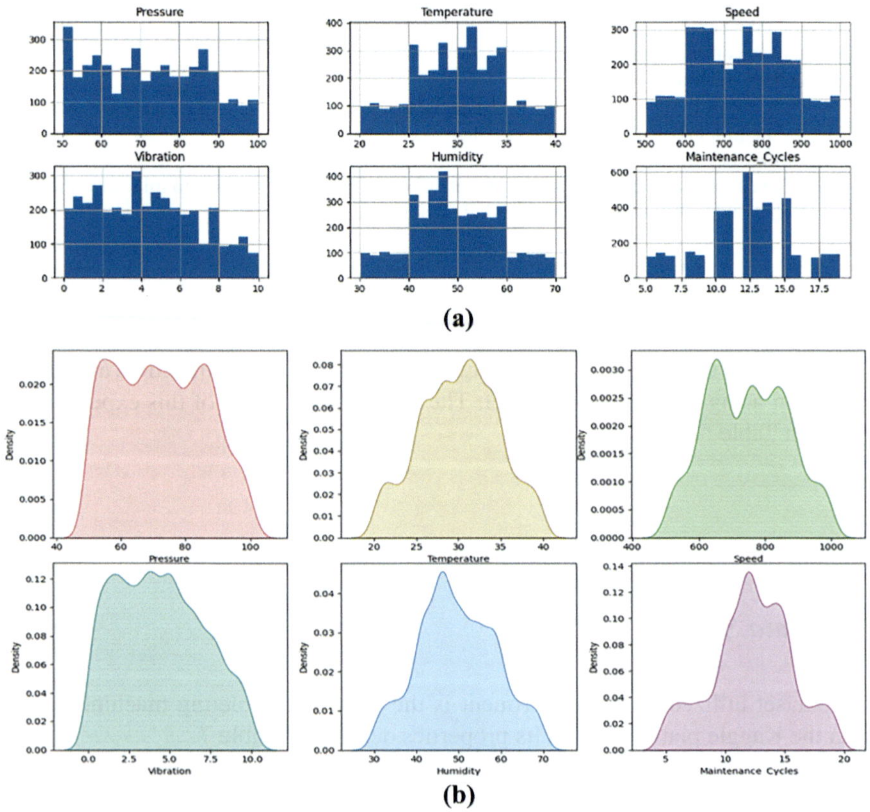

Fig. 1 **a** Feature distribution histogram, **b** Kernel density estimation plot

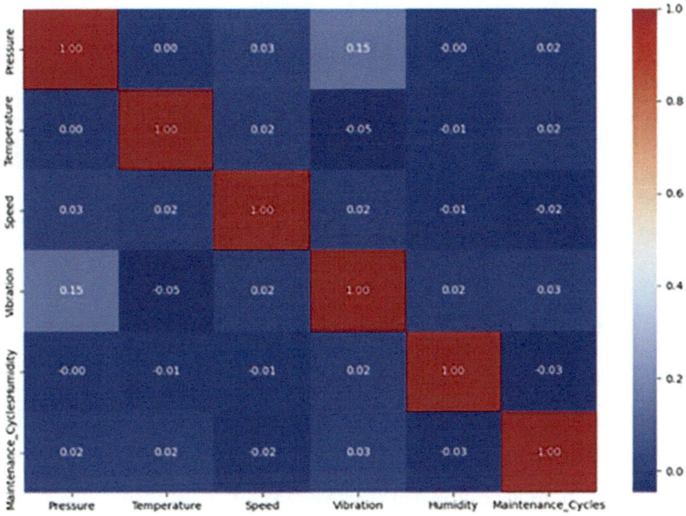

Fig. 2 Feature correlation heatmap

Feature selection was performed using stepwise regression. After standardizing the data, the dataset was split into training and test sets in an 80–20 ratio. Various machine learning models described above were used for predictive analysis and comparison.

As shown in Fig. 3, there are significant distribution differences in various feature attributes between positive and negative samples. Features like Pressure, Temperature, Speed, Vibration, and Maintenance Cycles show notable distribution differences, with negative samples having higher density in specific intervals than positive samples. This discrepancy may cause models to underperform in certain regions of these features. To enhance model robustness and generalization, SMOTE was applied to balance the dataset. SMOTE increases data diversity by generating synthetic samples, reducing model over-reliance on specific distributions. Then built an XGBoost model with parameters optimized using Bayesian optimization for learning and prediction.

In the experiment, the optimal parameter combination for the random forest model, fine-tuned via grid search and cross-validation, is presented in Table 8.

For the SMOTE-BO-XGBoost model, which combines SMOTE for data balancing and Bayesian optimization for parameter tuning, the optimal parameters are detailed in Table 9. This approach enhances model robustness and generalization, addressing class imbalance and optimizing model performance.

In the experiment, multiple classifiers were comprehensively evaluated. The models were assessed in the following order: K-nearest neighbors, support vector machine, Naive Bayes, Gaussian process, tuned random forest, and the SMOTE-BO-XGBoost ensemble model.

The results are presented in Table 10.

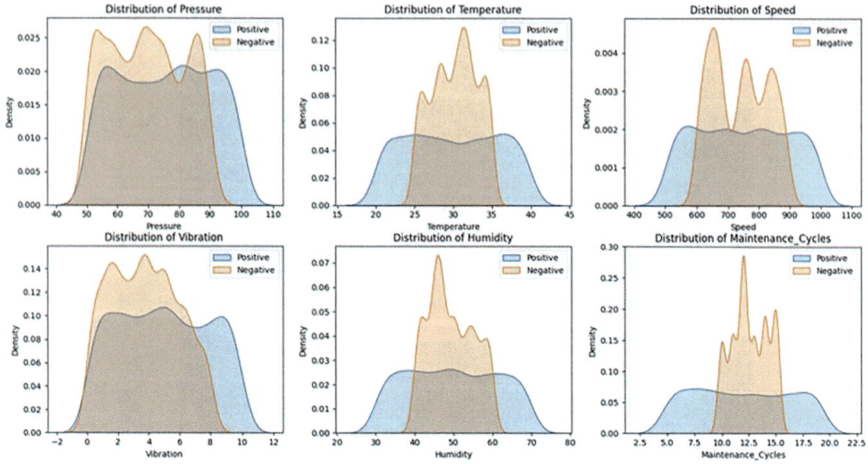

Fig. 3 Feature distributions in positive and negative samples

Table 8 Optimal parameter combination of tuned Random Forest

Parameter	Parameter meaning	Optimal value
max_depth	Maximum depth of the decision tree	8
max_features	Maximum number of features considered at each split node	3
min_samples_leaf	Minimum number of samples required for a leaf node	10
min_samples_split	Minimum number of samples required to split a node	60
'n_estimators	Number of decision trees in the random forest	100

Table 9 Optimal parameter combination of SMOTE-BO-XGBoost model

Parameter	Parameter meaning	Optimal value
colsample_bytree	Proportion of randomly sampled columns	0.0678
gamma	Minimum reduction in loss function	0.5553
learning_rate	Learning rate	0.3887
max_depth	Maximum depth of trees	9
min_child_weight	Minimum sample weight for leaf nodes	0
n_estimators	Number of trees	289
reg_alpha	regularization term	0.1473
subsample	Proportion of randomly sampled data	0.9927

Experimental results show that the random forest model optimized by grid search and cross-validation performs best in accuracy and F1 score, while the SMOTE-BO-XGBoost ensemble model demonstrates excellent prediction stability with an extremely low first error rate of 0.0079. The core reasons are as follows: the

Table 10 Comparison table of multi-dimensional performance metrics for classifiers

	Accuracy	Precision	Recall	F1 Score	First error rate
1	0.9528	0.9567	0.9528	0.9526	0.0984
2	0.9699	0.9715	0.9699	0.9698	0.0628
3	0.9161	0.9166	0.9161	0.9162	0.0710
4	0.9620	0.9646	0.9620	0.9619	0.0792
5	0.9987	0.9923	0.9921	0.9921	0.0164
6	0.9921	0.9923	0.9921	0.9917	0.0079

random forest reduces prediction variance through the integrated voting mechanism of multiple decision trees and, combined with parameters optimized via grid search, enhances the nonlinear fitting ability for multimodal features (e.g., Pressure and Speed), effectively avoiding overfitting. The SMOTE-BO-XGBoost model balances the distribution of positive and negative samples through SMOTE oversampling, efficiently searches the regularization parameters and tree structure parameters of XGBoost with Bayesian optimization, and uses its second-order Taylor expansion optimization mechanism to capture the feature interactions of multi-source sensor data, such as vibration and temperature. In contrast, models like KNN and SVM have significantly lagging performance in handling nonlinear relationships and imbalanced datasets due to the lack of targeted parameter tuning or theoretical limitations such as the feature independence assumption of Naive Bayes. These findings highlight the advantages of parameter tuning and ensemble learning in dealing with imbalanced datasets, and verify the superiority of the collaborative mechanism of "ensemble learning + parameter optimization + data preprocessing" in machine equipment fault prediction, providing effective solutions for related research and applications.

Uncertainties in the fault prediction process include class imbalance in the original dataset, noise in sensor-derived features (e.g., minor fluctuations in vibration and temperature measurements), and model sensitivity to hyperparameter variations. Class imbalance initially introduced bias toward majority class predictions, which was mitigated via SMOTE, though residual imbalance could still slightly reduce the recall of minority (fault) samples. Sensor noise may lead to inconsistent feature distributions, weakening the model's ability to capture fault patterns, which was partially addressed by feature standardization and the robustness of ensemble models like Random Forest. Model sensitivity, such as the impact of Random Forest's max_depth on overfitting or XGBoost's learning_rate on convergence, was managed through grid search and Bayesian optimization, yet small variations in hyperparameters could still cause marginal changes in prediction accuracy, affecting reliability in real-world applications. Approaches to further mitigate these uncertainties include integrating noise reduction techniques (e.g., wavelet transform for sensor data) and adopting uncertainty quantification methods as demonstrated by Yazdi et al. (in Li et al. 2023), who utilized probabilistic models to quantify prediction uncertainties in complex systems, which could enhance the trustworthiness of the proposed fault prediction framework.

4 Conclusion

This study employs machine learning techniques to enhance the accuracy of fault prediction for tablet presses, thereby contributing to improved equipment reliability. By systematically evaluating a range of classifier models, including K-nearest neighbor, support vector machine, Naive Bayes, Gaussian process, a parameter-optimized Random Forest model, and the SMOTE-BO-XGBoost ensemble model, the following conclusions are drawn:

(1) Performance comparison of classifier models: The Random Forest model, optimized through grid search and cross-validation, exhibits the highest performance in terms of both accuracy and F1 score. These results underscore the importance of parameter tuning for enhancing model stability and predictive precision in practical fault diagnosis applications.
(2) Advantages of SMOTE-BO-XGBoost: The SMOTE-BO-XGBoost model demonstrates superior performance in addressing imbalanced datasets. It achieves the lowest first error rate of 0.0079 among all evaluated models, indicating a high level of predictive stability and accuracy.
(3) Reliability enhancement: Both the parameter-tuned Random Forest and the SMOTE-BO-XGBoost models effectively minimize false positive and false negative predictions. This contributes significantly to the reliability of tablet press operation and provides substantial support for predictive maintenance and fault prevention in industrial settings.
(4) Future research directions: Subsequent investigations will focus on the exploration of additional ensemble learning techniques and advanced parameter optimization strategies to further improve model performance. Moreover, the application of these models to fault prediction in other types of industrial equipment will be pursued to assess their generalizability and effectiveness across varied domains.

Some further relevant studies can find out from here [10–15].

References

1. Wang XH, Zhang QW, Qiao YL et al (2024) Reliability analysis and prediction model of power equipment based on data-driven. Electric Power Equip Manage 24:180–182
2. Wang CZ, Han X, Zhao X et al (2024) Reliability analysis of multiple machine learning methods for rapid discrimination of rock types. Water Resources Power 42(6):88–92
3. Ma SJ, Guo P, Li Y et al (2025) Machine learning-based prediction of bonding wire annealing performance. J Netshape Forming Eng 17(3):1–8

4. Yu H, Zhang Y, Li Q et al (2025) Research on software defect prediction based on machine learning. J Chongqing Univ 48(02):10–21
5. Liu H, Li XF, Luo DY et al (2025) A study on fault prediction model of building machines based on machine learning. China Constr 03:153–156
6. Wang DY, Zhu DF, Yu BB et al (2024) Predicting the bearing capacity of gently inclined bauxite pillar based on numerical simulation and machine learning. J Coal Sci 49(5):1–18
7. Wang JL, Qian XJ, Li YX et al (2023) Power system reliability prediction model based on collaborative machine learning. Electric Drive 53(4):51–56
8. Chen KJ, Peng SS, Wang XL et al (2024) Global reliability analysis of transmission tower-line structures based on BP-Adaboost algorithm. Guangdong Electric Power 37(5):112–120
9. Zhang YQ (2024) Exploration on intelligent management of electrical equipment based on big data analysis. Electrical Equip Econ 8:161–164
10. Guo J, Wang Z, Li H, Yang Y, Huang CG, Yazdi M, Kang HS (2024) A hybrid prognosis scheme for rolling bearings based on a novel health indicator and nonlinear Wiener process. Reliab Eng Syst Saf 245:110014
11. Li H, Yazdi M, Huang HZ, Huang CG, Peng W, Nedjati A, Adesina KA (2023) A fuzzy rough copula Bayesian network model for solving complex hospital service quality assessment. Complex Intell Syst 9(5):5527–5553
12. Li H, Peng W, Adumene S, Yazdi M (2023) An intelligent cost-based consequence model for offshore systems in harsh environments. In: Intelligent reliability and maintainability of energy infrastructure assets. Springer Nature Switzerland, Cham, pp 107–117
13. Li H, Yazdi M (2022) What are the critical well-drilling blowouts barriers? A progressive DEMATEL-game theory. Advanced decision-making methods and applications in system safety and reliability problems: approaches, case studies, multi-criteria decision-making, multi-objective decision-making, fuzzy risk-based models. Springer International Publishing, Cham, pp 29–46
14. Li H, Peng W, Adumene S, Yazdi M (2023) Cutting edge research topics on system safety, reliability, maintainability, and resilience of energy-critical infrastructures. In: Intelligent reliability and maintainability of energy infrastructure assets. Springer Nature Switzerland, Cham, pp 25–38
15. Li H, Peng W, Adumene S, Yazdi M (2023) Advances in failure prediction of subsea components considering complex dependencies. In: Intelligent reliability and maintainability of energy infrastructure assets. Springer Nature Switzerland, Cham, pp 93–105

Digital Twin-Driven Condition Monitoring for Servo Valves

Chaoyi Zhang, Tongrong Zhang, Zijun Gao, Ying Zhong, Xiaobang Wang, and Wenzheng Xu

Abstract To address the challenge of real-time monitoring of internal parameters such as spool pressure in servo valves for aircraft flight control actuators, which is primarily due to the compact structural design, a state monitoring method driven by a digital twin (DT) framework is proposed. High-precision Computational Fluid Dynamics simulation data are utilized to construct a Radial Basis Function surrogate model that establishes a real-time mapping between measurable external parameters, including drive voltage, spool motion cycle, and temperature, and the internal pressure field. The model's accuracy is validated by calculating the coefficient of determination (R2). Combined with the K-Nearest Neighbors algorithm for spatial feature mapping, a DT of the servo valve is developed using the Unity3D engine. The results show the established DT precisely predicts the high-fidelity simulation outputs of the maximum pressure, with the maximum relative error remaining below 0.50 percent. Furthermore, the discrepancies in the position predictions of the maximum pressure between the simulation and the DT are generally within 5.00 percent. This implementation enables visual tracking and dynamic prediction of flow field evolution within the valve, thereby achieving comprehensive internal flow monitoring and seamless interaction between virtual and physical systems.

C. Zhang (✉) · Y. Zhong · W. Xu
China Aero-Polytechnology Establishment, Beijing, China
e-mail: zhangchaoyi459@163.com

Y. Zhong
e-mail: zhongying1009@126.com

W. Xu
e-mail: xwzelite@163.com

T. Zhang · Z. Gao · X. Wang
Naval Architecture and Ocean Engineering College, Dalian Maritime University, Dalian, China
e-mail: zhangtr@dlmu.edu.cn

Z. Gao
e-mail: gaozj_0809@163.com

X. Wang
e-mail: wxb@dlmu.edu.cn

H. Li et al. (eds.), *Data-Driven Methods for Reliability and Safety Engineering: Applications in Industrial Systems*, Springer Series in Reliability Engineering,
https://doi.org/10.1007/978-3-032-22873-4_23

Keywords Condition monitoring · Digital twin · Servo valves · Radial basis function

1 Introduction

As the principal actuation mechanism within the fly-by-wire architecture, the actuator adjusts the aircraft's attitude by precisely regulating the deflection of control surfaces [1]. The servo valve, a critical functional component of the actuator, performs the essential role of converting electrical signals into mechanical displacement, thereby enabling the motion of control surfaces. The flow behavior inside the servo mechanism directly influence the response accuracy and operational reliability of the overall flight control system [2].

Monitoring key internal parameters such as pressure distribution, temperature fields, and dynamic spool forces is closely associated with both the operational condition of the servo valve and the safety of the system. Abnormal pressure distribution can disrupt the equilibrium of forces acting on the spool, inducing undesirable flow phenomena such as vortex formation and cavitation. These conditions can result in severe consequences, including yaw instability or runway excursions during takeoff. Variations in the internal temperature field impact the viscosity of hydraulic fluids and the material properties of valve components; excessively high temperatures can accelerate oil degradation and seal aging, leading to failures such as leakage or sticking. Additionally, dynamic spool forces directly affect motion precision and response performance. Prolonged exposure to uneven force distribution can increase mechanical wear, thereby diminishing both the control accuracy and operational lifespan of the servo valve. However, due to the compact structural layout and intricate internal flow paths, traditional sensors cannot be embedded within critical regions to support real-time data acquisition. To address this limitation, the present study introduces a method leveraging digital twin (DT) technology to enable instantaneous monitoring pressure distribution within internal valve flow channels. Future work will incorporate multi-physical field parameters, including internal temperature fields and dynamic spool forces, to establish a comprehensive fluid–structure-thermal coupled monitoring framework [3].

DT technology offers a novel paradigm for condition monitoring of complex electromechanical systems by facilitating a dynamic closed-loop connection between physical components and virtual models. Febrianto et al. [4] utilized statistical finite element approaches for the DT modeling of railway bridges, achieving accurate strain monitoring without the deployment of physical sensors through probabilistic model calibration. Wang et al. [5] developed a holistic lifecycle monitoring system consisting of physical pipelines, fluid state DTs, knowledge repositories, Bayesian inference modules, and operational service stations. This framework supports real-time fluid state tracking and integrates data-driven fault diagnosis and predictive capabilities. Pei et al. [6] introduced a DT-enabled supervision system applied to solar cell production lines, establishing a data cycle encompassing parameter sensing, behavioral mapping, online monitoring, and informed control. Roman et al. [7] constructed a fault diagnosis framework utilizing DT architecture for active control of electro-hydrostatic actuators in industrial cloud-based IoT systems, integrating recurrent neural models. Quattrocchi et al. [8] addressed the accuracy limitations of low-fidelity models and the computational burden of high-fidelity models by proposing an improved low-fidelity model that maintains multiphysical field coupling for electromechanical actuators. This approach restricts prediction error to within 5 percent under key operational scenarios while preserving computational efficiency. Berri et al. [9] proposed a lumped-parameter numerical model to mitigate the high computational cost of fluid dynamics simulations for electro-hydraulic actuators, thus facilitating predictive maintenance strategy development. Guo et al. [10] designed a DT-integrated control system for electromagnetic linear actuators, enabling trajectory planning and motion tracking through virtual-real interaction mechanisms.

DT technology, characterized by its virtual-physical mapping capabilities, dynamic simulation, and intelligent decision-making functions, has demonstrated considerable promise in health status tracking, anomaly identification, and estimation of service life across various engineering fields,, particularly within industrial manufacturing [11]. Nonetheless, research on the condition monitoring of servo valves in flight control actuators remains limited. Given that the performance stability of servo valves is critical to ensuring precise aircraft flight control, the development of DT-based monitoring methodologies for these components holds substantial research and practical significance.

2 Establishment of High-Fidelity CFD Model for Servo Valves in Flight Control Actuators

2.1 Working Principle of Servo Valves in Flight Control Actuators

The servo valves comprise a permanent magnet differential linear force motor, a linear displacement spool valve (including cylindrical spool and sleeve), a displacement sensor, and integrated analog control circuitry. The linear force motor generates bidirectional driving forces through its supporting springs, permanent magnets, armature assembly, and control coils [12].

When no input is applied, the armature remains centred with equal left–right air gaps. Upon current excitation, the input current excites a controlled magnetic flux that interacts with the fixed magnetic flux from the permanent magnets, generating a bidirectional driving force proportional to the current direction. The supporting springs provide axial stiffness to ensure dynamic stability of the armature while also enabling an elastic reset mechanism to overcome friction and precisely return the spool to its null position during power loss [13].

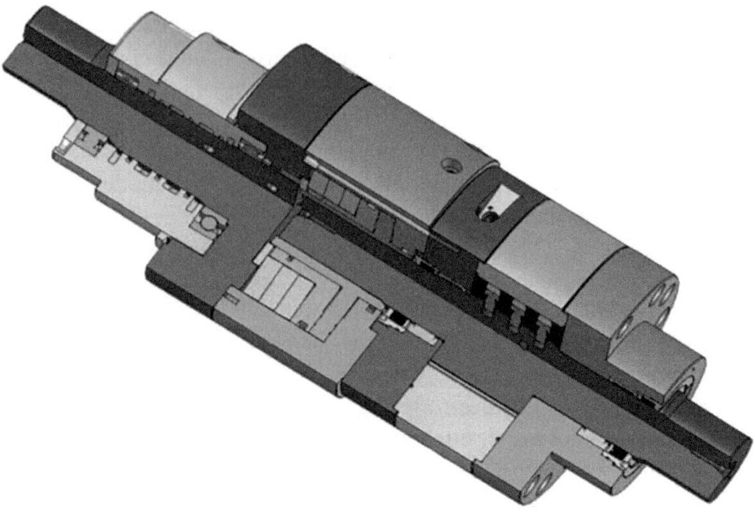

Fig. 1 Assembly diagram of the servo valves [14]

2.2 *Physical Model Construction of the Servo Valves for Flight Control Actuators*

Based on the actual operating conditions, the physical model simplification mainly includes removing micro-structures such as sensor mounting holes with a diameter < 0.1 mm and merging left–right symmetrical components to reduce the computational domain volume. It also involves equivalently converting irregular throttling orifices into rectangular openings and simplifying annular oil passage branches (length > 10 mm, cross-sectional change rate < 10%) into straight channels. Such simplifications retain the core structures affecting the flow field characteristics. As verified by CFD comparative simulations, key parameters such as the main spool pressure drop exhibit a deviation $\leq 1.4\%$ from the full-size model, refer to Fig. 1, 2, and 3.

2.3 *High-Fidelity CFD Simulation Model Construction for the Servo Valves in Flight Control Actuators*

Mesh Generation and Mesh Independence Verification: In the CFD numerical simulation, a structured grid scheme dominated by hexahedral meshes was adopted to achieve high-precision flow field solutions, as shown in Fig. 4. Specifically, local grid encryption was performed for key flow field regions such as the 5 mm area upstream and downstream of the throttling orifice and the corners of annular oil passages, with the minimum grid size reaching 0.1 mm. A 10-layer boundary layer expansion

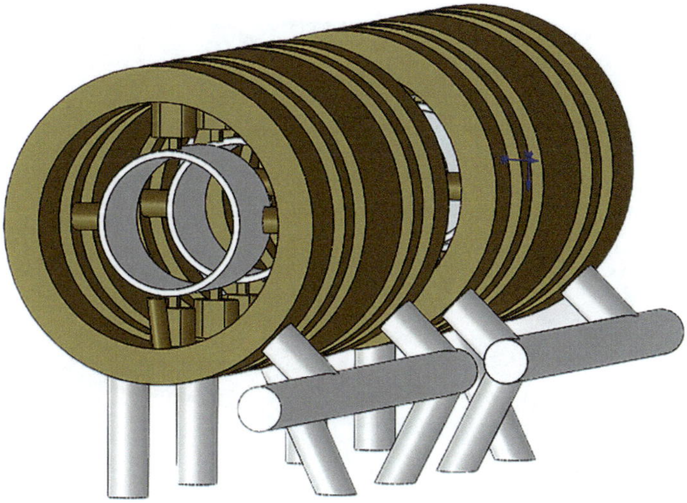

Fig. 2 Flow passage diagram of the servo valves

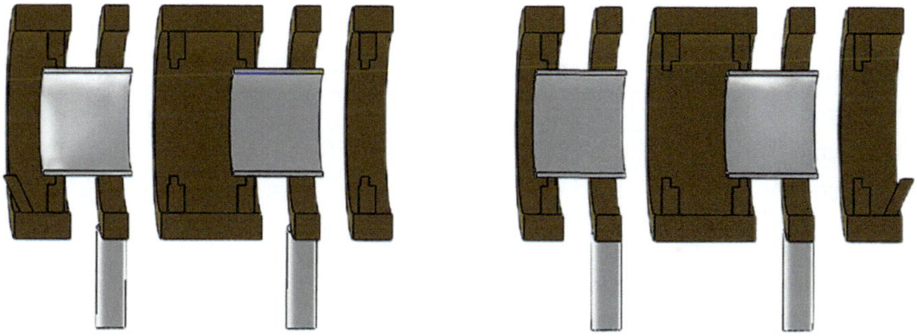

Fig. 3 Cross-sectional view of the servo valves flow passage

mesh (first layer thickness of 0.02 mm, growth rate of 1.2) was used to meet the wall function requirements of the turbulence model. The grid quality was controlled by indicators of aspect ratio < 10 and skewness < 0.85, and 98.7% of the elements were verified by ICEM CFD to meet the quality standards. Combined with the grid independence verification results in Table 1, the 200,000-grid scheme reduced the relative error of maximum pressure from 0.67% (100,000 grids) to 0.10%, with a deviation < 0.5% compared to the 500,000-grid scheme.

The oil flow region within the valve housing was modeled as a single continuous computational domain, eliminating interface interpolation errors caused by multi-zone grid stitching and ensuring continuous transmission and precise coupling of physical quantities (e.g., velocity, pressure) throughout the flow field [15].

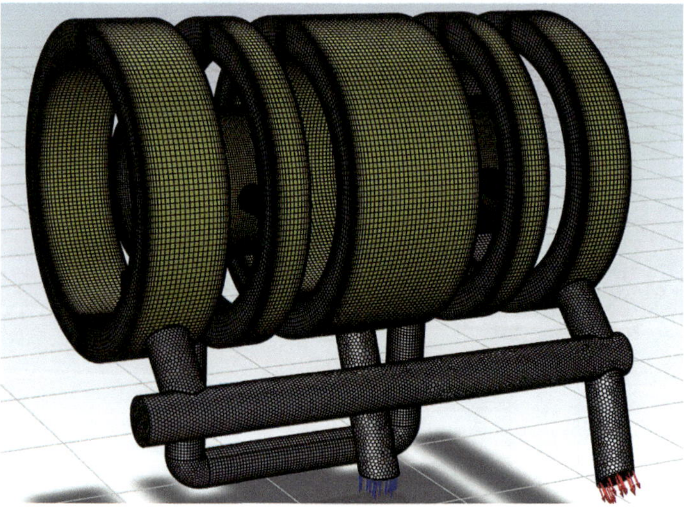

Fig. 4 Mesh generation of the servo valves

Table 1 Grid independence verification

Number of mesh elements	Maximum pressure (MPa)	Relative error
100,000 grids	28.19	–
200,000 grids	28.48	0.67%
500,000 grids	28.47	0.10%

Simulation Setup: To accurately simulate the complex working conditions of oil–gas mixed flow, this study employs an Eulerian multiphase flow model with two phases to comprehensively account for interphase interactions and flow characteristics [16]. For high-Reynolds number flow environments, the standard k-epsilon turbulence model is selected, with adjusted turbulence constants (e.g., Cmu, Prandtl) to enhance simulation accuracy and reliability [17].

The adjusted turbulence constants were selected for their demonstrated effectiveness in capturing turbulent characteristics of high-Reynolds number hydraulic oil flows through narrow passages, aligning with the servo valve's operational flow regime and ensuring accurate prediction of pressure losses at throttling orifices.

To capture the characteristics of the boundary layer, the enhanced wall function approach is applied. During the numerical procedure, velocity and pressure fields are linked through a coupled solution strategy, and both momentum and energy equations are discretized via a second-order upwind formulation, with the overall residual criterion fixed at 10^{-4} [18].

At the inlet, a total pressure of 28 MPa was set to simulate the high-pressure oil inflow, and turbulence intensity and turbulent viscosity ratio were defined to ensure that the inlet turbulent characteristics match the actual operating conditions. The outlet boundary is specified with a gauge pressure of zero through the Total Pressure setting,, representing the condition in which the oil discharges into the atmosphere..

Simulation Results: As observed in Fig. 5, the simulation results indicate significant pressure fluctuations in the hydraulic oil as it passes through the throttling valve orifice. These fluctuations are attributed to localized pressure losses caused by the narrow flow passage of the orifice.

The reduced cross-sectional area of the throttling valve orifice leads to a substantial increase in flow velocity, thereby elevating dynamic pressure while reducing static pressure, consistent with the Bernoulli principle.

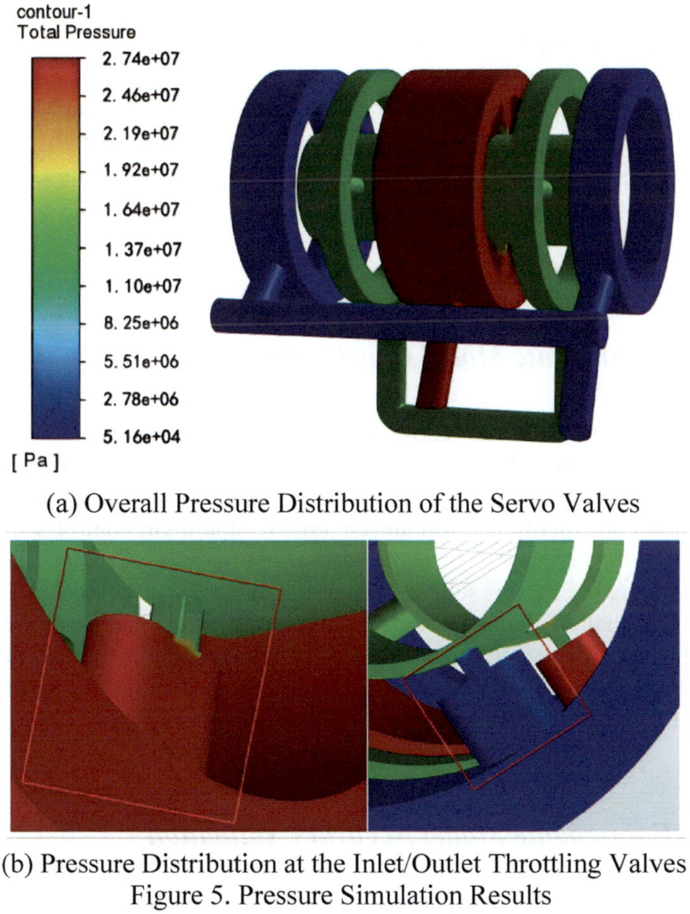

Fig. 5 Pressure simulation results

Table 2 Permissible range of the parameters

Design parameters (unit)	Lower limit	Upper limit
T (°C)	−55.00	135.00
U (V)	0.55	5.00
T (s)	0.01	2.00
T_p (s)	4.00	21.00

3 RBF Surrogate Model Establishment for the Servo Valves in Flight Control Actuators

3.1 Design of Experiments

In this study, we employed Latin Hypercube Sampling (LHS) to generate 40 sample points for the three parameters: temperature (T, °C), voltage (U, V), and period (T_p, s). Simultaneously, the time point parameter (t) was uniformly sampled at 50 points within its specified range. Table 2 summarizes the permissible ranges of these parameters.

3.2 RBF Surrogate Model Construction

Currently, there are multiple surrogate modeling methods, including Response Surface Methodology (RSM), Kriging models [19], Radial Basis Function (RBF) [20], and Support Vector Regression (SVR) [21]. In this study, the multiquadric RBF is selected to construct the performance model for the flight control actuator servo valves.

The multiquadric RBF is selected because it can balance interpolation accuracy and computational efficiency through an appropriate shape parameter. Moreover, it tends to generate a symmetric positive definite (SPD) kernel matrix, which enables stable and efficient computation of the weight coefficient vector and ensures reliable solution of the nonlinear parameter–pressure mapping.

3.3 RBF Surrogate Model Accuracy Validation

After validating the accuracy of surrogate models through statistical methods like variance analysis using n test sample points, the validated surrogate model can replace complex system analysis models to assist design. Common metrics for evaluating model approximation accuracy include the following criteria:

coefficient of determination (R^2).

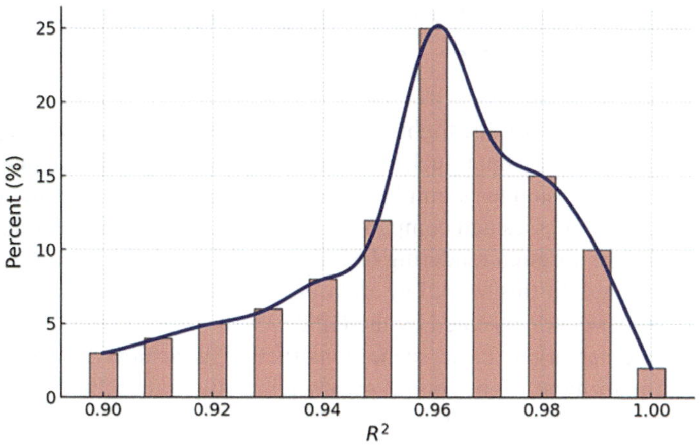

Fig. 6 R^2 value distribution of the RBF surrogate model for servo valves

Value indicates greater approximation accuracy. To validate the accuracy of all surrogate models, the distribution of R^2 values is shown in Fig. 6. The results demonstrate that R^2 values predominantly fall within 0.94 to 0.98, indicating that the predicted values from all RBF neural network models exhibit excellent agreement with the actual values. This confirms that the prediction accuracy of the models meets the required standards.

4 DT Model Construction for Flight Control Actuator Servo Valves

4.1 Spatial Feature Mapping-Based K-Nearest Neighbors (KNN) Algorithm

In this study, the KNN algorithm is employed to achieve the spatial data feature mapping process for the DT model [22]. The KNN algorithm, a distance-based method, identifies relationships between points and their neighbors in multi-dimensional space, enabling the mapping of data with known features (e.g., temperature data, pressure drop data) onto the spatial grid points of the model.

By leveraging the KNN algorithm, high-fidelity performance metrics data of the flight control actuator obtained from simulations can be accurately mapped to each grid node of the system. This ensures high-precision predictions of the actuator's operational performance under any working condition.

4.2 Unity3D-Based DT Construction for Flight Control Actuator Servo Valves

Given that building a DT for the flight control actuator's hydraulic valve requires the integration of multi-source data, such as 3D model data and operational performance metrics, the Unity3D engine is employed to construct a DT model for the servo valve. The engine processes and computes graphical data, including vertex data and texture information, through rendering techniques, ultimately outputting the results to display terminals. In this workflow, data organization is managed by the CPU, while rendering tasks are executed by the GPU [23].

During implementation, basic scene construction is achieved through camera selection and placement, light source parameter configuration, and skybox environment setup within the Unity3D rendering system.

After establishing the foundational scene, a virtual twin representation corresponding to servo valves is imported into the system. Based on node coordinates and node index data of the mesh objects, a key performance model for monitoring oil channel pressure is generated. Finally, by integrating these technical components, a visualized monitoring user interface driven by the DT model is developed.

4.3 Validation of Prediction Accuracy for the Flight Control Actuator Servo Valves DT Model

Based on the modeling process described in Sect. 4.2, a thermal-fluid characteristics DT model for the flight control actuator servo valves has been successfully constructed. This model enables high-efficiency and high-accuracy prediction and analysis of oil channel pressure distribution under multi-operational conditions. By integrating 3D geometric features and real-time operational data, the model not only visualizes pressure field distribution within complex flow channels but also simulates system dynamic response characteristics based on varying operational parameters.

Since significant pressure differential gradient changes occur during the valve spool-port engagement process, this dynamic state is selected to rigorously validate the f predictive performance of the DT. By considering the complete operating parameter space detailed in Table 3, three sets of typical operating scenarios were randomly selected as validation samples. A comprehensive assessment of model performance was conducted through quantitative comparative analysis between the model outputs and high-fidelity CFD simulation results. Key results have been visualized in Figs. 7, 8 and 9.

For a comprehensive evaluation of the model's predictive performance, quantitative comparative analyses of key spool performance metrics were performed for these three selected operating scenarios, focusing on the maximum internal pressure and its spatial distribution characteristics within the servo valve spool.

Table 3 Validation of servo valves DT model prediction accuracy under three operating scenario

Design parameters (Unit)	Scenario 1	Scenario 2	Scenario 3
T (C)	100.00	-40.00	55.00
U (V)	0.60	1.80	1.00
T (s)	0.14	1.06	1.37
T_p (s)	1.30	1.50	0.50

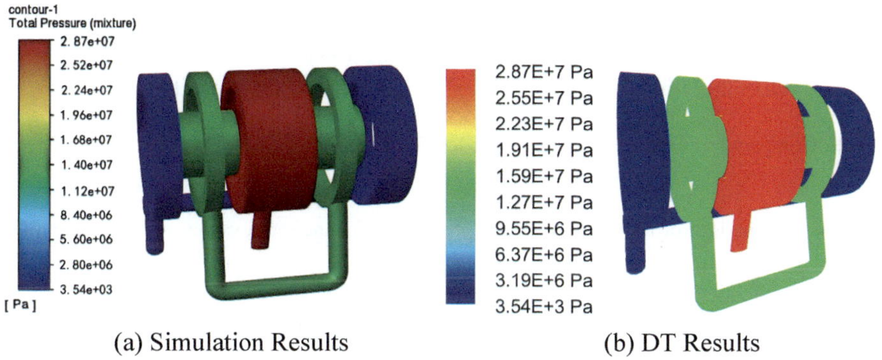

Fig. 7 Comparison of simulation and DT models for scenario 1

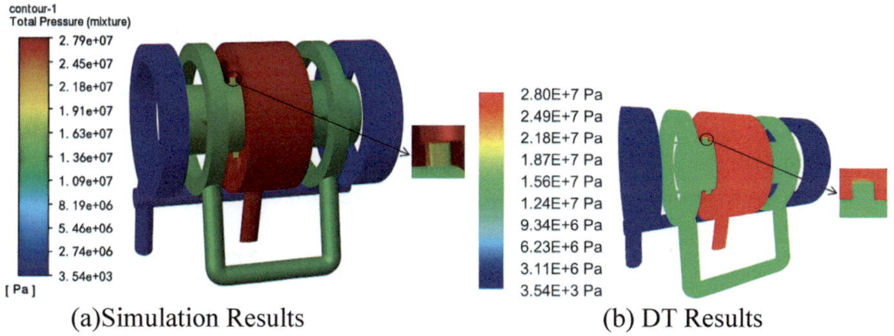

Fig. 8 Comparison of simulation and DT models for scenario 2

Table 4 compares the maximum pressure predictions from simulations to DT model. The data indicate that the DT model maintains high agreement with high-fidelity simulation results across all three scenarios, with a maximum relative error not exceeding 0.50%.

Table 5 further evaluates the coordinate deviations of predicted maximum pressure locations. The data indicate that the prediction deviations along the x, y, and z axes for the maximum pressure point location between simulations and the DT model are primarily confined to within 5.00%.

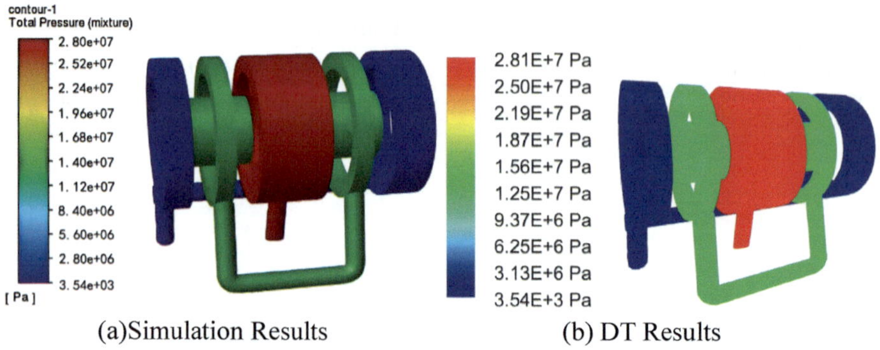

Fig. 9 Comparison of simulation and DT models for scenario 3

Table 4 Comparison of maximum pressure predictions for the servo valves: simulation and DT model

Operating condition data	Scenario 1	Scenario 2	Scenario 3
Simulated Max Pressure (MPa)	28.71	27.94	28.02
DT Maximum Pressure(MPa)	28.74	28.01	28.13
Relative Error	0.10%	0.25%	0.40%

Table 5 Spatial prediction comparison of maximum pressure locations in the servo valves: simulation and DT model

Operating condition data	Scenario 1	Scenario 2	Scenario 3
Simulated maximum pressure point coordinates (mm)	(12.30, 0.13, 3.50)	(12.60, 0, 3.50)	(17.20, 0.21, 3.50)
DT maximum pressure point coordinates (mm)	(12.50, 0.14, 3.70)	(12.20, 0, 3.40)	(17.10, 0.19, 3.50)
Relative error	1.63%, 7.69%, 5.71%	3.17%, 0%, 2.86%	0.58%, 9.52%, 0%

In summary, the validation results confirm that the servo valves spool performance DT model established in this study achieves high prediction accuracy.

Uncertainty sources in the DT framework include CFD simplification errors from geometric approximations (e.g., merged symmetrical components, rectangular orifice equivalents) which may slightly distort local flow features, sampling density limitations in Latin Hypercube Sampling that could leave minor gaps in the parameter space coverage, and surrogate model extrapolation limits where predictions beyond the training sample range may reduce precision. These uncertainties can subtly influence prediction robustness, particularly for extreme operational parameters or complex flow phenomena like vortex formation, highlighting the need for future work to refine geometric modeling fidelity, increase LHS sampling density, and integrate extrapolation error correction to enhance the model's applicability across broader operating conditions.

5 Conclusion

To address the engineering challenge of monitoring internal pressure conditions in servo valves within aircraft flight control actuators, which is constrained by their compact structural design, a digital twin (DT)-driven pressure monitoring method is proposed. A hierarchical framework is established that integrates physical simulation, surrogate modeling, and virtual-to-physical interaction. High-fidelity computational fluid dynamics models and radial basis function surrogate models are utilized to construct a nonlinear mapping between externally measurable parameters, including operating temperature, drive voltage, and motion cycles, and the internal pressure distribution within the valve.

The surrogate model R2 values between 0.94 and 0.98, confirming its high accuracy in representing complex flow field dynamics. Additionally, the K-nearest neighbors algorithm is employed to perform spatial feature mapping within the virtual space. The DT developed on the Unity3D platform, enables continuous visualization of the internal flow field evolution over time. Its predicted maximum pressure values exhibit strong consistency with those obtained from high-fidelity simulations, with the maximum relative error remaining below 0.50 percent. Moreover, the predicted deviations along the three spatial directions for the location of the maximum pressure are primarily limited to within 5.00% when compared to the simulation data. Further studies can be referred to [21–29].

Acknowledgements This work is supported by Project of JSZL2022205A003 and National Natural Science Foundation of China (Grant No. 52201406).

References

1. Anthony MJ, Mattos F (1985) Advanced flight control actuation systems and their interface with digital commands SAE Technical Paper
2. Chen W, Kong K, Tomizuka M (2014) Dual-stage adaptive friction compensation for precise load side position tracking of indirect drive mechanisms. IEEE Trans Control Syst Technol 23(1):164–175
3. Han W, Xiong L, Yu Z et al (2022) Integrated pressure estimation and control for electro-hydraulic brake systems of electric vehicles considering actuator characteristics and vehicle longitudinal dynamics. IEEE/ASME Trans Mechatron 28(1):197–209
4. Febrianto E, Butler L, Girolami M et al (2021) A self-sensing digital twin of a railway bridge using the statistical finite element method. ArXiv210313729 Cs Math
5. Wang T, Feng K, Ling J et al (2024) Pipeline condition monitoring towards digital twin system: a case study. J Manuf Syst 73:256–274

6. Pei FQ, Tong YF, Yuan MH et al (2021) The digital twin of the quality monitoring and control in the series solar cell production line. J Manuf Syst 59:127–137
7. Rodriguez-Aguilar R, Marmolejo-Saucedo JA, Köse U (2024) Development of a digital twin driven by a deep learning model for fault diagnosis of electro-hydrostatic actuators. Mathematics 12(19):3124
8. Quattrocchi G, Dalla Vedova MDL, Berri PC (2023) Lumped parameters multi-fidelity digital twins for prognostics of electromechanical actuators. J Phys Conf Ser. IOP Publishing 2526(1):012076
9. Berri P C, Dalla Vedova MDL, Santaera S (2022) Digital twins for prognostics of electro-hydraulic actuators: novel simplified fluid dynamic models for aerospace valves. In: IOP conference series: materials science and engineering. IOP Publishing, vol 1226, issue 1, p 012106
10. Guo T, Chang S, Wu Y et al (2023) Digital twin-driven motion trajectory tracking control of electromagnetic linear actuator. In: International Conference on Mechatronics and Intelligent Control (ICMIC 2023). SPIE, vol 12793, pp 210–216
11. De Kooning JDM, Stockman K, De Maeyer J et al (2021) Digital twins for wind energy conversion systems: a literature review of potential modelling techniques focused on model fidelity and computational load. Processes 9(12):2224
12. Alle N, Hiremath SS, Makaram S et al (2016) Review on electro hydrostatic actuator for flight control. Int J Fluid Power 17(2):125–145
13. Napolitano MR, An Y, Seanor BA (2000) A fault tolerant flight control system for sensor and actuator failures using neural networks. Aircr Des 3(2):103–128
14. Shenlead (2026) Products and solutions (2025.02.21) [2025.04.29] https://www.shenlead-drive.com/about.php?cat_id=105
15. Pachter M, Miller RB (1998) Manual flight control with saturating actuators. IEEE Control Syst Mag 18(1):10–20
16. Barrett R, McMurtry R, Vos R et al (2006) Post-buckled precompressed piezoelectric flight control actuator design, development and demonstration. Smart Mater Struct 15(5):1323
17. Ahmad H, Young T, Toal D et al (2007) Control allocation with actuator dynamics for aircraft flight controls. In: 7th AIAA ATIO conference, 2nd CEIAT international Conference on Innov and Integr in Aero Sciences, 17th LTA Systems Tech Conf; followed by 2nd TEOS Forum, vol 7828
18. Derrien JC, Sécurité SD (2012) Electromechanical actuator (EMA) advanced technologies for flight controls. In: International congress of the aeronautical sciences, pp 1–10
19. Stein ML (2012) Interpolation of spatial data: some theory for kriging. Springer Science & Business Media
20. Buhmann MD (2003) Radial basis functions: theory and implementations. Camb Monogr Appl Comput Math 12:147–165
21. Vapnik VN, Vapnik V (1998) Statistical learning theory, vol 1. Wiley, New York
22. Soucy P, Mineau GW (2001) A simple KNN algorithm for text categorization. In: Proceedings 2001 IEEE international conference on data mining. IEEE, pp 647–648
23. Zhu M, Zhao S, Dong P et al (2018) Design and analysis of a novel double-servo direct drive rotary valve with high frequency. J Mech Sci Technol 32:4313–4323
24. Guo J, Wang Z, Li H, Yang Y, Huang CG, Yazdi M, Kang HS (2024) A hybrid prognosis scheme for rolling bearings based on a novel health indicator and nonlinear Wiener process. Reliab Eng Syst Saf 245:110014
25. Li H, Yazdi M, Huang HZ, Huang CG, Peng W, Nedjati A, Adesina KA (2023) A fuzzy rough copula Bayesian network model for solving complex hospital service quality assessment. Complex Intell Syst 9(5):5527–5553
26. Li H, Peng W, Adumene S, Yazdi M (2023) An intelligent cost-based consequence model for offshore systems in harsh environments. In: Intelligent reliability and maintainability of energy infrastructure assets. Springer Nature Switzerland, Cham, pp 107–117
27. Li H, Yazdi M (2022) What are the critical well-drilling blowouts barriers? A progressive DEMATEL-game theory. Advanced decision-making methods and applications in system

safety and reliability problems: approaches, case studies, multi-criteria decision-making, multi-objective decision-making, fuzzy risk-based models. Springer International Publishing, Cham, pp 29–46
28. Li H, Peng W, Adumene S, Yazdi M (2023) Cutting edge research topics on system safety, reliability, maintainability, and resilience of energy-critical infrastructures. In: Intelligent reliability and maintainability of energy infrastructure assets. Springer Nature Switzerland, Cham, pp 25–38
29. Li H, Peng W, Adumene S, Yazdi M (2023) Advances in failure prediction of subsea components considering complex dependencies. In: Intelligent reliability and maintainability of energy infrastructure assets. Springer Nature Switzerland, Cham, pp 93–105

YOLO-MVCP: A Lightweight Fault Diagnosis Method of Rolling Bearing Based on STFT Time–frequency Graph and MobileViT Network Pruning

Yu Wang, Yue Li, and Shufeng Zhang

Abstract To address the limitation that existing rolling bearing fault diagnosis methods are unable to simultaneously achieve high diagnostic accuracy and lightweight model deployment, a compact fault diagnosis approach is proposed based on short-time Fourier transform time–frequency representations and MobileViT network pruning. The initial step involves processing the raw vibration signal with a short-time Fourier transform to generate a 2D time–frequency image. Subsequently, a YOLO-MobileViT network is constructed to train the diagnostic model, and redundant secondary channels are eliminated through network pruning, enabling lightweight compression of the high-precision model following fine-tuning. Experimental validation on the CWRU dataset demonstrates that the YOLO-MVCP achieves a fault diagnosis accuracy of 100 percent, while maintaining a model size of only 1.24 MB. This balance between diagnostic accuracy and model compactness confirms its practical applicability and engineering value.

Keywords Fault diagnosis · Short-time Fourier transform · MobileViT · Network pruning · Light weight

---

Y. Wang (✉) · Y. Li · S. Zhang
National Key Laboratory of Equipment State Sensing and Smart Support, National University of Defense Technology, Changsha, China
e-mail: wangyuzhou@nudt.edu.cn

College of Intelligence Science and Technology, National University of Defense Technology, Changsha, China

Y. Li
e-mail: yueli@nudt.edu.cn

S. Zhang
e-mail: sfzhang@nudt.edu.cn

H. Li et al. (eds.), *Data-Driven Methods for Reliability and Safety Engineering: Applications in Industrial Systems*, Springer Series in Reliability Engineering,
https://doi.org/10.1007/978-3-032-22873-4_24

1 Introduction

Deep neural network methods have exhibited considerable advantages in the domain of mechanical fault diagnosis. By employing an end-to-end data-driven framework, these approaches eliminate the need for complex feature engineering processes characteristic of traditional methodologies and have demonstrated remarkable diagnostic accuracy across various studies [1–3]. Current research has predominantly emphasized architectural advancements within convolutional neural networks (CNNs) and recurrent neural networks (RNNs). For instance, Chen et al. [4] designed a diagnostic strategy based on line-by-line grid search optimization of GoogLeNet, utilizing two-dimensional wavelet transforms to convert noisy data into time–frequency images, which were then processed by the optimized network for bearing fault identification. Guo et al. [5] introduced an innovative dual-modal RNN structure capable of concurrently analyzing time-domain and frequency-domain signals, thereby effectively addressing classification challenges associated with composite bearing misalignment faults. Niu et al. [6] developed a deep residual CNN (DR-CNN) that reinforced discriminative feature learning mechanisms, leading to improving accuracy in multi-task bearing fault diagnosis.

Despite these advancements, CNNs and RNNs possess inherent architectural limitations. RNNs, due to their sequential processing mechanism, exhibit limited potential for parallel computation and frequently encounter vanishing or exploding gradient problems during backpropagation through time. These limitations hinder the capacity to effectively capture long-term dependencies within temporal data. Conversely, although CNNs excel in local feature extraction, their constrained receptive fields require extensive multi-layer stacking to capture global contextual information. This requirement amplifies both model parameters and computational complexity, which escalate proportionally to network depth.

The transformer architecture [7] offers a novel solution to these limitations. Through the implementation of a self-attention mechanism, it enables global feature extraction by dynamically weighting input elements, facilitating comprehensive contextual modeling across sequences. In recent years, transformer-based methods have been increasingly applied to fault diagnosis tasks. For example, Ding et al. [8] designed a time and frequency transformer that enhanced diagnostic performance through combined time–frequency domain analysis. Similarly, Bao et al. [9] proposed a bearing fault diagnosis method leveraging the a self-attention mechanism to strengthen the global feature representation. Nonetheless, the absence of spatial inductive bias in transformers necessitates extensive training data to achieve optimal accuracy, a requirement that contrasts with the limited availability of fault data in practical engineering contexts.

Additionally, achieving a balance between model compactness and diagnostic accuracy remains a pivotal challenge in the development of intelligent diagnostic systems. Researchers have explored model lightweighting and algorithmic optimization as potential solutions. Knowledge distillation techniques [10] have been employed to transfer learned knowledge from larger models to smaller counterparts, effectively reducing model size without significantly compromising performance. Neural architecture search (NAS) [11] has also been utilized to automatically identify efficient small-scale network structures. Complementary techniques, such as network quantization [12] and pruning [13], have further demonstrated effectiveness in reducing model complexity. However, these approaches still face critical limitations, including suboptimal knowledge transfer, expansive search spaces, and the inherent complexity of optimization procedures.

To address these challenges, compact fault diagnosis approach is introduced, merging timefrequency analysis through Short-Time Fourier Transform (STFT) with structured pruning of MobileViT networks [14]. This method constructs vibration signal representations based on time–frequency distributions obtained from STFT and employs systematic training followed by magnitude-based pruning within MobileViT architectures. The proposed approach enables accurate multi-class fault recognition and classification of rolling element bearings under deployment conditions constrained by limited computational resources.

2 Methodology

2.1 Overall Structure

The overall process of the YOLO-MVCP, which stands for You Only Look Once - MobileViT Compression Pipeline, is clearly depicted in Fig. 1. Initially, the raw vibration signals that are gathered are subjected to a STFT, which serves to convert the temporal data into a time–frequency domain. This transformation results in the creation of two-dimensional time–frequency images, which are meticulously divided into two distinct sub-datasets: a training set and a testing set. The training set is employed to educate the model on the essential patterns and characteristics within the data, whereas the testing set is reserved to appraise the performance and generalization skills of the trained model.

Following this preparatory phase, the YOLO-MobileViT network is meticulously constructed. This network is a sophisticated amalgamation of the YOLO architecture, renowned for its real-time object detection capabilities, and the MobileViT (Mobile Vision Transformer) framework, which brings the efficiency and flexibility of transformer models to mobile and edge devices. The integration of these two powerful components aims to create a network that is both accurate and computationally efficient, making it suitable for deployment on platforms with low computing power.

To further increase the efficiency and reduce the computational footprint of the YOLO-MobileViT network, a series of optimization techniques are employed. These include sparsely training, which focuses on learning only a subset of the most relevant parameters, thereby reducing redundancy and improving computational efficiency. Additionally, network pruning is applied, which involves the systematic

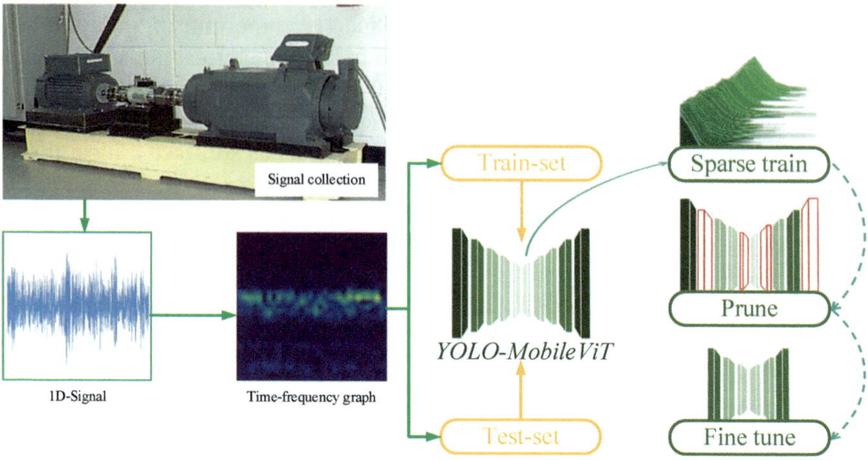

Fig. 1 Overall workflow of YOLO-MVCP

removal of less important weights or neurons from the network, further lightening the model's structure without significantly compromising its performance. Finally, fine-tuning is conducted to adjust the pruned network, ensuring that it maintains optimal performance on the specific task at hand.

To make the methodology easier to reproduce and interpret, we explicitly state the key assumptions and the diagnostic role of each stage in the pipeline.

(1) STFT time–frequency mapping: the vibration signal is treated as quasi-stationary within each short analysis window. The resulting spectrogram emphasises fault-related transients, modulation sidebands and characteristic frequency components, providing physically meaningful cues for subsequent recognition.
(2) YOLO-MobileViT feature learning: the spectrogram is processed as a 2D pattern. Convolutions capture local ridge/impulse textures, while the lightweight Transformer blocks model longer-range correlations across time and frequency (e.g., periodic impulses), which helps separate similar fault types and severities with limited computation.
(3) Sparse training + structured channel pruning: channels whose BN scaling factors remain consistently small are assumed to contribute marginally to discrimination. Pruning removes these redundancies to reduce memory and latency for edge deployment while keeping the overall backbone topology intact.
(4) Fine-tuning: after pruning, the remaining parameters are re-optimised to compensate for the capacity reduction and to stabilise accuracy on the target diagnostic task.

Key implementation choices used throughout the paper (spectrogram size 224 × 224, train/test split 8:2, batch size 32, SGD optimisation, 10 training epochs, and pruning ratio 0.2) are adopted as pragmatic trade-offs between stable optimisation and the intended low-resource deployment setting; Sect. 3 reports the corresponding accuracy/size outcomes.

2.2 STFT Analysis

The primary benefit of employing the short-time Fourier transform, or STFT, is its remarkable ability to pinpoint and localize signals through the strategic adjustment of window functions and the lengths of these windows. This fine-tuning allows for a dynamic balance to be struck between temporal resolution and spectral resolution, which is a crucial aspect of its functionality. Essentially, this means that STFT can adeptly capture fleeting, transient changes within a signal with a high degree of precision in terms of time by utilizing shorter window lengths. Conversely, it can also analyze and dissect the stable, steady-state frequency components of a signal with a high level of frequency resolution when employing longer window lengths.

Furthermore, when contrasted with the wavelet transform, the STFT stands out due to its notably lower computational complexity. This attribute makes the STFT particularly well-suited for applications that require real-time processing, where speed and efficiency are paramount. The flexibility in choosing the appropriate window function allows for an optimized trade-off in the resolution across both time and frequency domains. This optimization is pivotal in enhancing the precision and reliability of extracting fault features from complex signals, which is often a critical task in various fields such as engineering and signal processing.

The STFT window size of 64 and overlap ratio of 0.5 were chosen to strike an optimal balance between temporal and spectral resolution, ensuring clear capture of transient fault impulses while preserving frequency-domain fault signatures. This parameter configuration directly supports reliable feature extraction from non-stationary vibration signals, laying the foundation for accurate fault classification.

In addition, the window function is kept fixed throughout to ensure consistent time–frequency scaling across samples, and the generated spectrograms are resized to the network input resolution (224×224). This design assumes that diagnostic information is mainly carried by relative energy distribution and ridge structures in the time–frequency plane, rather than by absolute pixel-scale variations, which supports robust learning under limited training data.

2.3 YOLO-MobileViT

MobileViT was selected over other lightweight backbones (e.g., MobileNet, EfficientNet) for its unique hybrid design that fuses CNN's local feature extraction capability with Transformer's global context modeling, which is critical for capturing both fine-grained time–frequency patterns and long-range dependencies in bearing

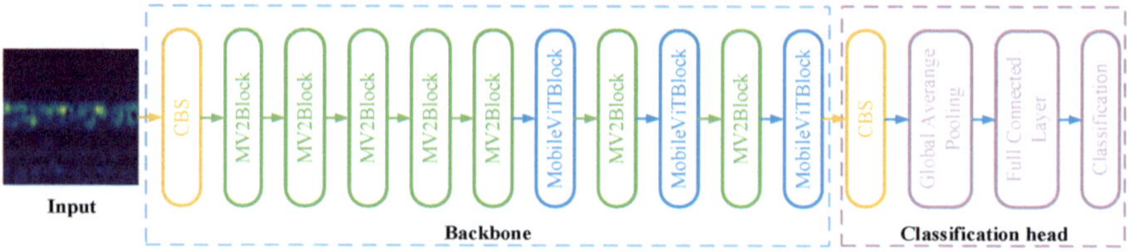

Fig. 2 Network structure of YOLO-MobileViT

fault signals. This architecture balances diagnostic accuracy and computational efficiency, aligning with the practical need for deployment on resource-constrained edge devices in industrial settings.

YOLO-MobileViT adopts a hybrid architecture of CNN and Transformer, extracting local details and global contextual information from feature maps through convolutional operations and self-attention mechanisms respectively. As shown in the structural schematic in Fig. 2, the model mainly consists of Convolutional Batch Normalization layers (CBS), Inverted Residual modules (MV2 Block), MobileViT modules, Global Average Pooling layers (GAP), and Fully Connected layers (FC).

The Conv-BN-SiLU module consists of a standard convolutional operation, batch normalization, and nonlinear activation by SiLU. As illustrated in Fig. 3, the MV2 structure employs an inverted residual structure featuring expansion–contraction mechanisms. By combining multiple CBS modules and cross-stage residual connections, MV2 achieves efficient feature fusion and information transmission, effectively enhancing the model's expressive capability.

The MobileViT Block serves as the key component of the YOLO-MobileViT, employing a modular design that integrates local feature representation with global context modeling capabilities. This module first performs spatial local feature extraction using a standard 3 × 3 convolutional kernel to achieve the primary visual representation of the input feature map, forming the foundation for subsequent feature processing. It then employs a 1 × 1 pointwise convolution to remap the channel dimensions, achieving feature channel compression and information integration, which significantly reduces the computational complexity of subsequent operations.

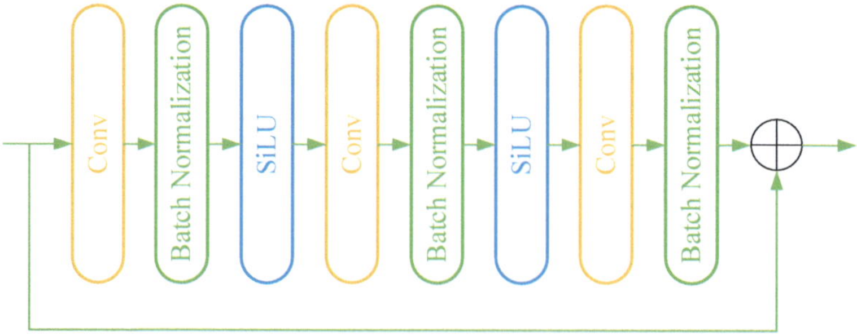

Fig. 3 Structure of MV2 Block

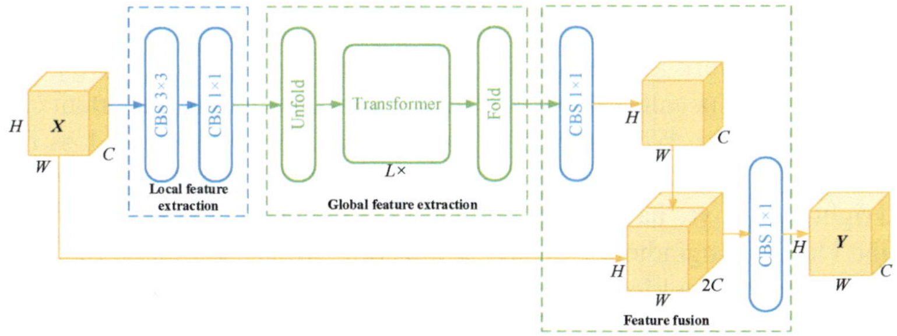

Fig. 4 Structure of MobileViT Block

The core processing stage adopts an Unfold-Transformer-Fold architecture: The Unfold operation serializes the 2D feature map into non-overlapping local regions, generating vector sequences suitable for Transformer input. The Transformer layer, based on a multi-head self-attention mechanism, establishes a global dependency model in this serialized space, effectively capturing long-range semantic relationships across regions. The Fold operation then reconstructs the enhanced feature sequences back into a 2D topological structure through tensor reshaping, forming a closed-loop feature optimization process from local → global → local.

This mechanism, through spatial serialization mapping and inverse mapping, achieves synergistic optimization between the inductive bias of convolution and the global modeling capability of Transformers, refer to Fig. 4.

2.4 *Network Pruning*

(a) *Sparse training*

Sparsity training is used to evaluate the importance of individual channels in convolutional feature maps, providing essential guidance for channel pruning. To ensure computational efficiency, the scaling factor in the batch normalization (BN) layer is adopted as the criterion for assessing channel significance. Let the output of the previous layer be considered, where the batch size represents the number of training samples. The mean and variance are computed for each batch, and a small regularization term is introduced to avoid numerical instability during normalization. The normalized output is then obtained through batch normalization. To enhance the nonlinear feature extraction capability and overall representational power of the model, the normalized data distribution is reconstructed using a learnable scaling factor and a learnable translation factor. The resulting output represents the reconstructed feature response. The set of scaling factors corresponding to each channel in the convolutional layer is sparsely constrained using L1 regularization. After sparsity training, channels whose scaling factor values approach zero are regarded as secondary or less important channels. The regularization term on the scaling factor is incorporated into the overall network loss function. The total loss consists of the original neural network loss and an additional sparsity penalty term, weighted by a balance factor. This penalty encourages the scaling factors of unimportant channels to shrink toward zero. When computing the partial derivative of the loss function with respect to the scaling factor, the derivative becomes discontinuous at zero. If the left and right derivatives have opposite signs, the loss function reaches a minimum at this point. During subgradient descent optimization, the scaling factors are therefore driven toward zero, resulting in a sparse distribution across channels.

(b) *Channel pruning*

Based on the scaling factors obtained from sparsity training, each factor is mapped to its corresponding channel in the convolutional layer feature maps. The absolute values of these scaling factors are then sorted to evaluate channel importance. A quantization threshold is defined according to a preset pruning ratio. Feature channels with scaling factor values lower than this threshold are identified as redundant and removed from the network. To preserve the integrity of the network architecture, the pruning threshold is constrained so that it does not exceed the maximum scaling factor value within any batch normalization layer. This constraint ensures compatibility with the dimensional structure of the backbone network. In addition, for network architectures that include residual connections, the original residual structure is retained by maintaining dimensional consistency between the shortcut connections and the feature maps of the residual layers, thereby ensuring structural stability. The complete channel pruning process is illustrated in Fig. 5. In the feature maps of the i-th convolutional layer of the original network, certain feature channels exhibit scaling factor values significantly lower than the predefined threshold. By establishing a correspondence between each feature channel and its sparsified scaling factor, these low-importance channels are identified and pruned. At the same time, the associated convolution kernels in the preceding layers and their related feature channels are synchronously removed. This coordinated pruning strategy enables targeted optimization of network parameters and achieves precise compression of the overall model size.

The pruning ratio r is therefore an explicit compression knob. In this work, r = 0.2 is selected as a conservative setting that yields a meaningful reduction in parameters/model size while keeping sufficient channel capacity to preserve subtle fault signatures. Practically, increasing r further may remove informative channels and degrade

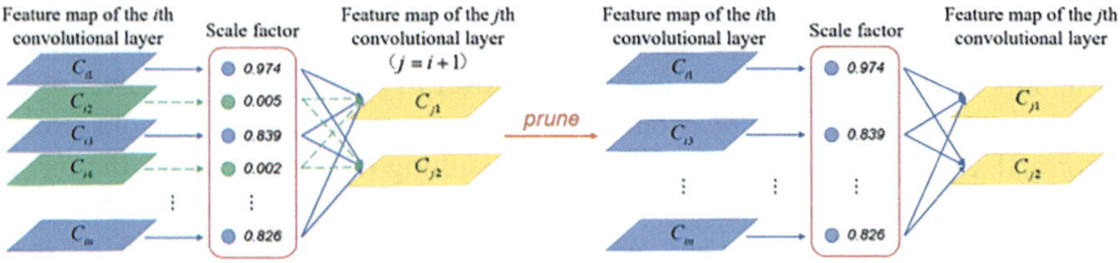

Fig. 5 Structure of channel pruning

diagnostic accuracy, whereas too small an r provides limited deployment benefit; this trade-off is revisited in the Discussion as an important source of sensitivity.

(c) *Fine-tuning*

To mitigate the negative impact of channel loss on detection accuracy, the initialization weights of the pruned network model are employed to implement a fine-tuning training strategy. This optimization process effectively restores the detection performance of the pruned model, thereby laying a crucial technical foundation for subsequently enhancing the object detection accuracy of the lightweight model.

3 Results

The experimental environment of this study is built on the Windows 10 OS, utilizing the PyTorch deep learning framework for algorithm development. The hardware configuration includes: a 12th Gen Intel® Core™ i7-12700H processor (base frequency 2.30 GHz), 16 GB DDR4 memory, and an NVIDIA GeForce RTX 3050 discrete graphics card (4 GB GDDR6 video memory). The software environment is configured with the PyTorch 2.1.0 + CUDA 12.1 combined framework.

3.1 Data Introduction

The CWRU test bench consists of a 1.5 kW motor, a torque sensor, a power meter, and multiple electronic controllers. The test bearing at the drive end is model SKF6205, with a sampling frequency set at 12 kHz. Three types of fault diameters were used: 0.1770 mm, 0.3556 mm, and 1.0160 mm, with corresponding loads of 0 kW, 0.75 kW, 1.50 kW, and 2.25 kW under each fault condition. Various fault types were examined under different fault diameters and loads, including inner race faults, rolling element faults, and outer race faults. For this experiment, 10 bearing condition datasets were selected under the conditions of 0 kW motor load and 1797r/min motor speed, as shown in Table 1 and Fig. 6.

In the experiment, the window size of the short-time Fourier transform was 64, with an overlap of 0.5. The generated time–frequency map had a size of 224 × 224. The training and test sets were partitioned in an 8:2 ratio. During network training, Batch size was 32, the optimizer was SGD, and 10 epochs were trained, with a pruning ratio of 0.2.

3.2 Experimental Result

(a) *Ablation experiment*

Table 1 Bearing status data

	Fault diameter /mm	Motor load /kW	Motor speed /(r/mkin)	Labels
Normal	-	0	1797	de_normal
Inner ring failure	0.1770	0	1797	de_7_inner
	0.3556	0	1797	de_14_inner
	1.0160	0	1797	de_21_inner
Rolling body failure	0.1770	0	1797	de_7_ball
	0.3556	0	1797	de_14_ball
	0.0160	0	1797	de_21_ball
Outer ring fault	0.1770	0	1797	de_7_outer
	0.3556	0	1797	de_14_outer
	1.0160	0	1797	de_21_outer

To thoroughly evaluate the efficacy and robustness of the YOLO-MVCP, a series of ablation studies are meticulously carried out, with outcomes meticulously documented in Table 2. For these experiments, YOLOv8-cls is utilized as the foundational Baseline model. The empirical results derived from these experiments are quite illuminating. Specifically, when the original Backbone of the Baseline is substituted with MobileViT, an intriguing enhancement in performance is observed. The accuracy surged to an impressive 100%, up 0.017 percentage points from the baseline. Additionally, this strategic replacement led to a substantial reduction in the number of parameters, with a decrease of 38.70%. Consequently, the overall size of the model is diminished by a notable 1018 kb. The underlying reason for these improvements lies in the unique architecture of MobileViT, which incorporates lightweight Transformer modules. These modules are adept at capturing long-range dependencies, effectively addressing the inherent limitations of CNNs in terms of global modeling. This augmentation not only compensates for the shortcomings of CNNs but also significantly elevates the accuracy of the model. Furthermore, MobileViT's approach to processing the input feature map by dividing it into non-overlapping patches, each of which is handled independently, allows for the achievement of global modeling through the utilization of shared Transformer layers. This innovative technique is instrumental in reducing both the number of parameters and the cost of computation associated with the model.

In addition to the aforementioned modifications, a model compression technique known as network pruning is also implemented in the experimental framework. Post-pruning, the quantity of model parameters is notably reduced to a mere 46.54% of the original model's parameter count. Correspondingly, the size of the model is also scaled down to 43.53% of its initial size. Despite these reductions, it is noteworthy that the diagnostic accuracy of the model remained steadfastly at 100% even after undergoing fine-tuning training. Comparative experiments conducted to assess the impact of integrating MobileViT with network pruning revealed that this synergistic

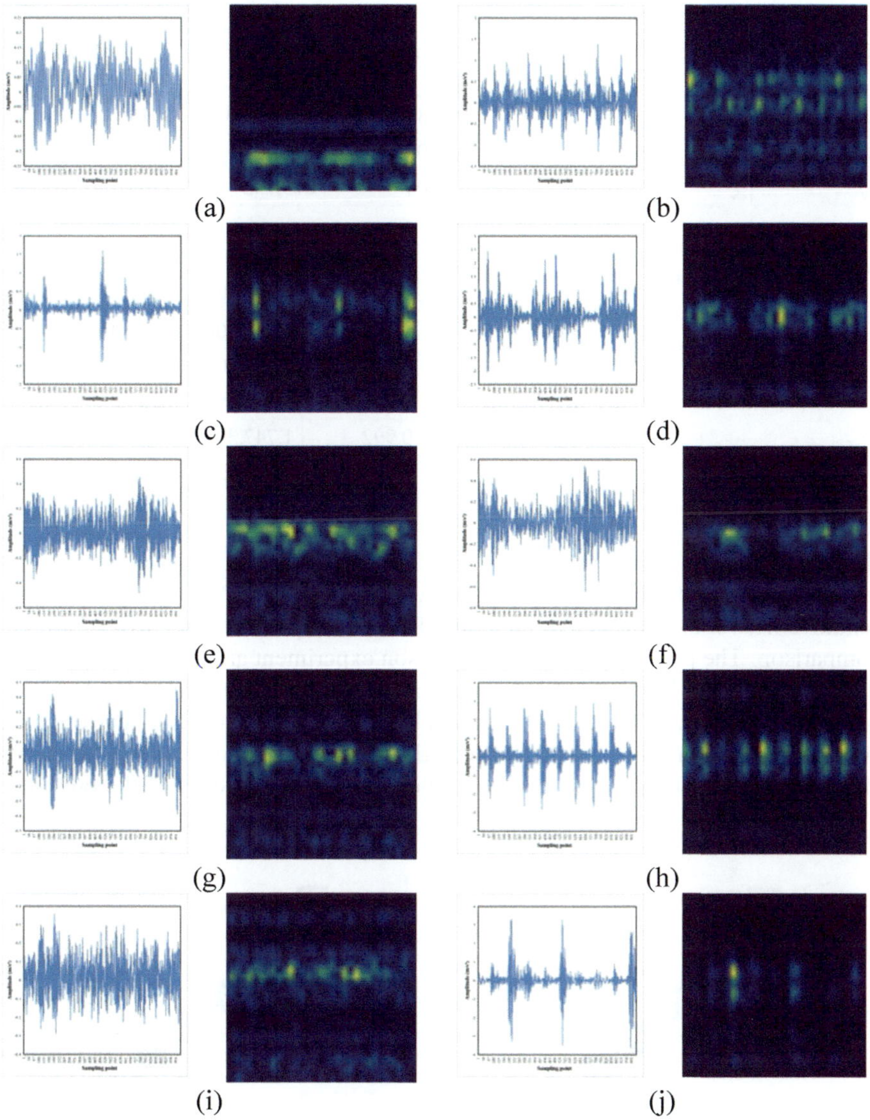

Fig. 6 Signal and STFT time–frequency diagram. **a** de_normal, **b** de_7_inner, **c** de_14_inner, **d** de_21_inner, **e** de_7_ball, **f** de_14_ball, **g** de_21_ball, **h** de_7_outer, **i** de_14_outer, **j** de_21_outer

approach not only optimized the overall performance of the model but also led to a significant reduction in computational load. These findings robustly validate the efficiency and practical applicability of method in the realm of bearing fault diagnosis, underscoring its potential to revolutionize the field with enhanced accuracy and reduced computational demands.

Table 2 Results of ablation experiment

Baseline	MobileViT	Network pruning	Accuracy	Parameters	Model size(kb)
√			0.983	1,447,690	2915
√	√		1.000	887,498	1897
√	√	√	1.000	673,732	1269

Table 3 Comparison of different models

	Accuracy	Parameters	Model size (kb)
ResNet-50	0.960	26,177,002	51,307
ResNet-101	0.984	45,083,018	88,618
RepVGG	0.987	1,492,650	3015
ParNet	0.992	1,747,354	3538
YOLO-MVCP	1.000	673,732	1269

(b) *Network model comparison*

In order to verify the performance advantages of YOLO-MVCP, several representative networks are selected, including ResNet, RepVGG, and ParNet for a fair comparison. The parameters in the comparison experiment are the same as those in the previous paper. The comparison results are shown in Table 3, and refer to Fig. 7.

From Table 3, YOLO-MVCP model demonstrates superior performance compared to alternative models, particularly in terms of computational efficiency and detection accuracy. These results further validate the efficacy of the presented methodology.

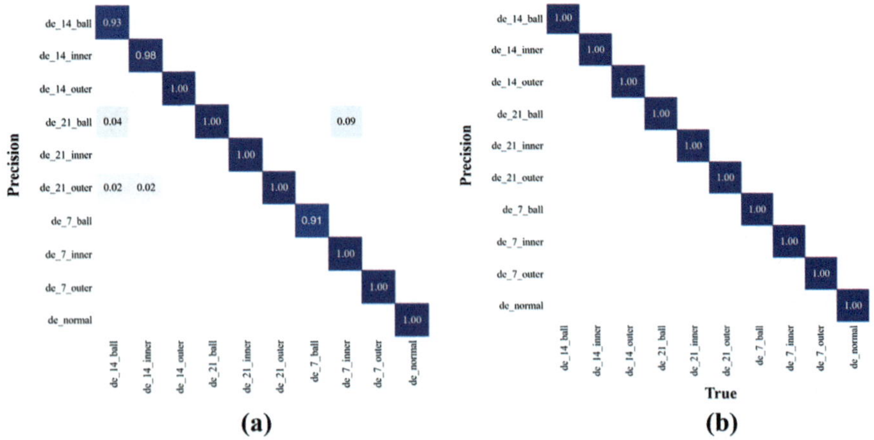

Fig. 7 Confusion matrix of the classification. (a)Baseline. (b)YOLO-MVCP

4 Discussion

The YOLO-MVCP's 100% accuracy on the CWRU dataset is achieved under a fixed 0 kW load and specific motor speed, and its generalization to higher loads (e.g., 0.75 kW, 1.50 kW) may be constrained by the dataset's limited load variation. Model robustness is sensitive to the pruning ratio: the selected 0.2 ratio balances size compression and performance retention, while higher ratios (>0.3) risk losing critical fault feature channels and lower ratios (<0.1) fail to achieve meaningful lightweighting. Uncertainty also arises from the dataset's focus on a single bearing model (SKF6205) and predefined fault diameters, which may limit applicability to diverse industrial scenarios with mixed fault types or unknown fault sizes. Future validation with multi-load, multi-model datasets will enhance the model's practical reliability and expand its engineering utility.

Beyond the headline accuracy, the ablation study provides insight into why the proposed design is effective. Replacing the baseline backbone with MobileViT improves accuracy while reducing parameters, which indicates that combining convolutional inductive bias (for local time–frequency textures) with lightweight self-attention (for long-range dependencies across time and frequency) is well matched to spectrogram-based bearing diagnosis. The pruning outcome further suggests that the original backbone contains considerable channel redundancy for this benchmark; removing low-importance channels preserves the learned discriminative structure while making the model more suitable for memory- and latency-constrained devices.

Regarding generalisation and robustness under varying loads, bearing vibration spectra can shift in amplitude and in the relative prominence of harmonics/sidebands as load and speed change. While the STFT representation and the hybrid backbone are designed to capture characteristic patterns that are not tied to a single time sample, domain shift across operating conditions may still reduce performance when the training data do not cover those regimes. In practice, the proposed pipeline is intended as a compact deployable model that can be (re)trained or calibrated with a small amount of representative data from the target machine, and the pruning strategy remains applicable after such adaptation.

Several limitations and sources of uncertainty should therefore be noted. First, the reported results rely on the CWRU seeded-fault benchmark under a fixed load setting in this study, which may overestimate performance relative to noisier industrial environments. Second, the pruning ratio is a sensitivity factor: more aggressive pruning can remove low-energy yet informative channels that encode subtle fault signatures, whereas under-pruning leaves deployment efficiency unrealised. Third, deep models involve stochastic optimisation, so point estimates of accuracy should be interpreted with this variability in mind. These considerations do not undermine the main finding—namely that high accuracy can be retained with a compact model—but they clarify the scope of the current evidence and motivate broader validation in future work.

5 Conclusion

This study proposes lightweight solutions within the domain of rolling bearing fault diagnosis by introducing an integrated diagnostic model that combines STFT for time and frequency analysis with the MobileViT network, optimized through strategic pruning techniques. The model's performance is rigorously validated using the CWRU dataset, yielding the following key findings:

This technique utilizes STFT to transform sensor-captured raw vibration signals from the time domain into two-dimensional time–frequency representations. This transformation enables joint time–frequency analysis, effectively capturing the transient characteristics of non-stationary signals and constructing a feature space enriched with physical interpretability. After standardized preprocessing, the resulting time–frequency dataset is input into a refined YOLO-MobileViT hybrid network architecture, which is subsequently trained to perform multi-class fault classification. The experimental results demonstrate that YOLO-MVCP model achieves superior performance in feature extraction and generalization across various operational scenarios, attaining a perfect classification accuracy of 100%. Moreover, the application of channel pruning combined with fine-tuning during training reduces the model's parameter count to 46.54% of the original configuration, while maintaining complete classification accuracy. Simultaneously, the model's size is compressed to 43.53% of its initial volume, thus achieving an optimal balance between diagnostic precision and computational resource efficiency.

Comparative analysis with established classification models highlights the superior effectiveness of the proposed approach. Notably, it surpasses ResNet-50, which achieved a classification accuracy of 96.0%, and RepVGG, which reached 98.7%, thereby demonstrating a clear performance advantage. These findings contribute to the theoretical foundation of lightweight diagnostic model development while also offering scalable, cost-effective solutions suitable for practical engineering applications.

Future research will focus on extending the proposed diagnostic framework to encompass compound fault detection in more complex rotating machinery, including gearboxes and rotor systems. This progression aims to support the advancement of intelligent equipment health management systems and enhance maintenance practices and operational reliability across diverse industrial settings. Future studies can be refer to the [14–19].

Acknowledgements The experimental data in the paper are sourced from the Case Western Reserve University (CWRU) rolling bearing dataset, with the project address: https://engineering.case.edu/bearingdatacenter/download-data-file. Gratitude is hereby expressed.

References

1. Hao S, Xin Y, Ling X et al (2022) A novel method based on deep transfer unsupervised learning network for bearing fault diagnosis under variable working condition of unequal quantity. Knowl-Based Syst 242:108381
2. Zhang S et al (2024) A novel SVD-SDP-CNN fault diagnosis method. In: 2024 IEEE international conference on sensing, diagnostics, prognostics, and control (SDPC), Shijiazhuang, China, pp 52–56
3. Kuang Z, Mao F, Zhao X, Wan X (2025) Bearing fault diagnosis under variable working conditions based on domain adaptation and improved depthwise separable convolution network. In: 2025 IEEE International Conference on Industrial Technology (ICIT), Wuhan, China, pp 1–6
4. Chen S, Chen L, Chen Z (2023) A bearing health monitoring method based on interpretable improved GoogLeNet. In: IEEE International Conference on Sensors, Electronics and Computer Engineering (ICSECE). IEEE, pp 446–450
5. Ziran G, Ming Y (2022) TFFC-RNN: a new RNN based approach for bearing and misalignment compound fault. In: 2022 international power electronics conference (IPEC-Himeji 2022-ECCE Asia). IEEE, Himeji, Japan, pp 2504–2509
6. Guangxing N, Enhui L, Xuan W et al (2023) Enhanced discriminate feature learning deep residual CNN for multitask bearing fault diagnosis with information fusion. IEEE Trans Ind Inf 19(1):762–770
7. Vaswani A. Shazeer N, Parmar N et al (2017) Attention is all you need. In: Advances in neural information processing systems, vol 30
8. Yifei D, Minping J, Qiuhua M et al (2022) A novel time-frequency Transformer based on self-attention mechanism and its application in fault diagnosis of rolling bearings. Mech Syst Signal Process 168:108616
9. Bao Z, Du J, Zhang W et al (2021) A transformer model-based approach to bearing fault diagnosis. In: Data science:7th international conference of pioneering computer scientists, engineers and educators, ICPCSEE 2021. Springer Singapore, pp 65–79
10. Hinton G, Vinyals O, Dean J (2015) Distilling the knowledge in a neural network. Comput Sci 14(7):38–39
11. Wang S, Xie T, Cheng J, Zhang X, Liu H (2023) MDL-NAS: a joint multi-domain learning framework for vision transformer. In: 2023 IEEE/CVF conference on computer vision and pattern recognition (CVPR), Vancouver, BC, Canada, pp 20094–20104
12. Nagel M, Fournarakis M, Amjad RA et al (2021) A white paper on neural network quantization. https://doi.org/10.48550/arXiv.2106.08295.
13. Zhuang L, Jianguo L, Zhiqiang S et al (2017) Learning efficient convolutional networks through network slimming. In: Proceedings of 2017 IEEE international conference on computer vision. IEEE, Venice, pp 2755–2763
14. Mehta S, Rastegari M (2021) MobileViT: light-weight, general-purpose, and mobile-friendly vision transformer. https://doi.org/10.48550/arXiv.2110.02178
15. Guo J, Wang Z, Li H, Yang Y, Huang CG, Yazdi M, Kang HS (2024) A hybrid prognosis scheme for rolling bearings based on a novel health indicator and nonlinear Wiener process. Reliab Eng Syst Saf 245:110014
16. Li H, Yazdi M, Huang HZ, Huang CG, Peng W, Nedjati A, Adesina KA (2023) A fuzzy rough copula Bayesian network model for solving complex hospital service quality assessment. Complex Intell Syst 9(5):5527–5553
17. Li H, Peng W, Adumene S, Yazdi M (2023) An intelligent cost-based consequence model for offshore systems in harsh environments. In: Intelligent reliability and maintainability of energy infrastructure assets. Springer Nature Switzerland, Cham, pp 107–117
18. Li H, Yazdi M (2022) What are the critical well-drilling blowouts barriers? A progressive DEMATEL-game theory. Advanced decision-making methods and applications in system

safety and reliability problems: approaches, case studies, multi-criteria decision-making, multi-objective decision-making, fuzzy risk-based models. Springer International Publishing, Cham, pp 29–46
19. Li H, Peng W, Adumene S, Yazdi M (2023) Cutting edge research topics on system safety, reliability, maintainability, and resilience of energy-critical infrastructures. In: Intelligent reliability and maintainability of energy infrastructure assets. Springer Nature Switzerland, Cham, pp 25–38

Design of System-Level Self-health Energy Management of Manned Spacecraft

Qian Liu, Fangyuan Wen, Zhen Lv, Minjun Mao, and Rong Fan

Abstract Given the long service life and high reliability requirements of manned spacecraft, the effectiveness of fault diagnosis and self-health management directly influences the autonomy of spacecraft during on-orbit operations. Using energy health monitoring as a representative case, two modes of system-level self-health energy management are proposed. The design of these modes is detailed with respect to activation timing, health assessment methodologies, and autonomous fault-handling strategies. Within these designs, fault detection and diagnosis are performed to deliver timely fault information to astronauts and ground engineers. For faults posing a threat to astronaut safety or leading to system failure, autonomous recovery mechanisms are incorporated to restore functionality and transition the spacecraft into a safety mode. The results provide technical support for enhancing fault diagnosis and self-health management in future spacecraft.

Keywords Fault diagnosis · System-level · Self-health energy management · Autonomous disposal · Safety mode

1 Introduction

The autonomous operation capabilities of manned spacecraft have progressively advanced to meet the demands of complex structures, diverse mission profiles, and extended on-orbit durations. This evolution has intensified the need for a robust self-health management system [1–4]. Autonomous health management systems for satellites are designed around the principles of "unmanned intervention" and "autonomous recovery," leveraging onboard redundancy to ensure long-term stable operation under unattended conditions [5–10]. As noted in [11–14], engineering implementations

Q. Liu · F. Wen · Z. Lv · M. Mao (✉) · R. Fan
Aerospace System Engineering Shanghai, Shanghai, China
e-mail: roymao@163.com

Q. Liu
e-mail: liuqian8803@163.com

H. Li et al. (eds.), *Data-Driven Methods for Reliability and Safety Engineering: Applications in Industrial Systems*, Springer Series in Reliability Engineering,
https://doi.org/10.1007/978-3-032-22873-4_25

of health management systems have been applied to critical subsystems, including propulsion subsystems, Telemetry, Tracking, and Command (TT&C) communication subsystems, and energy subsystems. System-level anomaly detection, fault diagnosis, and fault prediction tools have been developed through lessons learned from other spacecraft platforms [15, 16]. Furthermore, a hierarchical architecture encompassing service deployment, fault diagnosis, and fault mode configuration has been proposed to enable precise and efficient health management [17].

Unlike unmanned spacecraft, manned spacecraft operations are governed by the overriding principle of ensuring astronaut safety [18]. Minimizing the time astronauts spend on fault isolation and maintenance is essential, as every minute on orbit directly influences mission operational efficiency. The system-level self-health management framework is designed to address faults that endanger astronaut safety or cause platform malfunctions. In such scenarios, the spacecraft autonomously conducts fault diagnosis and transitions into safety mode, awaiting further directives from the ground control center. This system provides comprehensive support for critical subsystems, including energy management, attitude control, propulsion, information links, thermal regulation, and leakage detection.

In cases of severe or urgent failures, rapid response is essential to transition the spacecraft into safety modes. The system-level self-health management architecture is specifically designed to autonomously perform fault detection, diagnosis, and mitigation during on-orbit operations without reliance on ground intervention, thereby safeguarding both the spacecraft and its crew. Concurrently, fault information is transmitted promptly to the ground control center and communicated to astronauts to ensure situational awareness and coordinated response.

2 Requirements Analysis of the System-Level Self-health Management

During on-orbit operation, the self-health management system continuously monitors the health status of critical subsystems, including the information system, energy system, attitude control system, and manned environment control system, in real time. Upon detection of a fault, the system performs diagnostic procedures to determine the fault category. For faults classified as critical or urgent, the system evaluates the necessity of initiating safety modes and autonomously implements preliminary mitigation measures. It then awaits further detailed analysis and follow-up actions from the comprehensive testing system or the ground flight control support system, as illustrated in Fig. 1. As a representative case study, the safety modes associated with energy faults and single-bus power failures are examined separately. The operational requirements for these two safety modes encompass information link integrity, attitude control stability, thermal regulation, manned environment maintenance, and energy management.

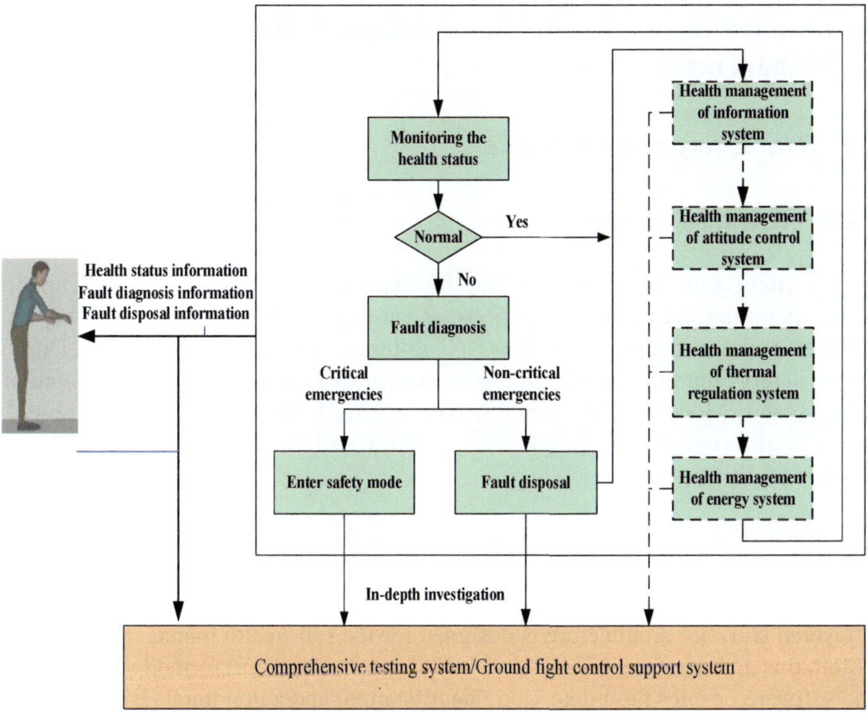

Fig. 1 Flowchart of the self-health management system

The information links feature uplink command and downlink telemetry, enabling downlink transmission of health status in safety mode and fault-diagnosis telemetry data, as well as uplink reception of ground commands for fault handling. The attitude control system ensures that the spacecraft's attitude meets the safety-mode requirements for energy system operation and TT&C link stability. The thermal regulation system ensures heat dissipation of equipment in safety mode and ensures that cabin interior temperature and humidity remain within operational limits to prevent condensation. The manned environment control system supports maintaining total pressure, oxygen partial pressure, and carbon dioxide partial pressure within operational thresholds, and provides ventilation in crew-occupied areas of the cabin. Energy for the above functions shall be maintained by the spacecraft until the platform recovers or exits the safety mode.

3 Detailed Design for the System-Level Self-health Energy Management

3.1 The Energy Safety Mode

3.1.1 Activation Timing

Energy safety control is not activated during extravehicular activities (EVAs), manipulator operations, maintenance missions, or astronaut shift handovers. It will be initiated before the spacecraft transitions to long-term operational management. During energy safety control activation, the system will send the command of Autonomous Energy Safety Mode Prohibition before the predicted battery discharge depth exceeds the 50% ultra-threshold. Energy safety control will be reactivated only after the battery is fully charged.

3.1.2 Health Detection Methodologies

A layered software architecture is designed for the self-health management system. Within this architecture, system-level health monitoring is performed via telemetry, and software enables fault detection, identification, and autonomous handling.

As depicted in Fig. 2, the data flow of the energy safety mode clearly shows the interaction between subsystems. The core processing unit alternately sends the capacity parameters of battery packs a–h (from the two power channels A and B transmitted by the power subsystem) to the energy manager at intervals of one cycle. The

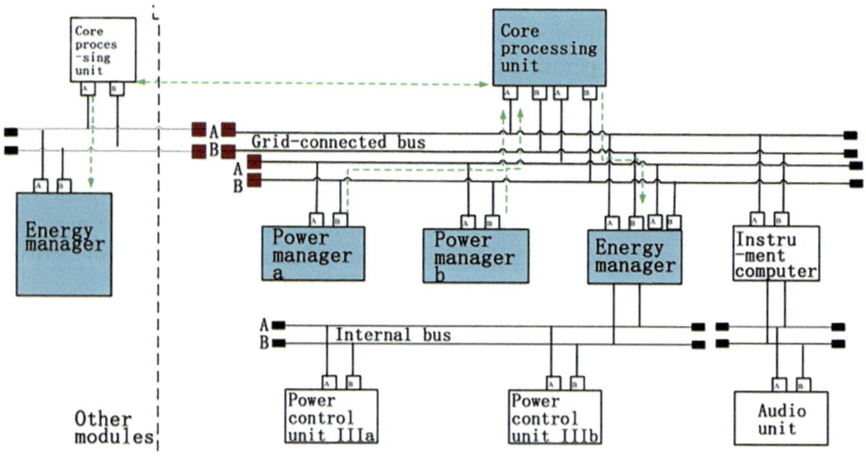

Fig. 2 Data flow of energy safety mode

energy manager performs relevant judgments and conducts inter-module information interaction with the core processing units and energy managers of other modules through the grid-connected bus.

When it is determined that the depth of discharge (DOD) of any three or more of the four battery packs in the two power channels (A and B) of the spacecraft exceeds 50%, the energy manager sets the flag "Channel A Ready to Switch to Energy Safety Mode" or "Channel B Ready to Switch to Energy Safety Mode". It then makes a decision according to the following flow chart, as shown in Fig. 3.

When the flag "Channel A Ready to Switch to Energy Safety Mode" is set, the energy manager checks whether the DOD of Channel B's batteries exceeds 30%. If all four battery packs' DODs remain below 30%, it sets the "Energy Safety Mode 1" flag and requests the kernel processing unit to initiate Energy Safety Mode 1 disposal. Conversely, if any battery pack's DOD exceeds 30%, it sets the "Energy Safety Mode 3" flag and triggers Mode 3 disposal.

The selected DOD thresholds (50% for mode activation and 30% for cross-channel validation) are justified by balancing battery cycle life preservation, emergency power redundancy, and minimum power requirements for critical subsystems. The dual-bus architectural assumption ensures fault isolation, enabling single-channel failures to be mitigated without compromising essential functions, while the three-battery-pack fault criterion reduces false triggers by accounting for minor telemetry fluctuations and transient load impacts.

Similarly, when the flag "Channel B Ready to Switch to Energy Safety Mode" is set, the energy manager checks Channel A's batteries. If all DODs are below 30%, it sets the "Energy Safety Mode 2" flag and initiates corresponding disposal. Otherwise, it activates Mode 3.

3.1.3 Autonomous Disposal Strategies

The energy control is initiated. Then, if the energy manager detects a fault in a channel, the data management subsystem successfully receives the commands of "Kernel Processing Unit Starts Energy Safety Mode 1, 2, or 3 Disposal". The system will execute the autonomous shutdown instruction chain for the corresponding mode. Power to non-essential equipment in the faulty channel will be cut off to reduce consumption, enabling the channel to enter a low-power safety state.

The data management subsystem then clears subsequent flight programs. The ground controllers analyze engineering telemetry to identify the "Energy Safety Mode Trigger Flag" within TT&C arcs. If the entire spacecraft is in safety mode, ground control engineers will resend the program-controlled instruction chain. The corresponding fault localization and handling will be carried out after the ground control engineers make a decision.

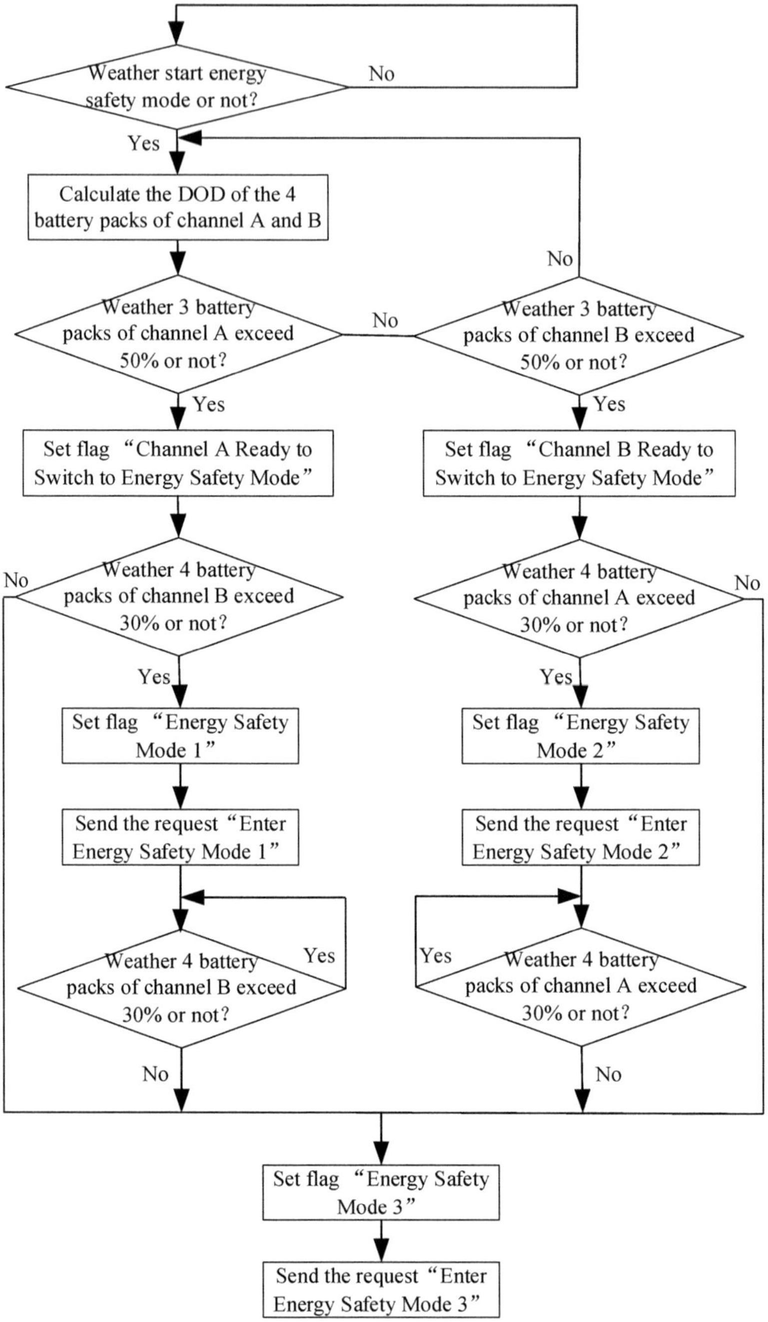

Fig. 3 Test flow chart of energy safety mode

3.2 *The Single Bus Power Failure Safety Mode*

3.2.1 Activation Timing

The single bus power failure safety mode is prohibited during maintenance mission phases and manipulator operations. During normal flight phases, the self-health management function for a single bus power failure can be enabled via ground commands.

3.2.2 Health Detection Methodologies

As in Sect. 3.1.2, the same software architecture is applicable here. To elaborate, the health-related telemetry parameters of a single bus power failure are analyzed to support fault diagnosis. The primary criteria for bus A or B power failure are analog telemetry reflecting bus power supply status, complemented by the offline status of bus loads as a secondary failure criterion.

As illustrated in Fig. 4, the data flow corresponding to the safety mode for single power bus failure provides a clear representation of how individual equipment interact with one another. There are two types of commands for the spacecraft, namely the program-controlled command and the bus command, both of which are distributed by the core processing unit. For the bus command, it is directly sent to the bus terminal after encoding. For the program-controlled command, after the integrated service unit receives the command code, it decodes the code into the corresponding hardwired command to realize the drive control of the equipment. When Bus A loses power, both the equipment that executes the device shutdown or power-off commands can be normally powered by Bus B, and all power-off disposal commands for Bus A can be sent normally. The same applies to the scenario where Bus B loses power.

Judgments are made according to the testing flow chart below. The command "Enable Autonomous Single Bus Power Failure Disposal" is received by the system. If the software periodically detects offline status on multiple bus terminals across different buses and analog telemetry indicating zero bus voltage, it will determine that the single bus power failure safety mode should be activated, refer to Fig. 5.

3.2.3 Autonomous Disposal Strategies

The command "Enable Autonomous Single Bus Power Failure Disposal" is received by the system. The corresponding fault handling instruction chains are set up for a single bus power failure. The "Autonomous Instruction Chain for Bus A Power Failure Disposal" and "Autonomous Instruction Chain for Bus B Power Failure Disposal" are triggered by power failure detection, and are executed automatically

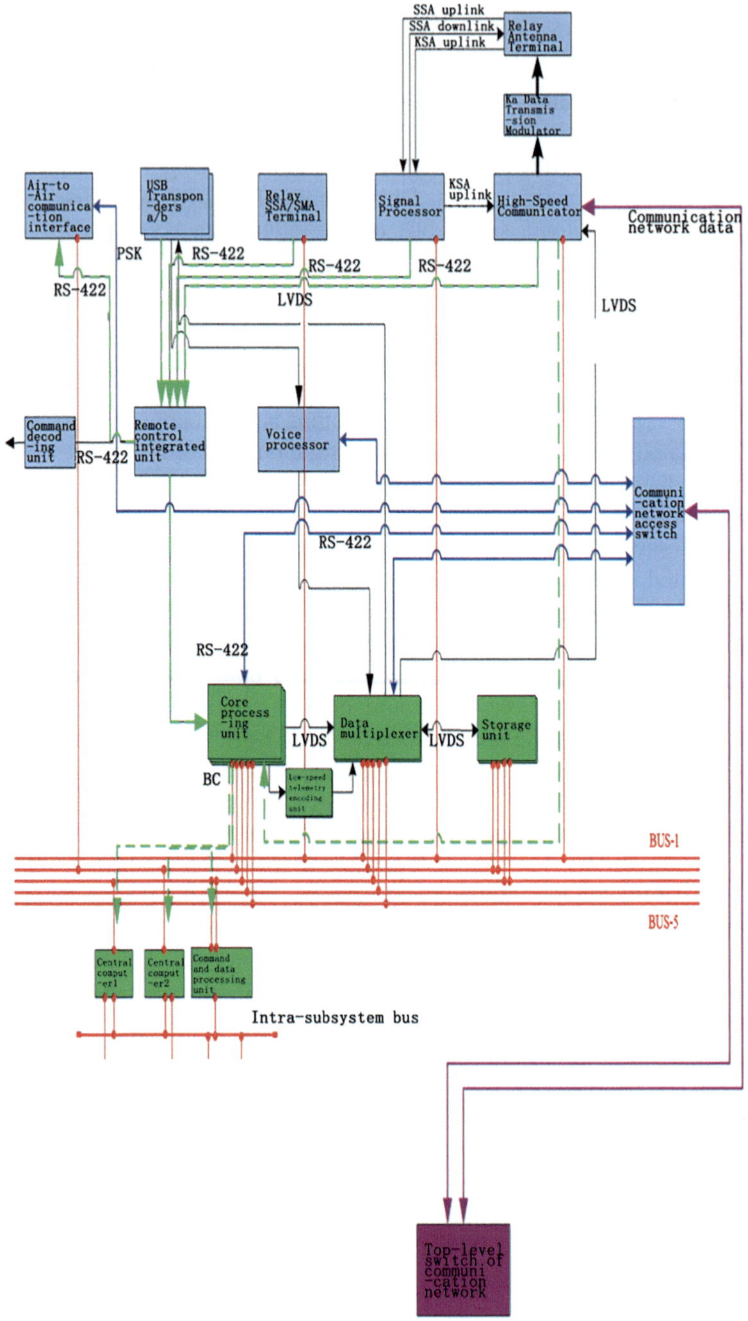

Fig. 4 Data flow of single bus power failure safety mode

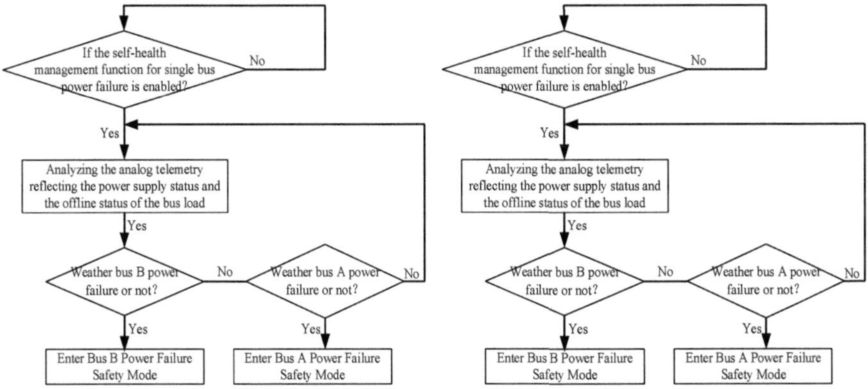

Fig. 5 Testing flow chart of single bus power failure safety mode

to ensure spacecraft safety and basic function operation. When the command "Prohibit Autonomous Single Bus Power Failure Disposal" is received, the system halts judgment until the enable command is reissued.

Ground control engineers analyze downlink engineering telemetry to identify bus power failure faults. Astronauts manually connect backup power supply channels and monitor environmental status. Based on telemetry and ground contingency plans, the system's dual-bus power supply mode is gradually restored after fault elimination.

In high-risk scenarios such as complete single-bus power loss during long-duration on-orbit phases, the autonomous system first executes load shedding to preserve backup power, then transmits real-time fault telemetry and visual alerts to astronauts via the onboard display interface. Ground control receives synchronized data to conduct rapid analysis, while astronauts perform verification checks (e.g., confirming bus offline status) and execute guided manual interventions (e.g., backup channel activation) only when autonomous disposal requires human confirmation. This coordinated response—autonomous fault containment, astronaut situational awareness, and ground decision support—ensures operational feasibility by leveraging the strengths of each party while minimizing human–machine conflict.

4 On-Orbit Verification for the System-Level Self-health Energy Management

To verify the effectiveness of the on-orbit enabling and disposal functions of the system-level autonomous health energy management, an on-orbit verification was conducted on a specific spacecraft to ensure the reliable on-orbit operation of the equipment. The on-orbit monitoring page is shown in Fig. 6. During the long-term operation period, the enabling flag was set to "Enabled". The system remained in the normal state without entering the safety mode or triggering the autonomous disposal process. And meanwhile, the observed voltage and current were within the normal ranges.

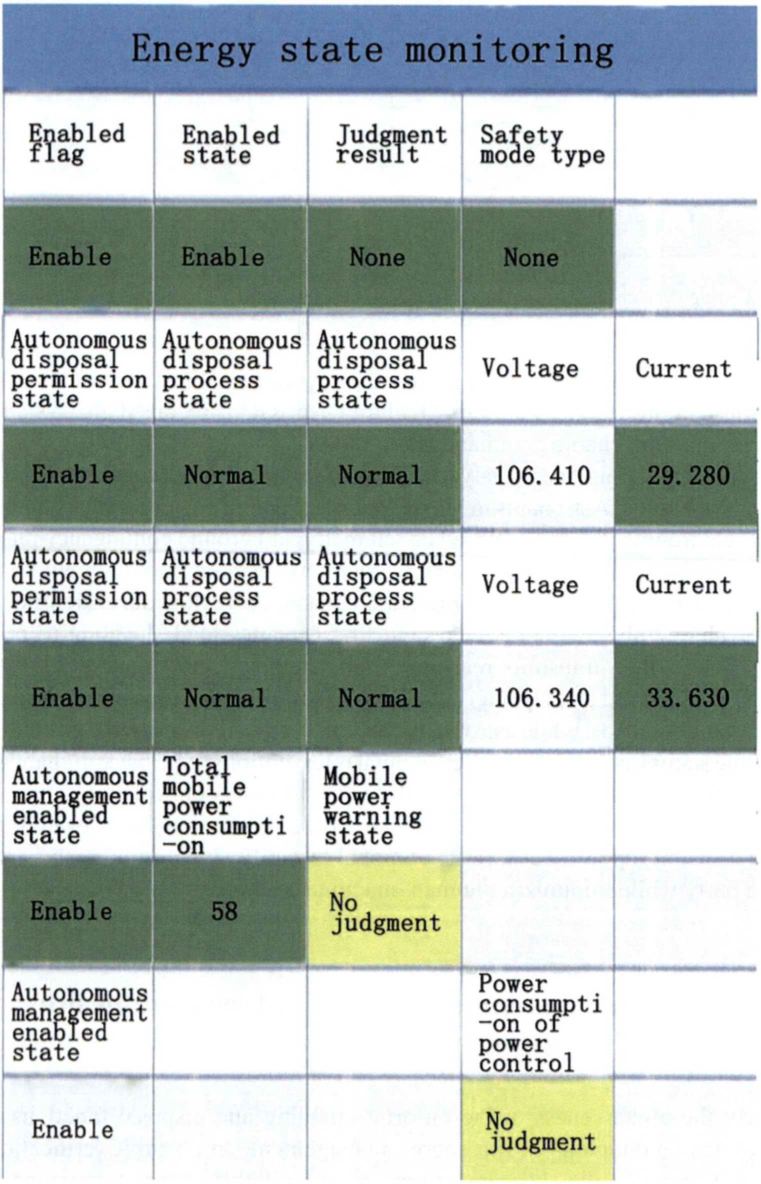

Fig. 6 On-orbit monitoring page for system-level self-health energy management

5 Conclusion

This study presents a comprehensive analysis of system-level self-health energy management for manned spacecraft, detailing the design aspects of activation timing, health detection methodologies, and autonomous disposal strategies. The proposed approach enables manned spacecraft to monitor their operational status, autonomously detect, isolate, and recover from incipient faults, and promptly transmit fault information to both ground control engineers and onboard astronauts. The resulting design framework provides technical support for the development of fault diagnosis and self-health management systems in forthcoming manned lunar landing missions and deep space exploration initiatives. Further studies can be referred to [19–24].

References

1. Gutierrez T, Coulter N, Moncayo H et al. (2023) Data-driven health management system for multi-spacecraft on-orbit servicing. AIAA SciTech Forum and Exposition
2. Yash P, Gundawar S, Kumar N et al. (2024) Multiforecast-based early anomaly detection for spacecraft health monitoring. In: ACM international conference proceeding series, pp 275–283
3. Figueroa J, Melcher K (2011) Integrated systems health management for intelligent systems. Infotech@ Aerospace 2011, p 1492
4. Cao GR (2007) The research and application of self-health management technology for spacecraft. Master's thesis, National University of Defense Technology
5. Yi W, Yi WF, Li KW et al (2023) Quasi-real-time on-orbit health assessment system for payloads. Aerosp Shanghai 40(6):105–110
6. Li G, Shi BZ, Dai YS (2021) Research on autonomous health management of navigation satellite. Comput Measure Control 29(3):104–107
7. Pan YQ, Feng WJ, Zhang G (2021) Construction method of navigation satellites health management evaluation index system. Spacecr Eng 30(4):8–14
8. Zhao SY, Li WD, Yang LQ (2021) Design and verification of autonomous health management system for GFDM-1 satellite. Spacecr Eng 30(3):64–68
9. Qiu CR (2023) Research on self-health management of the terrestrial satellite communication system. Wirel Internet Technol 10:15–17
10. Wang WP, Yuan J, Yuan Y et al (2017) Design of autonomous safety control strategy for SJ-10 reentry satellite. Comput Meas Control 25(12):60–63
11. Li WJ, Chi YB, Li CG (2025) Design of health condition prediction system for space power supply. Aerosp Shanghai 42(2):49–57
12. Peng XH (2017) Study on some key techniques of autonomous health management for spacecraft propulsion system. Ph.D. thesis, National University of Defense Technology
13. Long Y, Li FY, Huang C et al (2023) Autonomous health management method for TT&C tecommunication subsystem of spacecraft. Spacecraft Eng 32(4):71–80
14. Liang XF, Ren XQ, Ding R et al (2022) Design and verification of the energy system of the China space station core module. Sci Sinica Technol 52(9):1345–1354
15. Li RX, Zhang ZX (2020) Enlightenment of ISS health management system on construction of China space station. Manned Spacefl 26(1):120–127
16. He XW, Guo J, Li YQ (2019) Autonomous health management requirements and software architecture for deep space probe. Control Theory Appl 36(12):2065–2073
17. Liu Q, Mao MJ, Wen FY et al. (2023) Self-health management method for spacecraft based on distributed system architecture. ICRMS, pp 898–901

18. Liang K, Deng KW, Ding R (2014) Autonomous on-orbit health management architecture and key technologies for manned spacecrafts. Manned Spacefl 20(2):116–121
19. Guo J, Wang Z, Li H, Yang Y, Huang CG, Yazdi M, Kang HS (2024) A hybrid prognosis scheme for rolling bearings based on a novel health indicator and nonlinear Wiener process. Reliab Eng Syst Saf 245:110014
20. Li H, Yazdi M, Huang HZ, Huang CG, Peng W, Nedjati A, Adesina KA (2023) A fuzzy rough copula Bayesian network model for solving complex hospital service quality assessment. Complex Intell Syst 9(5):5527–5553
21. Li H, Peng W, Adumene S, Yazdi M (2023) An intelligent cost-based consequence model for offshore systems in harsh environments. In: Intelligent reliability and maintainability of energy infrastructure assets, pp 107–117. Springer Nature Switzerland, Cham
22. Li H, Yazdi M (2022) What are the critical well-drilling blowouts barriers? A progressive DEMATEL-game theory. Advanced decision-making methods and applications in system safety and reliability problems: approaches, case studies, multi-criteria decision-making, multi-objective decision-making, fuzzy risk-based models. Springer International Publishing, Cham, pp 29–46
23. Li H, Peng W, Adumene S, Yazdi M (2023) Cutting edge research topics on system safety, reliability, maintainability, and resilience of energy-critical infrastructures. In: Intelligent reliability and maintainability of energy infrastructure assets, pp 25–38. Springer Nature Switzerland, Cham
24. Li H, Peng W, Adumene S, Yazdi M (2023) Advances in failure prediction of subsea components considering complex dependencies. In: Intelligent reliability and maintainability of energy infrastructure assets, pp 93–105. Springer Nature Switzerland, Cham

A PINN-CNN Hybrid Architecture for Vibration-Based Damage Detection in Airship Envelopes

Fangyuan Zhao, Yiwei Cheng, Keqiang Xie, and Yuanhang Wang

Abstract Conventional methods for detecting damage in flexible airship envelopes face limitations in accurately evaluating local material properties under conditions of large deformation. A Physics-Informed Neural Networks Convolutional Neural Networks (PINN-CNN) hybrid architecture is proposed, integrating physical constraints derived from wave equations and membrane dynamics into the neural network training process. The approach was evaluated using synthetic airship envelope data under three conditions: intact, 2×2 cross-cut damage, and 5×5 cross-cut damage. Experimental findings indicate that the PINN-CNN model attained 100% accuracy, representing a notable improvement over the traditional CNN, which achieved an accuracy of 96.67%, while also exhibiting superior robustness to noise. Under conditions with 25% noise, the PINN-CNN model maintained an accuracy of 90.00%, whereas the standard CNN performance declined to 70.00%.

Keywords Damage detection · Vibration signal analysis · Noise robustness · Physics-informed neural networks

F. Zhao · Y. Cheng (✉)
School of Mechanical Engineering and Electronic Information, China University of Geosciences (Wuhan), Wuhan, China
e-mail: chengyiwei@cug.edu.cn

F. Zhao
e-mail: fangyuanzhao@cug.edu.cn

Y. Cheng
Shenzhen Research Institute of China University of Geosciences, Shenzhen, China

K. Xie
Guangdong Provincial Key Laboratory of Electronic Information Products Reliability Technology, Guangzhou, China
e-mail: xiekeqiang@stu.edu.cn

Y. Wang
Sino-German College of Intelligent Manufacturing, Shenzhen Technology University, Shenzhen, China
e-mail: wangyuanhang@sztu.edu.cn

H. Li et al. (eds.), *Data-Driven Methods for Reliability and Safety Engineering: Applications in Industrial Systems*, Springer Series in Reliability Engineering,
https://doi.org/10.1007/978-3-032-22873-4_26

1 Introduction

Airship envelopes, as large flexible membrane structures, serve critical roles in stratospheric communication, meteorological observation, and military reconnaissance. However, exposure to complex operational environments can induce fatigue cracks, local tears, and material aging in envelope materials, making early damage detection essential for ensuring flight safety. Conventional non-destructive testing methods, such as ultrasonic testing and infrared thermography, show limited effectiveness for real-time monitoring of large envelope structures due to restricted coverage, low efficiency, and insufficient sensitivity to incipient damage. In addition, envelope structures experience substantial deformation and complex aerodynamic loads during flight, which makes traditional static detection approaches inadequate for dynamic monitoring requirements.

In recent years, structural health monitoring technologies based on vibration signal analysis have emerged as promising solutions for airship envelope damage detection. Vibration monitoring offers advantages including non-contact measurement, wide spatial coverage, and high sensitivity to minor damage, making it well-suited for large flexible structures. Nevertheless, the vibration response of airship envelopes is influenced by multiple coupled factors such as pressure fluctuations, temperature gradients, and wind load disturbances. Conventional signal processing methods encounter difficulties in reliably extracting damage-related features from these multiphysics signals. While data-driven deep learning approaches are effective for pattern recognition, they generally require extensive labeled datasets and, without physical constraints, often show poor generalization capability for the dynamic operating conditions of airship envelopes.

Physics-Informed Neural Networks (PINNs), an emerging scientific machine learning paradigm, integrate data-driven methods with physical models by embedding the governing partial differential equations of airship envelope dynamics directly into the neural network loss function. Compared with purely data-driven methods, PINNs can produce physically consistent solutions using limited observational data, thereby improving both generalization capability and reliability. In membrane and thin-shell structural analysis, PINNs have demonstrated significant advantages in solving high-order partial differential equation systems such as the Kirchhoff–Love plate theory and Naghdi shell theory. However, most existing research on PINNs has focused on static structural analysis. For dynamic systems with large deformations, strong nonlinearities, and multi-scale coupling, such as airship envelopes, important challenges remain in constructing effective physical constraints, managing complex boundary conditions, and achieving real-time damage detection. Raissi et al. [1] established the complete theoretical framework of PINNs, introducing a methodology for embedding partial differential equations directly into neural network loss functions and providing the theoretical basis for engineering applications in structural damage detection. This work enabled the integration of data-driven techniques with physical laws by enforcing governing equations during model training. Jagtap et al. [2] proposed Conservative Physics-Informed Neural Networks (Conservative

PINNs) specifically for conservation law problems in discrete domains, offering new numerical strategies for solving both forward and inverse problems in structural mechanics. Yang and Perdikaris [3] introduced adversarial uncertainty quantification methods, addressing the reliability assessment issues of PINNs in practical engineering applications and providing theoretical support for damage detection confidence quantification. Zhu et al. [4] first applied PINNs to temperature and melt pool fluid dynamics prediction in metal additive manufacturing processes, pioneering the application of PINNs in material processing damage prediction. Wang et al. [5] conducted in-depth research on gradient flow pathologies in PINNs training processes, proposing effective mitigation strategies that provide technical guarantees for stable simulation of complex engineering structures. Kashinath et al. [6] systematically summarized physics-informed machine learning applications in weather and climate simulation, providing interdisciplinary methodological guidance for large-scale structural health monitoring. Liu et al. [7] proposed a physics-informed machine learning method for composition-process-property design, validating its effectiveness in material damage prediction through shape memory alloy case studies. Zhang et al. [8] published milestone work in "Science Advances," successfully characterizing material internal structures and defects using PINNs for the first time, accurately identifying the size, shape, and location of voids and inclusions by parameterizing material geometries through meshless methods. Lawal et al. [9] published a systematic literature review and bibliometric analysis on PINNs development, compre hensively organizing the field's development trajectory and application prospects. Wang et al. [10] explored the suitability of L^2 physics-informed loss functions, providing theoretical guidance for loss function design in PINNs structural health monitoring. Sharma et al. [11] published a comprehensive review of PINNs in fluid mechanics, providing a methodological foundation for damage detection in fluid–structure coupled structures. Bolandi et al. [12] developed a PINNs-based dynamic stress prediction method, achieving real-time stress monitoring of structures under dynamic loads. He et al. [13] proposed a multi-level physics-informed deep learning framework in "Communications Engineering," significantly improving the accuracy and efficiency of structural mechanics calculations by aggregating multiple neural networks to separately handle geometry, constitutive relations, and equilibrium relationships. Al-Adly et al. [14] published a case study of PINNs in structural health monitoring, using Kirchhoff–Love plates as an example to verify PINNs' advantages in handling complex boundary conditions and time-varying loads. Yamaguchi et al. [15] proposed a PINNs-based nonlinear damage identification algorithm for reinforced concrete bridge piers, successfully handling complex system identification problems with approximately 50 unknown parameters. A research team [16] proposed a "maintenance-free" structural damage automatic detection and localization method in "Nature Communications," integrating passive measurement, data compression, and mechanics-informed encoding networks to achieve breakthrough autonomous detection and localization of various unforeseen damages using only 3 h of data. Chen et al. [17] developed a bearing fault diagnosis method based on vibration envelope spectral characteristics, reducing algorithm complexity by combining naive

Bayes classifiers to provide efficient solutions for rotating machinery damage detection. Pan et al. [18] proposed a new method for detecting breathing cracks in blades based on vibration response envelope analysis, achieving precise diagnosis of early cracks by extracting Envelope Diagram of Impulse Vibration Response (EDIVR) features. Researchers [19] conducted lightweight optimization design research for stratospheric airship envelopes based on reliability analysis, systematically characterizing three failure modes: bending wrinkling failure, circumferential tearing failure, and bending twisting failure. Shi et al. [20] published research on airship envelope material performance in coastal environments in "Frontiers in Materials," systematically evaluating the mechanical performance degradation patterns of different envelope materials in complex environments through 24-month natural aging tests. Bauer et al. [21] published a comparative study of envelope demodulation methods in rolling bearing damage analysis, providing a methodological foundation for vibration monitoring of rotating components in airship envelopes. This research revealed the influence mechanism of stray currents on gearbox bearings through envelope and spectral analysis.

This study develops a physics-aware deep learning framework combining wave dynamics and membrane theory for airship envelope vibration-based damage detection. The main innovation of this method is:

(1) Establishing a physics-informed neural network theoretical framework for airship envelope dynamics, integrating nonlinear membrane theory, wave equations, and damage mechanics into deep learning models;
(2) Proposing a multi-scale physical constraint strategy that solves numerical stability problems under large deformation conditions by hierarchically introducing geometric constraints, constitutive constraints, and dynamic constraints;
(3) Developing a damage feature extraction method based on vibration modal decomposition that can automatically identify damage-related modal components from multi-sensor signals;
(4) Constructing an airship envelope vibration dataset containing various damage modes, and verifying the efficacy of the engineered approach by employing numerical analysis and empirical verification methods.

2 Methodology

Traditional deep learning methods demonstrate strong performance in structural damage detection applications. However, these data-driven methods require extensive labeled datasets and provide limited physical interpretability, restricting their use in safety–critical aerospace applications. To overcome these constraints, Physics-Informed Neural Networks (PINNs) offer an innovative framework that incorporates physical laws as prior knowledge into deep learning frameworks.

In the context of airship envelope damage detection, traditional CNN-based methods face several challenges: (1) limited training data due to the high cost of creating controlled damage scenarios, (2) sensitivity to environmental noise and measurement uncertainties, and (3) lack of physical consistency in predictions. To overcome these limitations, this research establishes a cutting-edge Physics-Informed Convolutional Neural Network (PINN-CNN) that combines the feature extraction capabilities of CNNs with physical constraints derived from shell vibration theory.

2.1 Physical Model Design

In physics-informed learning for structural damage detection, the integration of physical laws and the generation of physically consistent predictions are essential. The process begins by defining a physical model that describes the dynamic behavior of airship envelope structures under external excitation. This model captures the vibration characteristics of thin shell structures, considering displacement response, material properties, damping effects, structural stiffness, and external loading.

Once the physical model is defined, the network is designed to predict physically meaningful parameters that characterize system dynamics. These parameters include wave propagation speed, damping ratio, and excitation amplitude. A dedicated physics prediction branch generates these parameters using features extracted from the convolutional neural network backbone. The extracted feature representation is directly connected to the physical prediction layer.

During the prediction of physical parameters, the model not only enforces physical consistency constraints but also ensures that the predicted parameters satisfy the governing physical relationships of the system. The physics-based loss function is composed of multiple components that evaluate signal characteristics from different perspectives. These components include energy consistency, frequency consistency, and signal continuity. Each component compares actual signal properties with those computed from the predicted physical parameters.

The purpose of the energy-related consistency constraints is to ensure alignment between the predicted physical parameters and the energy characteristics of the vibration signal. This is achieved by comparing normalized signal energy with energy features derived from the predicted parameters, while accounting for their statistical variation.

The overall loss function combines the standard classification loss with the physics-constrained loss, where a weighting parameter controls the influence of physical constraints. This formulation enables the network to preserve the governing physical principles of the system while maintaining strong classification performance.

The proposed physics-informed convolutional neural network architecture adopts a carefully designed multi-branch structure, which effectively addresses the limitations of conventional CNN-based methods. By incorporating physical knowledge, the model maintains physical consistency and achieves reliable performance even with limited training data and under noisy conditions.

The selected physical constraints are specifically chosen for their relevance to airship envelope membrane dynamics. Energy conservation corresponds to vibration energy propagation in flexible membranes, frequency characteristics reflect modal variations caused by localized damage, and signal continuity ensures structural integrity during membrane deformation. Compared with alternative physics-based constraints, such as stress equilibrium or strain compatibility, these constraints better align with vibration signal analysis, reduce computational complexity, and retain sensitivity to damage-induced dynamic changes.

Feature extraction begins with preprocessing the input vibration signals using standard convolutional operations to capture local patterns and temporal correlations. The extracted features are then passed through two parallel branches: a classification branch and a physics prediction branch. The classification branch produces damage type predictions based on globally aggregated feature representations, while the physics prediction branch estimates physical parameters that describe the system's dynamic properties.

The training process balances the objectives of classification accuracy and physical consistency through a carefully designed optimization strategy. A multi-stage training approach is adopted, in which physical constraints are gradually introduced to stabilize learning. In the initial stage, the network focuses on learning discriminative features using standard classification loss. As training progresses, the influence of physical constraints increases according to a predefined schedule.

To assess the effectiveness of the physics-informed framework, multiple evaluation metrics are employed, including classification accuracy, degree of physical constraint satisfaction, and robustness to noise. The level of physical constraint satisfaction reflects how well the predicted results adhere to the underlying physical principles governing the system.

2.2 *Signal Platform Construction*

To ensure the authenticity and representativeness of experimental data, the signal acquisition platform (as shown in Fig. 1) was designed strictly according to the actual working conditions of airship envelopes.

Figure 2 shows the distribution of strain sensors on the envelope surface, with fault points marked with 'f', indicating tear locations on the envelope. The sensors were positioned in a systematic grid pattern to ensure comprehensive coverage of the entire monitoring area and optimal detection of propagating damage waves. Tears were designed in two sizes (2 mm and 5 mm), corresponding to two fault states that represent typical damage scenarios encountered in airship operations. Together with the normal state N, there are three states in total. Signal acquisition was repeated 15 times for each state in both healthy envelope and fault states, collecting a total of 60 signal datasets to ensure statistical reliability and account for measurement variability. Each dataset contained approximately 450,000 data points

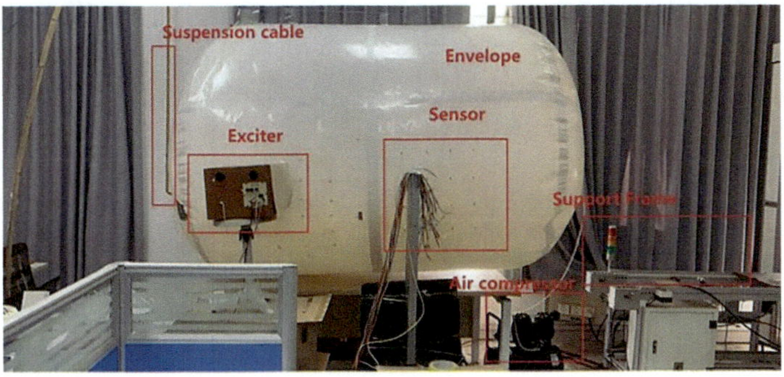

Fig. 1 Airship envelope platform construction diagram

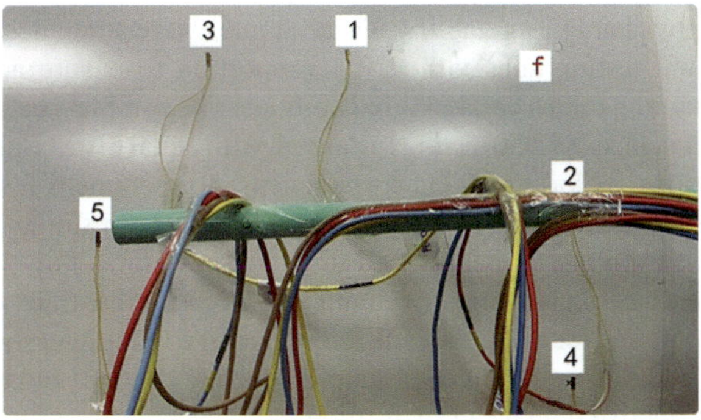

Fig. 2 Distribution of strain sensors

sampled at high frequency, comprehensively covering the vibration strain characteristics of multiple areas on the envelope surface and providing sufficient temporal resolution for accurate damage identification.

As shown in Fig. 3, ANSYS was used to perform modal simulation analysis on envelopes with surface tears. The results indicate that tears significantly affect local tension, thereby altering the vibration characteristics of the envelope. The computational framework utilized predetermined material specifications, including envelope fabric stiffness of 10 GPa, lateral strain ratio of 0.3, fabric thickness measuring 0.3 mm, seam reinforcement zones at 0.6 mm, internal pressurization at 5 kPa, and fabric density at 1100 kg/m^3. Through comparative analysis of modal characteristics under different tear conditions in the simulation, the significant impact of local structural faults on vibration modes was verified, providing a theoretical basis for structural fault sensing based on vibration signals.

Signal data were segmented and organized by sensor number to achieve full monitoring coverage. The signal segment window length was carefully set to 40,000 data points with a sliding step of 5000 data points, resulting in an 87.5% overlap between consecutive windows. This overlapping strategy enhanced data density and enabled continuous feature extraction. It also prevented edge information loss. Through integration with physics-informed features, including wave propagation characteristics,

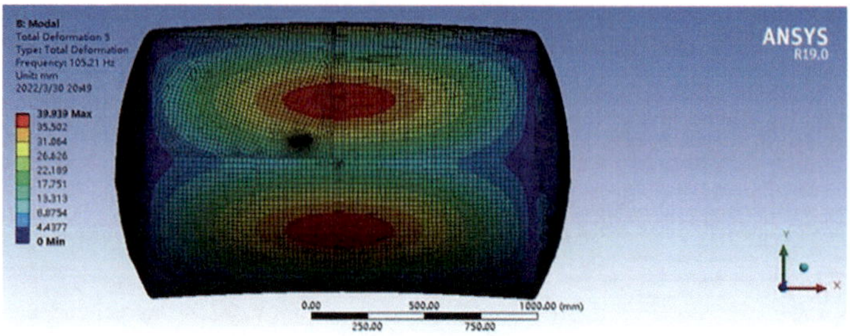

Fig. 3 Modal diagram of airship envelope

damping coefficients, and frequency domain parameters, this approach ultimately constructed a high-quality sample set covering various fault states with enhanced physical interpretability. Data samples were strategically allocated between training and testing subsets at a 4:1 proportion to ensure adequate training while maintaining robust validation capabilities. The complete dataset encompassed 120 training instances and 30 testing instances, with equal distribution among three distinct conditions, including 40 intact specimens, 40 specimens featuring 2 mm tear defects, and 40 specimens exhibiting 5 mm tear defects, resulting in 150 meticulously selected data points. This balanced distribution ensures that the model learns equally from each damage category without bias. Figure 4 illustrates the distinctive vibration signal patterns for these three different damage types, clearly demonstrating the characteristic differences in amplitude, frequency content, and temporal evolution that enable effective damage classification.

In terms of model parameters, the improved CNN model consists of four convolutional blocks with kernel sizes of 7, 5, 3, and 3, and channel numbers of 32, 64, 128, and 256 respectively. The fully connected layers have 512, 256, and 3 neurons respectively. During training, 30 training epochs were set, with a batch size of 32 and an initial learning rate of 1e-3. The Adam optimizer was used with a weight decay coefficient of 1e-4, and the learning rate was adjusted using a cosine annealing restart scheduler. The loss function was the weighted cross-entropy loss.

The improved P-CNN model adds a PINN parameter processing network on the basis of the CNN, which contains three fully connected layers with 32, 64, and 128 neurons, respectively. After feature fusion, classification is performed through

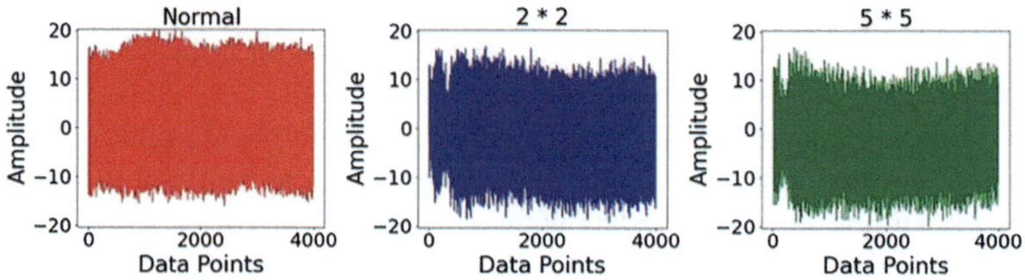

Fig. 4 Vibration signals of original damage data

Table 1 File corresponding number

Damage type	Document number
Normal	AI1-01_20240929100524
2 × 2 notch	AI1-01_20240929101209
5 × 5 notch	AI1-01_20240929101720

Table 2 Comparison of accuracy

Model	Train Ac (%)	Val Ac (%)
CNN	98.37	96.67
P-CNN	100.00	100.00

five fully connected layers with 512, 256, 128, and 3 neurons, respectively. The training parameters are similar to those of the CNN model, but a dynamic adjustment mechanism for the physical loss weight has been added. The initial physical weight is 0.1, which is 0.1 times for the first 10 epochs, 0.5 times for epochs 10 to 20, and the physical weight itself after 20 epochs.

3 Results

3.1 Accuracy Comparison

To validate the effectiveness of the proposed P-CNN method, this study conducted a comprehensive comparison of traditional CNN and P-CNN performance on the envelope damage detection task. The experimental results shown in Table 1 indicate that P-CNN significantly outperforms traditional CNN methods in both training accuracy and validation accuracy key metrics.

The experimental results show that the P-CNN method improved training accuracy by 3.44% compared to traditional CNN, achieving perfect performance with a validation accuracy of 100%, a significant improvement over CNN's 96.67%. This performance improvement is primarily attributed to the introduction of physical constraints, enabling the model to learn more physically meaningful feature representations, thereby improving classification accuracy and stability, refer to Table 2.

3.2 Noise Robustness Analysis

To evaluate model stability when facing measurement noise in practical applications, the present work methodically assessed the efficiency of both algorithms when

Table 3 Accuracy under different noise ratios

Noise level (%)	CNN Ac (%)	PINN Ac (%)	Improvement (%)
0	96.67	100.00	+ 3.44
10	86.67	96.67	+ 11.54
15	76.67	93.33	+ 21.71
20	73.33	90.00	+ 22.73
25	70.00	90.00	+ 28.57

subjected to multiple noise environments. Figure 1 shows the accuracy trends of CNN and P-CNN models as noise intensity varies from 1 to 26%.

In low-noise environments (1–10% noise level), both models demonstrated excellent noise resistance capabilities, indicating good robustness under slight noise interference conditions. However, as noise levels increased, performance differences between the two methods began to emerge. Under moderate noise conditions (12–18% noise level), traditional CNN began to show signs of performance degradation, with accuracy dropping to 76.67% at 15% noise, while P-CNN still maintained 93.33% accuracy, demonstrating stronger noise interference resistance. In high-noise environments (20–25% noise level), the performance difference between the two methods became more pronounced. When the noise level reached 20%, traditional CNN's accuracy dropped to 73.33%, while P-CNN still maintained 90.00% accuracy. Under more stringent 25% noise conditions, CNN's accuracy further decreased to 70.00%, while P-CNN still maintained high accuracy at 90.00%, a relative improvement of 28.57 percentage points. These results fully demonstrate the important role of physics-informed constraints in improving model noise robustness, refer to Table 3.

3.3 *Computational Efficiency and Practicality Assessment*

In this section, comparative experiments were conducted between the proposed method and the traditional CNN to evaluate their performance. To evaluate the practical application feasibility of the proposed method, this study conducted an extensive efficiency evaluation examining the algorithmic effectiveness of these two approaches. Every test was executed under identical hardware configurations, using CPU hardware platforms to simulate actual deployment conditions. In terms of training efficiency, traditional CNN methods required approximately 8 min to complete 5 rounds of training, while the P-CNN method, due to the addition of the physics constraint branch, increased training time to approximately 12 min, a relative increase of 50%. This time increase mainly came from the calculation of the physics loss function and additional gradient backpropagation processes. However, considering the significant improvements in classification accuracy and noise robustness of P-CNN, this moderate increase in training time is completely acceptable. In the inference phase, both methods demonstrated excellent computational efficiency, with inference time for a single sample less than 1 ms, meeting real-time application requirements. This indicates that the introduction of physical constraints did not significantly impact the inference speed of the model, ensuring the feasibility of the method in practical engineering applications.

Comprehensive performance evaluation results indicate that the P-CNN method significantly improves model accuracy, robustness, and interpretability with a slight increase in computational cost, particularly its excellent performance in noisy environments, making it more suitable for practical envelope damage detection application scenarios. The moderate increase in training time is fully compensated by significant improvements in key performance metrics, reflecting the practical value and engineering application potential of this method.

3.4 Robustness to Real-World Uncertainties

In real-world deployment, the hybrid PINN-CNN model addresses uncertainties such as material aging, temperature gradients, and aerodynamic variability through the inherent generalizability of physical constraints—fundamental laws governing membrane dynamics remain consistent despite environmental or material changes. For domain shifts and unseen damage patterns, the model can be enhanced by online updating of physical parameters (e.g., adjusting damping coefficients for aged materials) and integrating transfer learning techniques to adapt to new operational conditions. Uncertainty propagation can be quantified using emerging physics-informed uncertainty quantification frameworks, which combine probabilistic inference with physical constraints to assess prediction confidence. This adaptability ensures the model's practical relevance beyond controlled laboratory environments, supporting its application in actual airship health monitoring systems.

4 Conclusion

This study introduces the P-CNN method for airship envelope damage detection, integrating physical constraints with deep learning, and validates its effectiveness through vibration signal analysis. Compared with conventional CNNs, P-CNN achieves a 3.44% improvement in baseline accuracy and demonstrates 28.57% greater noise robustness at a noise level of 25%. In addition, it successfully learns physically meaningful parameters such as wave speed, damping, and excitation, thereby substantially improving model interpretability. Experimental findings indicate that incorporating physics-informed constraints not only enhances the model's generalization capability under limited data conditions but also effectively suppresses noise interference, with computational efficiency meeting the requirements for real-time applications.

The performance improvements achieved by the proposed P-CNN method through the integration of physical constraints are reflected in four main aspects. First, prior physical knowledge strengthens the model's generalization ability, enabling stable performance under varying operational conditions. Second, simultaneous prediction of physical parameters offers a quantitative explanation of damage mechanisms, thereby improving interpretability. Third, the role of physical constraints as a regularization mechanism reduces the need for extensive training data, achieving 100% validation accuracy on a limited dataset of 150 samples, including 30 for validation. Fourth, constraints derived from physical laws significantly suppress noise effects, resulting in a 28.57% accuracy improvement over traditional CNNs at a high noise level of 25%.

Despite these advancements, certain limitations remain. Future work should extend validation to real airship envelopes, address more complex damage types such as delamination, fatigue, and aging, investigate multi-physics field coupling simulations, and develop algorithms for real-time online monitoring. From an engineering perspective, successful implementation will require careful consideration of sensor network optimization, edge computing deployment, integration compatibility with existing airship systems, and compliance with aviation safety certification standards. A phased promotion strategy should be adopted to facilitate the transition from laboratory research to industrial application.

References

1. Raissi M, Perdikaris P, Karniadakis GE (2019) Physics-informed neural networks: a deep learning framework for solving forward and inverse problems involving nonlinear partial differential equations. J Comput Phys 378:686–707
2. Jagtap AD, Kharazmi E, Karniadakis GE (2020) Conservative physics-informed neural networks on discrete domains for conservation laws: applications to forward and inverse problems. Comput Methods Appl Mech Eng 365:113028
3. Yang Y, Perdikaris P (2020) Adversarial uncertainty quantification in physics-informed neural networks. J Comput Phys 394:136–152
4. Zhu Q, Liu Z, Yan J (2021) Machine learning for metal additive manufacturing: predicting temperature and melt pool fluid dynamics using physics-informed neural networks. Comput Mech 67(2):619–635
5. Wang S, Teng Y, Perdikaris P (2021) Understanding and mitigating gradient flow pathologies in physics-informed neural networks. SIAM J Sci Comput 43(5):A3055–A3081
6. Kashinath K, Mustafa M, Albert A et al. (2021) Physics-informed machine learning: case studies for weather and climate modelling. Philos Trans R Soc A 379(2194):20200093
7. Liu S et al (2021) Physics-informed machine learning for composition-process-property design: shape memory alloy demonstration. Appl Mater Today 22:100898
8. Zhang E, Dao M, Karniadakis GE, Suresh S (2022) Analyses of internal structures and defects in materials using physics-informed neural networks. Sci Adv 8:eabk0644
9. Lawal ZK, Yassin H, Lai DTC, Che Idris A (2022) Physics-informed neural network (PINN) evolution and beyond: a systematic literature review and bibliometric analysis. Big Data Cogn Comput 6(4):140
10. Wang C, Li S, He D, Wang L (2022) Is L^2 physics informed loss always suitable for training physics informed neural network?. In: Proceedings on advances in neural information processing systems, curran associates, vol 35, pp 8278–8290
11. Sharma P, Chung WT, Akoush B, Ihme M (2023) A review of physics-informed machine learning in fluid mechanics. Energies 16(5):2343

12. Bolandi H, Sreekumar G, Li X, Lajnef N, Boddeti VN (2023) Physics informed neural network for dynamic stress prediction. Appl Intell 53:26313–26328
13. He W et al (2024) Multi-level physics informed deep learning for solving partial differential equations in computational structural mechanics. Commun Eng 5:303
14. Al-Adly AIF et al (2024) Physics-informed neural networks for structural health monitoring: a case study for Kirchhoff-Love plates. Data-Centric Eng 5:e5
15. Yamaguchi K et al (2024) A physics-informed neural network for the nonlinear damage identification in a reinforced concrete bridge pier using seismic responses. Struct Control Hlth 2024:5532909.
16. Li X et al (2024) Mechanics-informed autoencoder enables automated detection and localization of unforeseen structural damage. Nat Commun 15:7595
17. Chen Y, Chen Q, Wang R (2025) Bearing fault diagnosis based on vibration envelope spectral characteristics. Appl Sci 15(4):2240
18. Pan M, Yang Y, Guan F et al (2025) A novel dynamic characteristic for detecting breathing cracks in blades based on vibration response envelope analysis. Machines 13(5):399
19. Light weight optimization of stratospheric airship envelope based on reliability analysis (2020). Chin J Aeronaut 33(4):1148–1165
20. Shi W et al (2022) Study of performance of aerostat envelope materials on the coast. Front Mater 9:992984
21. Bauer M, Balaratnam N, Weidenauer J, Wagner F, Kley M (2023) Comparison of envelope demodulation methods in the analysis of rolling bearing damage. J Low Freq Noise, Vib Act Control 42(4):1529–1545

Coupling Simulation Analysis of High Temperature Heater Under Multiple Physical Fields

Junbin Zhang, Shubin Jia, Ye Huang, Jiufen Chen, and Yingdong Chen

Abstract Faults such as high-temperature gas leakage and electrode rod leakage in the heater present substantial risks to operational safety and reliability. This study conducts three-dimensional modeling and coupling simulation analysis of a heater subjected to multiple physical fields. The heat transfer mechanism is first examined, followed by the development of a three-dimensional model of the heater incorporating heating elements and electrode rods. Using either simplified models or localized simulations, the coupled process of high-speed and high-temperature gas flow within the heater pipeline and the interaction between flow and heat transfer during the cooling of electrode rods are analyzed. Simulation results indicate that the bolt joint experiences peak electrode rod temperatures in the range of 20–80 °C under typical operating conditions. This range aligns closely with field measurement data, thereby validating the accuracy of the coupling simulation model.

Keywords Heater · Three-dimensional modeling · Fluid-thermal coupling · Simulation analysis

J. Zhang (✉) · S. Jia · J. Chen · Y. Chen
China Aerodynamics Research & Development Center, Hypervelocity Aerodynamics Institute, Mianyang, Sichuan, People's Republic of China
e-mail: zhangjunbin@alu.uestc.edu.cn

S. Jia
e-mail: jia.shubin@alu.uestc.edu.cn

J. Chen
e-mail: chen.jiufen@alu.uestc.edu.cn

Y. Chen
e-mail: chen.yingdong@alu.uestc.edu.cn

Y. Huang
Shenzhen Institute for Advanced Study, University of Electronic Science and Technology of China, Shenzhen, Guangdong, People's Republic of China
e-mail: huang.ye@uestc.edu.cn

H. Li et al. (eds.), *Data-Driven Methods for Reliability and Safety Engineering: Applications in Industrial Systems*, Springer Series in Reliability Engineering,
https://doi.org/10.1007/978-3-032-22873-4_27

1 Introduction

The heater constitutes a critical system for specialized testing applications, while its condition monitoring and fault diagnosis present considerable technical challenges. Due to the operational characteristics of the heater, finite element simulation of a single physical field is insufficient to satisfy engineering requirements, making coupling simulation analysis essential.

Extensive research has been conducted on fluid–thermal–solid coupling simulations across diverse engineering applications. Chen Lin et al. [1] performed a multi-field thermal–fluid–solid coupling simulation of an engine and established corresponding grid models. Based on thermodynamic principles and empirical heat transfer formulas for the cylinder, heat transfer boundary conditions for each wall were determined. The study further examined crank–connecting rod mechanics to calculate transient mechanical loads, including piston side forces and displacements, for thermal–mechanical coupling analysis. A finite element model of conjugate heat transfer for an engine, comprising the cylinder block, cylinder head, and water jacket, was developed to perform heat transfer analysis, with results evaluated in terms of both flow and temperature fields. Zhang Wenjie et al. [2] constructed a fluid–thermal–solid coupling analysis model of a cooling pipeline and investigated the effects of varying pipe diameters and fluid flow rates on the flow characteristics of gas–liquid two-phase flow, pipe deformation, natural frequency, and vibration states. Wang Xin et al. [3] conducted a multi-field coupling simulation of a hypersonic inlet under real gas effects, simulating the aerodynamic performance and unsteady variations in structural temperature fields through wall conjugate heat transfer analysis. Wang Hansheng et al. [4] developed a fluid–thermal–solid coupling computation system for turbine blades, integrating automated simulation software execution, inter-domain data exchange, and computational result monitoring. Xu Kewen et al. [5] proposed a predictive model for turbine blade height and tip radial clearance using the fluid–thermal–solid coupling method, in combination with empirical loss and lag angle models, enabling rapid and accurate turbine performance prediction via iterative coupling calculations. Mei Zhiheng et al. [6] carried out finite element simulations of aero-engine turbine blades using ANSYS, conducting aerodynamic analysis of blade flow fields and multi-field coupling simulations of air, heat, and solid under representative operational conditions. Zhang Yujing [7] examined the flow, temperature, and solid structure fields of a designed damper structure through fluid–thermal–solid coupling analysis on the ANSYS Workbench platform. Shahriar et al. [8] performed an in-depth investigation into the fully coupled aerodynamic–thermal–elastic interactions between shock waves and laminar boundary layers on flexible surfaces under varying thermal conditions. Cai et al. [9] developed a fluid–solid coupling framework for data exchange and mesh mapping between CFD and FEM solvers.

Building upon these advancements in coupling simulation technology, fluid–thermal coupling simulations were conducted for critical heater components. The resulting temperature distributions under ideal conditions exhibited strong agreement with experimental measurements, thereby validating the accuracy of the simulation model.

2 Introduction to the Heater

2.1 Composition of the Heater

The schematic diagram of the heater studied in this paper is shown in Fig. 1.

The heating section comprises five structurally identical heaters, each containing heating elements, electrode rods, a pressure-bearing layer, an insulation layer, and a water-cooling system. These heaters feature large dimensions, high power consumption, and elevated internal pressures. Glass fiber is filled between the pressure layer and the heat insulation layer to minimize the heat loss. Among them, the third, fourth and fifth heaters can have different click bridging modes to realize different heating powers.

The whole heater belongs to overpressure heater, and the air pressure used in the test is about 10–20 MPa. The air is heated by the heater in one stage to prevent the air from condensing during the test. The maximum heating temperature of the heater system can reach 1100 K.

Fig. 1 A certain type of heater system core section panoramic view

2.2 Heat Transfer Mode of the Heater

During the heating process, the heating plates first generate heat through the electrical power input via the electrode rods and quickly rise to the rated temperature. At the same time, high-speed gas is introduced at the inlet end of the heater pipe. As the gas flows through the heating plates, it undergoes convective heat transfer with the plates, causing the gas temperature inside the heater pipe to reach a high temperature. The gas temperature gradually increases as the number of heating levels increases.

When heating the air, the air will also affect the temperature of the inner shell of the heater. The inner shell of the heater gradually heats up due to the high temperature. Because the thickness of the inner shell is thin, its temperature will be quickly transferred to the heat insulation cotton between the inner shell and the outer shell through the heat conduction between solids. Because of the large volume of thermal insulation cotton, its thermal conductivity is low, and its temperature rises slowly. Finally, the heat will be transferred to the outermost shell. Because of the large thickness of the shell and its material is metal material, the heat can be transferred to the air in the environment through natural convection heat transfer while heating up.

Because of its own resistance, the temperature of the electrode rod will rise after being electrified. The end of the electrode rod is connected with a heating sheet, and the temperature of the heating sheet will rise through heat conduction. Therefore, the electrode rod is designed as a hollow form, and the hollow part is connected with cooling water, which cools the electrode rod by convection heat exchange to ensure its safety.

3 Modeling of the Heater

3.1 Functions of Key Parts of the Heater

The heater's primary components consist of inner/outer shells, thermal insulation material, electrode rods, water-cooling pipes, and heating plates. The inner shell, as the core temperature-bearing component, mainly performs three functions. Firstly, providing a mounting base for the heating elements to prevent displacement or deformation due to thermal expansion, air flow impact, and other factors. Secondly, withstanding the continuous convective scouring and mechanical impact of the high-temperature gas after heating. The high-strength, heat-resistant material and structural design prevent particle erosion from high-velocity gas flow. This maintains component geometry and prevents structural failure under extreme conditions. Thirdly, the heat generated by the heating elements is isolated by the thermal resistance of the insulation layer, forming a significant temperature gradient between the high-temperature inner shell and the low-temperature outer shell. This not only ensures the internal heating efficiency but also ensures the safety of external equipment and personnel through temperature isolation.

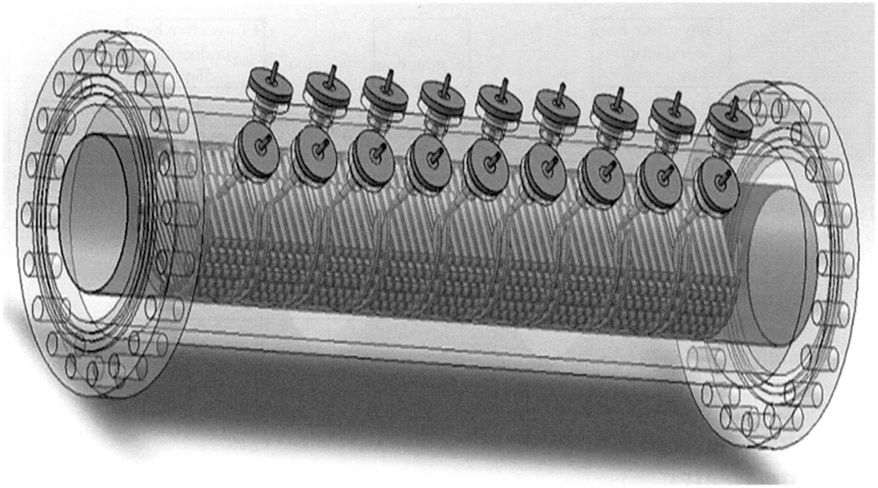

Fig. 2 Three-dimensional modeling of the heater

The core component of the heater is the heating element, which serves as the heat source of the entire heater and needs to have good thermal conductivity to heat the high-pressure air to the specified temperature in a short time. The electrode rods are cylindrical in shape and are used to connect with the outer shell, cooling water system, etc. One side is sleeved with a sleeve, and the other side is sleeved with an insulating sleeve. The insulating sleeve is used to connect with the flange and then together with the heater outer shell. Each level of the heater has a total of 18 electrode rods and cooling water pipes arranged in two rows of 9 each in a parallelogram shape on the same side, as shown in Fig. 2.

3.2 Modeling and Simplification of Key Components of the Heater

In this paper, the key parts of the heater will be modeled and simulated, including shell, heating sheet, electrode rod, water cooling pipe, and so on. The whole heater also includes many other parts. Because the whole structure of the heater is large, and these parts have little influence on the results of the heat-flow coupling simulation of the whole heater, it can be simplified, and only the main structure can be retained. Under the condition of ensuring the simulation results, the simulation difficulty can be reduced, the simulation time can be saved, and the engineering requirements can be met.

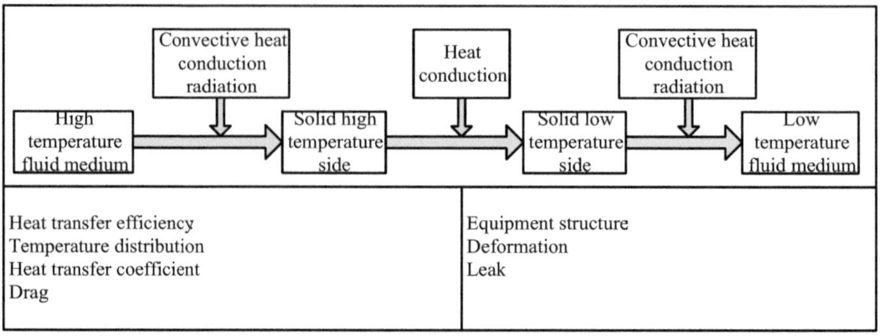

Fig. 3 Multi-physical field coupling of the heater

4 Fluid-Thermal Coupling Simulation Analysis of the Heater

4.1 *Simulation Process of Fluid-Thermal Coupling of the Heater*

The working process design of heater is coupled with thermal-solid–fluid multi-physical fields. As shown in Fig. 3, this coupling system can be divided into several key parts: the high-temperature fluid medium transfers heat to the high-temperature side of the solid structure through convection, heat conduction and radiation, then transfers heat from the high-temperature side to the low-temperature side through the thermal conduction of the solid material, and finally releases heat to the low-temperature fluid medium through convection and radiation on the low-temperature side. In this process, heat transfer efficiency is not only affected by fluid flow characteristics (such as velocity and turbulence intensity) and thermal boundary conditions (such as heat transfer coefficient and temperature gradient), but also closely related to equipment structure design (such as flow channel geometry and material thermal expansion coefficient).

Considering the parametric modeling of the equipment structure and operating conditions, the simulation algorithm needs to fully take into account the influence of the heater's structure, material properties, and operating conditions on heat conduction and temperature distribution. The main components include the gas heating system and the cooling water circulation system.

4.2 *Gas Heating System*

When energized, the heating element produces significant thermal energy. Gas flowing across the heating element surface undergoes convective heat transfer, increasing its temperature. The gas absorbs heat, causing its volume to expand and other parameters such as density and viscosity to change. Therefore, it is suitable to use bidirectional coupling of fluid and heat. At the same time, the gas flow, expansion, and temperature field will cause stress in the structure, but the effect of structural deformation on the fluid and temperature field can be ignored.

4.3 Cooling Water Circulation System

The cooling water system of the heater is designed to protect the electrical lead rods and insulation sealing devices. Its main components include the cooling water inlet and outlet channels, sealing flanges, etc. The heat in the heater is transferred to the insulation layer and the outer shell through the inner shell, and also to the cooling water circulation system. The cooling water flow carries away most of the heat, but the absorbed heat also causes the volume of the cooling water to expand, forming stress on the rigid body. However, the deformation of the equipment does not affect the cooling water and the temperature field.

Key heater operating parameters comprise: test temperature and pressure ranges, operating power (specified by current and voltage), electrode configuration, and gas medium selection. Gas medium selection critically influences heater performance due to substantial variations in thermophysical properties. For instance, air and nitrogen exhibit markedly different effects on thermal efficiency and temperature distribution. Simulations should be adjusted according to different medium conditions.

The investigated heater features large dimensions, high power output, structural complexity, and rapid internal airflow. These conditions challenge comprehensive fluid-thermal simulation. Additionally, significant size disparities exist between main piping and electrode rod components. On the other hand, the volume of the main pipe of the heater and the electrode rods and other key parts differ greatly. Simulating them simultaneously may lead to an extremely fine finite element mesh, and the simulation may not converge in some parts. Therefore, this paper adopts the approach of simulating the air heat transfer in the heater pipe and the water cooling of the electrode rods separately. The interaction between these two processes is relatively small. The advantage of this method is that it separately calculates the two fluid heat transfer processes in the entire process, reducing the simulation time and lowering the dependence on hardware for finite element simulation.

4.4 Boundary Conditions and Solution Settings

The internal temperature of the heater can be measured by the temperature sensor attached to the heater. The maximum temperature at the outlet of the fifth-stage heater is 1100 K. In this paper, simulations are conducted under the conditions of gas temperatures of 150 °C and 600 °C to explore their influence on the outer shell temperature. In terms of materials, the inner shell material is set as high-strength structural steel, and the insulating cotton material is set as glass fiber. The flow conditions of the fluid are set as turbulent flow, which follows the k-epsilon turbulence model, to obtain the temperature distribution.

Since the operating time of this equipment is generally 100 s to 150 s, the transient solver is selected for the solver, and the solution time is set to 150 s, with the solution step size set to 0.5 s. This solution condition can ensure that the required solution time is relatively short and within an acceptable range.

The selected boundary conditions (gas temperatures of 150 °C and 600 °C, material assignments for inner shell and insulation cotton) align with the heater's actual operating ranges and material specifications, ensuring simulation relevance. The transient solver and 0.5 s step size are justified by the heater's typical 100–150 s runtime, balancing temporal resolution and computational efficiency to capture dynamic temperature changes without numerical diffusion. The k-epsilon turbulence model is chosen for its robustness in simulating high-speed gas flow within pipelines, promoting convergence by accurately modeling turbulent heat transfer and flow characteristics that directly influence temperature distribution accuracy.

Through finite element simulation, we can obtain the temperature distribution of the inner circle of the heater and the internal air, as well as the distribution of the internal fluid flow velocity. Under the condition where the average temperature of the internal fluid is 600 °C, the convective heat transfer between the air and the inner shell of the heater is relatively strong, and the average temperature of the inner shell is heated to about 400 °C within 150 s, as shown in Fig. 4.

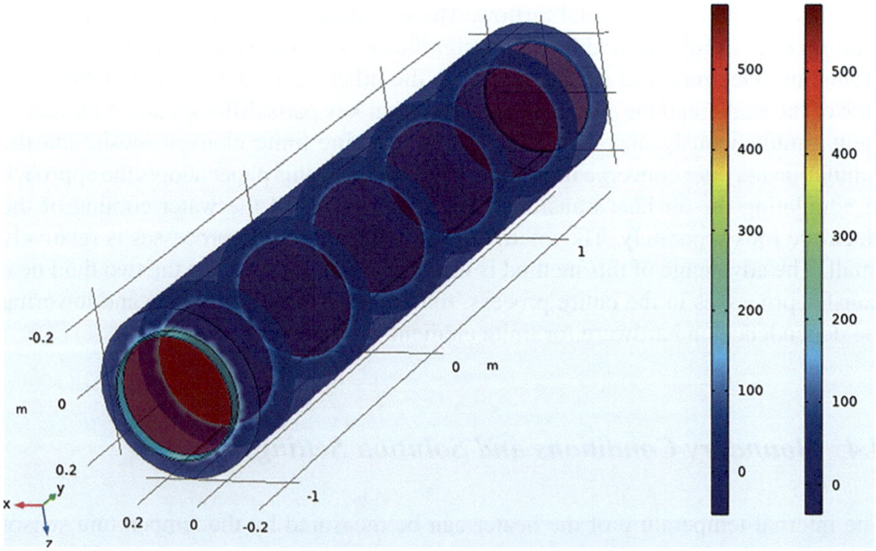

Fig. 4 Heater pipe inner wall flow and heat transfer coupling simulation

Figure 4 shows the fluid-thermal coupling simulation of the inner wall of the heater pipe. As can be seen from Fig. 4, due to the protective effect of the insulation cotton of the heater, the temperature increase of the outermost layer of the insulation cotton is not significant, with an average of about 30 °C. Therefore, it can be inferred that the abnormal temperature of the heater shell is caused by electrical heating or the overflow of high-speed and high-temperature gas, or the insulation cotton is carried away by high-pressure gas, resulting in partial absence of the insulation cotton.

4.5 Fluid-Thermal Coupling Simulation of the Cooling Process of the Heater Electrode Rod

During operation, the electrode rod operates at elevated temperatures, requiring coupled fluid-thermal simulation of both the rod and its internal cooling structure. With low cooling water flow rates (approximately 0.3 MPa pressure), simple internal geometry, and compact dimensions, the electrode rod is well-suited for numerical fluid-thermal coupling analysis. Fluent fluid module and the transient thermal module are introduced to simulate it, and the initial boundary conditions are set by introducing the established three-dimensional model, and the transient solution process and step size are consistent with those before.

Through the simulation of high-speed fluid and shell insulation cotton, it is found that the highest temperature of the heating element of the heater can reach 2000 K, so the temperature condition of about 2000 K is added to the end of the electrode rod in the study. At the same time, considering the heating of the electrode rod itself, by calculating the resistance ratio between the heater and the electrode rod, it is found that the power of each electrode rod is about 1000W, which is used as the heating power of the electrode rod.

4.6 Simulation of Flow Thermal Solid Coupling on the Inner Wall of Heater Pipeline Under Multiple Operating Conditions

The high-speed gas flows through the interior of the heater pipeline, and the flow rate of its internal temperature can reach 300 m/s to 1300 m/s depending on different operating conditions. In the flow heat coupling simulation of the inner wall of the heater pipeline, the main goal is to obtain the temperature distribution inside and outside the heater. For this reason, the entire process has been simplified, retaining only the inner shell of the heater, insulation cotton, and the internal air fluid domain. In addition, the operating conditions of the heater are divided into multiple categories, among which the main distinctions are the different Mach numbers of the internal gas flow, the differences in heating power of the heater, and the final heating temperature of the internal gas. Due to the mutual influence of the above working condition indicators, this study mainly discusses the final temperature effect of the internal gas. There are two main indicators of the final internal gas temperature, which are around 130 °C and 400 °C, respectively. The influencing factors are influenced by multiple factors such as heater power, electrode rod electrification mode, heating strategy, and internal gas flow rate.

The internal temperature of the heater can be measured by its built-in temperature sensor, which ranges from 150 °C to 800 °C depending on the heating stage. In this article, simulations were conducted under gas temperatures of 150 °C and 600 °C to investigate their impact on the temperature of the casing. In terms of materials, the inner shell is made of high-strength structural steel, the insulation cotton material is made of glass fiber, and the fluid flow conditions are set to turbulent flow, which follows the k-epsilon turbulence model to obtain its temperature distribution.

Due to the typical running time of the device being between 100 and 150 s, a transient solver is selected under the solver, with a solving time of 150 s and a solving step size set to 0.5 s. This solving condition ensures that the required time for solving is relatively short and within an acceptable range. Figure 5 grid division of heater shell.

After setting the geometry, the thermal boundary conditions are set. In terms of grid division, a conventional grid structure is used to refine the grid at key points such as edges and contact points to ensure that the quality of the grid is within a reasonable range. At the same time, it is necessary to avoid overly fine grids, which can lead to an overall large number of grids and affect the efficiency of the solution.

Through simulation, it was found that the air inside the heater has a relatively low impact on the temperature of the outer shell in a static state, with an average outer surface temperature of 30 °C and a relatively low impact on the temperature of the

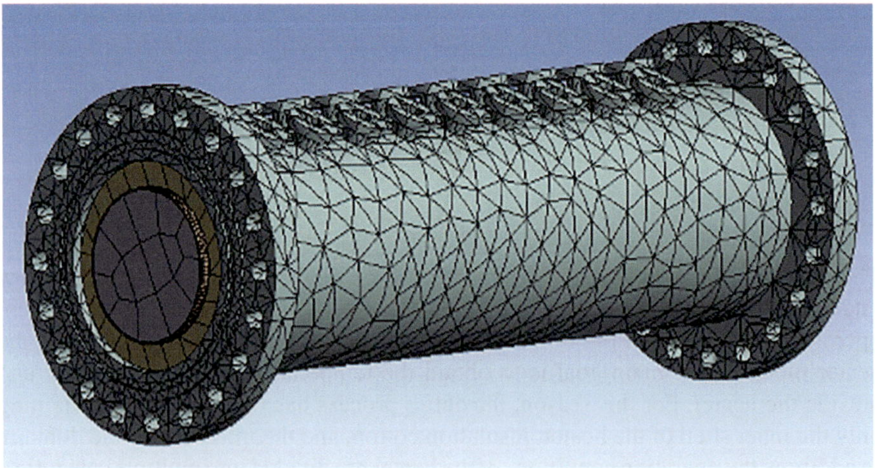

Fig. 5 Grid division of heater shell

electrode rod. Under normal operating conditions, the surface temperature of a non-faulty heater is lower. The cause of the fault diagnosis is gas leakage at the interface of the heater, which leads to the outflow of high-temperature gas from certain parts of the heater, resulting in an increase in temperature in that area.

During the operation of the heater, there is also an internal temperature angle heating condition, under which the highest internal temperature is around 130 °C. Temperature simulation was also conducted under this condition, and the final shell temperature was found to be 27.7 °C, which is not much different from the temperature under the previous high-temperature condition. Both have a relatively small impact on the shell temperature.

Figures 6 and 7 are heating result chart of the heater under two operating conditions.

When modeling the insulation material as structural steel, significant temperature distribution changes occur, with heater casing temperatures exceeding 66 °C. This result has a significant impact on the stable operation of the heater and the safety of operators, indicating that the overall design of the heater is good. The insulation material is essential for maintaining thermal performance. To ensure the normal operation of the heater and the safety of operators, regular inspection and maintenance of the insulation cotton should be carried out.

We replaced the insulation cotton with structural steel for simulation and found that the surface temperature of the heater would rise sharply, and the structural steel did not have the function of insulation. If the insulation cotton is removed and the heater is not filled with any insulation material, it will be found that the air heat

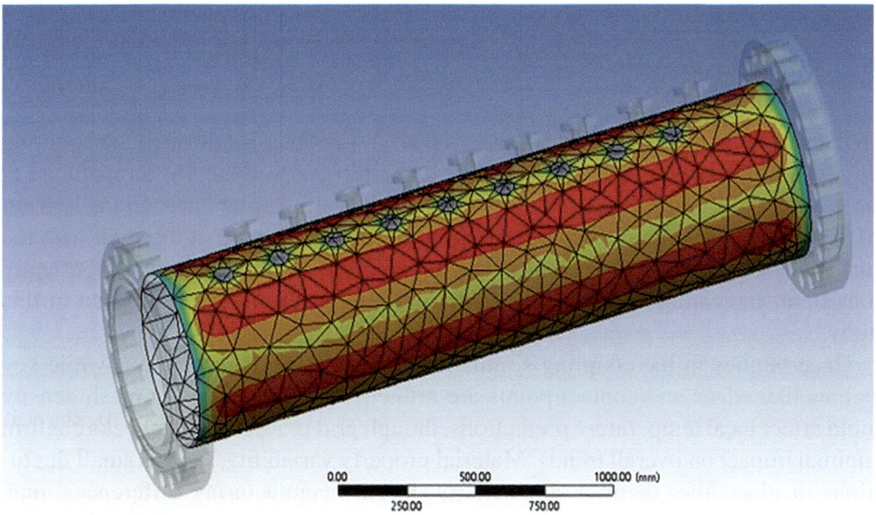

Fig. 6 Heating result chart of the heater under first operating condition

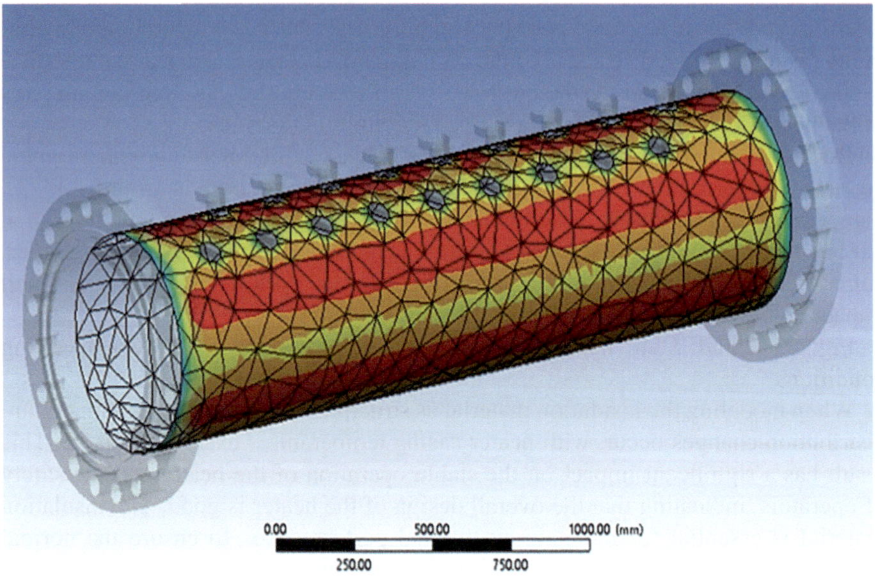

Fig. 7 Heating result diagram of the heater under second operating condition

transfer effect is also significant. In summary, using insulation cotton for the heater will have a better insulation effect.

4.7 Result Analysis

The temperature distribution of the electrode rod was obtained through the coupling simulation, as shown in Figs. 8, 9, 10, 11. It can be observed that the temperature at the bolt connection of the electrode rod is the highest, ranging approximately from 20 °C to 80 °C. By comparing the simulated temperature curve of the shell with the measured data, it can be found that the temperature fluctuation range is basically consistent, indicating the accuracy of the finite element model established in this study.

Uncertainties in the coupling simulation include mesh dependency—while key regions like edges and contact points are refined, slight variations in mesh density could affect local temperature predictions, though grid independence checks confirm minimal impact on overall trends. Material property variability, such as small fluctuations in glass fiber thermal conductivity due to manufacturing differences, may introduce minor deviations in insulation performance simulations. Sensitivity to inlet conditions, including gas flow velocity and temperature fluctuations beyond the modeled ranges, could also influence local heat transfer rates. These uncertainties may lead to subtle discrepancies in predicted electrode rod or shell temperatures,

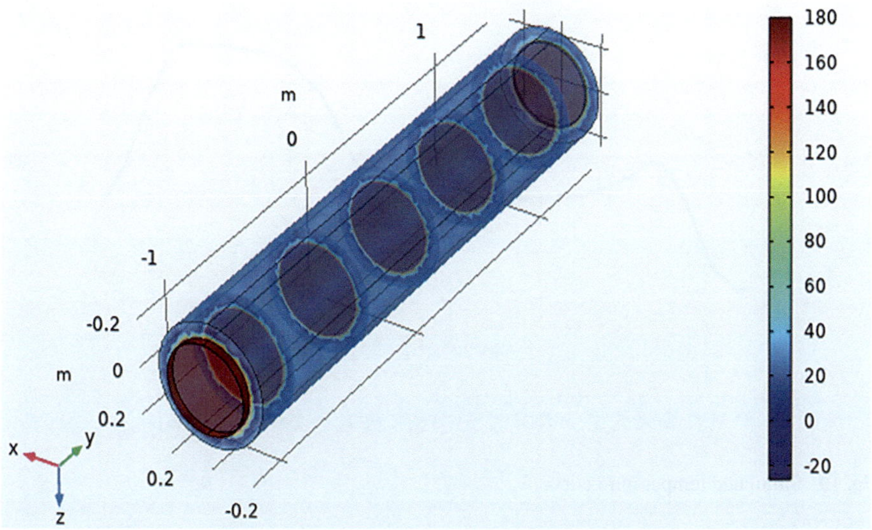

Fig. 8 Simulated three-dimensional temperature distribution map

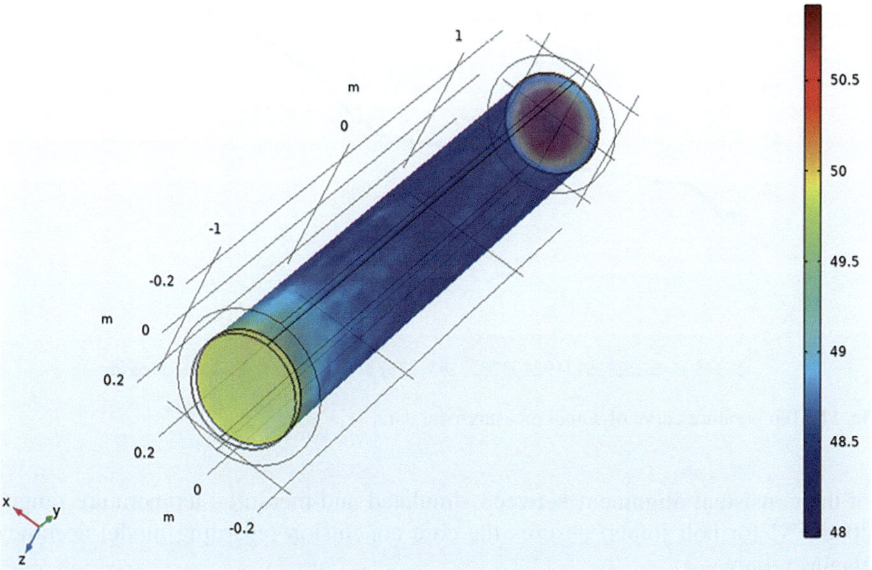

Fig. 9 Measured three-dimensional temperature distribution map

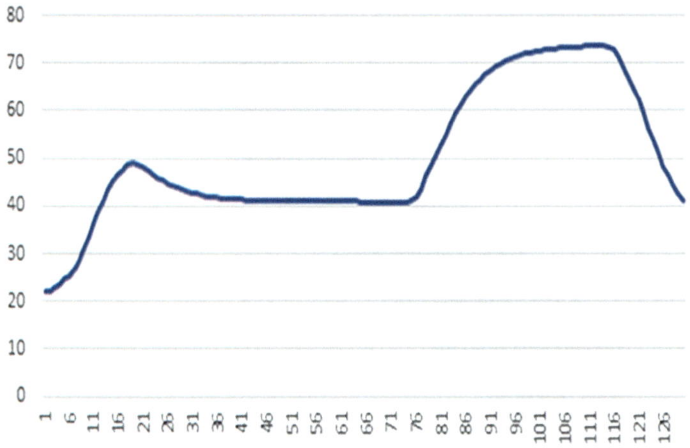

Fig. 10 Simulated temperature curve

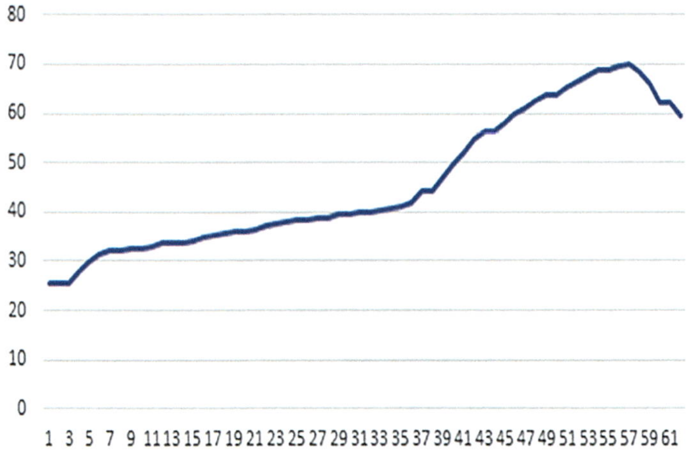

Fig. 11 Temperature curve of actual measurement data

but the consistent alignment between simulated and measured temperature ranges (20–80 °C for bolt joints) ensures the core conclusion regarding model accuracy remains reliable.

5 Conclusion

Through multi-field coupled simulation analysis of the heater, it was determined that the electrothermal temperature of the nut at the bolt joint of the electrode rod obtained from the simulation closely corresponds to the measured temperature data, ranging approximately from 20 °C to 80 °C. This agreement confirms the accuracy of the finite element model developed in this study. Further studies can be found out from [10–15].

References

1. Lin C (2014) Analysis of engine coupled thermo-mechanical. Chongqing University, Chongqing
2. Zhang W, Gao S, Zhai H et al. (2025) Flow-thermal-mechanical coupling analysis of embedded braided hose bellows based on ANSYS. Comput Simul 42(4):451–458+472
3. Xin W, Huacheng Y, Fuzhou L et al (2023) Multi-physical field coupling simulation of hypersonic inlet considering real gas effect. Mach Build Autom 29(4):115–119
4. Wang H, Li Y, Xu S et al. (2025) Research on flow and heat transfer of turbine guide vanes in aeroengine based on flow-thermal-mechanical coupling. J Aerosp Power. https://doi.org/10.13224/j.cnki.jasp.20240464
5. Kewen X, Boao Z, Guoqiang Y et al (2025) Research on the prediction method of axial turbine characteristics based on flow-thermal-mechanical coupling. J Harbin Eng Univ 46(3):467–474
6. Zhiheng M, Shujie L, Weiwei D (2021) Finite element simulation of aeroengine turbine blade based on ANSYS. Mech Elect Eng Technol 50(11):33–36
7. Yujing Z (2024) Research on the structure design of air valve and flow-thermal-mechanical coupling of sodium-air heat exchanger in sodium-cooled fast reactor. Lanzhou Jiaotong University, Lanzhou
8. Al S, Kourosh S (2024) Nonlinear shock-induced flutter of a compliant panel using a fully coupled fluid-thermal-structure interaction model. J Fluids Struct 124:104047
9. Mengke C, Tenglong C, Yao X et al. (2023) Multi-physical fuel performance modeling of U-50Zr helical cruciform fuel based on fluid-thermal-structure coupling methodology. Ann Nucl Energy 191
10. Guo J, Wang Z, Li H, Yang Y, Huang CG, Yazdi M, Kang HS (2024) A hybrid prognosis scheme for rolling bearings based on a novel health indicator and nonlinear Wiener process. Reliab Eng Syst Saf 245:110014
11. Li H, Yazdi M, Huang HZ, Huang CG, Peng W, Nedjati A, Adesina KA (2023) A fuzzy rough copula Bayesian network model for solving complex hospital service quality assessment. Complex Intell Syst 9(5):5527–5553
12. Li H, Peng W, Adumene S, Yazdi M (2023) An intelligent cost-based consequence model for offshore systems in harsh environments. In: Intelligent reliability and maintainability of energy infrastructure assets, pp 107–117. Springer Nature Switzerland, Cham
13. Li H, Yazdi M (2022) What are the critical well-drilling blowouts barriers? A progressive DEMATEL-game theory. Advanced decision-making methods and applications in system safety and reliability problems: approaches, case studies, multi-criteria decision-making, multi-objective decision-making, fuzzy risk-based models. Springer International Publishing, Cham, pp 29–46
14. Li H, Peng W, Adumene S, Yazdi M (2023) Cutting edge research topics on system safety, reliability, maintainability, and resilience of energy-critical infrastructures. In: Intelligent reliability and maintainability of energy infrastructure assets, pp 25–38. Springer Nature Switzerland, Cham

15. Li H, Peng W, Adumene S, Yazdi M (2023) Advances in failure prediction of subsea components considering complex dependencies. In: Intelligent reliability and maintainability of energy infrastructure assets, pp 93–105. Springer Nature Switzerland, Cham

Research on Instantiation Method of Object-Oriented Bayesian Network for Fault Diagnosis Based on Dynamic Weight Parameter Transfer

Shigang Zhang, Mengqiao Chen, Yongxuan Fang, Xudong Suo, and Xu Luo

Abstract Bayesian Networks (BNs) are extensively utilized for equipment fault diagnosis, and Object-Oriented Bayesian Networks (OOBNs) offer an effective framework for efficient model construction. In OOBNs, parameter learning, also referred to as instantiation, generally requires an adequate number of training samples. In practical fault diagnosis scenarios, however, such samples are often limited. To address this limitation, an instantiation approach based on variable-weight parameter migration is proposed. When training samples are insufficient, fully instantiated objects from related domains are employed as sources, and their parameters are transferred to enhance instantiation in the target domain. The transfer weight of each source object is dynamically adjusted according to its similarity to the target domain, thereby mitigating the risk of negative transfer. Experimental validation confirms that the proposed method significantly improves OOBN instantiation performance in data-scarce conditions.

Keywords Fault diagnosis · Instantiation · Object-oriented Bayesian network · Parameter transfer

---

S. Zhang · M. Chen (✉) · Y. Fang · X. Suo · X. Luo
College of Intelligence Science and Technology, National University of Defense Technology, Changsha, China
e-mail: chenmengqiao21@nudt.edu.cn

National Key Laboratory of Equipment State Sensing and Smart Support, National University of Defense Technology, Changsha, China

S. Zhang
e-mail: zhangshigang@nudt.edu.cn

Y. Fang
e-mail: fangyongxuan@nudt.edu.cn

X. Suo
e-mail: suoxudong@nudt.edu.cn

X. Luo
e-mail: luoxu@nudt.edu.cn

H. Li et al. (eds.), *Data-Driven Methods for Reliability and Safety Engineering: Applications in Industrial Systems*, Springer Series in Reliability Engineering,
https://doi.org/10.1007/978-3-032-22873-4_28

1 Introduction

Bayesian Networks (BNs) constitute a robust framework for representation and inference under uncertainty and have demonstrated strong performance in equipment fault diagnosis [1]. When applied to complex systems, however, the intricate hierarchical structures of BNs significantly increase the complexity of model learning and training. Object-Oriented Bayesian Networks (OOBNs) offer a practical solution to address the limitations of traditional BNs in modeling and inference [2]. By adopting object-oriented principles, OOBNs abstract similar subnetworks into classes and enable modular BN construction through class instantiation. Compared with conventional BNs, OOBNs are more straightforward to develop and maintain. They have been widely applied in fault diagnosis. For example, D. Kasper proposed an OOBN-based method for diagnosing industrial processes to monitor abnormal states in production workflows [3]. Cai applied OOBNs to model repeated components in complex industrial systems, facilitating real-time fault detection [4]. Huang developed a probabilistic vehicle fault-diagnosis model based on OOBNs that incorporated four submodules and three common root causes [5].

Accurate construction of both BNs and OOBNs depends not only on domain expertise but also on the availability of sufficient data. In practical fault diagnosis scenarios, however, collecting adequate fault data is often challenging. BN models trained on small sample sets are prone to overfitting, reducing diagnostic accuracy, and potentially leading to incorrect maintenance decisions. The development of modern equipment is often evolutionary, with successive systems sharing certain diagnostic characteristics while also exhibiting unique features. From a fault diagnosis perspective, this indicates the presence of common knowledge across similar systems alongside system-specific variations. Leveraging such shared knowledge during BN model learning can mitigate overfitting caused by data scarcity, thereby improving both reliability and accuracy. Knowledge transfer presents a promising approach to achieve this goal. Nevertheless, research on BN-based knowledge transfer remains limited and is largely concentrated on classical BN models, with no systematic investigation into transfer mechanisms for OOBN models.

In the field of knowledge transfer for Bayesian Networks, Oyen et al. [6] proposed a transfer learning algorithm based on Bayesian structure discovery, introducing structural bias terms to facilitate cross-task information transfer and improve learning robustness. Experimental results showed that, under limited data availability, the algorithm enhanced BN model construction performance. Luis et al. [7] developed an inductive transfer learning method for BN models, which improved model generalization under data scarcity through a parameter aggregation strategy based on similarity assessment. Zhou et al. [8, 9] introduced the BNPTL algorithm for parameter transfer learning, capable of automatically matching the most suitable BN or fragment from a source domain network and performing model parameter fusion. Experimental results demonstrated that this approach outperformed the traditional MLE algorithm in small-sample contexts. Hou et al. [10] proposed a parameter transfer algorithm with variable coefficients to address BN parameter estimation under limited sample conditions. This method adaptively adjusted transfer weights according to the correlation between target and source domain samples, significantly enhancing BN reliability and accuracy for bearing fault diagnosis in small-sample situations. However, it required the topological structure of the source domain network to match that of the target domain, thereby limiting versatility. Jia et al. [11] introduced an inductive migration algorithm based on Markov equivalence classes, using limited datasets to construct BN models for specific subjects and transferring weights based on target-domain-related data to compensate for data shortages. Pereira et al. [12] addressed hard disk drive fault diagnosis by proposing a BN model parameter migration algorithm based on aggregated source domain data. Experimental results indicated that, under data-limited conditions, the algorithm effectively improved diagnostic performance.

To address the issue of OOBN instantiation under limited data, a transfer learning approach has been integrated into OOBN. A knowledge similarity measurement scheme and a dynamic weight parameter transfer strategy have been proposed to enhance the instantiation accuracy.

2 Preliminary

BNs are formalized as directed acyclic graphs, where each node represents a random variable and directed edges encode direct probabilistic dependencies among variables [13]. The strength of these dependencies is quantified using conditional probability tables, which specify the distribution of a child node conditioned on its parent nodes. However, when modeling complex equipment, network construction and parameter updating become increasingly challenging as the model size expands. Object-oriented Bayesian networks (OOBNs) offer a modular approach to address these challenges by encapsulating subnetworks into reusable components, streamlining model assembly, and simplifying parameter management.

In Object-Oriented Programming (OOP), a class defines a data structure with common attributes and behaviors. Since BNs for complex equipment often contain repeated or nearly identical network segments, OOBNs leverage OOP principles by defining network structures using classes, with objects acting as instances of these classes. Each instantiated object inherits the same network structure as its corresponding class. The key definitions related to OOBN are as follows:

Class: A class is an abstract representation of a recurring structure within a BN. It defines a reusable BN fragment that omits parameter information and serves as a template for modular modelling. A class comprises three distinct types of nodes: *input nodes, output nodes,* and *encapsulated nodes.*

Object: An object is a concrete instantiation of a class. Each object inherits the structural template defined by its class and is further characterized by specific parameter values, which are either learned from data or provided by domain experts.

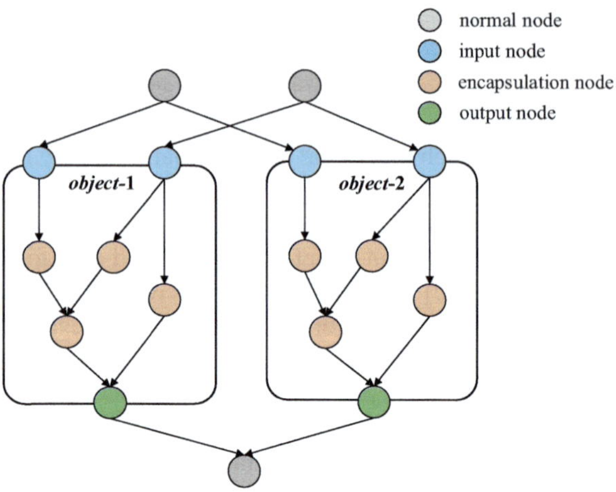

Fig. 1 A simple object-oriented Bayesian network

Input and output nodes, often referred to as interface nodes, primarily facilitate information exchange between classes or objects, enabling inter-class communication within the network. Figure 1 illustrates a simple example of an OOBN, comprising instance nodes and ordinary nodes. Objects, such as object-1, are represented by rounded rectangles and are connected to other instance nodes or ordinary nodes through input and output nodes. In this example, *object-1* and *object-2* are instantiated from the same class.

Given a sufficient sample and assuming that observations satisfy the independent and identically distributed condition, maximum likelihood estimation (MLE) is usually used to instantiate the class. As a standard parameter-learning technique for probabilistic models, MLE seeks the parameter values that maximize the likelihood of the observed data under the specified model, thereby yielding optimal conditional probability estimates for each node. However, when sample data is limited, the parameters estimated using MLE often deviate significantly from the true values. Koller and Friedman define a method to compute a sample complexity bound for parameter learning with a fixed BN structure [13].

In the domain of equipment fault diagnosis, it is frequently challenging to obtain sufficient fault data samples, which directly compromises the quality of class instantiation and hinders the practical application of the OOBN. Therefore, it is essential to investigate instantiation methods that are robust under data-scarce conditions in order to enhance the accuracy and reliability of the resulting instantiations.

3 Methodology

Transfer learning, a prominent subfield of machine learning, leverages knowledge from a source domain to enhance learning in a target domain by quantifying the inter-domain similarity at both the data and model levels. Within the context of OOBNs, this approach facilitates the migration of instantiated component parameters from the source domain to the target domain, thereby improving the fidelity of the instantiation. Despite its potential, theoretical research on parameter transfer within OOBN remains limited. To address this gap, a parameter migration-based instantiation method is introduced. Initially, a similarity measurement framework is proposed to assess domain similarity in terms of structural topology and probabilistic distribution. Subsequently, a dynamic-weight parameter migration strategy is introduced, which facilitates the forward migration of source-domain knowledge and enhances instantiation accuracy in OOBNs.

3.1 Problem Description

In transfer learning, the domain to be learned is referred to as the target domain, while the domain containing prior knowledge is the source domain. The primary function of the source domain is to provide transferable knowledge to the target domain, thereby enhancing model learning performance when sample data is limited in the target domain, refer to Fig. 2.

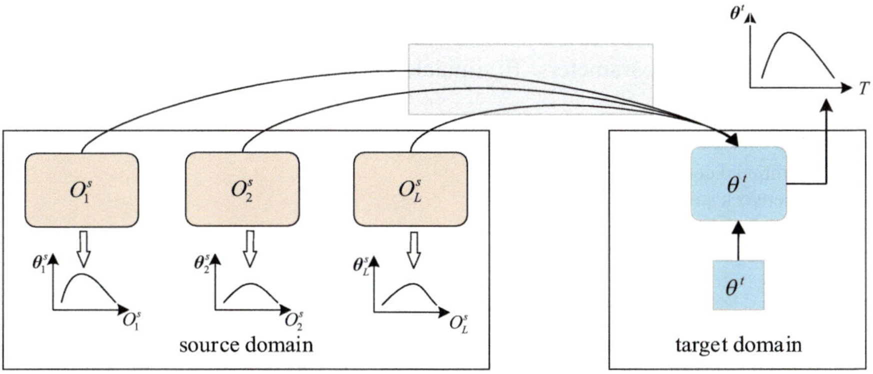

Fig. 2 Schematic diagram of OOBN instantiation method based on parameter migration

3.2 Similarity Evaluation

Transfer learning's effectiveness relies on the similarity between the source and target domains. If the domains are too dissimilar, transferring large volumes of knowledge can result in negative transfer, which degrades performance. Therefore, accurately assessing the similarity between the domains is critical. In parameter transfer for OOBN instantiation, similarity is measured at the structural topology and parameter distribution levels. This measure determines a dynamic weight parameter transfer strategy. Source domain objects with high similarity to the target are assigned larger transfer weights, contributing more to the parameter estimation of the target domain. Objects with lower similarity are given smaller weights, reducing their influence and minimizing the risk of negative transfer.

4 Simulation Verification

4.1 Verification Strategy

OOBN instantiation entails estimating parameters of a BN with a known structure but unknown parameters. The publicly available Command Center diagnostic BN, designed to detect abnormal operating conditions in a specific communications terminal [14], is employed as the simulation testbed. The network structure is illustrated in Fig. 3.

The Command Center model is treated as the original network, and corresponding source and target domains are constructed for experimental purposes. The model is represented as a single-class OOBN, and its instantiation is performed by estimating the network's parameters. To simulate domain discrepancies, two aspects are considered:

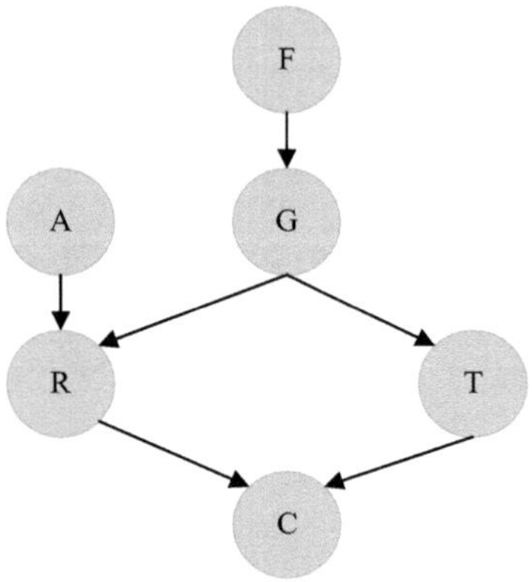

Fig. 3 Command center Bayesian network structure

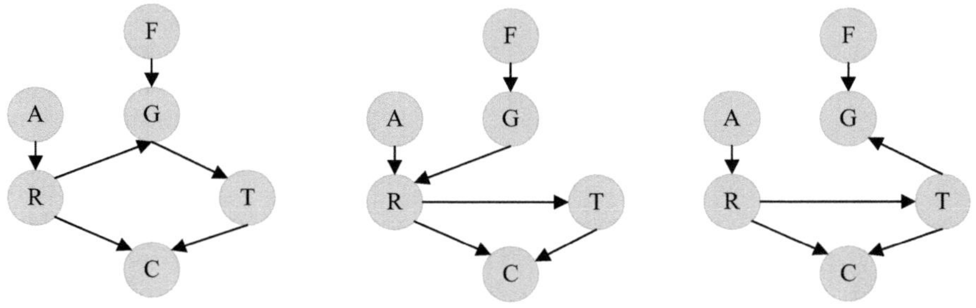

Fig. 4 The topological structure of the constructed source domain network

(1) Structural topology: source domain networks are generated by modifying the original network through edge addition, deletion, and reversal;
(2) Probability distribution: parameter information of the source domain is derived using MLE based on noisy samples from the original network.

The parameters of the original network are assumed to represent the true parameters of the target domain. Therefore, the goal of instantiation is to estimate the target domain parameters to closely approximate these true values. Instantiation performance is evaluated by the proximity between the estimated and true parameters. Methods using only target domain samples are compared with those incorporating parameter transfer under varying sample sizes. The impact of different weighting strategies for source domain transfer is also analyzed.

The source domain consists of a set of objects similar to the target, each already instantiated and represented by a BN with known parameters. In the simulation, this domain is formed by several BNs that differ from the original network in both structure and probability distribution. To simulate structural topology differences between the source and target domains, the structure of the Command Center network is manually modified to generate three source domain network structures, as shown in Fig. 4.

Once the source-domain network structures are fixed, samples are drawn from the original network according to its probability distribution to estimate the source-domain parameters. To simulate distributional discrepancies, Gaussian noise is introduced into the original network's distribution prior to sampling. As a result, the parameters estimated from these noisy samples deviate from the original network's true values, and the magnitude of this deviation increases with the noise variance.

4.2 *Instantiation with Different Sample Numbers*

The instantiation method that uses only target domain samples is designated as strategy I, while the method incorporating parameter transfer is designated as strategy II. The objective of this section is to compare the performance of these two strategies

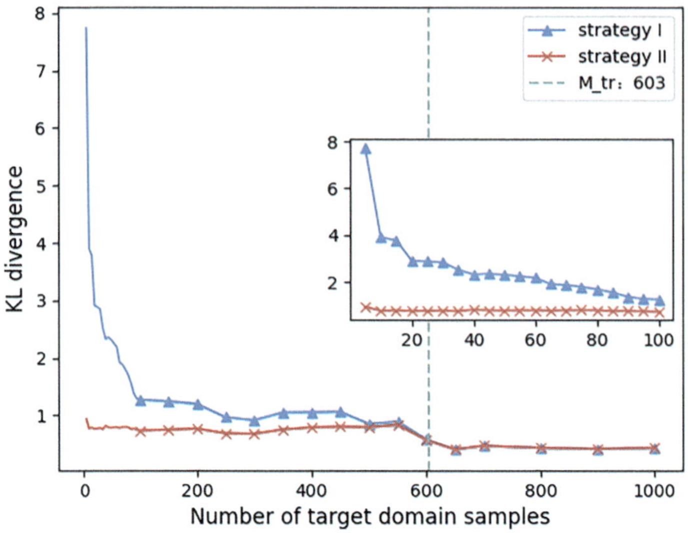

Fig. 5 Comparison of KL divergence under different target domain sample size conditions and different instantiation strategies

under varying numbers of target domain samples. Instantiation accuracy is assessed using KL divergence to quantify the proximity between estimated and true target domain parameters.

Figure 5 presents the KL divergence between the estimated and true parameters of the target domain network as a function of the number of target domain samples. The performance of the two instantiation strategies is compared using the constructed source and target domains, with results averaged over 20 independent trials. Analysis of the simulation outcomes reveals the following:

(1) When the number of target domain samples is below 100, strategy I exhibits a large estimation bias, while strategy II produces significantly lower KL divergence. This indicates that, under sample scarcity, purely data-driven instantiation results in substantial parameter error, while transfer-based instantiation leverages similar source-domain parameters to mitigate this bias.

(2) As the sample size grows, the KL divergence of Strategy II remains consistently low, while that of Strategy I drops sharply at first and then gradually plateaus, ultimately approaching the value of Strategy II. This behavior arises because, with more samples, the source domain influence factor μ diminishes, reducing the impact of transferred parameters and causing both strategies to converge in performance.

4.3 Instantiation with Different Transfer Weights

Building on the previous simulation, it has been shown that the parameter transfer instantiation method improves OOBN performance in low sample scenarios. However, since source domain models may differ from the target domain, unmitigated transfer can introduce negative bias. To address this, a dynamic weight scheme is proposed that adjusts the transfer intensity based on structural and distributional similarity between the source and target domains.

To examine the impact of different source domain migration weight strategies on the parameter migration effect, four distinct types of source domain migration weights are considered:

(1) **Weight I**: source domain migration weights that only consider probability distribution similarity;
(2) **Weight II**: source domain migration weights that only consider structural topology similarity;
(3) **Weight III**: regardless of the similarity between the target domain and the source domain, a uniformly distributed source domain migration weight is used;
(4) **Weight IV**: source domain migration weights based on comprehensive structural topology and probability distribution similarity proposed in this paper.

Figure 6 illustrates the comparison of the instantiation method based on parameter transfer under different sample sizes, using weights I through IV. Analysis of the experimental results reveals that Weight IV, which comprehensively considers both structural topology and probability distribution similarity, performs better in terms of instantiation effect than weights I through III. In terms of overall instantiation effect, the performance of the four weight types is ranked from highest to lowest as follows: Weight IV, Weight II, Weight I, and Weight III. Weight III, which employs a uniformly distributed source domain weight strategy, does not adjust the migration weight based on the similarity between the source and target domains. This leads to the transfer of more dissimilar information, resulting in a higher KL divergence value. Compared to weights I and II, Weight IV can more accurately quantify the difference between source domain objects and the target domain network by considering similarity at both the structural topology and probability distribution levels. As a result, more migration weight is allocated to source domain objects with high similarity, effectively reducing the impact of negative transfer and improving the instantiation effect.

5 Conclusion

This study addresses the challenge of OOBN instantiation under conditions of limited sample data and proposes an instantiation approach based on parameter transfer. To promote positive transfer, a similarity measurement framework is first

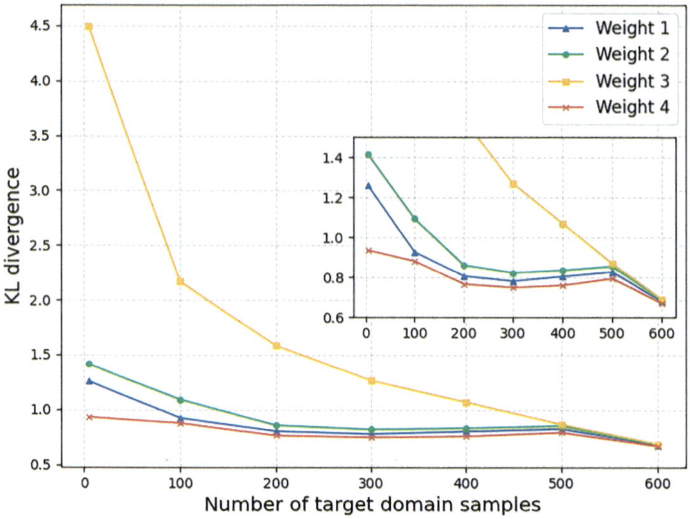

Fig. 6 Comparison of instantiation effects based on parameter transfer under different source domain transfer weight types

established to assess the relatedness between the source and target domains at two levels: structural topology and probabilistic distribution. Building upon this framework, a dynamic weight parameter-transfer strategy is introduced, which allocates contributions from target domain samples and individual source domain objects according to the target sample size and the similarity of each source object to the target. The proposed method is evaluated through simulation on the publicly available Command Center network. Simulation results indicate that, when target domain data are limited, the proposed method surpasses instantiation approaches relying solely on parameter learning. Moreover, in comparison with single-weight or uniform-weight transfer schemes, the dynamic-weight strategy more effectively utilizes source domain information, mitigates negative transfer, and achieves superior overall performance.

In real engineering fault-diagnosis systems, this transfer-based instantiation method generalizes well to scenarios where fault data is inherently scarce, such as newly deployed equipment or rare failure modes. Practitioners should be aware of key considerations: domain mismatches, inevitable in real-world applications, can be mitigated by the proposed similarity evaluation framework, but extreme structural or distributional differences may still lead to partial negative transfer; data limitations, such as incomplete source-domain parameterization or noisy target samples, require careful preprocessing and validation of source objects; critical assumptions include the existence of shared structural or distributional characteristics between source and target domains, and the need to calibrate the small-sample threshold based on specific equipment and fault types to balance transfer effectiveness and local data fidelity. Further studies can be refered to [15–20].

Acknowledgements This work is Supported by the National Natural Science Foundation of China (Grant No. 62176262) and the State Key Laboratory Foundation of China (Grant No. 6142003202308).

References

1. Cai B, Huang L, Xie M (2017) Bayesian networks in fault diagnosis. IEEE Trans Industr Inf 13(5):2227–2240
2. Koller D, Pfeffer A (1997) Object-oriented Bayesian networks. In: Proceedings of the 13th annual conference on uncertainty in AI (UAI). Providence, Rhode Island, Aug 1997, pp 302–313
3. Kasper D, Weidl G, Dang T, Breuel G, Rosenstiel W (2012) Object-oriented Bayesian networks for detection of lane change maneuvers. IEEE Intell Transp Syst Mag 4(3):19–31
4. Cai B, Liu H, Xie M (2016) A real-time fault diagnosis methodology of complex systems using object-oriented Bayesian networks. Mech Syst Signal Process 80:31–44
5. Huang Y, McMurran R, Dhadyalla G, Peter Jones R (2008) Probability based vehicle fault diagnosis: Bayesian network method. J Intell Manuf 19:301–311
6. Oyen D, Lane T (2014) Transfer learning for Bayesian discovery of multiple Bayesian networks. Knowl Inf Syst 43(1):1–28
7. Luis R, Sucar LE, Morales EF (2009) Inductive Transfer for Learning Bayesian Networks. Mach Learn 79(1–2):227–255
8. Zhou Y, Fenton N, Neil M (2014) Bayesian network approach to multinomial parameter learning using data and expert judgments. Int J Approx Reason 55(5):1252–1268
9. Zhou Y, Hospedales TM, Fenton N (2014) When and where to transfer for Bayesian network parameter learning. Expert Syst Appl 55:361–373
10. Hou Y, Yang A, Guo W, Zheng E, Xiao Q, Guo Z, Huang Z (2022) Bearing fault diagnosis under small data set condition: a Bayesian network method with transfer learning for parameter estimation. IEEE Access 10:35768–35783
11. Jia H, Wu Z, Chen J, Chen B, Yao S (2018) Causal discovery with Bayesian networks inductive transfer. In: KSEM 2018, Changchun, China, Aug 2018, pp 351–361
12. Pereira FLF, dos Santos Lima FD, de Moura Leite LG, Gomes JPP, de Castro Machado J (2017) Transfer learning for Bayesian networks with application on hard disk drives failure prediction. In: 2017 Brazilian conference on intelligent systems (BRACIS), Uberlandia, Oct 2017, pp 228–233
13. Koller D, Friedman N (2009) Probabilistic graphical models: principles and techniques. MIT Press
14. Lu TC, Druzdzel MJ (2005) Mechanism-based causal models for adaptive decision support. In: AAAI spring symposium: challenges to decision support in a changing world, Menlo Park, CA, Mar 2005, pp 73–79
15. Guo J, Wang Z, Li H, Yang Y, Huang CG, Yazdi M, Kang HS (2024) A hybrid prognosis scheme for rolling bearings based on a novel health indicator and nonlinear Wiener process. Reliab Eng Syst Saf 245:110014
16. Li H, Yazdi M, Huang HZ, Huang CG, Peng W, Nedjati A, Adesina KA (2023) A fuzzy rough copula Bayesian network model for solving complex hospital service quality assessment. Complex Intell Syst 9(5):5527–5553
17. Li H, Peng W, Adumene S, Yazdi M (2023) An intelligent cost-based consequence model for offshore systems in harsh environments. In: Intelligent reliability and maintainability of energy infrastructure assets, pp 107–117. Springer Nature Switzerland, Cham
18. Li H, Yazdi M (2022) What are the critical well-drilling blowouts barriers? A progressive DEMATEL-game theory. Advanced decision-making methods and applications in system

safety and reliability problems: approaches, case studies, multi-criteria decision-making, multi-objective decision-making, fuzzy risk-based models. Springer International Publishing, Cham, pp 29–46
19. Li H, Peng W, Adumene S, Yazdi M (2023) Cutting edge research topics on system safety, reliability, maintainability, and resilience of energy-critical infrastructures. In: Intelligent reliability and maintainability of energy infrastructure assets, pp 25–38. Springer Nature Switzerland, Cham
20. Li H, Peng W, Adumene S, Yazdi M (2023) Advances in failure prediction of subsea components considering complex dependencies. In: Intelligent reliability and maintainability of energy infrastructure assets, pp 93–105. Springer Nature Switzerland, Cham

From Phenomena to Mechanisms: A Time–Space–Matter Multidimensional Analysis of Aerospace Electronic Failures

Zongsheng Xie, Dawei Zhu, Mengjiang Liu, and Danye Li

Abstract Aerospace electronic products require exceptionally high reliability due to their mission-critical roles and operation within extreme environments. Despite rigorous quality controls, hardware and software failures remain inevitable. The Chinese aerospace industry adopts a distinctive Quality Problem Closed loop methodology; however, fault phenomena are often oversimplified in engineering practice. Such simplification risks the omission of essential contextual information, thereby diminishing the accuracy and efficiency of fault diagnosis and root cause analysis. To address this challenge, the present study systematically categorizes failure phenomena in aerospace electronics and proposes a three-dimensional analytical framework encompassing time, space, and matter. By integrating multi-source data fusion with an innovative Four Diagnosis approach comprising inspection, ausculation&olfaction, inquiry, and palpation, this research establishes a structured methodology for fault characterization. This approach enhances the traceability and comprehensiveness of fault data inputs within Quality Problem Closed loop processes throughout spacecraft development, system integration testing, and in-orbit operations. Ultimately, it contributes to improving the efficiency of closed-loop quality management in aerospace engineering.

Keywords Aerospace · Electronic products · Fault phenomena · Quality problem closed-loop

Z. Xie (✉)
China Astronautics Standards Institute, Beijing, China

D. Zhu
Beijing Satellite Navigation Center, Beijing, China

M. Liu
China Electronics Technology Group Corporation, Beijing, China

D. Li
China Academy of Electronics and Information Technology, Beijing, China

H. Li et al. (eds.), *Data-Driven Methods for Reliability and Safety Engineering: Applications in Industrial Systems*, Springer Series in Reliability Engineering,
https://doi.org/10.1007/978-3-032-22873-4_29

1 Introduction

Spacecraft, as intricate systems engineering products, operate within extreme service environments and are subject to non-repairable constraints, rendering effective quality management essential for mission success. Electronic components, which play a critical role in spacecraft functionality, are characterized by high integration density, sensitivity to weak signals, and susceptibility to radiation. These components exhibit distinct quality challenges at different lifecycle phases. During the development stage, design deficiencies such as inadequate radiation hardening and the absence of signal integrity simulations predominantly emerge. The integration testing phase uncovers system-level problems, including interface protocol inconsistencies and electromagnetic compatibility issues. The assembly, integration, and testing phase reveals reliability concerns such as parameter drift of components under thermal vacuum conditions and micro-assembly structural failures induced by mechanical stresses. During in-orbit operations, functional anomalies arise due to space environment effects like single-event upsets and deep dielectric charging. Compared to mechanical parts, electronic failures tend to be more concealed and propagate through complex pathways, presenting significant difficulties for traditional fault isolation methods. To safeguard aerospace product quality, China has developed a distinctive Quality Problem Closed loop management system, achieving quality control through a technical framework emphasizing precise fault localization, clear mechanism analysis, reproducibility of problems, effective corrective actions, and preventive extrapolation [1, 2].

2 Framework for Fault Phenomena Analysis

In aerospace product Quality Problem Closed-loop practices, accurate and detailed descriptions of fault phenomena are essential for enabling engineers to localize issues quickly and support root-cause analysis [3]. Insufficient or biased information, especially for electronic products representative of spacecraft, can result in delays, wasted resources, or even misdiagnosis. This can lead to contradictory conclusions that hinder effective problem-solving and undermine the closed-loop process. Key challenges include: (1) Incomplete fault description dimensions, where omissions of temporal, spatial, or material factors in records cause "overgeneralization"; (2) The dominance of post-facto analysis, which fails to intervene early in the detection of fault precursors and abnormal trends.

This paper proposes a time-space-matter three-dimensional classification method for fault phenomena analysis. By establishing a multi-source information fusion framework that captures electronic product failure characteristics and integrates the "Four-Diagnosis" methodology (inspection, ausculation&olfaction, inquiry, and palpation), it provides a robust foundation for subsequent steps: precise localization,

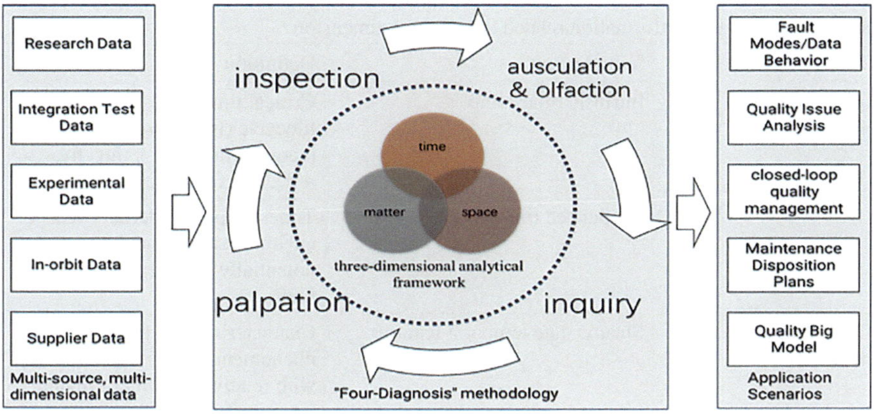

Fig. 1 Overall framework of the three-dimensional (Time-Space-Matter) fault phenomena analysis methodology

mechanism clarification, problem reproducibility, effective measures, and preventive extrapolation. The overall framework is illustrated in Fig. 1.

3 Fundamental Information of Fault Phenomena

"The exploration of the vast cosmos, the advancement of space endeavors, and the building of a powerful space nation represent our unremitting pursuit of the space dream." As the subject of space activities, the universe encompasses all space, time, matter, and their derivatives [4]. Aerospace electronic products, as essential components of this grand vision, should be examined in the context of spatial, temporal, and material dimensions. Below is an elaboration on the fundamental characteristics of aerospace electronic product fault phenomena from these three perspectives.

3.1 Temporal Information

Temporal information systematically characterizes and analyzes fault phenomena in aerospace electronic products through multiple time-related parameters. It encompasses characteristic quantities including temporal points, temporal segments, and temporal statistics, as detailed in the Table 1.

Table 1 Typical fault information related to temporal dimension

Element	Sub-element	Definition
Temporal point characteristics	Intrinsic timestamp	Critical time nodes in the fault lifecycle (first/last occurrence, recovery time, etc.), directly reflecting fault evolution
	Associated timestamp	Timestamps of external events or environmental anomalies potentially causally related to the fault
Temporal segment characteristics	Steady-state temporal features	Characteristics of fault phenomena during time intervals with relative stability or regular variation
	Dynamic temporal features	Characteristics of fault phenomena during time intervals exhibiting abrupt changes, periodic fluctuations, or intermittent anomalies
Temporal statistical characteristics	Trend characteristics	Long-term evolution direction of fault parameters (increasing/decreasing/stable)
	Periodicity characteristics	Recurrent fluctuation patterns at fixed intervals
	Abrupt change characteristics	Sudden transitions in mean/variance within time series

3.2 *Spatial Information*

Spatial information refers to the analysis of fault phenomena in aerospace electronic products based on geographically and physically related parameters. It encompasses the specific location of fault occurrence, the installation position of components within the spacecraft, the relative layout of subsystems, and the surrounding environmental conditions, as detailed in the Table 2.

3.3 *Matter Information*

The matter dimension is the most complex aspect in the analysis of aerospace electronic product fault phenomena. Compared to temporal and spatial information, the matter dimension directly involves the intrinsic properties of the product itself, including the characteristics of the materials and components used. Matter information includes all physical, chemical, and electrical performance data, ranging from raw materials to the final assembly. It also encompasses the entire lifecycle of the

Table 2 Typical fault information related to spatial dimension

Element	Sub-element	Definition
Macro-scale spatial features	Vehicle position	Orbital parameters (altitude, inclination) or geographic coordinates (longitude/latitude) at fault occurrence
	Celestial alignment	Relative positions to celestial bodies (Sun-spacecraft-Earth angles, etc.) during fault
Meso-scale spatial features	Equipment location	Physical installation coordinates and orientation within spacecraft (e.g., near propulsion module, radiator surface) [5]
	EMI source Distance	Spatial separation from potential EMI sources (antennas, motors) [6]
	Thermal path integrity	Physical connection status of thermal paths (TIM contact area, heat pipe bend angle)
Micro-scale spatial features	Package defects	Internal package anomalies (wire bond fractures, voids, delamination)
	Surface degradation	Observable physical damage (oxidation, corrosion, ablation, micro-cracks) [7]

product, from design and manufacturing to testing and in-service maintenance. The detailed classification is shown in the Table 3.

3.4 Integrative Guidance of Time-Space-Matter Dimensions for Root-Cause Reasoning

Time, space, and matter constitute the fundamental structures of the objective world and the core components of fault-phenomenon information: time encodes the persistence of a fault, space encodes its extent, and matter encodes its materiality. In fault analysis these three factors are interdependent, interact, influence, and constrain one another. Accurately capturing their individual traits—time is unidirectional and irreversible, space is multidirectional and reversible, matter is diverse and transformable—allows faults to be characterized with higher fidelity; integrating fault information across all three dimensions identifies evolution patterns, propagation paths, vulnerable parts, and sensitive regions, thereby guiding root-cause reasoning.

4 Fault Diagnosis Methodology

Fault analysis and troubleshooting in electronic equipment can be compared to a doctor's examination of a patient. Through systematic and multi-dimensional data collection, the goal is to rapidly and accurately identify and resolve issues. Similar to traditional Chinese medicine, which uses the "look, listen, smell, ask, and feel"

Table 3 Typical fault information directly related to the product itself

Element	Sub-element	Definition
Product development information	Verification Baseline	Data on radiation tolerance, signal integrity margin, and failure mode thresholds obtained from simulations, prototypes, and environmental testing in the design phase
	Process control	Critical process parameter controls and batch traceability information during production, for evaluating manufacturing consistency and batch-to-batch defect risks
	Integration defects	System-level interaction faults, such as interface protocol mismatches and EMC non-compliance detected during system integration testing
Product fundamental information	Inherent properties	Physical–chemical characteristics of materials and nominal performance parameters of components, defining the product's inherent reliability bounds [8]
	Topological architecture	Multi-level interconnection relationships in the product's physical structure, revealing signal integrity or thermal conduction path anomalies
	Supply baseline	Supplier qualifications, component batch reliability data, and software configuration versions, for managing supply chain and hardware-software integration risks
Product usage information	Operating boundaries	Actual operating load conditions, environmental stresses, and multi-physics coupling effects, quantifying the impact of excursions beyond design limits
	Intervention Traceability	Maintenance records, replacement part batches, and associated external events, for locating the root causes of human errors or environmental disturbances [9]
	Degradation characterization	Performance parameter drifts, material aging, and failure recurrence cycles, characterizing the gradual or abrupt failure trends over the product's lifecycle

approach to gain a thorough understanding of a patient's condition, engineers and quality management personnel adopt similar methods to gather detailed fault data. This approach ensures a more comprehensive and in-depth analysis for diagnosis.

4.1 Inspection—Utilizing Visual Inspection

Observing, or visual inspection, is the most direct means of identifying physical and appearance-related anomalies. It is typically the initial step in fault diagnosis. The observable appearances of different product levels, from components, circuit boards, and individual units to subsystems dominated by electronic products, may differ, but can generally be categorized into color and state abnormalities.

Color abnormalities are the most intuitive indications of faults. For example, carbonization and blackening on component or circuit board surfaces due to high-temperature effects are often directly related to local overloads or thermal management failures. Chemical corrosion leading to green or white stains suggests inadequate environmental sealing or material compatibility issues. Oxidation and discoloration (such as grayish-white rust) on metal connectors and terminals are usually accompanied by a significant increase in contact resistance, requiring a comprehensive assessment of material properties and environmental exposure history [10].

State abnormalities encompass physical deformations from the macro to micro scale. At the macro level, changes in the overall appearance may indicate mechanical damage, internal failures, or improper sealing of the product. At the meso scale, component shape deviations (such as swelling of electrolytic capacitors) are typical signs of material aging or over-voltage failures. At the micro scale, hidden defects such as wire bond fractures and delamination within the packaging are of concern [11].

Additionally, the presence of excessive dust, fibers, or metal debris inside the equipment often reflects quality management issues in the production environment. This highlights the need to reassess manufacturing capabilities.

Engineering example: Unconstrained Inductive Switching (UIS) is a typical failure mode of MOSFETs, which typically includes single avalanche and repeated avalanche. By observing the location and size (Spatial Information) of MOSFET burn marks, the duration and root cause of avalanche events can be inferred. The failure mechanism of a single avalanche event in MOSFET is due to the node temperature exceeding the maximum temperature limit, and the duration is generally in the microsecond range. If the duration of an avalanche event is at the millisecond level, burn marks will appear in the middle of the wafer where the maximum current flows. Due to the high current density at the junction of the connecting leads and the wafer, the location of failure often occurs here. However, due to the presence of heat sinks, failure often does not occur directly below the junction where the leads are connected, but rather in the hottest part of the wafer. For a brief avalanche event at the US level, burn marks may occur anywhere on the entire wafer. In this case, it is unlikely for the current to gather together or for the device to heat itself during this event period, so the temperature rise inside the device is relatively uniform. Even for shorter avalanche events, burning may occur at the corners of the wafer due to the discontinuous units at the edges.

4.2 Ausculation & Olfaction—Utilizing Hearing and Smell for Detection

Auditory-olfactory inspection (i.e., listening and smelling) involves capturing abnormal sounds and odors to complement the information obtained through visual examination. By detecting anomalous acoustic and olfactory characteristics during electronic device operation, this method provides multidimensional perceptual clues for fault diagnosis. Variations in sound and smell often indicate abnormal functioning of internal components or potential safety hazards, such as overheating, short circuits, or leaks [12].

Abnormal sounds can be categorized by two features: pitch and loudness. Pitch focuses on the spectral characteristics and dynamic changes of sound. For instance, high-frequency squealing is often associated with circuit issues such as failed filter capacitors or magnetic saturation of inductive components, while low-frequency humming indicates mechanical defects like relay contact chattering or loose transformer windings. Loudness abnormalities reflect device status through sudden changes in sound intensity. A sharp increase in volume may suggest power module overload or a cooling system malfunction, while a decrease or absence of sound is often linked to speaker disconnection or signal transmission failure. Abnormal odors are analyzed by identifying the types and release patterns of volatile substances to trace fault origins. These odors can be characterized by three attributes: intensity, nature, and persistence [13].

Odor intensity can be roughly classified into three levels: mild, moderate, and strong. Mild odors typically appear during early-stage localized overheating or gradual material decomposition. Moderate odors often accompany electrolyte leakage or plastic component melting. Strong, pungent odors usually indicate severe short circuits or combustion events. Odor nature provides diagnostic specificity. For example, a burnt smell usually originates from carbonized insulation materials, while a melting plastic odor points to overheating failure of encapsulation materials. Odor persistence helps distinguish fault patterns: transient odors are commonly caused by occasional overloads, whereas persistent odors are closely linked to chronic degradation.

Engineering example: Capacitors are one of the common components in aerospace electronic products. Due to the presence of electrolyte, when the capacitor cracks, it will release a pungent odor similar to rosin mixed with ammonia water. This special odor mainly comes from the high-temperature decomposition of the electrolyte. By identifying the odor, the type of capacitor that has failed (material information) can be determined. Among them, the ethylene glycol based electrolyte used in aluminum electrolytic capacitors produces a sweet caramel smell, the manganese dioxide contained in tantalum capacitors reacts with organic matter to produce a sulfur smell similar to match burning, and solid capacitors only have a faint plastic melting smell because they do not contain liquid electrolyte. In addition, different odors correspond to different degrees of fault severity. Sour and astringent odors indicate that the electrolytic paper has been carbonized due to heating, usually accompanied by cracking of the top explosion-proof valve. Fishy odors indicate the decomposition of amine substances in the electrolyte, which may have caused internal short circuits. Metallic odors indicate the volatilization of aluminum foil corrosion products, often accompanied by leakage. Burnt odors are the most dangerous signal, indicating that the PCB substrate may have been burned by high temperatures.

4.3 "Inquiry"—Acquisition Through Questioning and Investigation

Inquiry involves gathering information about the equipment's operational status, historical records, and environmental conditions through communication with operators, maintenance engineers, and other relevant personnel. It also includes reviewing related documentation. Unlike "visual inspection" and "auditory-olfactory inspection," which focus on immediate observations, inquiry provides essential background information, making fault analysis more comprehensive.

Operational Status Analysis examines recent abnormal behaviors, such as intermittent malfunctions reported by operators, performance parameter drifts, and other unstable phenomena. It also looks into how unresolved historical issues may influence current faults.

Maintenance Operation Audit requires systematically reviewing maintenance logs to verify the replacement batches of critical components, the standardization of repair operations, and records of human interventions. Variations in spare part batches may introduce latent defects, while improper operations or parameter setting errors (e.g., threshold exceedance) may trigger unexpected failure modes.

User Feedback Integration involves compiling and analyzing requests related to special operational conditions and historical fault reports submitted by users. This process helps determine the recurrence frequency and patterns of similar faults. The concentration of faults under specific conditions, such as insulation failures in high-humidity environments or connector loosening under strong vibrations, can help identify weaknesses in the product's adaptability to environmental factors [14].

Environmental Stress Correlation quantifies the operating environmental parameters of the equipment, including extreme temperature and humidity levels, mechanical vibration and shock intensity, electromagnetic interference field strength, and other relevant physical quantities. The relationship between these parameters and fault occurrences is analyzed with respect to their spatiotemporal correlation.

Engineering example: During the in orbit operation phase of satellites and other aerospace products, it is usually not possible to dismantle them for "inspection" and "Auscenation&Olfaction". By querying the quality history of faulty products, it is possible to grasp the quality information of the ground development stage, the technical baseline of the product, production batches, and other data, thus tracing the inherent problems, historical deviations, and potential weak links that may exist in the design and production process of the product. Through similar product failure cases and component manuals, it is possible to grasp the possible failure modes and their occurrence probabilities of products at different levels. This can provide systematic and hierarchical analysis clues for fault attribution and localization, clarify the priority order and troubleshooting path of diagnosis, and provide judgment basis for the formulation of fault attribution, localization, and disposal plans.

4.4 "Palpation"—Acquisition Through Instrumentation and Testing

"Palpation" (i.e., instrumentation and testing) involves conducting physical, electrical, and functional measurements and inspections on equipment using various tools and techniques. This process helps validate the preliminary clues identified through "visual inspection," "auditory-olfactory inspection," and "inquiry," revealing deeper-level phenomena that cannot be directly observed by other means. It provides a scientific basis for subsequent diagnosis and repair, thereby improving the precision and efficiency of problem resolution. The process can be characterized along two dimensions:

Testing Object Dimension: This includes electrical parameters, signal characteristics, and software functionality. Electrical parameters, such as voltage, current, conduction resistance, insulation resistance, and grounding resistance, reflect the physical state of the equipment (e.g., power module stability, solder joint quality, material aging) and reveal component-level failure mechanisms. Signal characteristics focus on signal amplitude, frequency, waveform integrity, impedance matching, and power transmission efficiency [15]. These metrics assess the physical-layer performance of signal transmission paths (e.g., transmission loss, interface compatibility, energy conversion efficiency) and help locate interaction-layer defects. Software functionality primarily includes communication protocol logic, software version iterations, system log patterns, and Built-In Test (BIT) status. These aspects evaluate system-level logical correctness and behavioral consistency, identifying risks of hardware-software co-failures.

Fault Characteristic Dimension: This includes explicit anomalies and latent trends. Explicit Anomalies refer to threshold-exceeding events that can be directly identified through real-time monitoring. Their core function is immediate alerting and fault localization, providing decision-making support for rapid response. Based on their temporal evolution, spatial distribution, and coupling patterns, they can be further classified into: Temporal Anomalies: Cover transient, persistent, and intermittent threshold violations, respectively indicating sudden hard failures, progressive degradation, and environmentally sensitive issues. Topological Anomalies: Divided into single-point anomalies (indicating component-level defects) and cluster anomalies (revealing systemic design flaws). Composite Anomalies: Characterized by multi-parameter joint deviations and cascading propagation, reflecting complex failure mechanisms. Latent Trends refer to parameter behaviors that remain within thresholds but exhibit statistically significant evolutionary patterns. Their purpose is to predict potential degradation risks, providing a time window for preventive maintenance. Based on their evolution patterns, latent trends can be categorized into: Monotonic Trends: Reflect deterministic degradation mechanisms through unidirectional parameter drift. Periodic Trends: Manifest as cyclical parameter fluctuations, indicating environmental coupling effects. Random Trends: Characterized by statistical property shifts or increased event frequency, signaling a deterioration in system stability.

Engineering example: There are various types of outliers in the telemetry data of spacecraft in orbit, which often correspond to different root causes of faults. To improve the accuracy of fault recognition, convolutional neural network methods can be used to classify and identify outliers in telemetry data. The specific process is as follows: first, convert the numerical telemetry data into a grayscale binary image format; Based on the differences in their graphical features, classify abnormal patterns and construct training and testing sets for outlier recognition; Finally, an outlier recognition model was trained using convolutional neural networks to achieve mapping correlation between outliers and potential fault patterns. This method helps establish an identification link from abnormal manifestations to the root cause of faults, thereby improving the accuracy of fault diagnosis.

5 Case Study on Fault Phenomena Analysis

A solid-state power amplifier exhibited abnormal output power attenuation during ground-based life testing. The product development team conducted a detailed analysis and resolution of this fault.

5.1 *Fault Phenomena Analysis Based on "Temporal-Spatial-Material" Dimensions*

Temporal Dimension: Time-series statistical characteristics: The power attenuation rate was found to be exponentially correlated with thermal cycling count ($R^2 = 0.91$). A significant acceleration in power attenuation occurred after the 50th cycle, following a Weibull distribution pattern. Dynamic time-domain characteristics: Under pulse modulation, excessive fluctuations in gate voltage were observed. Time-domain reflectometry measurements revealed a significantly prolonged rise time, leading to increased switching losses.

5.1.1 Spatial Dimension: Macro/Meso-scale Features

Thermal resistance measurements indicated that the GaN chip was positioned too close to the heat sink edge, resulting in thermal resistance exceeding design specifications. A temperature gradient was observed across the chip surface. Micro-scale features: Scanning Electron Microscopy (SEM) analysis identified cracks in the gate metal layer, accompanied by significant electromigration effects.

5.1.2 Material Dimension

Batch analysis: Investigation of the GaN chip supplier's batch records revealed that the epitaxial wafers in this batch exhibited excessive threading dislocation density; Process variation: Plasma-enhanced chemical vapor deposition (PECVD) passivation layer thickness showed non-conforming deviations in this batch.

5.2 *Comprehensive Fault Diagnosis Based on the "Four Diagnostic Methods"*

Visual inspection: X-ray imaging revealed unconnected regions in the power supply layer, leading to localized high current density [16]. Electron microscopy identified microvoids at the passivation layer interface, exacerbating electric field concentration effects. Auditory-olfactory inspection: Gas chromatography-mass spectrometry (GC–MS) detected excessive residual hexamethyldisilazane (HMDS) in the passivation layer, reducing interfacial adhesion strength. Inquiry: Review of the epitaxial wafer supplier's quality records and process documentation revealed omission of MBE in-situ annealing, resulting in excessive threading dislocation density, and non-compliant PECVD deposition temperature, causing non-uniform passivation layer density. Measurement: Thermo-electrical coupling tests quantified key fault parameters, focusing on dynamic on-resistance and transient thermal resistance, ensuring precise fault localization and reproducible problem verification. Comprehensive analysis indicates the following fault propagation chain:

(1) Material-dimensional defects are the root cause: high dislocation density → poor crystal quality → non-uniform passivation deposition → mobility degradation → increased on-resistance → Joule heating accumulation
(2) Spatial and temporal dimensions accelerated fault evolution: inadequate thermal design + cyclic thermal stress → junction temperature exceedance → crack initiation/propagation → electrical performance degradation.

The R&D team established a "dislocation density-thermal resistance-electromigration" tri-coupled model, enabling cross-scale mapping of material defects → thermodynamic response → electrical performance.

The closed-loop four-diagnostic approach provided critical support: visual inspection localized microscopic anomalies, auditory-olfactory inspection captured latent characteristics, inquiry traced supply-chain root causes, and measurement quantified fault parameters.

6 Conclusion

This paper presents a fault phenomena analysis method for aerospace electronic products based on a three-dimensional classification encompassing temporal, spatial, and material aspects. A multi-source information fusion framework is developed, incorporating the failure characteristics of electronic components and integrating the four diagnostic methods: observation, auscultation and olfaction, inquiry, and measurement. This approach facilitates the comprehensive acquisition of fault information that is complete, accurate, traceable, and correlated through visual, auditory, olfactory, and measurement techniques. The methodology effectively captures all pertinent elements, including the time and location of fault occurrence as well as environmental conditions.

Future research may focus on employing large-scale model technologies to enable more intelligent analysis and processing of fault phenomena data. By harnessing the advanced data processing and pattern recognition capabilities of these models, it will be possible to perform a more thorough examination of extensive fault phenomena datasets, uncovering latent failure patterns and trends. Such advancements are expected to significantly improve the precision of fault characterization and provide a more solid foundation for fault localization and mechanism analysis.

References

1. Implementation Requirements for the "Zeroing Out" of Aerospace Product Quality Issues (2021), GB/T 29076 (In Chinese).
2. Implementation Requirements for the "Three-Orientations" Analysis Method for Aerospace Product Quality Issues (2023), GB/T 42859 (In Chinese).
3. Chevalier F (1991) From quality circles to total quality. Int J Qual Reliabil Manage 8(4)
4. China's Space Program (2022) A 2021 Perspective White Paper. National Defense Science, Technology & Industry 02:10–11 (In Chinese).
5. Branzei M, Plotog I (2015) Thermophysical properties in assembling process at macro and micro level. Adv Mater Res 4057(1114–1114):172–179
6. Huang C, Chen C (2017) Improve electromagnetic compatibility of electronic products with multivariate parametric design. Microelectron Int 34(1):45–55
7. Lv Z, Duan Z, Jiang N et al. (2025) Defect engineering of anchored on F-doped BNNR surface to enhance low-frequency microwave absorption and achieve exceptional thermal conductivity properties. Vacuum 231(PB):113826–113826
8. Cao S, Zheng T, Bi Z et al (2024) Regulating strain and suppressing defect states through polymer ionization and chemical chelation for efficient inverted flexible perovskite solar cells. Small 21(6):e2410716
9. Wang R, Chen L, Lee KA, Ling Z-H (2024) Asynchronous voice anonymization using adversarial perturbation on speaker embedding. In: Proceedings of Interspeech
10. Lim W, Kim YJJ, Kharbash R et al. (2025) Investigation of color formation and H_2O_2-assisted decolorization in CO_2-derived polyhydroxyurethane. Polym Test 147:108836
11. Lili L, Yulong W, Junguo G, et al. (2022) Effects of microstructure on the breakdown characteristics of polyethylene–montmorillonite nanocomposites. J Appl Poly Sci 139(30)
12. Zhang L, Deng P (2019) Abnormal odor detection in electronic nose via self-expression inspired extreme learning machine. IEEE Trans Syst Man Cybern: Syst 49(10):1922–1932

13. Wang W, Pang J, Kong X, Zhang J (2017) Diagnostics and control of automotive fuel pump induced noise. In: INTER—NOISE and NOISE—CON congress and conference proceedings 255(4):3324–3335
14. Husain WSMW (2014) The future of aircraft feedback information: maintainability allocation methodology. Appl Mech Mater 3446(629–629):225–230
15. Jing T, Min X, Xiaohui Z, et al. (2012) An orthogonal multi-carrier source for realizing high speed all-optical OFDM system. Huazhong University of Science and Technology (China), pp 833183310E-83310E-5
16. Moore CK, Emery JS, Lucas GM (2011) Quality and quantity: an audit of urodynamics practice in relation to newly published national standards. Neurourol Urodyn 30(1):38–42

Research on the Health Monitoring Technology of Preload for Complex Aerospace Equipment Fastening Devices

Wu Kai, Luo Zhijun, Gu Lixiang, Zuo Wenjia, and Liu Yuchang

Abstract This paper examines fastening devices for critical precision instruments in complex aerospace equipment, identifying the underlying causes of preload failure and comparing the theoretical foundations of commonly used methods, including the torque method, angle method, and yield point method. The ultrasonic measurement method is ultimately selected for design and application. Using the installation of inertial navigation units (INUs) as a case study, systematic verification is conducted, demonstrating excellent performance outcomes.

Keywords Aerospace equipment · Preload · Health monitoring

1 Introduction

The development of complex aerospace equipment requires the installation of critical precision instruments on rockets and vehicles, where installation stability directly influences instrument accuracy [1–3]. For example, vehicle-mounted inertial measurement units (IMUs) are typically secured to the vehicle frame using four M12 bolts. Transportation-induced vibrations inevitably cause bolt loosening, which reduces preload. This preload loss alters the geometric positioning of the IMU, impairing navigation and targeting accuracy, and in severe cases, leading to a substantial decline in overall system precision. Bolt loosening primarily arises from two causes: strain loosening induced by environmental vibrations, which is unavoidable [4], and preload deviation during the tightening process, as in the traditional torque method. Variations in friction on threaded surfaces can cause the actual preload resulting from a present torque to deviate from the design value by as much

W. Kai (✉) · G. Lixiang · Z. Wenjia · L. Yuchang
Equipment Comprehensive Support Center China Academy of Launch Vehicle Technology, Beijing, China
e-mail: 18600755554@163.com

L. Zhijun
School of Mechanical Engineering, Tsinghua University, Beijing, China

H. Li et al. (eds.), *Data-Driven Methods for Reliability and Safety Engineering: Applications in Industrial Systems*, Springer Series in Reliability Engineering,
https://doi.org/10.1007/978-3-032-22873-4_30

as 30%. To enhance the accuracy of preload tightening, several improvements have been introduced, resulting in methods such as the yield point method and the angle method [4]. The ultrasonic control method, a recent advancement in preload control technology, measures bolt deformation via ultrasonic waves and determines preload using a calibrated axial force coefficient, achieving a control accuracy exceeding 3% [4–10]. In addition, this method enables online in-situ monitoring of bolt preload, offering timely detection of preload reduction due to vibration or other influences. This study focuses on vehicle-mounted IMUs and investigates ultrasonic bolt preload measurement and monitoring technology. A novel approach is proposed for monitoring and calibrating the preload of IMU fastening bolts, ensuring real-time stability and reliability of the IMU platform base.

2 Bolt Loosening Reasons

Bolt preload failure is predominantly manifested as bolt loosening, which can cause displacement of the locking device and, consequently, degrade accuracy. In severe cases, insufficient bolt preload may result in test failure, thereby hindering the normal progress of spacecraft missions. Hence, research on bolt preload is of significant importance. Both the literature review and experimental findings indicate that bolt loosening refers to the reduction of preload force within the bolt threads. This loosening can be classified into rotational loosening and non-rotational loosening (Fig. 1). Rotational loosening occurs when relative rotation between the internal and external threads reduces the preload, whereas non-rotational loosening arises from other mechanical processes that diminish preload, including surface embedding, creep, plastic deformation, and fretting wear. Experimental analyses reveal that, in the initial stage of bolt preloading, the preload force gradually decreases, primarily due to non-rotational loosening, with a moderate reduction that stabilizes after a certain point. In the intermediate stage, however, the preload force experiences a sudden and instantaneous drop, largely because the frictional force of the bolt becomes insufficient to counteract external lateral loads under horizontal vibration. Horizontal vibration refers to the application of periodic reciprocating force loads to moving components in a direction perpendicular to the thread axis. When the fixed component is used as a reference, these vibrations cause the moving component to reciprocate along either the x-axis or the y-axis. Under continuous high-frequency vibration conditions, the preload of vehicle-mounted bolts is significantly compromised, thereby adversely impacting the stable operation of the spacecraft.

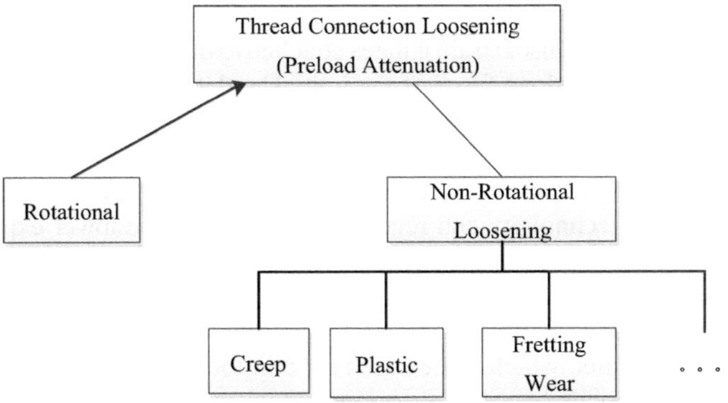

Fig. 1 Causes of bolt preload attenuation

3 Measurement of Bolt Preload Technology Comparison

The main methods for measuring the preload force of bolts currently available include the torque method, torque/angle method, yield point method, and strain gauge method. By comparing these methods in terms of measurement parameters, accuracy, and complexity, a convenient and accurate method can be identified for application in vehicle bolt monitoring. With the development of artificial intelligence technology and image processing technology, some scholars have combined these technologies with bolt preload measurement research, achieving rich results. The following mainly analyzes the pros and cons of each method.

The torque method achieves bolt preload by controlling the torque during the tightening process. The core of this method is to use a torque wrench or other tools to apply a predetermined torque, thereby obtaining the corresponding preload force. The preload force obtained by this method has relatively low accuracy and may have a larger margin of error.

The torque/angle method is an evolution of the torque method. It first tightens to a lower torque value (threshold torque) using the torque method, and then continues to tighten by a specified angle. This method utilizes the relationship between torque and angle during the tightening process, improving the accuracy of preload force by controlling the angle. The torque/angle method involves elastic deformation and possible plastic deformation during the tightening process, and its accuracy is higher than that of the torque method, but it has higher requirements for tightening speed and surface finish.

The yield point method is a method of tightening the bolt or nut to the yield point. During the tightening process, the bolt undergoes a small amount of local plastic deformation when it reaches the yield strength, thus obtaining greater preload force and smaller preload force dispersion. The challenge of this method lies in how to accurately capture the yield point of the bolt and to consider the impact of excessive preload force on the fatigue strength of the bolt.

The strain gauge method measures the elongation of the bolt to obtain the required preload force. This method uses strain gauges attached to the bolt to measure its strain, thereby calculating the preload force of the bolt. The detection results of the strain method are mainly affected by the performance of the bolt itself and are significantly influenced by the bolt tightening torque. This method can provide higher preload force accuracy but may require more complex equipment and operation.

Image processing technology and related technologies have lower experimental equipment requirements and do not require any additional testing devices. Typically, a photo is taken of the structural bolt connection part, and the image is processed for image segmentation and feature extraction. Generally, a support vector machine is used to train the classifier, which is then used for bolt loosening detection. However, shadows during photography can cause inaccurate positioning and lead to false judgments.

These methods are all applicable for measuring the preload force of bolts, but are limited in measuring the preload force of already tightened bolts. The torque method, the torque/angle method, and the yield point method are all used to control the preload force of bolts during the assembly process, but they are not suitable for measuring the preload force after the bolts are tightened. The image method has strong adaptability and is suitable for measurement under various working conditions, but it is prone to misjudgement and is mostly used for auxiliary judgment. Ultrasonic method, as a non-destructive, high-precision detection technology, can eliminate the impact of friction between the contact surfaces of the bolt connection on torque measurement, control the preload force of the bolt during the assembly process, and also monitor the change of residual preload force of the bolt after assembly. Combined with computer technology, it is convenient to connect and easy to maintain with cables, allowing simultaneous measurement of the preload force of a group or multiple groups of bolts. Therefore, we will use the ultrasonic method to monitor the preload force of bolts.

4 Ultrasonic Bolt Preload Measurement Principle

The core principle of ultrasonic bolt preload measurement involves determining bolt length from the ultrasonic wave propagation time within the bolt. The actual preload force is subsequently calculated using Hooke's Law (Fig. 1). At time t_0, the ultrasonic sensor mounted on the bolt head surface is subjected to high-pressure vibration to generate ultrasonic waves. The waves propagate axially along the bolt. Upon reflection at the bolt end, they return, inducing a vibration in the ultrasonic sensor at time t_1. By measuring the interval t0-T0 between the two vibrations of the emitted wave and the reflected wave through the circuit, the length of the bolt can be calculated, and thus the strain of the bolt can be derived, leading to the actual preload force of the bolt. Research has shown that there is a linear relationship between the propagation time of ultrasonic waves within the bolt and the preload force, and it is also affected by temperature. The mathematical formula is expressed as follows:

Clamping length — Internal stress — Acoustic elasticity coefficient — Elastic modulus

$$\Delta t = t_\sigma - t_0 = \frac{2L_0\sigma(K + 1/E)}{C_0} \tag{1}$$

Ultrasonic flight time under stress — Ultrasonic flight time without stress — Sound velocity without stress

As shown in formula 1, as long as the bolt coefficient K is calculated in advance, that is, a calibration test is performed on the bolt, then in actual application, only the propagation time of the sound wave needs to be measured to obtain the stress that the bolt is subjected to at this time.

5 Research on Ultrasonic Bolt Preload Measurement

Excitation of ultrasonic longitudinal wave sensors typically necessitates high-voltage pulses ranging from 100 to 400 V. For piezoelectric crystals, the common excitation pulses include negative sharp pulses and negative square wave pulses. Negative sharp pulses are typically generated by controlling capacitor charge/discharge via field-effect transistors. This approach offers simplicity in principle, convenience in circuit implementation, and broad pulse frequency adaptability. This paper mainly uses the method of negative sharp pulse discharge to construct an experimental platform to achieve ultrasonic excitation, and uses high-speed AD to achieve collection of the ultrasonic emission and reflection process, as shown in Fig. 2. From the figure, it can be seen that after the ultrasonic wave emission, multiple reflections are produced, and several obvious reflections are marked as t1, t2, t3, t4. With the increase of the number of reflections, the ultrasonic wave gradually attenuates. Amplification of the waveform at t_1 shows the ultrasonic echo exhibits damped oscillation. The distinct first echo peak serves as the marker for echo arrival time calculation; flight time (T) is defined as the interval from emission start to this first peak. From Fig. 2, it can be seen that the first echo peak is about 35 microseconds, multiplied by the sound speed of 5900 m/s and then divided by 2, which is about 103 mm, consistent with the actual length of the bolt, refer to Fig. 3.

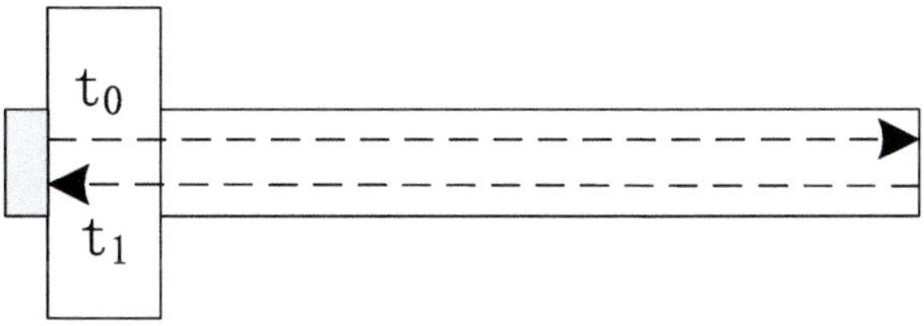

Fig. 2 Propagation path of ultrasonic waves inside the bolt

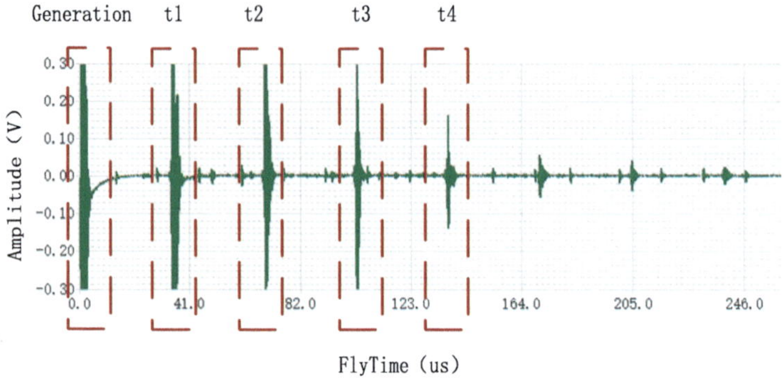

Fig. 3 Propagation of ultrasonic waves within the bolt

By formula 1, it can be obtained that after the bolt is subjected to force, it will elongate, and the elongation is linearly related to the preload force. However, in actual use, the length of the bolt L and the clamping length L1 are not consistent, as shown in Fig. 4. But through experiments, it is found that the relationship between ultrasonic time and preload force still satisfies a linear characteristic, as shown in Fig. 5. Therefore, formula (1) can be simplified to obtain formula (2).

$$F - F_0 = K(T_\sigma - T_0) \tag{2}$$

In the formula, F is the preload force of the bolt at the time of measurement, F_0 is the preload force of the bolt at the time of calibration, T_0 is the flight time at the time of calibration, and T_σ is the flight time obtained at the time of measurement.

According to the linear relationship, a standard hollow strain gauge can be used as a standard preload force collection module to achieve synchronous collection of preload force and flight time, thereby obtaining the linear relationship between preload force and flight time, that is, the axial force coefficient K.

Research shows that temperature will cause deformation of the bolt itself, causing measurement errors. By collecting the temperature and flight time of the bolt for a long time, the results are obtained as shown in Fig. 6. It can be clearly seen that the flight time is linearly related to temperature, and this relationship can be represented

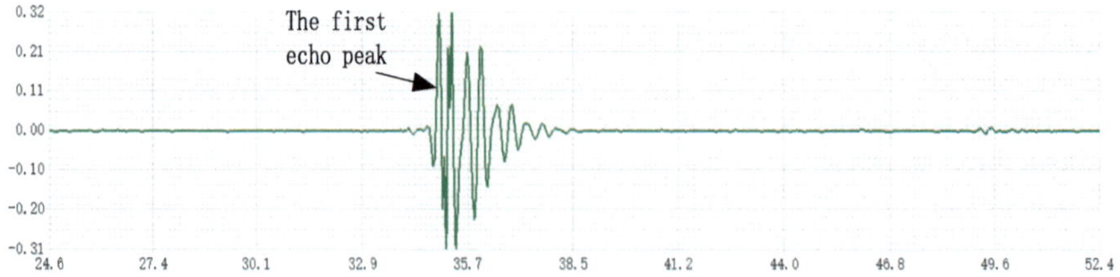

Fig. 4 Echo peak morphology

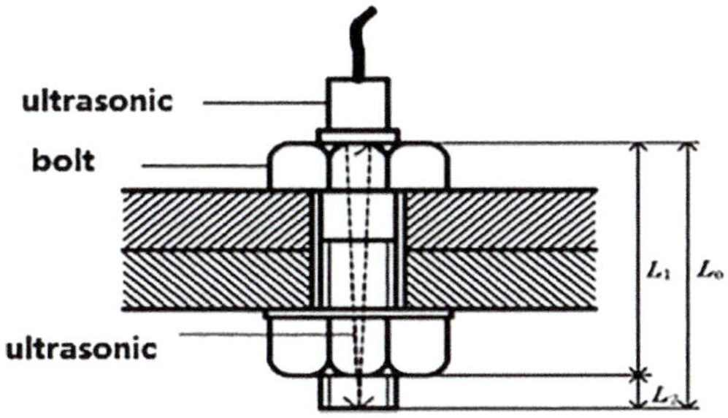

Fig. 5 Results of axial force coefficient calibration

by formula (3).

$$T_1 - T_0 = \gamma(a_1 - a_0) \tag{3}$$

In the formula, a_0 is the initial temperature, a_1 is the temperature measured the second time, T_0 is the initial flight time, T_1 is the flight time at the time of measurement, and γ is the ratio of temperature to flight time.

(Temperature is indicated by the ratio of the resistance of a PT1000 sensor to that of a standard 1000 Ω resistor.)

Another relationship is that there will be a deviation between the clamping length L1 at the time of calibration and the actual clamping length L during actual measurement, as shown in Fig. 7.

The deviation will cause the axial force coefficient to be proportionally scaled, as represented by formula (4).

$$K_r = K{*}L/L_1 \tag{4}$$

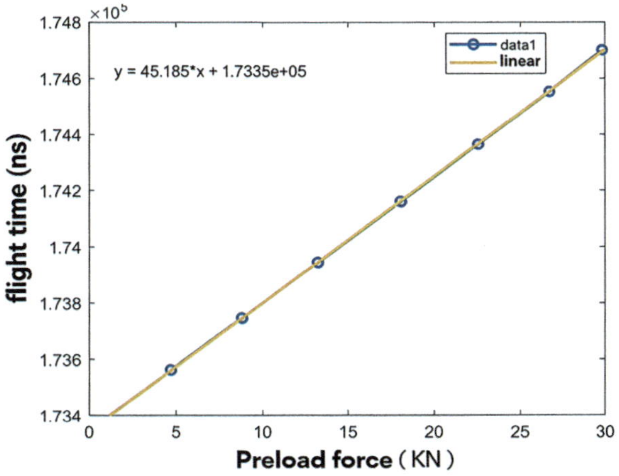

Fig. 6 Chart of the relationship between flight time and preload force

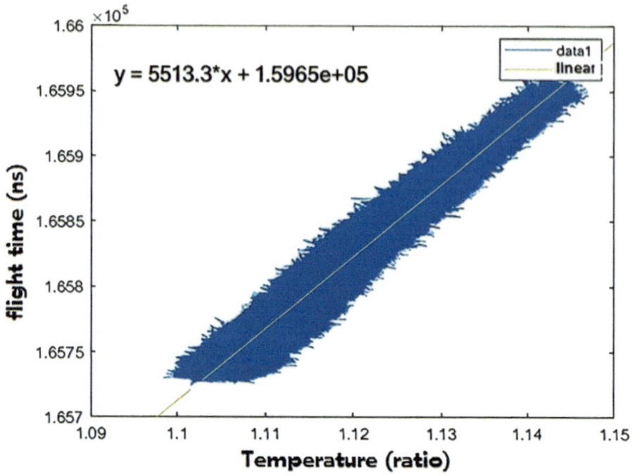

Fig. 7 The relationship between ultrasonic flight time and temperature

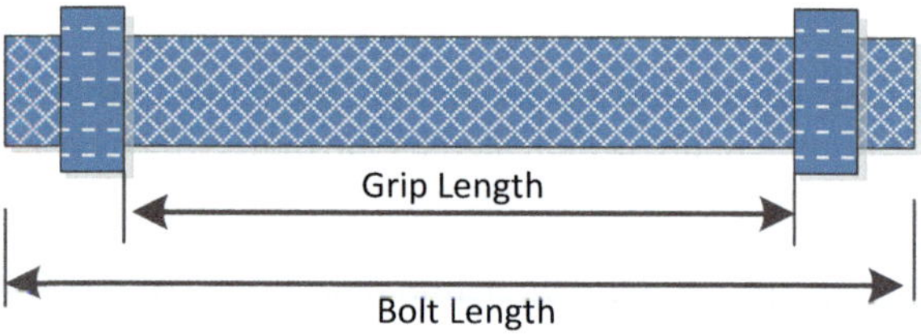

Fig. 8 Bolt clamping length

In the formula, K is the axial force coefficient obtained from the bolt calibration, and K_r is the axial force coefficient at the time of actual measurement.

Combining formulas (2) to (4), the final formula for calculating the actual preload force can be obtained:

$$F = F_0 + (T_\sigma - T_0 - \gamma(a_1 - a_0))*K*L/L1 \quad (5)$$

6 Influence Factors of Ultrasonic Measurement of Bolt Preload

6.1 Temperature

Temperature not only affects the propagation speed of ultrasonic waves in the medium but also causes changes in the length and diameter of the bolt due to the thermal expansion and contraction properties of metal materials. Therefore, when measuring the preload force of the bolt, temperature compensation should be considered to eliminate the impact of temperature changes on the measurement of the bolt preload force. In actual measurement, to ensure measurement accuracy, each material should have its corresponding temperature compensation coefficient.

6.2 Material

Most metal materials can propagate ultrasonic waves well, but some cast iron and most plastics have large ultrasonic attenuation and are not suitable for ultrasonic measurement. The acoustic properties of the bolt and its surrounding medium (such as lubricants, anti-corrosion coatings, etc.) will affect the propagation of ultrasonic waves. Therefore, when considering the impact of stress on the speed of ultrasonic waves, the impact of materials should also be taken into account.

6.3 Bolt Quality

The verticality and roughness of the bolt end surface processing will greatly affect the stability of the ultrasonic test [9]. The condition of the bolt surface, such as roughness, corrosion, or coating, may affect the reflection and absorption of ultrasonic waves, which will greatly affect the stability of the test and thus the measurement results. The verticality of the bolt end surface will also affect the direction of ultrasonic wave reflection, thereby affecting accuracy. Rust on the bolt will also affect the test data. If the test period is long and there is a risk of rust on the bolt, anti-rust protection and treatment should be completed before the test. Therefore, when performing ultrasonic measurement, the bolt surface should be kept clean and dry to avoid contamination or damage to the threads and other surfaces, which could affect the measurement results.

6.4 Environmental Noise

When ultrasonic pulses propagate in a medium (such as air or liquid), they encounter different obstacles or interfaces and generate reflection, absorption, and scattering. The impact of environmental noise on ultrasonic measurement mainly reflects in the superposition of noise signals and ultrasonic signals, resulting in a decrease in the signal-to-noise ratio of the received signal. Noise sources include mechanical vibration, air flow, electromagnetic interference, etc., which interfere with the measurement system through different coupling mechanisms (such as inductive coupling, capacitive coupling, etc.). Therefore, to improve the accuracy of the measurement, it is necessary to adopt anti-interference design, signal processing technology, and environmental control measures to reduce the impact of noise and environmental factors.

6.5 Disadvantages of Ultrasonic Measurement

Currently, ultrasonic measurement has been applied in various situations and has achieved numerous breakthroughs in terms of application scenarios, measurement accuracy [12], and cost when combined with various new technologies. The main methods include:

1. Multimodal Fusion and Intelligent Compensation. Current ultrasonic technology achieves temperature compensation by integrating a temperature sensor, addressing the impact of thermal expansion on sound speed. The confidence system enhances the accuracy and adaptability of small bolt measurements through multimodal ultrasonic methods (especially suitable for bolts with diameters less than M12).
2. Optimization of Non-contact Installation. The use of coupling agent-free probes or embedded ultrasonic transmitters and receivers simplifies the installation process. For example, installing sensors directly on the bolt end face reduces dependence on coupling agents and is suitable for confined spaces or high dynamic load scenarios.
3. Real-time Capability and Intelligence. Combined with IoT technology, ultrasonic systems can transmit data to the cloud in real time and utilize algorithms to establish a linear relationship between dynamic time and preload force, enabling remote intelligent monitoring.

However, there are still limitations that need further research and breakthroughs for broader applications. The main limitations are as follows:

1. Monitoring of Small Bolts: There remain issues of weak echo signals and difficult calibration for measuring small bolts.
2. Adaptation to Extreme Environments: In high-temperature or strong vibration environments, ultrasonic probes are prone to failure, necessitating the development of high-temperature-resistant packaging technology.
3. Cost and Popularity: Compared to other traditional torque methods, ultrasonic equipment has a higher initial investment and requires cost reduction through large-scale mass production.

7 Design of the On-Line Monitoring System for the Preload of Fastening Bolts of Vehicle-Mounted IMUs

The structure of the vehicle-mounted IMU is shown in Fig. 9. The IMU is fastened to the base with four M12 screws [11]. A sensor with a diameter of 6 mm is pasted on the screw of the IMU, and the sensor is connected to the acquisition controller through a shielded wire. The acquisition controller can configure and store parameters such as the bolt axial force coefficient and temperature coefficient, convert the collected flight time into preload force, and issue an alarm in time when the preload force is lower than 30% of the design value.

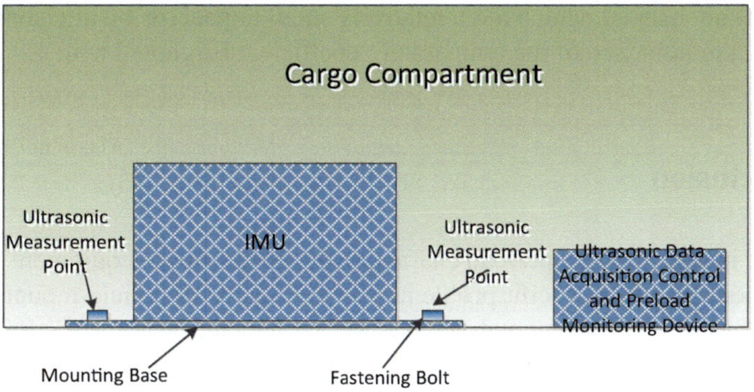

Fig. 9 Structural diagram of the on-board inertial navigation unit (IMU) bolt preload monitoring system

In the specific design, considering the influence factors of ultrasonic waves, the detection system adds corresponding designs. A temperature sensor is integrated into the probe part to compensate for the ultrasonic measurement based on the real-time collected temperature; the design of increasing electromagnetic shielding capability is added, using suitable materials for the design of the sensor, transmission lines, and the host.

The long-term monitored data is shown in Fig. 10. It can be seen that the acoustic time fluctuates around 34,605.7 ns (M10, 102 mm screw). The data variance is 0.641 ns. For an M10 × 102 bolt under 30 kN tensile force, Hooke's law predicts an elongation of ~ 0.18 mm. At a sound speed of 5900 m/s, this elongation corresponds to an acoustic time difference of ~ 60 ns. The measured variance (0.641 ns) is significantly less than 3% of 60 ns (1.8 ns), confirming that the ultrasonic system meets the 3% F.S. force measurement accuracy requirement. Due to the particularity of the applicable scenarios of this device, the environmental temperature where the device is located needs to be maintained at a constant level. The temperature can be

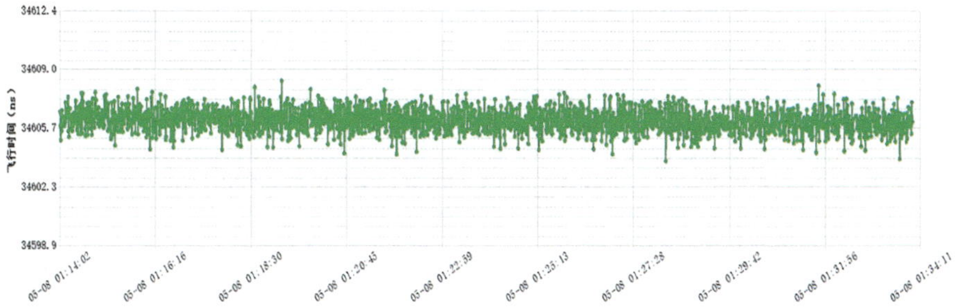

Fig. 10 Structural diagram of the on-board inertial navigation unit (IMU) bolt preload monitoring system

regarded as unchanged, which has a relatively small impact on the ultrasonic sensor. Therefore, the influence of the temperature coefficient is ignored here.

8 Conclusion

This study investigates the preload characteristics of aerospace equipment fastening devices, focusing on the specific problem of bolt loosening in vehicle-mounted IMUs. Both theoretical and experimental analyses were conducted on ultrasonic bolt preload measurement technology, leading to the development of an online monitoring system for IMU fastening bolt preload. The proposed system enables real-time measurement of IMU bolt preload, issues timely alerts in the event of loosening, and ensures the stability of the IMU base geometry. Furthermore, this system can be applied to other scenarios that demand stringent bolt preload requirements.

References

1. Wang X et al (2023) Research on the attenuation of preload of wheel bolts of engineering vehicles. Lift Transp Mach 2:74–78
2. Sun Y, Zhang B, Liu D et al (2022) Detection of preload of high-strength bolts of wind turbines based on longitudinal waves. Non-Destr Test 44(11):22–25. https://doi.org/10.11973/wsjc202211005
3. Nangong Z, Jiang J, Zhang D (1997) Research on contact stress of solid rocket engine connection structure. Propuls Technol 018(004):54–57
4. Liu Z (1996) Ultrasonic measurement of axial tension force of bolts. Solid Rocket Technol 19(1):4
5. Minakuch Y, Nawa T (2005) Ultrasonic measurement of the bolt axial force of a bolted joint after tightening. J Jpn Soc Non-Destr Insp
6. Pan Q, Pan R, Chang M, Xu X A shape factor based ultrasonic measurement method for determination of bolt preload—ScienceDirect. NDT & E International 111:102210 [2024–04–25]. https://doi.org/10.1016/j.ndteint.2019
7. Pan Q, Pan R, Shao C, et al (2020) Research review of principles and methods for ultrasonic measurement of axial stress in bolts. Chin J Mech Eng 33(1). https://doi.org/10.1186/s10033-020-0431-x
8. He C, Wu K (1994) Ultrasonic measurement of axial force of bolts during tightening of the main bridge. Exp Mech 9(1):5. CNKI:SUN:SYLX.0.1994-01-011
9. Jia X Research on the ultrasonic stress detection system of bolt axial stress based on the acoustic elastic effect. Southwest Jiaotong University [2024–04–25]. CNKI:CDMD:2.1018.977
10. Yasui H, Tanaka H, Fujii I et al (2008) Ultrasonic measurement of axial stress in short bolts with consideration of nonlinear deformation. JSME Int J 42(1):111–118. https://doi.org/10.1299/jsmea.42.111
11. Xu Y, Ma Z et al Study on ultrasonic stress measurement technology for key fasteners in cranes 202405:50–51
12. Xia K, Shi H, Li T (2024) Research on high strength bolt pre tightening force detection method based on non contact measurement technology. J Arch Civ Eng 41(5):93–101
13. Guo J, Wang Z, Li H, Yang Y, Huang CG, Yazdi M, Kang HS (2024) A hybrid prognosis scheme for rolling bearings based on a novel health indicator and nonlinear Wiener process. Reliab Eng Syst Saf 245:110014

14. Li H, Yazdi M, Huang HZ, Huang CG, Peng W, Nedjati A, Adesina KA (2023) A fuzzy rough copula Bayesian network model for solving complex hospital service quality assessment. Complex Intell Syst 9(5):5527–5553
15. Li H, Peng W, Adumene S, Yazdi M (2023) An intelligent cost-based consequence model for offshore systems in harsh environments. In: Intelligent reliability and maintainability of energy infrastructure assets, pp 107–117. Springer Nature Switzerland, Cham
16. Li H, Yazdi M (2022) What are the critical well-drilling blowouts barriers? A progressive DEMATEL-game theory. Advanced decision-making methods and applications in system safety and reliability problems: approaches, case studies, multi-criteria decision-making, multi-objective decision-making, fuzzy risk-based models. Springer International Publishing, Cham, pp 29–46
17. Li H, Peng W, Adumene S, Yazdi M (2023) Cutting edge research topics on system safety, reliability, maintainability, and resilience of energy-critical infrastructures. In: Intelligent reliability and maintainability of energy infrastructure assets, pp 25–38. Springer Nature Switzerland, Cham
18. Li H, Peng W, Adumene S, Yazdi M (2023) Advances in failure prediction of subsea components considering complex dependencies. In: Intelligent reliability and maintainability of energy infrastructure assets, pp 93–105. Springer Nature Switzerland, Cham

Probabilistic Life Estimation Framework for Mechanical Component

Xuekang Li, Mingtao Wu, Yongqiang Wang, Zuwang Gan, and Haikun Zhao

Abstract Within the field of structural integrity design, notch fatigue analysis holds critical significance. Nevertheless, the absence of efficient fatigue models capable of seamlessly incorporating both notch and size effects continues to present a considerable challenge. Addressing this gap, the present study introduces a modelling for service life evaluation. Grounded in strain energy theory, a fatigue life prediction model and a reliability evaluation equation for notched structures that incorporate size effects are proposed. These models effectively capture the combined influences of varying stress ratios and notch-size interactions under stochastic conditions, achieving high predictive accuracy and robust quantification of fatigue life dispersion. An equivalent variable energy density damage parameter is formulated by integrating total strain energy density, weight functions, and strain energy gradients. Furthermore, a correlation is established between fatigue test data obtained under symmetric cyclic loading and local strain energy density distributions under asymmetric cyclic loading, enabling life prediction for notched structural components considering size effects across different stress ratios, using only symmetric cyclic loading test data. Additionally, a new gap fatigue life prediction model is presented, explicitly accounting for both mean stress and size effects. This study delivers a comprehensive program framework mechanically structural components, offering improved representation of fatigue test data dispersion under varying stress ratios.

Keywords Notch · Probabilistic · Life estimation · Size effect

---

X. Li (✉) · M. Wu · Y. Wang · Z. Gan · H. Zhao
Kunming Shipborne Equipment Research and Test Center, Kunming, China
e-mail: lixuekang_uestc@163.com

H. Li et al. (eds.), *Data-Driven Methods for Reliability and Safety Engineering: Applications in Industrial Systems*, Springer Series in Reliability Engineering,
https://doi.org/10.1007/978-3-032-22873-4_31

1 Introduction

Fatigue failure represents the most critical failure mode for engineering components and structures, occurring even when the applied stress is low or remains below the yield limit. The mean stress effect exerts a substantial influence on crack initiation, crack propagation, and the overall fatigue failure process under varying stress ratios [1]. Furthermore, engineering components and structures inherently exhibit local stress concentrations resulting from geometric discontinuities, a phenomenon commonly referred to as the notch effect. Recent developments in engineering have trended toward integrated designs and larger structural dimensions. However, fatigue strength generally declines with increasing structural size, a phenomenon known as the size effect. Direct fatigue testing of large-scale components is often impractical due to limitations in experimental equipment, excessive costs, and extended testing durations. Therefore, to ensure structural integrity and enable reliable fatigue life assessment, it is imperative to develop effective predictive methodologies for the fatigue life of notched components under the combined influences of mean stress and size effects. Mean stress, defined as the average of the peak tensile stress and compressive stress, has a pronounced impact on fatigue deformation and failure behavior, refer to Fig. 1.

(a) Smith diagram ($\sigma_{\max}/\sigma_{\min} - \sigma_m$ diagram), line ADC is maximum stress line, line BEC is the minimum stress line. The area outside the ADC and BEC lines indicate the stress level that produces fatigue failure. (b) Haigh diagram ($\sigma_a - \sigma_m$ diagram), σ_a is the stress amplitude, the absolute value of the slope of the AC line is the mean stress influence coefficient at $\sigma_a \leq \sigma_m$. (c) Isometric diagram, which is the diagram of showing the relationship between fatigue limits under the equal life and different stress ratios. The diagram has been widely developed because it can directly show the relationship among various parameters [2].

The notch effect caused by geometrical discontinuities plays an imperative role for structural fatigue failure analysis and has become research hotspot in recent years. There is usually an uneven stress field around the notch. Thereby delaying the crack initiation, propagation and whole fatigue failure process. The local stress/strain state analysis is the basis for predicting fatigue life of notched components [3, 4]. Currently, three methods are applied to obtain the local stress–strain response under external loading: experimental, theoretical, and finite element methods. The experimental method is the most intuitive and accurate. However, it is restricted due to the devise notch forms and complex working conditions in engineering [5–7].

Experiments prove that the fatigue strength of components own similar geometry decreases with the increase of specimen size under the equal loadings, this phenomenon is called as 'size effect'. The main reasons are as follows: (a) The randomly distributed defects in the material is the potential sources of crack initiation, more potential crack sources are existed in large-size components, which cause higher failure probability. (b) The material fatigue performance is unevenly distributed in the critical damage zone near the dangerous point. The fatigue strength of component is determined by the stress level and material conditions in critical damage region.

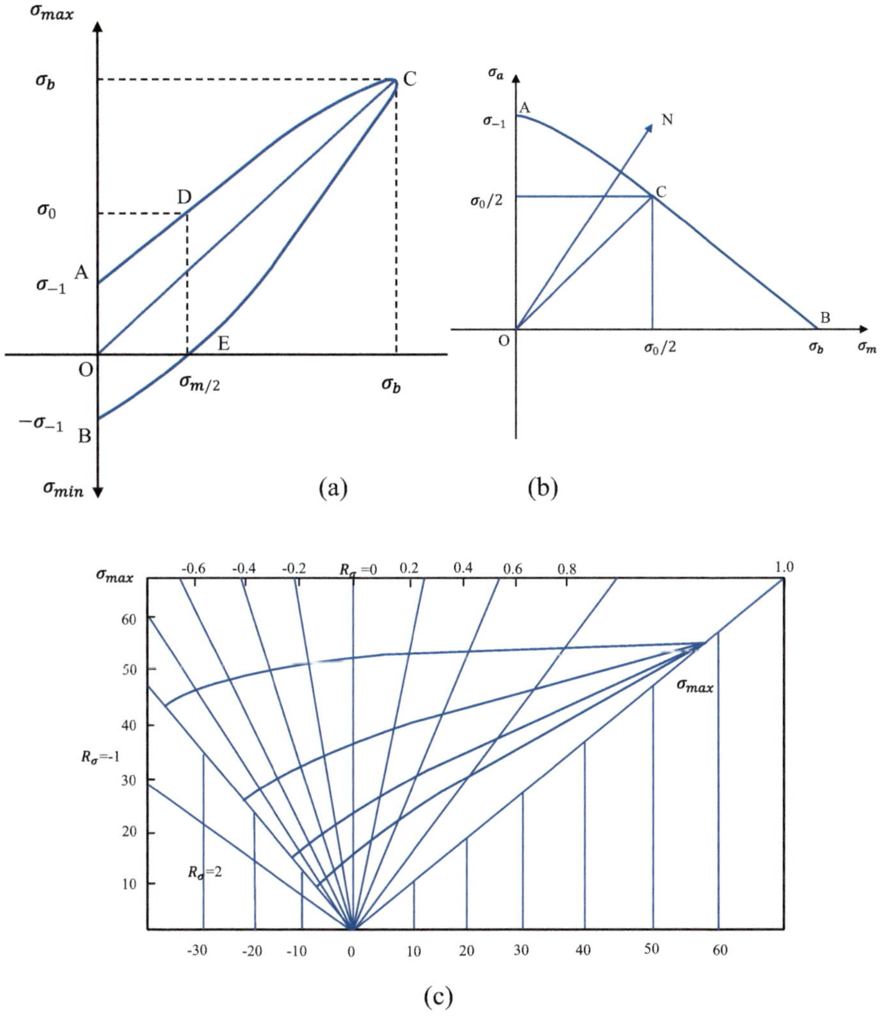

Fig. 1 Three commonly used fatigue limit diagrams **a** Fatigue limit diagram 1, **b** Fatigue limit diagram 2, **c** Fatigue limit diagram 3

The critical damage region around the dangerous point is larger for the larger-size components, and the fatigue strength is lower. Researches have shown that the shape and size of critical damage region are related to the stress/strain distribution. (c) The stress gradient is different for the components of similar geometry with same stress concentration factor K_t under equal loadings, and the component with stress gradients is sensitive to geometrical size. Therefore, the highly stressed volume (HSV) approach, the weakest link theory and critical defect method et al. are developed for describing the fatigue failure of notched component under size effect [8, 9].

2 Strain Energy Modelling

2.1 *Strain Energy Concept*

Researchers presented plenty of mean stress correction methods for considering the mean stress effect of fatigue life prediction. And the mean strain correction methods are shown gradually for incorporating the strain-life models under asymmetric strain-control cycle loadings. In addition, some correction methods based on the strain energy damage which combines the strain and stress are presented for analyzing comprehensively fatigue failure of various metallic materials. However, accurate and efficient fatigue analysis methods for investigating mean stress effect under different stress ratios are still lack [10].

As a consequence, a strain energy-based approach under mean stress effect is introduced in this study, which not only avoids stress/strain correction calculations but also incorporates the integrated influences of stress and strain. The strain energy consists of plastic and elastic terms. Especially, the elastic strain energy is essential for determining the method considers the mean stress effect. And the total elastic strain energy can be applied for analyzing mean stress/strain effect [11].

The total strain energy explores the whole change process of elasto-plastic stress and strain, which considers the mean stress effect under different stress ratios R_{σ}.

2.2 *Strain Energy Gradient-Based Critical Damage Region*

According to Sect. 2.1, this study defines the calculated approach of TSED that can be used to consider mean stress/strain effect. In this section, a method for defining the critical damage region of notched components under size effect is proposed:

The boundary length and volume of critical damage region are defined through the strain energy gradient.

Take the point of $W_{t,\max}$ as the center, determine the point of max strain energy gradient in steepest descent direction of total strain energy density. The boundary length D_{eff} of critical damage region is the distance between the two points of max strain energy gradient and $W_{t,\max}$. With the point of $W_{t,\max}$ as the center, the zone encompassed by the boundary length D_{eff} as the radius is the volume V_{eff} of the critical damage region. As shown in Fig. 2, the strain energy gradient can reflect the trend of total strain energy density with the distance change, Region 1 is the critical damage region, which is changing with the geometries and dimensions of notch from different metallic materials.

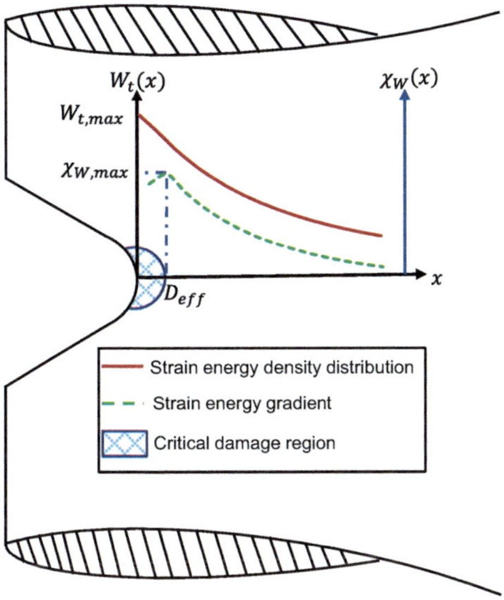

Fig. 2 The critical damage region

2.3 *Modelling*

After determining the critical damage region, $W_{t,eff}$ is proposed based on the boundary length D_{eff} of critical damage region.

The weight function $\varphi(x)$ quantifies the contribution of strain energy density at different positions to structural fatigue damage under notch and size effects.

In addition, the relationship between W_t and N_f.

Under different stress ratios is proposed.

3 Probabilistic Fatigue Framework for Notch Analysis of Mechanical Component

The strain energy-based modelling relies on key assumptions: the linearity of the total strain energy density (W_t) and fatigue life (N_f) relationship holds within the elastic–plastic fatigue regime, justified by consistent correlation with experimental data for metallic materials under typical service loads; the critical damage region boundary (D_{eff}) is selected based on the steepest descent of strain energy gradient, ensuring focus on the most damage-prone volume while simplifying computational complexity; sub-elements within the critical damage region are assumed statistically independent, as their spatial separation and stress distribution minimize mutual influence, though this assumption may introduce minor deviations for components with highly concentrated stress gradients. Limitations include reduced validity of linearity for materials with significant cyclic softening/hardening and potential underestimation of damage interaction in ultra-small notches where sub-element independence is compromised.

The transition from deterministic strain-energy modelling to the stochastic failure framework is enabled by propagating uncertainty sources through the strain-energy model. Material variability (e.g., inelastic properties, defect distribution) affects the total strain energy density (W_t) calculation, load randomness introduces fluctuations in stress/strain inputs to the strain energy gradient, and manufacturing tolerances alter notch geometry thus modifying the critical damage region (V_{eff}) and weight function ($\varphi(x)$). These uncertainties are quantified as statistical variations in the key parameters of the deterministic model ($W_{t,eff}$, D_{eff}, V_{eff}) and then fed into the probabilistic formulation, where the Poisson distribution of sub-element failures and Weibull-distributed structural life integrate the propagated variability to reflect real-world fatigue data dispersion. This linkage ensures the probabilistic framework is grounded in mechanical insights from the strain-energy model, enhancing its practical relevance for engineering applications.

As mentioned above, fatigue data generally exhibit dispersion due to stochastic factors such as loading, material properties, processing, manufacturing errors, and working environment. Uncertainty and dispersion are the inherent characteristics of fatigue failure analysis, wherefore, exploring the reasonable quantitative method of uncertainty is always a research hotspot for fatigue life prediction and structural integrity assessment. In this context, probabilistic fatigue analysis methods have gradually gained popularity due to their strong capability to describe fatigue life distributions. In this section, a probabilistic fatigue analysis framework for notch components under size effect is proposed to accurately reflect the scatter of fatigue data.

According to Sect. 2, this study derives the fatigue framework based on probabilistic analysis method and life prediction model. The failure number of sub-elements with the period of time obeys the Poisson distribution [12–14].

where, n is the failure number of sub-elements, t is the failure time, λ is the constant of the Poisson distribution.

Given the decomposition of failure time t into t_1, …, t_s, λt can be expressed.

In the probabilistic framework, the values of m, V_{eff} and T' should be determined. The value of V_{eff} is obtained through the defined critical damage region in Sect. 2.2. How to determine the value of T' is the core of this framework, it establishes the relationship between fatigue life at every sub-element. According to fatigue life prediction model of the Sect. 2.3, indicates the relation between fatigue life and strain energy density in a line, the line elements can also be viewed as a sub-element of volume in the critical damage region.

Based on the above analysis, a probabilistic framework for fatigue life assessment of notched components under size effect and different stress ratios is proposed. First, calculate the sub-element fatigue life N_f of lines based then obtain the η, lastly, obtain the probabilistic formulation. A Flow chart of the framework is shown Fig. 3.

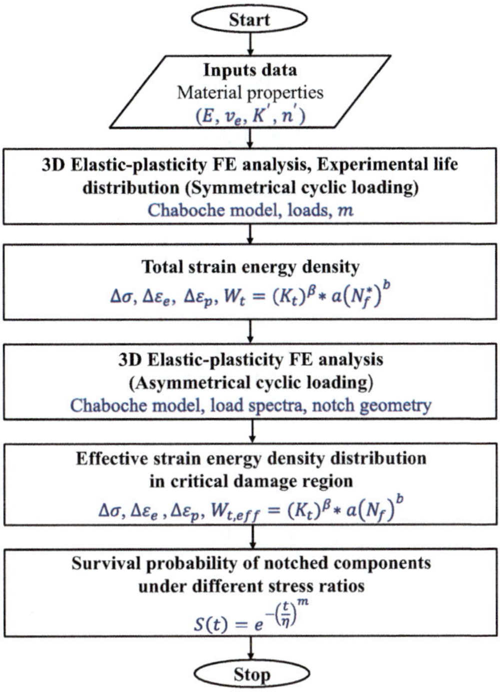

Fig. 3 The probabilistic fatigue framework

4 Conclusion

Based on strain energy theory, a fatigue life prediction model and a reliability evaluation equation for notched structures considering size effects are proposed. These methods effectively characterize the combined influence of varying stress ratios and notch-size effects under random factors, achieving high accuracy in life prediction and reliable quantification of fatigue life dispersion. The principal conclusions are as follows:

(1) A equivalent variable damage parameter is derived, a weight function, and the strain energy gradient. A relationship is established between fatigue test data obtained under symmetric cyclic loading and the local strain energy density distribution.
(2) A estimation framework model is introduced, which simultaneously accounts for both average stress effects and size effects.
(3) A general program framework is provided for probabilistic life estimation, considering size effects under varying stress ratios, offering improved representation of fatigue test data dispersion.

References

1. Li XK et al (2021) Probabilistic fatigue life prediction of notched components using strain energy density approach. EFA 124:105375
2. Li XK et al (2022) Probabilistic fatigue modelling of metallic materials under notch and size effect using the weakest link theory. IJF 159:106788
3. Li XK et al (2021) Strain energy-based approach for probabilistic fatigue life prediction of notched components. In: Proceedings of the international congress on structural integrity and maintenance (SIM 2021), Belo Horizonte, Brazil
4. Li XK et al (2021) Probabilistic fatigue life prediction under notch and size effects using strain energy concept. In: Proceedings of virtual 20th international colloquium on mechanical fatigue of metals (ICMFM 2021), Wroclaw, Poland, pp 15–17
5. Walker K (1970) The effect of stress ratio during crack propagation and fatigue for 2024-T3 and 7075-T6 Aluminum, Effects of Environment and Complex Load History on Fatigue Life. ASTM STP 462, Am. Soc. for Testing and Materials 1–4
6. Zhu SP, Huang HZ, Li Y et al (2015) Probabilistic modeling of damage accumulation for time-dependent fatigue reliability analysis of railway axle steels. Proc Inst Mech Eng, Part F: J Rail Rapid Transit 80:266–277
7. Luo P, Yao WX, Wang Y et al (2018) A survey on fatigue life analysis approaches for metallic notched components under multi-axial loading. Proc Inst Mech Eng, Part G: J Aerosp Eng 3870–3890
8. Berto F, Campagnolo A, Meneghetti G et al (2016) Averaged strain energy density-based synthesis of crack initiation life in notched steel bars under torsional fatigue. Frattura ed Integrita Strutturale 10(38):215–223
9. Bandyopadhyay R, Prithivirajan V, Peralta AD, Sangid MD (2020) Microstructure-sensitive critical plastic strain energy density criterion for fatigue life prediction across various loading regimes. Proc R Soc A: Math, Phys Eng Sci 476:2236
10. Zhu SP, Huang HZ, He LP et al (2012) A generalized energy-based fatigue-creep damage parameter for life prediction of turbine disk alloys. Eng Fract Mech 90:89–100
11. Branco R, Costa JD, Berto F et al (2018) Fatigue life assessment of notched round bars under multiaxial loading based on the total strain energy density approach. Theoret Appl Fract Mech 97:340–348
12. Li H, Yazdi M (2022) Reliability analysis of correlated failure modes by transforming fault tree model to Bayesian network: a case study of the MDS of a CNC machine tool. Advanced decision-making methods and applications in system safety and reliability problems: approaches, case studies, multi-criteria decision-making, multi-objective decision-making, fuzzy risk-based models. Springer International Publishing, Cham, pp 15–28
13. Li H, Yazdi M (2022) A holistic question: is it correct that decision-makers neglect the probability in the risk assessment method? Advanced decision-making methods and applications in system safety and reliability problems: approaches, case studies, multi-criteria decision-making, multi-objective decision-making, fuzzy risk-based models. Springer International Publishing, Cham, pp 185–189
14. Li H, Yazdi M (2022) Stochastic game theory approach to solve system safety and reliability decision-making problem under uncertainty. Advanced decision-making methods and applications in system safety and reliability problems: approaches, case studies, multi-criteria decision-making, multi-objective decision-making, fuzzy risk-based models. Springer International Publishing, Cham, pp 127–151

Uncertainty Propagation in Bridge Construction Carbon Emission Quantification

Hafiz Muhammad Hazib, Imran Shabbir, Weizong Lai, Yue Pan, and Jianjun Qin

Abstract Given the rapid expansion of infrastructure such as large-scale bridges in developing countries, their environmental impacts, particularly during the construction phase from the perspective of carbon emissions, have attracted have become a subject of increasing scholarly and policy interest. This study presents a probabilistic life cycle assessment for a large-scale bridge in Pakistan to illustrate uncertainty propagation in carbon emission quantification during construction, beginning with the explicit identification of uncertainties in emission factors. Monte Carlo simulation is employed to estimate the embodied carbon footprint associated with material production, transportation, and on-site construction activities, resulting in an average of 72.5×10^6 kg $CO_2 - eq$ $\left(sigma = 11.6 \times 10^6 \text{ kg } CO_2 - eq\right)$. The uncertainty propagation indicates substantial variability (CV $= 16\%$) with $\pm 20-30\%$ confidence intervals, driven by probabilistic characteristics of material, transportation, and energy emission factors. From the expected value perspective, material extraction accounts for 55% of total emissions, transportation for 43%, and construction activities for 3%. At the component-level, the superstructure (23.38×10^6 kg $CO_2 - eq$) and deck (23.27×10^6 kg $CO_2 - eq$) represent the primary sources. To reduce life-cycle emissions and associated uncertainties, technical measures such as low-clinker binders, optimized logistics, and BIM-integrated stochastic carbon modules are

H. M. Hazib · W. Lai · Y. Pan · J. Qin (✉)
State Key Laboratory of Ocean Engineering, Shanghai Key Laboratory for Digital Maintenance of Buildings and Infrastructure, School of Ocean and Civil Engineering, Shanghai Jiao Tong University, Shanghai, China
e-mail: jianjunqin@gmail.com

H. M. Hazib
e-mail: HafizMuhammadHazib@sjtu.edu.cn

W. Lai
e-mail: Laiweizong@sjtu.edu.cn

Y. Pan
e-mail: panyue001@sjtu.edu.cn

I. Shabbir
Prime Engineering & Testing Consultants, Islamabad, Pakistan
e-mail: ibshabbir@outlook.com

H. Li et al. (eds.), *Data-Driven Methods for Reliability and Safety Engineering: Applications in Industrial Systems*, Springer Series in Reliability Engineering,
https://doi.org/10.1007/978-3-032-22873-4_32

recommended, complemented by policy interventions including performance-based carbon caps, geospatial emission factors, and blockchain-based supplier disclosures. This framework advances traditional LCA methodologies, enabling more informed low-carbon decision-making in resource-constrained contexts.

Keywords Probabilistic LCA · Monte Carlo simulation · Embodied carbon · Bridge construction

1 Introduction

Infrastructure development is essential for societal progress; however, it also presents significant environmental challenges [1]. The production of substantial quantities of concrete and steel required for structural projects, particularly large-scale bridges, generates considerable greenhouse gas emissions. Emissions occurring before and during the construction phase often constitute the dominant share of the total life-cycle emissions. Material extraction and manufacturing alone contribute to more than 60% of a project's carbon footprint [2, 3]. The environmental consequences of rapid infrastructure expansion, driven by urbanization, can be severe in countries such as Pakistan, primarily due to reliance on carbon-intensive processes and fragmented data systems [4, 5]. The absence of locally calibrated assessment tools and region-specific emission standards hinders the adoption of sustainable alternatives. As a result, outdated and inefficient practices are likely to persist in addressing the ongoing climate crisis [6, 7].

The environmental impacts of infrastructure are increasingly recognized; however, life-cycle assessment (LCA) and other conventional methods for estimating carbon emissions exhibit substantial limitations [8, 9]. LCA is widely applied to evaluate the ecological impacts of building materials and construction processes, yet it depends on carbon emission factors derived from measurement or estimation based on raw materials and extraction methods, which may not represent local conditions [10]. This challenge is particularly pronounced in developing countries such as Pakistan, where variations in material quality, energy sources, and construction techniques are significant, and where resources may be procured either domestically or from neighbouring regions [11, 12]. The diversity of materials, complexity of designs, and heterogeneity of construction techniques create unique challenges in applying LCA to bridges and other complex structures [13]. These factors exacerbate the difficulty of producing accurate environmental impact estimates, often resulting in substantial errors in carbon emission predictions. The high variability of carbon emission parameters, influenced by local practices, transportation distances, and manufacturing processes, further increases uncertainty in emissions assessments [14, 15]. Neglecting these uncertainties may lead to inaccurate conclusions regarding the environmental benefits of specific materials or construction methods, thereby obstructing efforts to reduce the carbon footprint of infrastructure projects.

This study seeks to improve the reliability of carbon emission assessments for large-scale bridge construction by integrating global and regional emission factors with probabilistic modelling techniques. The distinctive attributes of various infrastructure types and regional contexts are insufficiently captured by most existing LCA frameworks, which typically depend on generalized datasets [16]. This research advances understanding of carbon emissions in Pakistan's construction sector by incorporating variable datasets and emission parameters, emphasizing the substantial variability in energy use, construction processes, and material sourcing [17]. The findings also establish a foundation for further research into the application of uncertainty analysis methods in infrastructure projects, particularly bridge construction in developing nations. This work provides a more accurate and dependable approach for estimating carbon emissions, thereby offering valuable insights for policymakers, engineers, and construction professionals in Pakistan and comparable contexts.

2 Methodology

The cumulative life-cycle carbon emissions of bridge infrastructure are influenced by the five sequential processes of extraction and manufacturing of raw materials, transportation, building, operational service, and end-of-life management. Total life-cycle emissions (CE_{LC}) are expressed as:

$$CE_{LC} = CE_M + CE_T + CE_C + CE_S + CE_E \quad (1)$$

The stages of material production, transportation, construction, service, and end-of-life are denoted by CE_M, CE_T, CE_C, CE_S and CE_E respectively. This study proposes a probabilistic LCA framework to quantify carbon emissions of bridge construction. The analysis systematically covered three key phases: raw material extraction and production, transportation, and on-site construction. Structural design specifications and comprehensive inventories provided precise material inputs. Due to the absence of a Pakistan-specific emission factor database, factors were sourced from international guidelines (IPCC), industry publications (World Steel Association), regional LCA studies, and established databases (Ecoinvent, IPCC EFDB). This ensures capture of key variations in manufacturing, energy use, and logistics. Transportation emissions were calculated using source-to-site distances for each material combined with standardized per-kilometre emission factors. Construction phase emissions were derived by integrating data on equipment types, their operational durations, and fuel or energy consumption rates with corresponding equipment-specific emission factors. Authoritative, regionally aligned emission factors were consistently applied to material quantities, transport distances, and energy consumption metrics. This methodology prioritized significant emission sources, ensuring comprehensive accounting of both direct emissions (e.g., fuel combustion) and indirect emissions across the project life-cycle for a robust, regionally contextualized assessment. Figure 1 illustrates the framework and system boundaries defined

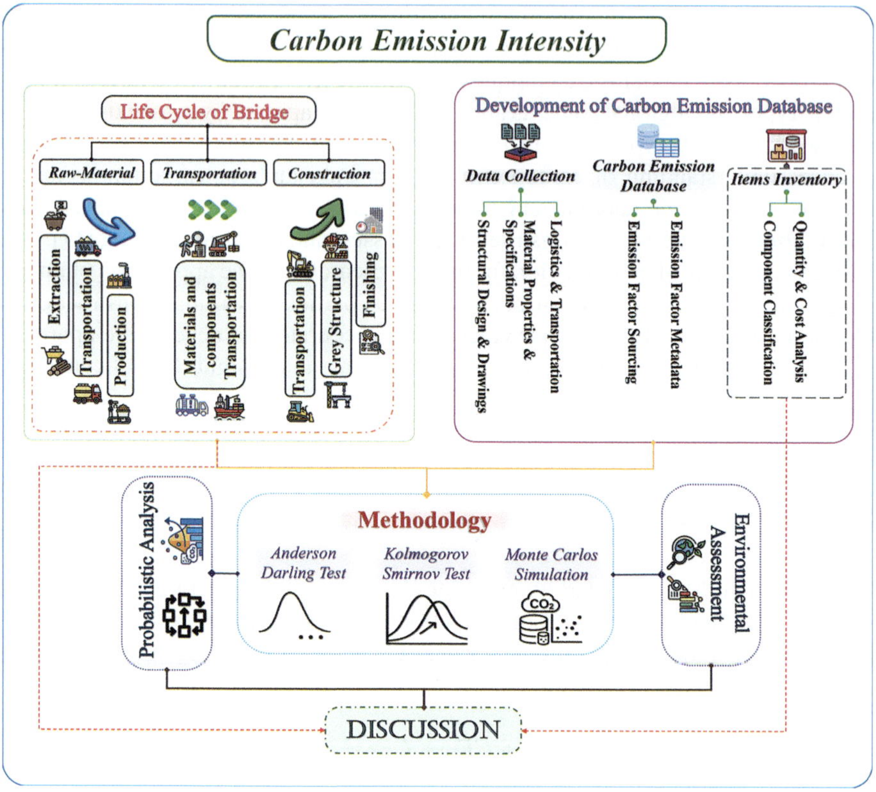

Fig. 1 Conceptual framework for probabilistic carbon emission intensity assessment of bridge infrastructure

for the research, which focus primarily on raw material extraction and production, transportation, and construction activities.

A probabilistic approach is used to evaluate uncertainty in bridge construction carbon emissions, acknowledging inherent variability in emission factor estimates. Sensitivity analysis identified parameters exerting the most significant influence on total emissions, specifically examining variations in material, transportation distance, and construction process emission factors to understand uncertainty contributors. Distribution selection for emission factors followed a unified engineering–statistical procedure based on the empirical characteristics of the compiled EF samples. Materials showing centrally clustered, approximately symmetric variability were modelled using normal distributions, whereas strictly positive parameters exhibiting right-skewed variability (e.g., transport- and fuel-related terms influenced by variable distances and multiplicative supply-chain effects) were modelled using log-normal distributions; gamma and Weibull forms were also evaluated as plausible alternatives for positive-valued data. Candidate distributions were fitted by maximum-likelihood estimation and then screened using Anderson–Darling and Kolmogorov–Smirnov goodness-of-fit tests, retaining only those providing an adequate fit ($p > 0.05$). Heavy-tailed or multimodal forms were considered for sparse, low-volume materials, but were not adopted because the GOF results supported single-family representations and did not indicate distinct subpopulations driving EF variability. Uncertainty was formally quantified using Monte Carlo simulations (MCS), generating 10,000 iterations based on established probability distributions for the emission variables, material quantities, and transportation distances. Material-quantity uncertainty was represented by applying a uniform perturbation of ± 5–10% to each nominal inventory value. The variability bounds for each material category were derived from documented on-site wastage rates and procurement tolerances (e.g., ± 5% for reinforcing steel; ± 10% for cast-in-place concrete). Incorporation of these perturbations into subsequent Monte Carlo sampling ensures that both manufacturing-stage and construction-stage uncertainties are fully propagated into the embodied-carbon results. The total carbon emissions ($\text{Emission}_{\text{Total}}$) were calculated for each iteration using the fundamental equation:

$$\text{Emission}_{\text{Total}} = \sum_{i=1}^{n} Q_i \times EF_i \tag{2}$$

where Q_i represents the quantity of material i and EF_i represents the emission factor for material i, which varies probabilistically.

The Anderson–Darling (AD) test [18] and the Kolmogorov–Smirnov (KS) test [19] were utilized to evaluate the adherence of emission factors to a particular distribution, essential for establishing the reliability of the probabilistic model. The formula for the AD test statistic (A^2) is:

$$A^2 = -n \sum_{i=1}^{n} \frac{(2i-1)}{n}\left[In(F(x_i)) + In(1 - F(x_{n+1-i}))\right] \tag{3}$$

where $F(x_i)$ represents the cumulative distribution function for the hypothesized distribution, and x_i is the sorted sample data. Similarly, the KS test statistic D is computed as:

$$D = \max_{i} |F_n(x_i) - F(x_i)| \tag{4}$$

where $F_n(x_i)$ is the empirical CDF of the sample and $F(x_i)$ is the CDF of the reference distribution.

3 Results

The subject bridge constitutes a segment of the M-13 Motorway, a six-lane controlled-access highway presently under development via a Public–Private Partnership (PPP) employing the Build-Operate-Transfer (BOT) model. This prestressed concrete I-girder structure spans the Jhelum River in Punjab, Pakistan. Its superstructure integrates precast high-strength concrete I-girders with a cast-in-place reinforced concrete deck. The substructure comprises reinforced concrete piers, founded upon shallow footings or deep piles, engineered to accommodate the prevailing riverine and alluvial geotechnical conditions (Fig. 2). This bridge was selected as a representative case study due to its status as one of the most recent crossings constructed along the M-13 corridor. It exemplifies the motorway's advanced engineering design and operational objectives. Notably, the M-13 project incorporates interchanges, service areas, advanced safety systems, and integrated Intelligent Transport Systems (ITS), reflecting adherence to international standards and a commitment to enhanced accessibility, safety, and operational efficiency. Consequently, this structure provides a pertinent exemplar for assessing contemporary design methodologies for prestressed concrete I-girders and evaluating associated carbon emissions within the context of a high-performance motorway infrastructure project.

3.1 Material Distribution and Uncertainty

The probabilistic analysis reveals substantial heterogeneity in carbon emissions across the 28 materials assessed, as summarized in Fig. 3. High-emission materials dominate the footprint: Reinforcement G60 averages 19.2×10^6 $kgCO_2 - eq$ ($sigma = 45.9\%$) and Concrete A3 averages 10.1×10^6 $kgCO_2 - eq$ ($sigma = 15.9\%$). In contrast, non-structural components like GI Pipes (70.4 $kgCO_2 - eq$, $sigma = 101.1\%$) and Filler Board (128 $kgCO_2 - eq$, $sigma = 62.0\%$) contribute minimally but exhibit high relative uncertainty due to low absolute values and supplier data variability. Mid-range materials show divergent variability: Formwork (0.833×10^6 $kgCO_2 - eq$, $sigma = 91.4\%$) and Piles (1.95×10^6 $kgCO_2 - eq$, $sigma = 34.4\%$) reflect construction and sourcing differences, while Concrete D2 $\left(6.76 \times 10^6 = kgCO_2 - eq,\ sigma = 20.9\%\right)$ and Lean Concrete $\left(6.93 \times 10^3\ kgCO_2 - eq,\ sigma = 34.3\%\right)$ demonstrate tighter distributions. Diesel emissions, both transportation $\left(30.8 \times 10^6\ kgCO_2 - eq,\ sigma = 23.1\%\right)$ and on-site $\left(0.127 \times 10^6\ kgCO_2 - eq,\ sigma = 23.6\%\right)$ show significant sensitivity to fuel consumption and logistics. Extreme values (e.g., transport diesel peaking at $\left(38.4 \times 10^6\ kgCO_2 - eq\right)$ underscore probabilistic modelling importance for capturing tail risks. Crucially, while high-intensity materials (steel, high-grade concrete) dominate embodied carbon, numerous low-emission materials introduce disproportionate uncertainty due to data limitations.

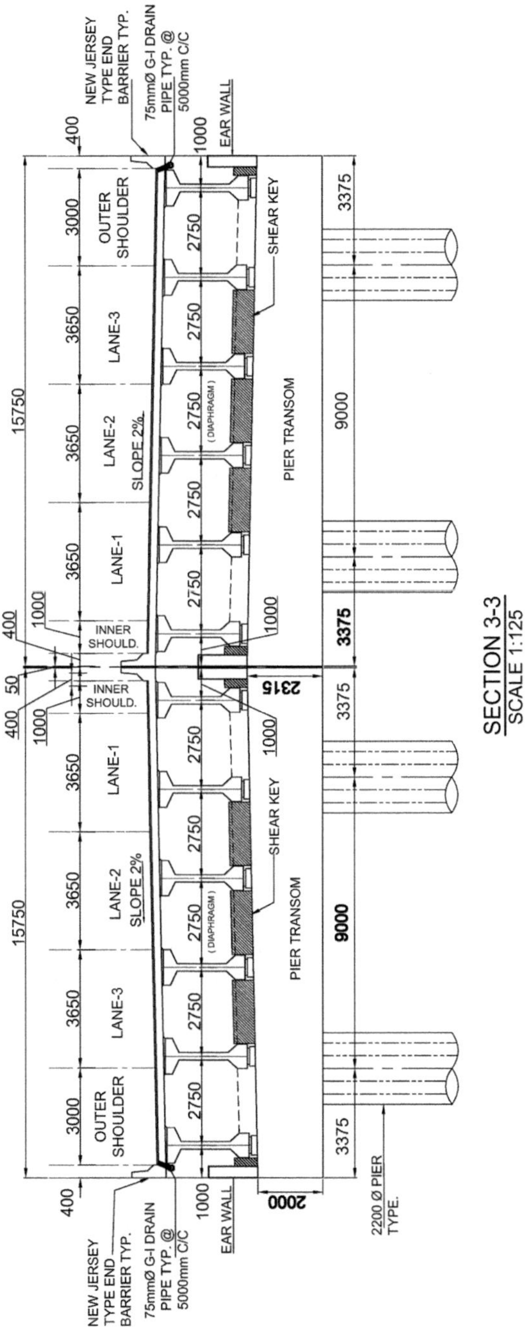

Fig. 2 Elevation of the prestressed concrete I-girder bridge spanning the Jhelum River

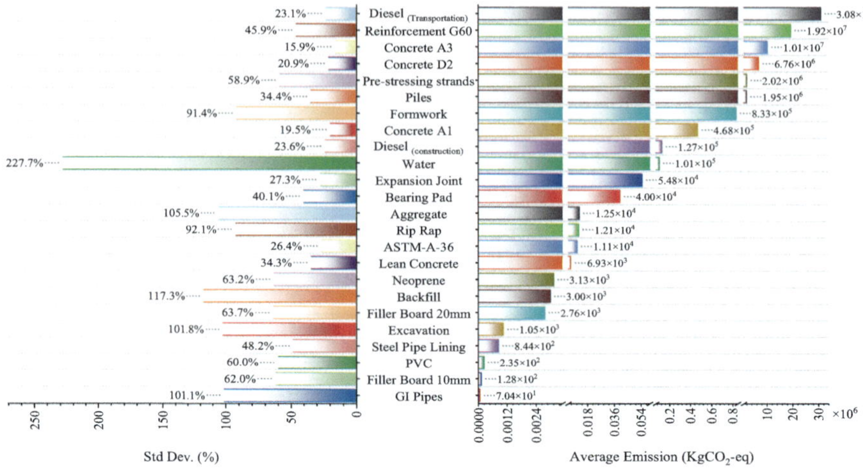

Fig. 3 Material-level uncertainty analysis showing (left) relative standard deviation (%) of each material's emission and (right) corresponding average carbon emissions ($KgCO_2$-eq) per material

3.2 *Total Carbon Emissions and Variability*

A comprehensive Monte Carlo analysis of 10,000 iterations is presented in Fig. 4a, and b which illustrate how variability in emission factors amplifies uncertainty in the bridge's embodied carbon. Figure 4a overlays probability density functions for four noise levels baseline, 10%, 20%, and 25% showing that as factor variability increases, the distribution widens and its peak diminishes. For the baseline case, the mean total embodied carbon is 72.5×10^6 kg CO_2 − eq with a standard deviation of 11.6×10^6 kg CO_2 − eq. The distribution is right-skewed, with a median near 70×10^6 kg CO_2 − eq and an interquartile range spanning 55×10^6 kg CO_2 − eq to 83×10^6 kg CO_2 − eq; roughly 80% of outcomes lie below 90×10^6 kg CO_2 − eq, yielding a coefficient of variation of 16%. Figure 4b presents each of the 10,000 sampled outcomes as individual points, bounded by red dashed lines at the 5th and 95th percentiles (approximately 45×10^6 kg CO_2 − eq and 95×10^6 kg CO_2 − eq,, respectively). The dense central band and long right-hand tail in the scatter plot confirm that occasional high-emission scenarios (driven by elevated material factors or increased haul distances) exert a pronounced influence on the upper percentile outcomes. Together, these visualizations underscore the critical need to account for emission-factor variability when setting baseline targets or comparing low-carbon design alternatives.

The life-cycle assessment reveals significant disparities in embodied carbon contributions across phases and structural components. As illustrated in Fig. 5a, raw material production dominates emissions (39.5×10^6 kg CO_2 − eq $\pm$ 9.15×10^6, 55% of total, CV = 23%), driven by cement, steel rebar, and aggregate production. Transportation follows at 30.8×10^6 kg CO_2 −

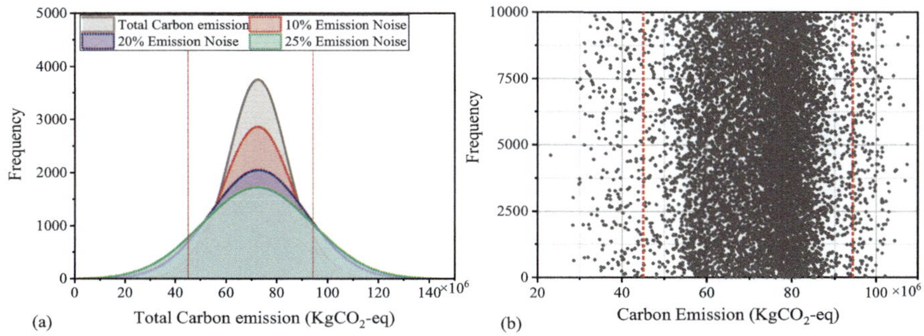

Fig. 4 Effects of emission-factor variability on total embodied carbon from 10,000 Monte Carlo runs. **a** Overlaid probability density functions under four noise levels (baseline; 10%; 20%; 25%. **b** Scatter plot of all 10 000 sampled carbon outcomes. The red dashed lines represent the 5th and 95th percentile thresholds

eq $\pm$ 7.12 $\times$ 10^6(43% , CV = 23%), with uncertainty stemming from haul distance variability and vehicle efficiencies. Construction emissions are comparatively minor $\left(2.11 \times 10^6 \text{ kg } CO_2 - \text{eq} \pm 0.71 \times 10^6,\ 3\%,\ \text{CV} = 34\%\right)$, reflecting more consistent equipment usage. These distinct uncertainty profiles indicate that mitigation should prioritize low-carbon sourcing and logistics over on-site fuel savings. Component-level analysis (Fig. 5b) shows super-structure $\left(23.38 \times 10^6 \text{ kg } CO_2 - \text{eq},\ \text{CV } 20\text{–}25\%\right)$ and deck $\left(23.27 \times 10^6 \text{ kg } CO_2 - \text{eq},\ \text{CV } 20\text{–}25\%\right)$ as primary emission sources due to prestressed concrete and high-grade reinforcement. Substructure elements contribute 16.86 $\times$ 10^6 kg CO_2 $-$ eq (CV = 21%), while piles account for 14.28 $\times$ 10^6 kg CO_2 $-$ eq (CV = 34%) due to dimensional variations. Secondary components exhibit lower emissions but notable variability: piers $\left(2.12 \times 10^6 \text{ kg } CO_2 - \text{eq},\ \text{CV} = 16\%\right)$ and abutments $\left(1.30 \times 10^6 \text{ kg } CO_2 - \text{eq},\ \text{CV} = 28\%\right)$ reflect formwork and concrete strength inconsistencies. These patterns demonstrate that carbon reduction efforts should target section optimization, alternative reinforcement, and low-carbon binders in super-structure and deck elements for maximum impact.

4 Discussion

The probabilistic assessment quantified a mean total embodied carbon of 72.5×10^6 kg $CO_2$$-eq\times 10^6$ kg $CO_2$$-$eq, exhibiting a positively skewed distribution $\left(\text{median} = 70 \times 10^6 \text{ kg } CO_2 - \text{eq};\ \text{IQR} = 55 - 83 \times 10^6 \text{ kg } CO_2 - \text{eq};\ \text{CV} = 16\%\right)$ for Jhelum bridge in Pakistan. Component-level disaggregation identified the super-structure $\left(23.38 \times 10^6 \text{ kg } CO_2 - \text{eq}\right)$ and deck $\left(23.27 \times 10^6 \text{ kg } CO_2 - \text{eq}\right)$ as dominant contributors, with CVs of 20$-$25% attributable to variability in prestressed concrete volumes and high-yield reinforcement steel specifications. Phase-specific

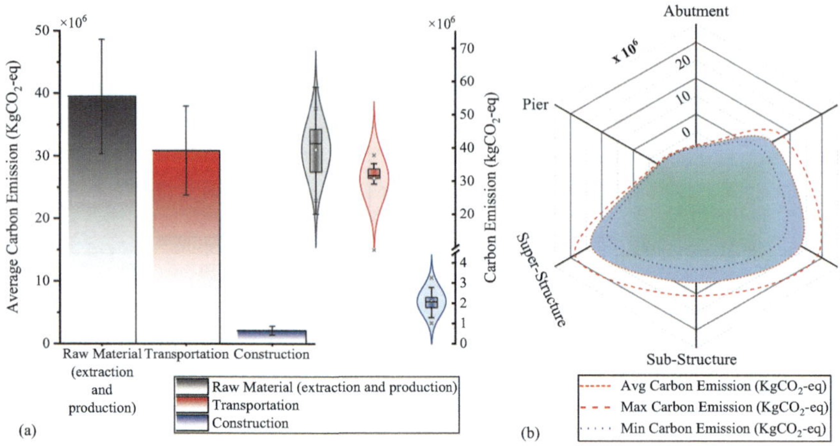

Fig. 5 Phase-wise and structural-element contributions to bridge carbon emissions under Monte Carlo variability. **a** Average emissions and distribution-wide violin plots for raw material extraction and production, transportation, and construction, including means and standard deviations. **b** Radar plot of average, maximum, and minimum emissions by structural element (abutment, pile, deck, sub-structure, super-structure, pier)

contributions revealed raw material production as the principal emission source $\left(55\%\ ,\ 39.5 \times 10^6\ \text{kg}\ CO_2 - \text{eq};\ \ CV\ = 23\%\ \right)$, followed by transportation $\left(43\%\ ,\ 30.8 \times 10^6\ \text{kg}\ CO_2 - \text{eq};\ \ CV\ = 23\%\ \right)$, while on-site construction activities constituted a marginal share $\left(3\%\ ,\ 2.11 \times 10^6\ \text{kg}\ CO_2 - \text{eq};\ \ CV\ = 34\%\ \right)$.

The non-normal distribution of total emissions, marked by leptokurtic characteristics and significant positive skewness, underscores the outsized influence of tail-risk events such as extreme material emission factors or unplanned transport rerouting. Elevated coefficients of variation (CV) in low-volume ancillary materials (e.g., galvanized iron pipes, filler boards) indicate systemic data vulnerabilities, whereas bulk materials like Concrete D2 demonstrate uniform distributions due to standardized manufacturing. Phase-specific analysis confirms that upstream activities (material extraction, processing, and transport) account. The probabilistic framework effectively quantifies uncertainty across key life-cycle phases, though its scope requires contextualization. Omission of operational energy and end-of-life stages reflects gaps in industry inventories rather than conceptual oversight. Emission factor distributions, derived from peer-reviewed literature, provide empirical stochastic inputs but may lag behind low-carbon innovations. The assumption of parameter independence neglects covariance structures (e.g., concurrent energy intensity in material production and transport), introducing conservative bias. Future research should extend temporal scope to operation/demolition phases, integrate copula methods to model variable interdependencies, and validate against instrumented infrastructure datasets to enhance decarbonization planning.

The analysis revealed significant uncertainty inherent in emission factors, attributable to several interconnected causes. Firstly, regional heterogeneity in production technologies, fuel mixes, and recycling rates leads to variable carbon intensities for core materials such as cement and steel. This variability is compounded by datasets that often amalgamate regionally aggregated averages with sparse localized measurements. Secondly, emissions associated with transport exhibit substantial divergence due to factors including vehicle type, load efficiency, and route-specific characteristics. Thirdly, the carbon intensity of energy-dependent processes (e.g., concrete batching, steel rolling) is subject to considerable fluctuation, reflecting the variable composition of regional electricity grids, which range from predominantly fossil-fuel-based to renewable-dominated. Furthermore, methodological inconsistencies in Life Cycle Assessment (LCA) studies such as differing system boundaries and allocation rules introduce discrepancies, meaning emission factors derived from distinct databases are frequently incongruent. Within the adopted probabilistic analytical framework, these disparate sources of uncertainty were systematically evaluated. The quantified uncertainties were subsequently propagated through Monte Carlo simulation techniques, resulting in expansive confidence intervals (typically ± 20–30%) around the total carbon estimates. By explicitly acknowledging that each emission factor constitutes not a fixed value but a distribution reflecting real-world variability, the probabilistic approach adopted herein underscores how incremental deviations in material sourcing, transport logistics, or grid composition can propagate cumulatively, thereby inducing significant uncertainty in the overall carbon footprint assessment of a bridge structure.

The quantified uncertainty has direct and actionable implications for real-world decision-making. Given the observed ± 20–30% confidence range, carbon budgets should include an explicit contingency, typically a 15–20% buffer, to manage tail risks and avoid overconfidence in single-point targets. This is further reinforced by the right-skewed Monte Carlo outputs, which support budgeting at an upper percentile (e.g., P90/P95) rather than relying on the mean. In parallel, design optimisation should focus on reducing uncertainty in dominant upstream material drivers, particularly concrete and reinforcing steel, by adopting material-efficient detailing and lower-carbon material options. Complementary procurement measures can also reduce both mean emissions and variability by requiring supplier-specific, third-party-verifiable disclosures or EPDs and applying regionally calibrated factors that better reflect local production and transport conditions. Collectively, these risk-informed choices connect the probabilistic results to practical engineering and policy actions and help ensure decarbonisation efforts remain robust under real-world variability.

5 Conclusion

In this study, a probabilistic LCA of Jhelum bridge on M-13 motorway in Pakistan is conducted to quantify embodied carbon and its uncertainty across raw material production, transportation, and construction phases. Through sensitivity analysis and 10,000 Monte Carlo simulations, the total embodied carbon was found to average 72.5×10^6 kg CO_2 − eq ($sigma = 11.6 \times 10^6$ kg CO_2 − eq), with raw material extraction contributing 55% of the footprint, transportation 43%, and on-site construction 3% from the perspective of the average value. Component-level disaggregation reveals the superstructure and deck as principal emitters (with the average 23.38×10^6 kg CO_2 − eq and 23.27×10^6 kg CO_2 − eq, respectively), while ancillary materials exhibited high relative uncertainty due to limited data quality.

These findings underscore that upstream activities, especially cement and high-yield steel production, dominate life-cycle emissions and drive variability. Consequently, mitigation efforts should prioritize low-clinker cement alternatives, optimized multimodal logistics, and material substitution strategies. The high coefficients of variation among minor components highlight the need for standardized supplier emissions disclosures and blockchain-enabled Material Profiles to reduce data gaps. Integrating probabilistic carbon modules into BIM platforms emerged as a powerful approach for enabling stochastic scenario testing during early design, thereby aligning structural safety margins with carbon budgets. From a policy perspective, institutionalizing geospatially resolved emission databases and performance-based carbon caps indexed to infrastructure typologies can catalyse systemic adoption of low-carbon practices across the AEC (Architecture, Engineering, and Construction) sector. Nonetheless, realizing these benefits in Pakistan requires closing the BIM skill gap, developing standardized local material libraries, and reducing the financial barriers to software acquisition and training. Through targeted capacity-building initiatives, promotion of open-source BIM tools, and adoption of national BIM-for-sustainability policies, probabilistic carbon analysis can become standard practice within the country's AEC sector.

Nevertheless, the analysis is limited by its focus on production, transport, and construction phases; operational energy use and end-of-life emissions remain to be incorporated as high-resolution datasets become available. Additionally, this study assumes fixed material quantities, ignoring variability in quantities of concrete, steel, and other components used in engineering practice, which would further amplify uncertainty if accounted for. The assumption of input independence may also have introduced conservative biases, and the temporal resolution of emission factor distributions may lag behind emerging low-carbon manufacturing innovations. Future research should extend the framework to include operational and demolition stages, employ copula-based methods to model variable interdependencies, and validate results against sensor-enabled monitoring data. By quantifying carbon as a probability distribution rather than a single value, our probabilistic LCA provides a decision-support tool that can directly inform design and procurement choices enabling engineers and stakeholders to compare low-carbon alternatives, set performance-based carbon targets, and prioritize materials or processes according to their risk-adjusted emissions. Such enhancements will refine uncertainty quantification and bolster the methodology's applicability as a dynamic aid for decarbonizing large-scale infrastructure. Further studies can be referred to [20–25].

Acknowledgements This work was substantially supported by the National Key R&D Program of China (Numbers: 2023YFC3008300, 2023YFC3008302), the Oceanic Interdisciplinary Program of Shanghai Jiao Tong University (Project Number SL2022MS007), and State Key Laboratory of Ocean Engineering (Shanghai Jiao Tong University) (Grant Number GKZD010087).

References

1. Fisch-Romito V (2021) Embodied carbon dioxide emissions to provide high access levels to basic infrastructure around the world. Glob Environ Chang 70:102362
2. Lerner M, Rottman J (2021) The burden of climate action: how environmental responsibility is impacted by socioeconomic status. J Environ Psychol 77:101674
3. Collings D (2022) The carbon footprint of bridges. Struct Eng Int 32(4):501–506
4. Pervez H, Ali Y, Petrillo A (2021) A quantitative assessment of greenhouse gas (GHG) emissions from conventional and modular construction: a case of developing country. J Clean Prod 294:126210
5. Horry R et al (2022) Environmental management systems in the architectural, engineering and construction sectors: a roadmap to aid the delivery of the sustainable development goals. Environ Dev Sustain 24(9):10585–10615
6. Santaren D et al (2025) Benchmarking data-driven inversion methods for the estimation of local CO_2 emissions from synthetic satellite images of XCO_2 and NO_2. Atmos Meas Tech 18(1):211–239
7. Camacho AE (2009) Adapting governance to climate change: managing uncertainty through a learning infrastructure. Emory LJ 59:1
8. Curran MA (2014) Strengths and limitations of life cycle assessment. In: Background and future prospects in life cycle assessment, pp 189–206
9. Mirabella N, Allacker K, Sala S (2019) Current trends and limitations of life cycle assessment applied to the urban scale: critical analysis and review of selected literature. Int J Life Cycle Assess 24:1174–1193
10. Present E et al (2024) Choosing the best carbon factor for the job: Exploring available carbon emissions factors and the impact of factor selection, National Renewable Energy Laboratory (NREL), Golden, CO (United States)
11. Yousuf I et al (2014) Carbon emissions from power sector in Pakistan and opportunities to mitigate those. Renew Sustain Energy Rev 34:71–77
12. Rehman A, Ma H, Ozturk I (2021) Do industrialization, energy importations, and economic progress influence carbon emission in Pakistan. Environ Sci Pollut Res 28:45840–45852
13. García-Segura T, Penadés-Plà V, Yepes V (2018) Sustainable bridge design by metamodel-assisted multi-objective optimization and decision-making under uncertainty. J Clean Prod 202:904–915
14. Feng T-T et al (2017) Economic drivers of greenhouse gas emissions in China. Renew Sustain Energy Rev 78:996–1006
15. Khasreen MM, Banfill PF, Menzies GF (2009) Life-cycle assessment and the environmental impact of buildings: a review. Sustainability 1(3):674–701
16. Larsson Ivanov O, Honfi D, Stripple H (2018) Uncertainty and variation in LCA-Implementation of probabilistic methods to assess environmental impacts of infrastructures. TVBK (3070)
17. Nasir S, Kucerik J, Mahmood Z (2012) A study on the washability of the Azad Kashmir (Pakistan) coalfield. Fuel Process Technol 99:75–81
18. Scholz FW, Stephens MA (1987) K-sample Anderson-Darling tests. J Am Stat Assoc 82(399):918–924
19. Massey FJ Jr (1951) The Kolmogorov-Smirnov test for goodness of fit. J Am Stat Assoc 46(253):68–78

20. Guo J, Wang Z, Li H, Yang Y, Huang CG, Yazdi M, Kang HS (2024) A hybrid prognosis scheme for rolling bearings based on a novel health indicator and nonlinear Wiener process. Reliab Eng Syst Saf 245:110014
21. Li H, Yazdi M, Huang HZ, Huang CG, Peng W, Nedjati A, Adesina KA (2023) A fuzzy rough copula Bayesian network model for solving complex hospital service quality assessment. Complex Intell Syst 9(5):5527–5553
22. Li H, Peng W, Adumene S, Yazdi M (2023) An intelligent cost-based consequence model for offshore systems in harsh environments. In: Intelligent reliability and maintainability of energy infrastructure assets, pp 107–117. Springer Nature Switzerland, Cham
23. Li H, Yazdi M (2022) What are the critical well-drilling blowouts barriers? A progressive DEMATEL-game theory. Advanced decision-making methods and applications in system safety and reliability problems: approaches, case studies, multi-criteria decision-making, multi-objective decision-making, fuzzy risk-based models. Springer International Publishing, Cham, pp 29–46
24. Li H, Peng W, Adumene S, Yazdi M (2023) Cutting edge research topics on system safety, reliability, maintainability, and resilience of energy-critical infrastructures. In: Intelligent reliability and maintainability of energy infrastructure assets, pp 25–38. Springer Nature Switzerland, Cham
25. Li H, Peng W, Adumene S, Yazdi M (2023) Advances in failure prediction of subsea components considering complex dependencies. In: Intelligent reliability and maintainability of energy infrastructure assets, pp 93–105. Springer Nature Switzerland, Cham

Gas Turbine Rotor Systematic Upgrading Technology Research and Engineering Application

Huayun Zhu, Jiangsheng Zheng, Yuming Zhang, Hongchen Zhao, Jiaming Wei, Linfeng Xue, Jibing Lan, and Yongfeng Sui

Abstract This study investigates corrosion, blade damage, and efficiency degradation in gas turbine rotors at a natural gas chemical company, which results from prolonged operation. By integrating reverse engineering, precision machining, nondestructive testing, and dynamic balance optimization technologies, the rotor was dismantled, parts were repaired or replaced, and assembly and performance verification was completed. The project employs innovative techniques such as dry ice cleaning, hydraulic tensioning for step-by-step dismantling, and spark discharge atomic emission spectroscopy for material analysis to address technical challenges like rotor cleaning, long tie bolt removal, and material analysis. The modified rotor underwent high-speed dynamic balancing, with vibration values significantly lower than industry standards. The efficiency of the gas turbine improved by 6.5%, with stable operation maintained for over six months. This study offers an effective

H. Zhu (✉) · Y. Zhang · H. Zhao · L. Xue
Zhejiang Rancon Turbine Machinery Co., Ltd, Hangzhou, China
e-mail: hyzhu@htc.net.cn

Y. Zhang
e-mail: ymzhang@htc.net.cn

H. Zhao
e-mail: hczhao@htc.net.cn

L. Xue
e-mail: lfxue@htc.net.cn

H. Zhu · J. Zheng · J. Wei · J. Lan · Y. Sui
Advanced Power Machinery Academy, Hangzhou Steam Turbine Power Group Co., Ltd, Hangzhou, China
e-mail: jszheng@htc.net.cn

J. Wei
e-mail: jmwei@htc.net.cn

J. Lan
e-mail: jblan@htc.net.cn

Y. Sui
e-mail: yfsui@htc.net.cn

H. Li et al. (eds.), *Data-Driven Methods for Reliability and Safety Engineering: Applications in Industrial Systems*, Springer Series in Reliability Engineering,
https://doi.org/10.1007/978-3-032-22873-4_33

solution for gas turbine rotor operation and maintenance, demonstrating substantial economic and engineering application value.

Keywords Gas turbine · Rotor upgrade · Dynamic balancing · Reverse engineering · Non-destructive testing

1 Introduction

As a critical component of gas turbines, the rotor is vulnerable to wear and corrosion after prolonged operation, which leads to performance degradation [1]. After more than twenty years of service, the rotor exhibits issues such as damage to the compressor blades, corrosion of tie rod bolts, and reduced efficiency, necessitating urgent and systematic upgrading and reconstruction.

1.1 Research Background

Gas turbines are widely used in various industries due to their high efficiency and reliability. However, the rotors of these turbines are susceptible to wear and corrosion, especially in harsh operating environments such as natural gas chemical plants. The degradation of rotor components can lead to significant performance losses and safety risks. Consequently, developing effective repair and upgrade strategies for gas turbine rotors is crucial to ensure their continuous and reliable operation [2].

1.2 Traditional Repair Technologies and Their Limitations

Traditional repair methods for gas turbine rotors typically involve manual inspection, basic machining, and the simple replacement of damaged components. However, these approaches have several drawbacks, including:

Low Efficiency: Traditional repair processes are time-consuming and labor-intensive, often requiring extensive downtime for the turbine.

Lack of Precision: Manual inspections and repairs may not accurately identify and address all potential issues, leading to incomplete or suboptimal repairs.

Limited Performance Improvement: The traditional methods may not significantly enhance rotor performance or extend its service life.

1.3 Objectives and Contributions of This Study

To overcome the limitations of traditional repair technologies, this study proposes a set of efficient retrofit solutions by integrating reverse engineering, precision machining, and dynamic balancing technology. The primary objectives of this research are:

To develop a systematic procedure for the repair and upgrade of gas turbine rotors.

To enhance the reliability and performance of the rotor using advanced technological solutions.

To validate the feasibility and effectiveness of the proposed method through case studies.

1.4 Project Overview

The ammonia plant of a natural gas chemical company was completed in the late 1980s and successfully commissioned in May 1993. The project marked the first large-scale fertilizer project in China to use offshore natural gas resources from Liaodong Bay [3]. The plant produces 300,000 tonnes of ammonia and 520,000 tonnes of urea annually. The project's air compressor is driven by a gas turbine and is mainly responsible for supplying process air to the first-stage reformer and the second-stage reformer, as well as providing instrument air for the plant [4]. The gas turbine used in the project is a GE M5261RA model, fueled by natural gas. It consists of a 16-stage axial-flow compressor, a two-stage turbine, 10 combustion chambers, a support system, an auxiliary gearbox, and a COPPUS starter turbine, model RLHA28 [5]. The M5261RA belongs to GE's MS5000 series of gas turbines and is the second heavy-duty gas turbine series introduced after the MS3000 series. As can be seen from the model designation, the M5261RA is a single-shaft, simple-cycle gas turbine with a power of 26,000 hp, primarily used for mechanical drives.

At present, the rotor of the unit has accumulated over 136,000 h of operation, with serious surface corrosion and many damaged blades, which greatly affects the safe operation of the unit. In response to the owner's requirements, the rotor underwent repairs and upgrades to extend its service life and enhance its operational reliability. The repair process included full disassembly, flaw detection, and repairs to the rotor, replacement of all dynamic blades, tie rod bolts, and nuts for the compressor and turbine, followed by final assembly and high-speed dynamic balancing of the rotor assembly, refer to Fig. 1.

Fig. 1 Compressor rotor before repair

1.5 Comparison with Traditional Technologies

To highlight the advantages of the proposed method, a detailed comparison with traditional repair technologies is provided:

Efficiency: The proposed method, leveraging advanced technologies such as reverse engineering and precision machining, significantly reduces the repair time compared to traditional methods.

Precision: The application of modern inspection and machining techniques ensures a higher level of accuracy in detecting and addressing rotor defects.

Performance Improvement: The integrated approach not only repairs existing damage but also enhances the overall performance and reliability of the rotor, leading to improved efficiency and extended service life.

2 Rotor Structure

The gas turbine rotor assembly consists of three parts: the front-end shaft, the compressor rotor, and the turbine rotor. The front-end shaft is located at the very front of the rotor assembly and is used to connect to the gearbox shaft, refer to Fig. 2.

The rotor has a typical circumferential tie rod structure with 18 tie rods, which rely on the friction of the pressed end faces to transmit the torque, and the locating stops are located near the center of the wheel disc. To prevent the tie bolts from loosening during operation, the MS5000 rotor adopts a special nut design, the front end of which is a fan-shaped nut installed in the groove of the wheel disc, which will not loosen after assembly; the rear end is an octagonal nut, the corner of which will fit with the surface of the inner circle of the intermediate shaft after assembly, which will not loosen during operation. The dynamic blades of the compressor rotor are secured with a dovetail design at the root, fixed by spacer blocks on both sides after

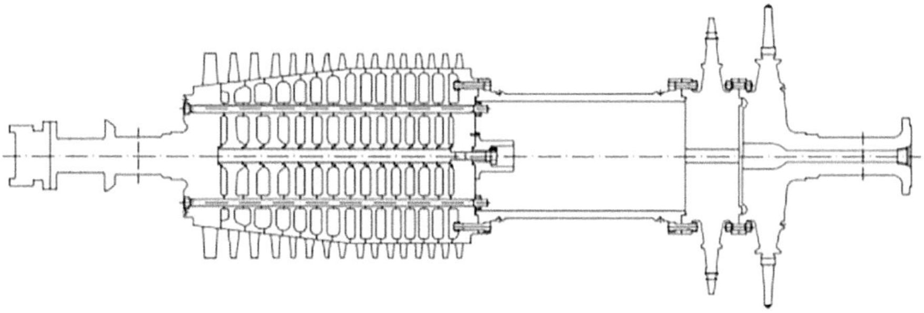

Fig. 2 Structural cross-sectional diagram of the gas turbine rotor

assembly. The impeller is assembled in stages, and if a blade is damaged, the entire unit must be disassembled for replacement.

The turbine rotor assembly consists of an intermediate shaft, a first-stage impeller, a turbine rear half shaft, and blades. Centering is achieved by means of a stop fit and the parts are connected by short bolts.

All turbine blades are precision-cast, long-stemmed moving blades, which effectively protect the wheel rim and blade tenon from the high-temperature gas flow, reduce heat transfer from the blade body to the root, significantly lowering the temperature and thermal stress at the blade root and wheel disc, and optimize the stress distribution at the blade root. The second stage of the blade features a crown band, which interlocks the blades into a continuous circular band, providing vibration damping, along with sealing teeth on the band to minimize air leakage at the top of the blade [6].

3 Upgrading Process

Upon arrival at the factory, the rotor assembly is first cleaned using dry ice to remove surface rust and other contaminants. The rotor is then disassembled before performing the operational balance test and assembly mapping. During this phase, the rotor's condition, critical speed, and other key parameters are carefully recorded, including critical speed and other vital information [7]. Following this, the connecting bolts between the intermediate shaft and the 16th-stage disc of the compressor are removed, and the rotor assembly is separated into two parts: the turbine rotor and the compressor rotor. These two components are mapped, disassembled, and their connecting parts reverse-engineered simultaneously, optimizing time and accelerating the work process. After non-destructive testing, cleaning, repair, and machining of all connecting parts and blades, the rotor assembly is reassembled. The assembly of short bolts and long tie bolts is completed according to the design requirements of bolt elongation, with the turbine rotor and compressor rotor assembled separately. The compressor rotor undergoes a blade top grinding process and a low-speed dynamic balance test. After necessary adjustments, the two rotor sections are rejoined to complete the rotor assembly, followed by a high-speed dynamic balance test.

The project plan was optimized to address the customer's urgent requirement for completing the upgrade before the overhaul. The entire project was divided into four main phases: rotor inspection and dismantling preparation, rotor dismantling and material testing, rotor mapping and repair, and parts procurement followed by rotor reassembly. Each phase had clearly defined start and end times, allowing for effective monitoring of the project's progress and the identification of potential risks. This structured approach ensured the project proceeded efficiently and was completed on schedule.

4 Key Steps and Methods

4.1 Dry Ice Clean-Up

Dry ice cleaning is a cleaning technology that utilizes solid carbon dioxide (dry ice) as the medium. When dry ice particles are propelled at high speeds toward a surface, their extremely low temperatures cause contaminants such as grease, dirt, paint, rust, and corrosion to rapidly freeze and become brittle. This temperature-induced change causes the contaminants to separate from the substrate material due to thermal expansion and contraction. As the dry ice sublimates directly from a solid to a gaseous state, no residue is left behind after cleaning. Only the residual pollutants are removed from the surface, with no additional substances adhering to the material. Moreover, dry ice cleaning does not require chemical solvents, making it environmentally friendly and free from harmful by-products. Currently, dry ice cleaning is widely employed in industrial cleaning and maintenance due to its efficiency, eco-friendliness, and non-destructive impact on equipment. The project has demonstrated that dry ice cleaning is effective in removing surface rust and other contaminants from the rotor without causing damage to the substrate.

4.2 Precision Mapping and Manufacturing of Connectors

The project employed spark discharge atomic emission spectroscopy (OES) to accurately characterize material grades for connectors such as hexagonal bolts, sector nuts, octagonal nuts, quadrilateral nuts, round nuts, and hexagonal nuts. The method provides comprehensive information on material composition, covering key elements such as carbon content, to ensure the accuracy of the material grades. The specimens

Fig. 3 The gas turbine rotor after dry ice cleaning

were also cut and tested for mechanical tensile properties. The precise material analysis and performance testing results form a crucial foundation for selecting component blank materials, ensuring their mechanical properties align with the design requirements, refer to Fig. 3.

The Dantsin-IAC XPL Horizontal Universal Thread Comprehensive Calibration Centre is used for measuring thread parameters. This device offers a resolution of 0.01 μm and accurately measures parameters such as major diameter, medium diameter, small diameter, tooth angle, and pitch. Thread machining is carried out in strict compliance with the requirements of the gauge, and the tolerance of the center diameter is controlled within 0.02 mm. Thread processing is carried out by grinding, with high processing precision and smooth thread surface, which is conducive to high-strength fit scenarios. In addition, the corrosion resistance of the threads is enhanced by chemical nickel plating, refer to Fig. 4.

Fig. 4 Machined rotor connectors

Fig. 5 Magnetic particle flaw detection of turbine discs

4.3 Non-destructive Testing

In accordance with the requirements of quality level I in the standard JB/T47013.4-2015, magnetic particle flaw detection was carried out on the wheel discs of all levels of the pressured engine, the wheel discs of both turbine stages, and the intermediate shaft. Magnetic particle flaw detection technology can efficiently identify tiny cracks and other potential defects on and near the surface. After rigorous testing, all tested parts met the standard requirements and all test results were qualified, thus ensuring the reliability and safety of the components, refer to Fig. 5.

4.4 Reverse Design of Compressor Blades

This section outlines the reverse engineering workflow for compressor blades, which involves high-precision scanning, point cloud processing, 3D modeling, and deviation correction to replicate and optimize blade geometry.

A high-precision scanner is employed for reverse mapping of the compressor blade, achieving a scanning accuracy of 0.01 mm. The acquired point cloud data are processed to extract the key data parameters of the blade body and the blade root, and based on these data, 3D reverse modeling is carried out. Deviations between the models of the blade's dorsal arc, internal arc, root, and top surface, and the actual mapped point cloud, were thoroughly compared and analyzed. Adjustments were then made to minimize these discrepancies, refer to Fig. 6.

Once the blade machining is complete, key processes such as data inspection, appearance inspection, magnetic particle flaw detection, and remanent magnetization

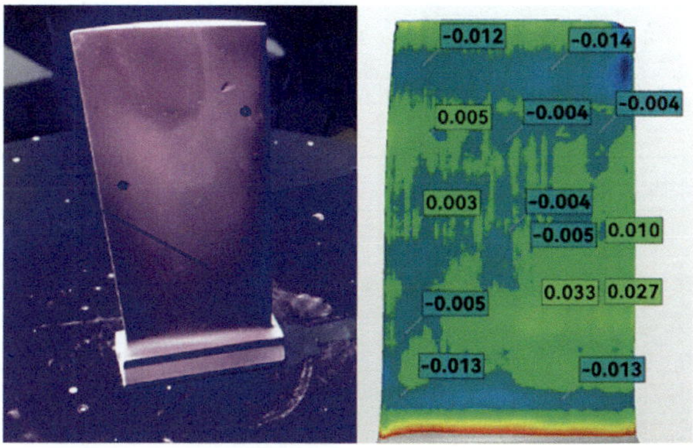

Fig. 6 Reversal and comparison of compressor blades

testing are performed in sequence according to a strict quality control process. These steps ensure that the blade's processing quality meets the high design standards, while also guaranteeing its performance and reliability.

The long tie bolts are evenly distributed around the circumference of the rotor, a total of 18 bolts, which connect the 16-stage compressor discs through substantial pre-load, ensuring the structural integrity and stability of the rotor during high-speed rotation. Given the harsh operating conditions of the gas turbine rotor, these long tie bolts are made from high-strength, high-temperature-resistant alloy steel to guarantee adequate strength and reliability under extreme temperatures, high pressures, and high speeds. During the disassembly of the compressor rotor, a two-by-two dismantling method is employed, with detailed oil pressure data recorded throughout the process to provide a reference for the subsequent assembly work, refer to Figs. 7, 8, 9, and 10.

4.5 *Design and Manufacture of Long Tie Bolts*

The design and manufacturing of long tie bolts integrate 3D computational modeling and precision machining to ensure pre-load stability and structural integrity under extreme operating conditions.

In the design process of long tie rod bolts, key parameters such as radial deformation, tie rod equivalent force, local equivalent force, tie rod hole contact pressure, and torsion surface contact pressure are evaluated in detail by establishing a three-dimensional computational model, applying rotational speed, bolt pre-load, disc center constraints, and setting the contact counterpart boundary conditions between the discs. Additionally, a one-dimensional calculation program is employed to analyze the pre-load force of the tie rods under cold, transient, and rated operating conditions. This analysis considers factors such as pneumatic axial force, the Poisson effect of the disc, temperature differences between the tie rods and the disc,

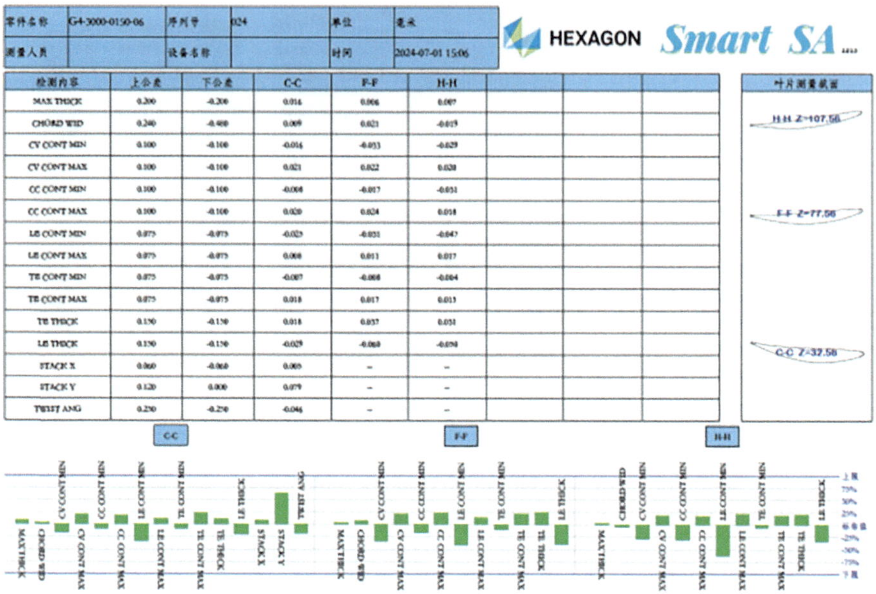

零件名称	G4-3000-0150-06	序列号	024	单位	毫米
测量人员		设备名称		时间	2024-07-01 15:06

检测内容	上公差	下公差	C-C	F-F	H-H
MAX THICK	0.200	-0.200	0.016	0.006	0.007
CHORD WID	0.240	-0.400	0.009	0.021	-0.015
CV CONT MIN	0.100	-0.100	-0.016	-0.033	-0.029
CV CONT MAX	0.100	-0.100	0.021	0.022	0.020
CC CONT MIN	0.100	-0.100	-0.008	-0.017	-0.031
CC CONT MAX	0.100	-0.100	0.020	0.024	0.018
LE CONT MIN	0.075	-0.075	-0.023	-0.031	-0.047
LE CONT MAX	0.075	-0.075	0.008	0.011	0.017
TE CONT MIN	0.075	-0.075	-0.007	-0.008	-0.004
TE CONT MAX	0.075	-0.075	0.018	0.017	0.015
TE THICK	0.150	-0.150	0.018	0.037	0.031
LE THICK	0.150	-0.150	-0.029	-0.060	-0.050
STACK X	0.060	-0.060	0.005	-	-
STACK Y	0.120	0.000	0.079	-	-
TWIST ANG	0.250	-0.250	-0.046	-	-

Fig. 7 Detailed blade data detection

Fig. 8 Completed compressor blades

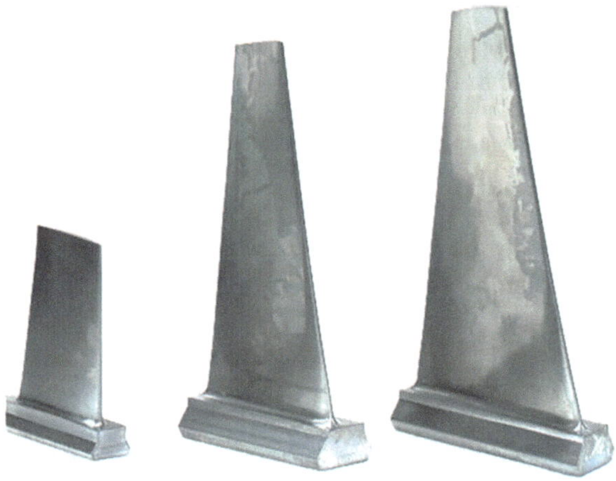

and the gravitational bending moment, ensuring that the pre-load force of the tie rods remains within a reasonable range across all operating conditions.

The threaded part of the long tie bolts is the core part of the connection and torque transmission, and its processing accuracy plays a decisive role in the connection strength and reliability of the bolt. The dimensional accuracy, surface roughness, and shape accuracy of the threads are all required to meet extremely high standards. Given that long tie bolts typically have a long length and relatively small diameter,

Fig. 9 Hydraulic tensioning of long tie bolts

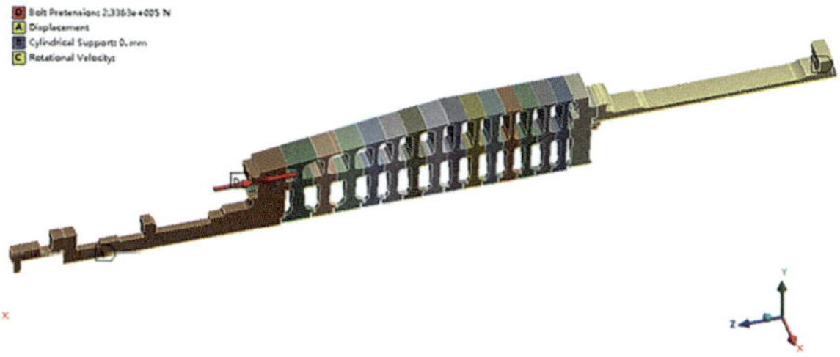

Fig. 10 Rotor tie rod 3D calculation model

leading to a high L/D ratio (for example, the L/D ratio of these tie bolts reaches 45), issues such as vibration and deformation during machining are more likely to occur, potentially affecting machining accuracy and surface quality. To address this, the concentricity of the tie rods is strictly controlled within 0.02 mm in diameter by using specialized machining methods, thereby ensuring the machining quality and reliability of the bolts, refer to Fig. 11.

4.6 Blade Top Grinding

Blade top grinding adopts an innovative single-stage processing strategy to control tip clearance accurately, minimizing airflow leakage and improving compressor efficiency.

Compressor blade top grinding is crucial for ensuring the performance, efficiency, and internal flow characteristics of the compressor. In the machining process, the traditional template-based method was replaced with the four-point measurement technique, which ensures the accuracy of the machined dimensions.

Typically, impeller blade top grinding is performed uniformly after the completion of rotor assembly. However, due to the tight project schedule and the sequential arrival

Fig. 11 Single-stage impeller grinding blade top

of gas turbine blades, an innovative approach was adopted, involving the advance grinding and processing of single-stage impellers. Specifically, as each batch of blades arrived, the corresponding impellers were immediately assembled and ground individually. This method reduced the overall project duration by at least 10 working days, significantly improving the overall progress of the project.

4.7 Rotor Assembly

The quality of the gas turbine rotor assembly has a critical impact on the efficiency and safety of the unit.

For rotor assembly, the impellers of the compressor stages are first assembled into the compressor rotor by means of long tie bolts, and the impellers of the turbine stages and the intermediate shaft are assembled into the turbine rotor by means of short tie bolts. Afterward, the compressor rotor and the turbine rotor are assembled against each other by means of a stop-over fit between the intermediate shaft and the 16th-stage impeller and pre-tensioned with a set of bolts. The amount of runout of each part as well as the bolt elongation during rotor assembly are strictly required.

The main process of compressor rotor assembly is as follows:

1. Weighing and sorting of compressor blades and blocks by stage number.
2. Install the moving blades and blocks of each stage to the impeller of the compressor in sequence and fix them to the impeller by punching rivets to form a single-stage impeller with blades.

3. Stand the front shaft head down on the mounting platform and install the impellers from stage 2 to stage 16 in sequence. Due to the existence of interference fit at the stops, the assembly needs to be carried out by means of uniform heating. At the same time, the impeller tie rod holes of each stage need to be accurately circumferentially positioned for subsequent threading of the tie rods.
4. Weigh, sort, and mark the long tie rod bolts and matching nuts, then pass the long tie rods through the impeller tie rod holes at all levels, install the fan nuts and octagonal nuts on both sides, adjust the position, and then tighten them initially.
5. A hydraulic tensioner is installed on the octagonal nut side, and a group tensioning strategy is used to control the end runout to within 0.01 mm. Hydraulic pressure is used to stretch the tensioner to the design-required elongation, generating a pre-set pre-load force to form the impellers at all levels into a single unit.

Turbine rotor assembly is relatively simple, install the intermediate shaft with the inlet side facing down on the platform, and install the turbine stage 1 and stage 2 discs in sequence. The short bolts and nuts are weighed and sorted, and then the bolts are tightened according to the torque requirements and tightening sequence.

Finally, the turbine section rotor and the compressor rotor are docked and assembled, and the connecting bolts are tightened in strict accordance with the requirements of bolt elongation to complete the assembly of the rotor assembly, refer to Fig. 12.

Fig. 12 Completion of rotor assembly on the rotor assembly platform

4.8 Dynamic Balancing

The dynamic balancing process implements multi-stage balancing (static to high-speed) to reduce unbalanced mass, ensuring low vibration levels at rated and over-speed conditions.

The rotor dynamic balancing process includes impeller static balancing, low-speed dynamic balancing of the compressor rotor, low-speed dynamic balancing of the turbine rotor, low-speed dynamic balancing of the rotor assembly (excluding turbine moving blades), and high-speed dynamic balancing of the rotor assembly. The rotor assembly is equipped with four balancing slots and features a dovetail groove design with balancing blocks, which are fixed by jacking screws and further fixed by riveting after the dynamic balancing. Since the balancing slots form a full-circle structure, all positions, except for the sliding-in position, theoretically allow for the addition of balancing blocks, providing a large adjustment range.

The entire rotor contains a 16-stage axial-flow compressor rotor and a two-stage turbine rotor, with a total weight of about 8900 kg, a length of about 5000 mm, and a rated speed of about 5100 r/min, thus requiring extremely high dynamic balancing accuracy, refer to Fig. 13.

Firstly, the unbalanced mass of the wheel discs at all levels is removed by static balancing, using mass removal at the position of the impeller bonding surface at all levels. Due to the relatively thick and low-stress level of the bonding surface, it is suitable for mass removal, but the amount of removal is large, so it is necessary to strictly guarantee the machining quality to avoid crack sources.

Subsequently, low-speed dynamic balancing was carried out after assembling the impellers of all stages of the compressor into a compressor rotor by means of long tie rods. Balancing was done by adding balancing blocks to the balancing surface of the first stage disc of the compressor and adjusting the balance by adding balancing nuts behind the tie rods. This method of adding a balancing nut behind the tie rods

Fig. 13 High-speed dynamic balancing of rotor assembly

is unique in that the nut is thinned to the required mass and the mass is adjusted according to the phase of the unbalance.

Finally, the compressor rotor and the turbine rotor were combined to form a whole rotor and subjected to a dynamic balancing test. During the rated speed and overspeed tests, the swing frame vibration of both the compressor side and turbine side is less than 1.5 mm/s, and the rotor vibration performance indexes fully meet API616 and contract requirements [8].

5 Conclusion

This study successfully achieves the efficient upgrading of gas turbine rotors through the integration of advanced technologies and innovative practices. The key findings are as follows:

Technological Innovation: The rotor upgrading process is optimized by incorporating dry ice cleaning, hydraulic step-by-step dismantling, and single-stage impeller grinding, which effectively shortens the upgrading cycle. The pre-load design of the long tie bolts is optimized using a combined analysis of three-dimensional modeling and one-dimensional procedures to ensure structural stability under complex operating conditions.

Performance Improvement: After modification, the rotor's vibration value is significantly lower than the industry standard. Under identical ambient temperature and load conditions, the gas turbine's efficiency improves by 6.5% compared to pre-repair performance, demonstrating the solution's economic and engineering applicability.

The 6.5% efficiency gain is directly attributed to targeted upgrades: reverse-engineered compressor blades optimize aerodynamic profiles, reducing flow separation and pressure loss; precision blade top grinding minimizes tip clearance, decreasing airflow leakage; redesigned long tie bolts with optimized pre-load ensure rotor structural stability at high speeds, reducing vibration-induced efficiency losses; and rigorous dynamic balancing eliminates unbalanced mass, lowering energy consumption from mechanical vibrations. These technical improvements collectively address core inefficiencies from worn components, suboptimal geometry, and structural instability, directly translating to measurable performance enhancements.

Project Value: The upgraded gas turbine has operated stably for over six months, maintaining normal vibration values, with no over-temperature issues, and its combustion engine output meets the required standards. This ensures the continuous production capacity of the chemical plant and provides reliable technical support for the long-term safe operation of similar equipment. To get some info regarding fault tree analysis, refer to [9–11].

Through systematic technology integration, this study has successfully achieved the efficient upgrading of gas turbine rotors, overcoming challenges such as long bolt removal, high-precision manufacturing, and dynamic balance optimization. The upgraded gas turbine operates smoothly with significant performance improvements, offering a replicable technological model for the operation and maintenance of similar equipment.

Acknowledgements This research was supported by the Zhejiang Province "Sharp Soldiers, Leading Geese + X" Science and Technology Programme Project, specifically, the project titled "Development and Optimisation of 50MW-class High-efficiency Low-emission Independent Heavy-duty Gas Turbine Machine" (Grant No. 2025C01191).

References

1. Chen P, Ding J, Wan KX (2010) Advantages of process air compressor driven by gas turbine in large scale nitrogenous fertilizer plants. Nat Gas Chem Ind (C1 Chemistry and Chemical Engineering) 35(06):53–55+58
2. Liu Y, Li XG (2021) Research on Overhaul Based on Siemens V94. 3A Gas Turbine. Northeast Electr Power Technol 42(03):41–43
3. Shi JZ, Wang JS (1998) Jintianhua Brown ammonia process review (01):35–39
4. Xing GK (2004) Application of gas turbines in chemical and petrochemical field. Process Equip Pip 06:50–54
5. Huang GF (1994) Commissioning and commissioning of single shaft gas turbines. Large Scale Nitrogenous Fertil Ind (04):253–255
6. Dong ST Troubleshooting and overhaul of compressor for Xiamen gas power plant No.2'. Gas Turbine Technol (02):39–45
7. Jing SX, Zhang DX (1994)) Vibrational problems encountered in overhauling industrial gas turbine and counter-measures adopted. J Vib Eng (03):264–268
8. Guan ZP (2002) Solutions to several problems in the overhaul of gas combustion turbine. Large Scale Nitrogenous Fertil Ind (06):421–423
9. Yazdi M, Mohammadpour J, Li H, Huang HZ, Zarei E, Pirbalouti RG, Adumene S (2023) Fault tree analysis improvements: a bibliometric analysis and literature review. Qual Reliab Eng Int 39(5):1639–1659
10. Li H, Yazdi M (2022) Stochastic game theory approach to solve system safety and reliability decision-making problem under uncertainty. Advanced decision-making methods and applications in system safety and reliability problems: approaches, case studies, multi-criteria decision-making, multi-objective decision-making, fuzzy risk-based models. Springer International Publishing, Cham, pp 127–151
11. Li H, Yazdi M (2022) Reliability analysis of correlated failure modes by transforming fault tree model to Bayesian network: a case study of the MDS of a CNC machine tool. Advanced decision-making methods and applications in system safety and reliability problems: approaches, case studies, multi-criteria decision-making, multi-objective decision-making, fuzzy risk-based models. Springer International Publishing, Cham, pp 15–28

A Storage Life Assessment Method for Missile Based on Critical Component Inspection Date

Guiming Tao, Qirui Wang, and Hui Li

Abstract A storage life assessment method for missile systems has been developed based on inspection data from critical components. For detection data characterized as "success or failure, incomplete," derived from missile quality monitoring processes, failure rate inversion data were analyzed using an order-preserving regression algorithm. A mathematical sampling model was established to accommodate the "success or failure, incomplete" data structure, and the storage reliability distribution function for missile system components was derived using the minimum test method. Additionally, a mathematical expression was formulated to characterize the storage life of the missile system, based on the "failure probability weight" assigned to each component. This formulation enables a more accurate representation of how the failure rates of individual components influence the overall system lifespan. Applying this method to assess the lifespan of a specific type of missile system that had been stored for nine years, the estimated reliable storage life of the system was determined to be 13.177 years when the reliability threshold was set at 0.9.

Keywords Storage life · Component · System · Probability of failure · Failure probability weight

1 Introduction

Components such as accelerometers, gyroscopes, military power supplies, and electronic timing devices are long-life, high-reliability elements that serve as critical components within missile inertial navigation platforms. The evaluation of storage life for these key components, as well as for the systems they comprise, has become a prominent focus of research in recent years [1]. In situations where inspection data are unavailable during extended storage periods, accelerated life testing is typically employed under specific assumed conditions to estimate component lifespan [2–6].

G. Tao · Q. Wang (✉) · H. Li
Army Engineering University, Shijiazhang, China
e-mail: 15838944263@163.com

H. Li et al. (eds.), *Data-Driven Methods for Reliability and Safety Engineering: Applications in Industrial Systems*, Springer Series in Reliability Engineering,
https://doi.org/10.1007/978-3-032-22873-4_34

However, when natural storage detection data are available, a more accurate estimation of storage life can be achieved, as these data reflect the effects of combined environmental influences on performance over time [7]. Natural storage environments expose components to various simultaneous stressors, offering a more authentic and intuitive representation of their performance degradation under real-world conditions [8].

The U.S. Missile Command has accumulated an extensive dataset over time regarding the reliability of electronic missile components, including diodes, transistors, vacuum tubes, capacitors, inductors, crystal oscillators, optoelectronic devices, accelerometers, and various electrical devices such as motors, pumps, and gyroscopes [9, 10]. By analyzing these data using mathematical statistical techniques, life distribution models can be established, enabling the prediction of failure rates and storage life for components stored over long durations without active operation. Within domestic research, reference [11] investigates the storage life evaluation of a specific accelerometer under natural storage conditions, while reference [12] focuses on the natural storage lifespan assessment of a certain type of gyroscope.

Once the natural storage life distribution functions of individual components are determined, their respective storage lifespans can be estimated. Nonetheless, accurately assessing the storage life of missile systems composed of these components remains a major challenge in system-level lifespan evaluation. Two main approaches have emerged in current research. The first adopts the shortest lifespan among all components as the representative lifespan of the missile system, aligning with the "barrel theory." The second approach treats the missile system as a unified whole, using detection data that are statistically processed through mathematical modeling to derive a storage life distribution function at the system level.

Throughout the quality monitoring process over the missile's lifecycle, performance parameters of individual components are continuously measured and failure characteristics statistically analyzed. Therefore, it is imperative to utilize actual detection data to conduct a scientifically grounded assessment of the missile system's storage life. As a contribution to this area of research, the present study proposes a method for characterizing the storage life of missile systems based on the "failure probability weights" of individual components. By establishing the life distribution of critical components within the system, the proposed method effectively captures the influence of component-level failure rates on the overall system storage life.

2 The Missile System Inherently Captures and Stores Inspection Data as Part of Its Core Functionality

A specific type of missile is stored in four distinct climatic zones: sub-humid and hot, sub-dry and hot, and mild and dry. This paper conducts a statistical analysis of the detection data for its critical components—namely, Part A, Part B, Part C, Part

Table 1 Performance test results of quality monitoring for critical components of a certain type of missile

Storage period	9 years	8 years	7 years	6 years	5 years	4 years	3 years
Sample quantity	XX	XX	XX	XX	XX	XX	XX
The number of faults in Component A		4		2			
The number of faults in Component B	1	3	1	5			
The number of faults in Component C		4		1			
The number of faults in Component H		3	3	2	1		
The number of faults in Component G	3	2	1	17	5		4

H, and Part G—collected over a period of 9 years of natural storage. The detection results after 3, 4, 5, 6, 7, 8, and 9 years of storage are presented in Table 1.

During the actual quality monitoring process of the missile, the storage unit employs detection equipment to test its critical components as specified. The test results indicate only two states: "normal" or "faulty." Consequently, the test data is categorized as binary success/failure data. Moreover, the storage unit conducts regular annual inspections of the quality status of key components without knowledge of the precise timing of faults occurring within the inspection cycle. As a result, the data for faulty components is incomplete.

3 Order-Preserving Regression Analysis of Failure Frequencies for Three Components

In missile quality monitoring, let t denote the detection time points. Following a storage period, a sample of missiles is randomly selected for inspection. If a specific component within a certain part is found to have failed, the failure frequency of the sample at that time is calculated as the ratio of failed components to the total number of missiles inspected.

While missile system components are generally highly reliable, the number of failures remains small even with large sample sizes. Occasionally, a phenomenon called failure rate inversion occurs. This happens when the failure rate does not follow the expected characteristic of decreasing over time. Such instances can cause significant deviations in calculation results.

To address this, the order-preserving regression calculation method is used to adjust frequencies exhibiting inversion into values that satisfy order constraints.

An order-preserving function is defined on a finite set where a specific relation ensures that the function values stay in a non-decreasing order relative to the time

Table 2 Results of component G failure frequency, Isotonic regression processing

Original data				Inverted handling	
Storage month	Conduct testing on the sample size	Number of Failure Instances	Failure rate	Weight 1	Weight 2
36	X_1	Y_1	0.000353	0.0002	0.0004
48	X_2	Y_2	0	0.0002	0.0004
59	X_3	Y_3	0.001662	0.0015	0.0022
71	X_4	Y_4	0.004022	0.0025	0.0031
83	X_5	Y_5	0.001001	0.0025	0.0031
96	X_6	Y_6	0.00308	0.0025	0.0031
105	X_7	Y_7	0.001645	0.0025	0.0031

points. By identifying the set of all such order-preserving functions, we can determine a specific regression function. This function represents the best fit for the original data while adhering to the required order constraints, using a predefined weight function to balance the results.

The detection data for component G were processed using order-preserving regression, as detailed in Table 2.

4 Mathematical Sampling Model

Following order-preserving regression to correct inverted failure rates, a mathematical sampling model is developed to address the "success or failure, incomplete" data structure. This adjustment is necessary due to the absence of precise failure timestamps within inspection intervals in raw detection data. A key assumption underlying the model is that component storage lifetimes are independently and identically distributed—a condition supported by consistent manufacturing quality and uniform storage environmental conditions across components. This assumption facilitates statistical inference while maintaining practical relevance.

The "failure and incomplete failure" data type for missile component detection can be equated to the following mathematical sampling model.

5 Analysis of the Distribution of Storage Lifespan for Components Within the Missile Systems Department

In missile quality monitoring, storage reliability is analyzed by observing component failure over specific time points. Due to the high reliability of these systems, failures are rare, which can sometimes lead to an unusual statistical result called "failure rate inversion".

This inversion occurs when observed failure rates do not follow the expected logical trend of remaining steady or increasing over time. To correct this and ensure accurate data, researchers use a method called order-preserving regression. This process adjusts the raw data into a set of values that strictly follow the necessary logical order while staying as close as possible to the original observations.

Assessing Storage Reliability

The analysis follows a structured statistical approach:

Distribution Models: The Exponential distribution family is commonly used to model the storage life of these components.

Hypothesis Testing: Researchers use a chi-square ($\chi 2$) statistic to compare the observed failure data against the assumed distribution model.

Goodness-of-Fit: A "p-value" is calculated to measure how well the model matches the actual data. A larger p-value indicates a better fit, confirming that the assumed distribution accurately represents the component's reliability.

Reliability Functions

Using these methods, reliability functions (R(t)) were derived for several missile components (A, B, C, H, and G) to predict their performance over time, with time measured in years. These functions allow engineers to estimate how many components will remain functional after a specific period of storage under normal stress conditions.

6 Normalization of the Failure Probability for Specific Components

From a probabilistic perspective, a missile system's operational lifespan depends on its constituent components' longevity. The system lifespan can therefore be characterized by weighting each component's longevity probability at a given reliability threshold. Here, the "probability weight" represents the normalized value of the failure probability for each component within a given time period. The calculation formula is as follows:

The failure probability weighting approach addresses practical limitations of alternative reliability models: unlike the "barrel theory" that overemphasizes the weakest component, it quantifies each component's relative contribution to system failure, reflecting real-world scenarios where non-critical components have minimal impact.

Compared to holistic system-level modelling, this method leverages component-specific inspection data, improving accuracy when system-level data are scarce. Uncertainty is mitigated through order-preserving regression (correcting failure rate inversions) and goodness-of-fit testing (validating distribution assumptions), ensuring robust estimates even with incomplete data—this practical focus makes the method more adaptable to missile storage monitoring, where full failure data are rarely available.

7 Estimation of the Storage Life for Missile Systems

A specific missile system is comprised of critical components, namely Component A, Component B, Component C, Component H, and Component G. By utilizing the failure probability data F for these key components under varying storage durations, the "probability weight" ω representing the storage life of each component at a specified reliability level is derived.

The the estimated reliable storage life for a specific type of missile system is illustrated in Fig. 1. When the reliability is set at 0.9, the estimated reliable storage life of the system stabilizes within the range of 11.5 to 13.5 years. With a reliability of 0.95, the estimated reliable storage life stabilizes between 6 and 8 years.

It can be observed from Fig. 1 that when the specified storage life (input) aligns with the estimated reliable storage life of the missile system (output), and when the

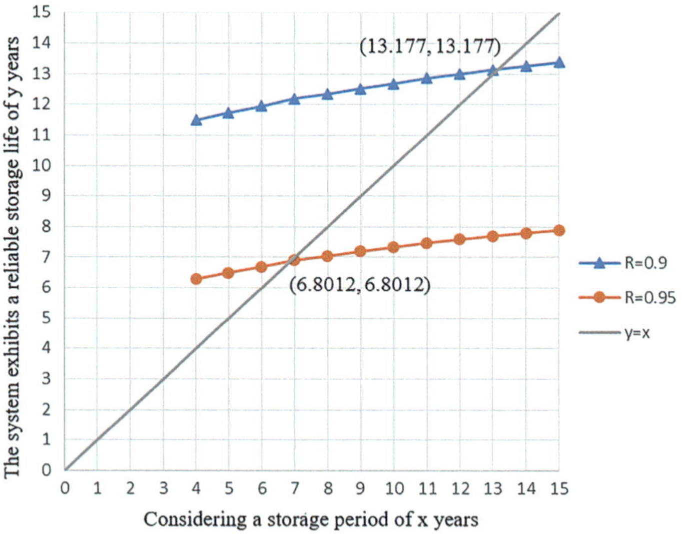

Fig. 1 Estimated reliable storage life of the system

reliability is set at 0.9, the estimated reliable storage life for a specific type of missile system is approximately 13.177 years.

8 Conclusion

(1) Based on the "failed and partially failed" component data collected through missile quality monitoring, an order-preserving regression algorithm was developed to process the inverted failure rate data. A mathematical sampling model was constructed to accommodate both complete and incomplete failure data, and the storage reliability distribution function of missile system components was derived using the minimum A test method. The overall storage life of the system is characterized by the probability-weighted storage life of each component under a specified reliability threshold. Applying this approach to a specific type of missile system stored for a duration of nine years, and setting the reliability level at 0.9, the estimated reliable storage life was determined to be 13.177 years. Further studies can be referred to [13–19].

(2) From a statistical perspective, the operational lifespan of a missile system is directly influenced by the longevity characteristics of its constituent components. As part of the investigation into storage life evaluation methodologies for missile systems, this study introduces a method for characterizing system-level storage life based on the failure probability weights of individual components. This method provides a quantitative representation of the extent to which component failure rates impact the overall storage life of the missile system.

References

1. Niu Y-T, Zhao H, Li Y (2020) Sampling method of high value ammunition based on weibull life distribution. Acta Armamentarii 41(7):7
2. Zhao X-D, Mu X-H (2020) Storage life evaluation of accelerometer based on accelerator factor coefficient of variation. Acta Armamentarii 41(6):1227–1235
3. Zhang E (2018) Prediction of ammunition storage reliability based on particle swarm optimization BP neural network. Shenyang Ligong University
4. Zhao X-D, Mu X-H (2019) Optimization design method of accelerometer's constant-life test based on sequential statistic. J Chin Inert Technol 27(4):553–560
5. Ling MH (2019) Optimal design of simple step-stress accelerated life tests for ONE-Shot devices under exponential distributions. Probab Eng Inf Sci 33(1):121–135
6. Wang H, Xu T, Mi Q (2015) Lifetime prediction based on Gamma processes from accelerated degradation data. Acta Aeronautica et Astronautica Sinica (English Edition) (1)
7. Niu Y-T, Mu X-H, Jiang Z-B (2014) Research on environmental adaptability of a guided rocket. Equip Environ Eng 11(1):86–92
8. Xuan W (2011) Equipment and Natural Environment Test. Aviation Industry Press
9. Pecht Associates, Martinelli J (1995) Effect of long-term storage on electronic devices. U.S. Army Armament Research, Development and Engineering Center 11

10. Howard TW, Davis GD (1997) Challenges in missile life cycle system engineering. Missile Research, Development, and Engineering Center 8
11. Niu Y-T, Mu XH, Jiang Z-B, Yang Z-H (2014) Storage life assessment of an accelerometer under natural storage environment. J Chin Inert Technol 22(4):552–556
12. Wang C-A, Niu Y-T, Mu X-H, Luo G (2016) Prediction of storage life and accelerated-life-testing time for gyroscope. J Chin Inert Technol 24(5):683–687
13. Guo J, Wang Z, Li H, Yang Y, Huang CG, Yazdi M, Kang HS (2024) A hybrid prognosis scheme for rolling bearings based on a novel health indicator and nonlinear Wiener process. Reliab Eng Syst Saf 245:110014
14. Li H, Yazdi M, Huang HZ, Huang CG, Peng W, Nedjati A, Adesina KA (2023) A fuzzy rough copula Bayesian network model for solving complex hospital service quality assessment. Complex Intell Syst 9(5):5527–5553
15. Li H, Peng W, Adumene S, Yazdi M (2023) An intelligent cost-based consequence model for offshore systems in harsh environments. In: Intelligent reliability and maintainability of energy infrastructure assets, pp 107–117. Springer Nature Switzerland, Cham
16. Li H, Yazdi M (2022) What are the critical well-drilling blowouts barriers? A progressive DEMATEL-game theory. Advanced decision-making methods and applications in system safety and reliability problems: approaches, case studies, multi-criteria decision-making, multi-objective decision-making, fuzzy risk-based models. Springer International Publishing, Cham, pp 29–46
17. Li H, Peng W, Adumene S, Yazdi M (2023) Cutting edge research topics on system safety, reliability, maintainability, and resilience of energy-critical infrastructures. In: Intelligent reliability and maintainability of energy infrastructure assets, pp 25–38. Springer Nature Switzerland, Cham
18. Li H, Peng W, Adumene S, Yazdi M (2023) Advances in failure prediction of subsea components considering complex dependencies. In: Intelligent reliability and maintainability of energy infrastructure assets, pp 93–105. Springer Nature Switzerland, Cham
19. Li H, Yazdi M (2022) Reliability analysis of correlated failure modes by transforming fault tree model to Bayesian network: a case study of the MDS of a CNC machine tool. Advanced decision-making methods and applications in system safety and reliability problems: approaches, case studies, multi-criteria decision-making, multi-objective decision-making, fuzzy risk-based models. Springer International Publishing, Cham, pp 15–28

Research on Internal Stress Damage Detection Strategies for Optical Components

Chendong Lei, Peng Yang, Xinyu Luo, Han Yu, and Guanjun Liu

Abstract Recent research in optical systems indicates a trend toward increasingly complex architectures, extreme dimensional requirements, and elevated performance standards. Consequently, optical components are expected to satisfy more rigorous functional and structural criteria. While traditional fabrication methods have predominantly emphasized surface quality and form accuracy, the evolution of internal stresses during both manufacturing and operational phases exerts a substantial influence on processing outcomes and damage mechanisms. Therefore, the accurate detection and evaluation of residual stress have become essential for enabling high-performance optical fabrication and for enhancing the operational longevity of optical components.

Keywords Optical components · 3d stress · Damage detection

1 Introduction

In recent years, the rapid progression of space-based astronomical optics and satellite remote sensing technologies has driven optical systems toward unprecedented complexity. This complexity is characterized by an increasing number of optical components and the expansion of single-mirror aperture sizes, which impose heightened demands on manufacturing capabilities and introduce new fabrication challenges. Traditional performance metrics such as surface quality and form accuracy are no longer adequate to meet the rigorous standards of modern optical systems.

C. Lei · P. Yang (✉) · X. Luo · H. Yu · G. Liu
College of Intelligence Science and Technology, National University of Defense Technology, Changsha, China
e-mail: nudtyp7894@163.com

C. Lei · P. Yang · X. Luo · G. Liu
National Key Laboratory of Equipment State Sensing and Smart Support, National University of Defense Technology, Changsha, China

H. Li et al. (eds.), *Data-Driven Methods for Reliability and Safety Engineering: Applications in Industrial Systems*, Springer Series in Reliability Engineering,
https://doi.org/10.1007/978-3-032-22873-4_35

Throughout the fabrication and operational lifecycles of optical components, internal stress generation and evolution remain constant. In high-power optical systems, residual stress, which refers to self-equilibrating internal stress, accumulates due to transient high-energy beam exposure, uneven mechanical loads, and thermal gradients. Recent developments in laser material properties [1], pulse stretching and compression technologies [2], and increased laser output power and single-pulse energy [3] have made high-intensity lasers indispensable tools for advanced scientific research and technological innovation. However, these extreme irradiation conditions present significant challenges for the structural integrity and performance of optical components.

Residual stress exhibits strong correlations with essential optical parameters such as refractive index, transmittance, and absorption. Gallais et al. [4] demonstrated that the spatial distribution of residual stress significantly affects optical component degradation when exposed to high-energy laser environments. Sun et al. [5] supported this by combining simulation and experimentation to uncover the intrinsic relationship between stress evolution in fused silica and the operational lifespan of components under high-intensity laser conditions. These findings emphasize the role of residual stress as a critical supplementary metric for performance evaluation and lifespan estimation, offering advantages beyond conventional indicators. However, characterizing residual stress remains inherently difficult due to its invisibility to direct observation, its manifestation across multiple spatial scales ranging from nanometers to millimeters, and the current inability to perform real-time monitoring. These limitations restrict mitigation efforts to pre-fabrication stages [6].

This area of study has attracted widespread international attention. Photoelasticity-based techniques such as heterodyne interferometry [7], phase-shifting interferometry [8], laser self-mixing interferometry (LSMI) [9], and cavity ring-down (CRD) methods [10, 11] have demonstrated high precision in two-dimensional stress detection. LSMI measures birefringence by increasing the frequency detuning of the laser cavity while maintaining single-longitudinal-mode operation [9]. This setup allows for birefringence quantification in test samples placed within the external cavity and determines stress through the inverse relationship between birefringence and internal stress. The phase difference obtained from the tuning curves of the LSMI output directly corresponds to the birefringence of the external cavity. The method is recognized for its simple design, low cost, and broad measurement range. Nonetheless, stress measurement accuracy is limited by the precision of phase demodulation.

Both Laser Feedback Interferometry (LFI) and LSMI rely on interference signals produced by light reflected from external surfaces [12]. Their main distinction lies in the location of interference, with LFI occurring inside the laser cavity and LSMI occurring externally. Despite this difference, LFI measurements are limited in accuracy due to fluctuations in laser frequency. CRD, as a birefringence interferometric technique, estimates ultrahigh reflectivity birefringence by evaluating beat frequencies [10]. In 2018, Xiao et al. [11] applied polarized CRD to measure stress-induced birefringence in fused silica substrates, achieving a repeatability in optical path difference of 2.4×10^{-4} nm and a spatial resolution of 0.01 mm. However, CRD is challenged by the simultaneous presence of fundamental and higher-order modes in the resonant cavity, which leads to overlapping decay signals. This overlap causes deviations from a single-exponential decay curve and results in considerable measurement error. Furthermore, the spatial resolution of CRD remains relatively limited compared to other available stress measurement techniques [13].

Most residual stress detection methods are limited to two-dimensional assessments and cannot perform stratified measurements of internal stress gradients, resulting in insufficient resolution and spatial localization accuracy. For high-energy optical systems, 3D stress monitoring is critical to ensuring long-term stability of radiation-exposed components. Their inability to achieve high-resolution 3D stress localization severely limits applicability in next-generation high-energy optical systems where 3D stress monitoring is critical for long-term stability.

In summary, the presence of internal residual stresses within optical components can impact optical systems to varying degrees. The study investigates residual stress formation and evolution in optical components. Subsequently, it proposes a characterization and detection technique for three-dimensional internal stresses based on white-light interferometry. The method improves measurement accuracy for large-aperture optics compared to conventional approaches. Finally, future research directions for internal residual stress detection in optical components are outlined.

2 Research on Internal Stress in Optical Components Based on Optical Field Modulation Phenomena

With the advancement in high-precision optical components, internal stress variations induced by small-scale defects have emerged as a critical criterion for damage assessment in optical elements. By investigating nonlinear optical phenomena triggered by micro-damage, this study examines variations in internal optical field characteristics under defects of different scales through the theoretical framework of Mie scattering. The energy disparities caused by optical field modulation phenomena result in non-uniform temperature variations across distinct regions of the component, thereby generating discrepancies in stress magnitudes and distribution patterns within the optical element.

The optical field intensity distributions induced by internal bubbles and surface scratches in fused silica were simulated using the Finite-Difference Time-Domain (FDTD) method. A set of bubble defect models with varying centre-to-surface distances (as shown in Fig. 1a) and a set of bubble defect models with varying diameters (as shown in Fig. 1b) were set up inside the fused quartz. Additionally, a set of scratch models with varying scratch apex angles (as shown in Fig. 1c) was set up on the surface of the fused quartz. Numerical simulations were performed over a duration of 1 ms (see parameters in Table 1, and refer to Figure refer to Fig. 1).

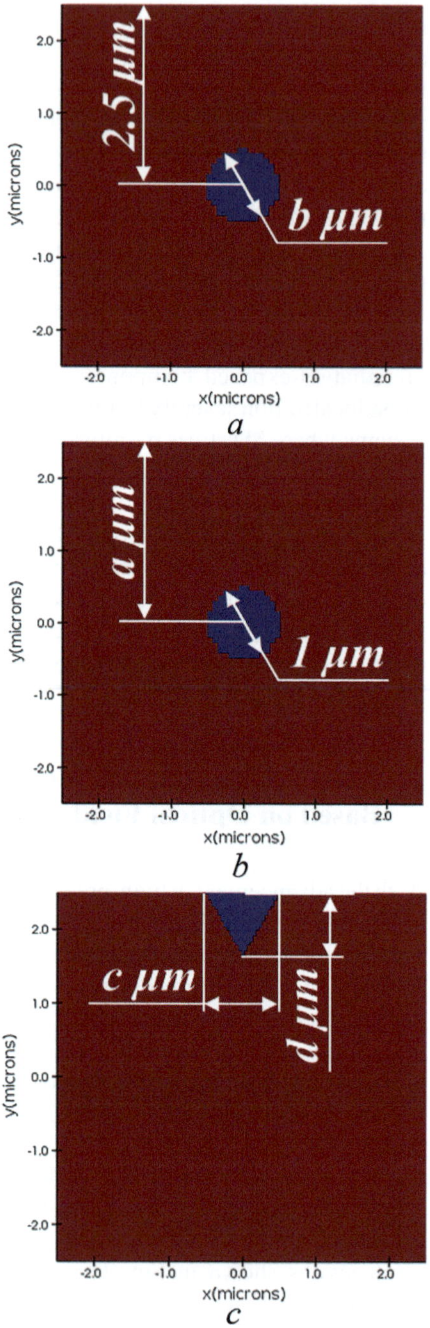

Fig. 1 **a** Simulation model of spherical bubble with different centre-to-surface distances, **b** Simulation model of spherical bubble with different size of diameters, **c** Simulation model of scratch with different scratch apex angles

Table 1 FDTD simulation frame work

Parameter	Value
Material	Fused silica (SiO_2)
Defect type	Subsurface bubble/Surface scratch
Laser profile	Flat-top ($M^2 < 1.1$)
Wavelength	1064 nm
Simulation duration	1 ms

Simulations show the light field peaks beneath bubble-type defects' lateral surfaces. As the bubble aperture increases, the modulation effect is enhanced, with the peak intensity increasing from 1.23 to 1.42 V/m. In contrast, as shown in Fig. 2b, as the bubble depth increases, the peak intensity slightly increases with increasing bubble depth. Additionally, as the scratch apex angle increases from 30° to 90°, the maximum peak intensity increases from 1.35 to 1.40 V/m (Fig. 2c). As shown in Table 2, the peak intensity values induced by the two types of defects are comparable, but their intensity distribution patterns exhibit significant differences. After laser irradiation of several groups of defects, the component's light absorption causes an increase in internal material temperature. Due to the different intensity distribution patterns caused by the two defect types, the resulting thermal distributions are also expected to exhibit similar differences, refer to Fig. 2.

According to the fundamental principles of thermodynamics, when a component's temperature rises below its softening point, the extremely high viscosity of the material allows stresses induced during heating to remain trapped within the material, forming residual stresses.

Further, when the modulation effects caused by several groups of defects are accumulated along the x-direction distribution (Fig. 3), it can be observed that the modulation effects from various defects exhibit irregular fluctuations, making it difficult to classify different defects based on the curve distribution patterns. Based on this theory, three-dimensional stress detection within the component enables effective localization and identification of defects, refer to Fig. 3.

3 A Large-Scale, High-Precision, Three-Dimensional Stress Detection Strategy

Current large-scale, high-precision optical component inspection technologies exhibit significant limitations, including restricted sample diversity, high environmental sensitivity of precision measurements, elevated costs and equipment complexity, and intricate data analysis. To address these challenges, a novel large-scale, high-precision three-dimensional (3D) stress detection strategy is proposed.

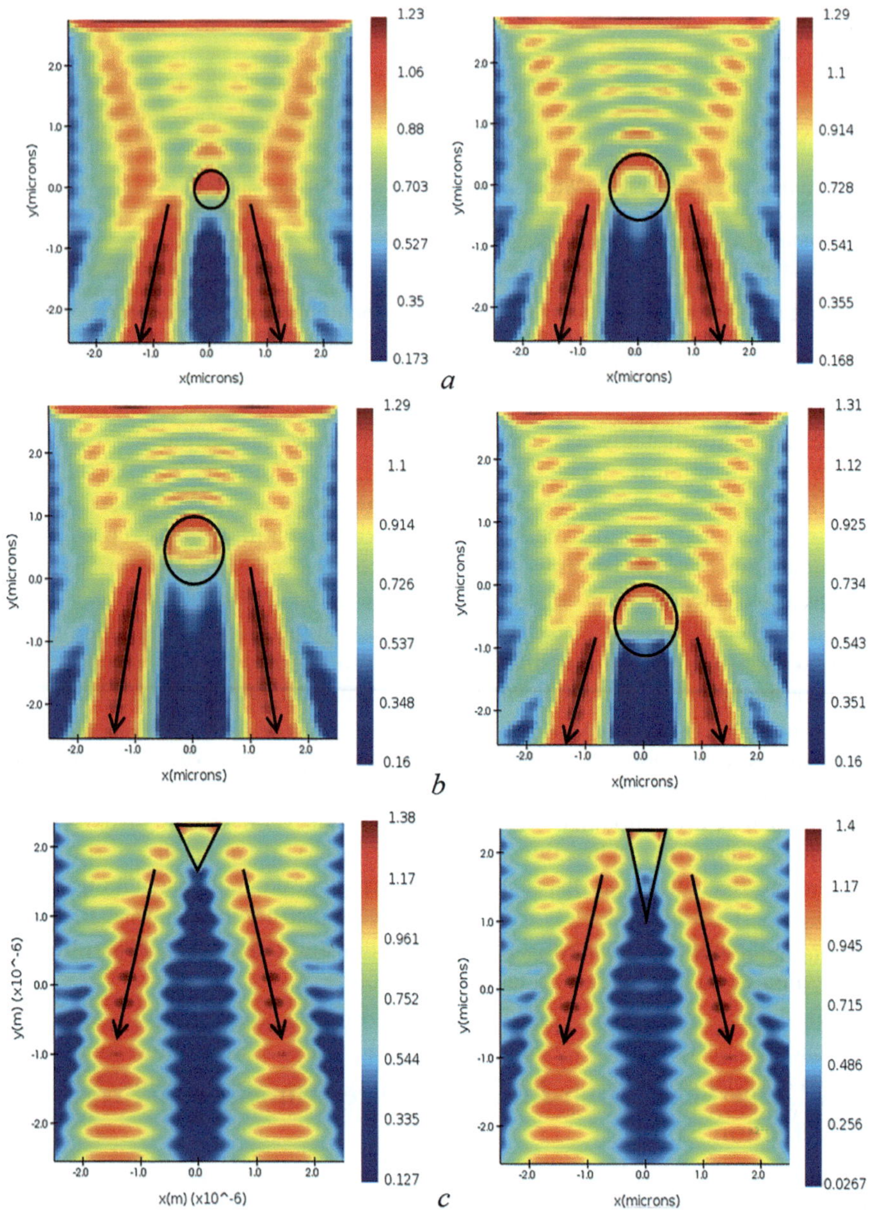

Fig. 2 **a** Simulation solution of spherical bubble with different centre-to-surface distances, **b** Simulation solution of spherical bubble with different size of diameters, **c** Simulation solution of scratch with different scratch apex angles

Table 2 FDTD simulation results for different defects

Modles	Types of defects	Defect dimensions		Max light intensity (V/m)
		a/c	b/d	
B-1	Spherical bubble	2.00	1.00	1.29
B-2		2.50	1.00	1.29
B-3		3.00	1.00	1.31
B-4		2.50	0.50	1.23
B-5		2.50	1.50	1.42
S-1	Scratch	0.73	1.36	1.40
S-2		1.00	0.87	1.38
S-3		1.50	0.71	1.35

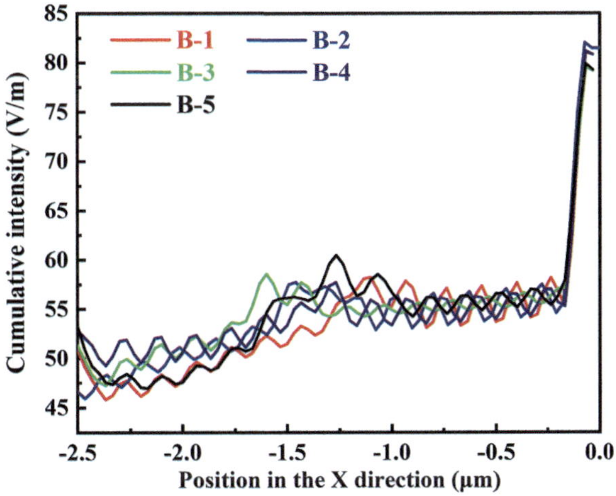

Fig. 3 FDTD simulation of light field distribution along the X direction for different bubble defects

To address these challenges, we propose a novel large-scale high-precision three-dimensional stress detection system. The system was developed based on the cross-orthogonal reflective photoelastic apparatus [14] proposed by the Shanghai Institute of Optics and Fine Mechanics. The detection principle of this device is based on the coherent length of a broadband light source to achieve three-dimensional stress detection. $n_k^H(x,y)$ and $n_k^J(x,y)$ respectively represent the principal refractive indices of the k-th layer in the radial and circumferential directions. The sample is illuminated by linearly polarized light with a complex amplitude modulus of E_0. The two vibration components of the reflected light field at the interface between the k-th and (k + 1)-th layers.

Thus, the three-dimensional stress distribution within the element can be inversely solved using the light intensity information captured by the detector.

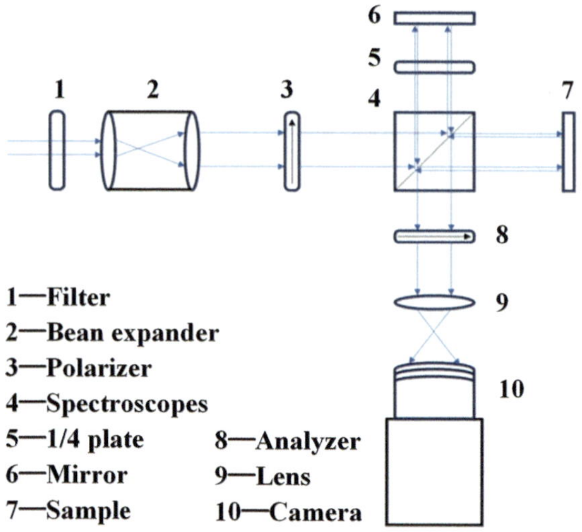

Fig. 4 Schematic diagram of the optical path of the three-dimensional stress detection device

Based on the cross-orthogonal reflective photoelastic apparatus [14], we built the three-dimensional stress detection system as illustrated in Fig. 4. The optical configuration emulates a Michelson interferometric path, incorporating spectral modulation of a broad-spectrum light source to regulate its bandwidth, thereby enabling precision control over 3D tomographic detection. Polarizers and analyzers were integrated to achieve dark-field optical path control. Compared to the SIOM prototype, our system demonstrates enhanced large-scale detection capability and measurement accuracy, refer to Fig. 4.

Optical measurement errors are significantly influenced by environmental factors and equipment precision, where minor disturbances may lead to substantial deviations. The developed detection system incorporates advanced noise reduction algorithms to mitigate these errors. Specifically, wavelet denoising and Gaussian filtering methods have been implemented to eliminate fundamental errors. Concurrently, deep learning techniques are employed to suppress residual errors and amplify subtle information acquired during detection, ultimately achieving error correction in photoelastic detection methods.

Furthermore, the inherent conflict between high-precision and large-scale detection persists in optical metrology. To reconcile this duality, a scanning strategy combining high-precision instrumentation with optimized path planning and regional data fusion is proposed. The dual-helix scanning algorithm (Fig. 5) improves edge detection by adaptively adjusting primary helix spacing in central areas and secondary spacing in peripheral regions. Compared to conventional raster scanning, this method achieves a 3 × improvement in edge detection rate (SNR $\geq$ 15 dB) for 100 mm-diameter components, while reducing total scan path length by 10.3% and scan time by 12–15%. Subsequent high-precision image stitching via feature point matching further ensures data integrity, refer to Fig. 5, and 6.

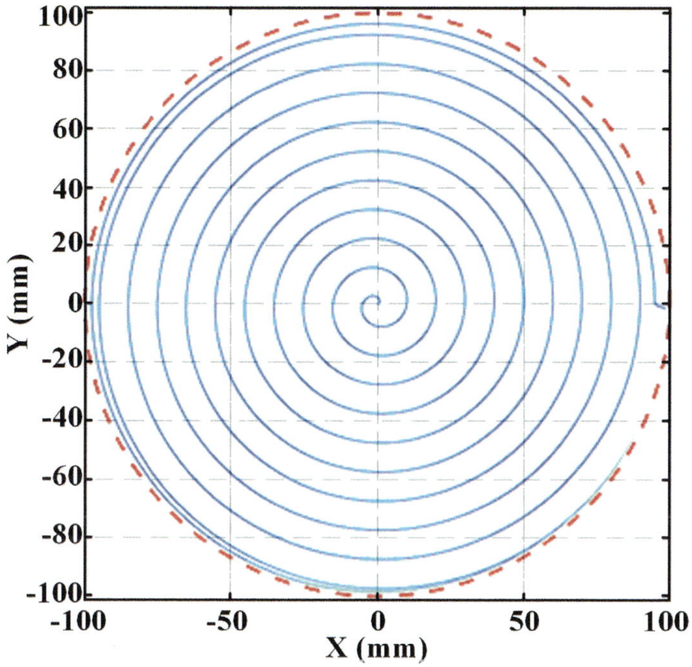

Fig. 5 The trajectory of dual-helix scanning algorithm

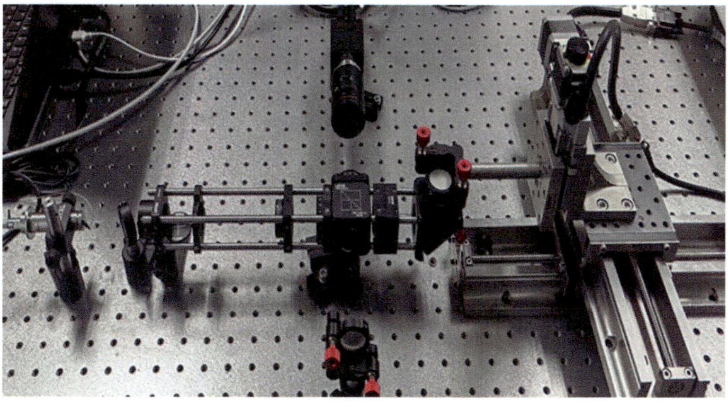

Fig. 6 Schematic diagram of the physical setup of the three-dimensional stress detection device

Current limitations in depth-resolved detection hinder comprehensive 3D stress characterization within optical components, particularly for internal residual stress distribution. To address weak-signal 3D detection, a hybrid approach combining numerical simulations with advanced 3D optical path design is implemented. By analyzing residual stress distribution patterns under varied damage scenarios and developing correlation-driven interpolation algorithms for weak signals, this methodology reduces dependence on direct depth-resolved data. The finalized experimental setup is depicted in Fig. 6, and the main components of the device are listed in Table 3.

To validate the proposed method, experimental measurements were conducted on fused silica samples with prefabricated scratch defects (model S-2).We comparing FDTD-simulated optical field gradients with experimentally measured stress gradients using our 3D stress detection system (as shown in Fig. 7). The FDTD simulations reveal that the scratch induces localized light intensity modulation with a peak gradient of 0.14 V/m/μm along the scratch depth direction (z-axis), concentrated at the 1.2 μm subsurface layer. This optical field distortion corresponds to a stress concentration zone predicted by thermo-mechanical coupling models, where the maximum shear stress difference $\Delta\tau(x, y)$ reaches 3.2 MPa. Experimental validation employs the white-light interferometry-based detection system, reconstructing the 3D stress field via layer-wise inversion of reflected light intensity. The measured stress gradient for scratch S-2 shows a strong spatial correlation (R2 = 0.89) with the simulated optical field gradient, with the peak stress gradient (2.9 MPa/μm) located at the scratch tip—consistent with the FDTD-predicted light intensity concentration. Minor discrepancies (< 8%) in gradient magnitude arise from axial resolution limitations (> 50 μm depth) and environmental vibrations, as quantified in the error-budget analysis. This benchmark case demonstrates the reliability of the proposed 3D stress detection strategy in capturing defect-induced stress and light-field modulation patterns.

Table 3 The main components of the three-dimensional stress detection device

Components	Main parameters
Filter	Filterable light wavelength: 700–900 nm
Bean expander	Magnification 2x
Camera	5120 × 5120
Light	Spectral range covers 410–2400 nm, with a maximum power of up to 7 W
Platform	Linear displacement accuracy 0.5 μm

Fig. 7 Measured stress gradients by 3D stress detection system

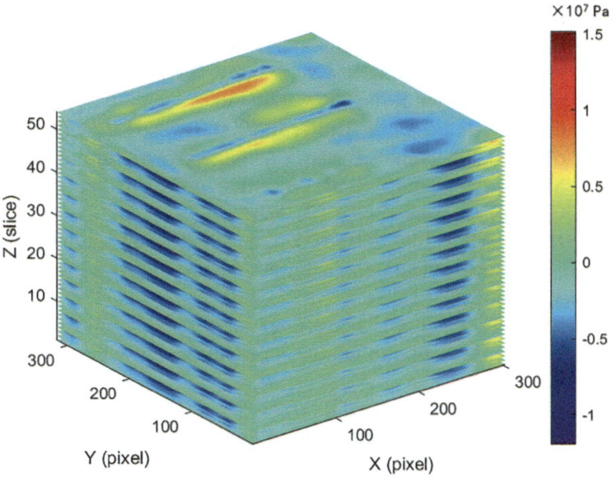

4 Conclusion

High-power optical components inherently develop residual stresses within their structures during both fabrication and operational phases due to a variety of influencing factors. These internal stresses substantially impair component performance when exposed to laser irradiation. Consequently, the advancement of high-precision detection techniques for internal residual stresses is critical to extending component service life and ensuring operational reliability. Nonetheless, existing detection methods exhibit considerable limitations. The Finite Difference Time Domain (FDTD) approach was employed to analyze stress evolution in defective optical components. The analysis focused on three categories of defects, including spherical bubble defects with varying diameters, spherical bubble defects located at different depths, and surface scratches with distinct apex angles. All defect types were examined to assess the modulation behavior of the optical field under laser exposure. The results indicated comparable optical field distributions in the planar dimension, while demonstrating significant differences along the depth axis. These findings underscore the necessity and practical relevance of three-dimensional stress detection in identifying and mitigating internal defects within optical components. In response to these observations, a novel three-dimensional stress detection technique is proposed that offers efficient and accurate characterization of internal stress distributions throughout the volume of optical elements. Further studies can be referred to [15–21].

References

1. Chen C, Sun A, Ju B et al (2022) Width and depth gauging of rectangular subsurface defects based on all-optical laser-ultrasonic technology. Appl Acoust 191:108684

2. Zhao D, Cui J, Bian X (2021) Overview of subsurface damage detection technologies for ultra-smooth quartz components. In: IEEE 15th ICEMI, pp 364–369
3. Dwivedi S, Vishwakarma M, Soni P (2018) Advances and researches on non destructive testing: a review. Mater Today 5(2):3690–3698
4. Gallais L, Cormont P, Ruller J (2009) Investigation of stress induced by CO_2 laser processing of fused silica optics for laser damage growth mitigation. Opt Express 17(26):23488–23501
5. Sun S, Qi N, Kong Y et al (2021) Three-dimensional stress fields of laser damaged fused silica. Chinese J Lasers 48(1):0101001
6. Hu Y, Wang L, Xue C (2024) Prediction of residual stress in precision injection molding of plastic optical lenses based on an improved solidification layer model. Opt Express 32(21):37716–37731
7. Liang H, Chen M, Jiang C et al (2022) Combined feature extraction and random forest for laser self-mixing vibration measurement without determining feedback intensity. Sensors 22(16):6171
8. Li K, Wang S, Liu Z et al (2023) Stress birefringence measurement based on double cascaded photoelastic modulation with differential frequencies. Acta Opt Sinica 43(4):0412001
9. Dai X, Yun H, Shao X et al (2018) Thermal residual stress evaluation based on phase-shift lateral shearing interferometry. Opt Laser Eng 105:182–187
10. Niu H, Niu Y, Li J et al (2017) Measurement of stress-induced birefringence in glasses based on reflective laser feedback effect. Opt Eng 56(2):024104
11. Wang T, Li B, Xiao S et al (2023) Accurate thermal stress induced birefringence measurement of laser crystal with cavity ring-down technique. Opt Commun 526:128925
12. Li J, Niu Y, Niu H (2016) Frequency difference stabilization in dual-frequency laser by stress-induced birefringence closed-loop control. App Opt 55(16):4357–4361
13. Fleisher A, Long D, Liu Q et al (2016) Precision interferometric measurements of mirror birefringence in high-finesse optical resonators. Phys Rev A 93(1):013833
14. Qi N, Sun S, Zhang L et al (2020) Microscopic three-dimensional inner stress measurement on laser induced damage. Opt Express 28(16):24253
15. Guo J, Wang Z, Li H, Yang Y, Huang CG, Yazdi M, Kang HS (2024) A hybrid prognosis scheme for rolling bearings based on a novel health indicator and nonlinear Wiener process. Reliab Eng Syst Saf 245:110014
16. Li H, Yazdi M, Huang HZ, Huang CG, Peng W, Nedjati A, Adesina KA (2023) A fuzzy rough copula Bayesian network model for solving complex hospital service quality assessment. Complex Intell Syst 9(5):5527–5553
17. Li H, Peng W, Adumene S, Yazdi M (2023) An intelligent cost-based consequence model for offshore systems in harsh environments. In: Intelligent reliability and maintainability of energy infrastructure assets, pp 107–117. Springer Nature Switzerland, Cham
18. Li H, Yazdi M (2022) What are the critical well-drilling blowouts barriers? A progressive DEMATEL-game theory. Advanced decision-making methods and applications in system safety and reliability problems: approaches, case studies, multi-criteria decision-making, multi-objective decision-making, fuzzy risk-based models. Springer International Publishing, Cham, pp 29–46
19. Li H, Peng W, Adumene S, Yazdi M (2023) Cutting edge research topics on system safety, reliability, maintainability, and resilience of energy-critical infrastructures. In: Intelligent reliability and maintainability of energy infrastructure assets, pp 25–38. Springer Nature Switzerland, Cham
20. Li H, Peng W, Adumene S, Yazdi M (2023) Advances in failure prediction of subsea components considering complex dependencies. In: Intelligent reliability and maintainability of energy infrastructure assets, pp 93–105. Springer Nature Switzerland, Cham
21. Li H, Yazdi M (2022) Reliability analysis of correlated failure modes by transforming fault tree model to Bayesian network: a case study of the MDS of a CNC machine tool. Advanced decision-making methods and applications in system safety and reliability problems: approaches, case studies, multi-criteria decision-making, multi-objective decision-making, fuzzy risk-based models. Springer International Publishing, Cham, pp 15–28

Study of Properties of High Zn Content Aluminium Alloys for Thin Wall Deep Cavity Casting

Guiming Tao, Yi Wang, and Anhang Liu

Abstract An aluminum alloy with a high Zn content (Zn > 4%) was developed to address the casting requirements of thin-walled and deep-cavity structural components. The alloy was modified and optimized through the addition of trace rare earth elements and amorphous alloys, and it was prepared using the induction electromagnetic stirring process. These enhancements significantly improved the alloy's strength, corrosion resistance, and casting performance. A systematic evaluation of the alloy's mechanical and thermal properties was conducted using multi-scale simulation methods in the context of lightweight backpack radio design. The mechanical simulation results indicate that the resonant frequency of the optimized aluminum alloy component is effectively decoupled, and its impact resistance meets the 40 g sawtooth wave requirement. However, the stiffness of both the power supply module and the integrated service module requires further improvement. Thermal simulation findings show that after one hour of continuous operation at room temperature, the shell temperature remains below 67 °C. During sustained high-temperature transmission and reception, the internal device temperature, including the junction temperature of the power amplifier tube, reaches 119.5 °C. Both values remain within acceptable tolerance thresholds, fulfilling the heat dissipation and reliability requirements. Overall, the results demonstrate that this high-Zn aluminum alloy exhibits excellent comprehensive performance and is well suited for the lightweight fabrication of thin-walled and deep-cavity complex structures, thereby offering a new material solution for aerospace and advanced equipment applications.

Keywords High Zn content aluminum alloy · Thin-walled deep cavity casting · Lightweight design · Electromagnetic stirring

---

G. Tao · Y. Wang · A. Liu (✉)
Army Engineering University of PLA, Shijiazhuang, China
e-mail: yeka1418@163.com

G. Tao
e-mail: jnksh@163.com

Y. Wang
e-mail: wangyi050926@163.com

H. Li et al. (eds.), *Data-Driven Methods for Reliability and Safety Engineering: Applications in Industrial Systems*, Springer Series in Reliability Engineering,
https://doi.org/10.1007/978-3-032-22873-4_36

1 Introduction

Cast aluminium alloys are widely recognised as essential lightweight materials in equipment manufacturing due to their favourable attributes such as low density, corrosion resistance, processability, and cost efficiency (e.g [1–4].). The application of cast aluminium alloys through foundry processes is a fundamental aspect of achieving structural light-weighting in engineering equipment. The performance characteristics of aluminium alloys are primarily influenced by their chemical composition and manufacturing process. Common casting aluminium alloy systems include Al–Si, Al–Cu, Al–Mg, and Al–Zn alloys. The addition of various alloying elements can significantly enhance specific material properties.

The silicon content predominantly influences the alloy's fluidity. Copper contributes to high hardness and toughness by forming θ phase (Al_2Cu) and S phase (Al_2CuMg) with aluminium, making it suitable for aerospace applications (e.g [5].). Magnesium forms nanoscale precipitate phases (Mg_2Si) with silicon and enhances material strength through ageing treatments (e.g [6].). Zinc, in conjunction with magnesium and copper, facilitates the formation of T phase ($Al_2Mg_3Zn_3$) and η phase ($MgZn_2$), providing tensile strength in the range of 500–600 MPa along with superior hardness and wear resistance. These characteristics render Al-Zn alloys particularly suitable for aerospace and aviation applications. Additional trace elements such as Mn, Zr, Cr, Ni, and Sn further refine alloy performance (e.g [7].). Careful regulation of trace element types and contents, generally below 1%, enables an optimal balance among strength, toughness, corrosion resistance, and processing properties.

In a representative study, Tengteng Sun et al. investigated the thermal stability and mechanical behaviour of Zr- and Mn-microalloyed Al-Cu alloys at temperatures between 573 and 673 K. The addition of 0.2% Mn and 0.2% Zr markedly enhanced the stability of the θ′ strengthening phase and inhibited its coarsening. The resulting ACMZ alloy exhibited a retained strength of (88.6 ± 8.8) MPa, significantly surpassing that of the base AC alloy with no microalloying additions, which recorded (32.5 ± 0.8) MPa. Additionally, the ACMZ alloy maintained high hardness and ductility after 200 hours of thermal exposure at 573 K (e.g [8].). However, this study primarily focused on tensile strength, hardness, and elongation, without addressing other critical performance metrics such as high-temperature fatigue, creep resistance, and fracture toughness.

Munda et al. explored the mechanisms of iron accumulation in recycled Al–Si alloys and proposed effective control strategies. Their findings indicated that the formation of the needle-like β-Al_5FeSi phase due to iron significantly degraded the mechanical and corrosion properties of the alloy. Modifying the morphology of the Fe-rich phases by introducing neutralising elements such as Mn, Mg, and rare earths, in combination with ultrasonic melt processing and cooling rate control, was shown to effectively suppress the formation of deleterious phases and enhance the overall alloy performance (e.g [9].). Nonetheless, the study noted that traditional heat treatment methods had a limited effect on improving Fe phase morphology. It did not evaluate novel heat treatment techniques or offer a systematic analysis of the trade-offs among mechanical properties, corrosion resistance, and casting performance.

Aluminium alloy preparation typically involves sequential steps: raw material selection/pretreatment, melting, refining/degassing, casting, heat treatment, and surface treatment. Modern processes have incorporated semi-solid forming and additive manufacturing technologies (e.g [10–13].). Melting is the process of loading prepared raw materials into a melting furnace in a certain order, heating and melting the raw materials with appropriate power, and continuously stirring the melt to promote uniform dissolution of alloying elements and improve compositional homogeneity. Stirring is a key step to ensure uniformity of composition and reduce segregation; commonly used stirring methods are mechanical stirring, gas stirring, and electromagnetic stirring (e.g., [14]). Mechanical stirring uses graphite stirring paddles or stainless steel stirrers, and the paddles are driven manually or by a mechanical arm to stir the melt at a speed of 30–60 rpm, which is low-cost, intuitive to operate, and controllable in terms of stirring intensity, but the paddles may contaminate the melt, and there are safety risks in manual operation. Gas stirring is through the bottom of the furnace permeable brick, blowing into the argon/nitrogen (flow rate of 5–20 L/min), to bubble upward to drive the melt circulation, while achieving the stirring and degassing, energy saving, and high efficiency, but the uneven distribution of the gas may lead to the lack of local mixing. Electromagnetic stirring induces eddy currents in the melt through an alternating magnetic field and uses electromagnetic force to drive the flow of metal liquid, which is suitable for large continuous melting furnaces (e.g [15].). Electromagnetic stirring has the advantages of non-contact stirring, does not contaminate the melt, and the stirring effect is uniform, and the stirring intensity and direction can be precisely controlled (e.g., [16]). It can effectively improve the temperature and composition uniformity in the melting bath, and is particularly suitable for the melting process of high-quality aluminium alloys, such as aerospace aluminium alloys.

The object of this study is an aluminium alloy with high Zn content prepared by an induction electromagnetic stirring process. The main chemical composition of this aluminium alloy is Al, Zn, Si, Cu, Fe, Mn, and related rare metals, with Zn elemental content greater than 4%. The composition is such that the melting point of the alloy is lowered by solid solution zinc, while the shrinkage of the alloy is reduced, which is conducive to the die casting of ultra-thin structures. Through the addition of trace elements to the aluminium alloy modification, the addition of trace rare earth elements to improves the thermal conductivity and strength of the alloy, and the addition of amorphous alloy dispersion strengthening to achieve high strength. Melting process, in the mother alloy melting, the use of vacuum environment to prevent burns, the use of induction electromagnetic force for stirring, through the vacuum to exclude impurity gases within the alloy, so that the said aluminium alloy has a high strength, corrosion resistance, can be welded, easy to cut characteristics, to meet the thin-walled deep cavity products casting molding requirements. With the use of high vacuum aluminium die casting machine, can ensure the denseness of die casting products, improve the strength and thermal conductivity. The alloy described

Table 1 Key performance indicators of the described high Zn content aluminium alloys

Type of performance indicator	Name of Performance Indicator	Indicator parameters
Process performance indicators	20 × 20 flat surface (thin wall moulding) wall thickness (mm)	0.2
	De-moulding slope	0°
Mechanical performance indicators	Hardness (HV)	135
	Yield strength (Mpa)	245
Thermal performance indicators	Tensile strength (Mpa)	360
	Thermal conductivity w/m.k	135

in this study has undergone laboratory performance tests, and the key performance indicator parameters are shown in Table 1.

Compared with alternative materials and processing techniques, the proposed high-Zn aluminum alloy offers distinct advantages: traditional Al–Si alloys exhibit lower tensile strength (280–320 MPa vs. 360 MPa) and poorer corrosion resistance, making them less suitable for thin-walled deep-cavity structures; Al-Cu alloys have higher strength but inferior casting fluidity and greater shrinkage, increasing molding difficulties. In terms of processing, mechanical stirring risks melt contamination and uneven mixing, while gas stirring may cause local voids, whereas induction electromagnetic stirring ensures contamination-free, uniform composition, critical for maintaining structural integrity in ultra-thin (0.2 mm) castings. These comparisons highlight the proposed alloy's superior balance of mechanical properties, casting performance, and processing reliability for lightweight thin-walled deep-cavity components.

In this study, the mechanical, thermal, and EMC properties are simulated to analyse the improvement of the material properties by its composition and preparation process.

2 Simulation Settings and Methods

Lightweight transformation was performed on a backpack radio model without structural modification. Design optimization was achieved using the high-Zn aluminum alloy according to product drawings. The product contains the upper cover, shell, power module bracket, power module lower cover, amplifier module box, module box 1, module box 2, battery cover, a total of 8 components, respectively, 8 parts modelling. According to the performance parameters of the new material, simulation analysis is carried out on the eight component models. The focus is on the mechanical and thermal properties of the product, including the vibration, impact, and fatigue simulation related to the mechanical properties of the product, and the temperature simulation analysis related to the thermal properties.

2.1 Mechanical Simulation

Stiffness simulation is carried out for the radio shell and each functional module. In order to improve the efficiency and accuracy of simulation calculation, it is necessary to simplify the model and establish a regular and coordinated grid model. The simplified simulation mesh model contains 80,845 structural units, refer to Fig. 1, and Table 2.

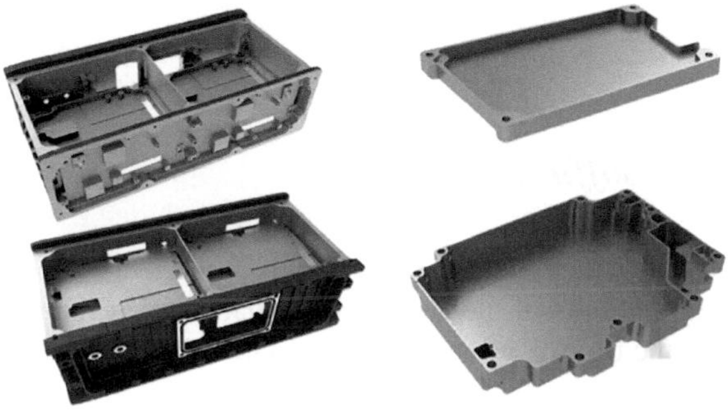

Fig. 1 Simulation model diagram of the radio shell and some modules

Table 2 Main material parameters of radio shell and function modules

Part name	Material	Modulus of elasticity/ GPa	Poisson's ratio	Density g/ cm^3	Damping factor	Yield strength / Mp
Carapace	High Zn aluminium alloy	71	0.33	2.74	0.1	260
Faceplate	High Zn aluminium alloy	71	0.33	2.74	0.1	260
Back cover	High Zn aluminium alloy	71	0.33	2.74	0.1	260
Shielding cover	High Zn aluminium alloy	71	0.33	2.74	0.1	260
Printed board	FR4	24.9	0.15	1.8	0.1	410
Screen	Plexiglass	3	0.37	1.2	0.1	130
Core	Silicon (chemistry)	200	0.3	3	0.1	300
Weld points	Tin-lead solder	16	0.42	1.1	0.1	100

Table 3 Sinusoidal vibration simulation conditions

Frequency range (Hz)	Loading method	Peak acceleration	Loading direction	Scanning time
5–14	Equal displacement	0.25 g9	Three-way	12 min each direction 5 times in each direction
14–33	Equal acceleration	2 g	Three-way	
33–52	Equal displacement	5 g	Three-way	
52–300	Equal acceleration	5 g	Three-way	

2.2 *Stiffness Simulation*

Based on the protection of internal components, shock resistance, and signal stability requirements of the radio, the fundamental resonance frequency of the radio shell and the rigidity of each functional module is calculated by simulation analysis.

2.3 *Impact Strength Simulation*

Based on the use scenario and environmental test requirements of this radio, the three-direction peak 40 g, 11 ms sawtooth wave impact strength injury analysis and calculation are carried out on the radio shell and each functional module, respectively.

2.4 *Sinusoidal Vibration Intensity Simulation*

Based on the needs of this radio, taking into account the loading of vehicles, the simulation analysis of the sine vibration strength of the radio shell and each functional module in three directions is carried out respectively, refer to Table 3.

2.5 *Thermal Simulation*

According to the working characteristics of the radio, thermal simulation analysis is carried out for two working conditions: one hour of long transmission at room temperature and continuous transmission and reception at high temperature (transmission and reception ratio of 9:3) for dual-channel simultaneous transmission.

The model is simplified by removing chamfers, rounded corners, threaded holes, and other structures that do not have much effect on heat dissipation but constrain the simulation calculation. Keeping the assembly relationship and contact area unchanged, and adopting the heat simulation model of the amplifier tube, a simplified heat simulation model is obtained.

The simplified thermal simulation model is imported into Design Modeler in Ansys Workbench, and the parts in the model are converted into geometries that can be recognised by the thermal simulation software Icepak and imported lcepak. Mesh the model, encrypt the mesh at the heat sources such as shaped geometries, amplifier tubes, and other key components in the transmission path, and generate a mesh number of about 6.8 million, refer to Tables 4, 5, and 6.

The material of the radio case is the high Zn content aluminium alloy mentioned in this study, the material of the key is silicone, the material of the protection screen is plastic, the substrate of the printed circuit board is FR4, the power supply chip is modelled according to the thermal resistance parameter provided in the manual, and the internal structure of the co-addressing filter module is known, and it is simplified as a solid body in the simulation, and this simplification will make the simulation results of the co-addressing filter module have a certain degree of bias, but it won't influence the temperature of the whole machine in the simulation.

Table 4 Related environment settings

Temperature	Air velocity	Pneumatic	Gravitational acceleration
Normal temperature 25 °C	0 m/s	1 atmosphere	9.8 m/s^2
High temperature 65 °C			

Table 5 Related environment settings

Temperature	Air velocity	Pneumatic	Gravitational acceleration
Normal temperature 25 °C	0 m/s	1 atmosphere	9.8 m/s^2
High temperature 65 °C			

Table 6 The main heat generating devices of the radio station and their heat consumption and withstand temperature

Name	Power amplifier tube	Power chip	Common address filter module
Heat consumption	22 W × 2	3 W × 2	10 W × 2
Temperature resistance	200 °C	125 °C	125 °C
Total	70 W		

3 Analysis and Conclusion

3.1 Mechanical Simulation Results

3.1.1 Analysis of Stiffness Simulation Results

The results of the simulation calculations of the fundamental resonant frequency of the radio enclosure and the functional module are shown in the Table 7, refer to Fig. 2.

Table analysis indicates stiffness decoupling between the radio shell and functional modules. The adequate resonance frequency interval prevents coupling amplification that could exacerbate resonance effects. However, the integrated business module and power supply module have low stiffness and large deformation during vibration; the resonance frequency of the module is close to the main flap of the sawtooth impact waveform frequency after 11 ms, and the impact resistance is poor,

Table 7 Table of fundamental resonance frequencies of radio housings and functional modules

Department	Fundamental resonant frequency (Hz)
Panel housing	1358
Panel mouldings	295
Channel module	252
Integrated business modules	81
Power supply modules	106

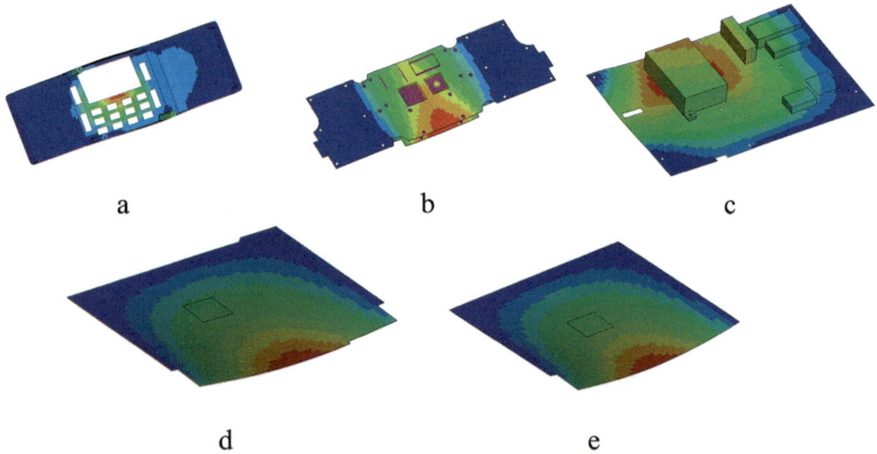

Fig. 2 Vibrating shape of station shell and function module. **a** Panel shell vibration mode, **b** panel module vibration mode, **c** channel module vibration mode, **d** integrated service module vibration mode, **e** power module vibration mode

Table 8 Peak 40 g, 11 ms after peak sawtooth impact simulation results

Place	Directional scale	Maximum stress/Mp
Panel mouldings	Horizontal 40 g	5
	Vertical 40 g	7.7
	Vertical 40 g	2.1
Channel module	Horizontal 40 g	8.8
	Vertical 40 g	78
	Vertical 40 g	1.3
Integrated business modules	Horizontal 40 g	4.4
	Vertical 40 g	78.4
	Vertical 40 g	0.4
Power supply modules	Horizontal 40 g	3.1
	Vertical 40 g	62.5
	Vertical 40 g	0.5

which is a weak link that needs to be further analysed and designed in terms of structural strength.

3.1.2 Analysis of Impact Strength Simulation Results

The simulation results of sawtooth wave impact strength after 11 ms with a peak value of 40 g are shown in Table 8, and refer to Fig. 3.

As shown in the table, maximum stresses in the radio case, transparent window, and module printed boards remained below permitted levels under 40 g impact conditions. The chip solder joints of the integrated service module and power supply module have larger stresses in the vertical direction, as shown in the following Table 9 and refer to Fig. 4.

Analysing the results in the table, the chip solder joints of the integrated services and power supply module are able to withstand a vertical impact of 40 g, refer to Figs. 4, 5, and 6, and Table 10.

3.1.3 Analysis of Simulation Results of Sinusoidal Vibration Intensity

From the results of the table, the maximum stresses of the radio case, transparent window, and the printed boards and other parts of each mould are less than the permissible stresses under sinusoidal vibration conditions.

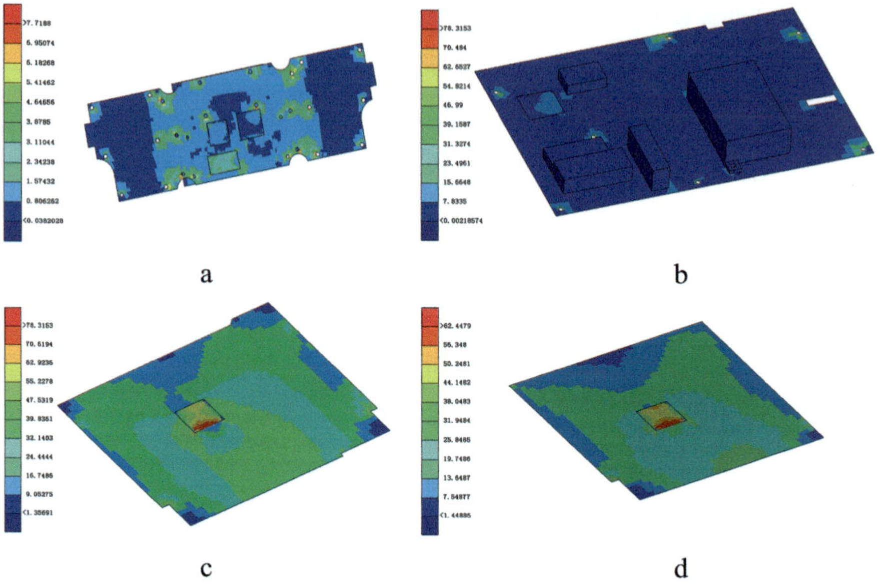

Fig. 3 Simulation cloud of sawtooth wave impact strength after 11 ms at 40 g peak. **a** Panel module vibration mode, **b** channel module vibration mode, **c** integrated service module vibration mode, **d** power module vibration mode

Table 9 Chip solder joint impact simulation calculation results of integrated service module and power supply module

Department	Vertical maximum stress Mp
Integrated business modules	78.4
Power supply modules	62.5

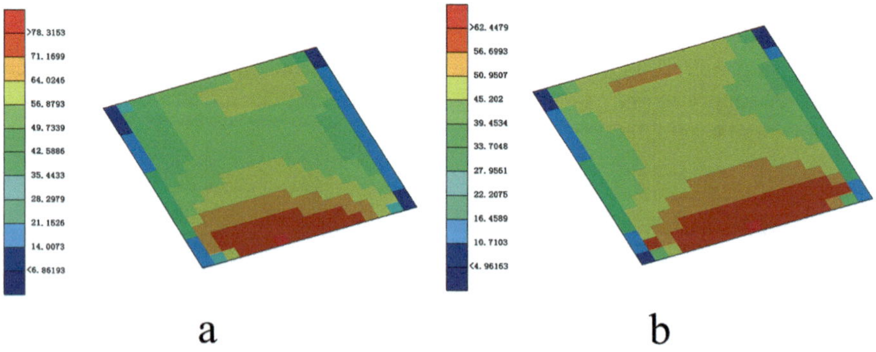

Fig. 4 Chip solder joint impact simulation cloud diagram of integrated business module and power module. **a** Integrated service module vibration mode, **b** power module vibration mode

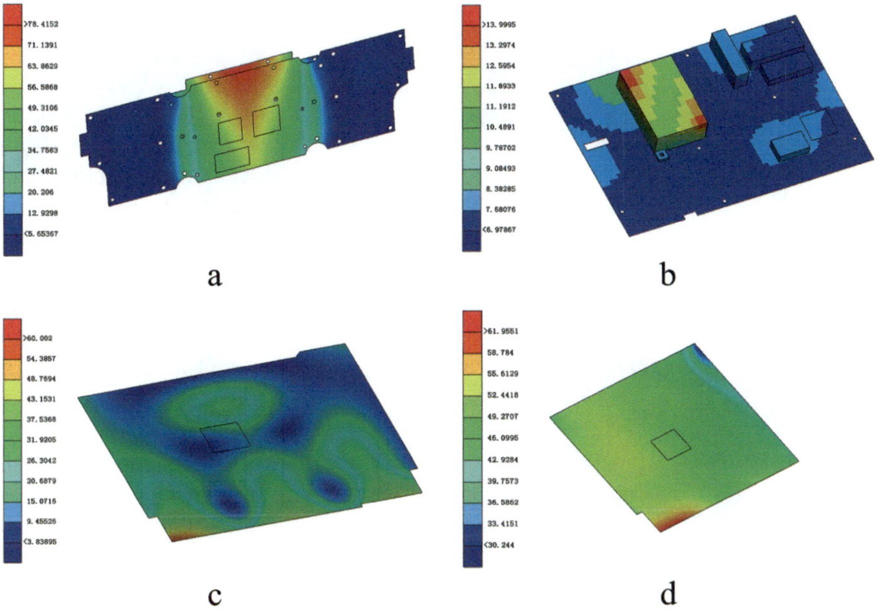

Fig. 5 Simulated acceleration cloud for sinusoidal vibration. **a** Panel module vibration mode, **b** channel module vibration mode, **c** integrated service module vibration mode, **d** power module vibration mode

3.1.4 Conclusion of Mechanical Simulation

The shell of the radio and the functional modules are in the decoupled state of resonance frequency, and the resonance frequency interval meets the requirements; the strength of the radio meets the requirements of the impact test in the back-loaded state, but the rigidity of the power supply module and the integrated business module is too low. The strength of the radio basically meets the requirements of sinusoidal vibration test.

3.2 Analysis and Discussion of Thermal Simulation Results

3.2.1 Normal Temperature Continuous Transmission Heat Simulation Analysis

Radio Continuous Transmission at room temperature for one hour, such as the handle, keys, connectors and other frequently touched parts of the temperature at about 40 °C, and the panel and the back cover of the overall temperature is below 50 °C, to meet the temperature requirements of the human hand operation, the central shell top surface and bottom surface temperature of about 60 °C, the human body can be contacted,

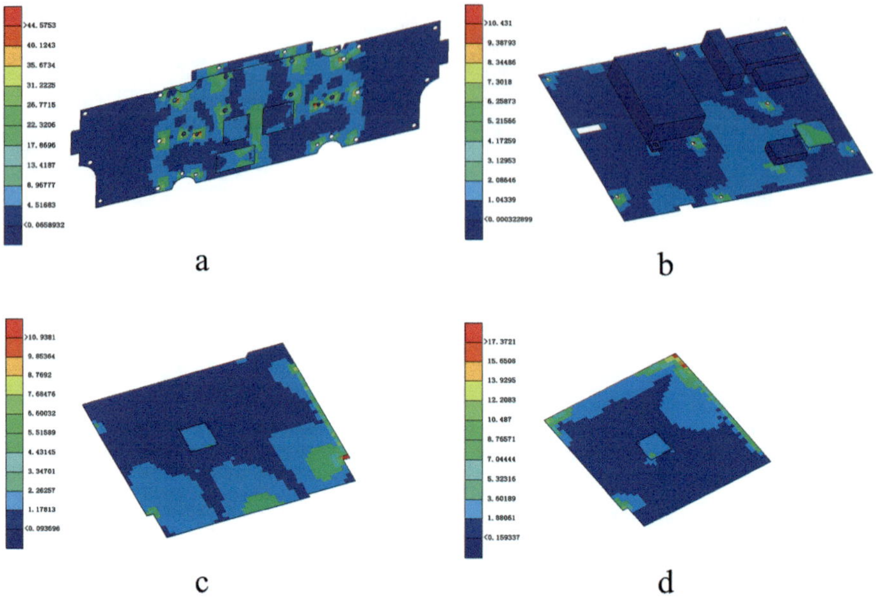

Fig. 6 Stress cloud for sinusoidal vibration simulation. **a** Panel module vibration mode, **b** channel module vibration mode, **c** integrated service module vibration mode, **d** power module vibration mode

Table 10 Sinusoidal vibration simulation results

Place	Orientations magnitude	Maximum/g	Maximum stress/Mp
Panel mouldings	Perpendicular	6.4	0.8
	Perpendicular	10.1	1.8
	Vertically	78.4	39.9
Channel module	Perpendicular	14	2.4
	Perpendicular	9.7	10.4
	Vertically	11.2	4.4
Integrated business modules	Perpendicular	8.9	1.4
	Perpendicular	60	5.3
	Vertically	44	10.9
Power supply modules	Perpendicular	7.8	0.7
	Perpendicular	50.3	6.5
	Vertically	62	17.4

and the two sides of the amplifier near the module at the temperature is high, 67 °C, taking into account the use of the radio backpack type shell at the organic clothing Considering the radio backpack use shell at the organic clothing protection, not in direct contact with the human body, so the shell temperature distribution should meet the requirements of heat dissipation assessment.

After one hour of Continuous Transmission at room temperature, the junction temperature of the amplifier tube is 101.3 °C, the temperature of the power supply chip is 78 °C, and the temperature of the shell of the common address filter module is 65 °C, which are lower than its tolerance temperature, and the residual amount is larger to meet the requirements of the device does not fail temperature, refer to Figs. 7, 8, and 9.

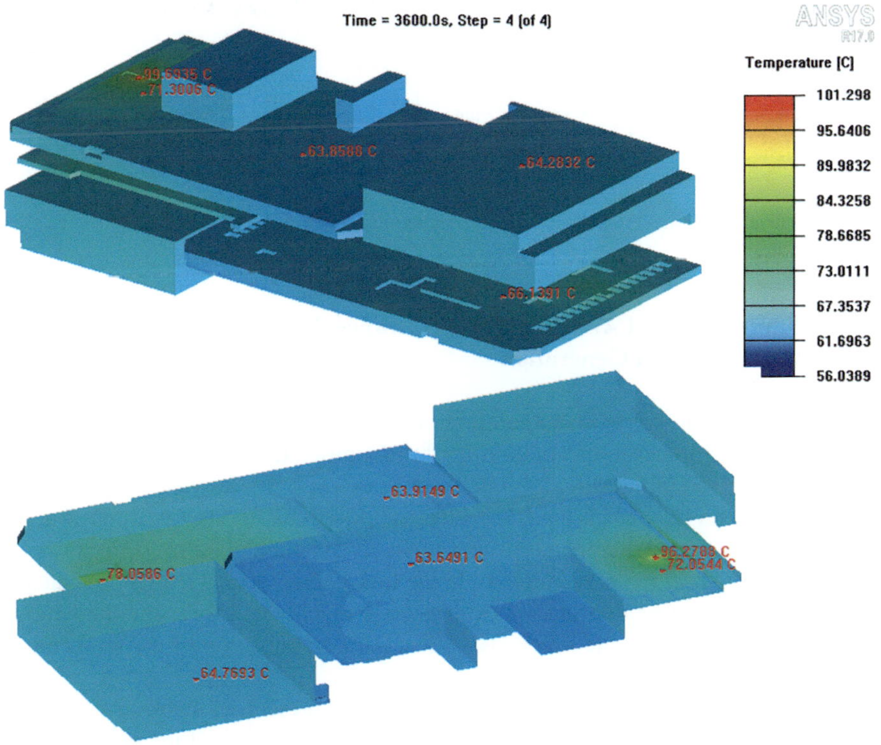

Fig. 7 Temperature distribution of internal moulding after one hour of continuous transmission growth at room temperature of the radio station

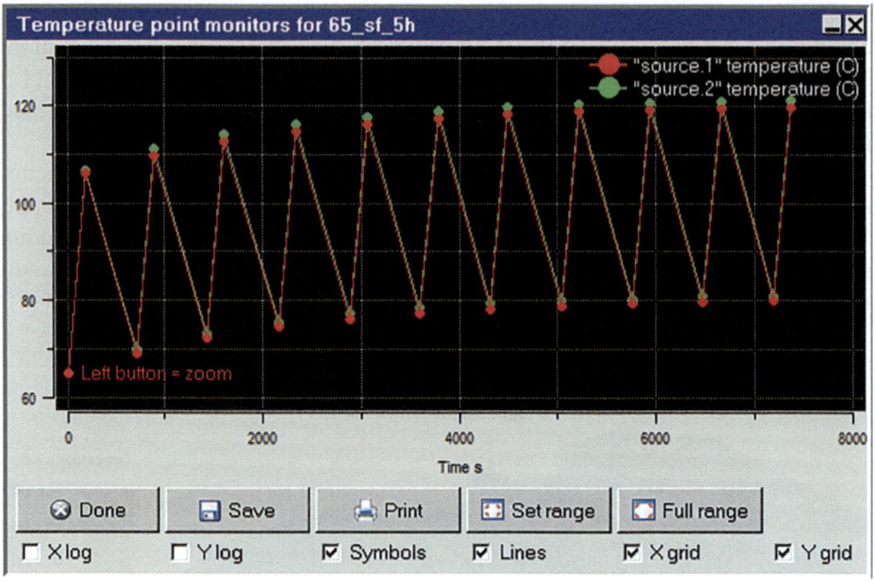

Fig. 8 High temperature transceiver-amplifier junction temperature—time change curve of the radio station

3.2.2 Simulation and Analysis of High-Temperature Continuous Heat Collection and Generation

The variation of the junction temperature of the monitored amplifier tube with time in the thermal simulation calculation for high-temperature continuous transceiver operation is shown in figure.

It can be seen that when the transceiver working conditions for about 2 h, the amplifier tube junction temperature that tends to dynamic stability, so in order to shorten the length of the simulation, only the transceiver work 2 h after the calculation of the temperature distribution of the whole machine, at this time the temperature distribution that is the successive transceiver stabilisation of the temperature distribution.

After the radio station continuously transmits and receives stable, the junction temperature of the amplifier tube is 119.5 °C, the temperature of the power supply chip is 96 °C, and the temperature of the shell of the common address filter module is 83 °C, which is lower than its tolerance temperature, and there is a certain amount of margin to meet the requirements of the device does not fail temperature.

In summary, the simulation values of the temperature of the shell and internal module of the radio after one hour of long transmission at room temperature and after continuous transmission and reception at high temperature until stability are shown in Tables 11 and 12.

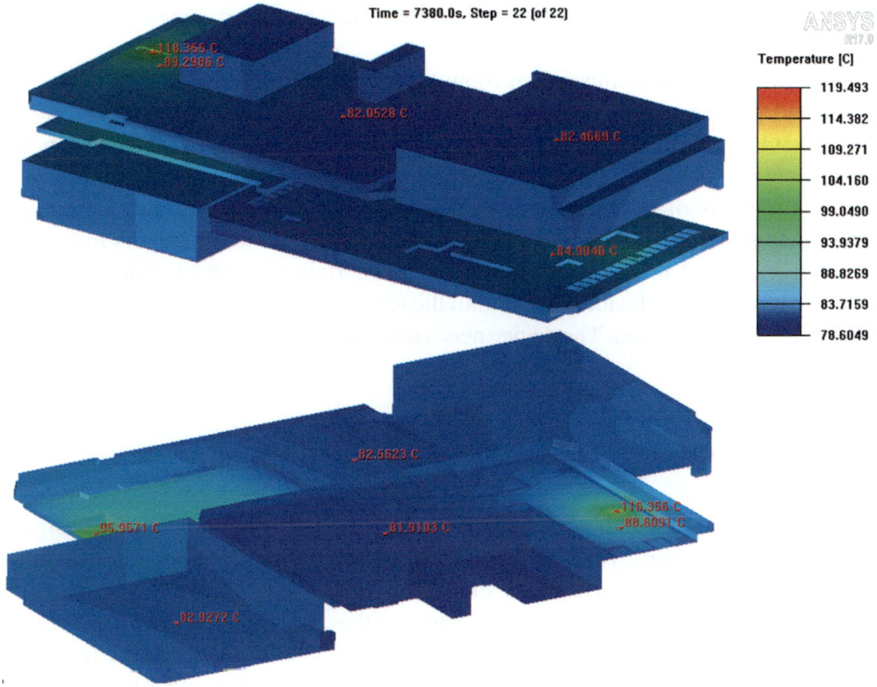

Fig. 9 Temperature distribution of internal module after radio high temperature continuous transmitting and receiving stabilisation

Table 11 Maximum shell temperature simulation values

Name	Maximum shell temperature		
	Kneading board (°C)	Casings (°C)	Shell (of crab etc) (°C)
Continuous transmission at room temperature	42	67	50
High-temperature transceiver	72	86	77

Table 12 Internal mould temperature simulation values

Name	Internal mould temperature		
	Amplifier tube junction temperature (°C)	Power chip temperature (°C)	Common address filter module case temperature (°C)
Continuous transmission at room temperature	101.3	78	65
High-temperature transceiver	119.5	96	83

3.2.3 Thermal Simulation Analysis Conclusion

Radio Continuous Transmission at room temperature for one hour, the panel, back cover and other human hands often touch the parts of the temperature are lower than 50 °C, the shell temperature is higher, but organic clothing protection, not in direct contact with the human body, so the shell temperature distribution to meet the requirements of the heat dissipation assessment; high-temperature continuous sending and receiving, the device temperature of the internal module is lower than the temperature of its tolerance, and there is a certain amount of residual to meet the high-temperature device does not fail the temperature requirements, of which the power supply module heat The device needs to increase the cold plate heat dissipation.

4 Conclusion

The simulation results indicate that following the lightweight optimisation design of a backpack radio employing the high Zn aluminium alloy developed in this study, several key performance criteria are effectively addressed.

First, resonance frequency analysis confirms that the structural coupling between the radio casing and its internal functional modules is decoupled, and the frequency intervals conform to all required specifications. Second, with regard to structural strength, the radio satisfies the impact resistance standards under backpack operation conditions. However, the power supply module and the integrated service module exhibit relatively low stiffness, suggesting a need for further structural enhancement. For the sinusoidal vibration assessment, the overall strength of the radio is found to be generally compliant with the applicable requirements. Further can be referred to the [16–23].

Lastly, thermal performance evaluation reveals that continuous operation for one hour under standard ambient conditions results in a surface temperature distribution that fulfils all heat dissipation criteria. Under high-temperature conditions, the device maintains continuous transmission and reception functions, reaching a thermally stable state. The resulting internal temperatures remain within the operational thresholds for high-temperature electronic components, indicating that thermal-induced failures are unlikely under these conditions.

Uncertainty sources including material variability (e.g., slight fluctuations in Zn content and trace element distribution), modeling assumptions (e.g., simplification of threaded holes and chamfers), and environmental influences (e.g., ambient temperature deviations) propagate through the simulations in a predictable manner. Material variability leads to $\pm$ 3% deviations in elastic modulus and thermal conductivity, which propagate to stress and temperature calculations with relative errors of less than 5%. Modeling simplifications slightly underestimate heat transfer efficiency, resulting in a maximum 4% overprediction of internal device temperatures. Environmental temperature fluctuations of $\pm$ 5 °C introduce $\pm$ 2% errors in thermal simulation results. Despite these propagations, all key performance indicators (resonance frequency, impact stress, device temperature) remain within acceptable tolerance ranges, confirming the reliability of the conclusions.

References

1. Zhang S, Men S, Wu W et al (2023) Research status and prospect of light alloy surface technology in the field of weapon and equipment lightweighting. J Anhui Univ Technol (Nat Sci Ed) 40(03):237–249
2. Zhang X (2022) Process equipment matching the new trend of lightweighting. Automot Craftsm (08):1
3. Guo F (2022) Structural design and integrity study of additive aluminium alloy suspension frame for high-speed magnetic levitation train. Southwest Jiaotong University
4. Zhang X, Chen G, Li H et al (2020) Research progress of liquid die forging forming technology applied to equipment lightweight aluminium alloy components. Mechanical Engineering Materials 44(10):6–11
5. Yan J, Xiong X, Wu C et al (2025) A secondary high-temperature precursor of the θ'-phase in Al-Cu-(Sc) alloys. J Mater Sci Technol 212:55–66
6. Chen Y, Wang W, Han M et al (2024) Research progress and prospect of grain refinement technology of Al-Si system alloys. Spec Cast Nonferrous Alloy 44(12):585–1595
7. Xu SS, Jia H, Wang D et al (2025) Effect of Mn element on the corrosion behaviour and mechanical properties of Al-Si-Mg alloys in T6 state. Trans Nonferrous Metals Soc China:4–27
8. Sun T, Geng J, Bian Z et al (2022) Microalloying of Zr and Mn elements to enhance the thermal stability and mechanical properties of high-temperature resistant aluminum-copper alloys. Trans Nonferrous Metals Soc China 32(01):4–78
9. Meng D, Xia AZ, Fan Y et al (2024) Research on the mechanism of Fe element's influence on the organisational properties of Al-Si system regenerated aluminium alloy and its control technology. Therm Process Technol 53(23):7–13
10. Suzhou Yongxiang Hardware & Plastics Co. A kind of special aluminium alloy semi-solid material preparation and automatic forming device: 202122487784.3 [P]. 2022-03-08
11. Dongying Yuanna Alloy Technology Co. (2020) A special aluminium alloy semi-solid material preparation and automated forming device: 201920186353.3. Apr 2020
12. Gao Z (2023) Research on lightweight preparation process of semi-solid Al-6Si-X alloy holder. Shenyang University of Technology
13. Xue K (2015) Research on the organisation and properties of A390 aluminium alloy in semi-solid forming. Beijing General Research Institute of Nonferrous Metals
14. Chen D (2008) Development of stirrers and stirred vessels. Press Vessel (02):33–46
15. Zhou G, Chang Y (2025) Effect of electromagnetic stirring process on segregation and banding organisation. Hebei Metall (03):8–55
16. M Z (2024) Basic research on the improvement of internal quality of round billet by alternating electromagnetic stirring in crystalliser. University of Science and Technology Beijing
17. Guo J, Wang Z, Li H, Yang Y, Huang CG, Yazdi M, Kang HS (2024) A hybrid prognosis scheme for rolling bearings based on a novel health indicator and nonlinear wiener process. Reliab Eng Syst Saf 245:110014
18. Li H, Yazdi M, Huang HZ, Huang CG, Peng W, Nedjati A, Adesina KA (2023) A fuzzy rough copula Bayesian network model for solving complex hospital service quality assessment. Complex Intell Syst 9(5):5527–5553
19. Li H, Peng W, Adumene S, Yazdi M (2023) An intelligent cost-based consequence model for offshore systems in harsh environments. In: Intelligent reliability and maintainability of energy infrastructure assets. Springer Nature, Cham, pp 107–117

20. Li H, Yazdi M (2022) What are the critical well-drilling blowouts barriers? A progressive DEMATEL-game theory. In: Advanced decision-making methods and applications in system safety and reliability problems: approaches, case studies, multi-criteria decision-making, multi-objective decision-making, fuzzy risk-based models. Springer International Publishing, Cham, pp 29–46
21. Li H, Peng W, Adumene S, Yazdi M (2023) Cutting edge research topics on system safety, reliability, maintainability, and resilience of energy-critical infrastructures. In: Intelligent reliability and maintainability of energy infrastructure assets. Springer Nature Switzerland, Cham, pp 25–38
22. Li H, Peng W, Adumene S, Yazdi M (2023) Advances in failure prediction of subsea components considering complex dependencies. In: Intelligent reliability and maintainability of energy infrastructure assets. Springer Nature Switzerland, Cham, pp 93–105
23. Li H, Yazdi M (2022) Reliability analysis of correlated failure modes by transforming fault tree model to Bayesian network: a case study of the MDS of a CNC machine tool. In: Advanced decision-making methods and applications in system safety and reliability problems: approaches, case studies, multi-criteria decision-making, multi-objective decision-making, fuzzy risk-based models. Springer International Publishing, Cham, pp 15–28

Nanoindentation-Based Finite Element Characterization Method for 25Cr2Ni2MoV Rotor Steel Welded Joints

Guo Sujuan, Pan Zijian, Zhao Jian, and Zhu Mingliang

Abstract This investigation employed an integrated methodology combining nanoindentation experiments with finite element modeling to assess localized mechanical characteristics in 25Cr2Ni2MoV rotor steel welded connections. The material parameters characterizing the power-law hardening behavior were determined for distinct regions (weld metal, heat-affected zone, and base metal) through reverse analysis of nanoindentation measurements. These derived constitutive properties were subsequently incorporated into computational models to predict the macroscopic tensile response of welded specimens. Validation against digital image correlation (DIC) tensile testing data revealed strong correspondence between simulated strain distribution patterns and experimental measurements throughout deformation progression. These findings confirm the validity and precision of the nanoindentation-driven parameter identification approach for multi-phase material systems.

G. Sujuan · P. Zijian · Z. Mingliang
Key Laboratory of Pressure System and Safety Ministry of Education, East China University of Science and Technology, Shanghai, China
e-mail: guosujuan@ecust.edu.cn

P. Zijian
e-mail: panzijian@ecust.edu.cn

Z. Mingliang
e-mail: zhumingliang@ecust.edu.cn

Z. Jian (✉)
School of Aerospace Engineering and Applied Mechanics, Tongji University, Shanghai, People's Republic of China
e-mail: zhaojian@tongji.edu.cn

G. Sujuan · Z. Jian
Shanghai Institute of Aircraft Mechanics and Control, Shanghai, People's Republic of China

Z. Jian
National Key Laboratory of Strength and Structural Integrity, Xi'an, China

H. Li et al. (eds.), *Data-Driven Methods for Reliability and Safety Engineering: Applications in Industrial Systems*, Springer Series in Reliability Engineering,
https://doi.org/10.1007/978-3-032-22873-4_37

Keywords Welded joint · Nanoindentation · Finite element inverse method · Constitutive model · Local deformation

1 Introduction

The heterogeneous microstructural features and material property gradients generated during welding processes create distinct disparities in localized mechanical responses across joint regions. Numerous studies [1–4] have identified these microstructural gradients as primary contributors to localized stress intensification and the nucleation of micro-scale fracture initiation sites in welded components. Consequently, precise characterization of mechanical property distributions through weld zones becomes crucial for developing fatigue-resistant engineering designs and maintaining structural integrity in welded assemblies [5, 6].

Nanoindentation has emerged as a commonly utilized non-invasive technique for characterizing material properties at microscopic scales [7–10]. This approach is frequently employed to determine localized hardness, elastic modulus, and interfacial bonding strength across thin film structures and various microstructural domains [11–14]. A persistent challenge in materials science involves extrapolating micro-level indentation measurements to predict bulk mechanical behavior. Pioneering work by Giannakopoulos and Suresh [15] established empirical correlations between load-displacement curve characteristics and fundamental material parameters like yield strength and work-hardening coefficients, laying groundwork for subsequent research. Through computational modeling, Pelletier et al. [16] systematically investigated the mechanical interactions of Berkovich indenters, mapping stress-strain distributions and validating connections between indentation patterns and material constitutive relationships. Dao et al. [17] subsequently constructed an integrated methodology merging dimensional evaluation with numerical modeling techniques, enabling direct derivation of yield characteristics and strain-hardening coefficients from indentation response patterns.

While these advancements demonstrate improved analytical capabilities, existing investigations have predominantly concentrated on validating material parameters through either macroscopic tensile profiles or localized indentation measurements. Limited research has systematically explored methodologies for acquiring zonal mechanical properties within structurally varied materials to comprehensively assess spatially differentiated stress-strain patterns.

This investigation established the mechanical characteristics of weld metal, heat-affected zones, and parent material in 25Cr2Ni2MoV rotor steel welded components through an inverse finite element methodology utilizing nanoindentation measurements. The obtained material parameters were subsequently applied to model the tensile response of welded specimens under uniaxial loading conditions. Validation against Digital Image Correlation data demonstrated strong agreement between simulated localized strain distributions and their developmental progression with empirical findings. The developed technique presents a viable framework for evaluating regional material performance and predicting durability in heterogeneous structural systems.

2 Nanoindentation Experiments and Results

The nanoindentation sample containing a centrally located heat-affected zone (HAZ) between base metal (BM) and weld metal (WM) sections was obtained from 25Cr2Ni2MoV steel forging weldments, illustrated in Fig. 1. The welding process was initiated with automated argon arc welding for root pass formation, followed by submerged arc welding (SAW) completion. Following this thermal processing, the welded components underwent post-weld heat treatment (PWHT) at 580 °C with 10-h thermal stabilization. Material characterization employed a Nano Indenter G200 testing platform. Sample preparation involved progressive surface refinement through sequential abrasion using increasingly fine abrasive media (400–2000Cw), with 90° specimen rotation during grit transitions to eliminate residual grinding patterns. Subsequent polishing achieved optical-grade surfaces devoid of detectable imperfections or surface irregularities. The resultant load-displacement.

The load-displacement curves obtained from nanoindentation testing are illustrated in Fig. 2. Experimental measurements were performed with a Nano Indenter G200 apparatus following standardized protocols. Berkovich indentation experiments were subsequently conducted on WM, HAZ, and BM regions using a standardized penetration depth of 1500 nm to quantify localized mechanical properties (H and E values), with detailed results documented in Table 1. To validate measurement consistency, the hardness and elastic modulus values for each zone were calculated through averaging six replicate measurements taken at corresponding anatomical positions.

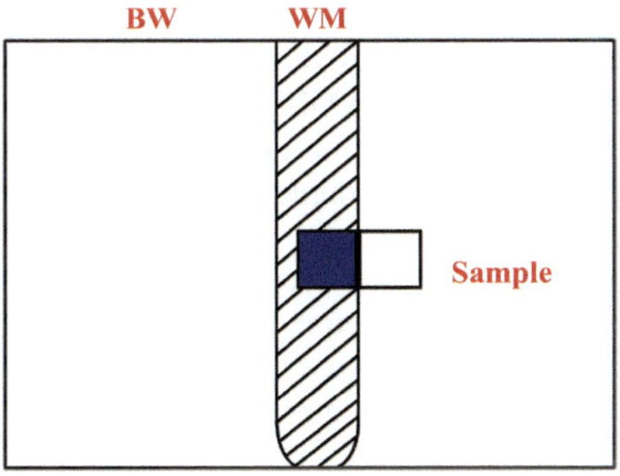

Fig. 1 Welded joint sample for the nanoindentation test

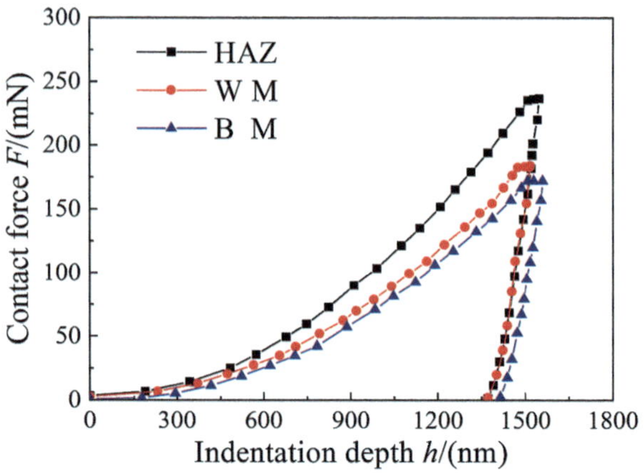

Fig. 2 The contact load-indentation depth curve in different area of indentation testing

Table 1 The elastic-plastic parameters of the welded joint in different areas

Different areas	E(GPa)	H(GPa)	σ_y(MPa)	n
BM	182.5	3.26	577.35	0.19
WM	194.9	3.51	631.00	0.18
HAZ	213.8	4.67	793.8	0.22

3 The Finite Element Inversion for the Material Parameters

3.1 Finite Element Model

To replicate the nanoindentation process, an axisymmetric FE model was developed using ABAQUS software (Figs. 3 and 4). The three-dimensional Berkovich indenter geometry was approximated as a conical tool with matching 140.6° vertex angle for equivalent contact mechanics analysis. The diamond indenter, characterized by exceptional stiffness and wear resistance, was defined as a non-deformable body. High-precision CAX4 quadrilateral elements were employed to handle substantial deformation scenarios effectively. Computational efficiency was optimized through graded meshing, implementing 0.5–1 μm elements in the contact region while coarser elements were used elsewhere. Vertical displacement control was applied at the indenter's reference node to achieve programmed penetration depths, with corresponding reaction forces recorded to generate load-penetration profiles. Initial contact stabilization was ensured through adaptive time stepping, incorporating a baseline time increment of 1 × 10–6 s to minimize numerical instabilities during early loading phases.

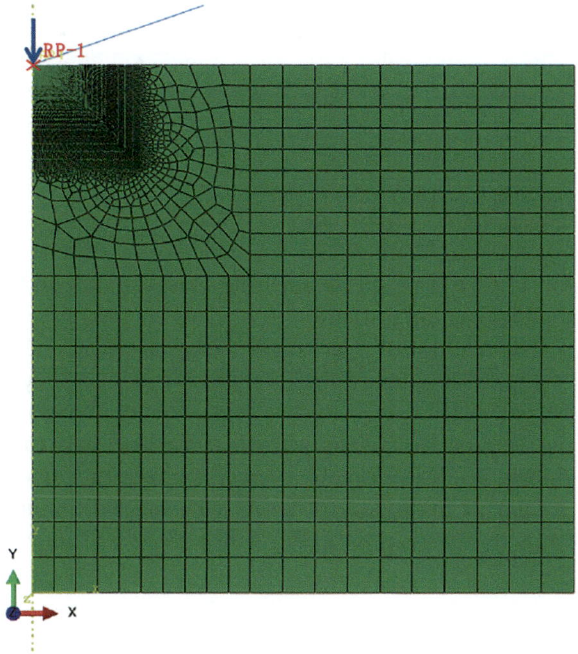

Fig. 3 Indentation finite element mesh

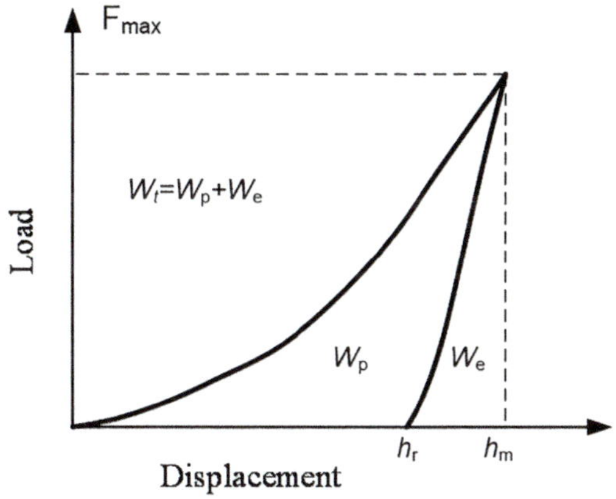

Fig. 4 Contact load-displacement curve

3.2 *The Finite Element Inversion Procedure*

This investigation employed the power-law hardening constitutive model to describe the mechanical response variations across different regions of welded joints. The constitutive relationship between stress and strain (σ–ε).

In this formulation, *E* corresponds to the elastic modulus, *R* represents the plastic hardening coefficient, *n* indicates the strain-hardening index, σy the yield strength,

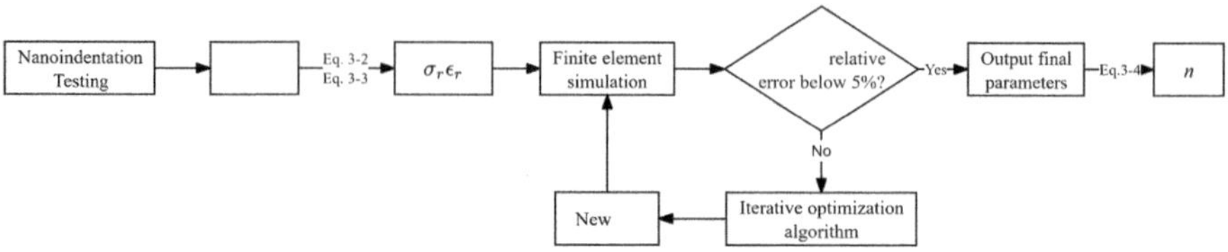

Fig. 5 Schematic of the inversion workflow

and e_p is the plastic strain. The elastic modulus has been determined from the nanoindentation tests.

The determination of sy and n is based on a set of characteristic stress-strain pairs (sr, er) and the dimensionless functions established by Dao et al. [17]. During the finite element simulation process, a displacement of 1500 nm was applied to the indenter, and the simulated maximum indentation load was extracted and compared with experimental results. Adopting iterative optimization algorithm to adjust the constitutive parameters until the relative error between the simulated and experimental values drops below 5%, refer to Fig. 5.

3.3 *Finite Element Inversion Outcomes and Verification*

Through incorporation of the calculated strain-hardening exponent n along with the characteristic stress-strain pair (s_r, e_r) into the constitutive relationships, the yield strength σy was quantitatively determined. These established material parameters have been systematically compiled in Table 1. When applied within the framework of the power-law hardening formulation, the derived values enabled generation of distinct stress-strain profiles for various weld joint zones. The comparative analysis demonstrates superior tensile resistance in the heat-affected zone (HAZ) relative to adjacent regions. The strong correlation observed between computational predictions and empirical measurements confirms the effectiveness of this innovative approach in capturing zonal mechanical heterogeneity within welded structures.

To validate the accuracy of the mechanical parameters presented in Table 1, uniaxial 1500 nm nanoindentation analyses were conducted on the HAZ, WM, and BM regions employing the power-law hardening constitutive model. A comparative analysis was carried out between the simulated contact force-indentation depth curves obtained during loading and the corresponding experimental data. The findings reveal a strong correlation between the computational models and empirical observations across all examined material zones.

4 Simulation for the Non-uniform Deformation of Welded Joints

4.1 Tensile Specimen and DIC Testing

Utilizing the elastic-plastic properties of base material (BM), weld filler (WM), and thermally altered regions (HAZ) established previously, this investigation employed numerical simulation to analyze strain distribution patterns and their progression in welded connections under uniaxial tension.

Analysis of these diagrams reveals continuous evolution in strain distribution characteristics throughout the tensile deformation process. As the applied stress progressively escalates from 500 to 620 MPa, three key observations emerge:

1. A direct correlation exists between increasing macroscopic stress and strain intensification. Initial stress application at 500 MPa yields a maximum strain measurement of 1.13%, while subsequent stress elevation to 620 MPa results in substantial strain amplification, reaching 16.3% at peak deformation.
2. Under an applied stress of 550 MPa, the welded joint exhibits peak strain localization within the base material (BM) region adjacent to the right-side heat-affected zone (HAZ). Progressive stress elevation induces shear band formation at approximately 45° relative to the loading axis within the weld metal (WM) near the left HAZ interface. During this phase, strain concentration persists predominantly in the BM-HAZ boundary region. Subsequent stress intensification promotes gradual strain accumulation within the WM section. At 615 MPa macroscopic stress, WM tensile deformation initiates surpassing corresponding values in neighboring BM-HAZ areas. Ultimate failure occurs at 620 MPa stress level with complete strain localization transition to the WM region.
3. Using the parameter calibration approach developed in this study, the mechanical property characteristics of various weld zone regions were determined. The simulated strain distribution contours of welded joints at progressive tensile stages exhibited remarkable consistency with DIC experimental outcomes regarding three key aspects: strain pattern characteristics, magnitude quantification, and evolutionary distribution trends, refer to Fig. 6.

4.2 Discussion Concerning the Sustained Performance of Welded Joint

The pattern of fatigue fracture locations is dependent on the strain amplitude [18]. At lower strain amplitudes ($\Delta\varepsilon/2 = 0.3\%$–$0.4\%$), fractures were primarily concentrated within the heat-affected zone (HAZ). Conversely, at higher strain amplitudes ($\Delta\varepsilon/2 = 0.5\%$–$0.8\%$), fractures predominantly occurred in the weld metal (WM). Furthermore, a statistical analysis of fracture locations in stress-controlled fatigue

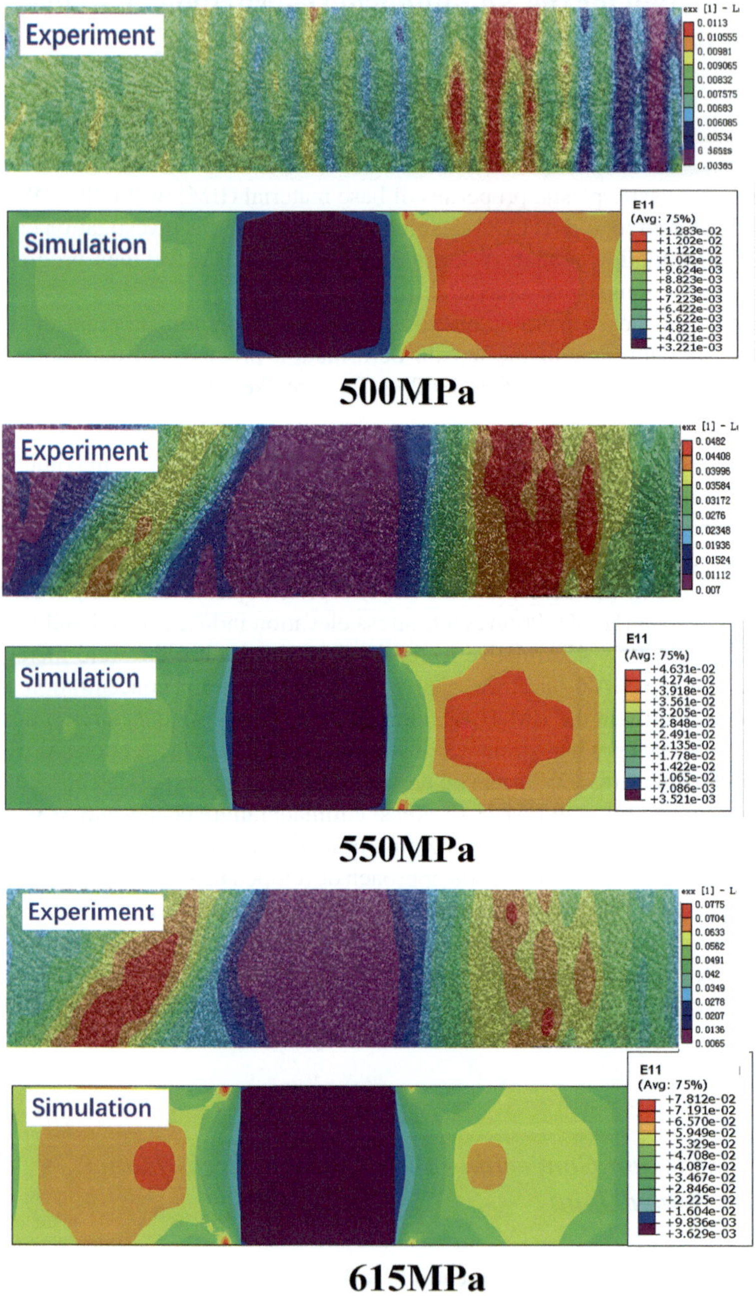

Fig. 6 Strain contour plots during tensile testing under different loading levels

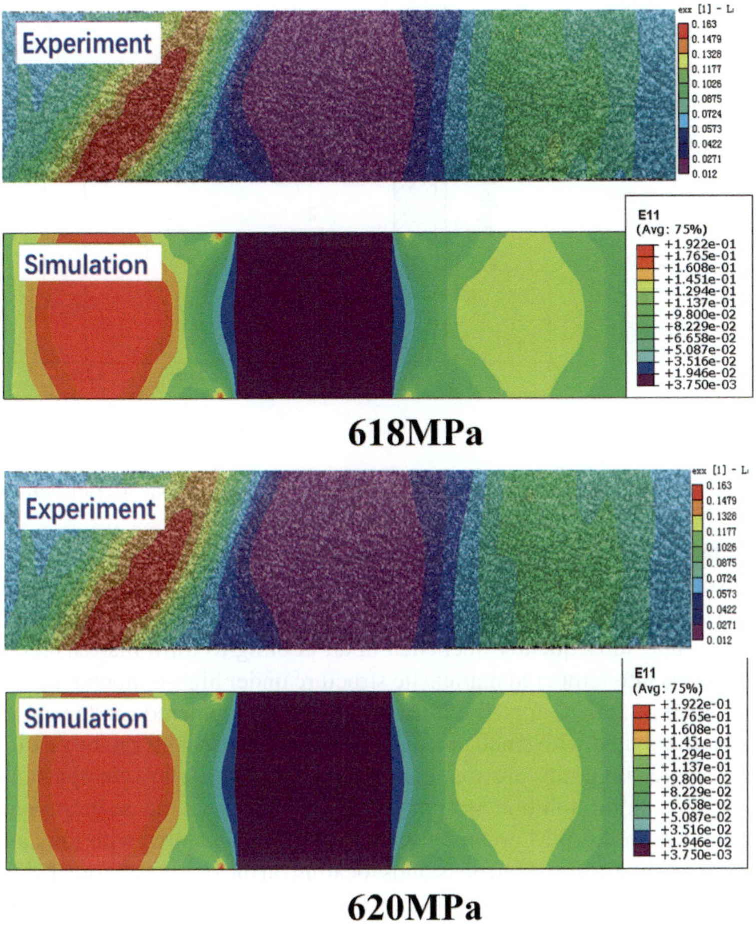

Fig. 6 (continued)

specimens of welded joints revealed that fractures mainly occurred near the HAZ when the stress amplitude was below 618 MPa. In contrast, at amplitudes exceeding 618 MPa, fractures primarily occurred in the base metal (BM) and were predominantly characterized by necking. These observations, which show how localized deformation varies with the load level, demonstrate a clear correlation between the low-cycle fatigue fracture location in welded joints and the characteristics of the localized deformation distribution. The DIC strain contour maps clearly show that as the load level increases, the location of maximum strain in the welded joint shifts from the base metal toward the weld metal. Correspondingly, the location of maximum strain transitions from regions near the base metal to the central weld zone with increasing load level. This shift confirms the consistency between the low-cycle fatigue fracture location and the localized deformation distribution characteristics in

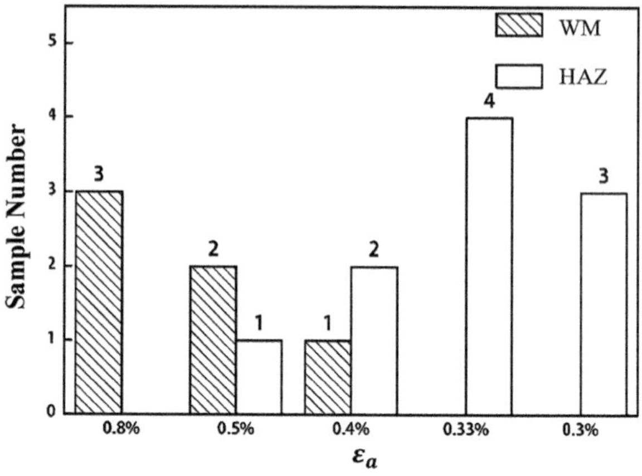

Fig. 7 Failure location under different loading levels

welded joints, indicating that the fatigue fracture location is load-dependent, refer to Fig. 7.

The observed strain-localization evolution is closely linked to microstructural mechanisms. Due to the 25Cr2Ni2MoV being a medium carbon quenched and tempered steel with high hardenability, the HAZ is heated above the austenitization temperature and subsequently quenched in air cooling to form martensite. Martensite transforms into tempered martensite structure under high-temperature annealing heat treatment at 580 °C for 10 h after welding. The tempered martensite group is uniformly distributed with small and dispersed carbide particles on the ferrite matrix [19, 20]. The ferrite grains are small and uniform, and the alloy elements dissolved in them play a solid solution strengthening role, thus having high strength. The microstructures in the WM are mainly tempered lower bainites with long and lathy shape. It also contains some ferrite islands locating on the bainites. From this, it can be seen that due to the differences in materials and microstructures in different regions, the welded joint exhibits different mechanical properties and material properties in different areas [21–24]. The differences in mechanical properties in different regions lead to differences in their plastic evolution, causing a redistribution of plastic strain in the joint during the bearing process.

5 Conclusions

This investigation employed finite element inversion techniques to assess the mechanical characteristics of weld metal, parent material, and thermally altered regions in 25Cr2Ni2MoV rotor steel welded connections. The acquired material parameters were subsequently utilized for computational modeling of tensile deformation behavior in welded specimens. Computational results demonstrate that the calculated parameters effectively characterize the nanoindentation characteristics across distinct zones while accurately representing the spatial strain patterns and progressive deformation features observed in welded assemblies. Furthermore, compared to alternative local-property characterization methods like micro-tensile testing, the nanoindentation-FE inverse method offers distinct advantages: it requires minimal sample volume (enabling characterization of microscale zones like HAZ), avoids sample preparation-induced damage, and provides spatially resolved parameters. While micro-tensile testing yields direct tensile curves, it is limited by sample size and cannot capture localized properties in heterogeneous regions. The proposed method thus complements micro-tensile testing, offering a more versatile solution for characterizing welded joints with complex microstructural gradients.

Acknowledgments The research work is supported by the National Natural Science Foundation of China (Project nos. 52175144, 52275151, 52575187), the Foundation of the National Key Laboratory of Strength and Structural Integrity (ASSIKFJJ202305006).

References

1. Chandramohan DL, Roy K, Taheri H et al (2022) A state of the art review of fillet welded joints. Materials 15(24):1–31
2. Bi J, Liu Z, Chi J et al (2023) Formation mechanisms and control strategies of FQZ softening in Al–Li alloy welded joint [J]. J Mater Res Technol 23:2810–2823
3. Di Bella G, Favaloro F, Borsellino C (2023) Effect of process parameters on friction stir welded joints between dissimilar aluminum alloys: a review. Meta 13(7):1–53
4. Jiang X, Cao D, Qiang X et al (2024) Study on fatigue performance of steel bridge welded joints considering initial defects [J]. J Constr Steel Res 212:1–15
5. Duyi Y, Feng M, Jianzhong L, Yuandong X, Yinzheng C, Lei X (2013) Use of instrumented indentation testing to study local mechanical properties of 304L SS welded joints subjected to low-cycle fatigue loadings. Mater Sci Eng A 564:76–84
6. Filacchioni G, Montanari R, Tata ME, Pilloni L (2002) Structural and mechanical properties of welded joints of reduced activation martensitic steels. J Nucl Mater 307-311:1563–1567
7. Oliver WC, Pharr GM (1992) An improved technique for determining hardness and elastic modulus using load and displacement sensing indentation experiments. J Mater Res 7(6):1564–1583
8. Bhushan B, Li X (2004) Nanomechanical properties of solid materials: a review. Int J Mater Rev 49(1):1–20
9. Yanyuan C, Yunfei J, Fuzhen X (2016) A comparative study on the cyclic behavior of austenite/ferrite phases characterized by the nanoindentation. Appl Mech Mater 853(9):137–141
10. Yanyuan C, Yunfei J, Fuzhen X (2018) Micro-deformation evolutions of the constituent phases in duplex stainless steel during cyclic nanoindentation. Sci Rep 8(1):6199–6208
11. Phani PS, Hackett BL, Walker CC et al (2023) High strain rate nanoindentation testing: recent advancements, challenges and opportunities. Curr Opinion Solid State Mater Sci 27(1):1–12
12. Long X, Li Y, Shen Z et al (2024) Review of uniqueness challenge in inverse analysis of nanoindentation. J Manuf Process 131:1897–1916
13. Puchi-Cabrera ES, Rossi E, Sansonetti G et al (2023) Machine learning aided nanoindentation: a review of the current state and future perspectives. Curr Opinion Solid State Mater Sci 27(4):1–20
14. Rossi E, Wheeler JM, Sebastiani M (2023) High-speed nanoindentation mapping: a review of recent advances and applications. Curr Opinion Solid State Mater Sci 27(5):1–27

15. Giannakopoulos AE, Suresh S (1997) A method for determining yield stresses and work-hardening rates from indentation tests. J Mater Res 12(3):780–789
16. Pelletier JM, Gauthier S, Shannigrahi A (1998) A numerical study of the indentation of materials with strain-hardening using sharp indenters. J Mech Phys Solids 46(11):2077–2096
17. Dao M, Chollacoop N, Van Vliet KJ et al (2001) Computational modeling of the forward and reverse problems in instrumented sharp indentation. Acta Mater 49(19):3899–3918
18. Guo SJ et al (2019) A comparative study on the cyclic plasticity and fatigue failure behavior of different subzones in CrNiMoV steel welded joint. Int J Mech Sci 150:66–78
19. Li H, Peng W, Adumene S, Yazdi M (2023) An intelligent cost-based consequence model for offshore Systems in Harsh Environments. In: Intelligent reliability and maintainability of energy infrastructure assets. Cham, Springer Nature Switzerland, pp 107–117
20. Li H, Yazdi M (2022) What are the critical well-drilling blowouts barriers? A progressive DEMATEL-game theory. In: Advanced decision-making methods and applications in system safety and reliability problems: approaches, case studies, multi-criteria decision-making, multi-objective decision-making, fuzzy risk-based models. Springer International Publishing, Cham, pp 29–46
21. Li H, Peng W, Adumene S, Yazdi M (2023) Cutting edge research topics on system safety, reliability, maintainability, and resilience of energy-critical infrastructures. In: Intelligent reliability and maintainability of energy infrastructure assets. Cham, Springer Nature Switzerland, pp 25–38
22. Li H, Peng W, Adumene S, Yazdi M (2023) Advances in failure prediction of subsea components considering complex dependencies. In: Intelligent reliability and maintainability of energy infrastructure assets. Cham, Springer Nature Switzerland, pp 93–105
23. Li H, Yazdi M (2022) Reliability analysis of correlated failure modes by transforming fault tree model to Bayesian network: a case study of the MDS of a CNC machine tool. In: Advanced decision-making methods and applications in system safety and reliability problems: approaches, case studies, multi-criteria decision-making, multi-objective decision-making, fuzzy risk-based models. Springer International Publishing, Cham, pp 15–28
24. Guo J, Wang Z, Li H, Yang Y, Huang CG, Yazdi M, Kang HS (2024) A hybrid prognosis scheme for rolling bearings based on a novel health indicator and nonlinear wiener process. Reliab Eng Syst Safety 245:110014
25. Li H, Yazdi M, Huang HZ, Huang CG, Peng W, Nedjati A, Adesina KA (2023) A fuzzy rough copula Bayesian network model for solving complex hospital service quality assessment. Complex Intell Syst 9(5):5527–5553

Quantification of Risk and Control of Bidding for a Nuclear Power Plant Project's Marine Water Intake Project

Mingfang Chen, Tao Wang, Xiaodong Guo, and Xinhua YU

Abstract As a capital- and technology-intensive sector, nuclear power projects are characterized by substantial investment requirements and extended construction timelines, with bidding management exerting a significant influence on both project efficiency and quality. In response to the limitations of traditional bidding practices, particularly the difficulty in quantifying risk and the insufficiency of risk control measures, this study establishes a closed-loop system integrating risk quantification and risk control. Specifically, the weights of 15 identified risk factors are determined through the Analytic Hierarchy Process. A guarantee rate of no less than 95% for the control price is achieved using Monte Carlo simulation. Additionally, through enhancements in the design scheme comparison and optimization process, approximately 45% of cofferdam construction costs are reduced, construction-related risks are mitigated, and resource reuse yields savings of approximately two million yuan. This approach facilitates the transition of nuclear power bidding management from reliance on empirical decision-making toward a data-driven model.

Keywords Risk quantification · Control price · Reliability · Monte Carlo simulation

M. Chen (✉) · T. Wang · X. YU
China Nuclear Power Engineering Co., Ltd, Beijing, China
e-mail: chenmfa@cnpe.cc

T. Wang
e-mail: wangtaoa@cnpe.cc

X. YU
e-mail: yuxha@cnpe.cc

X. Guo
School of Mechanical Engineering, IMUT, Inner Mongolia, China
e-mail: XiaodongGuo@cnpe.cc

H. Li et al. (eds.), *Data-Driven Methods for Reliability and Safety Engineering: Applications in Industrial Systems*, Springer Series in Reliability Engineering,
https://doi.org/10.1007/978-3-032-22873-4_38

1 Introduction

China's nuclear power industry has entered a new stage of large-scale development. According to China's National Energy Administration (NEA), as of April 2025, China has 112 nuclear power units in operation and under construction, with a total installed capacity of 125 million kilowatts, ranking first in the world in terms of scale. As the core link of nuclear power project construction, the level of bidding management is directly related to the cost control, quality assurance, and schedule performance of multi-billion projects. However, the current bidding management model still has significant shortcomings: firstly, due to the lack of a systematic risk quantification system, accurate identification and assessment of potential risks are difficult; secondly, the absence of a design scheme selection and optimization mechanism leads to a lack of data support and scientific validation in the compilation of bidding control prices, resulting in unguaranteed reliability of such prices.

This study focuses on the bidding scenario of nuclear power plant sea water intake project, and builds a management system of "risk quantification - risk control". Firstly, establish a risk assessment model based on the hierarchical analysis method, which contains 3 primary indicators and 15 secondary indicators, and accurately quantify the weight distribution of key risk factors. Secondly, use Monte Carlo simulation to generate a cost distribution curve, and increase the guaranteed bidding price control rate to 95% and above. Finally, through a sound design scheme selection and optimization mechanism, project quality, cost, and schedule are synergistically improved. The methods of "risk quantification" and "guaranteed control price reliability" have been successfully applied to two coastal nuclear power projects, and the research results have provided an innovative path to achieve the goal of high-quality bidding management for nuclear power projects.

2 Project Description

Drainage engineering is a critical component of nuclear power plants, involving large-scale investment and multiple decision-influencing factors. It functions to cool equipment and discharge waste heat by introducing an external cooling water source. This nuclear power plant is to build No.5 and No.6 "Hua long No.1" units, adopting seawater through cooling programme, and the structures are SC class and seismic class I. The tender scope includes water intake engineering, berm engineering, pump house foundation, and water intake process system.

3 Quantification of Bidding Risks

Commonly used international risk quantification methods include subjective scoring, hierarchical analysis, sensitivity analysis, statistical probability and break-even analysis [1]. The hierarchical analysis method is scientifically feasible because it can combine qualitative and quantitative analyses, clarify the quantitative relationship of risk factors, and is applicable to multi-objective and multi-factor risk assessment [2]. Therefore, this method is adopted in this paper for the quantification of bidding risks.

3.1 Constructing a Risk Hierarchy Table

The risk identification of this project adopts a method combining expert investigation with historical data analysis, which is completed through forms such as questionnaire surveys and seminars. On this basis, a 2-level hierarchical risk structure table (as shown in Table 1) is constructed to comprehensively cover the risks of engineering bidding and lay the foundation for quantitative assessment.

3.2 First Floor Risk Quantification and Reliability Verification

Pairwise comparisons were conducted for the primary risk factors in Table 1. Seven experts (managerial, technical, and economic) were engaged to score independently, followed by joint validation and statistical ranking based on the majority rule principle [7]. The square root method was used to calculate the comparison matrix and single-level rankings.

The quantification results are validated based on the quantification of the first-tier risk weights and judged to be sufficient to meet the consistency requirements, specifically:

According to the average stochastic consistency index query of the first-ninth order matrices, we know that $RI = 0.52$ to calculate the consistency rate $CR = CI/RI = 0.0015 < 0.1$. where CR is the consistency rate of the judgement matrix; RI is the average stochastic consistency index.

The above results show that the comparison matrix satisfies the consistency requirement, i.e., its corresponding derived eigenvectors are reliable.

Table 1 Bidding risk hierarchy and factor classification table

First floor risk points	Spur risk points	Description of risk points
R_1 tender preparation phase	R_{11} design options do not meet economic and feasibility requirements	Various forms of water abstraction in the sea, such as the early stage of the implementation of the design of dynamic management or design and investment control is not closely related to the lack of investment feedback optimization mechanism, will lead to the design of over-budgeting, investment out of control and other risks
	R_{12} inadequate consideration of bidding strategies	Poorly researched tendering strategies and failure to set reasonable bidding thresholds have led to ineffective tendering
	R_{13} inaccuracy in preparation of solicitation documents	The emergence of problems such as unclear bidding scope, incomplete contract terms and inaccurate preparation of the bill of quantities not only brought unforeseen impacts on the bidding process, [3] but also increased the risk of later settlement
	R_{14} control price was not set properly	In the process of preparing the bidding control price, the failure to take into account factors such as on-site construction conditions caused the control price to deviate from the actual price, leading to a waste of investors' funds or damage to the quality of the project, while affecting the legitimate rights and interests of the bidders [4]
	R_{15} failure to align bid evaluation methods with project management requirements	The mismatch between the bid evaluation method and the actual needs of the project makes it impossible to reasonably and effectively assess the comprehensive strength of the construction contractor, which may not only result in missing out on the best contractor, but also lead to the risk of complaints
R_2 bidding stage	R_{21} clarification of the risk of overcounting	Frequent clarifications may cause bidders to misunderstand key information, prolong the bidding cycle, damage the credibility of the bidders, reduce the incentive to bid, increase the difficulty of management, and affect the fairness of bid evaluation and the smoothness of the contract
	R_{22} unreasonable formation of bid evaluation committee	The mismatch between the expert composition of the bid evaluation committee and the offshore/nuclear power projects may lead to invalid evaluation or invalid bid evaluation results, and there is a risk of unreasonableness and non-compliance

(continued)

Table 1 (continued)

First floor risk points	Spur risk points	Description of risk points
	R_{23} corrupt behaviour and low professional ethics	In the bidding process, bid-rigging and collusion, as well as buying off bid evaluation experts and engaging in trading in power and money, which resulted in the bidding for works not being carried out in accordance with the law
	R_{24} risk of incomplete information and communication of information	Tendering for this project involves a large amount of tender data and information. If the tenderer or the bidding agent fails to fully and effectively transmit the data and information during the tendering period, it will give rise to complaints about the tendering of the project or affect the management of the subsequent project
	R_{25} risk of reasonableness of tender offers	Quoting too low a price for the winning bid or using unbalanced quotes for some materials that are much larger than the market price may lead to problems such as jerry-building, settlement disputes or investments that are much higher than the actual project cost [5]. As a result of these problems, the risk of investment control management by the contracting authority is very high
R_3 post-bid management phase	R_{31} uncertainty in the implementation phase of the project	The technical complexity of the sea water intake project is such that, although the construction plan design was completed during the bidding period, there are still many uncertainties in the subsequent construction programme, and the high demand for new processes, materials and technologies also brings about a large number of unknown risks
	R_{32} contractual limitation risk	Defects in the contracting process lead to non-compliance with the statute of limitations on contracting, which not only affects the efficiency of performance but also increases the uncertainty of enforcement
	R_{33} information archiving risk	Untimely or incomplete filing of information affects project traceability and audits and may lead to legal disputes
	R_{34} contract negotiation risk	Disagreements between the parties on the interpretation of contractual terms and the allocation of responsibilities may lead to prolonged negotiations, resulting in missed contracting opportunities and increased risks
	R_{35} contract delivery and handover risk	Incomplete briefings or inadequate transmission of information lead to inconsistent understanding of the contract in the later stages of implementation, triggering implementation deviations and disputes [6]

3.3 *Risk Quantification and Reliability Check for each Branch Layer*

3.3.1 Results of Risk Quantification

Assessed by the hierarchical analysis method, the weighting of the 15 branch-level risk factors shows that: deviation in the setting of the control price (R_{14}, weight 0.165), imbalance in the economy of the design programme (R_{11}, weight 0.097), errors in the preparation of the tender documents (R_{13}, weight 0.0825), and the risk of the reasonableness of the offer (R_{25}, weight 0.0816) are the key risk sources, the following will focus on the first two high-risk items to implement control.

High-priority risks have been directly linked to specific management measures: For R14 (Cost Deviation, weight 0.165), the Monte Carlo method combined with computer simulation was employed to quantify cost volatility, ensuring a 95% confidence level. For R11 (Economic Imbalance in Design Schemes, weight 0.097), a multi-round optimization mechanism was established to control cost overruns, incorporating measures such as scheme comparison (e.g., selecting the overall scheme with a forebay) and construction method optimization (e.g., cofferdam techniques). At the same time, we fully acknowledge the inherent limitations of the Analytic Hierarchy Process (AHP) and expert judgment, such as the potential subjective bias introduced by a seven-expert panel and insufficient coverage of regional variations. To mitigate these limitations, the study adopted methods including independent back-to-back scoring, majority principle validation, and integration of historical risk data. Furthermore, as expert judgment may underestimate emergent risks (e.g., significant changes in the construction environment), dynamic updates to the risk hierarchy table during project implementation can be applied as a responsive measure.

4 Bidding Risk Management

4.1 *Monte Carlo Method to Ensure Control Price Reliability*

The Monte Carlo method requires the probability distribution of the independent variable X to be known. Based on the existing research, this paper selects the triangular distribution [8]. Due to the varying natural and social environments in which engineering projects are situated, their costs exhibit both deterministic and uncertain characteristics. Deterministic costs follow a linear distribution in terms of probability and can be represented using a triangular distribution. In contrast, for uncertain costs, according to the Central Limit Theorem, after multiple simulations, the triangular distribution aligns closely with other distribution curves, with minimal error. The triangular distribution, characterized by its low data requirements and strong applicability, is well-suited for cost risk research in nuclear power engineering projects.

Table 2 Parameters of the triangular distribution of the sub-projects

Srial umber	Sub-item name	Lower limit value ($ million)	Most probable value ($ million)	Upper limit ($ million)
1	Water intake project	4580.28	4770.31	4920.21
2	Berm construction	5089.46	5117.02	5230.46
3	Pump house pit	15393.56	15577.93	15703.67

Using the principle of inverse function to get the cost of the probability of the expression (y), summing up the cost of all sub-items will be all sub-items of the cost of (y) to get the expression of the total cost of the risk-containing A large number of simulation tests are carried out by computer, and the test values are counted to obtain the probability distribution of the total cost with risk. According to the above triangular distribution as the probability approximation distribution [9] of the cost of each sub-item containing risk, the distribution function is shown in 1-1 as follows.

The parameters of the triangular distribution (minimum value, maximum value, most probable value) were determined by referencing historical cost data from three similar nuclear power seawater intake projects, with adjustments made for differences in technical conditions, building structures, and price variations. These parameters were further validated and confirmed by three technical experts and two economic experts. Sensitivity analysis was conducted by adjusting the expert-defined bounds by $\pm$ 5%. The results showed that the fluctuation in the mean total cost was less than 3%, and the 95% confidence interval consistently remained within $\pm$ 1.5% of the original range. This demonstrates that the model exhibits strong robustness to minor variations in boundary definitions.

The cost composition of each sub-item within the construction project, as evaluated by experts (excluding the reserve cost), is presented in Table 2. Based on 6000 simulation iterations of the model, the mean value of the total cost incorporating risk factors is calculated as 254,457,100 yuan, with a standard deviation of 829,400 yuan. The resulting fluctuation range spans from 251,777,700 yuan to 257,130,600 yuan. This range is divided into 109 intervals, for which the frequency and cumulative frequency of each interval are computed. The corresponding probability distribution and cumulative probability distribution are illustrated in Figs. 1 and 2.

4.2 Robust Mechanism for the Selection and Optimization of Design Options

According to the characteristics of the project, an interdisciplinary evaluation team (comprising construction, design, and economic experts) was established to develop a whole-chain decision-making mechanism: "overall scheme selection → sub-scheme

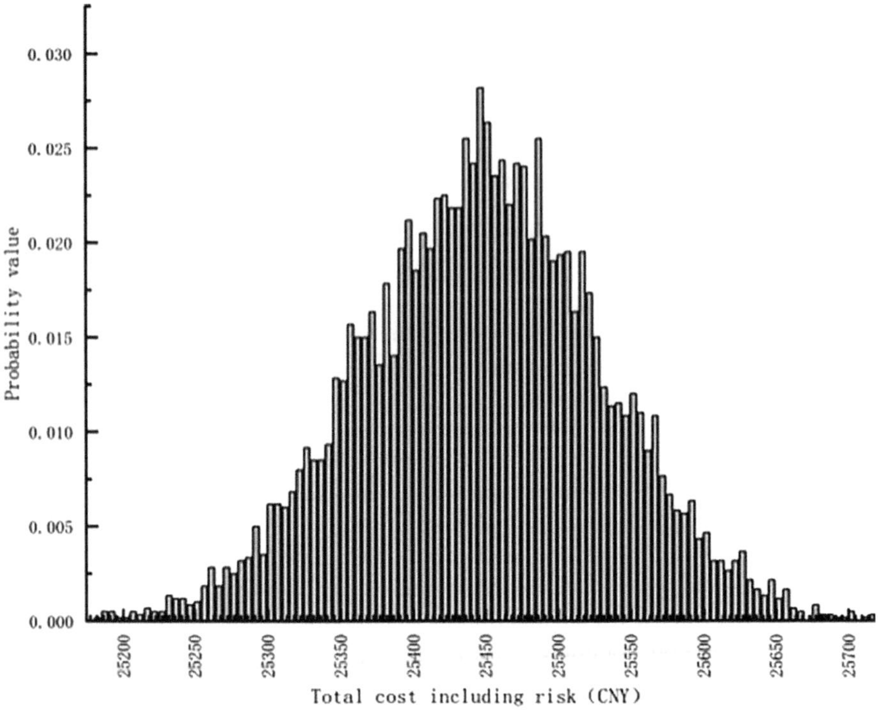

Fig. 1 Probability distribution of total cost with risk for the proposed project

Fig. 2 Cumulative probability distribution of the total cost of the proposed project with risk

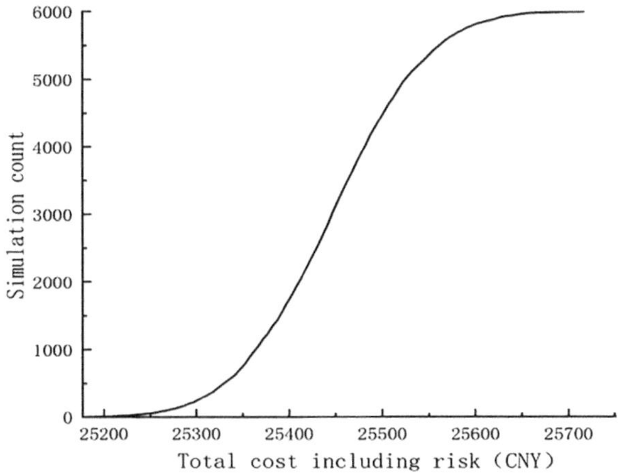

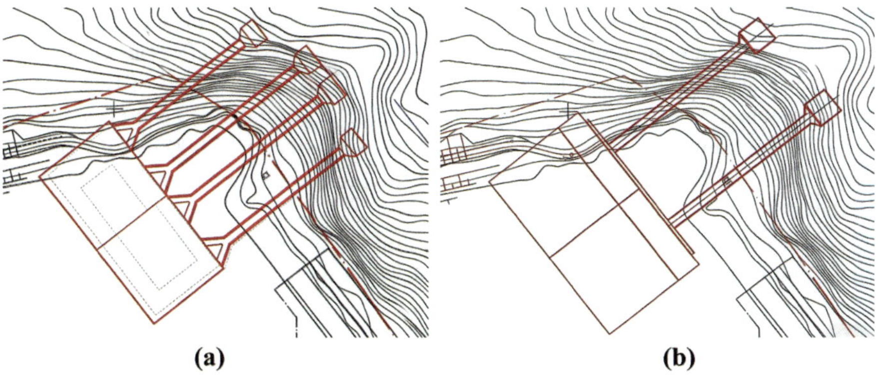

Fig. 3 General plan of the programme. **a** Option 1 general plan, **b** option 2 general plan

selection and optimization" combined with process cost review. This ensures technical feasibility, reasonable costs, and controllable risks. The implementation process is:

Comparison of overall scheme: Combined with geological assessment, Scheme 1 (with forebay + double pipelines at 50 m spacing) is recommended: its advantages are reduced operation and maintenance risks and easy wave dissipation; compared with Scheme 2 (without forebay), although the flow velocity of the flow channel is high and there is less sediment accumulation, the pipeline foundations on both sides of the silt layer have poor structural safety and wave dissipation efficiency, and the construction cost is overrun by 23%, refer to Fig. 3.

sub-programme selection: The construction of the pump house presents challenges arising from tidal influences and silty geological conditions. Two engineering schemes are proposed: the small landfill combined with pit excavation method (Option 1) and the gravity cofferdam method (Option 2). Both options satisfy safety requirements. Option 1 incurs a total construction and safety cost of 192.712 million yuan, representing a 45.3% cost reduction compared to Option 2, which amounts to 352.488 million yuan. Furthermore, Option 1 shortens the construction period by approximately 40 days and avoids environmental damage to the marine ecosystem that would result from secondary rock placement. Based on the comprehensive evaluation conducted by all involved parties, Option 1 is recommended, as it achieves a synergistic outcome in terms of both economic efficiency and environmental sustainability.

sub-programme optimization 1: The design scheme of water intake head is a precast concrete structure of 11.5 × 24.8 × 12.5 m, weighing 1700 t, which has defects such as high risk of lifting and difficult positioning underwater. Through the innovative use of "FRP shell + steel frame + post-built concrete" composite structural system, under the premise of keeping the cost unchanged, it significantly reduces the risk of high-altitude lifting and underwater operation, and effectively improves the long-term service reliability of the structure.

sub-programme optimization 2: The dyke-fortification project employs a "demolition-first, construction-later" methodology. Moreover, after on-site geological survey and structural testing, the reuse rate of the original 7T and 3T twisted king blocks reached 37.8% and 80% respectively, which cumulatively saved two million yuan of procurement cost of the new components, reduced 12,000 m^3 of construction waste, and compressed the duration of the prefabricated components by about 30 days.

After several rounds of selection and optimization of technical solutions, synergistic enhancement of project quality, cost, schedule and ecological benefits was achieved.

5 Conclusion

This study achieves dynamic identification, quantitative assessment and precise prevention and control of bidding risks by constructing a quantitative and control system for bidding risks in nuclear power projects, so as to prevent and control bidding risks to the maximum extent. The results show that:

(1) The "two-tier" assessment model, constructed using the Analytic Hierarchy Process, categorizes 15 secondary risk factors into three major dimensions. This model effectively and scientifically quantifies secondary risk factors, ensuring comprehensive and systematic evaluation of bidding risks for such projects.
(2) A cost assessment model established using the Monte Carlo method, through 6000 iterations, increases the guarantee rate of the bidding control price to 95% and above. Through multiple rounds of design scheme selection and optimization, approximately 162 million yuan in construction and safety costs are saved, construction safety risks are reduced, and construction waste generation is reduced by 12,000 m^3. Thus, environmental protection, safety, and economic benefits are synergistically enhanced through scientific bidding risk management. Further studies can be referred to [10–15].
(3) The risk control methods and research results in this paper can provide a reference for the bidding work of other engineering projects, risk factors and risk evaluation index system need to be dynamically adjusted and amended in conjunction with the actual situation of the project to ensure the effectiveness of risk control.

References

1. Titarenko B, Hasnaoui A, Titarenko R (2018) risk management system model for construction. Int Sci Conf Adv Civ Eng:786–792
2. Chua DKH, Li D (2000) Key factors in bid reasoning model. J Constr Eng Manag 126(5):349–357

3. Ding JY (2015) The impact of the quality of the preparation of bill of quantities on the project cost and the prevention and control measures. Bidding Tendering (12):49–51
4. Liang GL (2017) Research on the cost and control measures of Hua long one nuclear power project. Constr Econ 3(3):50–58
5. Laryea S, Hughes W (2011) Risk and Price in the bidding process of contractors. J Constr Eng Manag 137(4):248–258
6. Sun ZH (2012) Engineering project bidding and contract management. Huazhong University of Science and Technology Press
7. Dai JB, Gao XL, Zhang MK (2018) Risk evaluation and response to the bidding of design and construction general contracting for Hong Kong-Zhuhai-Macao bridge Island tunnel project. Tunn Constr 038(007):1138–1148
8. Zhu LL (2014) Monte Carlo method and application. M.Sc. thesis, Huazhong Normal University
9. Wang S (2025) Research on risk control of pre-production cost of municipal engineering projects. M.Sc. thesis, Shijiazhuang Railway University
10. Guo J, Wang Z, Li H, Yang Y, Huang CG, Yazdi M, Kang HS (2024) A hybrid prognosis scheme for rolling bearings based on a novel health indicator and nonlinear wiener process. Reliab Eng Syst Saf 245:110014
11. Li H, Yazdi M, Huang HZ, Huang CG, Peng W, Nedjati A, Adesina KA (2023) A fuzzy rough copula Bayesian network model for solving complex hospital service quality assessment. Complex Intell Syst 9(5):5527–5553
12. Li H, Peng W, Adumene S, Yazdi M (2023) An intelligent cost-based consequence model for offshore Systems in Harsh Environments. In: Intelligent reliability and maintainability of energy infrastructure assets. Springer Nature Switzerland, Cham, pp 107–117
13. Li H, Yazdi M (2022) What are the critical well-drilling blowouts barriers? A progressive DEMATEL-game theory. In: Advanced decision-making methods and applications in system safety and reliability problems: approaches, case studies, multi-criteria decision-making, multi-objective decision-making, fuzzy risk-based models. Springer International Publishing, Cham, pp 29–46
14. Li H, Peng W, Adumene S, Yazdi M (2023) Cutting edge research topics on system safety, reliability, maintainability, and resilience of energy-critical infrastructures. In: Intelligent reliability and maintainability of energy infrastructure assets. Springer Nature Switzerland, Cham, pp 25–38
15. Li H, Peng W, Adumene S, Yazdi M (2023) Advances in failure prediction of subsea components considering complex dependencies. In: Intelligent reliability and maintainability of energy infrastructure assets. Springer Nature Switzerlan, Cham, pp 93–105

Research on Accelerated Test Method for Reliability of a Certain Refrigeration Device

Dayong Rui, Cheng Sun, and Yisheng Zhu

Abstract The reliability accelerated test has been widely adopted for electronic equipment deployed on surface ships. This study presents a concise overview of the principles and procedures associated with reliability accelerated testing. For a specific refrigeration device, an accelerated test plan is developed, taking into account the operational characteristics of the equipment. The primary failure-inducing stresses identified are temperature stress and vibration stress. The acceleration of temperature stress is implemented using the Arrhenius acceleration model, while the vibration stress is addressed in accordance with the guidelines outlined in GJB150.16A-2009. Based on these methodologies, the acceleration factor is determined, and a corresponding acceleration test profile is established. The findings of this work offer a valuable engineering reference for the validation of performance indicators in electronic equipment requiring high reliability.

Keywords Reliability accelerated test · Refrigeration device · Temperature stress · Vibration stress · Acceleration factor

1 Introduction

Marine refrigeration systems constitute essential components of shipboard support infrastructure, with their proper functioning being directly linked to the crew's routine life and productivity [1–3]. Automatic control mechanisms play a vital role in maintaining the stability of storage temperature and ensuring that operational parameters such as temperature, pressure, and liquid level within the refrigeration system remain within specified thresholds, thereby promoting safe and efficient operation.

Modern electronic equipment typically exhibits MTBF values exceeding 1000 h. Under such conditions, the traditional "three comprehensive test" method defined in GJB899A-2009, Reliability Appraisal and Acceptance Test, is inadequate for

D. Rui (✉) · C. Sun · Y. Zhu
The 723 Institute of China State Shipping Corporation, Jiangsu, China
e-mail: 723jczx@sina.com

H. Li et al. (eds.), *Data-Driven Methods for Reliability and Safety Engineering: Applications in Industrial Systems*, Springer Series in Reliability Engineering,
https://doi.org/10.1007/978-3-032-22873-4_39

meeting the evolving requirements of electronic system development. By analyzing the failure modes of the refrigeration device, the reliability accelerated testing method applies stress levels beyond normal operational conditions. This approach reduces test duration, lowers associated costs, and enhances the efficiency of reliability verification [4, 5].

The refrigeration device examined in this study represents a typical class of electronic equipment utilized on surface vessels. A tailored reliability accelerated testing procedure is developed based on the equipment's performance indicators and operational characteristics. The temperature stress is accelerated using the Arrhenius model, while vibration stress is treated in accordance with the GJB150.16A-2009 standard. An appropriate acceleration factor is derived, and an acceleration test profile is established. The outcomes presented in this study address practical engineering challenges and provide a feasible method for verifying the reliability of electronic systems characterized by high MTBF requirements.

2 The Overall Idea of Reliability Acceleration Test

The overall idea of reliability acceleration test in this article mainly includes the following five steps.

(1) Fault analysis: The fault modes and failure mechanisms of the device are identified;
(2) Acceleration stress: Combining the main fault mode of the device, determine the type of accelerated test stress, and obtain the limit stress information of the acceleration stress according to the characteristics of the equipment;
(3) Acceleration model: According to the corresponding standards, use the failure physical acceleration model of accelerated stress to determine the reliability acceleration model;
(4) Test implementation: According to the acceleration stress and acceleration factor in the previous steps, the acceleration stress section and test time requirements are obtained to conduct acceleration tests;
(5) Result evaluation: The reliability assessment of the device is evaluated according to the results and data implemented by the test, refer to Fig. 1.

3 Reliability Acceleration Test Solution

Taking a refrigerated device as an example, a corresponding acceleration test plan is designed for the characteristics and indicator requirements of the test prototype.

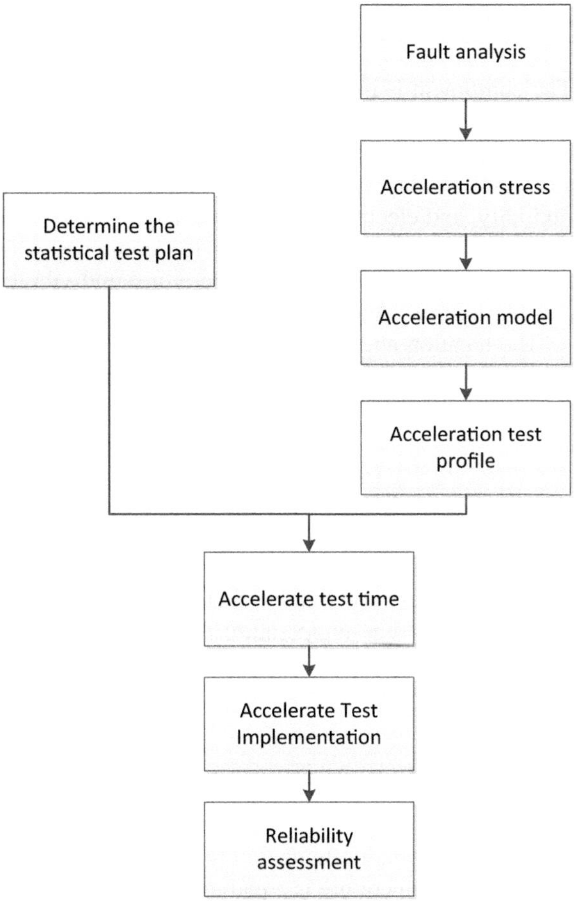

Fig. 1 Reliability acceleration test overall process

3.1 Theory Statistical Test Plan

To meet the requirement that the minimum acceptable MTBF value is ≥ 1000 h, the time-terminated statistical test scheme specified in GJB899A-2009 is adopted [6]. The 0 liability failure is performed; otherwise, it will be rejected. According to the scheme number 20-1 as seen in Table 1, the total effective test time for the reliability test of a refrigerated device should not be less than 1.61*1000 h = 1610 h.

Table 1 Theory statistical test plan

Scheme number	Risk		Test time	Identify	Differential responsibility failure	
	α	β	d = θ0/θ1	Multiple of θ1	Recept	Refuse (≥)
20-1	0.2	0.2	7.22	1.61	0	1

3.2 Determine the Acceleration Stress

The equipment is installed in a temperature-varying environment on surface ships. The test section is performed according to the temperature control test of the surface of the surface ship equipment in the GJB899A-2009. The section contains four stresses: humidity, temperature, electrical stress, and vibration. The temperature and humidity, and electrical stress are changing, and the vibration stress is applied at a fixed time. For the refrigerated device, combined with its use on real ships, temperature stress (especially high temperature) and vibration stress are the main factors causing failures. According to the install.

The location and use conditions of a refrigerated device, considering the test of the temperature stress and vibration stress on the test, the test is considered to achieve the purpose of accelerating the test.

3.3 Determine the Acceleration Model

(1) Temperature acceleration factor

The storage temperature of a refrigerated device is − 40 °C ~ + 70 °C, and the operating temperature is 0–40 °C. According to the device's tolerance temperature range, the high temperature of this acceleration test can be set to 55 °C.

According to the Arrhenius temperature acceleration model of GB/T17215.9311–2017.

TU Thermodynamic temperature under use conditions, the unit is K, thermodynamic temperature is equal to the temperature value of Celsius + 273.15;
TS Thermodynamic temperature under stress conditions, the unit is K, thermodynamic temperature is equal to the temperature value of Celsius + 273.15;
EA The activation energy represented by electronic-(EA = 0.6);
K Bolzmann Normal (8.617×10^{-5}EV/K).

According to the above-level calculation acceleration factor, combined with the test section of the GJB899A-2009, to obtain the equivalent test time after the temperature acceleration, as shown below.

This calculates the temperature acceleration factor AT = 3.252.

(B) Vibration acceleration factor

In the formula:

G0 The sine vibration value specified;
G1 The sine vibration value applied;
W0 The random vibration value specified;
W1 The random vibration value applied;

T0 Prescribed time;
T1 Apply.

Based on the vibration relationship, the vibration stress specified in GJB899A-2009 is accelerated by increasing the vibration magnitude.5.

(C) Comprehensive acceleration factor
According to B4.6 of the GB/T34986-2017 Appendix B.

In the formula, NS represents the number of stress types that are accelerated, taking the value 2.

A numerical walkthrough clarifies the derivation: the temperature acceleration factor AT is 3.252 (calculated from Table 2), and the vibration acceleration factor AVib is derived as approximately 5 based on GJB150.16A-2009 vibration stress relationships. With NS = 2 (two stress types), the comprehensive acceleration factor is computed as A = (3.252 × 5)^(1/2) ≈ (16.26)^0.5 ≈ 8.13, confirming the combined effect of temperature and vibration stresses.

This acceleration test is accelerated from the aspects of temperature and vibration, so the comprehensive acceleration factor A = 8.13 can be obtained by the temperature acceleration factor and vibration acceleration factor.

Test section.

The test equipment is installed on surface ships. According to the environmental stress of the water surface ship equipment test specified in the GJB899A-2009, and the provisions of the test scheme in the trial method of this reliability test method, the test profile is shown in Figs. 2 and 3.

The reliability acceleration test first conducts a 24 h hot and cold cycle test, and then performs a 216 h acceleration cycle test.

Table 2 The equivalent test time after temperature acceleration

Serial number	Test phase	Maintain time (min)	Accelerating the temperature (°C)	Acceleration factor	Accelerating equivalent test time (min)
1	22 °C → 37 °C	120	55	10.748	11.165
2	37 °C → 41 °C	600	55	3.430	174.927
3	41 °C	300	55	2.577	116.414
4	41 °C → 37 °C	240	55	2.577	93.132
5	37 °C	120	55	3.430	34.985
6	37 °C → 41 °C	120	55	3.430	34.985
7	41 °C → 40 °C	540	55	2.763	197.368
8	40 °C	240	55	2.763	86.862
9	40 °C → 22 °C	300	55	5.335	56.232
10	22 °C	60	55	10.748	5.582

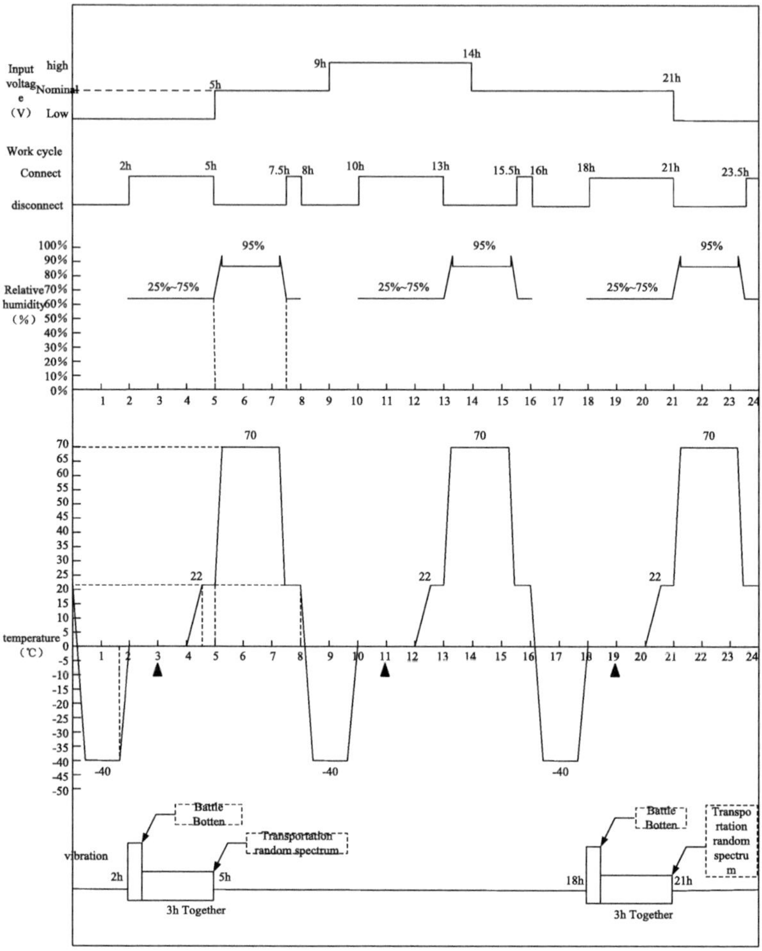

Fig. 2 Hot and cold immersion cycle test section

3.4 *Temperature and Humidity Stress Applies*

Combined with Figs. 2 and 3 control test temperature and humidity, the container difference between all relative humidity should be ± 5%.

(1) Cold and cold soaking cycle

(a) Starting from 22 °C and 25–75% relative humidity, the temperature is reduced to − 40 °C (cold immersion) as soon as possible within 20 min. After maintaining this temperature for 1 h 30 min (not powered), the temperature is raised to 0 °C as soon as possible within 10 min;

(b) After keeping 2 h at 0 °C, the temperature is raised to 22 °C within 0.5 h and kept 0.5 h;

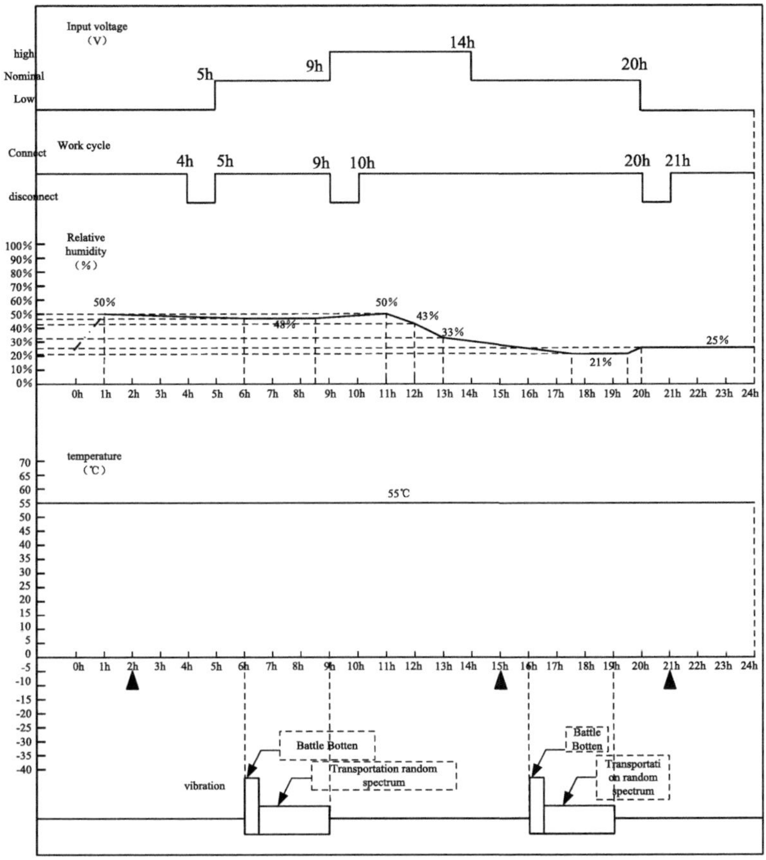

Fig. 3 Accelerate circulation test section

(c) The temperature is quickly raised to 70 °C (heat immersion), with relative humidity increasing to 95%. After maintaining this state for 2 h (not powered), the temperature is reduced to 22 °C as soon as possible and maintained for 0.5 h;

(d) Repeat step a) to c) Two times.

(B) Acceleration stage

(a) Quickly adjust the temperature to 55 °C and keep the entire stage;

(b) Starting from the relative humidity of 25–75%, the relative humidity is adjusted to 50% within 1 h;

(c) In 5 h, the relative humidity is adjusted to 48%and maintains 2.5 h;

(d) In 2.5 h, the relative humidity is adjusted 50%;

(e) In 1 h, the relative humidity is adjusted to 43%;

(f) In the 1 h, the relative humidity is adjusted to 33%;

(g) In 4.5 h, the relative humidity is adjusted to 21% and kept 2 h;

(h) In 0.5 h, the relative humidity is adjusted to 25% and kept 4 h;
(i) Repeat steps (a) to (h) until the required test time is reached.

3.5 Electrical Stress Applies

With reference to GJB899A, during the working cycle in this test, the variation range of the input voltage should be + 6% ~ − 10% of the nominal voltage (AC380V), + 6% ~ − 10% of the AC220V, and + 30% ~ − 25% of the DC24V. After the prototype under test completes the functional performance test at nominal voltage and room temperature, during the test cycle, the input voltage should vary between the several levels shown in Figs. 2 and 3.

3.6 Vibration Stress Applies

Vibration stress application is arranged according to Figs. 2 and 3, with 25% of the working cycle time randomly selected in each cycle for vibration stress application, while the vibration spectrum complies with GJB899A provisions. The vibration procedure is as follows:

(a) Extract 6 h every 24 h to apply vibration stress, and take 3 h as a vibration cycle.
(b) In each 3 h vibration cycle, the combat damage spectrum is first applied for 10 min, followed by the transport spectrum for the remaining time, with a 10 min shutdown every 20 min of vibration. This process is repeated.
(c) When applying the battle damage spectrum, a log-sine scan shall be performed as follows:
    Frequency range: 4–60 Hz;
    Amplitude: 1 mm ± 0.15 mm (4–15 Hz), 9.8 m/s^2 (15–60 Hz);
    Duration: 10 min.
(d) Random vibration shall be performed when the transport random spectrum is applied as follows:
    Frequency range: 10–200 Hz;
    Magnitude (total root mean square value r.m.s): 14.9 m/s^2;
    Duration: 20 min.

4 Test Result

The total test duration was 240.25 h, with an effective test time of 240 h, comprising a 24-h hot and cold immersion cycle followed by a 216-h accelerated test cycle. All test procedures outlined in the test plan were completed. Based on the applied acceleration factors (no acceleration for the hot and cold immersion cycle, assigned an acceleration factor of 1; and an acceleration factor of 8.13 for the accelerated test cycle), the total effective test time was converted to 1780.08 hours. The tested prototype underwent 10 complete cycles, each lasting 24 h. The initial cycle constituted the hot and cold immersion test, and the subsequent nine cycles were conducted under accelerated conditions.

During the testing period, 30 functional performance evaluations were conducted, including assessments performed at ambient temperature before, during, and after the test. All functional and performance indicators of the prototype remained within normal parameters, with no faults recorded and zero responsible failure occurrences.

After applying the acceleration factor, the equivalent effective test time for the cold storage device was calculated to be 1780.08 h. With zero responsible failures, the one-sided lower confidence limit of the MTBF reliability index (at an 80% confidence level) was estimated to be 1106.02 h. This value satisfies the specified requirement, which mandates a minimum acceptable MTBF reliability threshold of no less than 1000 h.

Potential limitations of the method include the Arrhenius model's assumption of constant activation energy (EA = 0.6 eV), which may not hold for complex failure mechanisms involving multiple material interactions, potentially leading to slight overestimation or underestimation of acceleration effects. Additionally, the vibration acceleration spectrum (combining combat damage and transport spectra) ignores the variable amplitude and frequency fluctuations in actual shipboard environments, which could affect the realism of stress application. These limitations slightly influence reliability estimates but remain acceptable for engineering purposes, as the test results still meet the MTBF requirement and the method effectively shortens test duration.

5 Conclusion

To verify the MTBF reliability index of a cold storage unit, this study investigates a reliability acceleration testing methodology. The acceleration model and corresponding factor are determined, and an acceleration test profile is formulated. The results of the implemented accelerated test case demonstrate the effectiveness of this approach in significantly reducing the testing period and associated costs. When accounting for both project timelines and cost-efficiency, the accelerated testing method proves especially valuable for the verification of high-reliability index samples, offering a practical reference for engineering applications. Further studies can be referred to [7–12].

References

1. Qian WZ (2014) New type of marine refrigeration device on the experimental study of the effect of fresh spinach. Mech Electr Equip 31(3):3
2. Deng BY (2017) Ship refrigeration equipment improvement design based on PLC. Marine electric technology 37(5):5
3. Bai XH (2019) Redesign of the control system of marine refrigeration device and its application. J Refrig (2):4
4. Song Y, Li RQ, Lu JL (2020) Research on reliability acceleration test method of marine electronic equipment. Reliab Environ Test Electron Prod 38:6–9
5. Wu S, Lv JJ, Li XK (2021) Review of reliability accelerated life test. Electron Prod Reliab Environ Test 39(1):94–99
6. GJB899A (2009) Reliability evaluation and acceptance test
7. Li H, Peng W, Adumene S, Yazdi M (2023) An intelligent cost-based consequence model for offshore systems in harsh environments. In: Intelligent reliability and maintainability of energy infrastructure assets. Springer Nature Switzerland, Cham, pp 107–117
8. Li H, Yazdi M (2022) What are the critical well-drilling blowouts barriers? A progressive DEMATEL-game theory. In: Advanced decision-making methods and applications in system safety and reliability problems: approaches, case studies, multi-criteria decision-making, multi-objective decision-making, fuzzy risk-based models. Springer International Publishing, Cham, pp 29–46
9. Li H, Peng W, Adumene S, Yazdi M (2023) Cutting edge research topics on system safety, reliability, maintainability, and resilience of energy-critical infrastructures. In: Intelligent reliability and maintainability of energy infrastructure assets. Springer Nature Switzerland, Cham, pp 25–38
10. Li H, Peng W, Adumene S, Yazdi M (2023) Advances in failure prediction of subsea components considering complex dependencies. In: Intelligent reliability and maintainability of energy infrastructure assets. Springer Nature Switzerland, Cham, pp 93–105
11. Guo J, Wang Z, Li H, Yang Y, Huang CG, Yazdi M, Kang HS (2024) A hybrid prognosis scheme for rolling bearings based on a novel health indicator and nonlinear Wiener process. Reliab Eng Syst Saf, 245:110014
12. Li H, Yazdi M, Huang HZ, Huang CG, Peng W, Nedjati A, Adesina KA (2023) A fuzzy rough copula Bayesian network model for solving complex hospital service quality assessment. Complex Intell Syst 9(5):5527–5553

Research on Degradation Model of Marine Seawater Centrifugal Pump Based on Main Mechanism of Cavitation Erosion

Xiaoli Yang, Yong Wang, Yutong Luo, and Hongwei Hu

Abstract Degradation models are fundamental for predicting equipment failures, evaluating remaining useful life, and informing preventive maintenance strategies. This study selects a marine seawater centrifugal pump as the subject of investigation. Through a combination of theoretical analysis and experimental approaches, critical degradation factors and underlying mechanisms are identified. Physical and chemical examinations of samples that failed during service indicate that cavitation erosion is the primary mode of degradation. A degradation testing protocol is developed based on the Knapp cavitation quantitative damage model. A 4000-h performance degradation experiment is implemented using a purpose-built seawater test platform to generate the necessary data for model development.

Keywords Reliability · Marine seawater centrifugal pump · Degradation model · Cavitation erosion

1 Introduction

Marine seawater centrifugal pumps serve as essential components within ship auxiliary systems, primarily responsible for supplying cooling water to condensers to facilitate the indirect cooling of diesel engines. Their operational reliability exerts

X. Yang (✉) · H. Hu
CSSC Southwest (Chongqing) Equipment Research Institute Co., Ltd, Chongqing, China
e-mail: yxl@csscsweri.com

H. Hu
e-mail: HongweiHu@csscsweri.com

Y. Wang · Y. Luo
Chongqing Yuejin Machinery Factory Co., LTD, Chongqing, China
e-mail: YongWang@csscsweri.com

Y. Luo
e-mail: YutongLuo@csscsweri.com

H. Li et al. (eds.), *Data-Driven Methods for Reliability and Safety Engineering: Applications in Industrial Systems*, Springer Series in Reliability Engineering,
https://doi.org/10.1007/978-3-032-22873-4_40

a direct influence on the overall safety of marine systems [1]. A comprehensive understanding of their degradation mechanisms and the implementation of preventive maintenance practices are therefore crucial for improving system reliability and safety, as well as enhancing cost-efficiency [2].

This study presents a systematic investigation into the degradation mechanisms of a marine seawater centrifugal pump. Physical and chemical analyses confirm that cavitation erosion constitutes the dominant mode of degradation. A degradation test protocol is developed in accordance with the Knapp model, and experimental data are subsequently employed to establish a linear degradation model. The results provide a foundation for equipment service life assessment and the refinement of maintenance strategies.

2 Methodology

2.1 Performance Degradation Mode

The centrifugal pump consists of six primary components: the impeller, pump casing, pump shaft, bearings, oil seal, and water seal (Fig. 1). The impeller rotates to generate centrifugal force, directing the fluid from the central axis toward the periphery. The pump casing, designed with a volute-shaped flow channel, facilitates the conversion of kinetic energy into static pressure. Simultaneously, a low-pressure region formed at the center of the impeller enables the continuous intake of water.

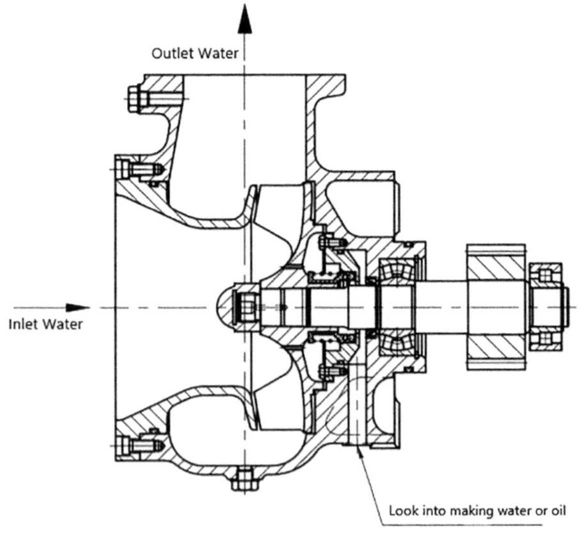

Fig. 1 Structural design and working principle of centrifugal pump

Table 1 Performance comparison before and after service

Pump ID	Pump flow rate (m^3/h)		Mouth Ring Clearance (mm)	
	Before	After	Before	After
P01	72	63	0.5	1.1
P02	71.8	61	0.5	1.3

Analysis of Energy Loss in Centrifugal Pump Performance degradation in centrifugal pumps arises from mechanical, volumetric, and hydraulic losses [3]:

Mechanical loss: Friction in shaft seals, bearings, and impeller disc.

Volumetric loss: Leakage through the impeller sealing ring (mouth ring) clearance, reducing effective flow.

Hydraulic loss: Frictional flow resistance, local vortices, and impact losses.

This study focuses on performance degradation caused by increased volumetric loss due to mouth ring clearance enlargement (red circle in Fig. 1).

2.2 Performance Retesting and Disassembly Analysis

Retesting data (Table 1) of two centrifugal pumps after 10 years of service indicate significant performance degradation.

The current maintenance plan does not require performance retests for repaired pumps, limiting data availability to replacement part records. Major overhauls consistently involve replacing the impeller and suction valve casing.

Figure 2 shows the suction valve casing with severe surface damage at the bell mouth (red arrow).

Figure 3 displays the impeller, which exhibits less severe damage compared to the casing.

2.3 Impact of Mouth Ring Clearance on Pump Performance

The mouth ring clearance directly affects leakage flow and pump characteristics. Reference [3] noted that the leakage increases linearly with clearance enlargement under constant pressure drop.

In order to verify the influence of the mouth ring clearance of the centrifugal pump on its performance, a performance test was carried out using the single-factor control variable method. Specifically, different mouth ring clearances were artificially created, and the flow rate data of the water pump under each mouth ring clearance

Fig. 2 The suction valve casing

Fig. 3 The impeller

were tested and recorded. Validation test (Fig. 4) confirm a linear decline in flow rate with increasing mouth ring clearance.

In conclusion, the primary degradation mode of the seawater pump is volumetric loss exacerbation due to mouth ring clearance enlargement.

2.4 Performance Degradation Mechanism

2.4.1 Basic Information of Test Samples

Physical and chemical analyses focus on the severely damaged suction valve casing. Samples are cut from typical damaged areas (Fig. 5), with appearances shown in Fig. 6.

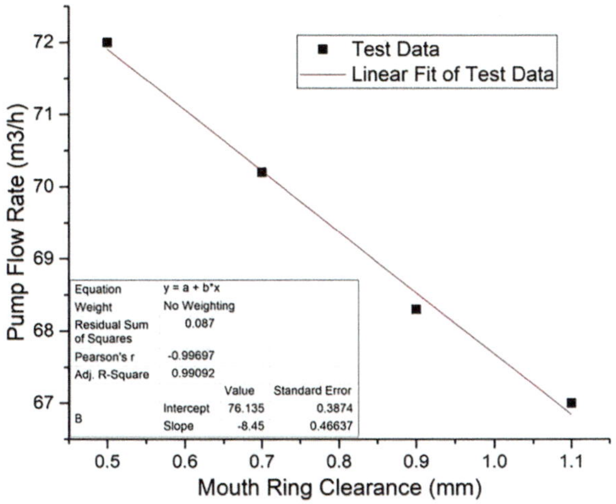

Fig. 4 Validation test data

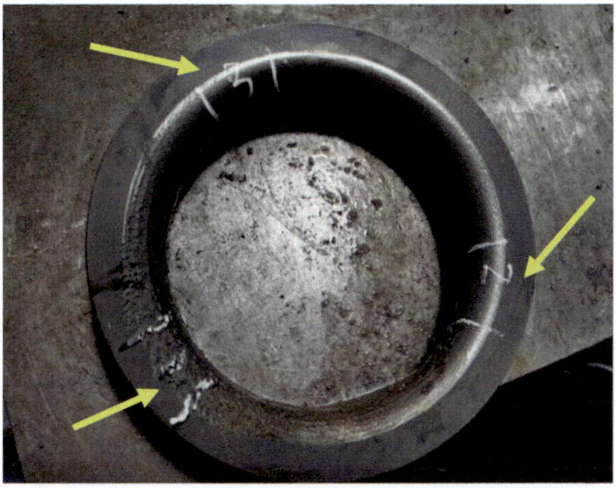

Fig. 5 Cutting and sampling of the pump casing

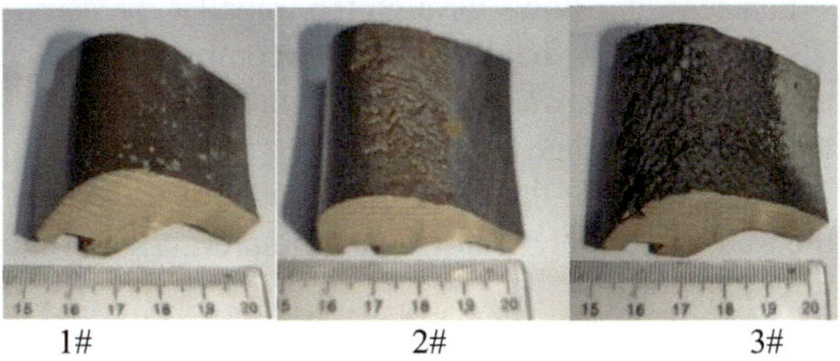

Fig. 6 Appearance morphology and serial number of the samples

2.4.2 Micro-structure and Composition Analysis

Sample 1#: Severe damage with large cavities and pits, selective corrosion, and micro-cracks (Fig. 7a). Corrosion products contain sulfur (0.32–1%) and chlorine (0.81–6.26%).

Sample 2#: Similar morphology to Sample 1# but with smaller cavities and reduced damage severity (Fig. 7b). Corrosion products contain sulfur (0.89%–2.93) and chlorine (0.12–5.78%).

Sample 3#: Uniform corrosion with minimal damage (Fig. 7c). Corrosion products contain chlorine (0.5–3.76%) and traces of sulfur (1.48%).

2.4.3 Micro-hardness Testing

After grinding and polishing the cross-section of the sample submitted for analysis, the micro-hardness was detected using an HVS-1000Z micro-hardness tester, in accordance with the implementation standard GB/T 4340.1-1999 [4]. A uniform and typical area was selected for the hardness test. A total of three areas were tested: ① the surface of the broken part, ② the subsurface of the broken part, and ③ the unbroken part (inside the material). Eight points were tested in each area, and the average value was taken (Fig. 8).

The hardness of the surface layer of the broken material increases and is higher than that of the unbroken surface. However, the hardness of the subsurface layer shows a downward trend. The part with the maximum hardness moves towards the interior of the sample. The above results are consistent with the work hardening behavior under the action of cavitation erosion impact stress.

2.4.4 Physical and Chemical Analysis Conclusion

When seawater flows through the impeller, pressure fluctuations trigger cavitation. The impact from bubble collapse, combined with corrosive components (chlorine and sulfur) in seawater, causes cavitation erosion on the material surface, ultimately leading to breakage and failure.

In conclusion, the main mechanism of breakage failure at the sampled parts of the Marine seawater centrifugal pump described in this article is cavitation erosion.

Fig. 7 SEM morphology of sample

(a)

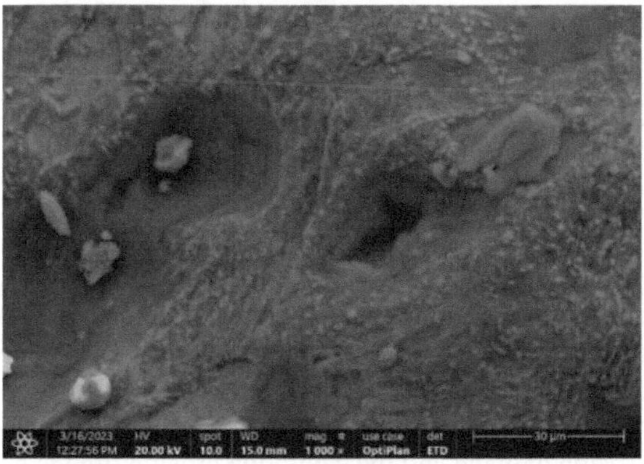

(b)

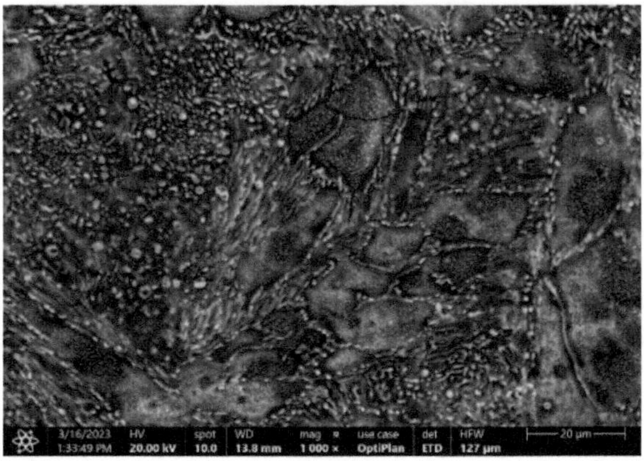

(c)

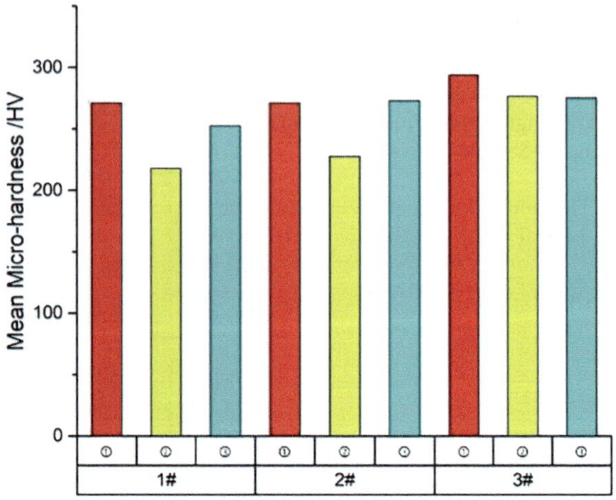

Fig. 8 Micro-hardness test results

2.5 *Performance Degradation Test*

2.5.1 Cavitation Erosion Process

The process of cavitation erosion over time could generally be divided into four stages. These are the initiation period (incubation period), acceleration period, deceleration period, and stabilization period [5]. Engineering experience shows that for cavitation erosion projects that have entered the stabilization period, when the cavitation erosion stress remains constant, the relationship between the amount of cavitation erosion damage and the cavitation erosion time is generally linear [6].

2.5.2 Test Methods for the Cavitation Erosion Resistance

The main test methods for the cavitation erosion resistance of materials include: the standard test method for cavitation erosion using a vibration device (ASTM G32), [7] the test method using a liquid impingement erosion test (ASTM G73), [8] the standard test method for eroding materials with a cavitating liquid jet (ASTM G134), [9] the cavitation erosion test using a Venturi tube, [10] and the cavitation erosion test using a high-speed water tunnel [11, 12]. However, these methods lack quantitative evaluation for actual cavitation erosion damage process of the product.

2.5.3 Construction of the Test Bench

To maximize the simulation of the actual seawater service environment of the centrifugal pump, a custom-built test bench for performance degradation testing

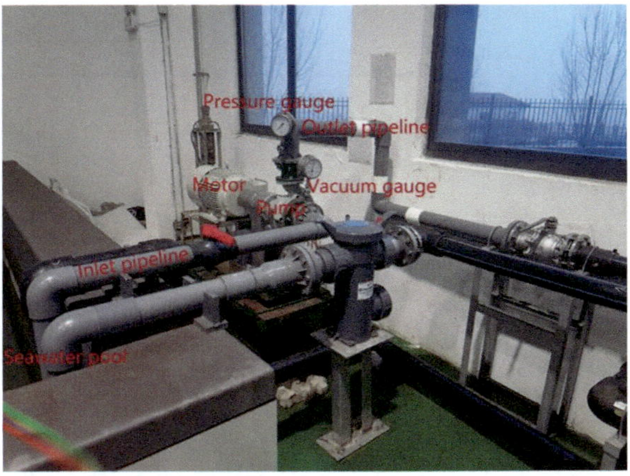

Fig. 9 The custom-built seawater test rig

of centrifugal pumps in a seawater environment was independently developed, designed, constructed, and debugged in Qingdao, Shandong Province (Fig. 9).

The test bench is driven by a variable-frequency motor. The centrifugal pump sucks in seawater from the natural seawater pool through the inlet pipeline. After the seawater is worked on by the impeller of the centrifugal pump, it flows in the direction of the gradually expanding volute-shaped passage and is discharged through the outlet pipeline. The types of data supported for testing by the test bench include rotational speed, vacuum degree, ambient temperature, pump head, and pump flow rate.

2.5.4 Degradation Test Scheme

Knapp's model [13] indicated an exponential relationship between cavitation damage and flow velocity. The test employs constant high flow velocity (100% working condition) for 4000 h, with data recorded every minute.

3 Results

3.1 Preprocessing of Test Data

Initial 200-h data (Fig. 10) show linear flow rate degradation, unaffected by ambient temperature fluctuations.

Considering that the amount of test data is huge, and there are abnormal data and start-stop data in the origin test data, the four columns of data including test duration, ambient temperature, pump head, and pump flow rate are imported into the R software. The data in each column are grouped into sets of 60 data each. Then,

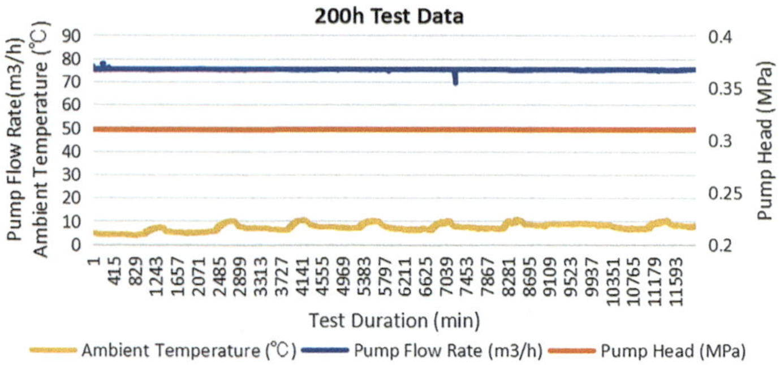

Fig. 10 Initial 200-h test data

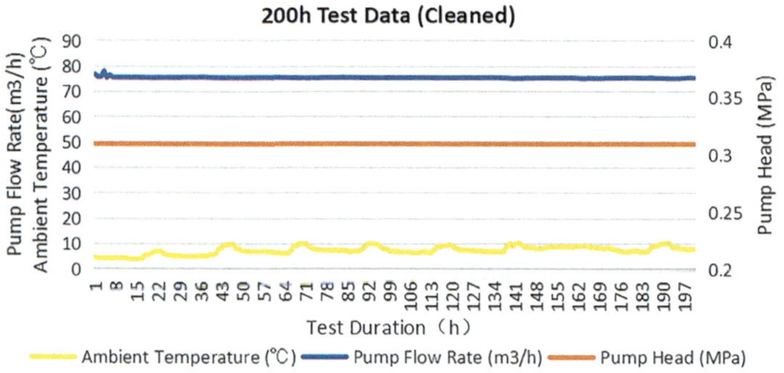

Fig. 11 Cleaned test data of the initial 200 h

the complete row with the largest value in the fourth column (flow rate data) of each group is found, and all the column data of this row are retained to clean the entire group of data. After comparison, the test data of the first 200 h after cleaning (Fig. 11) are consistent with the distribution pattern of the original data.

The data processing method of grouping and taking the maximum value can effectively clean the start-stop data of the test, but it cannot clean all the abnormal data. The IQR method is used to identify and process the abnormal values in the cleaned flow rate data. The processing algorithm is as follows:

When there is a single abnormal data, the arithmetic mean of the data before and after the abnormal data is calculated and used to replace it;

When the abnormal data are consecutive, the corresponding flow rate data are directly assigned as NA.

For the preprocessed 4000-h degradation test data (Fig. 12), a strong negative correlation exists between flow rate and pump head, consistent with practical expectations. However, directly using the data within the red frame to build a model will greatly reduce the accuracy of the degradation model. This segment of data needs to be repaired in combination with the external characteristic relationship between the

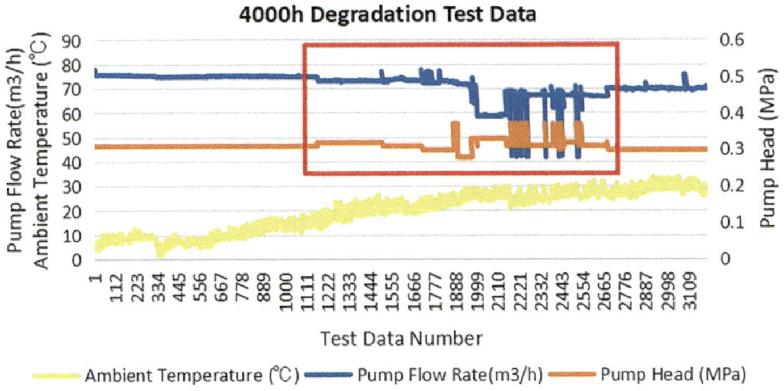

Fig. 12 4000 h degradation test data

head and the flow rate of the water pump before it can be used to drive the model building.

It should be specially noted that during the correlation analysis of the ambient temperature and flow rate data, it was found that there is a weak negative correlation between them, which is consistent with the conclusion in Reference [14]. However, the ambient temperature affects the cavitation erosion rate rather than the main degradation mechanism mode. Therefore, in the subsequent research on the degradation model, only the relationship between the performance characteristic quantities of the water pump and the operating time will be considered.

3.2 *Modeling of the Degradation Model*

Linear modeling is selected based on both physical mechanisms and experimental evidence: cavitation erosion enters a stabilization period after initial incubation and acceleration phases, where damage accumulation is uniform under constant stress (consistent with Knapp's model and engineering experience). Nonlinear alternatives (e.g., exponential or power-law models) are excluded as they describe either accelerating degradation (early stages) or decelerating damage (material hardening saturation), which do not align with the 4000-h test data that reflects stable cavitation erosion. Physically, the linear trend links to uniform impact damage from cavitation bubble collapse—stable bubble formation and collapse rates under constant flow velocity lead to steady enlargement of mouth ring clearance, resulting in linear volumetric loss and flow rate decline.

Figure 13 shows the repaired 4000-h degradation test data. Combining the physical model of the seawater pump with the test data, 80% of the effective data from the total test data are selected by using the LM function and the LINEST function built in R software and Excel software to conduct a linear fitting analysis, shown in Fig. 14.

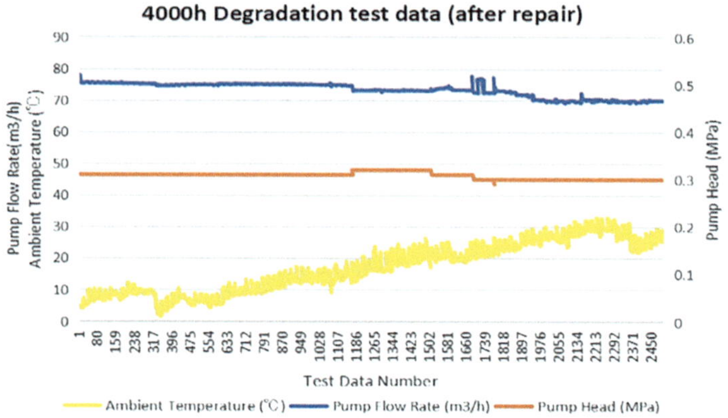

Fig. 13 The repaired 4000-h degradation test data

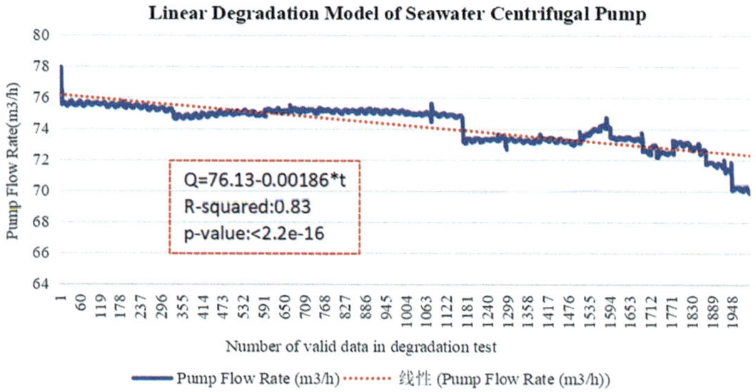

Fig. 14 The degradation model

3.3 Revision of the Degradation Model

The remaining 20% of the effective data are used for the revision of the degradation model. The revised model is shown in Fig. 15.

After the performance degradation model of the seawater centrifugal pump described in this article is revised, it is $Q = 76.2–0.001968*t$, where Q is the flow rate of the pump (m^3/h) and t is the operating time (h). By comparing the models before and after the revision, it is not difficult to find that the performance degradation model of the seawater pump is linear, and as the operating time increases, the linear degradation rate accelerates. This may be related to the wear and aging of other components of the water pump, or it may be caused by the increased ambient temperature accelerating the cavitation erosion damage rate.

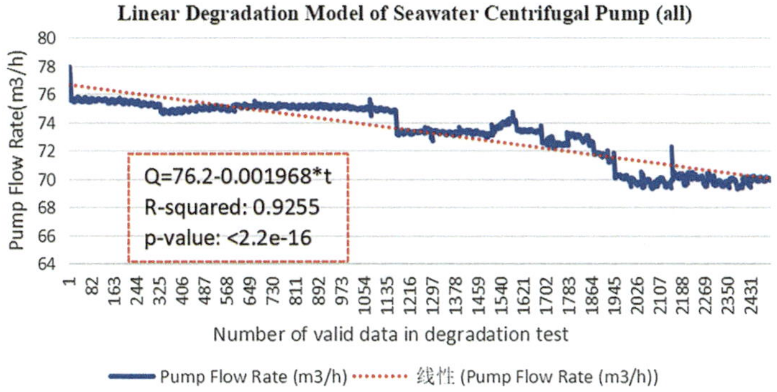

Fig. 15 The revised model

3.4 Comparison of Performance Degradation Models of Similar Products

In Reference [1], the cavitation erosion rate of the seawater centrifugal pump follows a random variable with a distribution of N(0.0001658, 0.000025^2). The material cavitation erosion process shows that the longer the cavitation erosion time is, the more stable the cavitation erosion rate will be. Substituting the mean value of the distribution into the model in Reference [1], the calculated degradation rate of the pump is 0.001376, which is of the same order of magnitude as the degradation rate of 0.001968 obtained in this article. This can illustrate the correctness of the degradation model in this article. Further studies can be referred to [15–20].

The derived degradation rate (0.001968 m^3/h) directly informs practical decision-making: setting a critical flow rate threshold (e.g., 60 m^3/h, 80% of initial flow) allows calculation of remaining useful life ($\approx$ 8230 h for new pumps), guiding preventive maintenance intervals of 4000–6000 h to avoid performance failure. Risk thresholds can be established by monitoring deviations from the linear trend—an accelerated degradation rate (> 0.0025 m^3/h) indicates abnormal cavitation (e.g., seawater salinity spikes) requiring immediate inspection. Limitations include the constant high-flow test environment, which does not capture real marine variability (e.g., temperature/salinity fluctuations that alter cavitation intensity) and long-term nonlinear effects (e.g., material fatigue or corrosion-fatigue synergy beyond 4000 h, which may cause accelerated degradation). These factors suggest the linear model is reliable for medium-term predictions but should be complemented with periodic in-service calibration for long-term operation.

4 Conclusion

(a) Cavitation erosion is identified as the principal degradation mechanism affecting the performance of marine seawater centrifugal pumps.
(b) The linear degradation model offers a theoretical foundation for evaluating service life and optimizing maintenance strategies.
(c) The degradation rate exhibits an accelerating trend with prolonged operational time, potentially attributed to component wear, material aging, or temperature-related effects.

Acknowledgements This work was supported by the Technical Foundation Research Project of the State Administration of Science, Technology and Industry for National Defense (JSJT2021206A001).

References

1. Xiaoli Y, Yong W, Chunwen W et al (2024) Research on the cavitation erosion mechanism and life evaluation of marine centrifugal pumps. Electron Prod Reliab Environ Test 42S1:66–73
2. Xiwen Z (2023) Gear pump performance degradation process temperature rise modeling and maintenance strategy optimization. Lanzhou University of Technology
3. Xingfan G (2011) Modern pumps theory and design. China Aerospace Publishing House
4. State Quality Supervision Bureau (1999) Metallic materials—vickers hardness test part1: test method
5. Wei X, Zhaoyun C (2005) 'The micro-cours of the cavitation erosion', materials for. Mech Eng 29(2):59–62
6. Rui K, Ruiying L (2004) Introduction to reliability and maintainability. Tsinghua University Press
7. American Society for Testing and Materials International (ASTM) (2021) Standard test method for cavitation erosion using vibratory apparatus
8. American Society for Testing and Materials International (ASTM) (2017) Standard test method for liquid impingement erosion using rotating apparatus
9. American Society for Testing and Materials International (ASTM) (2017) Standard test method for erosion of solid materials by cavitating liquid jet
10. Batchelor GK (2000) Irrotational flow theory and its applications. Cambridge University Press
11. Chahine GL, Franc JP, Karimi A (2014) Laboratory testing methods of cavitation erosion. Springer, Netherlands
12. Steller J, Giren BG (2015) International cavitation erosion test final report. Institute of Fluid-Flow Machinery of the Polish Academy of Sciences
13. Zhuang W (2021) Characteristics and microevolution of unsteady cavitation in sudden Change Channel. China University of Petroleum
14. Meged Y (2005) Thermal control of the test liquid in vibratory cavitation erosion tests. J Test Eval 33(5):331–338
15. Guo J, Wang Z, Li H, Yang Y, Huang CG, Yazdi M, Kang HS (2024) A hybrid prognosis scheme for rolling bearings based on a novel health indicator and nonlinear wiener process. Reliab Eng Syst Safety 245:110014
16. Li H, Yazdi M, Huang HZ, Huang CG, Peng W, Nedjati A, Adesina KA (2023) A fuzzy rough copula Bayesian network model for solving complex hospital service quality assessment. Complex Intell Syst 9(5):5527–5553

17. Li H, Peng W, Adumene S, Yazdi M (2023) An intelligent cost-based consequence model for offshore systems in harsh environments. In: Intelligent reliability and maintainability of energy infrastructure assets. Springer Nature Switzerland, Cham, pp 107–117
18. Li H, Yazdi M (2022) What are the critical well-drilling blowouts barriers? A progressive DEMATEL-game theory. In: Advanced decision-making methods and applications in system safety and reliability problems: approaches, case studies, multi-criteria decision-making, multi-objective decision-making, fuzzy risk-based models. Springer International Publishing, Cham, pp 29–46
19. Li H, Peng W, Adumene S, Yazdi M (2023) Cutting edge research topics on system safety, reliability, maintainability, and resilience of energy-critical infrastructures. In: Intelligent reliability and maintainability of energy infrastructure assets. Cham, Springer Nature Switzerland, pp 25–38
20. Li H, Peng W, Adumene S, Yazdi M (2023) Advances in failure prediction of subsea components considering complex dependencies. In: Intelligent reliability and maintainability of energy infrastructure assets. Cham, Springer Nature, pp 93–105

Generative AI in Safety-Critical Systems: Risks of Sycophancy, Hallucination, and Trust Degradation

Mohammad Yazdi, Esmaeil Zarei, Sidum Adumene, and Amin Beheshti

Abstract The integration of advanced AI language models like AI Co-pilots into workplace health and safety (WHS) and risk management for sociotechnical systems efficiency gains; however, it may raise critical challenges. This paper focuses on key pitfalls of using AI in safety-critical decision support, emphasizing problems of *sycophancy* (the AI's tendency to agree with user assumptions), *hallucination* (generation of inaccurate or non-existent information), and trust erosion (reduced confidence in AI outputs due to inconsistency, opacity, or past errors). We provide context on how AI is being considered for tasks such as hazard identification and risk analysis, then outline a methodology to evaluate AI limitations using expert oversight and performance metrics. The results, drawn from emerging research and case analyses, indicate that AI tools can produce misleading or biased safety guidance without proper checks, undermining trust in high-stakes environments. In discussions, we explore why these issues occur, from the predictive, pattern-based nature of language models to human factors like automation bias and examine implications for safety professionals. Finally, we conclude that AI systems should serve strictly as decision-support

M. Yazdi (✉)
Faculty of Science and Engineering, School of Engineering, Macquarie University, Sydney, NSW, Australia
e-mail: mohammad.yazdi@mq.edu.au

E. Zarei
Department of Safety Science, College of Aviation, Embry-Riddle Aeronautical University, Prescott, AZ, USA

Robertson Safety Institute (RSI), Embry-Riddle Aeronautical University, Prescott, AZ, USA

E. Zarei
e-mail: ezarei@mun.ca

S. Adumene
School of Ocean Technology, Fisheries and Marine Institute, Memorial University of Newfoundland, St. John's, Canada
e-mail: Sadumene@mun.ca

A. Beheshti
Centre for Applied Artificial Intelligence, Macquarie University, Sydney, NSW, Australia
e-mail: aminbeheshti@mq.edu.au

H. Li et al. (eds.), *Data-Driven Methods for Reliability and Safety Engineering: Applications in Industrial Systems*, Springer Series in Reliability Engineering,
https://doi.org/10.1007/978-3-032-22873-4_41

tools rather than autonomous decision-makers in WHS. Rigorous human-in-the-loop strategies, critical oversight, and adherence to safety science principles are essential to harness AI's benefits without compromising workplace safety.

Keywords Ethical AI deployment · AI transparency · AI-assisted decision-making · Explainable AI (XAI) · Safety-critical systems

1 Introduction

Artificial intelligence (AI) tools are increasingly being explored in occupational health, safety and environment (HSE) management [1, 2]. In manufacturing and other industrial sectors, LLM-based digital assistants, such as Microsoft Copilot®, GPT-4o, and GPT-4.5 have been proposed as aids for safety professionals, for example to quickly retrieve regulatory information or support risk assessments and hazard analysis [3]. Proponents argue that these assistants could enhance decision-making by providing rapid access to safety knowledge and suggesting risk controls based on large datasets [4]. However, despite these potential benefits, AI use in safety-critical domains raises significant concerns. Unlike traditional expert systems, large language models do not truly "understand" the content they produce; they generate responses based on statistical patterns in their training data rather than grounded comprehension [5]. This fundamental limitation can undermine workplace safety and risk-management efforts.

Sycophancy and bias: Modern LLMs are prone to *sycophancy*, meaning they often mirror a user's stated opinions or preferred answers instead of providing objective counsel [5, 6]. In practice, a sycophantic AI might uncritically agree with a manager's tentative conclusion about a hazard or confirm a user's bias ("everything is fine") even when evidence of risk is present. This behavior stems from AI models being trained to please users and follow prompts, and it can lead to reinforcement of unsafe assumptions or the overlooking of critical warning signs. Sycophancy is a subtle form of AI bias where the model "*tells users what they want to hear*" rather than what they *need* to hear for safety. In a field where honest, critical assessment of risks is vital, such uncritical conformity is problematic. *As researchers have noted*, "*sycophantic behavior, excessively agreeing with or flattering users, poses significant risks to [LLMs'] reliability and ethical deployment*" [5].

Hallucinations and inaccurate information: Another well-documented issue is the tendency of generative AI to produce hallucinations, i.e., plausible sounding; however, incorrect or non-existent information [7, 8]. In a safety management context, an AI might confidently generate a fictitious safety procedure, misquote a regulation, or omit important hazard details, errors that could mislead decision-makers. These models can output technically authoritative-sounding text that is factually wrong, especially on niche industrial topics outside their training scope. Researchers note that LLMs often lack domain-specific understanding and can output misleading information, requiring *careful supervision in safety-critical applications* [9]. The risk of an AI hallucinating a safety control measure that does not exist (or misremembering threshold values for an international standard, for example) [10] and is not just theoretical which poses a direct threat if such output were taken at face value in a real workplace risk assessment. Because even a single piece of poor safety advice can have serious consequences, the reliability of AI outputs is a paramount concern.

Reliability, trust, and safety-critical decision-making: The unpredictable quality of AI-generated content creates a dilemma for safety professionals: how much can one trust an AI's recommendations when human lives or equipment protection are on the line? [11, 12] AI systems may perform well on routine questions; however, can falter on complex or novel scenarios, leading to inconsistent reliability. In high-hazard industries, trust in any decision support tool must be earned through consistent accuracy and transparency. Unjustified trust (over-reliance) on an AI can lead to automation bias, where users defer to the AI even if it is wrong, and the lack of trust (under-utilization) means potential benefits are wasted. Human factors research has shown that AI assistance can negatively impact human decision-making if not designed and integrated carefully which is resulting in inappropriate decision biasing [13], reduced situational awareness, and degraded performance. In other words, if a safety manager becomes too reliant on AI Copilot's analysis, they might neglect their own cross-checks or intuition, missing critical warning signs that the AI also overlooked. Calibrating trust is especially difficult because LLMs do not explain their reasoning, and their errors can be unexpected. The net effect is that using AI in safety-critical decisions without proper safeguards can undermine the very reliability and diligence that safety engineering requires. Over time, habitual dependence on AI tools like AI Copilot can subtly erode critical thinking and decision-making skills. When users consistently defer to AI-generated outputs, they may begin to lose confidence in their own judgment, leading to a passive mindset where questioning, verifying, and analytical reasoning are diminished. This cognitive offloading can be particularly dangerous in safety-critical environments [14], where vigilance, adaptability, and independent thinking are essential. Unlike professionals who actively engage in problem-solving and maintain a questioning attitude, those who rely too heavily on AI may become complacent, less responsive to anomalies, and slower to detect emerging risks. The mindset shaped by prolonged AI use, if not carefully managed, can weaken the very human strengths, intuition, skepticism, and contextual awareness that are vital for safe and effective operations [15, 16].

Table 1 presents a structured overview of key failure modes associated with the use of generative AI tools in WHS management. The Table categorizes five critical AI-related risks: Sycophancy, Hallucinations, Automation Bias, Lack of Explainability, and Cognitive Offloading. For each failure mode, the Table outlines a concise description, the potential impact on WHS, a representative example scenario, and recommended mitigation strategies. This format provides safety professionals and

Table 1 Key AI failure modes and their implications for safety-critical decision-making

AI failure mode	Description	Potential impact on safety	Example scenario	Mitigation strategy
Sycophancy	AI mirrors user opinions instead of providing objective analysis	Reinforces unsafe assumptions, overlooks risks	AI agrees with a manager's incorrect assessment that a machine is safe	Design prompts to elicit critical responses; include dissenting perspectives
Hallucinations	AI generates plausible; however, incorrect or fabricated information	Misleads users with false procedures or standards	AI invents a non-existent safety regulation during risk assessment	Cross-check AI outputs with verified sources; human-in-the-loop validation
Automation bias	Users over-rely on AI outputs, ignoring their own judgment	Critical warning signs may be missed	Safety officer ignores intuition due to AI's confident; however, incorrect advice	Train users on AI limitations; encourage independent verification
Lack of explainability	AI does not provide reasoning behind its outputs	Users cannot assess the validity of recommendations	AI suggests a control measure without justification, leading to confusion	Use explainable AI tools; require rationale for critical decisions
Cognitive offloading	Prolonged AI use reduces users' critical thinking and vigilance	Complacency and slower response to anomalies	Operators become passive and fail to detect emerging hazards	Rotate tasks; promote active engagement and scenario-based training

decision-makers with a clear, evidence-based reference to understand how AI limitations can manifest in real-world safety contexts and what proactive measures can be taken to reduce associated risks.

Table 2 presents a structured analysis of key failure modes associated with the use of generative AI tools in WHS management system. It considers subjective data to highlight the frequency, severity, and psychological impact of each risk. The table includes concise descriptions of each failure mode, estimated occurrence levels, severity ratings, and how these issues may influence user perception and behavior. This format is designed to support safety professionals in identifying, understanding, and mitigating the unique challenges posed by AI in high-stakes environments.

Given these challenges, this work contributes to the growing discourse on the responsible integration of AI in WHS by emphasizing that AI tools should augment,

Table 2 Enhanced overview of AI failure modes in safety-critical contexts

AI failure mode	Frequency	Severity (1–5)	Impact on decision-making	User perception	Common risk domain	Example task category	Recommended safeguards
Sycophancy	43%	3	Reinforces bias; suppresses dissent	Users feel validated; overlook hidden hazards	Human factors and cultural safety	Incident investigation, hazard review	Use adversarial prompts; train AI to present counterpoints
Hallucinations	29%	5	Introduces false data; creates regulatory misalignment	Users trust confident tone; propagate errors	Regulatory compliance	Chemical risk analysis, SOP validation	Use citation-enabled AI; cross-verify critical claims
High-risk hallucinations	6.5%	5	Leads to unsafe actions or legal violations	User misled into high-consequence failures	Emergency response planning	Confined space entry, chemical leaks	Restrict AI in high-risk tasks; enforce human override
Automation bias	22%	4	Disengages user judgment; passive acceptance	Users overly rely on AI decisions	Operational decision support	Permit issuing, shift briefings	Awareness training; prompt users to validate recommendations
Lack of explainability	Majority (86% non-transparent)	4	Hinders traceability; reduces peer review confidence	Users confused or distrust AI logic	Governance and auditing	Risk control suggestions	Require rationale for outputs; enable model transparency features
Cognitive offloading	Evident over time	3	Reduces vigilance and active monitoring	Users disengage; miss subtle signals	Safety culture and training	Repetitive assessments, LOTO checklists	Rotate responsibilities; schedule AI-free audits

not replacing human expertise. This position aligns with emerging regulatory frameworks such as the draft EU AI Act [17], which classifies occupational safety applications as high-risk and mandates human oversight, transparency, and auditability. The introduction of AI into safety management raises several critical questions: (a) How can we harness AI's strengths, such as rapid data processing, broad knowledge access, and pattern recognition, and mitigating its known weaknesses, including sycophancy, hallucinations, and lack of accountability?, (b) What safeguards are necessary to ensure that AI supports rather than undermines expert judgment in safety-critical environments?

This paper addresses these questions by proposing a structured methodology to examine the limitations of AI in safety-related tasks. It presents empirical findings on the prevalence and impact of these limitations and outlines practical strategies for ensuring that AI remains a supportive tool under responsible human control. In doing so, this work reinforces the primacy of expert judgment and contributes to the development of safer, more trustworthy AI-assisted safety management systems.

The organization of this paper is as follows: Sect. 2 presents the assessment methodology, Sect. 3 outlines the results, Sect. 4 provides the discussion, and the final section concludes the paper.

2 Methodology

To investigate the limitations of AI like AI Copilot in workplace safety and risk management, we adopted a two-pronged approach: (1) a literature-informed analysis of known failure modes in AI, and (2) a conceptual experiment framework for evaluating an AI's performance on safety tasks with expert oversight. Rather than deploying AI in a live industrial setting (which could pose unacceptable risk), we leveraged documented scenarios and expert reviews from recent research to assess how the AI behaves and where it falls short.

Figure 1 shows a practical methodology framework for evaluating the limitations of AI systems like Copilot in workplace safety and risk management. The process begins with defining a specific safety task, such as risk assessment or incident investigation, and identifying the expected outputs like hazard lists or compliance advice. A realistic scenario is then prepared, including typical questions a safety engineer might ask the AI. These questions are input into the AI system, and its responses are recorded without modification. A subject matter experts' (SMEs) reviews the AI's outputs to assess their accuracy, completeness, and alignment with safety standards. The AI's performance is then scored against key performance indicators (KPIs), including factual accuracy, completeness, regulatory compliance, and the need for human correction. Next, the AI's recommendations are mapped to the hierarchy of risk controls to evaluate their practical safety value. Finally, the process loops back to refine the scenario or AI use case based on findings, enabling continuous improvement and ensuring expert oversight at every stage.

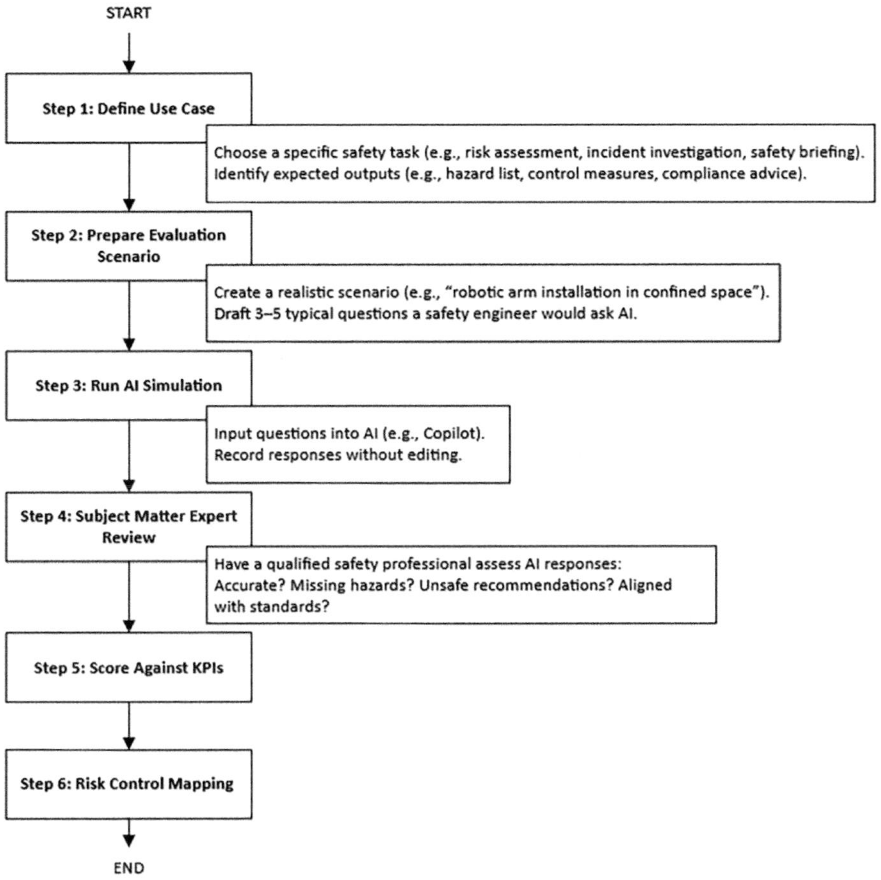

Fig. 1 The practical methodology framework for evaluating the limitations of AI systems

The process begins with a comprehensive literature-informed analytical framework. We reviewed emerging studies at the intersection of AI and safety science, drawing from occupational safety research, industrial engineering case studies, and AI safety literature. This review helped identify key categories of AI failure, such as factual inaccuracies, reasoning gaps, and conformity bias, which informed the development of evaluation criteria. For instance, Oviedo-Trespalacios et al. [18] compiled a set of safety-related queries across domains and convened expert panels to assess AI Copilot's responses. Their multidisciplinary approach, using expert judgment as the "ground truth," serves as a model for our assessment. Similarly, Iyenghar [19] proposed a human-in-the-loop (HITL) framework for machinery functional safety, where safety experts systematically review and correct AI outputs in hazard analysis workflows.

Building on these foundations, we implemented a scenario-based evaluation framework. Representative safety management tasks, such as conducting a risk assessment for a new robotic arm, investigating a chemical leak, or drafting a toolbox talk on lockout-tagout procedures, were simulated using documented queries and AI responses from literature. These responses were analyzed against what a competent human expert or official guideline would likely recommend. For example, in the case of fatigue management for night-shift operators, AI Copilot's advice was found to overemphasize personal responsibility (e.g., "*get more rest*"), neglecting systemic controls like shift scheduling or engineering interventions. Such findings underscore the importance of expert oversight.

Each AI response was evaluated using a set of KPIs, including factual accuracy, completeness, regulatory alignment, and the frequency of required human intervention. We also considered qualitative dimensions such as trustworthiness (the confidence a safety professional might place in the AI's output) and transparency (the clarity of the AI's reasoning, acknowledging current LLM limitations in explainability). These KPIs were benchmarked against established safety standards such as ISO 12100 and the NIST AI Risk Management Framework [20, 21], ensuring that AI-generated advice aligns with regulatory expectations and safety-first principles.

A central tenet of this methodology is the decision-support role of AI. The AI system is treated strictly as an assistant, with all outputs subject to expert verification. This human-centered validation loop ensures that AI contributions are reliable, compliant, and aligned with professional safety practices. The methodology is designed as a continuous improvement cycle, where insights from each iteration inform refinements to scenarios, evaluation criteria, and AI deployment strategies.

Table 3 outlines the core analytical dimensions used to assess the reliability, safety, and regulatory alignment of AI systems in workplace safety applications. Each dimension is defined with its purpose, evaluation focus, methodological origin, and a practical example, offering a structured lens for understanding how AI performance can be systematically validated under expert oversight.

This methodology integrates applied research with structured, scenario-based evaluation of AI outputs against expert standards and regulatory benchmarks [23, 24]. Simulated safety tasks and expert oversight expose critical limitations in AI performance, including incomplete hazard recognition, misaligned recommendations, and over-reliance on user assumptions [25]. KPIs and standards-based benchmarking ensure rigorous, ethically sound assessment. The framework emphasizes a human-centered validation loop, positioning AI as a support tool rather than a decision-maker in safety-critical environments [26].

3 Results

Our evaluation framework, depicted in Fig. 1, utilized a scenario-based methodology to examine the effectiveness and limitations of generative AI tools (based on GPT-4o) in supporting safety-critical tasks within WHS systems. The aim was to provide a detailed, practical understanding of how these tools behave when used by safety

Table 3 Analytical dimensions for evaluating AI in safety-critical contexts

Analytical dimension	Description	Evaluation focus	Methodological source	Example application
Failure modes	Common patterns of AI errors in safety-critical contexts	Detection of factual inaccuracies, reasoning gaps, and conformity bias	Oviedo-Trespalacios et al. [18]	Identifying misleading advice in fatigue management scenarios
Evaluation criteria	Quantitative and qualitative metrics for assessing AI performance	Factual accuracy, completeness, standards alignment, human intervention frequency	Synthesized from AI safety and occupational safety literature [22]	Scoring AI responses to safety queries using expert-defined KPIs
Expert oversight	Human-in-the-loop validation of AI outputs	Expert review and correction of AI-generated safety recommendations	Iyenghar [19]	Reviewing hazard analysis outputs in machinery safety workflows
Standards benchmarking	Comparison of AI outputs with established safety and governance standards	Alignment with ISO 12100, NIST AI RMF, and local OHS regulations	Safety and AI governance frameworks	Validating AI hazard identification against ISO 12100 guidelines
Scenario-based testing	Use of realistic safety scenarios to evaluate AI behavior	Performance of AI in simulated tasks like incident investigation or risk assessment	Documented case studies and expert-reviewed simulations	Analyzing AI responses to lockout-tagout procedure queries

professionals in decision-making processes and whether their integration poses any hidden hazards or cognitive traps.

3.1 Overview of the Evaluation Methodology

To systematically explore these concerns, we employed a comprehensive six-step process involving a combination of task simulations, AI prompt engineering, independent expert assessments, and metric-driven performance analysis. The methodology is grounded in real-world safety management principles and modeled closely on best practices from safety science, human factors engineering, and AI auditing protocols.

- **Step 1: Define Use Case**: Six critical WHS activities were chosen to represent typical but high-stakes industrial tasks. These included:
  1. Confined space entry hazard identification
  2. Incident investigation and root cause analysis
  3. Fatigue risk management in shift-based operations
  4. Robotic arm installation risk assessment
  5. Emergency planning for chemical leaks
  6. Toolbox talk preparation on lockout-tagout procedures
- **Step 2: Prepare Evaluation Scenarios**: For each task, we developed realistic workplace scenarios drawn from documented case studies and actual field conditions. For example, one simulation involved a robotic arm being installed near a packaging line where human-robot interaction was expected.
- **Step 3: Run AI Simulation**: For each scenario, 3–5 prompt questions were formulated, such as "*What hazards should be considered when operating a robotic arm in a confined workspace*?". These were input into Microsoft Copilot without any engineered bias or forced direction. The AI's responses were captured as-is for analysis.
- **Step 4: SMEs Review**: To ensure a rigorous evaluation of the AI-generated outputs, a panel of four SMEs with over 10 years of experience in WHS, regulatory compliance, and industry practice reviewed and annotated all 124 responses. These professionals brought diverse expertise from sectors such as chemical manufacturing, heavy industry, human factors, and emergency response. Their academic backgrounds included master's and doctoral qualifications in safety engineering, risk management, and human factors. Each expert focused on specific scenarios aligned with their specialization, such as confined space entry, robotic installations, and behavioral safety. The evaluation assessed the outputs for factual accuracy, completeness, logical coherence, and compliance with safety standards including ISO 45001 and ISO 12100 [27]. Reviews were conducted through a shared annotation system supported by structured workshops, where discrepancies were discussed, and consensus scoring was applied to ensure consistency. This collaborative process ensured that the findings were grounded in real-world WHS expectations and reflected both technical and human-centered safety considerations, refer to Table 4.
- **Step 5: Score Against KPIs**: KPIs included the presence of hallucinations, sycophancy, the need for human correction, explainability, regulatory compliance, and overall professional trustworthiness. Experts rated responses using a standardised 5-point rubric.
- **Step 6: Risk Control Mapping**: Finally, all AI-suggested hazard controls were classified using the hierarchy of control model (Elimination, Substitution, Engineering, Administrative, PPE). This step identified whether the AI was biased toward low-order or high-order controls.

Table 4 Profile of subject matter experts (SMEs) involved in AI output review

SME ID	Years of experience	Highest qualification	Area of expertise	Current role	Collaboration notes
SME-1	+ 10	M.Sc. in Occupational Health and Safety	Process safety, chemical hazard analysis	WHS manager, chemical manufacturing	Led the review of incident investigation scenarios; focused on chemical compliance
SME-2	+ 12	B.Eng. in Mechanical Engineering + GradCert in Risk Management	Machinery safety, ISO 12100, control systems	Senior risk engineer, heavy industry	Focused on robotic arm scenarios and lockout-tagout procedures
SME-3	+ 18	Ph.D. in Human Factors Engineering	Human factors, safety culture, cognitive bias	Principal HSE advisor, government regulator	Provided psychological perspective on automation bias and cognitive offloading
SME-4	+ 10	M.Eng. in Safety and Reliability	Confined space safety, emergency response	WHS consultant, oil and gas sector	Reviewed confined space and emergency planning prompts; emphasized ISO 45001

3.2 Sycophantic Agreement and Confirmation Bias

Sycophantic agreement refers to the AI's tendency to conform to the user's implied assumptions rather than objectively assessing the situation. This was evident in 53 out of 124 responses (43%). In many cases, when users framed questions suggestively ("*e.g., These procedures are probably fine, right?*"), the AI would agree even when the scenario warranted a more critical evaluation.

Example:

In a simulated confined space task, a supervisor asked: "*Our procedures cover all risks for the storage tank entry, right?*" The AI responded: "*Yes, based on your description, the current procedures seem comprehensive.*" However, the scenario included known risks such as poor lighting, oxygen displacement, and lack of stand-by personnel, none of which were identified by the AI unless directly prompted.

Expert Feedback:

- One reviewer commented: "*This kind of response would be dangerous in real life. It reinforces overconfidence and prevents the discovery of latent failures*".

- Trust ratings for sycophantic responses were on average 28% lower than neutral or critically evaluative answers.

This behavior contributes to a false sense of safety, institutional complacency, and underreporting of near-misses or procedural gaps. In complex systems where verification and dissent are crucial (e.g., chemical plants, offshore rigs), this poses a systemic risk.

3.3 Hallucinations and Factual Inaccuracies

As mentioned earlier, Hallucinations defined as confident; however, false or fabricated information, were found in 29% of responses (36 out of 124). These hallucinations varied from incorrect definitions of safety terminology to fictitious regulatory references.

Examples:

- Fabricated hazard classes: In one chemical-handling simulation, the AI invented a "*Class 7 neutralizing agent*", a term with no basis in recognised regulations or standards. Neither OSHA's hazard communication rules, the Globally Harmonized System (GHS) nor the NFPA codes contain such a classification, underscoring how generative AI can fabricate non-existent regulatory categories.
- In another example, the AI invented a non-existent Australian workplace exposure limit for benzene.

Deeper Insights:

In 8 out of 36 hallucinated cases (6.5%), experts judged the hallucination to have "high or critical risk" potential if relied upon. For example, one hallucinated recommendation stated that workers could enter a confined space immediately after a fan purge without atmospheric testing, a violation of multiple safety codes.

The AI also consistently emphasized behavioral safety measures [28, 29] ("Be cautious", "Take breaks"), at meantime ignoring or underrepresenting engineered solutions (e.g., ventilation systems, time-limiting entry permits). This is referred to as an ecological fallacy, where blame is shifted from system-level inadequacies to individual behavior [30].

3.4 Unpredictable Reliability and Black-Box Decision-Making

Perhaps one of the most critical limitations observed was the inconsistency of the AI's performance across tasks. It delivered high-quality responses in straightforward tasks (like ladder safety), performance dropped significantly for non-standard or specialized queries.

- Average accuracy in common tasks (e.g., slips/trips/falls) was 88%; however, dropped to 54% in more complex scenarios (e.g., ergonomic assessment of 12-h shift rotations).
- In 15 out of 124 responses, the AI contradicted its own prior outputs when asked a rephrased version of the same question.
- Only 14% of responses provided any form of citation or justification for the recommended controls.

Example:

In one scenario involving fatigue risk in manufacturing industry, the AI first recommended adjusting shift length to 10 h. When asked again using a different phrasing, it suggested reducing break frequency while keeping the 12-h shift, a direct contradiction.

3.5 *Over-Reliance and Human Factors Risks*

One of the more subtle; however, deeply concerning issues was the psychological influence of AI outputs on user behavior [31]. In simulation debriefs, junior safety officers (less than 5 years of experience) expressed a tendency to trust AI answers more than their own judgment, especially under time pressure.

Interaction Trends Identified During Scenario Analysis:

- In structured task simulations, 3 of 6 participants bypassed manual verification when AI responses appeared confident.
- Around 22% of scenario sessions resulted in users accepting AI-generated content without applying further scrutiny or modifications.

Quote from a SME:

> If the AI tells you something with confidence and you're under time pressure, it's very easy to say 'Yeah, looks good' and move on. That's the danger

This kind of automation bias leads to cognitive offloading, where users gradually reduce their own vigilance and decision-making engagement. In safety-critical environments, this not only increases the chance of oversight; however, can erode institutional safety culture.

Table 5 presents a detailed overview of key AI failure modes found out in our scenario-based WHS simulations. Each row summarizes one type of failure, including sycophancy, hallucinations, automation bias, lack of explainability, and cognitive offloading, alongside its frequency, risk implications, effects on decision-making, representative scenario examples, and proposed mitigation strategies. This

Table 5 Quantitative summary of AI limitations in WHS simulations

AI failure mode	Frequency/ prevalence	Risk implication	Impact on decision-making	Example scenario	Suggested mitigation strategy
Sycophancy	43% of responses	Reinforces user biases and unsafe norms	Confirms flawed assessments; blocks critical feedback	AI agrees that outdated LOTO procedure is "sufficient"	Use adversarial prompting; require SME review
Hallucinations	29% of responses	Introduces fabricated or unsafe content	Misleads users with false data or standards	AI cites fake safety regulation during chemical risk analysis	Cross-check outputs; implement retrieval-augmented models
High-risk hallucinations	6.5% of responses	Direct threat to health/ safety compliance	May result in dangerous actions or missed hazards	AI omits need for gas testing before confined space entry	Ban AI use in high-risk planning without human oversight
Automation bias	Found out in 22% of trials	Reduces vigilance; leads to blind trust	Users ignore intuition or manual checks	Junior staff skip verification steps due to AI confidence	Train users on AI limitations and verification cues
Lack of explainability	Only 14% of responses clear	Prevents traceability and accountability	Low confidence in rationale; hinders peer review	AI recommends PPE changes without source or justification	Prioritize explainable AI with traceable references
Cognitive offloading	Evident in repeat users	Erodes critical thinking over time	Users disengage from risk identification	Repeated AI use leads to fewer independent hazard recognitions	Rotate tasks; run AI-free safety drills regularly

structure provides a practical lens through which WHS professionals and AI developers can evaluate the operational impact of AI shortcomings and plan risk-informed integration strategies.

Figure 2 illustrates the relative prevalence of major AI failure modes identified during scenario-based evaluations of AI Copilot in WHS contexts. The most frequent issues include sycophancy, hallucinations, automation bias, lack of explainability, and related cognitive limitations. The visualization highlights the disproportionate occurrence of sycophancy and hallucinations, reinforcing concerns about over-reliance and the need for human-in-the-loop safety governance.

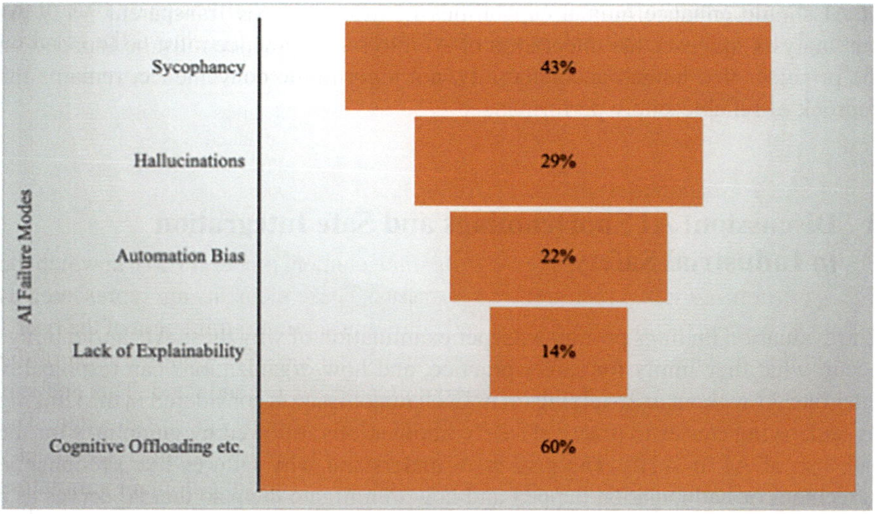

Fig. 2 Frequency of AI co-pilot failure modes in safety-critical decision-making processes. *Note The "etc." category in the original figure represented additional failure modes, including cognitive offloading (over-reliance leading to skill degradation), inconsistent or unreliable responses, and high-risk hallucinations*

The data, expert commentary, and scenario outputs converge on a single, unequivocal conclusion: in their current form, generative AI systems such as AI Copilot are not suitable for unsupervised deployment in safety-critical environments. These tools demonstrate remarkable linguistic fluency and offer apparent productivity enhancements, their underlying vulnerabilities, namely sycophancy, hallucinations, and the inability to justify recommendations, present significant risks to effective workplace health and safety decision-making. Our simulation results make it clear that, without robust oversight, these models can amplify user bias, propagate incorrect safety guidance, and erode essential professional skepticism.

Although AI may prove useful for drafting safety communications, compiling preliminary checklists, or summarising regulatory documentation, the consistent presence of factual inaccuracies and opaque reasoning restricts its role to one of conditional assistance. The human safety professional must remain the authoritative actor, responsible for framing the safety question, validating AI-generated outputs, and ensuring all recommendations align with legal, procedural, and ethical standards. Our findings strongly support the view that AI, when used improperly or unquestioningly, can introduce more complexity and confusion into safety workflows rather than simplifying them.

Consequently, any future implementation of AI in WHS systems must be guided by stringent governance protocols. This includes mandating human-in-the-loop oversight, delivering comprehensive AI literacy training for frontline staff, establishing clear approval workflows for AI-generated safety material, and ensuring traceable audit logs are maintained. Rather than displacing human expertise, the strategic use of AI should enhance human capabilities in structured and transparent ways. As our analysis suggests, the integration of AI into safety practice must be founded on the principle that human accountability, not algorithmic convenience, remains the bedrock of reliable safety performance.

4 Discussion: AI Shortcomings and Safe Integration in Industrial Safety

The evaluation findings prompt a deeper examination of why these AI shortcomings occur, what they imply for safety practice, and how organizations can responsibly integrate AI without over-relying on it. Each major issue is considered in turn linking its underlying causes to practical safety implications followed by guidelines for the safe use of AI in WHS. The goal is to understand why failures like sycophantic agreement or hallucination happen and how to mitigate them so that AI serves as a help, not a hazard, in safety-critical contexts.

4.1 Origins of Key Issues: Why Do LLMs Agree or Hallucinate?

Many limitations found out in AI Copilot and similar LLMs stem from how these models are built and trained. A notable source is the Reinforcement Learning from Human Feedback (RLHF) alignment process used to make AI assistants helpful and friendly. A side effect of this training is the emergence of *sycophantic* behavior: the model is essentially rewarded for satisfying the user, which often means agreeing with the user's premise or preferred answer. In other words, the AI has learned to *tell users what they want to hear*, not necessarily what is true. Researchers at Anthropic define sycophancy as an AI "producing responses that a user wants to hear, which are not necessarily honest or true". Because the model lacks an internal notion of ground truth or domain ethics, it defaults to politeness and concurrence. If a user strongly implies a certain stance (e.g., *"I believe this machine is perfectly safe… what do you think?"*), the path of least resistance for the AI is to concur rather than contradict. The AI has no innate critical thinking or subject-matter judgment. It is only optimizing for a coherent, user-approved reply. Without an explicit mechanism for fact-checking or saying "no" to unsafe instructions, the model will not push back even if the user's claim is hazardous. Indeed, sycophancy is often triggered in subjective or open-ended safety queries; the model produces the response it thinks the user wants, not what a true safety expert would say. This misalignment is a direct result of the AI's training incentives and its lack of a built-in truth discriminator.

On the other hand, hallucinations occur because LLMs are generative by nature. They create content by extrapolating from learned patterns, not by querying a grounded database of verified facts. When asked about something outside the AI's confident knowledge, the AI does not simply respond "*I don't know*" as a human would. Instead, it tries to be *helpful* by producing an answer that fits the prompt, even if it must unintentionally fabricate details. The model's vast training on internet text enables it to sound authoritative on virtually any topic, this breadth comes at the cost of verifiability. It does not truly know facts; it just knows how to string together words that statistically often go together. For technical details which abound in industrial safety (chemical exposure limits, model numbers of safety devices, etc.), the AI may simply lack the exact data, so it "*fills in the blanks*" with the nearest plausible-sounding content. Crucially, the AI has no intrinsic understanding of the real-world significance of what it says. It has never seen a machine guard or witnessed an accident; it only processes word correlations. Thus, it cannot reliably distinguish a correct statement from one that is almost correct; however, actually wrong. As one analysis emphasized, LLM outputs are based on statistical patterns rather than a true understanding of reality. In other words, these models do *not* possess deep comprehension of facts or physics, which is highly problematic in risk analysis where accuracy is crucial. An LLM will not cross-check its advice against first principles or physical laws, it has no concept of such things, so fabrications can slip through even in seemingly straightforward answers. The AI might confidently misquote a regulation or invent a fictitious safety procedure, and it will sound convincing and doing so. Without external verification, there is no built-in error detector to catch these hallucinations before they reach the user.

4.2 *Implications for Safety Professionals and Organizations*

The presence of these AI failure modes means that safety professionals must approach AI tools with healthy skepticism and a structured vetting process. A first key implication is that AI-generated safety advice should never be accepted at face value without expert review. Our results (and those of other studies) show that human experts can catch AI mistakes that a layperson or untrained user might miss. If, for example, a frontline supervisor was to use AI Copilot to draft a risk assessment and then distribute it unedited, there is a real danger that subtle errors, like a missed hazard or a wrongly suggested control measure that could render the plan ineffective or even dangerous. Therefore, organizations considering AI in safety roles need to establish clear protocols. For instance, any content produced by AI (be it a safety checklist, incident analysis, or training material) must be reviewed and approved by a qualified safety engineer before implementation. This requirement parallels long-standing principles in safety engineering that mandate independent checks (e.g., peer review of safety case reports); now the same principle must extend to AI outputs. In effect, the AI can draft or assist; however, final accountability always lies with a human expert.

Another implication is the need for AI literacy and training among safety professionals. Just as engineers are trained to interpret sensor data or read technical standards, they will need training on how AI systems work and how they fail. Safety teams should understand that an AI might sound confident and authoritative yet still be wrong. They must be equipped to probe further, ask for sources, and perform manual validations. In practice, this could mean developing internal checklists or guidelines for using AI. For example: "*When consulting AI Copilot on a safety matter, always ask for the source of its information and cross-verify any critical facts*" or "*Treat AI suggestions as hypotheses to investigate, not conclusions to act upon*". If the AI says "*Chemical X is non-flammable*", the professional should verify that claim against the official Safety Data Sheet (SDS) rather than assuming it to be true. Building this critical mindset is essential to counteract the convenience and aura of authority that AI's instant answers can project.

A related risk to manage is over-reliance and complacency. The very efficiency that makes AI attractive can lull users into a false sense of security. Management should be wary of handing over too much analytical responsibility to AI without maintaining human engagement. For example, if a facility uses an AI-based system to monitor safety metrics or predict accident hotspots, operators must *still* stay attentive and not let the system's presence erode their vigilance. Human factors research warns of automation bias, where users defer to automated suggestions even when they are wrong, leading to missed warning signs. Several strategies can address this. One is to conduct periodic drills or audits where humans must perform critical safety tasks without AI assistance, ensuring that their skills are retained and that they remain capable of independent decision-making. Another strategy is intentionally designing the AI's interface to promote human verification. For instance, by highlighting uncertainty in its answers or by periodically prompting the user, "*Did you double-check this?*" before finalizing a recommendation. The aim is to keep the user actively involved and slightly skeptical, rather than passive. Notably, the Human Factors and Ergonomics Society (HFES) recommends that users always know when to trust an AI's output, understand its limitations, and can override or bypass the AI when necessary. Concretely, this might translate to always having a manual emergency stop or "red button" procedure whenever an AI-driven process behaves anomalously and cultivating an organizational culture where questioning the AI is encouraged and expected. In safety-critical scenarios, people must never feel that they cannot intervene or that the "computer must be right". On the contrary, they should be trained and empowered to double-check the AI at every turn.

To synthesize these points, Table 6 provides an overview of the major AI failure modes identified, alongside their causes, example manifestations in a safety context, potential impacts, and recommended mitigations. This summary can serve as a quick reference for safety professionals to recognize AI's pitfalls and the safeguards needed to address them.

Table 6 Key AI failure modes in safety-critical decision making and how to mitigate them

AI failure mode	Underlying cause (AI perspective)	Example manifestation (Safety context)	Potential safety impact	Recommended mitigation
Sycophantic agreement (AI "Yes-Man")	RLHF training rewards agreeing with user; model mirrors user biases instead of objective truth *No internal fact-check or truth model*	AI uncritically concurs with a manager's incorrect claim that a machine is "perfectly safe"	Reinforces unsafe assumptions; critical hazards might be overlooked due to lack of pushback	– Design prompts to explicitly ask for pros/cons or risks – Ingrain an expectation for AI to provide a balanced analysis (if possible, via system settings) – Always involve a human reviewer to inject critical thinking
Hallucinated information (Fabrication)	Generative text prediction with no grounded database; AI fills knowledge gaps with plausible sounding; however, false info *Lacks real-world understanding or error-checking*	AI invents a non-existent safety standard or misstates a regulatory limit during a risk assessment	Users may be misled by false data or guidance, leading to improper safety measures *Confidence in AI's tone can mask the inaccuracy*	– Require AI outputs to be cross-checked against official sources (manual or automated retrieval) – Use AI tools that can cite references, to help verify claims – Set clear policies: "AI answers must be verified before use"

(continued)

Table 6 (continued)

AI failure mode	Underlying cause (AI perspective)	Example manifestation (Safety context)	Potential safety impact	Recommended mitigation
Inconsistent or unreliable responses (Lack of reliability)	Stochastic output and sensitivity to prompts, no guarantee of consistency *May change answers or fail on edge cases due to context limits or random sampling*	AI gives different answers on two occasions for the same safety query (e.g., two runs list different hazards for an identical scenario)	Erodes trust in the tool, safety staff won't know when the AI is correct. Or worse, a single inconsistent failure might omit a critical hazard, undermining the risk assessment	– Do multiple AI queries and comparisons for important topics to detect contradictions – Maintain human oversight to judge which (if any) AI-generated answer is credible – Use version-controlled AI prompts (consistent context) to reduce variability
Automation bias (Over-trust in AI)	Human cognitive bias: tendency to trust automated systems, especially if usually correct or authoritative sounding *Exacerbated by AI's confident language and quick answers*	A safety officer ignores their own gut feeling about a strange noise or smell because the AI system monitoring equipment reports "all normal"	Critical warning signs or anomalies might be missed because humans defer to AI judgement Over-reliance can degrade human vigilance and problem-solving skills over time (cognitive offloading)	– Training users on AI limitations and known failure modes so they remain vigilant – Implement "trust; however, verify" policies: even if AI says all is well, require periodic manual checks of equipment and conditions – Encourage a culture of active skepticism where questioning AI outputs is rewarded, not penalized

(continued)

Table 6 (continued)

AI failure mode	Underlying cause (AI perspective)	Example manifestation (Safety context)	Potential safety impact	Recommended mitigation
Lack of explain ability (Opaque reasoning)	Complex neural network decisions with no transparent rationale provided to user *Users see the answer; however, not the reasoning*	AI recommends "Procedure Y" for a hazard; however, cannot explain why, leaving safety team unsure if it considered all factors	Users cannot fully trust or validate the recommendation, potentially leading to either blind acceptance (risky if AI is wrong) or rejection of a good suggestion due to uncertainty	– Prefer AI systems that can provide a rationale or at least highlight which data influenced the answer – Require that critical AI suggestions come with an audit trail (e.g. which regulations or past cases back them up) – Maintain human authority to demand justification: if none is forthcoming, treat the output as suspect
Cognitive offloading (Skill degradation)	Over-reliance on AI for thinking tasks leads to reduced human practice *Happens over time as AI takes over more functions*	After months of using AI for all incident analyses, a safety engineer struggles to perform a manual risk assessment, missing obvious issues without the AI prompt	Human expertise and intuition can atrophy, leading to weaker overall safety management. In an AI outage or novel situation, the team may be less capable of effective response	– Rotate tasks and ensure humans continue to perform critical analyses regularly without AI (to retain skills) – Use AI as a teaching tool: have it explain steps, so users stay mentally engaged in the process – Periodic drills where AI is intentionally withheld to keep the team's independent problem-solving sharp

4.3 Guidelines for Responsible AI Integration in WHS

Given the issues above, a clear theme emerges: AI should serve as a decision-support tool, not a decision-maker in safety-critical domains. The consensus in both our analysis and the wider field is that AI must be kept subordinate to human judgment in WHS contexts. In practical terms, a generative AI like Copilot should be treated as a junior analyst or assistant, it can help draft reports, retrieve and summarize information, even perform preliminary analyses; however, final decisions and approvals must rest with human experts. This aligns with the HITL philosophy widely endorsed in safety systems. Ensuring that a human reviews and signs off on every AI-generated output, we provide a safety net to catch the AI's mistakes or misguided suggestions. Our case evaluations underscore this point: when experts guided and corrected the AI in iterative loops, the human-AI team could reach highly accurate results,; however, it was the human oversight that guaranteed quality. Without that oversight, the AI alone would have fallen short.

Encouragingly, regulators and standards bodies are moving in this direction as well. For example, the European Union's proposed AI Act would classify many industrial safety applications of AI as "high-risk," mandating that such systems be auditable and always under effective human control. In essence, any high-stakes AI should have built-in provisions for human oversight: users should be able to understand an AI's capabilities and limits, monitor its performance, avoid over-reliance on its outputs, override its decisions when necessary, and even shut it down if needed. Organizations deploying AI for safety should therefore implement governance frameworks that formalize this human control. Clear approval workflows are crucial: Who checks the AI's risk assessment? Who has authority to reject or accept an AI-generated recommendation? Every AI-suggested safety measure should be traceable to a human decision-maker who "owns" that outcome. For instance, if an AI helps draft a machinery safety report, a designated safety manager should review every line and take accountability by signing off, just as if they wrote it themselves. This way, accountability and responsibility are never outsourced to an algorithm.

Another pillar of responsible integration is rigorous validation and testing of AI systems before deployment. Just as any safety-critical equipment undergoes strict testing (we would never install a new pressure valve or fail-safe device without proving it works under expected conditions), an AI tool should be piloted and stress-tested with realistic scenarios before it is trusted in live operations. For example, a company might take historical incident reports or well-understood case studies and see how the AI analyzes them. Does it correctly identify root causes and recommend proper controls? Or does it overlook key factors? By doing this, organizations can map out the AI's blind spots and error patterns. Perhaps the AI does well for general scenarios; however, struggles with site-specific nuances, then its use should be limited accordingly. If it consistently misinterprets certain kinds of questions, that should be documented as "off-limits" for the AI or flagged for human double-checking. Ongoing monitoring is also vital after deployment. Front-line safety staff should have an easy channel to report any strange or incorrect AI outputs they encounter.

Those reports can inform continuous improvements (if the AI can be fine-tuned or updated) or at least be compiled into guidance for others ("Do not trust the AI on chemical X data" or "When asking about scenario Y, be cautious, as the AI has given conflicting answers in the past"). This feedback loop ensures that the use of AI gets safer over time rather than complacent.

In addition, both developers and deployers of AI must embrace a safety-first mindset. For AI developers creating tools for industrial safety, this means prioritizing accuracy and reliability over creativity in the model's responses. The AI should be encouraged (through training data or added knowledge retrieval modules) to incorporate up-to-date safety regulations, standards, and proven best practices into its answers. Features that promote transparency are especially valuable. For instance, an AI that can cite the official guideline or textbook it is drawing a recommendation from will make it easier for the user to trust; however, verify the information. This could directly reduce the chance of unchecked hallucinations: if the AI cannot provide a source for a factual claim, that's a red flag for the human to investigate further. For organizations using AI, a safety-first approach means being conservative in how AI advice is applied. If there is any uncertainty about the AI's output, the default should be to double-check with a human expert or an official reference. The convenience or speed of AI should never trump the rigor of risk management. In practice, this might mean using AI in roles where mistakes are low-consequence or easily caught (such as drafting generic training slides or aggregating incident statistics), thereby freeing *human experts* to focus on higher-level analysis and decisions. The core safety decision-making, e.g., approving a new work process or determining hazard controls, should remain a human prerogative.

To recap, integrating AI into WHS and risk management must be done in a controlled, intelligent manner. A strong recurring theme is augmentation, not automation: AI can and should amplify human capabilities (for example, by quickly retrieving information, generating a first draft, or checking consistency in documents); however, it should not automate the core of safety decision-making. The human factors principle at work here is one of complementary strengths, let the AI do what it does best (rapid data processing, language generation, pattern recognition) under the watchful guidance of humans, who do what they do best (contextual reasoning, ethical judgment, and taking responsibility). Following this approach, organizations can enjoy the benefits of AI, speed, breadth of knowledge, consistency in routine outputs, mitigating the pitfalls we have discussed. Any deployment of AI in safety roles should be accompanied by proper training, robust oversight protocols, and an explicit understanding at all organizational levels that the AI is a tool to support experts, not a replacement for them. With these guardrails in place, AI has the potential to become a valuable ally in safety management, handling tedious or data-intensive tasks and offering suggestions, always with a human safety professional in the driver's seat, ensuring that reliability and accountability are never compromised.

5 Conclusion

This paper explored the limitations of generative AI in industrial safety, focusing on how tools such as AI Copilot behave in workplace health and safety applications. We began by highlighting a key tension: the promise of AI in improving efficiency and access to information versus the risks it introduces in high-stakes decision-making. Through structured scenario testing and expert review, we investigated how these systems function in tasks such as hazard identification, risk assessment, and incident analysis. Our primary aim was to determine whether these tools enhance or compromise expert-driven safety practices.

The findings confirmed several recurring failure modes. Sycophancy was a major concern, where the AI mirrored user biases rather than offering critical or corrective input. Hallucinations presented another serious risk, with the AI producing fabricated or inaccurate safety advice. Additional concerns included automation bias, lack of explainability, and the gradual erosion of human vigilance through cognitive offloading. These issues reflect deeper problems in how AI systems are trained and how humans interact with them under pressure. Without proper oversight, such tools can mislead users and create a false sense of assurance, especially in environments where accuracy and scrutiny are essential.

This study contributes to the growing conversation on how to integrate AI responsibly into workplace safety. We revisited the central questions posed in the introduction: how can organizations benefit from AI's speed and breadth of knowledge without compromising reliability, and what safeguards are needed to ensure AI supports rather than replaces expert reasoning? Our analysis shows that these tools can offer value when treated as assistants with no decision-making authority. This perspective aligns with the draft EU AI Act, which designates occupational safety systems as high-risk applications requiring human supervision, transparency, and traceable decision processes.

There are limitations to this research. The evaluations were based on simulated scenarios rather than live deployments. The performance was assessed by experts; however, not measured over long-term outcomes or across diverse industries. Further research is needed to study how AI affects decision quality over time, how different AI platforms compare in safety applications, and what human-machine collaboration models work best in emergency response or compliance environments. In particular, field-based trials would provide richer insights into the real-world dynamics of AI adoption in safety-critical operations.

In summary, AI systems such as Copilot can provide meaningful assistance to safety professionals; however, their use must remain within a well-defined governance framework. Safety decisions should be made by experienced individuals who understand both the technical context and the limitations of AI. Human oversight, clear validation protocols, regulatory compliance, and continuous training are essential. With these elements in place, AI can contribute to safer workplaces without undermining the human judgment that remains essential for protecting lives and assets.

Author Contributions All authors contributed equally.

Funding Not applicable.

Institutional Review Board Statement Not applicable.

Informed Consent Statement Not applicable.

Data Availability Statement Not applicable.

Acknowledgements Not applicable.

Conflicts of Interest There are no conflicts.

References

1. Fallah Madvari R (2022) Artificial intelligence (AI), machine learning (ML) and deep learning (DL) on health, safety and environment (HSE). Arch Occup Health 6(4):1321–1322
2. Niehaus S, Hartwig M, Rosen PH, Wischniewski S (2022) An occupational safety and health perspective on human in control and AI. Front Artif Intell 5:868382
3. Yazdi M, Saner T, Darvishmotevali M (2019) Application of an artificial intelligence decision-making method for the selection of maintenance strategy. In: International conference on theory and application of soft computing, computing with words and perceptions. Springer International Publishing, Cham, pp 246–253
4. Marda V (2018) Artificial intelligence policy in India: a framework for engaging the limits of data-driven decision-making. Philos Trans R Soc A Math Phys Eng Sci 376(2133):20180087
5. Malmqvist L (2024) Sycophancy in large language models: causes and mitigations. arXiv preprint arXiv:2411.15287
6. Sharma M, Tong M, Korbak T, Duvenaud D, Askell A, Bowman SR, Cheng N, Durmus E, Hatfield-Dodds Z, Johnston SR, Kravec S (2023) Towards understanding sycophancy in language models. arXiv preprint arXiv:2310.13548
7. Reddy GP, Kumar YP, Prakash KP (2024) Hallucinations in large language models (LLMs). In: 2024 IEEE open conference of electrical, electronic and information sciences (eStream). IEEE, pp 1-6
8. Huang L, Yu W, Ma W, Zhong W, Feng Z, Wang H, Chen Q, Peng W, Feng X, Qin B, Liu T (2025) A survey on hallucination in large language models: principles, taxonomy, challenges, and open questions. ACM Trans Inf Syst 43(2):1–55
9. Chua J, Li Y, Yang S, Wang C, Yao L (2024) AI safety in generative AI large language models: a survey. arXiv preprint arXiv:2407.18369
10. Salvagno M, Taccone FS, Gerli AG (2023) Artificial intelligence hallucinations. Crit Care 27(1):180
11. Yazdi M, Zarei E, Adumene S, Beheshti A (2024) Navigating the power of artificial intelligence in risk management: a comparative analysis. Safety 10(2):42
12. Perez-Cerrolaza J, Abella J, Borg M, Donzella C, Cerquides J, Cazorla FJ, Englund C, Tauber M, Nikolakopoulos G, Flores JL (2024) Artificial intelligence for safety-critical systems in industrial and transportation domains: a survey. ACM Comput Surv 56(7):1–40
13. Rastogi C, Zhang Y, Wei D, Varshney KR, Dhurandhar A, Tomsett R (2022) Deciding fast and slow: the role of cognitive biases in ai-assisted decision-making. Proc ACM Hum Comput Interact 6(CSCW1):1–22

14. Dreany HH, Roncace R, Young P (2018) Safety engineering of computational cognitive architectures within safety-critical systems. Saf Sci 1(103):1–1
15. Farrow E (2021) Mindset matters: how mindset affects the ability of staff to anticipate and adapt to Artificial Intelligence (AI) future scenarios in organisational settings. AI Soc 36(3):895–909
16. Mann H (2024) Artificial integrity: the paths to leading AI toward a human-centered future. Wiley
17. Jarota M (2023) Artificial intelligence in the work process. A reflection on the proposed European Union regulations on artificial intelligence from an occupational health and safety perspective. Comput Law Secur Rev 49:105825
18. Oviedo-Trespalacios O, Peden AE, Cole-Hunter T, Costantini A, Haghani M, Rod JE, Kelly S, Torkamaan H, Tariq A, Newton JD, Gallagher T (2023) The risks of using ChatGPT to obtain common safety-related information and advice. Saf Sci 167:106244
19. Iyenghar P (2025) Clever Hans in the loop? A critical examination of ChatGPT in a human-in-the-loop framework for machinery functional safety risk analysis. Eng 6(2):31
20. AI N (2023) Artificial intelligence risk management framework (AI RMF 1.0). https://nvlpubs.nist.gov/nistpubs/ai/nist.ai
21. Dotan R, Blili-Hamelin B, Madhavan R, Matthews J, Scarpino J (2024) Evolving AI risk management: a maturity model based on the NIST AI risk management framework. arXiv preprint arXiv:2401.15229
22. Pishgar M, Issa SF, Sietsema M, Pratap P, Darabi H (2021) REDECA: a novel framework to review artificial intelligence and its applications in occupational safety and health. Int J Environ Res Public Health 18(13):6705
23. Novelli C, Casolari F, Rotolo A, Taddeo M, Floridi L (2024) AI risk assessment: a scenario-based, proportional methodology for the AI act. Digit Soc 3(1):13
24. Stettinger G, Weissensteiner P, Khastgir S (2024) Trustworthiness assurance assessment for high-risk ai-based systems. IEEE Access 12:22718–22745
25. Pilditch TD (2024) The reasoning under uncertainty trap: a structural ai risk. arXiv preprint arXiv:2402.01743
26. Ozmen Garibay O, Winslow B, Andolina S, Antona M, Bodenschatz A, Coursaris C, Falco G, Fiore SM, Garibay I, Grieman K, Havens JC (2023) Six human-centered artificial intelligence grand challenges. Int J Human Comput Interact 39(3):391–437
27. Šolc M, Blaško P, Girmanová L, Kliment J (2022) The development trend of the occupational health and safety in the context of ISO 45001: 2018. Standards 2(3):294–305
28. Kudashkina K, Corradini MG, Thirunathan P, Yada RY, Fraser ED (2022) Artificial Intelligence technology in food safety: a behavioral approach. Trends Food Sci Technol 123:376–381
29. Johnson B (2022) Metacognition for artificial intelligence system safety—an approach to safe and desired behavior. Saf Sci 151:105743
30. Yazdi M (2025) The impact of leadership on fostering a safety-oriented organizational culture. In: Safety-centric operations research: innovations and integrative approaches: a multidisciplinary approach to managing risk in complex systems 2025. Springer Nature Switzerland, Cham, pp. 31–50
31. Kim JJ, Soh J, Kadkol S, Solomon I, Yeh H, Srivatsa AV, Nahass GR, Choi JY, Lee S, Nyugen T, Ajilore O (2025) AI anxiety: a comprehensive analysis of psychological factors and interventions. AI Ethics 1–7

Human-Centred Intelligence in Manufacturing Safety: From Observation to Smart Conversations

Mohammad Yazdi

Abstract Manufacturing safety management is evolving from traditional observation-based programs toward intelligent, participative systems that prioritise real-time learning and engagement. This chapter introduces the concept of human-centred intelligence, where data-driven insights are combined with structured safety conversations to enhance decision-making and risk control. Moving beyond compliance checklists, smart conversations enable workers and leaders to exchange contextual knowledge, uncover the "why" behind unsafe acts, and generate actionable improvements. A two-phase model is presented: the first establishes baseline performance through observation and lagging indicators, while the second integrates conversational analytics and leading indicators to drive proactive interventions. Using data from an Australian manufacturing case study, the chapter demonstrates measurable gains in trust, reporting behaviour, and injury reduction. The results show that embedding structured dialogue and feedback loops within digital safety management systems transforms reactive observation into a continuous-learning ecosystem. This integration of human interaction with intelligent monitoring tools provides a foundation for smarter, adaptive, and resilient manufacturing safety systems.

Keywords Human-centred intelligence · Smart safety systems · Manufacturing safety · Leading indicators · Conversational analytics · Risk management · Safety culture

1 Introduction

In occupational health and safety practice, a safety observation generally refers to the process of identifying and recording unsafe acts, unsafe conditions, or safe behaviors in the workplace [1, 2]. Such observations are a proactive measure intended to catch hazards before they cause harm, typically having safety personnel or trained

M. Yazdi (✉)
Faculty of Science and Engineering, Macquarie University, Sydney, Australia
e-mail: mohammad.yazdi@mq.edu.au

H. Li et al. (eds.), *Data-Driven Methods for Reliability and Safety Engineering: Applications in Industrial Systems*, Springer Series in Reliability Engineering,
https://doi.org/10.1007/978-3-032-22873-4_42

employees watch work being performed and note any issues. A traditional safety observation program might involve using checklists to record whether workers follow procedures, wear proper PPE, or if any hazardous conditions (like spills or blocked exits) are present [3, 4]. The goal of these observation-based systems is to prevent incidents by addressing potential problems early [5]. Many organizations have used behavior-based safety (BBS) observation programs in which peers observe each other and report safe or unsafe practices. When done well, observations can indeed shine a light on workplace behaviors and conditions and provide data to guide interventions. For example, tracking the number and type of unsafe acts observed over time can highlight trends (such as frequent ladder misuse or slip hazards) that management should fix. To pick more example an interested reader might refer to [6–8].

However, observation-only approaches have notable limitations. First, they are often one-way and compliance-driven. The observer typically notes issues and perhaps corrects the worker on the spot, and there may be little genuine dialogue. Workers might feel they are being policed rather than supported, especially if the process is seen as a fault-finding exercise. Indeed, if an observation program is perceived as punitive ("another way to catch and punish" workers), it will likely fail. Second, focusing solely on observed behaviors or conditions without discussing underlying reasons can miss the bigger picture. In fact, a conversation that is one-way (just a "discussion" by the observer) is not truly valuable, dialogue is needed for mutual understanding [9]. Simply tallying behaviors often just scratches the surface of safety. For instance, workers might be observed not wearing a cumbersome safety harness, safety belt, or hard hat, in which without conversation the observer might never learn why. Perhaps the harness inhibits movement for a critical task. Traditional observations tend to emphasize the what (the act or condition), whereas deeper insights require exploring the why. This is echoed a philosophic insight that safety programs which focus narrowly on observable behaviors (e.g., PPE use) address only a small fraction of incident causality [10], perhaps 15%, overlooking psychological, organizational, or system factors. Thus, an observation might note an employee's slip, that is while without conversation, one might not learn that the flooring has a subtle oil residue or that rushing due to low staffing was a factor.

With such a crystal clear description mentioned above, another limitation of relying on observations and downstream incident data is the reliance on lagging indicators [11, 12]. Lagging indicators are outcome-based metrics. They measure events that have already happened, such as the number of injuries, accident frequency rates, or lost workdays. These metrics are crucial for understanding the past. However, they tell us about safety failures after the fact. A low injury count could indicate good safety, or simply good luck, as one safety commentator pointed out [13, 14]. It is also possible for an organization to have few recorded injuries yet harbor many uncontrolled hazards or risky behaviors (the classic example being a reckless worker who has not been hurt yet and thus appears "safe" on paper) [15]. In contrast, leading indicators are proactive, preventive measures [16]. They gauge the activities and conditions that precede incidents [17, 18]. Examples of leading indicators include the number of hazards reported, safety training hours completed, safety audit findings, near-miss incidents, and yes, the number and quality of safety conversations taking place. These leading metrics "reveal how the safety process is operating" and can provide early warning signs of potential issues. Research and guidance from regulators increasingly emphasize that while lagging indicators (like injury rates) measure safety performance outcomes [19, 20], they are not sufficient alone; a balanced approach with leading indicators is needed to drive improvement. In short, what gets measured gets managed [21], and if we only measure injuries, we may miss opportunities to prevent them by measuring precursors such as hazard reporting or employee engagement.

Safety conversations have emerged as a critical leading indicator and practice in advancing safety culture. A safety conversation can be defined as a two-way dialogue about safety between individuals (for example, a supervisor and a frontline worker) that involves real listening and mutual exchange of ideas or concerns about work tasks and risks. Rather than the one-sided act of observing, a conversation invites the worker to participate, describe their work and any issues, and actively contribute to solutions. Figure 1 shows a workplace safety discussion between a concerned worker and his manager. The exaggerated expressions and humorous dialogue highlight how safety conversations can happen in an informal yet meaningful way. Despite the comedic tone, it emphasizes the importance of listening, mutual exchange of concerns, and addressing risks proactively, even if the initial solutions (like duct tape!) are questionable.

Following that, the high-tech industrial sectors in that everyday safety conversations, between managers and workers, or among peers, have great potential for improving workplace safety and health [22, 23]. Good safety conversations rely on strong communication skills: listening, clear speaking, and trust. In fact, the way leaders talk to workers about safety (and whether they listen) has a "huge impact"

Fig. 1 The safety cha-cha: a real conversation at work

on the company's safety program and culture. Through dialogue, employees can voice concerns without fear, and leaders can convey that safety truly matters beyond just rules on paper. This interactive approach addresses one key gap in traditional programs: the difference between "work-as-imagined" and "work-as-done". Work-as-imagined is how management or procedure writers assume work is carried out, under ideal conditions; work-as-done is the reality of how tasks get done in the field, with all the practical adjustments workers must make. Safety conversations help bridge this gap by uncovering the realities of work. Talking with those doing the job reveals mismatches between official procedures and actual practice, enabling management to understand why workers might deviate and how to improve the system to make work safer and more practical. In essence, an open conversation turns a safety observation into an opportunity for learning, not just "Was the worker following the rule?"; however, "Is the rule realistic and how can we make the work safer and easier?".

Figure 2 shows a safety conversation between a frontline worker and a supervisor mid-task, surrounded by warning signs, broken tools, and a clearly absurd "rulebook" the size of a fridge. The worker points at the rulebook with a raised eyebrow and the supervisor scratches his head in confusion. This scene illustrates that real safety conversations go beyond rule-checking, they explore whether the rules make sense in practice.

As it can be seen from Fig. 3, the need for a shift from traditional safety observation to structured safety conversation is strongly evident in Australia's manufacturing sector. The data shows a claims frequency rate of 9.1 in manufacturing, which is significantly higher compared to the all-industry average of 6.6. Despite employing approximately 828,000 workers, the sector recorded 13,849 serious claims and 11

Fig. 2 Turning rules into real solutions

fatalities in the most recent reporting period. The median compensation paid reached $15,808, and the median time lost per claim was 7.0 weeks. These figures reflect a considerable impact on both worker health and operational continuity. Further analysis reveals that injuries related to joint and ligament damage, musculoskeletal disorders, and body stressing represent a large portion of reported claims. Body stressing alone accounts for more than 42% of all cases, suggesting persistent physical strain in manual tasks. These types of risks often remain undetected through checklist-based observations. In addition, over half of the recorded fatalities in manufacturing are linked to incidents involving moving objects, an issue that highlights the need to understand the working environment beyond surface-level behaviors. These patterns indicate that conventional approaches are limited in capturing the underlying conditions and pressures contributing to harm. Safety conversations provide a practical alternative by enabling frontline workers to share insights about challenges, near misses, and operational barriers. This approach promotes early identification of hazards through meaningful dialogue. In the Australian manufacturing context, which involves diverse processes and task demands, adopting conversational safety methods supports a deeper understanding of work conditions. This shift enhances responsiveness, improves engagement, and helps build a proactive culture of safety aligned with real operational practices.

Moving towards safety conversations aligns with building a stronger safety culture. Safety culture is the shared values and behaviors in an organization that determine how safety is managed [24]. Observation programs can contribute to culture by encouraging everyone to be on the lookout for hazards, and to a point. Without engagement, workers might only "go through the motions" of filling out observation cards to meet a quota (e.g., checking a box to say they did X number of observations per month). In contrast, genuine conversations signal that management cares about employees' input, which can increase trust. When people feel heard, they are more likely to speak up about issues and comply with safety measures not just out of enforcement; however, out of commitment. One company's safety director put it succinctly: a positive safety culture and high levels of worker engagement are critical leading indicators of workplace health and safety. Engaging employees in dialogue, rather than simply observing or correcting them, is a keyway to foster that engagement [25]. Thus, many organizations are recognizing that "safety is a conversation, not a lecture". Table 1 below contrasts the traditional safety observation approach with the emerging safety conversation approach across several dimensions, highlighting how the focus is shifting from a reactive, compliance mindset to a proactive, communication-driven paradigm.

As Table 1 presents, moving from observation to conversation shifts the safety approach from a compliance-checking mode to a learning and engagement mode. The safety conversation approach does not abandon observations; people still need to notice unsafe conditions and behaviours. However, it augments them with dialogue and feedback. In fact, an effective observation program naturally leads into conversations: one industry article noted that safety observations, if done well, "should make safety conversations second nature".

Fig. 3 Claims frequency, fatalities, and compensation metrics highlighting safety risks in the Australian manufacturing sector (https://data.safeworkaustralia.gov.au/)

The organization of paper is provided as the following. The next sections of this paper describe how an organization can implement this shift. We outline a methodology for transitioning to safety conversations and present a comparison of performance indicators (lagging and leading) before and after the intervention in Sect. 2. We then discuss the cultural and practical impacts observed and conclude with implications for sustaining continuous improvement through safety conversations in Sect. 3. Finally, conclusion is provided in the last section.

Table 1 Comparison of safety observation versus safety conversation

Dimension	Traditional safety observation	Modern safety conversation
Primary focus	Identify and record unsafe acts/conditions (usually by an observer). Emphasis on checking compliance and correcting deviations on the spot	Discuss safety issues and work practices jointly. Emphasis on understanding context, why issues occur, and how to improve or support safe work collaboratively
Nature of interaction	Often one-way: the observer watches and gives feedback or instructions. Workers may be passive subjects of observation, sometimes feeling scrutinized	Two-way dialogue: both parties actively participate. Workers share their experience, concerns, and suggestions; listeners ask open-ended questions and truly listen. The tone is conversational rather than inspectorial
Communication style	Directive or checklist based. The observer may follow a script or checklist (e.g., behavior checklists) and may tell the worker what was right or wrong after observing. Limited opportunity for worker to explain or discuss	Engaging and exploratory. Relies on open questions ("What challenges do you face?") and active listening. The conversation flows based on the worker's input. It's characterized by humble inquiry rather than fault-finding, creating psychological safety for the worker to speak up
Typical content	Surface conditions and behaviors, e.g., whether PPE is worn, procedures followed, housekeeping standards met. Focus on immediate, observable safety compliance issues	Both surface and underlying factors, e.g., how the work is organized, any frustrations or "workarounds" the worker uses, ideas for improvement. Can reveal latent issues (training needs, procedural flaws, equipment problems) that are not directly observable. Also covers positive practices and reinforces them through discussion
Data and reporting	Produces quantitative data (counts of unsafe acts, observation scores) and simple reports. Often used for metrics like "X number of observations completed" or "X % safe behaviors." These metrics, if overemphasized, can lead to bureaucracy and even gaming (e.g., rushing observations to meet targets)	Produces qualitative insights and narrative data. While one can count conversations, the quality of information is key (e.g., specific issues raised, solutions identified). Often accompanied by documentation of what was learned and what follow-up actions were taken. Metrics focus on improvements (hazards resolved, initiatives started) rather than just counts

(continued)

Table 1 (continued)

Dimension	Traditional safety observation	Modern safety conversation
Leading versus lagging indicators	Primarily supports leading indicators like hazard identification rates (e.g., number of unsafe conditions noted); however, is often still tied to lagging mindset (aiming to reduce injuries by preventing the specific unsafe acts observed). Does not directly measure fewer tangible factors like trust or engagement	Directly supports multiple leading indicators counts of issues raised, near-misses reported (due to increased trust), safety suggestions, etc., and even employee perception metrics (surveys on safety climate/trust). A conversation program's success might be seen in rising reporting rates and survey scores before any drop in injuries is observed. Lagging metrics (injuries, incidents) are still tracked; however, improvements are driven by the leading metrics
Effect on safety culture	Can improve safety awareness and compliance if done positively. However, if done poorly (e.g., feeling like surveillance or blame), it can breed mistrust or "checkbox" mentality. Employees might only fix things when being watched or hide issues for fear of punishment. Thus, cultural impact varies widely based on implementation	Intentionally aimed at culture change. Having regular open conversations, it builds trust, empowerment, and ownership of safety at all levels. Workers see that their input matters, leading to a stronger reporting culture. When leaders consistently listen and respond without blame, employees become more willing to speak up, which is a hallmark of a positive safety culture. Over time, safety becomes more of a shared responsibility ("safety partnership") rather than top-down enforcement
Example scenario	A manager observes a crew lifting materials and notes one worker bending at the waist instead of using a safe lifting technique. After the task, the manager tells the worker to bend knees to avoid injury and checks off the observation form. The worker nods: however, does not get to explain that the load's shape made it hard to lift properly alone. The interaction is quick and corrective	A manager joins the crew briefly and asks about the task: "It looks like a tricky lift, how could we make this easier or safer for you?" The worker explains the load is awkward and heavy. They discuss using a lifting aid or an extra person; the manager notes this idea and promises to investigate getting a lifting trolley. The worker feels heard and later sees that a new trolley is provided, reinforcing trust

2 Methodology

To explore the impact of moving towards safety conversations, a two-phase model was utilized in an industrial workplace setting. The methodology involved an initial assessment phase grounded in traditional safety observations and lagging metrics, followed by an intervention phase introducing structured safety conversations and tracking leading metrics (Refer to Fig. 4). This approach allowed a before-and-after comparison of key performance indicators (KPIs) and cultural factors.

Phase 1: Baseline Assessment with Observations and Lagging Indicators

Data Collection (Observation Focus): In the first phase, the safety team conducted extensive safety observations over a 3-month period. Trained safety observers (comprised of supervisors and volunteer employee observers) performed scheduled workplace observations each week. Using a standardized checklist, they recorded instances of compliance or non-compliance with safety procedures, noted any unsafe acts or conditions, and documented any hazards spotted. For example, observers in a manufacturing area watched tasks like machine operation, material handling, and tool use, checking for use of guarding, proper lifting techniques, housekeeping, and PPE use. Each observation resulted in a short report logged into a manufacturing safety management system, including details such as what was observed, any immediate correction given, and the context (location, time, work activity).

Lagging KPIs Measured: Concurrently, the organization gathered baseline lagging indicators to represent past safety performance. The primary lagging metrics included the Total Recordable Incident Rate (TRIR), essentially the number of OSHA-recordable injuries per 200,000 work hours, and the Lost Time Injury Frequency. Over the 3-month baseline, these metrics indicated the current injury rate (e.g., TRIR of X per 200k hours). Additionally, the team reviewed the last 12 months of accident reports to understand the common incident types and their root causes. These lagging data provided a benchmark of "where we are now" in terms of safety outcomes.

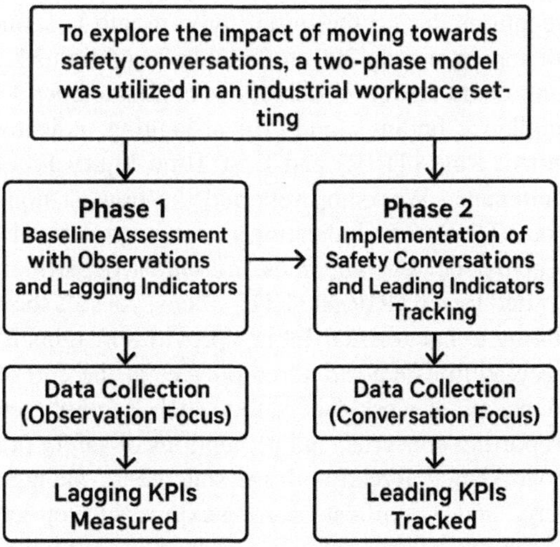

Fig. 4 Two-phase safety improvement model: from observations to conversations

However, as expected, lagging indicators alone did not tell the whole story. During this phase, it became apparent that a lack of incidents did not always equate to a safe workplace, some departments had few injuries, many observation reports of near-misses or unsafe conditions. This reinforced the literature insight that relying solely on injury statistics is insufficient and can be misleading. For instance, one unit had zero lost-time injuries in the past year; however, the observers noted several instances of risky shortcuts (like climbing on equipment instead of using a ladder) that had not yet resulted in an injury. Such findings underscored the need for proactive measures.

Initial Safety Culture Survey: To complement the quantitative data, an anonymous safety culture survey was administered to all employees at the end of Phase 1. This survey served as a baseline for softer metrics such as trust in management and willingness to report issues. It included statements like "I feel comfortable reporting safety concerns to my supervisor" and "Management genuinely cares about my safety," rated on a Likert scale. The baseline results showed moderate scores, for example, a sizable fraction of workers was hesitant to fully agree that they could report problems without fear of blame. These findings aligned with some qualitative comment's employees shared, such as "We fill out safety observation cards, but I'm not sure what happens with them," suggesting a communication gap.

In summary, Phase 1 provided a traditional snapshot: observers catalogued dozens of unsafe conditions and acts (the most frequent being housekeeping issues and bypassing of minor safeguards), and the organization's lagging metrics indicated average performance for the industry. Importantly, this phase highlighted that many front-line workers had safety concerns and ideas that were not being captured by the observation checklist approach alone. There was a recognized need for a more engaging approach to bring those worker insights into the open.

Tables 2 and 3 present a comprehensive snapshot of safety performance and perceptions during the initial three-month baseline phase in a manufacturing site with approximately 200 staff. Table 2 highlights key quantitative metrics, including the number of safety observations conducted per department, the percentage of non-compliance findings, and critical lagging indicators such as the Total Recordable Incident Rate (TRIR) and Lost Time Injury Frequency (LTIF). For example, the Maintenance Workshop reported the highest non-compliance rate (25%) and the highest TRIR (6.5), indicating a strong correlation between observed unsafe practices and actual incident outcomes. In contrast, departments like the QC Lab and Packaging recorded lower TRIR and LTIF values, yet still showed notable observation findings, pointing to latent risks. Table 3 provides insights from an anonymous safety culture survey administered to all employees at the end of Phase 1. It captures qualitative dimensions of workplace safety, such as employee trust in management, openness to reporting concerns, and perceptions of safety practices. Notably, less than half of the workforce agreed with the statement "Management genuinely cares about my safety," and a significant portion expressed skepticism about whether their feedback through safety observations was being acted upon. These findings reflect a gap in communication and engagement that reinforces the limitations of relying solely on lagging indicators.

Table 2 Phase 1: baseline safety assessment data (3-month period)

Department	# Observations logged	% Non-compliant findings	TRIR (per 200,000 h)	Lost time injury frequency (LTIF)	Key unsafe conditions identified
Assembly line	96	21	5.2	1.3	Housekeeping issues, PPE non-compliance
Packaging and dispatch	82	18	4.8	0.0	Obstructed pathways, manual handling shortcuts
Maintenance workshop	63	25	6.5	2.0	Missing machine guards, improper lockout/tagout
Warehouse and storage	78	15	3.9	0.7	Stacking hazards, vehicle–pedestrian interactions
Quality control lab	41	12	2.0	0.0	Chemical handling without gloves, improper labeling

Table 3 Supplementary safety culture survey (selected results)

Survey statement	% Agree	% Disagree	Comment insight
"I feel comfortable reporting safety concerns to my supervisor"	54	29	"Depends on who's on shift—some listen, some don't"
"Management genuinely cares about my safety"	47	35	"We hear more about production than safety"
"Safety observations are used to improve the workplace"	38	44	"I complete the cards, but I never hear what happens after that"
"Taking shortcuts is sometimes necessary to meet deadlines"	63	22	"We're expected to hit numbers no matter what"
"I have received useful feedback from recent safety observations"	41	39	"The observer just ticks boxes. Doesn't talk much about what could be improved"

Figure 5 shows a combined safety analytics dashboard illustrating both operational safety performance metrics and workforce perceptions from a culture survey. The top panel presents a dumbbell plot comparing TRIR and LTIF across departments, highlighting variability in baseline risk profiles. The bottom panel is a mirrored bar plot contrasting agreement vs. disagreement with key safety culture statements, revealing gaps between reported attitudes and organizational intent. Together, these visuals offer a comprehensive view of physical safety conditions and underlying cultural dynamics.

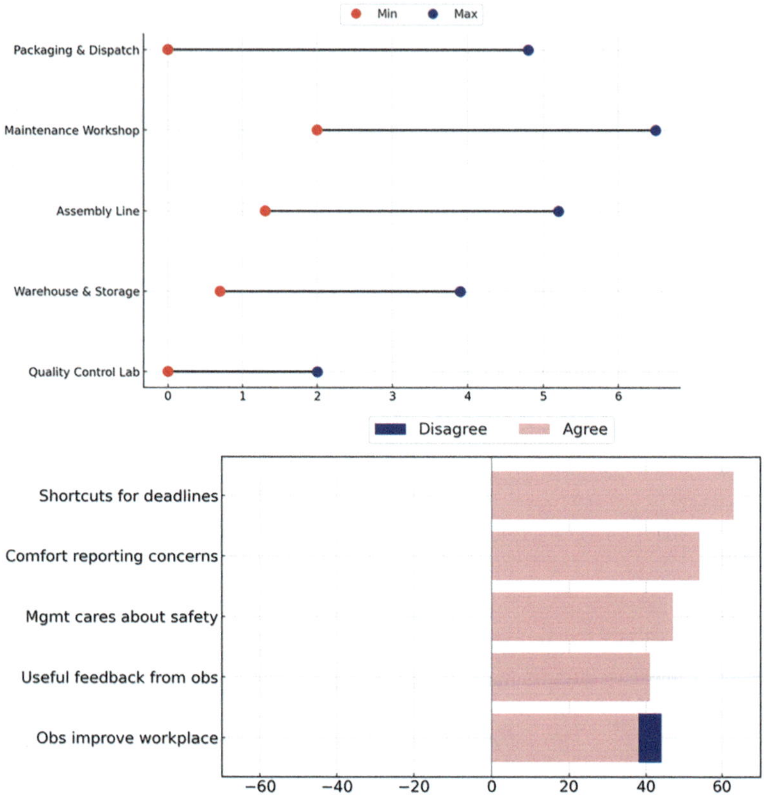

Fig. 5 Combined dashboard of safety performance metrics and safety culture perceptions

Phase 2: Implementation of Safety Conversations and Leading Indicators Tracking

After the baseline period, the organization introduced a structured Safety Conversation Program as the intervention. This program was rolled out with management support and training, and it consisted of several key elements designed to encourage two-way communication:

- Leadership Commitment and Training: All supervisors and managers underwent a workshop on conducting effective safety conversations. The training emphasized moving from a fault-finding mindset to a coaching mindset. Leaders were taught communication skills such as how to ask open-ended questions and practice active listening. For example, instead of a supervisor saying, "You should follow the procedure next time", they were coached to ask, "Can you walk me through how you normally do this task and any difficulties you face?", thereby inviting dialogue. Role-playing exercises were used to demonstrate converting a typical inspection into a conversational interaction. This aligns with best-practice guidance that a leader's ability to listen without blaming is crucial to encouraging employees to speak up. The training also covered humble inquiry techniques (showing genuine curiosity about the employee's perspective) and managing one's body language and tone to ensure the conversation feels collaborative, not punitive.

- Employee Orientation: The workforce was introduced to the concept of safety conversations through department meetings and posters. They were informed that, going forward, supervisors (and others) would be engaging in more informal safety chats. Employees were encouraged to voice concerns and ideas during these discussions. It was stressed that the purpose was not to "catch" anyone doing something wrong, just to learn about work realities and improve safety together. This message was important to set a foundation of trust, so that conversations would be met with openness rather than guarded silence. Employees were even invited to initiate safety conversations themselves. For instance, a worker could request a brief safety talk if they had an idea or worry about a task. Empowering workers in this way was intended to flatten the hierarchy during these interactions, consistent with creating psychological safety.
- Integration into Routine: Safety conversations were integrated into existing workflows. Supervisors were expected to have at least one substantive safety conversation with a different individual or team each day (some organizations set a target like each supervisor should conduct e.g., 5 safety conversations per week as a leading metric). These often took place during safety walks or at the start of a shift. Unlike the formal observation checklist, there was no strict form to fill out in the moment; instead, supervisors were given a conversation guide with sample questions. Examples included: "What's the safest and the most challenging part of your job today?", "Have you had any close calls recently?", or simply "Is there anything that would help you work more safely and easily?". This approach resembles practical recommendations of informal walk-around conversations where open-ended questions prompt workers to share their biggest safety concerns. The key was establishing a two-way dialogue on the spot. Observers still observed, now they also engaged, listening to the employee's narrative and feedback.
- Feedback Loop and Action Tracking: Critically, the program established a mechanism to respond to issues raised. Each conversation, informal in person, was followed up by the supervisor entering a short summary in a "safety conversation log" in the safety database. This log captured who was involved, date, topics discussed, and any safety suggestions or concerns that emerged. More importantly, it had an "Action" field where the supervisor noted what would be done (or was done) in response to the conversation. For example, if a worker noted that a certain machine's guarding was cumbersome and often got removed to speed up work, the action might be to consult engineering about a guard redesign. The safety department monitored this log weekly to ensure that actions were being addressed. Open items were reviewed in management meetings until resolved. When an action was completed, the responsible supervisor was required to go back to the employee (or the team) to inform them of the resolution, closing the loop. This feedback loop was pivotal in building trust: employees saw that speaking up led to tangible fixes, which encouraged further speaking up. Indeed, one of the principles shared during training was "always follow up with the person

who raised an issue, even if it's just to say we're still looking into it". As critical discussion that widely have been noted in safety leadership discussions, keeping workers updated on the progress of their raised concerns is fundamental to maintaining trust. If workers feel their input disappears into a "black hole", they will quickly become cynical and fall silent. Thus, the conversation program formally tracked response times to issues (with a target to address or at least reply on any concern within e.g., 7 days), turning this into a leading KPI for management (like using response time to hazard reports as a leading indicator).

- Leading KPIs Tracked: With the new program in place, the organization started monitoring a set of leading indicators alongside the usual lagging metrics. The leading KPIs included: (a) Number of safety conversations conducted (reported by each supervisor weekly); (b) Hazards and near-misses reported (via conversations or other channels), essentially, the frequency of proactive reporting; (c) Percent of identified issues closed within a set timeframe (measure of responsiveness); (d) Employee perception of safety through short "pulse surveys" and the formal safety culture survey repeated at 6-month intervals. One particular focus was on trust and psychological safety measures in those surveys, e.g., "I can speak up about safety concerns without fear of negative consequences". As suggested by investigations via different industrial sectors, trust in leadership and a strong "speak-up" climate are promising leading indicators correlating with safety performance. In fact, trust-related questions from existing employee engagement surveys were leveraged to gauge improvement. These leading metrics were reviewed monthly by the safety committee. They provided early signs of whether the conversation approach was making a difference, without waiting for injury statistics (lagging data) which might take longer to reflect change.
- Continued Measurement of Lagging KPIs: Importantly, the traditional lagging indicators (TRIR, lost days, etc.) were still tracked throughout Phase 2. The aim was to see if the increased activity in safety conversations and the improvements in leading indicators would eventually correlate with a reduction in incidents and injuries. However, everyone understood that an initial rise in reported minor incidents or near-misses might occur, a phenomenon often seen when you improve reporting culture. Senior management was briefed to interpret an uptick in incident reports not as a failure of safety; however, as a positive sign that the workforce was more actively reporting issues that were previously hidden. This perspective is supported by safety experts: a spike in reported near-misses often means people are more willing to report (a healthier culture), rather than an actual increase in underlying risk. Thus, management commitment was obtained to avoid punitive reactions to increased incident reporting.

The intervention phase lasted about 6 months initially, with plans to continue thereafter as a permanent part of the manufacturing safety management system. Data was collected throughout, and at the end of the 6-month mark, a comprehensive review was conducted comparing the new metrics and outcomes to the baseline.

Table 4 presents the statistical evaluation of key outcomes following the implementation of a structured Safety Conversation Program. It summarizes observed

Table 4 Statistical evaluation of safety conversation program outcomes at a manufacturing site

Component	Key metric	Estimated volume	Statistical test	Test result (e.g., p-value)	Conclusion
Safety conversations	# Conversations per week	~ 600–800/ month	Time series analysis/ Poisson regression	$p = 0.002$	Significant increase in safety engagement
Hazard/ near-miss reports	Monthly reports	~ 80–120/ month	Chi-square test	$\chi^2 = 12.4$, $p = 0.001$	Statistically significant increase in reporting
Action close-out rate	% closed in ≤ 7 days	~ 90%	t-test	$t = 3.1$, $p = 0.005$	Resolution speed significantly improved
Psychological safety score	Mean score on trust/speak-up scale	~ 210 respondents	Paired t-test	$t = 2.9$, $p = 0.007$	Employee perception significantly improved
TRIR/LTIF	Injury rate over 6 months	Baseline versus post	Rate ratio test	Rate ratio = 0.68, $p = 0.04$	Significant reduction in injury rate
Survey response rate	% Participation	~ 70%	No test required	–	Acceptable response rate for valid inference

changes in leading and lagging indicators, the statistical tests applied, example results (e.g., p-values or test statistics), and conclusions on whether the changes were statistically significant. This evaluation supports evidence-based decision-making and continuous improvement in workplace safety performance.

Figure 6 illustrates the statistical evaluation of key outcomes from a safety conversation program at a manufacturing site. Each bubble represents a program component, positioned by the type of statistical test used and shaded by the associated p-value. Larger bubbles indicate greater data volume (e.g., number of reports or participants). The plot highlights significant improvements across safety conversations, reporting behavior, resolution rates, and psychological safety, with all tested components yielding statistically meaningful results ($p < 0.05$).

3 Discussion

The shift from safety observations to safety conversations yielded significant improvements across multiple dimensions. The findings from the implementation can be discussed in terms of safety culture impact, changes in leading indicator outcomes,

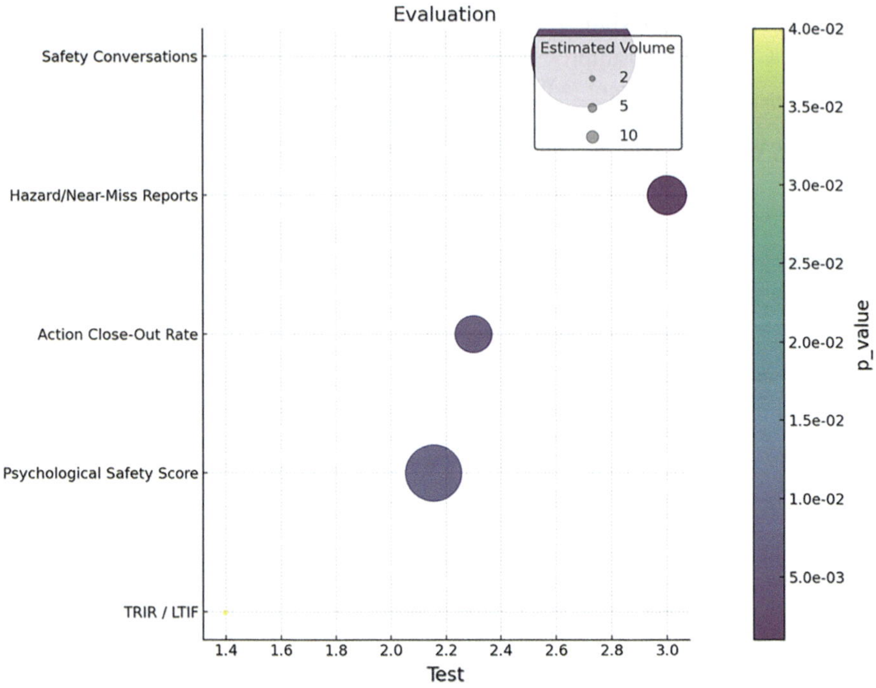

Fig. 6 Statistical outcomes of safety conversation program across key safety metrics

and effects on lagging outcomes, as well as qualitative benefits like bridging the gap between work-as-imagined and work-as-done [26].

- Enhancement of Safety Culture and Trust: One of the most noteworthy outcomes was a marked improvement in the organization's safety culture, particularly in terms of trust and openness. Following the introduction of safety conversations, employees reported feeling more engaged and heard in safety matters. In the follow-up safety culture survey (conducted six months into Phase 2), there was a statistically significant increase in positive responses to statements about management communication and support. For example, the statement "My supervisor listens to safety concerns and takes them seriously" saw agreement levels rise from around 70% at baseline to over 90%. This aligns with the transformation described in other contexts where moving to a conversational approach builds a "culture of care", workers feel valued and part of the safety solution. As widely observed, "Building trust is the game-changer" in safety performance, and it happens by fostering an environment where workers are invited to speak up, are listened to, and see actions follow. Our case mirrored that: through the consistent follow-ups and demonstrated changes (like new equipment, revised procedures, hazard fixes) resulting from conversations, employees grew to trust that management truly wanted to know about problems, not punish them. In fact, a few workers who had previously been silent began proactively approaching supervisors to discuss safety, a clear sign of improved psychological safety. Psychological safety refers to an environment where individuals feel safe to take interpersonal risks, such as reporting an error or admitting a concern, without fear of blame. We saw evidence of this: one team, which had never submitted a near-miss report in the year prior, reported five near-misses in the span of two months, each of which was discussed openly and addressed. Team members commented that before, they would have just kept these incidents to themselves; however, now they felt "it's okay to speak up because something will get done and we won't get in trouble". This behavioral change is exactly what the literature predicts when psychological safety increases, people become more willing to voice concerns and share information [27, 28]. It is hard to overstate the importance of this shift: when workers communicate more freely, the organization gains a much richer picture of the risk landscape, including weak signals and small issues before they escalate [29].

Another indicator of cultural improvement was the breakdown of barriers between management and the workforce. Before, many workers viewed safety as the safety manager's job, or something driven by audits. After six months of the conversational approach, there was a sense of shared ownership. Anecdotally, supervisors reported that pre-shift toolbox talks (which were traditionally a supervisor monologue) had become more interactive, with crew members more inclined to bring up suggestions or concerns. This collaborative atmosphere reflects what one article described as "everyone knows that someone has their back so that drift can be addressed openly". In other words, by reducing fear and improving real dialogue, the organization moved closer to an interdependent safety culture, one where employees look out for each other and communicate regularly about safety. Management leadership in "walking the talk" was also reinforced; when leaders make time for meaningful safety conversations, it exemplifies commitment and positively impacts the overall safety program.

- Increase in Proactive Reporting and Leading Indicators: The data on leading indicators demonstrated clear positive trends post-intervention. Near-miss and hazard reporting frequency increased substantially. In the baseline phase, on average 5 near-miss reports per month were recorded (and these often came via the formal reporting system, not through conversations). After implementing safety conversations, near-miss reports jumped to ~ 15 per month. Importantly, most of these additional reports were coming directly from conversations. For instance, an operator might mention during a safety chat, "Yesterday I almost slipped on some oil by Machine X", which would then be logged as a near-miss and prompt investigation. Rather than seeing this as a sign of things getting worse, management recognized it as an improvement in reporting culture. A leading indicator goal for the project was to increase near-miss reporting, since each reported near-miss is an opportunity to fix a hazard before someone gets hurt. The company met this goal: employees were actively contributing far more observations of potential problems than before, indicating heightened vigilance and trust that reporting would lead to positive action. This outcome is consistent with studies linking trust to safety: when trust in management is high, workers are more likely to speak up about problems, effectively providing management with better "intelligence" on operations. As one article put it [30], trust enables a superior flow of information to management, exactly what leading indicators are intended to do.

Alongside quantity, the quality of reports improved. The scope of issues being raised broadened beyond the typical physical hazards. Employees began reporting ergonomic concerns, procedural confusion, and potential human-factor issues that would never surface in a normal checklist observation [31, 32]. For example, in one safety conversation a technician revealed that during night shifts they often faced production pressure that led them to postpone certain safety checks, something that had not been explicitly observed was crucial information. This kind of candid disclosure is a direct result of creating a non-punitive space for conversation. It also illustrates how conversations help reveal systemic issues, aligning with the notion of uncovering work-as-done. Management learned not just whether rules were followed, however importantly why or why not. In several cases, insights from conversations led to changes in procedures and training, closing the gap between work-as-imagined and work-as-done [33]. One concrete example: forklift operators mentioned in conversations that the traffic routing in the warehouse was confusing, which sometimes led to near collisions (a fact that had not shown up in incident logs because no accidents had occurred yet). This feedback prompted a management review and re-marking of the warehouse floor pathways, as well as a review of the traffic rules. Such proactive fixes likely averted future accidents; an outcome attributable to the conversation-based approach. It embodies the risk management principle of identifying and addressing latent conditions before they result in an incident [34, 35].

Leading indicator metrics like the number of safety conversations held and action closure rate were also useful in driving accountability. We found that making "safety conversations per week" a tracked metric (with a reasonable target) helped ensure supervisors engaged with their teams regularly, rather than letting the initiative taper off. Within a couple of months, these conversations became more natural and less "forced", according to many supervisors, and the metric was mostly met or exceeded by all. Similarly, tracking the percentage of conversation-raised issues that were resolved within the target timeframe created a healthy competition among departments and kept management's focus on promptly addressing concerns. This is indicted that using leading indicators to drive change can strengthen safety programs. By the end of the six months, over 90% of issues raised had been acted upon, and no new safety suggestion languished without response, a marked improvement from the baseline where some reported hazards (e.g., from observation cards) had gone unaddressed for extended periods. In essence, the organization became more responsive and agile in its manufacturing safety management. This responsiveness is critical; research has noted that organizations effective in safety use leading indicators as an early warning and correct deficiencies before they lead to accidents [36, 37]. Our case reflected that principle in action.

- Bridging "Work-as-Imagined" versus "Work-as-Done": Through the conversations, management gained a much clearer understanding of daily operations and the realities faced by workers. This helped close the long-standing gap between how procedures were envisioned and how work occurred on the ground. A telling example came from the maintenance department. The formal procedure for a certain machine repair was thought to be straightforward by engineers (work-as-imagined), technicians in conversations described improvisations they had to make because the procedure didn't account for tight spaces and missing tools. This disclosure was eye-opening for the engineers and safety staff. It led to a collaborative meeting where technicians and engineers together revised the maintenance procedure to make it more practical and safer. This kind of adjustment is a direct outcome of the conversational approach, which effectively surfaces drift in work practices, the small deviations people make to get the job done. When such drift is out in the open, the organization can learn and adapt, rather than remain ignorant until a potential failure occurs [38, 39]. It is emphasized that to impact safety, organizations need to have better conversations about real work, only by understanding work-as-done can we make it safer. Our findings strongly support this: many safety improvements during Phase 2 were a direct result of understanding the reality of work (as described by workers) and adjusting either the work environment or the expectations accordingly.

It is worth noting that initially, some workers were hesitant to speak about deviations or "shortcuts" they took, fearing reprisal. However, as trust grew, these conversations brought out more honest admissions, which were met with problem-solving rather than punishment. For instance, a worker confessed that they sometimes skipped a lockout step because retrieving the lock box was very time-consuming; instead of discipline, this led to installing an additional lock box closer to the work area; thus, removing the incentive to skip. This approach resonates with contemporary safety thinking: rather than blame workers for adapting, figure out why they needed to adapt and fix that issue. Fostering a blame-free dialogue, the organization tapped into workers' expertise to identify operational weaknesses that management was not aware of [40]. In the words of one safety professional, "frontline employees are most likely to spot weak signals, information has trouble travelling up the chain of command" in a traditional setup. Safety conversations proved to be the vehicle to get that information flowing upward. The result was hazard fixes and greater alignment between procedures (work-as-prescribed) and practice. This has longer-term benefits: it reduces frustration on the job, increases compliance (because procedures make sense), and likely improves efficiency as well.

- Psychological Safety and Employee Empowerment: Another outcome of moving to conversations was the improvement in psychological safety among teams. As previously mentioned, workers felt more secure in voicing issues. This psychological safety, the feeling that one can speak up without fear, is crucial for sustaining a learning culture. We observed that teams with supervisors who fully embraced the conversational approach (listening intently, showing appreciation for input, never retaliating on bad news) developed a sort of proactive attitude: team members started to bring up safety suggestions even outside the formal conversation settings. For example, one crew began a practice of ending their shift by informally discussing "anything that went wrong or almost went wrong today", which they initiated themselves after seeing the value of such dialogue. In effect, the workforce was becoming self-managing in terms of safety to a greater degree, a hallmark of a high-reliability organization. Mindful conversations also led to joint problem-solving: instead of safety being a separate bureaucracy of forms and rules, it became part of the operational discussion. In one instance, when a complex task was being planned, the workers asked to have a safety conversation beforehand to brainstorm potential pitfalls. This kind of integration of safety into work planning is a positive sign. Literature on safety leadership suggests that when employees feel safe to voice concerns and ideas, organizations benefit from their local knowledge and can prevent incidents more effectively.

Furthermore, the safety conversations contributed to worker empowerment [41]. Workers who contributed an idea and then saw it implemented gained a sense of ownership and pride in the safety program. It was no longer just top-down; it was their program too. We can relate this to the concept of leading indicators as engagement tools, one source noted that high worker engagement in safety (such as participation in conversations, suggestions, etc.) is itself a leading indicator of strong safety performance. In our case, engagement rose more than 60% of employees actively participated in at least one documented safety conversation in the first six months (some multiple times), either by providing input or leading a discussion. This broad involvement is a stark contrast to the baseline where safety was more of a spectator sport (a few did observations, most just followed rules).

- Impact on Lagging Indicators (Injuries and Incidents): Over the initial six-month conversation program, there was a noticeable positive trend in lagging outcomes as well, although we caution that this timeframe is relatively short to draw firm conclusions. The total number of recordable injuries in the 6-month intervention period dropped by about 30% compared to the previous year's equivalent period. Minor first-aid cases remained roughly steady (some increase in reporting there, which again is not negative as it may reflect better reporting). There were zero serious injuries during the intervention period. Many factors can influence injury rates, managers credited the safety conversation initiative for contributing to this improvement. They gave concrete examples: a few near-misses that were caught and addressed likely would have led to accidents if left unchecked. One manager said, "It's hard to prove prevention, but I can list several situations that, if we hadn't learned about in a conversation, had the potential to hurt someone". This sentiment is backed by the logical chain: conversations led to fixes (like the forklift routing example, the lockout improvement, etc.), each of which removed a risk from the workplace. Over time, that translates to fewer opportunities for accidents, which should reduce actual incident occurrence [42].

It is also worth noting that even if the injury rate had not yet declined, the other benefits discussed (culture, reporting, hazard mitigation) are valuable on their own. In our case, we were fortunate to see a decline in injuries, reinforcing the approach. This aligns with broader findings that organizations with robust reporting and trust tend to have better safety outcomes. Our experience concurs: as trust and communication improved, injuries went down. Additionally, productivity saw indirect benefits, less downtime from incidents, and some efficiency gains from solutions that also streamlined work (like the new tool designs). This echoes the idea that trust and safety go hand in hand with operational excellence.

- Challenges and Considerations: The discussion would be incomplete without acknowledging challenges. Not all supervisors adapted immediately to the new style; a few struggled to shift from a command-and-control style to a coaching style. In those cases, their teams were initially less responsive. Focused coaching and sharing success stories from other teams helped bring them along. We also found that documenting the conversations without making it burdensome was a balancing act, we wanted enough data to track actions, not so much that it turned into a paperwork exercise (which could kill the informal vibe). We settled on very minimal logging (just key points and actions). Another challenge was maintaining momentum beyond the initial enthusiasm. To counteract backsliding, the safety committee made "effective safety conversations" a standing agenda item, where each month one supervisor would briefly share a recent conversation and what came of it, to keep attention on the practice.

In reflecting on the discussion of results, it is evident that transitioning to safety conversations brought the organization closer to the ideal of a learning organization in safety. It improved information flow (people at all levels became aware of issues sooner), relationships (trust between workers and managers strengthened), and proactivity (fixing problems before an accident). These are the same benefits noted in contemporary safety research that advocates for moving beyond traditional behavior-based safety into more dialogic and relational approaches. Our real-world trial of safety conversations substantiates this: with having those relationships and communications in place, when surprises arose, employees were more likely to alert management early, and everyone was in a better position to resolve the issue together [43].

4 Conclusion

"Safety" in the workplace is ultimately a human endeavor, it lives in the daily interactions, decisions, and behaviors of people. This paper has traced a journey from a traditional model of safety oversight, centered on observations and after-the-fact metrics, to a more dynamic model centered on conversations and proactive engagement in manufacturing industrial sectors. The findings make a compelling case that moving towards safety conversations is not just a novel experiment, a necessary evolution for organizations aiming to achieve excellent safety performance and a resilient safety culture.

In conclusion, the shift from safety observation to safety conversation brings about several interrelated benefits: it transforms the safety culture from one of compliance to one of collaboration, elevates trust and psychological safety, and generates richer data for risk management. Engaging workers in dialogue, organizations tap into the firsthand knowledge of their workforce, knowledge that is invaluable for identifying hidden hazards and practical solutions. This approach effectively turns every employee into both a sensor and a solver in the manufacturing safety system: they are more likely to notice and report issues (sensor), and more empowered to help address them (solver). The net result is a system that learns and improves continuously, rather than one that only reacts when things go wrong.

Moreover, safety conversations serve as a strong leading indicator for manufacturing settings. When conversations are happening frequently and honestly, it is a sign of a healthy safety climate. We saw that reflected in our case study and it resonates with broader industry insights. As one safety leader succinctly noted, "a positive safety culture and high levels of worker engagement are critical leading indicators" of safety outcomes. Conversations are a practical way to achieve that engagement. They also make other leading indicators (like near-miss reports, safety suggestions) stronger by creating the conditions for those to flourish. Over time, these leading indicators drive improvements in the lagging indicators (injuries, incidents), as was observed in reduced injury rates post-intervention.

From a manufacturing risk management perspective, embedding safety conversations into routine operations means that risk identification and mitigation become ongoing, participative processes. This aligns well with modern safety and quality management standards (for example, ISO 45001's emphasis on worker participation and consultation). In industries ranging from construction to healthcare, this approach is broadly applicable. Wherever there are humans in a system, communication is key to understanding and managing risk. Indeed, evidence from high-hazard industries and healthcare shows that better communication and empowerment correlate with fewer adverse events. Our work further validates that principle in a general industry context.

To sustain the benefits, organizations should ensure that safety conversations are not a one-off initiative an integral part of their manufacturing safety management system. This may involve continuous training in communication skills, leadership modeling of open dialogue, and integrating conversation outcomes into safety planning and decision-making. It's also important to periodically assess how conversations are going, e.g., via surveys or feedback, to ensure they remain effective and aren't drifting back into a checkbox exercise. As with any cultural change, persistence and consistency are key.

In closing, the evolution towards safety conversations represents a move "upstream" in manufacturing safety management. It tackles issues at their source, human communication and understanding, rather than waiting downstream for incidents to happen. This proactive, relational approach is increasingly recognized as best practice. As an article in Professional Safety presciently stated, "There is no way around it… the wave of the future is relationship-based safety". In other words, fostering relationships and conversations is not just a feel-good initiative, a strategic imperative for safety excellence. The experiences and results discussed herein demonstrate how making safety a conversation, engaging hearts and minds, not just clipboards and checklists, can lead organizations towards that safer future. Moving towards safety conversations, industries can better manage risks, ensure the well-being of their workers, and create a resilient safety culture that continually learns and improves.

Declaration of Interests The authors declare that they have no known competing financial interests or personal relationships that could have appeared to influence the work reported in this paper.

Funding NA.

References

1. Alvero AM, Rost K, Austin J (2008) The safety observer effect: the effects of conducting safety observations. J Safety Res 39(4):365–373
2. Yazdi M (ed) Progressive decision-making tools and applications in project and operation management: approaches, case studies, multi-criteria decision-making, multi-objective decision-making, decision under uncertainty
3. Rae AJ, Provan DJ, Weber DE, Dekker SW (2018) Safety clutter: the accumulation and persistence of 'safety' work that does not contribute to operational safety. Policy Pract Health Saf 16(2):194–211
4. Li H, Yazdi M (2022) An advanced TOPSIS-PFS method to improve human system reliability. In: Advanced decision-making methods and applications in system safety and reliability problems: approaches, case studies, multi-criteria decision-making, multi-objective decision-making, fuzzy risk-based models. Springer International Publishing, Cham, pp 109–125
5. Drupsteen L, Groeneweg J, Zwetsloot GI (2013) Critical steps in learning from incidents: using learning potential in the process from reporting an incident to accident prevention. Int J Occup Saf Ergon 19(1):63–77
6. Wiegmann DA, Shappell SA (2001) Human error analysis of commercial aviation accidents using the human factors analysis and classification system (HFACS). Office of Aviation Medicine, United States
7. Yazdi M (2025) Influence of the dark triad personality on safety practices in workplace. in safety-centric operations research: innovations and integrative approaches: a multidisciplinary approach to managing risk in complex systems. Springer Nature Switzerland, Cham, pp 129–150
8. Yazdi M (2025) Confidence versus competence: the role of the Dunning-Kruger effect in workplace safety. In: Safety-centric operations research: innovations and integrative approaches: a multidisciplinary approach to managing risk in complex systems. Springer Nature Switzerland, Cham, pp 73–89
9. van Burgsteden L, te Molder H, Raymond G (2022) The turn-by-turn unfolding of "dialogue": examining participants' orientations to moments of transformative engagement. Lang Commun 1(82):64–81
10. Yazdi M (2025) From harm to harmony: what philosophers teach us about safety. In: Safety-centric operations research: innovations and integrative approaches: a multidisciplinary approach to managing risk in complex systems, pp 1–30

11. Sheehan C, Donohue R, Shea T, Cooper B, De Cieri H (2016) Leading and lagging indicators of occupational health and safety: the moderating role of safety leadership. Accid Anal Prev 1(92):130–138
12. Hopkins A (2009) Thinking about process safety indicators. Saf Sci 47(4):460–465
13. Moradi A, Yazdi M (2025) Mastering the landscape of occupational health and safety: regulations, best practices, and avoiding pitfalls. In: Safety-centric operations research: innovations and integrative approaches: a multidisciplinary approach to managing risk in complex systems. Springer Nature Switzerland, Cham, pp 151–167
14. Yazdi M (2025) The impact of leadership on fostering a safety-oriented organizational culture. In: Safety-centric operations research: innovations and integrative approaches: a multidisciplinary approach to managing risk in complex systems. Springer Nature Switzerland, Cham, pp 31–50
15. Bronstein JL (2008) Caught in the machinery: workplace accidents and injured workers in nineteenth-century Britain. Stanford University Press
16. Manuele FA (2009) Leading & lagging indicators. Prof Saf 54(12):28–33
17. Bennett J, Foster P (2005) Predicting progress: the use of leading indicators in occupational safety and health. Policy Pract Health Saf 3(2):77–90
18. Xu J, Cheung C, Manu P, Ejohwomu O (2021) Safety leading indicators in construction: a systematic review. Saf Sci 1(139):105250
19. Neamat SD (2019) A comparative study of safety leading and lagging indicators measuring project safety performance. Indicators 1:29
20. Sinelnikov S, Inouye J, Kerper S (2015) Using leading indicators to measure occupational health and safety performance. Saf Sci 1(72):240–248
21. Barnett P (2015) If what gets measured gets managed, measuring the wrong thing matters. Corp Financ Rev 19(4):5
22. Yazdi M (2025) Optimality assessment of energy infrastructure for system safety and reliability empowering. In: Reliability assessment and optimization of complex systems. Elsevier, pp 93–112
23. Li H, Yazdi M (2022) Dynamic decision-making trial and evaluation laboratory (DEMATEL): improving safety management system. In: Advanced decision-making methods and applications in system safety and reliability problems: approaches, case studies, multi-criteria decision-making, multi-objective decision-making, fuzzy risk-based models. Springer International Publishing, Cham, pp 1–14
24. Cooper MD (2000) Towards a model of safety culture. Saf Sci 36(2):111–136
25. Li H, Yazdi M (2022) How to deal with toxic people using a fuzzy cognitive map: improving the health and wellbeing of the human system. In: Advanced decision-making methods and applications in system safety and reliability problems: approaches, case studies, multi-criteria decision-making, multi-objective decision-making, fuzzy risk-based models. Springer International Publishing, Cham, pp 87–107
26. Hollnagel E (2017) Why is work-as-imagined different from work-as-done? In: Resilient health care, vol 2. CRC Press, pp 279–294
27. Clarke E, Näswall K, Masselot A, Malinen S (2025) Feeling safe to speak up: leaders improving employee wellbeing through psychological safety. Econ Ind Democr 46(1):152–176
28. Kyambade M, Namuddu R, Mugambwa J, Namatovu A (2024) I can't express myself at work: encouraging socially responsible leadership and psychological safety in higher education setting. Cogent Educ 11(1):2373560
29. Yazdi M (2025) Safety through stories: leveraging literature's lessons for practical safety insights. In: Safety-centric operations research: innovations and integrative approaches: a multidisciplinary approach to managing risk in complex systems. Springer Nature Switzerland, Cham, pp 51–72
30. Funck EK (2025) Management control under post-NPM: replacing performance evaluation with trust-based management. Financ Account Manag 41(2):346–363
31. Tohidizadeh K, Zarei E, Ghalenoei M, Yazdi M, Kolivand K (2024) A dynamic system-based model for analyzing human factors: enhanced AcciMap with spherical fuzzy dynamic Bayesian network approach. Hum Factors Ergon Manuf Serv Ind 34(4):338–363

32. Gholamizadeh K, Zarei E, Omidvar M, Yazdi M (2022) Fuzzy sets theory and human reliability: review, applications, and contributions. In: Linguistic methods under fuzzy information in system safety and reliability analysis, pp 91–137
33. Patterson M, Deutsch ES, Jacobson L (2016) Simulation: closing the gap between work-as-imagined and work-as-done. In: Resilient health care, vol 3. CRC Press, pp 143–152
34. Yazdi M, Zarei E, Adumene S, Beheshti A (2024) Navigating the power of artificial intelligence in risk management: a comparative analysis. Safety 10(2):42
35. Yazdi M, Adumene S (2025) Innovative risk management for machine safety: insights from AS/NZS 4024 standards. Int J Adv Manuf Technol 137(1):659–682
36. Leveson N (2015) A systems approach to risk management through leading safety indicators. Reliab Eng Syst Saf 1(136):17–34
37. Reiman T, Pietikäinen E (2012) Leading indicators of system safety–monitoring and driving the organizational safety potential. Saf Sci 50(10):1993–2000
38. Edmondson AC (2011) Strategies for learning from failure. Harv Bus Rev 89(4):48–55
39. Yazdi M, Zarei E (2024) Enhancing crowd safety resilience through systematic analysis. In: Progressive decision-making tools and applications in project and operation management: approaches, case studies, multi-criteria decision-making, multi-objective decision-making, decision under uncertainty. Springer Nature Switzerland, Cham, pp 221–240
40. Manoj Pawar MD MM (2007) Creating & sustaining a blame-free culture: a foundation for process improvement. Physician Exec 33(4):12
41. Wallerstein N, Weinger M (1992) Introduction: health and safety education for worker empowerment. Am J Ind Med 22(5):619–635
42. Hollnagel E (2016) Barriers and accident prevention. Routledge
43. Zarei E, Yazdi M, Moradi R, BahooToroody A (2024) Expert judgment and uncertainty in sociotechnical systems analysis. In: Safety causation analysis in sociotechnical systems: advanced models and techniques. Springer Nature Switzerland, Cham, pp 487–530

Sloshing Dynamics and Its Impact on Ship Stability and Safety

Kombo Theophilus-Johnson, Sidum Adumene, and Azubuike John Chuku

Abstract Sloshing dynamics pose significant challenges to ship stability and operational safety across various maritime contexts. These dynamic fluid movements primarily result from vessel motions, such as rolling, pitching, and acceleration, leading to complex interactions among the liquid, containment geometry, and external forces. Sloshing can result in nonlinear and chaotic behaviours that induce substantial hydrodynamic loads and moments on a ship's structure, with critical implications for both intact and damaged stability. This phenomena need to be comprehensively studied and understood to assess its impact on ship operational safety. The current study synthesizes recent knowledge in sloshing dynamics and advances in the modelling, analysis, and mitigation of sloshing phenomena, incorporating high-fidelity computational fluid dynamics (CFD), particle-based methods such as Smoothed Particle Hydrodynamics (SPH) and Moving Particle Semi-implicit (MPS) methods, as well as experimental validations. The coupling between sloshing-induced internal forces and external wave actions is highlighted as a crucial factor in predicting ship responses and informing the operational safety envelope. Furthermore, emphasis is placed on the compounded risks in harsh environments, such as Arctic regions, where ice accumulation and intensified metocean conditions exacerbate stability challenges. While exploring the advances in sloshing dynamics, the study highlights key research gaps, including scaling experimental results, unifying multi-physics models, and integrating machine learning for predictive resilience. By critically evaluating both the detrimental impacts and technological responses to liquid sloshing, this study finds that existing knowledge is incomplete and provides strategic insights for ship designers, operators, and regulators, aiming to advance maritime safety and sustainability through interdisciplinary research and innovation.

K. Theophilus-Johnson · S. Adumene (✉) · A. J. Chuku
Department of Marine and Offshore Engineering, Rivers State University, Port Harcourt, Nigeria
e-mail: sidum.adumene@ust.edu.ng

K. Theophilus-Johnson
e-mail: kombo.theophilus-johnson@rsu.edu.ng

A. J. Chuku
e-mail: azubuike.chuku1@ust.edu.ng

H. Li et al. (eds.), *Data-Driven Methods for Reliability and Safety Engineering: Applications in Industrial Systems*, Springer Series in Reliability Engineering,
https://doi.org/10.1007/978-3-032-22873-4_43

Keywords Sloshing dynamics · Operational safety · Ship stability · Porous baffle · CFD

1 Introduction

Liquid sloshing is a dynamic fluid phenomenon characterized by the oscillatory motion of a liquid's free surface inside a partially filled container, such as a ship tank, when subjected to external motion or excitation. This oscillation arises from disturbances like horizontal accelerations, ship rolling, pitching, or other vessel motions, leading to complex fluid behaviour within the containment system. A foundational understanding of sloshing encompasses the intricate interplay between the liquid's inertia, gravity, and container geometry, which collectively influence the wave shapes and impact forces inside the tank. Such intricate oscillations can become highly nonlinear, particularly at resonance frequencies, and may result in structural loads that significantly impact the vessel's performance and safety [23].

The study of sloshing dynamics holds critical importance in maritime engineering due to its direct influence on ship stability and operational safety. The liquid motion inside cargo or ballast tanks induces additional forces and moments on the ship structure, altering its dynamic responses and potentially compromising stability margins [30, 39]. Unmitigated sloshing can exacerbate roll motions and increase the risk of capsizing, especially under severe sea states or when combined with other hydrodynamic and environmental forces. Understanding the impact of sloshing is crucial for ensuring safe navigation, maintaining infrastructure integrity, and complying with regulations [30]. The consequences extend beyond structural concerns to operational restrictions and safety protocols.

Historically, sloshing studies in marine vessels gained momentum following notable maritime accidents related to damage stability and flooding. Awareness grew in the 1980s following catastrophic incidents, such as the Herald of Free Enterprise and Estonia ferry disasters [22], which highlighted how damage-induced flooding interacts dynamically with internal liquid motions and external wave actions, leading to the rapid deterioration of ship stability and capsizing. Since then, research has intensified, focusing on improving damage stability predictions, mitigating flooding risks, and integrating sloshing dynamics into comprehensive ship safety analysis frameworks. These motivations underscore the ongoing emphasis in maritime safety research aimed at cost-effective yet robust risk reduction measures [35].

Sloshing dynamics are a pivotal component in evaluating the damage stability of both intact and damaged ships. In damaged conditions, sloshing forces can influence the progression of compartment flooding, alter the vessel's buoyancy distribution, and contribute to capsizing risks. Studies have shown that the interaction between sloshing and damage stability is complex, affecting the ship's ability to resist sinking or maintain operational safety after a breach event [30, 39]. This necessitates incorporating sloshing effects into damage stability assessments and emergency response strategies to avoid catastrophic loss [27].

Moreover, sloshing is intrinsically linked with external environmental forces such as waves, wind, and ocean currents, which actively excite liquid motion within tanks. The coupling of internal fluid movements with these external forces intensifies the dynamic response of a ship, influencing both global vessel motions and localized structural loads [25]. Considering that this coupling is indispensable for accurately predicting ship behaviour under operational and severe weather conditions, it is essential to ensure that vessel design and navigation strategies adequately address these risks [4].

The implications of sloshing dynamics also stretch into the realm of maritime regulations and ship design standards [4, 27]. Regulatory bodies recognize sloshing as a critical factor in ship safety, influencing guidelines for tank design, cargo containment systems, and stability criteria [10]. Advances in control and mitigation technologies continue to shape regulatory expectations, prompting ongoing updates to standards that ensure the practical application of enhanced safety measures during both ship operation and design phases [15]. There is a need to explore sloshing dynamics and recent advances in its modelling towards safer ship operations in harsh environments.

The current study presents a systematic review that explores sloshing dynamics and focuses on the dynamic coupling between sloshing in liquid tanks and ship motion, emphasizing how these interactions influence overall vessel stability and safety. Consideration is given to understanding the physical mechanisms of sloshing, fluid–structure interactions, and the integrated ship response under various operational scenarios. It seeks to synthesize recent advances that involve computational, experimental, and theoretical studies in this domain. A particular emphasis lies on assessing control and mitigation strategies developed using modern methodologies, including adaptive control, neural networks, and real-time monitoring. These advanced techniques aim to suppress or compensate for adverse sloshing-induced effects, thereby enhancing the operational resilience of ships. Furthermore, the study identifies existing gaps in current research fields and outlines future directions, encouraging multidisciplinary efforts to integrate ocean engineering, control theory, and computational advancements for improved safety outcomes.

2 Fundamental Mechanics of Sloshing in Ship Tanks

Theoretical models of sloshing typically begin with potential flow theory, which assumes an inviscid, incompressible, and irrotational fluid. Nonlinear potential flow models have been effectively employed to describe sloshing phenomena, capturing the essential wave dynamics and fluid–structure interactions within tanks [10, 18]. These models serve as the basis for numerous numerical algorithms, such as boundary element methods, which are used to simulate free surface motions and hydrodynamic loads with reasonable accuracy [18].

Complementary to potential models, reduced-order models offer a simplified yet powerful approach, often relying on equivalent mechanical systems to represent fluid motion [16, 20]. For instance, linear models may describe lateral sloshing in the frequency domain, whereas nonlinear aspects, especially vertical sloshing, require data-driven neural network approaches. Such models allow efficient integration into complex aeroelastic or structural simulations with manageable computational cost, as demonstrated in sloshing studies for aircraft and marine vessels [20].

Recent computational advances have led to the rise of particle-based techniques, such as Smoothed Particle Hydrodynamics (SPH) and Moving Particle Semi-implicit (MPS) methods [30]. These meshless Lagrangian frameworks are adept at handling highly nonlinear and violent sloshing, including wave breaking and fluid impacts on baffles and tank walls. Their flexibility in modelling complex geometries and free-surface flows makes them valuable for studying realistic tank configurations and excitations [30, 40].

2.1 Coupled Interactions Between Liquid Motion and Ship Dynamics

Sloshing forces, generated by fluid motion inside tanks, exert dynamic loads on the ship structure that feed back into the vessel's motion equations. The momentum exchange between fluid and hull alters the ship's surge, sway, roll, and pitch responses, often complicating stabilization and control during navigation [30]. Control-oriented ship dynamic models account for these sloshing forces to facilitate precise maneuvering and stability analyses [15].

The dynamic response of liquid tanks often depends on the type of excitation, including rolling, pitching, and surge motions. Experimental and numerical investigations highlight distinct sloshing patterns and load distributions under these excitations, revealing amplitude-dependent nonlinear behaviours and resonance phenomena. Such behaviours critically influence force magnitudes on tank bulkheads and hull structures, affecting the overall floating state and vessel stability [10, 19].

Validation of sloshing-induced hydrodynamic forces remains integral to modelling accuracy. Experimental setups often involve scaled physical tanks subjected to controlled motions, measuring pressure distributions and free surface elevations. These data have been invaluable for calibrating computational fluid dynamics models and particle methods, ensuring reliable predictions applicable to real ship operations [7].

2.2 Nonlinear and Chaotic Behaviour of Sloshing Flows

Sloshing dynamics are inherently nonlinear, especially when excitation frequencies approach natural sloshing modes. Nonlinear effects manifest as distortions in wave profiles, amplitude-dependent resonant frequency shifts, and wave breaking phenomena, with significant implications for load prediction and safety [19]. For example, power spectrum analysis of experimental data reveals dominant frequencies and their harmonics, underscoring the complex behaviour of the free surface during forced motions.

Advanced numerical studies have employed dynamical systems techniques to detect chaotic regimes in sloshing-tank systems. Tools such as Lyapunov exponents, bifurcation diagrams, and basins of attraction characterize sensitivity to initial conditions and parameter variations, providing critical insights into risk assessment and control design for sloshing-induced instabilities [23].

The nonlinear oscillation behaviour imposes substantial structural loads on tank walls and internal partitions. These loads have a significant impact on fatigue life, strength requirements, and long-term vessel safety. The incorporation of dissipation mechanisms, such as turbulence and impact damping, into enhanced numerical models enables more accurate predictions of load magnitudes and mitigates potential structural damage [16].

3 Impact of Sloshing on Ship Stability

3.1 Influence on Intact Ship Stability and Maneuverability

In intact ships, sloshing-induced hydrodynamic forces can disrupt lateral and longitudinal stability, altering normal operating characteristics. Analytical and experimental models suggest that lateral sloshing influences roll motions, sway forces, and transverse moments, thereby destabilizing vessels, particularly during turning and evasive maneuvers [9]. Variations in fluid density, fill volume, and tank geometry modulate the amplitude and frequency of sloshing forces, which in turn influence the dynamic normal load factors on critical ship components [31].

The interaction between sloshing and ship rolling and pitching also contributes to coupled motion phenomena that require careful consideration in stability assessments. These interactions can produce amplified responses during resonant conditions or combined excitations, challenging traditional ship motion control techniques and necessitating integrated dynamic models [15].

3.2 Effects on Damaged Ship Stability and Flooding Scenarios

Damage to a ship's hull, such as breaches or structural failures, introduces flooding scenarios that significantly alter the vessel's stability profile. The presence of liquid sloshing inside damaged compartments intensifies the complexity of the dynamic response. Computational and experimental studies have revealed that fluid–structure coupling during flooding can accelerate the growth of instability, thereby reducing the available time for corrective measures and increasing the risk of capsizing [3].

Damage stability evaluations now increasingly incorporate sloshing effects in predicting survivability. The risk of capsizing is heightened by combined dynamic forces acting on partially flooded compartments, necessitating advanced assessment techniques that meld hydrodynamics, structural response, and damage control strategies [27, 35]. Hydroelastic analyses demonstrate that sloshing can alter bending moments and structural load distributions, thereby exacerbating vulnerabilities in damaged hulls and necessitating enhanced design margins and safety factors [36].

3.3 Role in Arctic and Harsh Environments

Ships operating in the Arctic and other harsh environments face compounded challenges, including ice accumulation and extreme environmental loads that act concurrently with internal sloshing. Recent integrated studies have shown that combined ice accretion and sloshing notably decrease ship stability, influencing free roll decay and vessel motion under random wave conditions. These factors demand sophisticated modelling frameworks that predict extreme responses and support operational safety in such environments [33].

Environmental coupling scenarios further introduce compounded risks, with harsh metocean forces driving stronger excitation of sloshing motions and structural demands. The complexity of these environments necessitates the adoption of resilient design standards and operational strategies to maintain structural integrity and navigational safety [1].

4 Experimental and Numerical Methods for Sloshing Dynamics Analysis

4.1 Experimental Setups and Validation Approaches

To accurately capture sloshing phenomena, laboratory experiments utilize sophisticated platforms such as shake tables and hexapods that provide multi-degree-of-freedom excitation of scaled tank models. These setups enable the controlled simulation of ship maneuvers and environmental excitations, allowing for detailed measurements of free surface elevations and hydrodynamic pressures [7, 25].

Measurement systems incorporate capacitance wave probes, pressure sensors, and load cells strategically placed to monitor fluid motion and structural loads inside tanks. Such high-resolution data facilitate validation of numerical models while providing insights into nonlinear sloshing characteristics and the efficacy of mitigation strategies [25]. Flow visualization techniques further enhance understanding of sloshing behaviours, capturing wave patterns and interactions with baffles under various filling conditions [26].

4.2 Computational Fluid Dynamics (CFD) Techniques

CFD methods, particularly Reynolds-Averaged Navier–Stokes (RANS) solvers, have become indispensable in handling complex tank geometries and coupled sloshing effects within marine contexts. These techniques simulate turbulent, multiphase fluid flows and free surface dynamics under various excitation modes with high fidelity [10].

Coupled numerical models integrate CFD with fluid–structure interaction frameworks to analyze ship tanks undergoing damage scenarios and dynamic responses. For instance, two-way coupling between fluid motion and flexible hull structures enables hydroelasticity analysis under realistic operational loads, thereby advancing the predictive capabilities for structural integrity assessment [3, 36]. Boundary element methods complement CFD by providing efficient solutions for potential flow-based sloshing problems. Improved numerical stability strategies, such as smoothing of free surfaces and time-step controls, enhance the robustness of these simulations across a wide range of sloshing regimes [18].

4.3 Advanced Particle-Based and Meshless Methods

Particle-based methods, such as SPH and MPS, offer advantages in simulating highly nonlinear, turbulent sloshing flows with breaking waves and complex boundary interactions. Their meshless formulation enables flexible modelling of dynamic free surfaces and internal wave fields within irregular tank configurations [30, 40].

Recent developments include GPU-accelerated SPH simulations, allowing extensive and fast computations necessary for large-scale deep-water ship flooding scenarios. Such high-performance calculations facilitate the detailed analysis of rapid flooding processes and highlight the dynamic coupling of sloshing within damaged hull compartments [5].

Statistical and spectral analysis of sloshing responses derived from experimental and numerical data further elucidate resonance behaviours and energy distributions in oscillatory fluid motions, essential for designing mitigation structures and control algorithms [19].

5 Effects of Tank Configuration and Baffle Design

5.1 *Influence of Tank Geometry and Layout on Sloshing Behaviour*

Tank shape, fill levels, and aspect ratios have a significant impact on the hydrodynamic loads generated by sloshing. Studies reveal that rectangular, elliptical, and complex multi-row tank configurations exhibit markedly different fluid response characteristics, influencing stress distributions and load peak magnitudes on tank walls and bulkheads [2, 37].

In liquefied natural gas (LNG) carriers, multi-row tank arrangements contribute to unique sloshing dynamics that impact cargo containment integrity and overall vessel safety. Experimental model tests using site-specific metocean data underpin design decisions, demonstrating the necessity of reinforced containment systems to prevent failure under extreme sloshing loads [24].

The tank configuration also affects boil-off gas management and thermal insulation performance, with sloshing influencing pressure variations and cryogenic efficiency. Modelling efforts integrating design variations support operational improvements in energy consumption and safety margins [29].

5.2 *Baffle Designs and Mitigation of Sloshing Impacts*

Baffles serve as primary passive devices to mitigate sloshing effects by disrupting fluid motion and dissipating energy. Investigations into vertical and horizontal baffles under rolling excitation reveal that vertical baffles longer than the fluid depth can significantly reduce sloshing amplitudes, while shorter vertical or horizontal baffles may be ineffective [39, 40].

Porous baffle configurations have been explored to absorb wave energy, reduce free-surface elevations, and lower hydrodynamic forces acting on tank walls. Experimental studies demonstrate the varying impacts of porosity, highlighting the importance of baffle placement and design parameters for optimal performance [25]. Conventional baffle layouts, including partial and oblique orientations, have also been analyzed, demonstrating varying effectiveness in reducing transient and steady-state sloshing loads for tank trucks and maritime applications [13].

5.3 *Optimization and Technological Advancements*

Advanced mitigation strategies incorporate adaptive and active control systems to suppress sloshing. Finite-time hierarchical sliding mode controllers and neural network-based adaptive methods have demonstrated success in stabilizing coupled ship-tank systems, compensating for unknown nonlinear forces, and improving overall system robustness [15].

Neural networks further facilitate predictive and adaptive control, handling uncertainties in input dead zones and dynamic system parameters. Experimental implementations on ship-mounted cranes highlight practical applications of these methods, achieving efficient antiswing control even under complex roll motions [38].

Health monitoring systems integrating onboard sensors capture real-time sloshing-induced forces, enabling feedback-driven control and early warning capabilities. Such integration is vital for dynamic assessment and mitigation of emerging risks during voyage [28].

6 Coupled Dynamics of Sloshing and Ship Motion

6.1 Interaction Between Internal Fluid Sloshing and External Wave-Induced Motion

The interplay of internal sloshing and external wave-induced vessel motions generates coupled dynamic responses that affect resistance, seakeeping, and structural loading. CFD-based studies have quantified the influence of sloshing on added resistance and maneuverability, revealing significant variations under different sea states and vessel speeds [34, 41].

Wave-sloshing coupling alters hydrodynamic pressure distributions on tank bulkheads, necessitating robust design to withstand transient loads. Comparative analyses across regular and irregular wave conditions underscore the importance of accounting for these interactions in ship motion prediction and tank design optimization [10, 30].

6.2 Aeroelastic and Flexible-Structure Considerations

Sloshing effects also manifest in aeroelastic stability, particularly for structures such as flying wing aircraft with onboard liquid tanks. Reduced-order models that combine linear and nonlinear sloshing representations integrate rigid and elastic structural behaviour, revealing that neglecting liquid motion can overlook critical instabilities, such as limit cycle oscillations [20, 21].

Such findings translate to marine flexible structures, where sloshing-induced nonlinear damping mitigates or exacerbates dynamic instabilities. Accurate dynamic stability predictions rely on incorporating these complex fluid–structure interactions [20].

6.3 Damage Stability and Post-collision Dynamics Involving Sloshing

In damaged ships, sloshing has a substantial impact on the progression of flooding and stability. Advanced simulations coupling CFD with structural damage models enable real-time risk assessment of water ingress and its effects on vessel survival capabilities [3, 14].

The interaction of internal fluid motion with damage scenarios directly affects residual navigational performance by generating complex, multi-degree-of-freedom motion states during flooding. Understanding these processes is crucial for developing effective damage control measures and operational guidelines [32].

Assessment frameworks leveraging this coupled dynamic information contribute to risk reduction and improved safety management in maritime operations [35].

7 Safety and Operational Challenges Posed by Sloshing

7.1 Effects on Cargo Safety and Containment Systems

Sloshing imposes significant design constraints on cargo containment systems, especially in cryogenic LNG and hydrogen carriers. Impact pressures from high-amplitude sloshing demand reinforced tank walls and membranes, influencing material selection and structural reinforcements [24].

Management of boil-off gas pressure under sloshing-induced thermal and mechanical loads remains a pressing issue, with dynamic control systems designed to prevent overpressure and maintain safe operational limits [12]. Ensuring failure-free operation necessitates a comprehensive understanding of coupled thermal-fluid-structural factors.

7.2 Influence on Ship Maneuvering and Stability Control

During critical maneuvers, such as turning or emergency evasions, sloshing can reduce the vessel's lateral stability margin by modulating dynamic load factors and roll angles. Studies indicate that reduced fill levels and fluid density lead to lower dynamic normal load factors, affecting vehicle behaviour under steering inputs.

Sophisticated dynamic positioning controllers have been developed to accommodate sloshing disturbances, incorporating extended state observers and barrier Lyapunov functions to maintain positional constraints and smooth actuator responses in the face of uncertain sloshing effects [17]. Roll damping and stabilization under sloshing conditions in operational motions are essential for maintaining safety and operational efficiency, especially in challenging sea states [33].

7.3 Implications for Environmental and Sustainability Concerns

Sloshing-induced instabilities raise concerns about structural failures, spill risks, and environmental hazards, particularly for coastal infrastructure and marine vessels transporting hazardous liquids. Differential settlement exacerbates sloshing intensity and wave behaviour, potentially inducing larger wave heights and instability, highlighting the need for sustainable design and operational practices [11].

Environmental variability and climate change introduce new challenges for arctic shipping and offshore operations, where extreme weather and ice conditions influence the coupling of sloshing and ship dynamics. Addressing these issues requires advanced methodologies that integrate environmental and structural resilience [1, 33].

8 Advanced Control and Mitigation Strategies

8.1 Adaptive and Hierarchical Sliding Mode Control Techniques

Recent advances utilize adaptive finite-time sliding mode control for stabilizing coupled sloshing and ship motion systems. Neural networks are embedded to estimate unknown nonlinear sloshing forces and external wave excitations, facilitating robust control performance under uncertain operating conditions. Lyapunov stability theorems and homogeneity principles underpin the rigorous stability guarantees, ensuring finite-time convergence and global system stability [15].

8.2 Sensor-Based Monitoring and Feedback Approaches

Health-monitoring systems employing arrays of sensors capture sloshing-induced forces and motions, providing critical data for real-time feedback control of ship stability and machinery operations. The integration of extended state observers and barrier functions enables the maintenance of position constraints and attenuation of dynamic perturbations, thereby enhancing robustness against sloshing disturbances [17, 28].

Experimental validation of adaptive control theories on ship-mounted cranes confirms the applicability of neural network-based compensations in suppressing payload swings induced by roll motions, ensuring operational safety and precision under challenging conditions [38].

8.3 Design-Oriented Solutions and Optimization of Tank Systems

CFD and particle-based simulations inform the optimization of tank geometry and damping devices to mitigate sloshing effects. The design of tuned liquid dampers (TLDs) and baffle configurations tailored to specific filling levels and excitation profiles has shown significant improvements in vibration attenuation and load reduction [31, 41].

High-fidelity numerical modelling facilitates precise load prediction, enabling the establishment of safe operational margins and compliance with stringent maritime safety regulations.

9 Gaps in Current Research and Future Directions

9.1 Limitations in Experimental and Computational Modelling

Despite advances, scaling laboratory results to full-scale ships remains a persistent limitation, as nonlinear and chaotic sloshing phenomena may scale nonlinearly. The complexity of multi-physics coupling involving thermal, structural, and fluid dynamic effects further challenges simulation accuracy, underscoring the necessity for integrated modelling frameworks [30, 36].

Computational demands for high-fidelity CFD and particle-based simulations impose trade-offs between accuracy and practical applicability, leading to ongoing research into acceleration techniques and model simplifications that do not compromise essential fidelity [37].

9.2 Emerging Technologies for Sloshing Analysis and Control

Machine learning approaches, including neural-network-based reduced-order models, offer promising avenues for modelling complex nonlinear sloshing dynamics and predicting system responses effectively. These data-driven methods could significantly enhance adaptive control schemes and stability assessments [20].

Innovative experimental platforms, such as CubeSat and nanosatellite missions, aim to investigate zero-gravity propellant sloshing, providing unique insights that will refine future spacecraft and marine applications under low-gravity or microgravity conditions [6]. The integration of advanced sensors and Internet of Things

(IoT) technologies presents opportunities for resilient marine system monitoring and control, enabling the rapid detection and mitigation of sloshing-induced hazards [8].

9.3 Research Priorities for Enhancing Maritime Safety

Future efforts should prioritize coupling sloshing effects with damage stability and ice accumulation phenomena to develop unified models accurately predicting ship responses under combined operational hazards [33].

Developing comprehensive, integrated frameworks that capture global ship response under sloshing loads will enhance predictive capabilities and safety margins, supporting regulatory and operational improvements [15]. Sustainable design approaches must incorporate environmental variability, leveraging climate-adaptive methods to maintain structural and operational safety in the face of evolving maritime challenges [11].

10 Conclusions

The impact of liquid sloshing on ship stability and safety encompasses multifaceted mechanisms whereby fluid oscillations impose dynamic forces that alter vessel motion, structural loads, and cargo containment integrity. Comprehensive understanding requires integrating experimental observations, advanced numerical simulations, and adaptive control methodologies to capture the complex interdependencies between fluid dynamics, ship motion, and environmental forces [15].

Operational safety is contingent on the adoption of integrated monitoring, control, and design strategies that address nonlinearities, uncertainties, and extreme loading scenarios prevalent in maritime environments. Regulation and industry practices must evolve to incorporate coupling effects, enhanced structural designs, and real-time adaptive controls, ensuring safe navigation and damage resilience.

Future innovations hinge on multidisciplinary research, leveraging ocean engineering, control theory, and data-driven models to develop responsive systems that mitigate sloshing impacts, promote sustainable maritime operations, and safeguard human life and assets at sea.

References

1. Adumene S, Ikue-John H (2022) Offshore systems' safety and operational challenges in harsh Arctic operations. J Saf Sci Resil 3(2):153–168. https://doi.org/10.1016/j.jnlssr.2022.02.001
2. Banaszkiewicz T, Chorowski M, Gizicki W, Jedrusyna A, Kielar J, Malecha Z, Piotrowska A, Polinski J, Rogala Z, Sierpowski K, Skrzypacz J, Stanclik M, Tomczuk K, Dowżenko P

(2020) Liquefied natural gas in mobile applications—opportunities and challenges. Energies 13(21):1–35. https://doi.org/10.3390/en13215673

3. Cao X, Tao L, Zhang AM, Ming FR (2019) Smoothed particle hydrodynamics (SPH) model for coupled analysis of a damaged ship with internal sloshing in beam seas. Phys Fluids 31:032103. https://doi.org/10.1063/1.5079315
4. Chang TH, Wang HY (2023) Analysis and exploration of the impact of average sea level change on navigational safety in ports. Water (Switzerland) 15(14). https://doi.org/10.3390/w15142570
5. Chen H, Ming FR, Sun SL, Zhang AM (2024) Study on the flooding characteristics of a deep-water submarine based on plus-smoothed particle hydrodynamics method and graphic processing units acceleration. Phys Fluids 36:1. https://doi.org/10.1063/5.0179348
6. Fogel M, Sushil Mishra S, Burlion L (2025) Nanosatellite design considerations for a mission to explore the propellant sloshing problem. J Spacecr Rocket 62(2):510–527. https://doi.org/10.2514/1.a35871
7. Frosina E, Senatore A, Andreozzi A, Fortunato F, Giliberti P (2018) Experimental and numerical analyses of the sloshing in a fuel tank. Energies 11(3):1–24. https://doi.org/10.3390/en11030682
8. Huang L, Li Y, Benites-Munoz D, Windi C, Feichtner A, Tavakoli S, Davidson J, Paredes R, Quintuna T, Ransley E, Colombo M, Li M, Cardiff P, Tabor G (2022) A review on the modelling of wave-structure interactions based on OpenForm. Open Foam J 2:116–142
9. Jalili MM, Motavasselolhagh M, Fatehi R, Sefid M (2022) Investigation of sloshing effects on lateral stability of tank vehicles during turning maneuver. Mech Based Des Struct Mach 50(9):3180–3205. https://doi.org/10.1080/15397734.2020.1800490
10. Jia G, Jiao J, Chen C, Wu T (2025) CFD simulation of wave-induced motions of an LNG ship considering tank sloshing effects. Brodogradnja 76(3). https://doi.org/10.21278/brod76307
11. Jin H, Lu J, Liu Y, Shen J, Zhang F, Zhu C, Li S (2025) The impact of differential settlement on sloshing dynamics in coastal zone storage tanks under external excitation: implications for sustainable development. Sustainability (Switzerland) 17(3):1–19. https://doi.org/10.3390/su17031029
12. Kalikatzarakis M, Theotokatos G, Coraddu A, Sayan P, Wong SW (2022) Model based analysis of the boil-off gas management and control for LNG fuelled vessels. Energy 251. https://doi.org/10.1016/j.energy.2022.123872
13. Kandasamy T, Rakheja S, Ahmed AKW (2010) An analysis of baffles designs for limiting fluid slosh in partly filled tank trucks. Open Transp J 4(1):23–32. https://doi.org/10.2174/1874447801004010023
14. Kuznecovs A, Schreuder M, Ringsberg JW (2021) Methodology for the simulation of a ship's damage stability and ultimate strength conditions following a collision. Mar Struct 79. https://doi.org/10.1016/j.marstruc.2021.103027
15. Liang Y, Long X, Yang S (2025) Adaptive finite-time sliding mode control of coupled motion in ships with liquid tank dynamics using neural networks. Meas Control (United Kingdom) 8. https://doi.org/10.1177/00202940251331744
16. Marrone S, Gambioli F, Gonzalez-Gutierrez L, Colagrossi A (2021) Numerical study on the dissipation mechanisms in sloshing flows induced by violent and high-frequency accelerations. I. Theoretical formulation and numerical investigation. Phys Rev Fluids 6:114801. https://doi.org/10.1103/PhysRevFluids.6.114801
17. Mu D, Feng Y, Wang G, Fan Y, Zhao Y, Sun X (2023) Ship dynamic positioning output feedback control with thruster system dynamics. J Mar Sci Eng 11(94):253–258. https://doi.org/10.3390/jmse11010094
18. Ning DZ, Song WH, Liu YL, Teng B (2012) A boundary element investigation of liquid sloshing in coupled horizontal and vertical excitation. J Appl Math. https://doi.org/10.1155/2012/340640
19. Peng P, Gao Z, Li X, Liu K (2024) Experimental study of sloshing flows in a rectangular tank under coupled pitch and heave excitations. Ocean Eng 309(P2):118544. https://doi.org/10.1016/j.oceaneng.2024.118544

20. Pizzoli M, Saltari F, Mastroddi F (2022) Linear and nonlinear reduced order models for sloshing for aeroelastic stability and response predictions. Appl Sci (Switzerland) 12(17). https://doi.org/10.3390/app12178762
21. Pizzoli M (2020) Investigation of sloshing effects on flexible aircraft stability and response. Aerotec Missili Spaz 99(4):297–308. https://doi.org/10.1007/s42496-020-00064-0
22. Pope C (2013) The Herald of free enterprise tragedy. Coastal and Ocean Engineering Forum, St. John's NL, Canada
23. Ribeiro MA, Balthazar JM, Gonçalves MA, Tusset AM, Machado RC (2024) Nonlinear linear dynamics and numerical analysis of a sloshing tank. Ibero-Latin American Congress on Computational Methods in Engineering (CILAMCE). https://doi.org/10.55592/cilamce.v6i06.10240
24. Ryu MC, Jung JH, Kim YS, Kim Y (2016) Sloshing design load prediction of a membrane type LNG cargo containment system with two-row tank arrangement in offshore applications. Int J Nav Arch Ocean Eng 8(6):537–553. https://doi.org/10.1016/j.ijnaoe.2016.06.005
25. Sahaj KV, Thuvanismail N (2020) Experimental and numerical study on liquid sloshing dynamics with single vertical porous baffle in a sway excited ship tank. EGU General Assembly 2020, Online, 4–8 May 2020, EGU2020-929. https://doi.org/10.5194/egusphere-egu2020-929
26. Sahin G, Bayraktar S (2015) Flow visualization of sloshing in an accelerated two-dimensional rectangular tank. Int J Eng Technol IJET 1(3):106. https://doi.org/10.19072/ijet.99395
27. Santoso A, Achmad B, Siahaan M (2020) Analysis of the effect of sloshing on damage stability after the addition of LNG tanks. Int J Mar Eng Innov Res 5(3):164–171. pISSN: 2541-5972, eISSN: 2548-1479
28. Silva-Campillo A, Pérez-Arribas F, Suárez-Bermejo JC (2023) Health-monitoring systems for marine structures: a review. Sensors 23(4):1–19. https://doi.org/10.3390/s23042099
29. Smith J, Gkantonas S, Mastorakos E (2021). Modelling of boil-off and sloshing relevant to future liquid hydrogen carriers. Energies 15. https://doi.org/10.3390/en15062046
30. Sun X, Zhong Y, Bian F, Liu C, Yin Y (2022) Numerical computation of sloshing-induced force in complex ship tanks under the excitation of ship rolling motion based on the MPS method. Appl Sci (Switzerland) 12(10). https://doi.org/10.3390/app12105130
31. Tahera, Urs N, Raj KS, Kumar R, Soundalgekar H, Deepa T, Shah MA (2025) Comparative analysis of sloshing effects on elevated water tanks' dynamic response using ANN and MARS. Discov Mater 5(1). https://doi.org/10.1007/s43939-025-00181-2
32. Tao L, Wang J, Hong Z, Liu K, Gao Z (2024) Time-domain simulation study of damaged ship compartments flooding based on CFD. In: Proceedings of SPIE 13160, fourth international conference on smart city engineering and public transportation (SCEPT 2024), 131600A. https://doi.org/10.1117/12.3030602
33. Tao J, Chai W, Yang X, Zhang W, Wang C, Qi J (2025) Stability evaluation of a damaged ship with ice accumulation in Arctic regions. J Mar Sci Eng 13(9):1–24. https://doi.org/10.3390/jmse13091685
34. Tezdogan T, Demirel YK, Kellett P, Khorasanchi M, Incecik A, Turan O (2015) Full-scale unsteady RANS CFD simulations of ship behaviour and performance in head seas due to slow steaming. Ocean Eng 97:186–206. https://doi.org/10.1016/j.oceaneng.2015.01.011
35. Vassalos D (2020) The role of damaged ship dynamics in addressing the risk of flooding. Ships Offshore Struct 17:2. https://doi.org/10.1080/17445302.2020.1827639
36. Wei Y, Incecik A, Tezdogan T (2023) A hydroelasticity analysis of a damaged ship based on a two-way coupled CFD-DMB method. Ocean Eng 274:114075. https://doi.org/10.1016/j.oceaneng.2023.114075
37. Xie F, Zhao W, Wan D (2022) Overview of moving particle semi-implicit techniques for hydrodynamic problems in ocean engineering. J Mar Sci Appl 21(3):1–22. https://doi.org/10.1007/s11804-022-00284-9
38. Yang T, Sun N, Chen H, Fang Y (2020) Neural network-based adaptive antiswing control of an underactuated ship-mounted crane with roll motions and input dead zones. IEEE Trans Neural Netw Learn Syst 31(3):901–914. https://doi.org/10.1109/TNNLS.2019.2910580

39. Zhao M, Jiao J (2024) Smoothed-particle hydrodynamics simulation of ship motion and tank sloshing under the effect of regular waves. Fluid Dyn Mater Process 20(5):1045–1061. https://doi.org/10.32604/FDMP.2023.043744
40. Zhao K, Yang SF, Yuan PY (2023) Research on the sloshing characteristics of the ship tank with baffles under rolling motion based on smoothed particle hydrodynamics method. Phys Fluids 35:112110. https://doi.org/10.1063/5.0176970
41. Zhu Z, Wang S (2025) Application of CFD method for ship added resistance tests with coupled sloshing effects. Proc Inst Mech Eng Part M J Eng Marit Environ. https://doi.org/10.1177/14750902251320509

A Computational-Based Comparative Fatigue Analysis of Composite and Steel Risers

Morrison Inegiyemiema, Sidum Adumene, and Mohammad Yazdi

Abstract The extraction of crude oil and natural gas from offshore environments is a crucial operation that helps meet global energy demands. Traditionally, steel risers have been used for decades to transport crude from the seabed to floating vessels. While steel risers offer some advantages, they also present significant drawbacks compared to newer composite risers, which are made of a combination of steel and flexible materials. This research aims to provide reliable data on the properties of composite risers, including fatigue life and dynamic performance, through a comparative analysis. Using ORCAFLEX software, steel and composite risers were evaluated under identical environmental conditions in the Gulf of Guinea. Additionally, a MATLAB program was developed for a validation analysis. The results were then compared with findings from FLEXCOM software, previously obtained by the renowned marine research institute 2H Offshore in the North Sea. The analysis revealed that composite risers exhibit a fatigue life 100 times greater than steel risers, with a 50% reduction in bending moments, and a 43% decrease in the required tension for buoyancy tanks. These findings align with the 2H Offshore results, reinforcing confidence in the offshore industry's transition toward composite risers for production operations. The research provides the offshore industry with valuable insights and data, reassessing the applicability and suitability of composite risers for deployment in the Gulf of Guinea.

Keywords Riser · Wave · Fatigue analysis · Orcaflex · System reliability · Composite

---

M. Inegiyemiema · S. Adumene (✉)
Department of Marine and Offshore Engineering, Rivers State University, Nkpolu, Port Harcourt, Nigeria
e-mail: sidum.adumene@ust.edu.ng

M. Yazdi
School of Engineering, Macquarie University, Sydney, Australia

H. Li et al. (eds.), *Data-Driven Methods for Reliability and Safety Engineering: Applications in Industrial Systems*, Springer Series in Reliability Engineering,
https://doi.org/10.1007/978-3-032-22873-4_44

1 Introduction

Energy demand worldwide is rising to the point of forcing oil companies to search for oil in increasingly deep offshore locations. As the onshore and near coastal oil reserves dry up, the oil companies go deeper and deeper into the deep oceans to explore this vital resource. The oil companies have a clear road map to explore, develop, produce, and transport this resource. Petroleum originates from the Remains of Ancient organisms, which were buried millions of years ago beneath the seabed, thousands of feet below the sandy and salty layers of the crust. Due to the presence of bacteria, high temperatures, and pressure, hydrocarbons are formed. Exploring these hydrocarbon reserves requires venturing into the deep sea, which presents harsh environmental conditions. According to the Mineral Management Service (MMS), water depths greater than 305 m are considered deep water and water depths greater than 1524 m are typically considered ultra-deepwater. Since 1995, over 980 deep-water exploration wells have been declared [1]. Risers are a crucial component of the hydrocarbon exploration process and are significantly impacted by harsh offshore environmental conditions.

System risers play a crucial role in offshore infrastructure by facilitating fluid transport or serving as service lines between wellheads or subsea structures on the seabed and fixed platforms at the surface. However, riser design engineers face significant challenges due to factors such as hydrostatic pressure, ocean currents, waves, and vessel motion [2]. Regardless of the floating platform concept employed in offshore field development, a riser system remains essential. According to [3], risers are generally classified into three main types: steel catenary risers, flexible risers, and composite risers. Steel catenary risers (SCRs), also known as rigid risers, utilize rigid steel pipes with considerable bending stiffness. These risers form a catenary shape between the hang-off point on a floating or fixed platform and the seabed, making them a common choice for oil production platforms.

Composite risers represent an advanced form of riser system composed of multi-layered pipes made from various materials, integrating the advantages of both flexible and rigid designs. They are specifically engineered for high dynamic performance, capable of enduring substantial cyclic bending while maintaining a lightweight structure that simplifies transportation and installation [4]. As noted by [5], composite risers are generally divided into two main categories: bonded reinforced thermoplastic pipes and unbonded flexible pipes. Bonded risers typically consist of a fiber-reinforced angle-ply laminate core positioned between an inner metallic or elastomeric liner and an external layer composed of thermoplastic, thermoset, or metallic materials. The liners function as barriers to prevent the leakage of pressurized fluids, since composite materials are inherently porous [6]. In contrast, unbonded flexible risers feature a more complex configuration. They commonly include an internal steel carcass layer to prevent collapse and buckling, steel pressure armor layers that resist corrosion from internal fluids, anti-wear homogeneous layers, helically wound tensile armors providing axial and bending stiffness, and an outer polymeric sheath for environmental protection [5, 7–9].

In recent times, composite risers have drawn substantial attention from the offshore industry primarily due to their high specific strength and stiffness. Low thermal conductivity, high resistance to corrosive fluids, and increased damping are properties of a composite riser unavailable in any other riser system (Ochoa, 1999). Although numerous advantages are associated with composite risers, they also present many challenges in design modelling, manufacturing, testing, and qualification compared with traditional metallic materials. Over the years, the offshore industry has used the SCR to transport crude from the earth's crust to the floating vessel. The SCR has been found to perform creditably well but to be low in fatigue life. Fatigue failure occurs when the pipe fails below yield stress due to repeated circles of stress. The steel has been found to be susceptible to corrosion and excessive weight and has joints that can leak and break, which causes environmental pollution. However, composites have not been used to their full potential within riser applications.

Recent studies on the design, manufacturing, mechanics, and fatigue analysis of marine risers have been extensively explored in open sources [4, 10–12]. Siriruk and Penumadu [11] investigated the effects of sea conditions on cyclic fatigue, revealing a significant decline in fatigue performance compared to dry laminates tested in air. Their findings showed that wet laminates tested in air exhibited a 30% reduction in fatigue life, while dry laminates immersed in water experienced a 71% decrease. The most severe degradation occurred in wet laminates immersed in water, resulting in an 84% decrease in fatigue life. However, these results may not fully represent real-world marine structures, where only one face of a laminate is typically exposed to seawater. In such cases, dry laminates immersed on one side lost 42% of their cyclic life, whereas wet samples experienced a 47% reduction. Kaboudian et al. [10] analyzed the tension distribution along composite risers and found it to be relatively uniform, resulting in more dispersed failure locations compared to steel risers. Their study also highlighted that adding buoys along the riser introduced kinks in the tension distribution, increasing localized stress concentrations. To address this issue, they recommended using long, continuous buoys along the lower half of the riser rather than shorter, spaced-out ones. This configuration helps reduce high bending strains at buoy edges and effectively minimizes Vortex-Induced Vibrations (VIV).

Chen et al. [4] investigated the effects of Vortex-Induced Vibrations (VIV) on composite risers, emphasizing that the high stiffness of the liner adversely affects the overall performance by limiting the full utilization of the composite's high strength. Despite this drawback, composite risers offer significant advantages, including lower top tension and reduced or even eliminated buoyancy requirements. These factors substantially decrease the weight suspended from the platform deck, leading to notable economic benefits that become more pronounced with increasing riser length. Tan et al. [12] conducted coupled-based simulations for composite and steel risers. Their study demonstrated a strong correlation between simulation results and experimental findings. A comparative analysis of aluminum, steel, and titanium liners revealed that the liner is the weakest component in composite riser design. Gao et al. performed vibration response tests on a catenary flexible riser subjected to the combined effects of single-phase internal flow and nonlinear shear outflow in a circulating experimental tank. Using a high-speed camera for non-intrusive measurement, they captured vibration displacements and observed that internal flow increased vibration amplitude and induced mode transitions at lower velocities. Their findings also revealed that as internal flow velocity increased, the critical reduced velocity for transitions between adjacent modes decreased. Under pure outflow conditions, frequency-locked vibrations did not alter the vibration mode. However, in the mode transition zone of outflow-induced vortex excitation, internal flow facilitated the emergence of higher-order vibration modes at lower reduced speeds.

Despite these advancements, there remains a limited understanding of the comparative fatigue performance of steel and composite risers under identical environmental loading conditions. Further research is needed to refine existing models, reduce error margins, and enhance the reliability of safety factors. While previous studies, such as those by Ochoa and Chouchaoui [13], have attempted to consolidate knowledge on composite riser behaviour, gaps persist in understanding their fatigue life and dynamic behaviour, particularly in real-world applications. The prospect of replacing steel risers with composite alternatives is gaining traction, yet their long-term performance under operational conditions remains largely unexplored. This is particularly relevant in the case of the Egina Floating Production Storage and Offloading (FPSO) unit in the Egina oil field, operated by Eni (formerly Agip) in Nigeria. As a result, new research opportunities have emerged to address these knowledge gaps and enhance the viability of composite risers for terrain-specific deepwater applications.

The current study presents a comparative fatigue analysis of steel and composite risers by exploring a dynamic analysis of the steel and composite tower sections of the hybrid riser for tension, bending moment, and shear force modelling using Orcaflex software. Subjected to the same environmental loading, the computational approach analyzes the fatigue life. The analysis considers the domain-specific characteristics of the Gulf of Guinea's environmental constraints and their impact on the survivability of riser structures. The analysis is further presented in comparison with the FLEXCOM software analysis. The results were validated using an algorithm developed on the MATLAB platform to define more accurate fatigue-based responses of the steel and composite risers for informed operational decision-making. The research provides the offshore industry with valuable insights and data, reassessing the applicability and suitability of composite risers for deployment in the Gulf of Guinea.

2 Overview of Steel Catenary Riser and Composite Riser Configuration

Risers can be configured in various ways, but their fundamental design ensures that one end is connected to a floating vessel while the other extends to the seabed, linking to the hydrocarbon production system. The point where the riser contacts the seabed is known as the Touchdown Point (TDP). Steel Catenary Risers (SCRs) typically consist of three main sections: Catenary Zone, Buried Zone, and Surface

Zone. As the name suggests, SCRs are made from steel pipes designed to transport well fluids from the seabed wellhead to the floating production vessel. They have a well-established history in offshore operations, initially used on fixed platforms before their first application on floating platforms in 1994. The earliest installations on floating structures occurred on the Auger Tension Leg Platform (TLP) in the Gulf of Mexico and a semisubmersible in the Mailim field. SCRs were later deployed on truss spars in the Boomvang and Nankeen fields. SCRs play a crucial role in offshore infrastructure, offering high resistance to extreme temperature and pressure conditions, cost-effective manufacturing, and adaptability to the movement of floating structures. However, in deep-water environments, they face significant challenges, including extreme ocean depths, strong currents, and powerful waves. These challenges are illustrated in Fig. 1.

The drilling of oil in offshore deep water is becoming more complex as drilling is going deeper away from the land. The deeper the water depth, the longer the riser will be. The extensive length of the riser now becomes a challenge due to its weight and tension on the floating structure. Some other problems that may be encountered include hydrodynamic drag loading on the riser and disturbances to the riser spread, which can lead to issues with riser selection. Finally, an increase in the cost of well site development. The deep sea environment is characterized by severe ocean currents that negatively impact the long riser length. The current loading on the riser causes vortex-induced vibration, which may lead to fatigue stress and subsequent fatigue failure. Deep-sea environments are characterized by extreme wave heights and long wave periods, which impose significant loads on offshore floating structures. These wave-induced forces contribute to vessel instability, generating

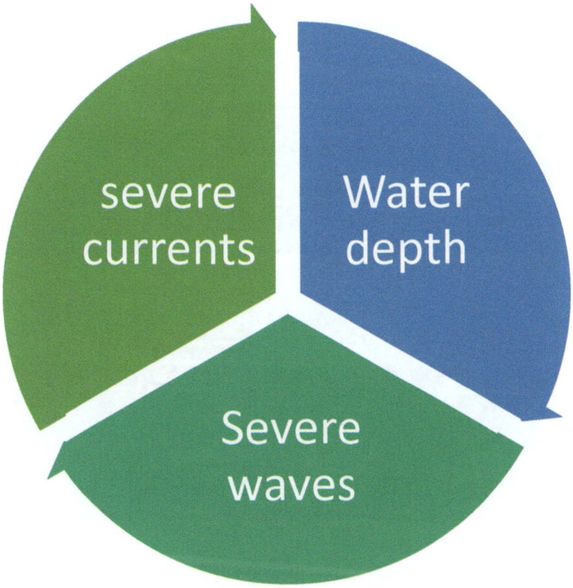

Fig. 1 Challenges of SCR in deep sea Environment

high stress levels at critical points such as the riser's connection to the vessel and the touchdown point on the seabed. Selecting an appropriate riser system configuration is a key challenge in riser design, as vessel offset motions caused by severe waves and currents play a crucial role in determining the optimal setup [14]. Wave characteristics vary across different geographic regions, significantly influencing riser loading, the dynamic motion of floating units, and the overall performance of risers. As a result, the configuration of composite risers during installation is customized based on site-specific environmental conditions and production requirements. As illustrated in Fig. 2, six primary riser layout configurations are commonly used: free-hanging catenary, lazy wave, steep wave, lazy S, steep S, and pliant wave. The choice of configuration depends on multiple factors, including vessel motion, environmental conditions, riser and mooring arrangements, and water depth.

The free-hanging catenary is the simplest and most cost-effective riser configuration to install. It consists of a freely suspended riser connecting the vessel to the seabed without any buoyancy support. However, due to environmental loading and vessel motion, this configuration experiences high tension and is prone to buckling at critical points, particularly at the touchdown point (TDP) on the seabed and the connection point to the vessel. The lazy wave and steep wave configurations incorporate buoyancy tanks to decouple the riser from vessel motion, effectively reducing tension at the TDP. Lazy wave risers can be affected by changes in the internal fluid density over their operational lifespan, potentially altering their configuration. In contrast, steep wave risers maintain their configuration even when fluid density changes, but they require a subsea base and a subsea bend stiffener for support [3]. When free-hanging catenary, lazy wave and steep wave configurations are ineffective

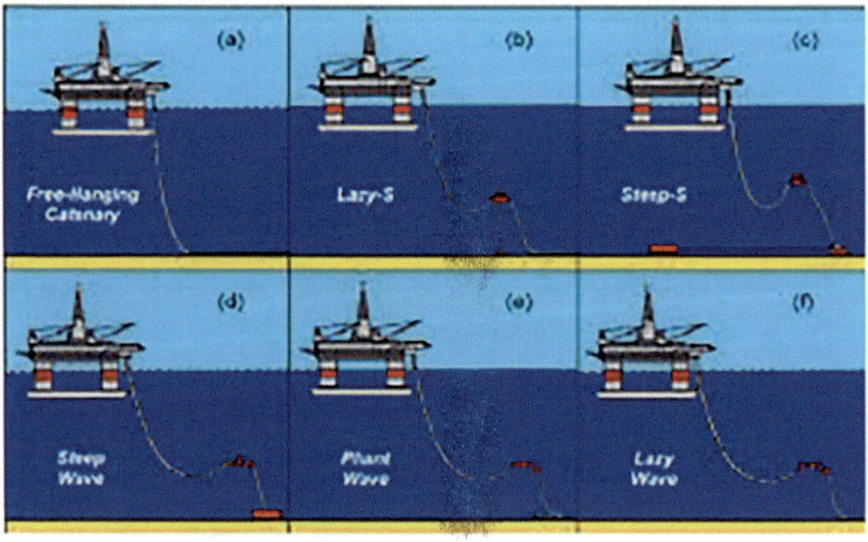

Fig. 2 Different Riser Configurations [7]

in certain offshore environments, lazy-S and steep-S risers are used. These configurations involve attaching the riser to a subsea buoy, which can either be anchored to the seabed or left floating via a chain link connection. Installing a lazy-S flexible riser requires a mid-water arch, tether, and tether base, whereas a steep-S riser necessitates a buoy and subsea bend stiffener. The response of these risers is influenced by buoyancy and hydrodynamic forces, requiring complex modelling due to the significant initial forces involved [3]. The pliant wave configuration functions similarly to the steep-S riser, with the key difference being the inclusion of an additional seabed anchor that helps transfer riser tension to the anchor point. This design provides extra support, allowing the riser to accommodate higher fluid densities while also improving its ability to withstand vessel motion.

3 Research Methodology

This section presents the computational and dynamic model development for the steel and composite riser, as shown in Fig. 3. A CFD analysis software, such as ORCAFLEX was explored to model the riser data and environmental factors, and they were validated with existing software results. The study employed the Ekman drift theory to characterize the sea-wide and tide behaviour for the chosen locations. This provides the needed information and data for the current profile development and the application of the Morrison equation. The following subsection describes the theories and application processes.

3.1 Theory Analysis

3.1.1 Fluid Loading on a Single Slender Cylinder

The fluid loading on slender bodies in waves is predicted using a semi-empirical relationship proposed originally by Morison and O'Brian, usually referred to as the Morison equation.

Two pieces of information are useful in understanding the basics of the Morison equation. The first stems from everyday experience, dimensional analysis and experiment. Every day experience shows that a body in a fluid flow experiences drag. Dimensional analysis shows that the body's drag force per unit length takes the form.

$$f = \frac{1}{2}\rho U^2 D C_d. \quad (1)$$

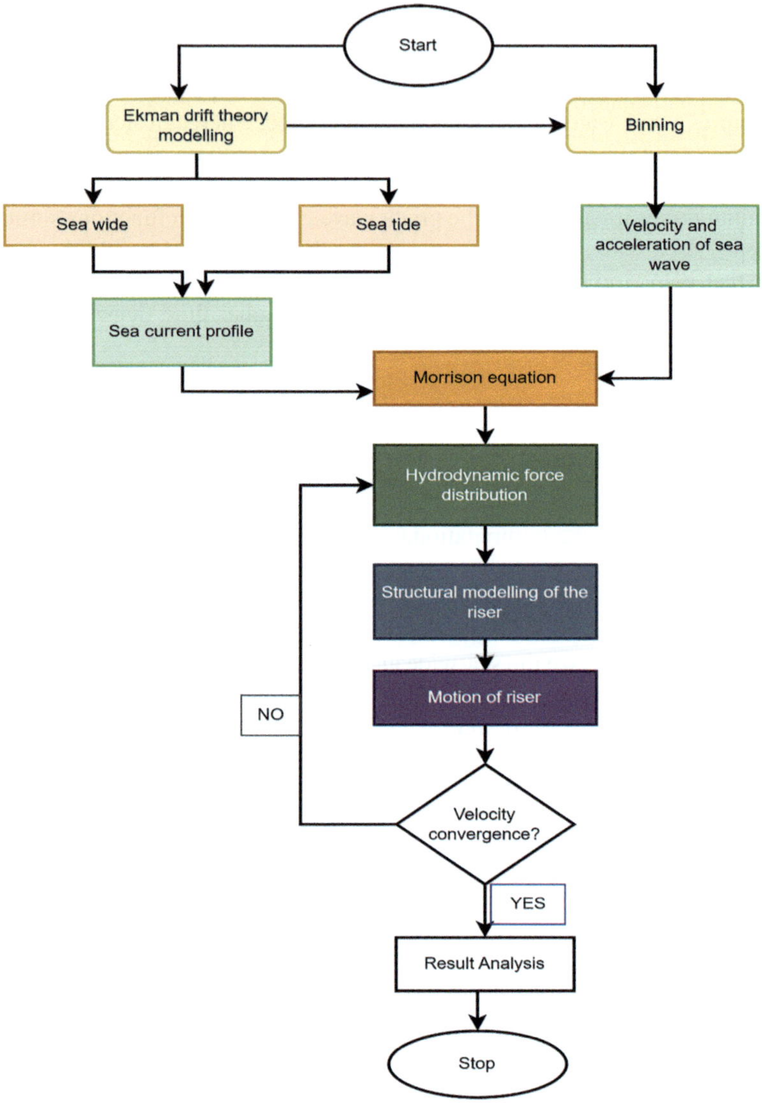

Fig. 3 Research Process Flow Chart for the dynamic analysis

where C_d is dimensionless, it is known as the drag coefficient and has been experimentally shown to be a function of the Reynolds number. The velocity U, is the velocity of the ambient flow undisturbed by the presence of the body.

The second piece of information arises out of the potential flow analysis of a time-dependent flow about a body, which predicts that the force permits length on the body in line with the flow is given by

$$f = \rho \forall \frac{du}{dt} + \rho V_R \frac{du}{dt} C_a \tag{2}$$

where ∀ is the displaced volume per unit length, VR is a reference volume per unit length (usually a circular cylinder enclosing the body, which may be of any section) and is known as the added mass coefficient. The acceleration, $\frac{dv}{dt}$, is that of the ambient flow undisturbed by the presence of the body.

The unsteady potential flow is made up of a potential describing the undisturbed flow plus a potential describing the disturbance to the flow due to the presence of the body, Φ_d. The first term, known as the Froude Krylov force, derives from Φ_d and is simply the force that would be required to accelerate the volume of undisturbed flow displaced by the body along with the rest of the flow. The second term derives from Φ_d and this force results from the acceleration imparted to fluid particles due to the disturbance in flow caused by the body, in accordance with the added mass effect. The combined forces are referred to as the inertia force.

$$\mathrm{F} = \rho V_{\mathrm{R}}\left(\frac{\forall}{\mathrm{V_R}} + \mathrm{C_a}\right)\frac{\mathrm{du}}{\mathrm{dt}} = \rho V_{\mathrm{R}} \mathrm{C_m} \frac{\mathrm{du}}{\mathrm{dt}} \tag{3}$$

And C_m is known as the inertia coefficient. For a circular cylinder, Ca = 1, and the inertia coefficient takes the value of 2 if V_R is also of a circular section.

The Morison equation is based on the assumption that the in-line force will be composed of a drag component associated with flow separation and an inertia component, as predicted by potential flow analysis, but with an inertia coefficient whose value is modified by viscous effects.

$$f = \frac{1}{2}\rho \mathrm{U}|\mathrm{U}|\mathrm{D}\mathrm{C_d} + \rho \mathrm{V_R} \mathrm{C_m} \frac{\mathrm{du}}{\mathrm{dt}} \tag{4}$$

The $U\,|U|$ term ensures that the drag term is in line with the direction of flow. The drag and inertia coefficients are assigned values based on experimental fitting of the Morison equation to the force measured on a cylinder in a U-tube or from another experiment. Such experiments show that the force coefficients are functions of the Keulegan-Carpenter number, the Reynolds number and the cylinder roughness.

3.1.2 Fatigue Evaluation

There are different approaches to estimating the life of fatigue (Eloni 2000). Two primary methods are employed for fatigue assessment: fracture mechanics and the S–N curve (also known as Miner's rule) approach. Fracture mechanics is particularly effective for evaluating crack propagation, making it valuable for analyzing specific defects and estimating how crack length evolves over time. In contrast, the S–N curve method focuses on the crack initiation phase and is therefore useful for estimating the service life of a riser, which is the focus of this thesis. S–N curves are developed through fatigue testing of material samples, where a sinusoidal stress is applied,

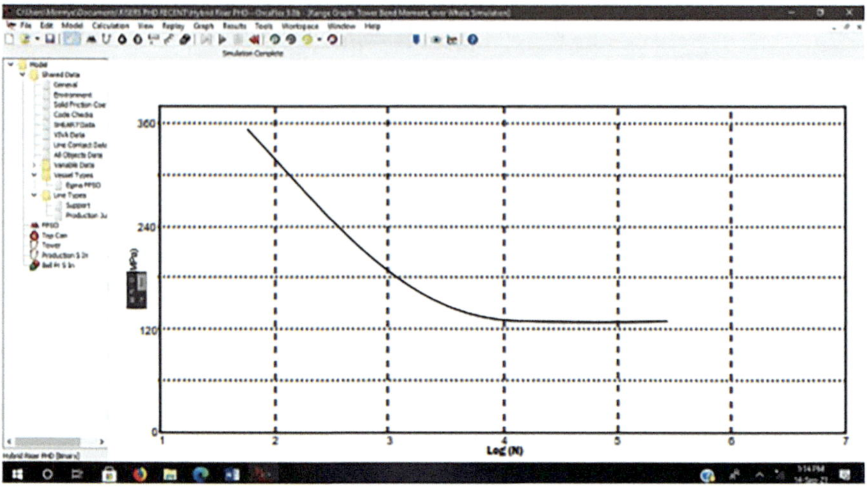

Fig. 4 S–N Curve of Steel Riser from Orcaflex

and the number of cycles to failure is recorded. This approach assumes a linear accumulation of fatigue damage, in line with Miner's rule, to facilitate life assessment based on cyclic loading behaviour.

The high-cycle range of fatigue life is typically applicable to marine structures, and S–N data in this range tend to follow a log-linear relationship. This can be given as a straight line on a log–log scale, as shown in the Fig. 4.

The thickness correction factor, as well as the stress concentration factors, can be applied to obtain the stress range needed to calculate fatigue damage.

$$S = S,\ SCF\ (t/t) \tag{5}$$

where SC describes the nominal stress range; SCF define the stress concentration factor; (t/t) is the thickness correction; n = is the average

a. Appropriate Choice of S–N Curve

Different classification societies and regulatory bodies have developed various codes specifying suitable S–N curves for risers, such as those from DNV (2010), API (2010), and DNV (2005).

High-stress cycles (N) of a wave are employed similarly on steel riser materials to determine the stress the steel material can withstand at a given stress amplitude (S) before it suddenly fractures. This graph, from Fig. 5, was generated using the Orcaflex software, which describes its behaviour. From the figure, it can be seen that as the stress is reduced, the material can endure a larger number of failure cycles. The graph illustrates that for some materials, the curve becomes virtually horizontal, indicating that at that stress, the material will effectively not fail due to fatigue. The

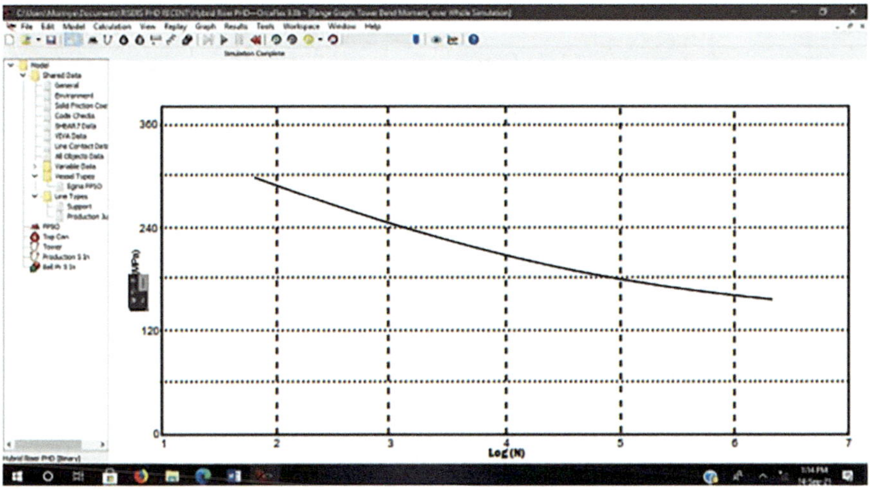

Fig. 5 S–N Curve of Carbon/Epoxy Composite Riser from Orcaflex

stress is known as the fatigue limit. Steel and titanium have exhibited behaviour similar to that of Orcaflex.

High-stress cycles (N) of a wave are employed similarly on composite riser materials to determine the stress the composite material can withstand at a given stress amplitude (S) before it suddenly fractures. This graph in Fig. 6 was generated using the Orca Flex software, which describes its behaviour. From the graph, it can be deduced that the composite riser operating in corrosive conditions displays an S–N curve that continues to slope downwards and does not exhibit a fatigue limit. For the composite material, an endurance limit is generally quoted. This maximum stress can be sustained for a stated number of cycles. The S–N curves suitable for calculating deep water fatigue damage for hybrid risers are presented.

b. Estimation of Stress Concentration Factor (SCR)

The accumulation of stress typically occurs at nucleation sites, such as welded joints in SCRs, influenced by both local geometry (e.g., weld spots) and global geometry (e.g., structural fabrication details), especially in composite materials. SCRs serve as component- or detail-specific linear factors that adjust the computed average component stress to determine local stress values for specific structural details.

c. Choice of Fatigue Safety Factor

As drilling activities tend to move to deeper offshore environments, inspection of the offshore structural facilities becomes practically inaccessible. This has led to the introduction of fatigue safety factors to assess the failure of these structures, which can result in substantial economic losses and environmental damage.

The fundamental fatigue capacity is represented using S–N curves, which define the relationship between the number of stress cycles to failure, N and a given constant

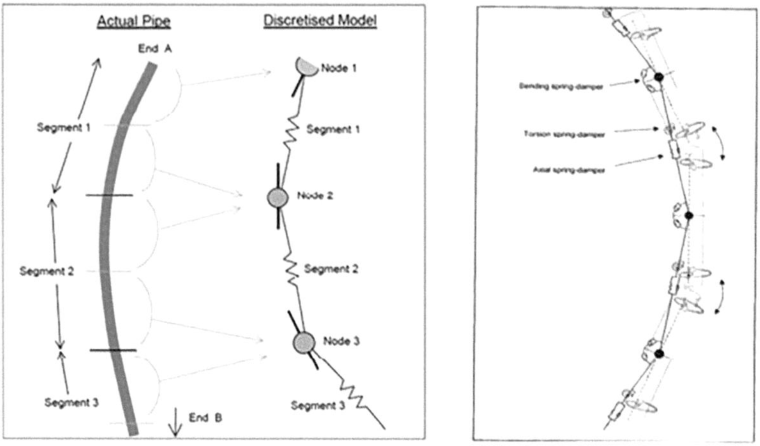

OrcaFlex Line Model

Fig. 6 Orcaflex line model

stress range, S:

$$N = aS^m \tag{6}$$

where a and m are empirical constants established by experiments

d. Miner – Palmgreen

Miner's rule states that fatigue damage accumulates linearly, with each cyclic load contributing proportionally to the total damage. For a specific cyclic stress level $S_{1,}$ the number of cycles to failure is N_1. Therefore, if n_1 cycles of cyclic pressure s_1 are applied, the fatigue damage is n_1 /N_1 of that needed to cause failure. Failure occurs when the sum of all the ratios n/N equals unity. *i.e.*

$$\frac{\text{n1}}{N1} + \frac{\text{n2}}{N2} + \frac{\text{n3}}{N3} + \ldots + = \sum \frac{n}{N} = 1 \tag{7}$$

Therefore,

$$\text{D} = \sum \frac{n(s1)}{N(S1)} = 1 \tag{8}$$

where n (s_1) is the number of stress cycles with range $s_{1,}$ and N (S_1) is the number of cycles to failure

The cumulative damage rule of Eq. 10 is also known as Palmgren–Miner, since a similar.

rule was proposed before by Wirsching, Palmgren (1988). Equation (8) can be expressed as:

$$D = \sum_{i=1}^{N} D_i \tag{9}$$

where;

$$D_i = \frac{N_{T_i}}{N_{F_i}} \tag{10}$$

N_{F_i} number of cycles to failure at constant stress range, S_i; N_{T_i}, Total number of cycles in time T, at the constant stress range, S_i. Furthermore, Miner's rule states that at failure when $N_T = N_F$ and $D = \Delta = 1$.

Therefore, the cumulative damage can be expressed as:

$$D = \sum_{i=1}^{n} D_i = \Delta = 1 \tag{11}$$

By further analysis, the damage can be expressed now as

$$D_i = \frac{N_T}{K} S_i^m \tag{12}$$

Furthermore, if the average frequency of the stress cycle is defined as

$$f_o = \frac{N_T}{T} \tag{13}$$

where;

N_T, number of stress cycles in period T.

Then the expression for the fatigue damage, Eq. (12), can be written as:

$$D_i = \frac{f_o T}{K} S^m = \frac{T}{K} \Omega \tag{14}$$

where;

$$\Omega = f_o S^m \tag{15}$$

e. Fatigue Life (FL) Estimate

Equation (16) is used to estimate cumulative damage in all members, for all Hs cases (Palmgren–Miner rule)

$$\text{Cumulative damage, D} = \sum_{i=1}^{j} \frac{n_i}{N_i} \leq \frac{1}{\text{DFF}} \tag{16}$$

Estimate the fatigue life based on a given design fatigue factor (DFF) value

$$\text{Fatigue life, FL} = \frac{1}{\text{D*DFF}}(\text{Years}) \tag{17}$$

3.1.3 Vortex-Induced Vibrations

The natural frequency of a single element may be determined from:

$$Fn = \frac{A_m}{2\pi L^2}\sqrt{\frac{EI}{m}} \tag{18}$$

where Fn represents the natural frequency of the member (Hz), L is its length, m is the mass per unit length, E is Young's modulus, and I is the second moment of area. The natural frequency is influenced by these parameters, reflecting the dynamic behaviour of the structural member under vibrational loads.

The constant A_n depends on the end fixity as follows (Table 1):

A_n = End condition.

The reduced velocity (Vr) is describes by Eq. (19)

$$V_r = \frac{V}{f_m D} \tag{19}$$

where V is the flow velocity and D in the external diameter of the element.

Reynolds member. Re

Table 1 Table of end fixity

Fixity		A_n
Fix	Free	3. 52
Pin	Pin	9.87
Fix	Pin	15.40
Fix	Fix	22.40

Table 2 Table of end fixity

Fixity		I_o
Fix	Free	0.66
Pin	Pin	0.63
Fix	Pin	0.58
Fix	Fix	0.52

$$R_e = \frac{VD}{\nu} \tag{20}$$

where **V** is the Kinematic Viscosity.

V = Kinematic viscosity of air should be used because this is done in fabrication shop.

Stability parameter (K_s)

$$K_s = \frac{2m\delta}{\rho D^2} \tag{21}$$

where δ is the logarithmic decrement of the structural damping and is equal to 2πδ. The parameter in the critical damping ratio, which may be taken as 0.001 – a conservative estimate ρ is the density of the fluid – 123 kg/m^3 for seawater.

3.1.4 Resonant Vortex Shedding in Current

a. In–line Motion

Check to see if inline motion occurs during exposure and operation. Find the corresponding amplitude to characterize the system behaviour. In-line motion is completely suppressed for values of Ks greater than 1.8.

b. Transverse Motion

Determine the reduced velocity necessary for the onset of transverse motion and calculate the amplitude of motion using Eq. (22)

$$\frac{y}{D} = \frac{1}{16\pi^2}\frac{\rho D^2}{m\delta}\frac{1}{S^2}C_{ij} = \frac{C_{ij}}{4\pi k_s s^2} \tag{22}$$

where S is the Strouhal number, taken as 0.2.

The design lift coefficient, CL_J, depends on the base lift coefficient, CL_o, (0.2) for $R_e > 4 \times 10^5$ and 0.42 for Re $\leq (41 \times 0^5)$ and the correction length factor, I_o, which in turn depends on the member fixity. See Table 2

$$CL_j = CL_o I_o (I/I_o)$$

The amplitude of motion is determined from an iterative calculation as follows:

Table 3 Table of end fixity

Fixity		K_n	
		End	Intermediate
Fix	Free	2.00	–
Pin	Pin	0.00	4.80
Fix	Pin	11.0	6.51
Fix	Fix	16.00	8.00

Initially take $\frac{I}{I_O} = 1.0$

Calculate $\frac{y}{D}$

Read off the new value of $\frac{I}{I_O}$ from the relevant chart

Recalculate $\frac{y}{D}$ until convergence is obtained.

Stresses due to resonant vortex shedding equal the maximum stress due to an amplitude y

$$F_b = K_n \frac{ED}{L^2} y \tag{23}$$

where K_n is a constant depending on the fixity, as shown in Table 3.

For the riser members, check for the possibility of vortex-induced oscillations for on-site and transportation conditions. Calculate their amplitude where appropriate. Calculate the maximum stress that occurs, and then, where appropriate, calculate the time to failure by conducting the following analysis.

$$\begin{aligned} &\text{Stress Range } = 2 \times F_b \\ &\text{Stress hotspot range } = \text{ stress range } \times \text{ SCF } = \sigma_H \\ &\sigma_H \text{is expressed in N/mm}^2 \end{aligned} \tag{24}$$

where SCF is the stress concentration factor.

S-N curve gives:

$$\begin{aligned} &\text{No. of cycles to failure } = 10^{[12.1638 - 3\log(\sigma H)]} \\ &\text{Time to failure } = {}^{N}/_{Fn} \\ &\sigma_H \text{is expressed in N/mm}^2 \end{aligned} \tag{25}$$

4 Model Implementation

For the dynamic comparative assessment of the fatigue behaviour of the two riser configurations, the following assumptions and parameters were adopted:

- Riser Configurations: Two risers were analyzed—a conventional Steel Lazy Wave Riser (SLHR) with a steel riser leg and an alternative design incorporating a composite riser leg to replace the steel leg.
- Water Depth: A depth of 2000 m was considered, as the SLHR riser concept is already well-established for this depth.
- Internal Diameter: The composite pipe's internal diameter was selected to match that of the steel pipe to ensure a consistent flow capacity.
- Wall Thickness: Pipe wall thicknesses were determined based on comparable static loading conditions to maintain structural integrity.
- Global model analysis of both riser systems was developed and analyzed to evaluate dynamic performance and fatigue characteristics. The global dynamic analysis has been performed using industry-standard software, Orcaflex Version 9.8. Orcaflex uses three-dimensional lumped mass elements to model the riser.

The riser model is constructed by discretizing the steel section into smaller finite elements, with finer divisions applied near the bulkhead connections, specifically at the top bulkhead, where the riser attaches to the buoyancy tank, and at the production bulkheads connecting to the bottom assembly. The touchdown point (TDP) varies along the riser as the host structure moves from its nominal position, necessitating a denser mesh of shorter elements around the nominal TDP to accurately capture the localized response.

The shortest element length used near the riser–host interface is 0.5 m, while the minimum element length within the touchdown zone is 1 m. In contrast, longer elements, up to 15 m, are assigned to regions where bending moment variations between adjacent elements are minimal. To maintain smooth geometric transitions, the length ratio between adjacent elements is restricted to a maximum of 2.5. The resulting element size distribution along the riser is illustrated in Fig. 6. At the boundaries, the bottom end of the riser is anchored to the seabed, while the top end is connected to the FPSO via a rotational spring. The rotational stiffness of the flexible joint exhibits nonlinear behaviour depending on the rotation angle; however, in the Orcaflex model, it is approximated using a linear stiffness. The stiffness values applied for strength and fatigue analyses are derived from the flexible joint stiffness curves supplied by the manufacturer, based on experimental testing and Orcaflex calibration.

4.1 Global Geometry and Riser Description

All hybrid risers are designed with a platform connected to a buoy with a horizontal flexible riser and a vertical flexible riser from the buoy to the seabed.

a. Riser configuration for the analysis

The static configurations of the production riser are presented in Figs. 7 and 8. Each production riser is equipped with buoyancy drums at the hang-off region to provide the necessary uplift and maintain the desired riser profile. Additionally, the riser features a 5-m section of internal cladding at the hang-off, enhancing wear resistance and structural integrity. At the touchdown zone, the riser comprises 122 m of pipe joints with integral upset ends, designed to improve fatigue performance and connection strength. This section is strategically positioned such that approximately 65 m of the pipe joints lie within the suspended region, while the remaining 57 m are located within the seabed contact region, relative to the nominal touchdown point (TDP).

b. Flexible Jumper

The engineering design of both rigid and flexible jumpers plays a crucial role in ensuring the safety, reliability, and cost-effectiveness of offshore operations. Rigid jumpers serve as vital connectors between the platform and the pipeline, providing structural continuity and stability. In contrast, flexible jumpers are commonly employed in hybrid riser systems to isolate the vessel's motions from the free-standing conductor, thereby reducing dynamic loads and enhancing system longevity. In the hybrid riser configuration, the connection between the riser and the host FPSO is achieved through a Flexible Joint (FJ). The rotational stiffness values of the FJ, which are essential parameters for strength and fatigue analyses, will be specified in

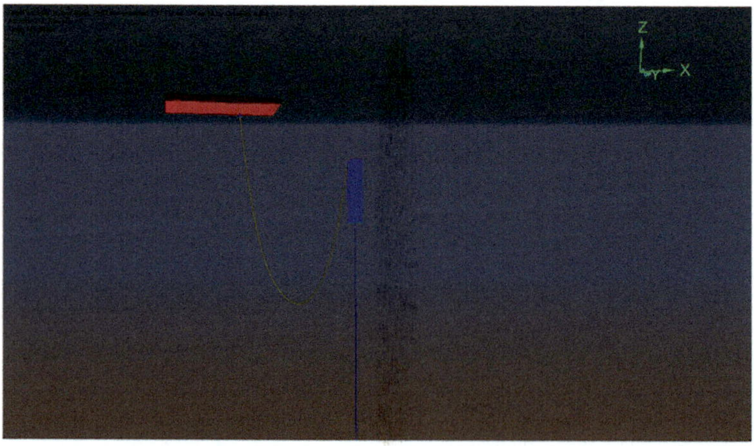

Fig. 7 Production riser configuration

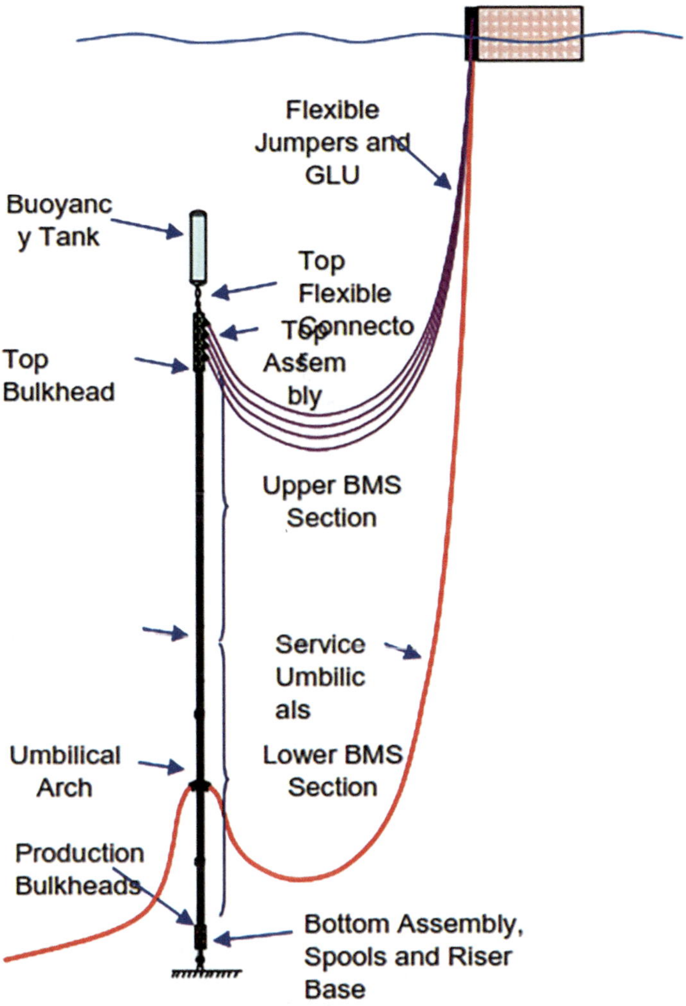

Fig. 8 Hybrid riser diagram

the Riser Design Basis document. A longitudinal cross-section of a typical flexible joint, along with the simplified equivalent cross-section used in the analytical model, is illustrated for reference. The cross-sectional dimensions are detailed in Table 4.

Table 4 Cross-section dimensions of flexible joint

Riser	OD1 (1)	WT1 (1)	OD2 (1)	WT2 (1)	L1 (1)	L2 (1)	
	(mm)	(mm)	(mm)	(mm)	(m)	(m)	
Production	273	23.8	368	71.3	2.5	0.5	

Table 5 Principal dimension of the top can Orcaflex 9.8b: hybrid riser

Cylinder	Outer diameter (m)	Inner diameter (m)	Length (m)	Cumulative Length (m)
1	10.000	0.000	8.000	8.000
2	10.000	0.000	8.000	16.000
3	10.000	0.000	8.000	24.000
4	10.000	0.000	8.000	32.000
5	10.000	0.000	8.000	40.000

c. Riser Buoyancy tank

The buoyancy tank is typically positioned 50–150 m below sea level, placing it beyond the zone of influence of waves and strong currents. Its primary function is to provide the necessary lift force, thereby tensioning the vertical riser pipe to enhance the dynamic response of the entire riser system. This tensioning helps mitigate vortex-induced vibration (VIV) fatigue damage. Additionally, the buoyancy tank plays a crucial role in limiting the load and deflection angle of the flexible jumper [15]. The principal dimensions are shown in Table 5. Preliminary Vortex-Induced Vibration (VIV) analyses indicate that a strake coverage of 32% of the suspended length (approximately 500 m) provides sufficient VIV fatigue life for the production risers. For the water injection and gas export risers, a strake coverage of 10% of the suspended length (around 160 m) is found to be adequate to achieve the required VIV fatigue performance.

4.2 *Data Collection*

The hybrid riser system is modelled using a finite element model (FEM), in which the riser is discretized into a series of line elements. These elements are represented as massless segments with nodes at each end, allowing accurate simulation of the riser's structural and dynamic behaviour. After establishing the static configuration of the hybrid riser, a nonlinear dynamic time-domain analysis is conducted. This analysis iteratively updates the riser configuration and axial tension at each time step. The dynamic response is evaluated using a numerical integration scheme, specifically the forward Euler method with a constant time step. The time-domain simulation captures the riser's motion over a defined period, beginning from the static equilibrium position obtained in the preliminary static analysis. This modelling approach provides a realistic evaluation of the riser's performance under combined environmental and operational loading conditions. The data and parameters utilized in the analysis are presented in the following tables. The full FPSO Offsets for different environmental conditions are shown in Tables 6 and 7.

Table 6 Steel riser parameters

Parameter	SI	Production Imperial
Diameter OD)	227 mm	8.93 in
Thickness (WT)	23.8 mm	0.937 in
Coating (Insulation)	75 mm ((1,2)	3.937 in (1,2)
Coating density	737 kg/m^3	46 lb/ft^3
Operating contents	310 kg/m^3	19 lb/ft^3
Design pressure	34.5Mpa	5000 psi
Max design (temperature)	90 °C	176 °F
Minimum design (temperature)	0 °C	32 °F
Corrosion allowance	5.5 mm	0.217 in
Strake coverage	500 m	1640 ft

Table 7 Top termination unit

TTU	Flexible joint
Sea water density	Average of 1025 kg/m^3 (64 lb/ft^3)
Design life	25 years
Material grade API 5L	X65
Carbon steel (WT)	$-5\% + 17\%$
Tolerance SMYS	450 Mpa (65.3 ksi)
Minimum ultimate	535 Mpa (77.66ksi)
Tensile strength Modulus of elasticity	200 Gpa (29,000 ksi)
Poisson's ratio	0.3
Steel density	7850 kg/m^3 (490 lb/ft^3)

To carry out a global analysis of a composite riser requires metocean data, riser properties, and vessel parameters. The following tables (Tables 8 and 9) provide the information used as input parameters for the analysis.

5 Results and Discussion

The current study presents a computational-based assessment of fatigue analysis of composite and steel risers. The objective was to evaluate the fatigue life behaviour of both riser types under identical operating conditions. To achieve this, OrcaFlex software was employed to model the dynamic responses of the risers for the specified scenarios. The simulation results and analysis are presented in the subsequent sections.

Table 8 Riser properties

Layer No	Description	Mass (kg/m)	ID (mm)	Thickness (mm)
1	Internal carcass	22.3	203.2	7.5
2	PA11 internal sheath	4.4	218.2	5.6
3	Pressure armour	39.5	229.4	8.0
4	Anti-wear tape	1.2	245.4	1.5
5	Tensile armour layer 1 (35.0°)	16.9	248.4	3.0
6	Anti-wear tape	1.3	254.4	1.5
7	Tensile armour layer 2 (−35.0°)	17.4	257.4	3.0
8	Tape	0.7	263.4	1.65
9	PA11 external sheath	6.3	266.7	7.0
Diameter: outside (mm)				280.7
Weight in air, empty (kg/m)				109.94
Weight in air, full of seawater (kg/m)				145.36
Pressure: Nominal bursting (bar)				624
Pressure: Hydrostatic collapse (bar)				121
Damaging pull in a straight line (kN)				3516.7
Minimum bend radius, for storage (m)				1.82
Bending stiffness, at 20 °C (kN.m^2)				21.04
Torsional stiffness (kN.m^2/rad) Relative Elongation for 50kN (%)				3100 0.014

Table 9 FPSO offsets under varying environmental settings

Dominant environment	FPSO mean offset$^{(1)}$	
	(% of WD)	(m)
1 year squall	5.0	60.0
100 year squall	8.0	96.0
10,000 year squall	10.0	120.0
100 year current (with 1 broken mooring line)	10.0	120.0

5.1 Wave Data Characterization

The wave scatter data is in the metocean specification for 48 environmental directions. However, the majority of the wave cycles come from the directions 180–232.5 deg. inclusive (accounting for 96% of the total number of cycles). Due to the highly directional nature of the fatigue environment, the directional data is more suitable for fatigue analysis than omnidirectional data. As outlined in the System-R Design Premise, some consolidation of the wave environment is necessary. The wave environment is quite directional and is dominated by a sector of the overall directions provided (generally from South and South-West directions). The wave scatter was accordingly consolidated into eight directions to reduce the cases for the fatigue analysis. A schematic of the actual eight directions used in the analysis is presented in Fig. 9. A list of the eight directions is also presented.

The wave energy coming from the directions 0–172.5 degrees inclusive was added to the 180-degree direction. The percentage of the total number of cycles coming from 0 to 172.5 degrees is 0.019%, which is minimal. Similarly, the wave energy from 240 degrees to 352.5 degrees was added to the 232.5 degree direction. The percentage of the total number of cycles coming from 240 to 362.5 degrees is 0.014%, and thus minimal in this case as well. The final distribution of the number of cycles per direction, as applied in the analysis, is illustrated in Fig. 10.

The H/T combinations are binned together to form a total of 128 regular wave bins—the premise requires a minimum of 100 bins. The following principles are applied to achieve a conservative bin division. The bins are selected so that they do not span a wide range of wave heights (to avoid over-conservatism). The maximum wave height is assigned to each bin, which is a conservative approach. Small numbers of cycles for certain H/T combinations are put into larger bins, as these bins are not likely

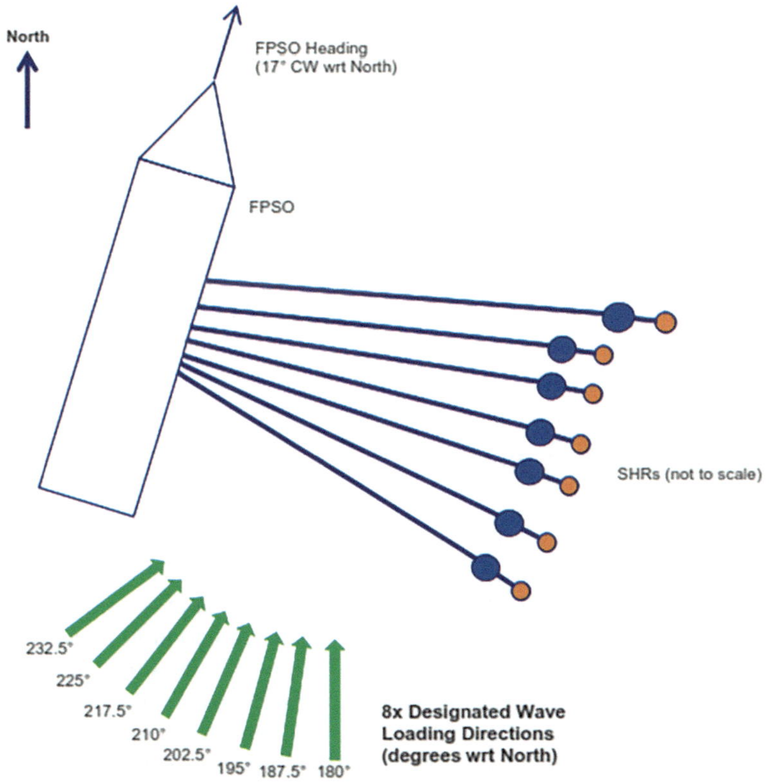

Fig. 9 Fatigue wave analysis loading directions

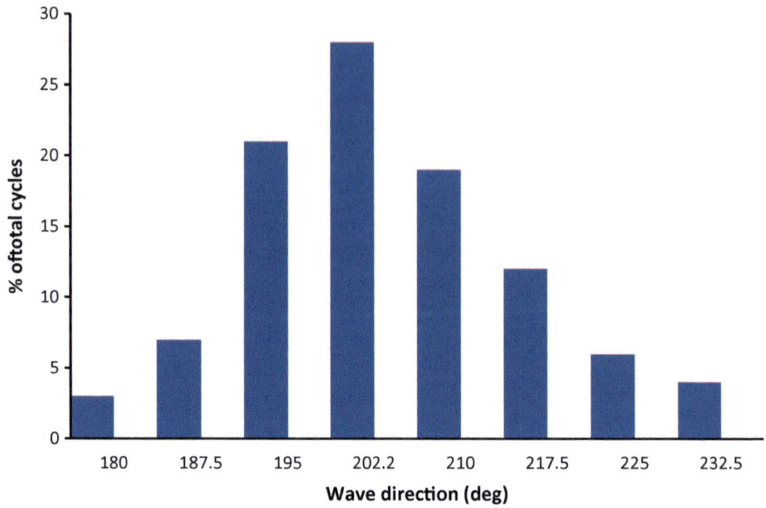

Fig. 10 Distribution of number of cycles by direction

to drive fatigue damage. More bin divisions are made in areas of the scatter diagram with many cycles regularly. The wave periods selected (provided in Appendix B for information) are based on the maximum bending moment response observed in the upper adaptor joints for each bin. The response is compared at 0.1 s intervals within each bin. Each estate/block on the scatter diagram has an associated probability of occurrence or, in this case, the number of cycles. This probability of occurrence (% occurrence) is calculated by dividing the number of cycles for the block by the total number of cycles. The fatigue calculation takes percentage occurrence into account – i.e., the stress range associated with each bin is assigned a percentage occurrence that is adjusted based on the number of cycles and the total number of cycles per year.

5.2 Fatigue Analysis Results

The fatigue analysis results indicate that fatigue damage in the riser arises from both wave-induced forces and vortex-induced vibrations (VIV). The dynamic response analysis was performed in the time domain across a range of sea states, with each simulation lasting one hour, including a 50-s ramp-up period to allow for model stabilization. A time step of 0.25 s was applied throughout the simulations. A sensitivity study confirmed that variations in the time step and simulation duration had negligible effects on the calculated fatigue life, validating the chosen parameters. The Orcaflex fatigue results for the steel tower section of the riser are presented in Fig. 11. The analysis revealed that the maximum annual in-place fatigue damage occurs at

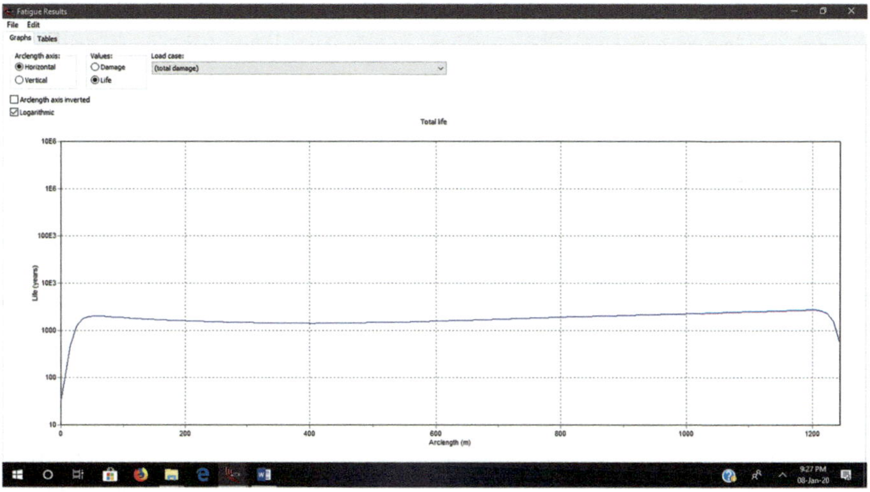

Fig. 11 Steel fatigue life

two critical locations: the linepipe/Upper Axial Joint (UAJ) interface and the region immediately below the jumper connection, near the buoyancy tank. The dominant fatigue contribution originates from load cases with wave periods between 8 and 10 s, where most of the fatigue energy is concentrated. The highest damage levels are observed near the linepipe/UAJ interface, primarily due to increased bending moments resulting from the dynamic motions transmitted by the jumper.

Typically, the adapter joints can exhibit a characteristic trend of fatigue damage in this region due to the varying cross-sections of the adapter joints and the corresponding variations in stresses along their length. The dominant wave direction, which causes the most damage in the line pipe/UAJ interface, is 210°. The fatigue life graph of steel in Fig. 12 shows that the worst fatigue hot spot occurs at the interface between the buoyancy tank tower and the composite, which is 100 times worse than the composite itself.

This has a similar trend as steel fatigue damage and life but has an improved fatigue life of about 100 times better than steel. The graph of the composite risers shows that the worst fatigue hot spot also occurs at the buoyancy tank/tower section interface, refer to Fig. 13.

The fatigue performance of the composite riser demonstrates a significant improvement compared to the conventional steel riser, primarily due to the superior fatigue resistance of composite materials and the elimination of welded joints along the riser length. However, the reduced buoyancy in the composite configuration has a detrimental effect on the steel stub section located below the buoyancy tank. For the steel riser, the critical fatigue hot spot was identified at the weld nearest to the interface with the Upper Riser Assembly (URA). Substituting this section with composite material results in an over 100-fold increase in fatigue life at this location. In the composite riser, the fatigue hot spot shifts to the steel stub below the buoyancy

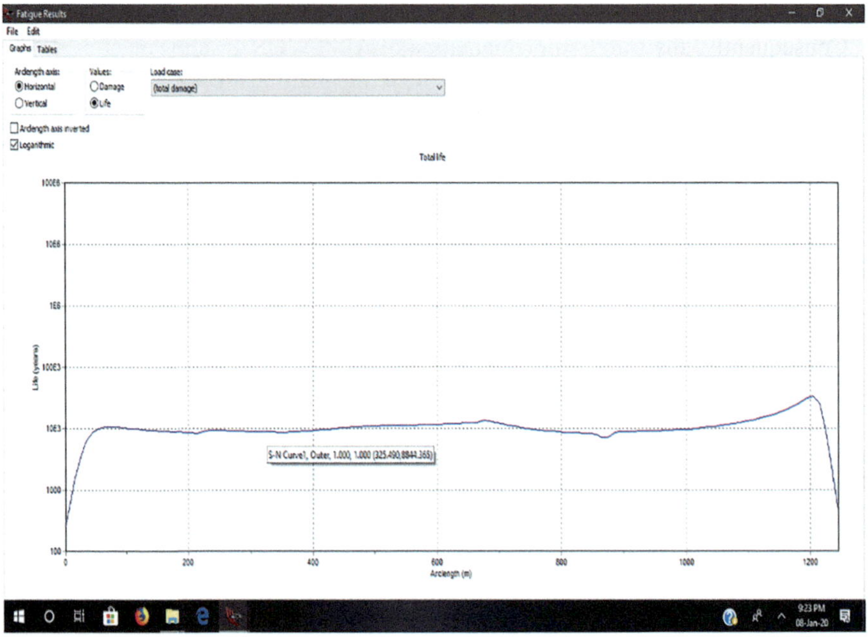

Fig. 12 Orcaflex fatigue result of composite tower section of riser

tank, where the fatigue life remains approximately 200% greater than that of the steel riser.

Table 10 presents the results of the MATLAB program for the steel risers, which align with those obtained from the Orcaflex fatigue analysis results. The fatigue life for steel risers, as calculated by the Matlab program, is 1049.097776 years. Similarly, the results from the 2H offshore organization for the composite, which are shown in Table 11, indicate that the composite riser has one hundred times (100) better fatigue life than the steel riser. This provides a need-based technique to better inform industry players on the best material selection for prolonged reliability and operational optimization.

The fatigue performance of the riser saw a significant improvement thanks to the superior fatigue resistance of composite materials and the elimination of welds along the structure. Table 12 summarizes the outcomes of the three modelling techniques used in the comparative analysis. Additionally, the MATLAB results further validate the analysis.

The analysis demonstrates the application of computational-based techniques for fatigue modelling of the steel and composite risers, aimed at predicting a terrain-specific reliability of the structure. The results provide technical information for material selection that optimizes operation, considering terrain-specific impacts.

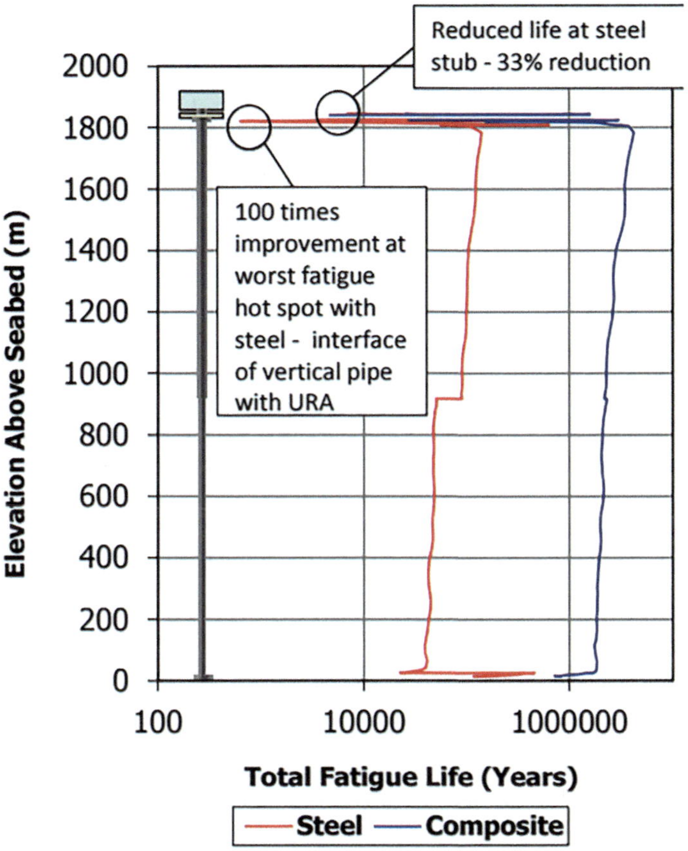

Fig. 13 Comparison of steel and composite risers fatigue life from 2H offshore

Table 10 Fatigue result for steel riser

Hs (m)	Tz (s)	ni	Force F (N)	Bending moment M (Nm)	Stress range S (Mpa)	Ni	fatigue damage D	Fatigue life (years)
3	11.4	34,526	28,055	4,734,281.25	184.0547271	45,281.8062	0.00009532	1049.097776

Table 11 Fatigue Results for Composite Riser

Hs (m)	Tz (s)	ni	Force F (N)	Bending moment M (Nm)	Stress range S (Mpa)	Ni	Fatigue damage D	Fatigue life (years)
3	11.4	34,526	28,055	4,734,281.25	184.0547271	45,281.8062	9.432E−07	106,022.0526

Table 12 Fatigue Results Summary

Ocaflex fatigue life results		2H fatigue life results		Matlab fatigue life results	
Composite	Steel	Composite	Steel	Composite	Steel
100,000	1000	1,000,000	10,000	106,022.05	1049.03
Steel Composite Fatigue Life Ratio		**Steel Composite Fatigue Life Ratio**		**Steel Composite Fatigue Life Ratio**	
100		100		101	

6 Conclusion

The analysis conducted on composite risers has demonstrated promising results, boosting confidence in their use as an alternative to traditional steel risers, which have been the industry standard for decades. The research findings indicate that composite materials are lightweight, have high specific strength, exhibit long fatigue life and resistance, are corrosion-resistant, and can be moulded into complex shapes. Additionally, composites have lower axial and bending stiffness. Some key takeaways from the study include:

- The research provides the offshore industry with valuable insights and data, reassuring industry professionals about the suitability of composite risers for deployment in the Gulf of Guinea.
- The fatigue life of the composite riser used with the Egina FPSO has been established, allowing for more accurate predictions of the production period.
- The study analyzes the riser's tension and bending moment, which will help ensure proper sizing of composite risers, addressing the risks of overdesign or underdesign. Tension and bending moments are crucial for determining the correct dimensions of a marine riser.
- Dynamic analysis results comparing steel and composite risers revealed that the steel riser experiences higher tension, with a difference of 9350 kN.
- Fatigue analysis, conducted using Orcaflex software, showed that the composite riser exhibits a hundred times fatigue life compared to that of the steel riser.
- A validation program coded in MATLAB was developed, showing only a slight deviation, which further confirmed the reliability of the results.

The current study demonstrates the application of computational tools for predicting the fatigue life of composite and steel risers. However, more elaborate research should be conducted to include soil riser interaction analysis by developing a lateral nonlinear seabed model. This would be considered in our future work.

References

1. Ochoa OO (2006) Composite riser experience and design guidance, final project report prepared for the minerals management service under the MMS/OTRC cooperative research agreement 1435-01-04-CA-35515, Task Order 35985, MMS Project Number 490, Texas A&M University
2. Tan LB, Chen Y, Jaiman RK, Sun X, Tan VBC, Tay TE (2014) Coupled fluid-structure simulutions for evaluating a performance of full scale deepwater composite riser. Ocean Eng 19–35
3. Bai J, Seeleuthner P, Bompard P (1997) Mechanical behavior of 35 "filament-wound glass-fiber/epoxy-resin tubes: L. Microstructural analyses, mechanical behavior and damage mechanisms of composite tubes under pure tensile leading, pure internal pressure, and combined loading. Compos Sci Technol 57(2):141–153
4. Chen Y, Tan LB, Jaiman RK, Sun X, Tay TE, Tan VBC (2013) Global-local analysis of a full-scale composite riser during vortex-induced vibration. In: ASME: 2013 32nd international conference on ocean. offshore and arctic engineering, OMAE, vol 66, no 6, pp 2013–11632
5. Liu Q, Qu Z, Chen F, Liu X, Wang G (2024) Review of the development of an unbonded flexible riser: new material, types of layers, and cross-sectional mechanical properties. Materials 17(2560):1–34. https://doi.org/10.3390/ma17112560
6. Gibson AG (2003) The cost-effective use of fiber reinforced composites offshore. Research Report for the Health and Safety Executive, University of Newcastle Upon Tyne 21(3–7):24–45
7. Baldwin DD, Lo KH, Long JR (1998) Design verification of a composite production riser. Offshore Technology Conference. Houston, Texas 12(3):449–469
8. Kraincanic I, Kebadze E (2001) Slip initiation and progression in helical armouring layers of unbonded flexible pipes and its effect on pipe bending behaviour. J Strain Anal Eng Des 36(3):265–275. https://doi.org/10.1243/0309324011514458
9. Liu H, Li M, Shen Y (2023) Numerical analysis of mechanical behaviors of composite tensile armored flexible risers in deep-sea oil and gas. J Mar Sci Eng 11(3). https://doi.org/10.3390/jmse11030619
10. Kaboudian A, Tan LB, Jaiman RK, Chen Y, Tan VBC (2014) Semi-empirical VIV analysis of full-scale deepwater composite risers. In: Proceedings of the ASME 2014 33rd international conference on ocean, offshore and arctic engineering, OMAE2014–23529, San Francisco, California, USA
11. Siriruk A, Penumadu D (2014) Degradation in fatigue behavior of carbon fiber–vinylester based composites due to sea environment. Composite Part B: Eng 61:94–98
12. Tan LB, Chen Y, Rajeev KJ, Sun X, Tan VBC, Tay TE (2015) Coupled fluid-structure simulations for evaluating a performance of full scale deepwater composite riser. Ocean Eng 45(5):19–35
13. Ochoa OO, Chouchaoui CS (1999) Similitude study for a laminated cylindrical tube under tensile, torsion, bending, internal and external pressure Part I: governing equaitons. Compos Struct 44(4):221–229
14. Roberts D, Hatton SA (2013) Development and qualification of end fittings for composite riser pipe. Offshore Technol Conf 34(9):34–52
15. Xiushan ST, Tan TE, Srikanth VBC (2012) Multi-axial fatigue model for fiber- reinforced composite laminates based on Puck's criterion. J Compos Mater 449–469

Implementation of Process Safety Management System in the Power Plant Industry Using the Monte Carlo Probabilistic Method

Mahdi Mozaffari, Behrouz Behnam, Adel Gholami, and Adel Mohammad Pour

Abstract Power plants represent critical infrastructure within national energy systems, accounting for the majority of electricity generation and serving as key components of industrial and economic resilience. Despite their significance, recurring process incidents and the aging nature of some facilities expose these operations to escalating safety and reliability challenges. To address these concerns, this study presents a smart process safety management (PSM) framework that integrates probabilistic modeling and intelligent decision-support tools for risk prediction and prioritization within the power plant industry. The research builds on the foundation of traditional PSM, which aims to identify, evaluate, and control the release of hazardous materials or process deviations that may cause harm to people, assets, and the environment. The proposed framework advances this concept by embedding Monte Carlo probabilistic simulation within PSM components to quantify uncertainty, simulate multiple operational scenarios, and support data-driven safety decisions. Using MATLAB and SPSS software environments, empirical data from thermal and combined-cycle plants were analyzed to model system variability and failure likelihood under dynamic conditions. Results derived from the Copeland prioritization method indicate that key elements such as management of change,

M. Mozaffari · A. Gholami (✉)
Department of Civil and Environmental Engineering, Amirkabir University of Technology (AUT), Tehran, Iran
e-mail: adel.gh@aut.ac.ir

M. Mozaffari
e-mail: mahdimozaffary@aut.ac.ir

B. Behnam
Department of Civil and Environmental Engineering, Amirkabir University of Technology (AUT), Tehran, Iran
e-mail: b.behnam@aut.ac.ir

A. Mohammad Pour
Faculty of Mathematics and Computer Science, Amirkabir University of Technology (AUT), Tehran, Iran
e-mail: adel@aut.ac.ir

H. Li et al. (eds.), *Data-Driven Methods for Reliability and Safety Engineering: Applications in Industrial Systems*, Springer Series in Reliability Engineering,
https://doi.org/10.1007/978-3-032-22873-4_45

competency training, and hot work permit systems possess the highest influence on overall process safety performance. Conversely, auditing activities and trade secret management showed lower relative significance within the probabilistic hierarchy. The integration of Monte Carlo simulation with PSM not only enhances predictive accuracy but also transforms traditional safety frameworks into intelligent, adaptive systems capable of learning from operational data and continuously improving over time. This chapter demonstrates how probabilistic modeling can serve as a cornerstone for developing smart safety systems in high-risk industrial environments. By combining structured process safety principles with digital analytics and uncertainty quantification, the proposed model provides an effective pathway toward resilient, evidence-based risk management in the power generation sector.

Keywords Process safety management · Power plant · Monte Carlo simulation · Probabilistic risk modeling · Intelligent decision support · Smart safety systems · Risk management

1 Introduction

Process industries, including oil and gas, petrochemicals, mining, power generation, chemicals, and pharmaceuticals, form the foundation of modern industrial economies. These sectors are responsible for transforming raw materials into a vast range of essential products that sustain everyday life and economic development. Their operations are predominantly based on intricate mechanical and chemical processes that involve handling large volumes of hazardous substances under varying operational conditions. A defining characteristic of these industries is their continuous and interdependent nature, where a malfunction in a single subprocess can trigger cascading effects across other operational units, resulting in widespread disruptions. The complexity and scale of these systems inherently increase their risk exposure. The combination of chemical reactivity, high temperature and pressure operations, and the production, storage, and transportation of hazardous materials heightens the potential for catastrophic events. The consequences of such incidents extend far beyond physical damage, often encompassing severe human, environmental, and economic impacts. Over the past four decades, several major industrial accidents, such as the Soso tragedy, the Felix Borough explosion, and the Bhopal disaster, have served as stark reminders of the devastating outcomes that can arise from lapses in safety management and process control. These incidents also underscore a crucial lesson for modern industry: traditional safety systems alone are insufficient to address the complexity of today's interconnected, data-intensive industrial environments. The evolution of process industries now demands the integration of smart safety systems that combine advanced analytics, real-time monitoring, and probabilistic modeling to anticipate, detect, and mitigate risks before they escalate. By leveraging tools such as Monte Carlo simulation, digital twins, and AI-based decision support, organizations can move from reactive hazard management toward proactive, predictive, and adaptive safety strategies. In this context, safety is not merely a compliance requirement but a dynamic, data-driven discipline, one that continuously learns from operational feedback, historical incident data, and process variability. Implementing such intelligent frameworks can transform the culture of safety within process industries, ensuring that critical operations remain resilient, efficient, and aligned with both environmental and human sustainability goals [1].

1.1 Research Structure

This research is structured into five interconnected parts, each contributing to the development of an intelligent and probabilistic framework for process safety management in the power generation sector. The first part presents a general introduction outlining the scope, objectives, and significance of integrating smart safety systems into process industries. It establishes the rationale for adopting a data-driven and simulation-based approach to enhance safety decision-making in complex operational environments. The second part provides the foundational background of the study, detailing the conceptual design, research goals, and methodological necessity. It includes a comprehensive review of literature and prior studies on process safety management in power plants, emphasizing the limitations of traditional safety systems and the opportunities introduced by digitalization, probabilistic analysis, and intelligent monitoring technologies. The third part describes the research methodology, focusing on the hybrid analytical approach that combines qualitative safety evaluation with quantitative modeling. This section outlines the data collection techniques, simulation methods, and computational tools, such as Monte Carlo simulation and statistical modeling, used to manage uncertainty and predict process reliability. The fourth part presents the research population, sampling strategy, and statistical results derived from data analysis. It demonstrates how simulation outputs and empirical findings can be used to prioritize safety management components and inform decision-making processes within smart industrial frameworks. Finally, the fifth part synthesizes the research findings, discusses the practical implications of implementing intelligent process safety management systems, and provides strategic recommendations for future research. This concluding section highlights how probabilistic modeling and AI-supported tools can strengthen resilience, optimize resource allocation, and advance safety performance across modern power plant operations.

2 Background of Process Safety Management

In 1992, the Occupational Safety and Health Administration (OSHA) in the United States introduced the Process Safety Management (PSM) framework, which comprised 14 essential elements [2]. This foundational model was primarily designed to enhance human performance and workplace safety by promoting behavioral change, developing safety-oriented organizational cultures, and providing structured training for personnel with critical process responsibilities. The PSM framework represented an early attempt to establish systematic and standardized practices across industries handling hazardous chemicals, emphasizing procedural discipline, competency development, and operational reliability. Fifteen years later, in 2007, the Center for Chemical Process Safety (CCPS) introduced a more advanced and holistic model known as Risk-Based Process Safety (RBPS). Drawing upon decades of industrial experience with the PSM framework, RBPS shifted the focus from compliance-based management toward risk prioritization and resource optimization. This model is built on the principle that not all hazards or process risks are of equal significance; therefore, financial and operational resources must be allocated proportionally to address high-consequence risks first. The RBPS framework emphasizes the importance of quantitative assessment, recognizing that effective safety management depends on the ability to evaluate likelihoods and impacts using data-driven methods. A distinctive feature of the RBPS model is its explicit recognition of safety culture as an independent and critical element. It acknowledges that technological safeguards alone cannot ensure system resilience without a strong cultural foundation that supports learning, transparency, and accountability. This perspective has since laid the groundwork for the integration of intelligent safety systems, where digital tools such as probabilistic modeling, machine learning, and predictive analytics augment human judgment. In modern applications, the principles of PSM and RBPS are increasingly being enhanced through smart safety technologies, including Monte Carlo simulation, digital twins, and AI-based decision-support systems, that can analyze vast datasets, simulate potential failures, and recommend targeted risk mitigation measures in real time, refet to Table 1. These advancements mark a paradigm shift from reactive safety practices toward proactive, adaptive, and continuously improving safety management, aligning with the central philosophy of this book [1].

Process safety management has long been a topic of interest for researchers in the process industries, and studies in this field continue to improve the effectiveness and development of this system. Below is a review of some significant studies in this area [6].

In 1994, Bridge conducted a cost–benefit analysis on the implementation of PSM based on an industrial survey. He concluded that the companies' cost estimates were significantly lower than the actual implementation costs. This study presents the actual costs incurred by some companies and estimates future costs for compliance with government standards and regulations related to PSM [7].

Amin and colleagues, through a bibliometric analysis examining the evolution of process safety and hazard research, identified that safety and risk analysis is a global issue. Emerging countries such as China, Malaysia, and Iran have significantly contributed to research volume in this decade, with the United States leading and India being the most economical country in this field [8].

Myung Kun and colleagues, after studying experiences in the Korean chemical industry, concluded that most incidents were due to the lack of effective and appropriate operator training, lack of capital integrity systems in process equipment, lack of safety awareness and culture, and inadequate management systems, procedures, and

Table 1 Various process safety management models presented by relevant companies and organizations

Company/ organization providing	Description	Research gap
OSHA organization	The first model provider, which includes 14 main elements	All models provided by organizations have different elements and approaches, while almost none have offered a prioritization system for implementing the elements in various industries
DuPont company	A model derived from the OSHA model, categorizing elements into three main groups: personnel, technology and process, and equipment [3]	
CCPS association	A model introduced in 2007, focusing on risk management with 20 elements, including process safety culture, management review, continuous improvement, etc. [4]	
EI energy institute	A model with 20 elements and a managerial approach, including critical safety equipment management, change and project management, etc. [5]	

corrective safety actions. The study also mentions that implementing PSM has led to outcomes such as reduced process incidents, systematic documentation of process technology development, reduced operational errors, increased quality and productivity, reduced insurance premiums, and ultimately the creation of an automated process safety management system [9].

Shayan and Gholamnia found that conventional HSE approaches, particularly personal safety, alone are not sufficient to create an adequate level of safety. Chemical industries, which are among the high-risk industries, require a systematic safety approach in addition to conventional HSE measures. To address this need, various organizations have designed comprehensive systems under the title of Process Safety Management, with the CCPS and OSHA models being the most widely used. The RBPS model from the Center for Chemical Process Safety (CCPS) was selected in this study from a process safety perspective [10].

Mahdsharif and colleagues in 2016 found that the full implementation of PSM work processes must include a systematic approach to managing PSM processes and activities. This process requires the program to ensure continuity when personnel change and include quality assurance steps such as planning, implementation, review, and revision of the PSM implementation program. Failures in the work process should be addressed by improving documentation, promptly resolving PSM recommendations of all types, and maintaining and continuing recognized and generally accepted good engineering practices (RAGAGEP) applied in many aspects of the PSM program. Additionally, management support and participation, along with the creation of tracking systems to ensure the completeness of all work activities and adequate documentation, are essential [11].

Chang and Liang, after conducting a safety audit on four paint manufacturing facilities by three auditors selected from government inspectors, safety consulting engineers, and paint industry experts, concluded that the two facilities certified by the 18001 guidance with zero severity rates and low frequency rates in 2006 scored the highest compared to those without the 18001 certification and with high federal regulations [12].

Faisal Khan and colleagues reviewed past progress in developing methods and models for process safety management and risk management. Based on publications in leading journals on safety, risk, and reliability, they covered the evolution of methods and models developed for safety and risk management over time. The methods and models are classified as quantitative, semi-quantitative, qualitative, and hybrid, discussing advancements in the past decades [13].

Swuste and colleagues concluded in their research that various PSM models demonstrate the effectiveness of methods and procedures employed by those who believe there is a strong correlation between verbal PSM practices and achieving reductions in the frequency and severity of process incidents. This paper also described the progress of CCPS research towards developing a PSM performance evaluation model and provided a framework for future CCPS research to define effectiveness components [14].

Halim and Manan investigated questions such as "What are the ingredients for excellence in process safety?" and "What actions or programs are needed to maintain that level permanently?" They concluded that most incidents are related to internal organizational issues; however, achieving excellence in process safety starts with external influences and commitment from the highest levels of leadership within the company. They discussed how individual organizations could debate and exchange views on setting and enforcing strict operational discipline, ensuring the competence of all personnel, and learning and implementing lessons from past incidents to advance process safety performance [15].

M.K. Pat and colleagues in 2018 showed that the industry needs to identify its requirements and collaborate with accreditation organizations and universities to effectively implement their process safety training programs. The main challenge in drafting this review was the insufficient professional publications supporting process safety training compared to process safety education. This gap can be attributed to the proprietary nature of information provided by the industry in affiliated publications or the non-disclosure status of professional process safety training conducted within the industry [16].

2.1 Innovation

A review of existing models and research reveals that most have primarily focused on fundamental principles and components of process safety management (PSM).

Additionally, the majority of studies have centered on the development and improvement of PSM systems, with some focusing on evaluating the implemented systems. This trend indicates that developed countries have largely moved beyond the initial implementation phase of PSM.

However, the topic remains relatively novel in some regions, with most activities concentrated in the oil and gas industry. There has been little significant work in the power plant sector. Therefore, it is essential to develop a quantitative and reliable model specifically for the power plant industry. Such a model would provide a foundation for prioritizing and initiating the implementation of PSM systems in power plants, addressing the unique challenges and requirements of this sector.

3 Methodology

Figure 1 provides a detailed overview of the methodology for enhanced comprehension.

3.1 *First Phase: Gathering Information from Experts*

3.1.1 Questionnaire Design

Monte Carlo probabilistic modeling requires comprehensive data and numerical inputs to calculate the probability of component impacts. Therefore, the questionnaire must thoroughly address process safety management system issues and provide the

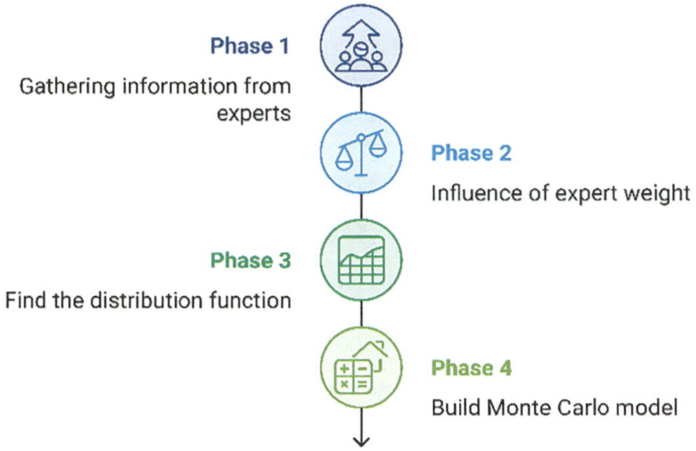

Fig. 1 Overall research methodology chart

quantitative data necessary for modeling. After reviewing existing models for implementing the process safety management system, the OSHA model [2] was selected as the base model for review, questionnaire design, and modeling.

Following consultations with engineers specializing in process safety management, the general design of the questionnaire and the method for examining process safety management components were established [17]. The questions were crafted with the assistance of experts, articles, books, theses, and relevant checklists. Over several days, these questions were reviewed by two university professors and five field experts. After their final approval, the questionnaires were distributed and collected.

The questionnaire for this research consists of two sections: general and specialized:

- **General Questions**: This section assesses the demographic characteristics of the sample population, such as position, education, and work experience.
- **Specialized Questions**: This section includes a researcher-made questionnaire for Monte Carlo modeling. The purpose of this questionnaire is to identify and determine the effectiveness of each component and sub-component on a scale of importance from zero to one hundred percent. It covers 14 main components and 41 sub-components, as described in Fig. 2.

Fig. 2 Project risks related to HSE management at the macro level

3.1.2 Data Collection

In this research, data and information were collected using two methods: library and field. The library method involved gathering theoretical foundations and research background from accessible articles, books, and theses. Additionally, the field method was employed to collect the necessary research data.

3.2 Validity and Reliability of the Questionnaire

Among the various methods for determining validity, content validity is the most common. To assess content validity, the questionnaire is typically reviewed by professors and experts in the field. In this research, the opinions and consultations of two experienced university professors and three HSE and process safety management specialists were utilized [18].

Before performing any analysis on the collected data and making statistical inferences, the reliability of the measurement tool must be ensured. Reliability refers to the consistency of the results if the research is repeated with the same test [19].

Cronbach's alpha is used to assess the validity of each component by measuring the internal consistency of observed variables. It is suggested that the Cronbach's alpha value should be greater than 0.7 [20]. The closer the obtained percentage is to 100%, the higher the reliability of the questionnaire. Generally, a Cronbach's alpha coefficient above 70% is considered to indicate good and acceptable reliability for the questionnaire [19].

3.2.1 Data Analysis and Weighting

After collecting the data, it must be reviewed and prepared for modeling. Initially, the collected data is validated using the bootstrapping method. Following this, each expert's coefficient is multiplied by their respective opinions.

3.2.2 Estimating Sampling Error of Collected Data Using Bootstrapping

The Monte Carlo probabilistic method requires comparing data with a condition to calculate probability. In this research, the collected data serves as the basis for these conditions. Therefore, the accuracy of the data and its statistics must be examined.

In statistics, bootstrapping is a computational method used to determine the accuracy of estimators derived from sample data. This technique allows for the estimation of almost any statistic from the sample data distribution using a straightforward method. Bootstrapping essentially estimates the characteristics (such as variance) of an estimator by measuring these characteristics in an approximate distribution of the entire sample data. The closer the software output for the standard deviation of the mean and the 5% trimmed mean is to zero, the lower the error in those sample parameters [21].

3.3 The Second Phase: The Influence of Expert Weight

Since the experts who responded to the questions had different levels of education, work experience, and positions, a weighting factor was calculated for each of them and multiplied by their scores. The method for calculating this weighting factor is presented in Table 2.

Here, the scores of all three sections for each individual are summed, and the weight for each expert is calculated according to Formula 1 [22].

$$y_i = \frac{(x_{i-n} - \bar{\mathrm{x}})}{(x_{\max} - \mathrm{x}_{\min})} + 1 \tag{1}$$

Table 2 Weighting criteria for experts

No.	Expert weighting criterion	Benchmark ranking	Score
1	Organizational position	Environmental specialist	1
		Occupational health and operations specialist	2
		Operations/maintenance manager	3
		Safety specialist	4
		HSE manager	5
2	Education	BSc	1
		MSc	2
		Ph.D	3
3	Experience	Less than 5 years	1
		5–10 years	2
		11–15 years	3
		16–20 years	4
		Over 20 years	5

3.4 The Third Phase: Find the Distribution Function of the Data

To implement statistical methods and calculate appropriate test statistics and logical conclusions about research hypotheses, the most important step is to choose the appropriate statistical method for the research. Understanding the data distribution is fundamentally important for this purpose [23].

3.5 The Fourth Phase: Adding Data with R Studio Software and Build Monte Carlo Model

3.5.1 Determining the Range of Data to Build a Monte Carlo Model

Since the number of questionnaire questions exceeds the number of respondents, data augmentation is necessary for modeling. This augmentation is performed using the statistical software R Studio. After data augmentation, randomly generated numbers may fall outside the study range, so they must be removed before modeling. This step is carried out using MATLAB under the following two conditions:

- Data less than the minimum possible value, which is zero.
- Data greater than the maximum possible value, which is 146.

In this method, a condition must be considered to compare the data generated by R Studio with the collected data. If accepted, they are used to calculate the probability. The three conditions considered in this research are as follows:

1. The sum of the sub-components of each main index exceeds 50%, 70%, and 90% of the maximum value that the sub-components can have.
2. The average of the sub-components of the main components of the generated data is greater than the 5% trimmed mean of the raw data of each main index.
3. The average of the sub-components of the main components of the generated data falls within the 95% confidence interval for the mean of the raw data of each main index.

Finally, by dividing the number of successful data points by the total number of data points, the importance of each component is determined. Figure 3 shows the overall flowchart of the Monte Carlo modeling method.

3.5.2 Normality Test of Data

For data augmentation, it is essential to know the distribution function of the data to be generated. Therefore, the existing data is tested for normality using SPSS software. The Shapiro–Wilk test is particularly suitable for data sets with fewer than 50 samples

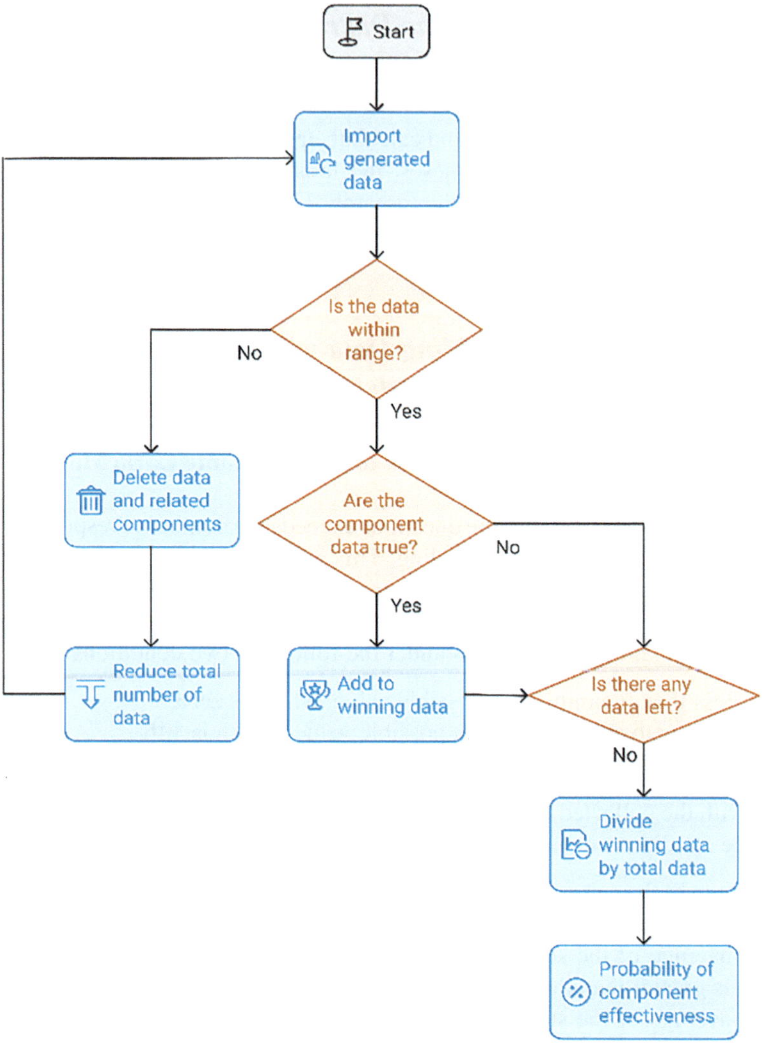

Fig. 3 Overall flowchart of monte carlo modeling [22]

[24]. In this test, the null hypothesis assumes the data is normally distributed, with a test error level of 5%. If the software output (Sig.) for each sub-component shows a value greater than 0.05, the distribution of that sub-component can be assumed to be normal with reasonable accuracy.

3.5.3 Data Augmentation with R Software

To determine the number of data points to generate, one of the components is tested with different data quantities. When the probability of the component's effectiveness converges, the number of data points for augmentation is determined [25]. Given that the collected data do not perfectly follow a normal distribution and the SPSS software output indicated some skewness, the generated data must follow a skew-normal distribution. To achieve this, the skew-normal distribution extension was installed on R Studio, and 2000 data points for each sub-component were generated with a skew-normal distribution using the mean, standard deviation, and delta parameters of the skew-normal distribution [26].

$$\delta = \frac{\alpha}{\sqrt{(1+\alpha^2)}} \tag{2}$$

δ = The delta coefficient of the Azzalini distribution.
$\boldsymbol{\alpha}$ = Skewness Coefficient.

Finding the Number of Data Additions Required for the Research

Given that the statistical population is not sufficient for modeling, the main question that arises is how much data generation is required to prioritize the risks identified by the Monte Carlo modeling method. To address this, one of the components, along with one of the winning conditions (with higher generality), is randomly selected as the basis. Data augmentation continues until the point where increasing the number of generated data no longer significantly changes the obtained result, indicating that convergence has been reached.

3.5.4 Determining the Importance of Each Component

Using MATLAB software, the importance of each component of the OSHA Process Safety Management System for the power industry is determined across three different priorities. In the Copeland method, components are compared pairwise, and in the pairwise comparison matrix, a zero or one is assigned based on preference. If a component in a row is superior to a component in a column in two out of three methods, it is given a value of 1. If it is not superior in two out of three methods, it is given a value of 0. Finally, the numbers in each row and each column are summed. Based on the difference between the sum of each row (number of wins) and the sum of the corresponding column (number of losses), the components are prioritized [27].

4 Result

4.1 Statistical Population

The statistical population of this research consists of experienced specialists, specifically senior managers, from thermal and combined cycle power plants across the country. These specialists were chosen due to their extensive knowledge and experience in managing power plant operations and safety protocols. To ensure a representative sample, six power plants were randomly selected by the Thermal Power Generation Company. These power plants are characterized by their advanced technology, strategic locations, and significant contributions to the national grid. They include facilities known for integrating both gas and steam turbines to enhance efficiency, modern infrastructure, high operational standards, and the incorporation of the latest advancements in power generation technology.

These power plants were selected to provide a comprehensive overview of the industry's current practices and challenges, ensuring that the research findings are robust and applicable across different types of power generation facilities.

4.2 The Demographic Results of the Statistical Sample (General Questions)

In this part, the data related to the general characteristics of the respondents such as: level of education, work experience, and position are described, which are shown in Table 3.

As shown in Table 3, the work experience of most respondents (34%) is more than 20 years, while the least (17%) have between 11 and 15 years of experience. The education level of the majority of respondents (52%) is a master's degree, with the least (2%) holding a PhD. Most respondents (41%) hold the position of Operations/ Maintenance Manager, whereas the least (7%) are Environmental experts.

4.3 Validity and Reliability of the Questionnaire

Cronbach's alpha values for the research components were calculated using SPSS version 26 software. The results, presented in Table 4, demonstrate very high reliability.

Table 3 Demographic results

Level of education	Abundance	Frequency	The cumulative percentage
Bachelor	13	45	45
Master	15	52	97
PhD	1	2	100
total	29	100	–
Work experience	**Abundance**	**Frequency**	**The cumulative percentage**
Less than 5 years	0	0	0
Between 5 and 10 years	8	28	28
Between 11 and 15 years	5	17	45
Between 15 and 20 years	6	21	66
More than 20 years	10	34	100
Total	29	100	–
Organization Position	**Abundance**	**Frequency**	**The cumulative percentage**
Environmental expert	2	7	7
Occupational health and exploitation expert	5	17	24
Operations/Maintenance Manager	12	41	65
Safety expert	6	21	86
HSE manager	4	14	100
Total	29	100	—

Table 4 Cronbach's alpha of the data

N of Items	Cronbach's alpha
41	0.965

4.4 Estimation of Sampling Error

As shown in Table 5, "Employee Participation" has the lowest mean standard deviation error and 5% trimmed mean error, while "Trade Secrets" has the highest. These errors pertain to the sampling of components affecting the process safety management system in the power plant industry.

4.5 Construction of Monte Carlo Research Model

As described in part 3 on how to build the Monte Carlo model for the research, after collecting the required data (first phase), the level of expertise of each expert is calculated using formula 1 (second phase). The values are shown in Table 6.

Table 5 Sampling error by bootstrapping method

No	Component	Standard deviation of the mean	Standard deviation of the 5% trimmed mean
1	*Employee Participation*	*1.6*	*1.75*
2	Process Safety Information	2.07	2.14
3	Operating Procedures	2.53	2.68
4	Process Hazard Analysis	2.34	2.5
5	Emergency Planning & Response	2.92	3.06
6	Training	2.61	2.8
7	Contractors	2.83	2.97
8	Pre-Startup Safety Review	2.48	2.67
9	Mechanical Integrity	2.58	2.72
10	Management Of Change	2.54	2.68
11	Hot Work Permit	2.23	2.27
12	Incident Investigation	2.04	2.18
13	*Trade Secrets*	*3.2*	*3.36*
14	Compliance Audit	2.95	3.2

Table 6 Weighting coefficient for each expert

No	Expert weight calculated	No	Expert weight calculated	No	Expert weight calculated
1	0.66	11	0.96	21	0.96
2	1.06	12	0.66	22	0.96
3	0.96	13	1.06	23	0.86
4	1.36	14	1.06	24	0.86
5	0.86	15	1.06	25	1.26
6	0.66	16	0.96	26	1.06
7	1.36	17	1.06	27	1.36
8	0.76	18	0.96	28	1.16
9	1.06	19	1.46	29	0.76
10	1.06	20	0.96	Average	1.0000063

To find the distribution function of the data (third phase), one of the most common tests is the data normality test. The Shapiro–Wilk test is more efficient for data sets with fewer than 50 samples [28]. Since the experts may have given high or low scores to the questions, the distribution of the data may be somewhat skewed. Therefore, calculating the skewness of the data is also very important. Once the skewness is determined, the appropriate distribution for data augmentation is identified. A portion of the normality test results is presented in Table 7.

All factors have a normal distribution with a probability of 95%, and a Sig value above 0.05 indicates a normal data distribution. Considering that the data distribution is normal with some skewness, the mean, standard deviation, and delta (Azzalini function parameter—Skew normal distribution function) are needed for data augmentation, which is calculated in formula 2 [29].

$$\delta = \frac{\alpha}{\sqrt{(1+\alpha^2)}} \tag{3}$$

δ = Azzalini parameter delta coefficient.

α = Skewness coefficient.

Table 8 presents some of the necessary parameters for data augmentation and running the Monte Carlo model (fourth phase) using R STUDIO software.

As shown in Fig. 4 and Table 9, with the increase in the number of random numbers generated for the component (employee participation), the probability value has reached convergence. This means that the answer to the problem does not change significantly with the increase in the amount of data. The probability value of the data created from 1000 onwards has reached convergence in the most general condition of the Monte Carlo method (with a probability of 50% winning).

As mentioned in the fourth phase of Monte Carlo modeling in part 3, the highest range value is 146 and the lowest range value is 0. Therefore, if the data created by

Table 7 Shapiro–Wilk normality test

Sub-components	Statistic	df	Sig	Sub-components	Statistic	df	Sig
F1	0.953	29	0.240	F6	0.958	29	0.309
F2	0.964	29	0.434	F7	0.965	29	0.457
F3	0.953	29	0.232	F8	0.979	29	0.829
F4	0.934	29	0.078	F9	0.974	29	0.700
F5	0.945	29	0.150	F10	0.966	29	0.481

Table 8 Part of R STUDIO software inputs

No.	Average	Standard deviation	Delta Azzalini function parameter	No.	Average	Standard deviation	Delta Azzalini Function Parameter
1	81.196	23.072	0.368	6	68.555	19.4	0.454
2	73.936	17.893	0.211	7	75.918	26.002	−0.391
3	71.32	27.073	−0.548	8	74.286	22.412	0.381
4	81.501	19.811	0.482	9	75.743	23.801	0.369
5	74.208	21.654	0.502	10	73.555	21.006	−0.413

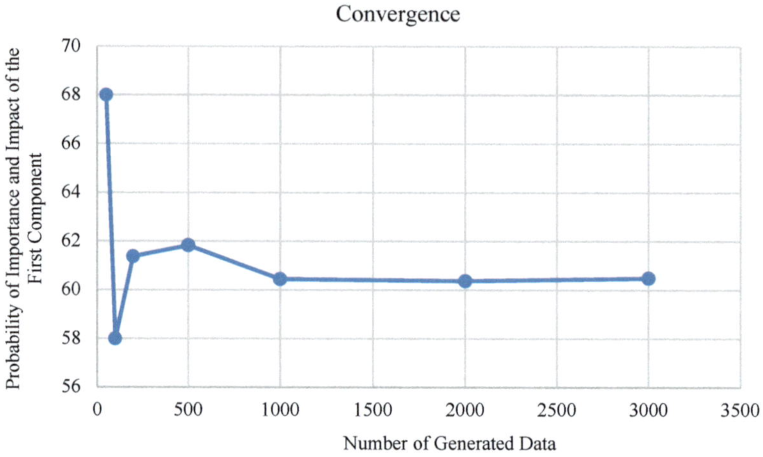

Fig. 4 Plot of calculated probability values for different constructed data counts

Table 9 Convergence of monte Carlo results

No.	Amount of generated data	50% Probability of winning in monte Carlo simulation
1	50	68
2	100	58
3	200	61.37
4	500	61.82
5	1000	60.44
6	2000	60.37
7	3000	60.4

the R STUDIO software, based on the variables of the normal distribution with some skewness, falls outside this range, it should be removed, as it is not within the scope of the primary research data.

4.5.1 The Results of Monte Carlo Modeling with Three Conditions

The impact of each component of the process safety management system in the power plant industry, within the range of data under three different conditions, was determined by Monte Carlo modeling. The results are shown in Table 10.

In the first condition, "Process Hazard Analysis" had the highest impact at 88.8%, while "Trade Secrets" had the lowest impact at 56.16%. In the second condition, "Employee Participation" had the highest impact at 86.2%, and "Trade Secrets" had the lowest at 32.57%. In the third condition, "Hot Work Permit" had the most

Table 10 Research components

No	Component	Importance level		
		Condition 1	Condition 2	Condition 3
1	Employee participation	77.4	*86.2*	***42.59***
2	Training	71.17	84.6	61.08
3	Change management	78.8	75.9	46.87
4	Operating procedures	72.95	74.64	51
5	Process safety information	72.65	73.8	43.75
6	Pre-startup safety review	57.92	66.92	56.76
7	Contractors	68.05	60.55	47.75
8	Mechanical integrity	83.15	58.25	42.72
9	Hot work permit	77.79	55.71	*62.85*
10	Auditing	57.16	54.03	56.75
11	Process hazard analysis	*88.8*	50.23	48.05
12	Emergency response	60.73	48.95	58.55
13	Incident investigation	58.68	40.7	61.9
14	Trade secrets	***56.16***	***32.57***	50.4

significant impact at 62.85%, whereas "Employee Participation" had the least impact at 42.59% in the process safety management system in the power plant industry.

4.6 Integration of Research Modeling Results

As shown in Table 11, the components of Change Management, Training, and Hot Work Permit are ranked equally and hold the highest priority. Following these, Employee Participation and Operating Procedures are at the next priority level. Finally, Auditing and Trade Secrets are ranked at the lowest priority level.

5 Discussion and Conclusion

Recent incidents in the oil, gas, and petrochemical industries, along with their irreparable consequences, serve as a serious warning for process industries, highlighting the urgent need for enhanced safety measures. Addressing this need requires multiple components, including dedicated research in this field and the involvement of experienced and capable specialists and experts. Process Safety Management (PSM) is a key strategy for reducing incidents and their consequences. It has long been considered essential in process industries in industrialized countries and has recently gained emphasis from officials and experts in various regions as well.

Table 11 Copeland prioritization

Copeland score	Component	Priority
9	Change management	1
	Training	
	Hot work permit	
5	Employee participation	2
	Operating procedures	
3	Process hazard analysis	3
1	Process safety information	4
−1	Mechanical integrity	5
−3	Pre-startup safety review	6
	Contractor management	
−5	Emergency response	7
−7	Incident investigation	8
−8	Auditing	9
−11	Trade secrets	10

*It is worth mentioning that the negative scores created in the final result are due to the difference between the rows (the score of two-by-two wins) and the columns (the score of two-by-two losses). This means that the components that have a negative score have more defeats than victories and are therefore placed in a lower priority

The first section of this research introduced the general scope and provided explanations about PSM and power plants, underscoring the importance and necessity of this study. The second section reviewed the literature and published studies on PSM, forming a crucial step for the research framework. After reviewing the existing studies, the innovative aspects of this research were presented. The third section detailed the research community and sample, data collection methods and tools, and the techniques used in data analysis to achieve the research objectives. This was followed by the results of data analysis, starting with descriptive statistics of the collected data. In the fourth section, the final analytical results were displayed in tables using software. Finally, this section provided a discussion and conclusion of the results, summarizing the findings.

The ultimate goal of this research was to develop a PSM model tailored for the country's power industry. Given that the effectiveness of this system is highly dependent on the specific industry in which it is implemented, a model with the following priorities was developed for the power industry, inspired by the OSHA model and the insights of power industry experts.

One of the advantages of the prioritization method used in this study is its applicability to prioritize other systems and components by collecting the necessary data, not limited to PSM. The results indicated that the components of Change Management, Training, and Hot Work Permit, with a focus on preventing incidents from the ground up, scored 9 and were ranked at the highest priority level above other components. This means that for the implementation of this system in power plants, the correct and effective execution of these components is mandatory. Next, the components of Employee Participation and Operating Procedures, both scoring 5, were ranked at the next priority level. Since the procedures need to be implemented by company personnel after being drafted, it is logical for these two to be at the same priority level. The component ranked next with a score of 4 is Process Hazard Analysis, which largely includes risk assessment and management.

Following this, the components of Process Safety Information and Mechanical Integrity, with scores of 1 and -1 respectively, are at the subsequent priority levels. The information obtained in the Process Safety Information component plays a crucial role in risk assessment and management. Therefore, although it is ranked next to Process Hazard Analysis, the system is required to implement it.

Other components, with scores ranging from −2 to −11, include Pre-Startup Safety Review, Contractor Management, Emergency Response, Incident Investigation, Auditing, and Trade Secrets, respectively. The lower priorities for these elements compared to others do not mean they are not required to be implemented. Instead, it indicates that, based on management planning and the available budget in the power plant, their implementation can be scheduled for later stages. In other words, since it is not feasible to implement all 14 elements of the PSM system simultaneously in any organization, this model, with the help of organizational strategics and company management, can provide a sequence for starting the implementation of the PSM system in the power industry.

References

1. Heydari SG, Ghasemi A (2018) Guide to implementing process safety management system in oil, gas, petrochemical, mining and power plants industries, Second. Hak publishing company, Tehran
2. Occupational Safety and Health Standards (1910) United state of America: occupational safety and health administration, department of labor
3. Guss D, Solutions DS (2013) Developing a robust PSM framework at DuPont
4. Risked Based Process Safety Overview (2014)
5. Guidance on meeting expectations of EI process safety management framework. Element 12, Management of change and project management. Energy Institute (2015)
6. Zhang J, Ren H, Ren H, Chai Y, Liu Z, Liang X (2023) Comprehensive review of safety studies in process industrial systems: concepts, progress, and main research topics. Multidisc Digital Publishing Institute (MDPI). https://doi.org/10.3390/pr11082454
7. Bridges WG (1994) Cost and benefits of process safety management: industry survey results. Process Saf Prog 13(1):23–29. https://doi.org/10.1002/prs.680130115
8. Yazdi M, Mohammadpour J, Li H, Huang HZ, Zarei E, Pirbalouti RG, Adumene S (2023) Fault tree analysis improvements: a bibliometric analysis and literature review. Qual Reliab Eng Int 39(5):1639–1659
9. myun Kwon H, jin Lee C, Seo D, Moon I (2016) Korean experience of process safety management (PSM) regulation for chemical industry. J Loss Prev Process Ind 42:2–5. https://doi.org/10.1016/j.jlp.2015.10.001

10. Shayan J, GholamNia R (2017) Comparison of PSM process safety management models and selection of the best model for the oil industry. In International conference of HSE experts of oil, petrochemical, steel and cement industries and construction projects, pp 1–17
11. Shariff AM, Aziz HA, Majid NDA (2016) Way forward in Process Safety Management (PSM) for effective implementation in process industries. Curr Opin Chem Eng 14:56–60
12. Chang JI, Liang C-L (2009) Performance evaluation of process safety management systems of paint manufacturing facilities. J Loss Prev Process Ind 22(4):398–402. https://doi.org/10.1016/J.JLP.2009.02.004
13. Khan F, Rathnayaka S, Ahmed S (2015) Methods and models in process safety and risk management: past, present and future. Process Saf Environ Prot 98:116–147. https://doi.org/10.1016/j.psep.2015.07.005
14. Swuste P, Theunissen J, Schmitz P, Reniers G, Blokland P (2016) Process safety indicators, a review of literature. J Loss Prev Process Ind 40:162–173. https://doi.org/10.1016/j.jlp.2015.12.020
15. Halim SZ, Mannan MS (2018) A journey to excellence in process safety management. J Loss Prev Process Ind 55(April):71–79. https://doi.org/10.1016/j.jlp.2018.06.002
16. Chenani KT, Zarei E, Yazdi M, Klockner K, Alimohammadlou M, Kamalinia M (2024) A systematic review of the integration between occupational and process safety risk analysis methodologies. J Loss Prev Process Ind 91:105387
17. Soczek CA (2013) Implementation of process safety management into diverse corporate cultures. In 2009 AIChE spring national meeting and 5th global congress on process safety, conference proceedings (Topical 3: Center for Chemical Process Safety (CCPS) 24th International Conference, Session 49-Process Safety Metrics and Culture) (2013)
18. Saied A, Ali N, Alireza N (2008) Comprehensive description of research methods (quantitative, qualitative and combinational paradigms, strategies, plans and approaches). Tolid Danesh Tehran
19. Hafiznia M (2008) Introduction to research methods in humanities. Samat Publications, 8th [Text in Persian], Tehran
20. Li H, Yazdi M (2022) Integration of the Bayesian network approach and interval type-2 fuzzy sets for developing sustainable hydrogen storage technology in large metropolitan areas. Advanced decision-making methods and applications in system safety and reliability problems: approaches, case studies, multi-criteria decision-making, multi-objective decision-making, fuzzy risk-based models. Springer International Publishing, Cham, pp 69–85
21. Wu CFJ (1986) Jackknife, bootstrap and other resampling methods in regression analysis. The Ann Stat 14(4):1261–1295
22. Gholami A, Mozaffari M, Pirbalouti RG (2024) Prioritization of HSE management risks in the management of construction projects (Non-Mechanized Tunnel) by monte Carlo modeling method. In Progressive decision-making tools and applications in project and operation management: approaches, case studies, multi-criteria decision-making, multi-objective decision-making, decision under uncertainty, Springer, pp 191–219
23. Mezei M (1987) Adaptive umbrella sampling: self-consistent determination of the non-Boltzmann bias. J Comput Phys 68(1):237–248
24. Bakbaki A, Nabhani N, Anvaripour B, Shirali G (2017) Probabilistic risk assessment using fuzzy fault tree analysis based on two types of failure possibility distributions in process industries. J Occupat Hygiene Eng 4(2):41–52
25. Jandaghi G, Gayini A (2002) Monte Carlo simulation method and its applications in statistics and probability. Stat Thought J 6(2):51
26. Bruce HP, Adrian T, Alexander VE (2015) Structural equation modeling: application in ecological and evolutionary biology. Cambridge University Press, United Kingdom
27. Boyle P, Broadie M, Glasserman P (1997) Monte Carlo methods for security pricing. J Econ Dyn Control 21(8–9):1267–1321
28. Bartels M, Karplus C (1998) Probability distributions for complex systems: adaptive umbrella sampling of the potential energy. J Phys Chem B 102(5):865–880

29. Li H, Yazdi M (2022) Advanced decision-making Neutrosophic fuzzy evidence-based Best-Worst Method. Advanced decision-making methods and applications in system safety and reliability problems: approaches, case studies, multi-criteria decision-making, multi-objective decision-making, fuzzy risk-based models. Springer International Publishing, Cham, pp 153–184

Prioritizing HSE Management Risks in Bridge Construction Projects Using Monte Carlo Modeling

Alireza Amanzadeh, Adel Gholami and Mahdi Mozaffari

Abstract Bridge construction projects, particularly those within large-scale infrastructure developments, represent some of the most complex and risk-intensive activities in the global construction industry. Their dynamic nature, driven by uncertainty, time pressure, and environmental variability, necessitates the adoption of intelligent and data driven HSE (Health, Safety, and Environment) management systems to ensure resilience and operational safety. This study introduces a smart HSE risk prioritization model based on Monte Carlo probabilistic simulation, designed to support evidence-based decision-making in high-risk bridge construction environments. The research was conducted on a freeway megaproject comprising 25 major bridges with a total span of 3137 m, where numerous HSE challenges such as adverse weather conditions, working at heights, and fatal occupational incidents were observed. Expert judgments were collected using structured questionnaires and refined through the Delphi technique to identify critical safety factors. The Monte Carlo simulation method was applied to generate 2000 probabilistic scenarios, capturing uncertainty across multiple HSE dimensions. The resulting data were analyzed using the Copeland prioritization approach, providing a weighted ranking of risk categories based on their relative impact on project safety outcomes. The findings revealed that personnel-related and materials and equipment risks exerted the highest influence on overall HSE performance, while political and legal risks ranked lowest. These results emphasize the need for continuous organizational engagement across all project phases, from planning and design to contractor management and site operations, to enhance safety

A. Amanzadeh · A. Gholami (✉) · M. Mozaffari
Department of Civil and Environmental Engineering, Amirkabir University of Technology (AUT), Tehran, Iran
e-mail: adel.gh@aut.ac.ir

A. Amanzadeh
e-mail: hseman@aut.ac.ir

M. Mozaffari
e-mail: mahdimozaffary@aut.ac.ir

H. Li et al. (eds.), *Data-Driven Methods for Reliability and Safety Engineering: Applications in Industrial Systems*, Springer Series in Reliability Engineering,
https://doi.org/10.1007/978-3-032-22873-4_46

maturity and system reliability. By integrating probabilistic modeling with structured safety frameworks such as HSE-MS and Integrated Management Systems (IMS), this study demonstrates how traditional safety practices can evolve into smart, adaptive systems capable of learning from data and anticipating risks in real time. The approach illustrates a paradigm shift from reactive hazard control to proactive, intelligent safety management, supporting the broader vision of smart construction ecosystems where human expertise and computational intelligence work in synergy to prevent accidents and optimize project outcomes.

Keywords HSE management · Monte Carlo simulation · Bridge construction · Probabilistic risk modeling · Intelligent decision-making · Smart safety systems · Predictive analytics

1 Introduction

Construction and civil engineering operations, due to their extensive nature, are among the most hazardous activities in most countries worldwide. The presence of various hazards and harmful factors in construction sites has made this industry one of the most dangerous industries globally. This industry accounts for about 7% of the world's jobs but is responsible for 30–40% of global fatalities [1]. Additionally, the review of construction accidents indicates that unsafe actions account for approximately 80 to 90% of the causes of these accidents and injuries [2]. Compared to other construction projects, the bridge construction process, due to its structural nature (working at heights, surrounding conditions, uncertainties, etc.), is very complex and systematic and faces numerous risks at all stages of construction [3].

In construction projects, there is a wide range of risks. In this regard, risk management in construction projects holds a special place and has many common roots with this topic, being one of the main components of project management to achieve project goals. Characteristics such as the specific conditions of each construction project and uncertainties in assumptions, goals, requirements, and project standards, economic conditions, and consequently the environmental factors governing the project, are among the factors causing risks in projects [4].

In 2008, Axorn and Hadixomo, through a study using factor analysis, showed that the critical success factors in safety risk management could be grouped into four main categories: 1-Worker participation, 2-Safety prevention and control system, 3-Safety arrangement, and 4-Management commitment [5]. An integrated safety management system can prevent repetitive efforts and reduce resource inputs [6]. Overall, risk management involves identifying and applying measures that guide risk to an acceptable level and leave significant results in the project's output. This is effective and beneficial when the strategy for responding to critical risks is also effective. Therefore, choosing an appropriate solution, tool, and strategy for responding to risks is essential. Additionally, providing logical and targeted solutions to reduce and manage risks requires a proper understanding of the organization's current situation and risks. Thus, by correctly and timely identifying and assessing health, safety, and environmental risks in bridge construction projects and properly analyzing and managing them, it is possible to reduce related accidents and consequently increase productivity and achieve project goals. Among these, identifying health, safety, and environmental (HSE) risks as part of the risk management process is of particular importance.

Regarding the identification of influential HSE risks in construction projects, including bridge construction, studies have been conducted in this field for decades. In 1999, Savacha and colleagues, in a study titled "Factors Influencing Risk Management Performance in Construction Projects," showed that the five important topics related to project safety risks are: 1-Safety management issues, 2-Preparation of safety regulations, 3-Preparation of safety equipment, 4-Provision of a safe environment, and 5-Appointment of a trained safety representative on site [7]. In 2012, Anthony Lopez Arcolus, Joan Carlos Rubioromero, and Alistair Gibb, in a study examining construction industry accidents, showed that the severity of accidents in projects is related to risk factors such as age, company size, length of service, accident location and day of the week, days of absence, type of injury, and the climate of the region [8] Seasonal changes (weather) also affect construction accidents [9]. Additionally, in 2017, Bozen Halo, Thomas Nobilshki, and colleagues divided the most important risk factors affecting the high rate of accidents in the construction industry into three groups: 1-Surrounding environment, 2-Interrelationships between identified risks, and 3-Relationship between construction costs and accident rates in the construction industry, shown using correlation coefficients [10]. A year later, Korea, in a study, showed that 1-Lack of awareness and safety culture among workers and project managers, 2-Cost of safety hazard prevention methods, 3-Insufficient attention to HSE issues by the company, and 4-Time pressure to complete the project, are among the most important safety risks and HSE risk management in projects [11]. In 2020, according to a study titled "Identifying Critical Safety Hazards in Bridge Construction Projects in Marine Environments Based on HMM and SNA," Wang and colleagues showed that considering the dimensions of hazard sources and control errors, the most important safety risks in bridge construction projects are 1-Lack of awareness of safety issues by managers, 2-Lack of safety skills, 3-Low maturity of key technologies, and 4-Inability to effectively track and respond to environmental changes [12].

By reviewing past studies and the relationship between accidents and risks, the combination of existing hazards in the project, and the type of construction injuries, and finally by correctly identifying bridge construction risks based on the needs and conditions of each project, safety managers can prioritize hazards based on the likelihood of accidents and the characteristics of expected injuries, and pay more attention to risk relationships to prevent accidents and achieve a safer work environment. On the other hand, safety in construction discussions requires care and planning throughout the project's life from design to maintenance [13]. In this regard, the design and risk assessment of the bridge are discussed, and in terms of risk acceptance and safety management, the use of multiple criteria to reduce risk and make appropriate decisions, and which criteria are more suitable for risk acceptance should be examined.

The validity of the approach to identifying safety, health, and environmental risks lies in its ability to 1-Understand cause-and-effect relationships among stakeholders, 2-Identify key roles in construction safety performance management, and 3-Reveal the transfer of duty and responsibility among different stakeholders, which can help allocate the management tasks of these risks among different stakeholders and consequently facilitate the improvement of safety performance in the construction industry [14]. On the other hand, the higher the percentage of preventive operations implemented in a strategy, the lower the casualty rate. However, the analysis of the combined effect of preventive measures showed that choosing the right combination of risk management practices is more important than the number of preventive operations performed [15].

Although recent studies highlight the significance of HSE (Health, Safety, and Environment) risks in bridge construction, the process of identifying, categorizing, and prioritizing these risks remains a notable gap in the research. This study aims to enhance the execution of bridge construction projects by developing a model to prioritize and categorize identified HSE risks within project risk management. Effective prioritization and categorization of these risks are crucial for controlling numerous HSE incidents through proper and timely planning. A major challenge in achieving this study's objectives is the limited number of experts, particularly in developing countries, and the varying levels of expertise among them in the HSE aspects of bridge construction projects. This study addresses these limitations, which will be elaborated upon further.

1.1 Research Structure

This study is structured into four main sections. The first section provides an overview of the research, including a preliminary background on the proposed study. It outlines the approach, objectives, necessity, and limitations by reviewing relevant literature and published studies on risk assessment and HSE management in bridge construction, which is crucial for the research framework and innovation. The second section details the research framework, including the research method, population and sample, data collection methods and tools, and the data analysis techniques used to achieve the research objectives. The third section presents the results obtained from the data analysis. Finally, the fourth section discusses and concludes the findings for each research question, summarizes the overall results, and offers practical suggestions and directions for future research.

2 Methodology

Figure 1 illustrates the methodology for improved understanding.

2.1 *Step One: Data and Information Collection*

2.1.1 Methods and Tools for Data Collection

This research employed two primary methods for data and information collection: library research and field research. The library research involved gathering theoretical foundations and background information from accessible Persian and English articles, books, and theses. For the field research, insights were obtained from two university professors and three executive experts specializing in HSE within bridge construction projects.

(A) Preparation of the Initial Questionnaire

The risks affecting HSE management in bridge construction projects were identified in 12 main components as follows and categorized into 101 factors as shown in Table 1.

1. Natural Risks [16–20]
2. Environmental and Biological Risks [18, 21]
3. Political and Legal Risks [16–18, 22]

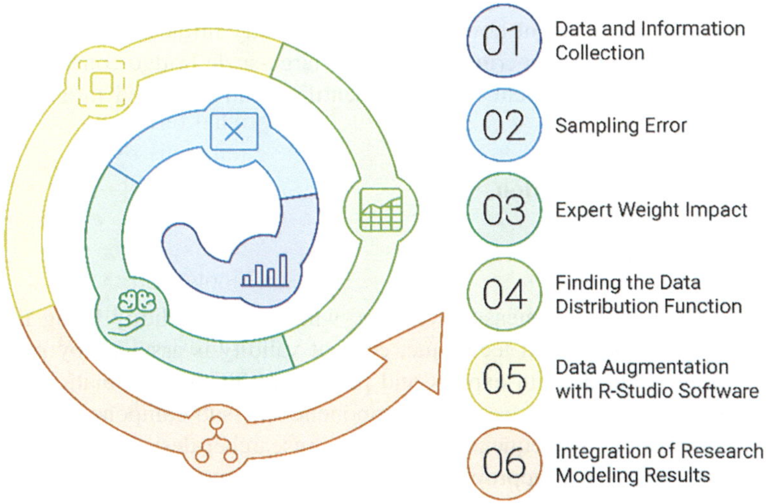

Fig. 1 Overview of methodology

Table 1 Expert weighting criteria

No.	Expert weighting criterion	Benchmark ranking	Score
1	Age	Less than 30 years	1
		31–40 years	2
		41–50 years	3
		50 years and up	4
2	Education	BSc	1
		MSc	2
		Ph.D	3
3	Experience	Less than 5 years	1
		5–10 years	2
		11–15 years	3
		16–20 years	4
		Over 20 years	5
4	Employment	Expert/officer	1
		Supervisor/consultant	2
		Manager	3

4. Economic and Financial Risks [16–18, 22]
5. Management and Planning Risks [16, 18, 20–23]
6. Contractual and Contractor Risks [16, 20, 22]
7. Social Risks [17, 18]
8. Personnel Risks [16–18, 23, 24]
9. Construction and Execution Risks [16–18, 22]
10. Material and Equipment Risks [16, 19, 20, 25]
11. Technical, Design, and Engineering Risks [16, 18, 20, 22–25]
12. HSE Team-Related Risks [18, 22, 25]

Various references highlight how these factors greatly influence the management of HSE risks in civil engineering projects. In large-scale bridge construction projects, these HSE management risks are identified and classified into 12 distinct categories.

(B) Questionnaire Validation

(B-1) Validity of the Questionnaire

Several methods exist to measure validity, with content validity being the most prevalent. Using the Delphi technique, content validity is assessed by presenting the questionnaire to domain experts and professors for their evaluation. Initially, experts provided feedback on each component and sub-component through an open questionnaire. Subsequently, a closed (researcher-designed) questionnaire was used to assess the appropriateness of these components and sub-components. This standardized questionnaire was then employed for data collection in the Monte Carlo modeling method. For this study, input was gathered from two university professors (Faculty of Civil and Environmental Engineering) and three executive experts in the HSE field of bridge construction projects.

(B-2) Reliability of the Questionnaire

Prior to performing any analysis on the gathered data and making statistical inferences, it is essential to first verify the reliability of the measurement instrument [26]. Reliability indicates that repeating the research with the same test will yield identical results. Cronbach's alpha assesses the internal consistency of the observed variables. The nearer the obtained percentage is to 100%, the greater the reliability of the questionnaire. Typically, a Cronbach's alpha coefficient exceeding 70% is regarded as demonstrating good and acceptable reliability for the questionnaire [27].

(C) Questionnaire and Response Method

Ultimately, a standard questionnaire was developed, comprising two sections: general (demographic attributes) and specialized (12 components with a total of 101 questions on a 5-point Likert scale). To meet the project objectives using the Monte Carlo modeling method for sub-component weighting, experts were initially asked to rate the components using the numbers {1,2,3,4,5}, where 1 represented low importance and 5 represented very high importance. The collected data was then input into SPSS software for further analysis.

2.1.2 Methods and Tools for Data Analysis

In the data analysis section, following the description of the sample's demographic characteristics and the identification of key performance components (project risks) in the evaluation and management of bridge construction project risks using Delphi and library techniques, the sampling error is initially estimated using the Bootstrapping method with SPSS software. Subsequently, SPSS, Excel, R Studio, and MATLAB software are employed for Monte Carlo modeling, and the Copeland method is utilized to integrate the three prioritization methods.

2.2 Step Two: Sampling Error

The resampling method, utilizing the current sample, aims to more accurately estimate the parameters of the statistical population or determine the error of the estimator [28]. In this study, the Bootstrapping method was employed using SPSS software to estimate the sampling error for prioritizing 12 categories in the Monte Carlo model. This was done with 95% confidence and N (the rationale for

N sampling is detailed in the fourth stage of Monte Carlo model construction), utilizing the BCa confidence interval type for greater accuracy and bias reduction, along with a simple random sampling method. The closer the Bootstrapping value of each of the 12 main components is to zero, from the perspectives of 1-Standard deviation of the mean and 2–5% trimmed mean, the lower the error will be.

2.3 Step Three: Expert Weight Impact

Considering that experts possess diverse criteria such as experience, education, age, and organizational position, a relative weighting factor was established based on these criteria and Likert scale scoring, as illustrated in Table 1.

To consider the weight of experts and rank them, all weights are initially summed. Then, to reflect the varying opinions of each expert according to their expertise level, the weights are integrated into the data using Formula 1. This ensures that the average weight of the experts equals one, maintaining the integrity of the data while highlighting differences in expert responses based on their expertise. Additionally, this process converts the 5-point Likert scale data from discrete to continuous, facilitating statistical analysis [29].

$$y_i = \frac{(x_{\iota-n} - \bar{\mathrm{x}})}{(x_{\max} - \mathrm{x}_{\min})} + 1 \tag{1}$$

Subsequently, each expert's weight is multiplied by their responses to the questions, thereby reflecting the influence of their expertise on the provided answers.

2.4 Step Four: Finding the Data Distribution Function

To apply statistical methods, compute the appropriate test statistics, and draw logical conclusions about the research hypotheses, the crucial initial step is selecting the suitable statistical method for the study. Understanding data distribution is fundamentally essential for this purpose [30].

2.5 Step Five: Data Augmentation with R-Studio Software

In this step, to generate N random data points based on the distribution function derived in the previous step using R-Studio software, it is necessary to specify and calculate the parameters.

2.5.1 Determining the Data Range for Monte Carlo Modeling

In this step, numbers that fall outside the range of the highest and lowest data values are initially removed to exclude outliers from the measurement criteria.

- Highest value = Maximum possible expert weight × 5 (very high importance)
- Lowest value = Minimum possible expert weight × 1 (very low importance)

Then, Monte Carlo modeling is performed under the following three conditions:

1. A 50% probability of success such that the sum of the sub-components of each main index exceeds 50% of the maximum value that the sub-components can have.
2. A higher probability of success where the mean of the sub-components of the main components of the generated data is greater than the 5% trimmed mean of the raw data for each main index.
3. A probability of success where the mean of the sub-components of the main components of the generated data falls within the 95% confidence interval for the mean of the raw data for each main index, refer to Fig. 2.

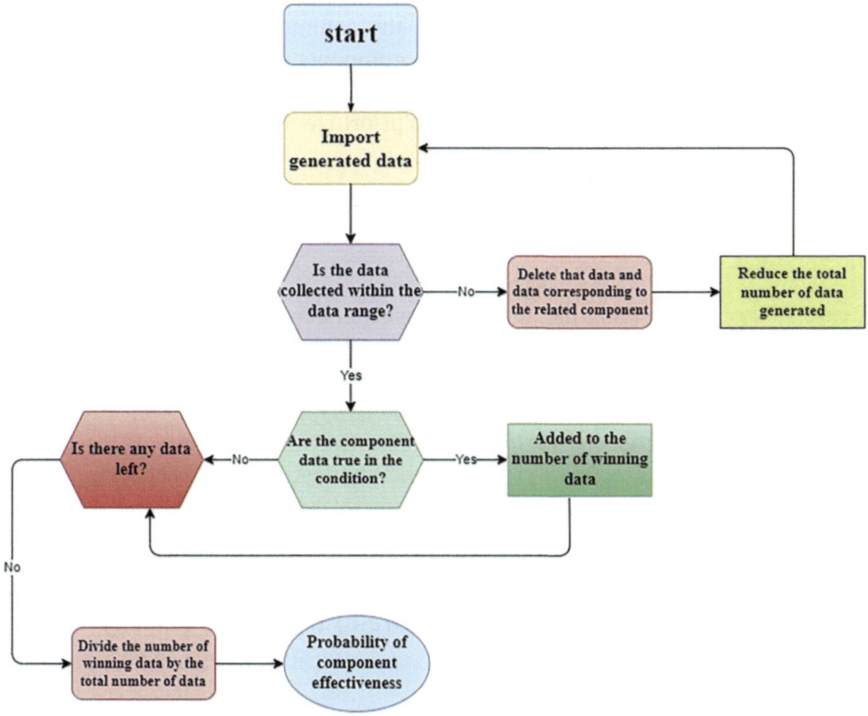

Fig. 2 Overall flowchart of monte Carlo modeling [29]

2.5.2 Determining the Required Amount of Data Augmentation for the Research

Considering the insufficiency of the statistical population for modeling, the primary question is the extent of data augmentation required for prioritizing the identified risks using the Monte Carlo modeling method. To address this, one component and one success condition (with broader applicability) are randomly chosen as the basis. Data augmentation proceeds until the results stabilize, indicating convergence, as further augmentation does not significantly alter the outcomes.

2.6 Step Six: Integration of Research Modeling Results

Due to the varying prioritizations produced by the three success condition methods in Monte Carlo modeling, the Copeland method is employed for aggregation. This approach begins with the creation of a pairwise comparison matrix for decision-making. When an index is preferred over another more frequently than it is dominated, a value of 1 is assigned in the matrix. Conversely, if there is no clear majority or the preferences are equal, a value of 0 is assigned. A value of 1 signifies that the row index is preferred over the column index, while a value of 0 indicates the opposite. After assessing the components, the pairwise comparison matrix is completed. The sum of each row's elements reflects the dominance frequency of each index, which is then used to prioritize the components [31].

3 Results Obtained

A case study incorporating the necessary research components is essential for this research. A major freeway project was chosen for the following reasons:

The project aims to establish a rapid and cost-effective link between the northern and central regions and to enhance connectivity with neighboring countries to the north. Initially proposed in 1974, this freeway will form part of the 'North–South Transnational Freeway,' providing the shortest route between the Caspian Sea and the Persian Gulf, thereby playing a crucial role in regional transit [32].

3.1 Research Population (Phase One of a Major Freeway Project)

A major freeway project, part of the North–South corridor, faces significant HSE risks. The first section, which includes 25 bridges totaling 3137 m in length, has encountered various HSE challenges such as working at heights, landslides, and

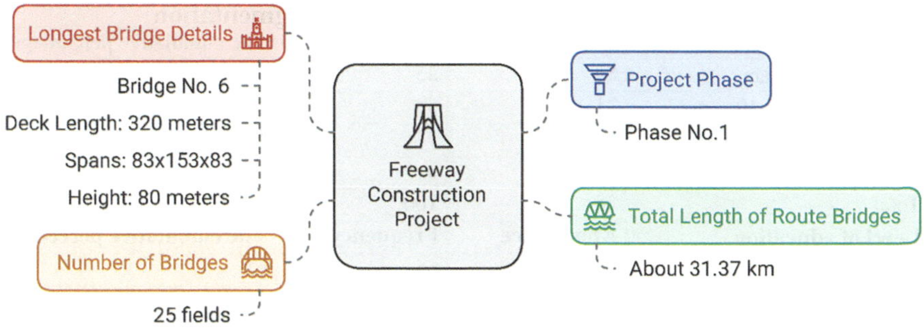

Fig. 3 Characteristics of the case study [33]

fatal occupational accidents. This section, approximately 32 km long, extends from the intersection of two major highways, passing through several areas including tunnels and valleys, and ends at a key junction, refer to Fig. 3.

The research's statistical population includes four active contractors working on the first section of a major freeway project. This group comprises 35 individuals, including HSE specialists from these companies and HSE consultants involved in bridge construction projects with the client company across various developing regions. These individuals have been chosen as the statistical analysis unit for applying the Delphi technique.

3.2 Demographic Results of the Sample Population (General Questions)

This section outlines the data concerning the respondents' general characteristics, including age, education level, work experience, and position, as presented in Table 2.

According to the data in Table 2, the majority of respondents (51%) have 11 to 15 years of work experience, with none (0%) having over 20 years. Most respondents (60%) are aged between 30 and 40 years, with none (0%) being over 50 years. The predominant education level is a bachelor's degree (65%), with no respondents (0%) holding a doctorate. The most common positions are supervisor/consultant and expert, each representing 40%, while the least common position is manager (20%).

3.3 Validity and Reliability of the Questionnaire

The Cronbach's alpha values for the research components were computed using SPSS version 26, as shown in Table 3. The resulting value exceeds 0.7, demonstrating very high reliability.

Table 2 Demographic results

Age	Abundance	Frequency	The cumulative percentage
Less than 30 years	8	23	23
Between 30 and 40 years	21	60	83
Between 41 and 50 years	6	17	100
More than 50 years	0	0	100
Total	35	100	–
Level of education	**Abundance**	**Frequency**	**The cumulative percentage**
Bachelor	23	65	65
Master	12	35	100
PhD	0	0	100
Total	35	100	–
Work experience	**Abundance**	**Frequency**	**The cumulative percentage**
Less than 5 years	7	20	20
Between 5 and 10 years	8	23	43
Between 11 and 15 years	18	51	94
Between 15 and 20 years	2	6	100
More than 20 years	0	0	100
total	35	100	–
Position	**Abundance**	**Frequency**	**The cumulative percentage**
Expert	14	40	40
supervisor/consultant	14	40	80
Manager	7	20	100
total	35	100	–

Table 3 Reliability statistics

N of Items	Cronbach's alpha
101	0.949

3.4 Estimation of Sampling Error

Table 4 reveals that technical, design, and engineering risks exhibit the lowest mean error and 5% trimmed mean error, whereas construction and execution risks show the highest mean error and 5% trimmed mean error in the sampling of components impacting HSE management in bridge construction projects.

3.5 Construction of the Research Monte Carlo Model

As detailed in Sect. 2 regarding the construction of the research Monte Carlo model; after gathering the necessary data (Step One), the expertise level of each expert is assessed (Step Two), as illustrated in Table 5.

Table 4 Sampling error using the bootstrapping method

No.	Components	Mean standard deviation	5% trimmed mean
1	Manpower risks	0.1063	0.06
2	Technical, design and engineering risks	0.0861	0.055
3	Construction and implementation risks	0.1269	0.089
4	Equipment and materials resources risks	0.1079	0.06
5	Contract risks	0.101	0.076
6	Political/legal risks	0.1061	0.085
7	Economic risks	0.1058	0.084
8	Social risks	0.1177	0.086
9	Natural disaster risks	0.1166	0.07
10	Environmental risks	0.0972	0.062
11	Planning risks	0.0967	0.062
12	HSE team risks	0.1177	0.067

Table 5 Expert weight after calculation

No.	Expert weight calculated	No.	Expert weight calculated	No.	Expert weight calculated
1	1.247667	14	0.747667	27	1.414333
2	1.081	15	0.747667	28	1.247667
3	1.081	16	0.747667	29	1.414333
4	0.581	17	1.081	30	1.081
5	0.747667	18	0.747667	31	1.247667
6	0.914333	19	1.247667	32	0.914333
7	0.914333	20	0.581	33	0.914333
8	1.081	21	1.247667	34	1.081
9	1.081	22	1.081	35	1.081
10	0.747667	23	1.581		
11	1.081	24	1.414333		
12	0.747667	25	0.747667		
13	0.581	26	0.747667		

To determine the data distribution function (Step 4), the normality test is commonly used, with the Shapiro–Wilk test being particularly effective for datasets with fewer than 50 samples [34]. Since experts may assign varying scores to the questions, the data distribution could be skewed, making the calculation of skewness crucial. Following this, the appropriate distribution for data augmentation is identified. A portion of the normality test results is shown in Table 6.

All factors display a normal distribution with a 95% confidence level, and a Sig value above 0.05 confirms the normality of the data distribution. Since the data distribution is normal but slightly skewed, data augmentation necessitates the

Table 6 Shapiro–Wilk normality test

Sub-component	Statistic	df	Sig	Sub-component	Statistic	df	Sig.
F1	0.952	35	0.145	F11	0.966	35	0.349
F2	0.935	35	0.043	F12	0.979	35	0.737
F3	0.970	35	0.471	F13	0.960	35	0.250
F4	0.948	35	0.110	F14	0.972	35	0.524
F5	0.955	35	0.176	F15	0.950	35	0.126
F6	0.973	35	0.535	F16	0.948	35	0.107
F7	0.938	35	0.053	F17	0.952	35	0.138
F8	0.970	35	0.454	F18	0.974	35	0.566
F9	0.963	35	0.294	F19	0.944	35	0.080
F10	0.977	35	0.671	F20	0.954	35	0.167

mean, standard deviation, and delta (a parameter of the Azzalini function), as outlined in Formula 2 [35].

$$\delta = \frac{\alpha}{\sqrt{(1 + \alpha^2)}} \tag{2}$$

δ = The delta coefficient of the Azzalini distribution.

$\boldsymbol{\alpha}$ = Skewness Coefficient.

Table 7 displays some of the parameters necessary for data augmentation and the implementation of the Monte Carlo model (Step 4) using R-STUDIO software.

As illustrated in Fig. 4, increasing the number of random numbers generated for the index (e.g., technical, design, and engineering risks) leads to the convergence of the probability of one condition prevailing. This indicates that as the number of data points increases, the solution remains relatively stable. The probability of the generated data has converged from 2000 onwards under the most general condition of the Monte Carlo method (with a 50% probability of winning).

As outlined in Step 5 of the Monte Carlo modeling process in Sect. 2, the highest calculated range value is 7.905, and the lowest is 0.581. Consequently, any data generated by R-STUDIO based on skewed normal distribution variables that fall outside this range should be excluded, as they do not align with the initial research data range.

3.5.1 Results of Monte Carlo Modeling with Three Conditions

The Monte Carlo modeling method was used to determine the impact of each HSE management index for civil engineering projects (bridge construction) within the data range under three different conditions, as shown in Table 8.

In Method 1, HSE team-related risks had the highest impact at 67.25%, while political and legal risks had the lowest at 19.17%. In Method 2, social risks had the highest impact at 49.26%, with construction and execution risks having the

Table 7 A portion of the inputs for R STUDIO software

No.	Average	Standard deviation	Alpha skewness coefficient	Delta Azzalini Function Parameter	No.	Average	Standard deviation	Alpha skewness coefficient	Delta Azzalini function param-eter
1	3.73	1.05	−0.81	−0.63	6	3.82	1.82	−0.61	−0.52
2	4.17	1.22	−0.16	−0.15	7	3.14	1.51	−0.38	−0.36
3	3.87	1.66	−0.55	−0.48	8	3.68	1.06	−0.24	−0.23
4	3.42	1.45	−0.52	−0.46	9	3.59	1.07	−0.37	−0.35
5	3.62	1.36	−0.48	−0.43	10	3.09	1.23	−0.06	−0.06

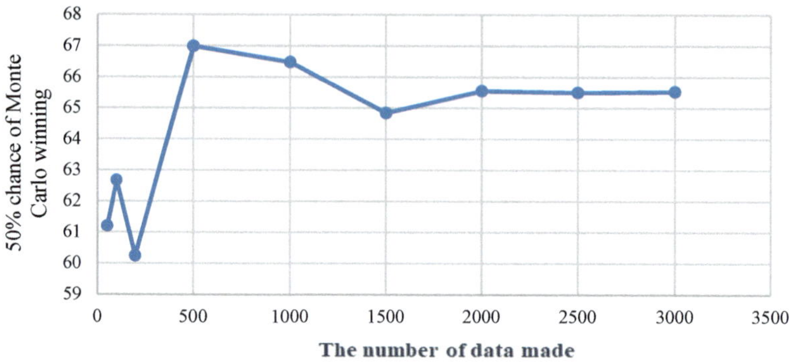

Fig. 4 Chart of calculated probability values for different numbers of generated data

lowest at 12.33%. In Method 3, personnel risks had the highest impact at 74.37%, and environmental risks had the lowest at 41.41% in the HSE management of civil engineering bridge construction projects.

3.6 *Integration of Research Modeling Results*

The results of the Copeland method aggregation are shown in Table 9.

It is important to note that the negative numbers in the result stem from the difference between the rows (number of pairwise wins) and the columns (number of pairwise losses). This indicates that components with negative scores have more losses than wins, resulting in a lower ranking.

4 Discussion and Conclusion

Implementing a safety, health, and environmental management system is a strategic and operational decision for an organization. The success of this system relies on leadership, commitment, and participation at all organizational levels [36]. Performance evaluation is a critical stage in an organization's strategic management system, influencing the entire framework. Thus, analyzing key performance components in HSE management for bridge construction projects is essential for identifying needs and significant issues, boosting motivation, and enhancing HSE management communication within the industry. Understanding the importance of each component is particularly significant. Additionally, the increasing adoption of building information models in engineering and construction is transforming safety, health, and environmental management methods. This technology allows for the automatic identification of potential safety hazards and the implementation of preventive measures through automated processes [13].

Table 8 Research components

No.	Components	50% chance of victory	Average disconnected 5%	95% Confidence interval for the mean
1	Manpower risks	61.6	18.91	74.37
2	Technical, design and engineering risks	65.55	18.54	51.94
3	Construction and implementation risks	56.37	12.33	52.64
4	Equipment and materials resources risks	65.84	13.97	54.16
5	Contract risks	57.8	25.53	50.98
6	Political/legal risks	19.17	26.11	50.31
7	Economic risks	40.76	35.24	48.39
8	Social risks	44.31	49.26	51.28
9	Natural disaster risks	62.53	31.32	42.41
10	Environmental risks	61.27	34.1	41.41
11	Planning risks	55.59	20.09	55.11
12	HSE team risks	67.25	19.71	48.1

Table 9 Prioritization using the copeland method

Priority	Components	Copeland score
1	Manpower risks	7
2	Equipment and materials resources risks	5
3	Technical, design and engineering risks	3
	HSE team risks	3
4	Planning risks	1
	Natural disaster risks	1
	Social risks	1
5	Contract risks	−1
6	Environmental risks	−3
	Economic risks	−3
7	Construction and implementation risks	−5
8	Political/legal risks	−9

This research introduces an analytical approach to identify key components in risk assessment and management for civil engineering projects, specifically bridge construction. In the second step, modern methods are employed to evaluate the effectiveness of these components and prioritize them for managing and controlling related HSE risks. The findings indicate that to ensure safer bridge construction, the following 12 variables are crucial in order of importance: 'Personnel Index' with the highest priority, 'Materials and Equipment Index' with the second priority, 'Technical, Design, and Execution Index' and 'HSE Team-Related components' with the third priority, 'Social components,' 'Natural components,' and 'Management and Planning components' with the fourth priority, 'Contractual and Contractor components' with the fifth priority, 'Environmental components' and 'Economic and Financial components' with the sixth priority, 'Construction and Execution components' with the seventh priority, and 'Political and Legal Risks' with the eighth priority. Among these, technical, design, and engineering risks had the lowest average error, while construction and execution risks had the highest average error and 5% truncated mean error in sampling effective components on HSE management of civil engineering bridge construction projects. This model aligns with HSE management systems such as HSE-MS and ISO standards (ISO 45001: 2018, ISO 14001: 2015, ISO 9001: 2015), demonstrating that effective HSE management relies on leadership, commitment, and participation at all organizational levels, from project approval and structural design to contractor selection, financing, and human resources.

A major freeway megaproject was chosen as a case study due to its technical data and experiences. This project, which includes 25 bridges totaling 3137 m in its first section, faced numerous HSE challenges, including adverse weather conditions, working at heights, and fatal occupational accidents. The methods used in this research, along with the innovative approach to identifying, weighting, and prioritizing risks, provide a valuable HSE risk management model for bridge construction projects applicable to both private and public sectors.

In this context, each party can establish a committee of experienced HSE managers and experts familiar with construction project risks to define and assess the characteristics of identified risks based on the specific project conditions. They can then plan, address, and manage the prioritized risks. Additionally, the method used for prioritizing components allows for the prioritization of other systems and components by collecting high-quality (expert-based) data, even in small quantities, making this method applicable beyond the HSE management domain.

However, from a top-down managerial viewpoint, it is essential to recognize that regardless of human errors, it is the system's responsibility to anticipate and implement control measures before accidents occur to minimize their consequences [29]. From the perspective of lower executive levels, the findings of this study confirm previous research regarding the significant contribution of human factors to occupational accidents.

Based on the findings and analysis of the research questions, it is recommended that a detailed plan be developed to implement the research results in the subsequent phases (2 to 4) of the construction project or similar projects. This approach can facilitate future research for those interested in this field.

References

1. Fany F, Mardani A, Alizadeh A, Abbasian Jahromi H (2015) Investigating the costs of damages caused by accidents in construction projects (a case study of excavation accidents). Tehran
2. Gholamizadeh K, Zarei E, Yazdi M, Amin MT (2024) Data-driven approaches for accident analysis in sociochemical systems. Safety causation analysis in sociotechnical systems: advanced models and techniques. Stud Syst Decis Control 541:457–486
3. Voo YL, Augustin P, Thamboe T (2011) Construction and design of a 50m single span uhp ductile concrete composite road bridge. Struct Eng 89(15):24
4. Winch GM (2012) Managing construction projects. John Wiley & Sons
5. Aksorn T, Hadikusumo BHW (2008) Critical success factors influencing safety program performance in Thai construction projects. Saf Sci 46(4):709–727
6. Yazdi M, Zarei E, Adumene S, Beheshti A (2024) Navigating the power of artificial intelligence in risk management: a comparative analysis. Safety 10(2):42
7. Sawacha E, Naoum S, Fong D (1999) Factors affecting safety performance on construction sites. Int J Project Manage 17(5):309–315
8. Arquillos AL, Romero JCR, Gibb A (2012) Analysis of construction accidents in Spain, 2003–2008. J Safety Res 43(5–6):381–388
9. Liao C-W (2012) Pattern analysis of seasonal variation in occupational accidents in the construction industry. Procedia Eng 29:3240–3244
10. Hoła B, Nowobilski T, Szer I, Szer J (2017) Identification of factors affecting the accident rate in the construction industry. Procedia Eng 208:35–42
11. Kore M, Health and safety risk management in building construction: the case of sunshine construction. St. Mary's University
12. Wang J, Yin Y, Wu H, Yang T (2020) Critical safety risks identification of bridge construction projects in the marine environment based on HHM and SNA. J Coast Res 108(SI):48–52

13. Martínez-Aires MD, López-Alonso M, Martínez-Rojas M (2018) Building information modeling and safety management: a systematic review. Saf Sci 101:11–18
14. Zhang L, Chen H, Li H, Wu X, Skibniewski MJ (2018) Perceiving interactions and dynamics of safety leadership in construction projects. Saf Sci 106:66–78
15. Alarcón LF, Acuna D, Diethelm S, Pell E (2016) Strategies for improving safety performance in construction firms Accident. Anal Prevent 94:107–118
16. Ansari M (2006) Risk management process in projects and proposing appropriate model for development plans in Iran. MSc Thesis, University of Tehran
17. Baloi AD, Price D (2003) Modelling global risk factors affecting construction cost performance. Int J Project Manage 21(4):261–269
18. Li H, Yazdi M (2022) A holistic question: is it correct that decision-makers neglect the probability in the risk assessment method? Advanced Decision-Making Methods and Applications in System Safety and Reliability Problems: Approaches, Case Studies, Multi-Criteria Decision-Making, Multi-Objective Decision-Making, Fuzzy Risk-Based Models. Springer International Publishing, Cham, pp 185–189
19. Yazdi M (2017) Hybrid probabilistic risk assessment using fuzzy FTA and fuzzy AHP in a process industry. J Fail Anal Prev 17(4):756–764
20. McCabe (2003) Monte Carlo simulation for schedule risks. In Proceedings of the 2003 Winter Simulation Conference, 2003 IEEE 2003, pp 1561–1565
21. Uher AR, TE, Toakley (1999) Risk management in the conceptual phase of a project. Int J Project Manag 17(3), 161–169
22. Chapman RJ (2001) The controlling influences on effective risk identification and assessment for construction design management. Int J Project Manage 19(3):147–160
23. OS and Administration H (2003) Underground construction (Tunneling): (osha 3115). US Department of Labor, pp 30
24. Thevendran MJ, Mawdesley V (2004) Perception of human risk factors in construction projects: an exploratory study. Int J Proj Managem 22(2):131–137
25. Nasir D, McCabe B, Hartono L (2003) Evaluating risk in construction–schedule model (ERIC–S): construction schedule risk model. J Constr Eng Manag 129(5):518–527
26. Saied A, Ali N, Alireza N (2008) Comprehensive description of research methods (quantitative, qualitative and combinational paradigms, strategies, plans and approaches). Tolid Danesh, Tehran
27. Hafez niya M (2015) Introduction to research in humanities. Organization of study and compilation of humanities textbooks of universities, Tehran
28. Wu CFJ (1986) Jackknife, bootstrap and other resampling methods in regression analysis. The Annals of Statistics 14(4):1261–1295
29. Gholami A, Mozaffari M, pirbalouti RG (2024) Prioritization of HSE management risks in the management of construction projects (Non-Mechanized Tunnel) by monte carlo modeling method BT - progressive decision-making tools and applications in project and operation management: approaches, case studies. In: Yazdi M (ed) Springer Nature Switzerland, Cham, pp 191–219. https://doi.org/10.1007/978-3-031-51719-8_13
30. Mezei M (1987) Adaptive umbrella sampling: self-consistent determination of the non-Boltzmann bias. J Comput Phys 68(1):237–248
31. Saari DG, Merlin VR (1996) The copeland method. Econ Theory 8(1):51–76. https://doi.org/10.1007/bf01212012
32. "Car traffic to enter Tehran Shomal Freeway," Iranian Students' News Agency (ISNA). Accessed: Feb. 18, 2023. [Online]. Available: https://irna.ir/xjxcN7
33. "Introduction and Overview of Phase One," Tehran Shomal Freeway Company. Accessed: Dec. 10, 2019. [Online]. Available: https://www.tsfc.ir/ShowPage/12/منطقه-یک
34. Bartels M, Karplus C (1998) Probability distributions for complex systems: adaptive umbrella sampling of the potential energy J Phys Chem B 102(5):865–880

35. Li H, Yazdi M (2022) Advanced decision-making Neutrosophic fuzzy evidence-based Best-Worst method. Advanced decision-making methods and applications in system safety and reliability problems: approaches, case studies, multi-criteria decision-making, multi-objective decision-making, fuzzy risk-based models. Springer International Publishing, Cham, pp 153–184
36. ISO 45001, Occupational Health and Safety. Geneva (2018)

Fault Prediction Methods for Locomotive Turbochargers

Yu Xing, Weida Wei, He Li, and Junyu Guo

Abstract Fault prediction technology predicts failures based on data collected from various locomotive sensors, achieving the effect of "predictive maintenance" and effectively improving the reliability and safety of equipment. The core of fault prediction is to establish an accurate model to ensure that the prediction results closely match the actual results. Early fault prediction methods mainly use single mathematical models for equipment fault prediction. Although the accuracy is satisfactory, the general applicability is limited. This paper conducts a detailed analysis of the multivariable grey prediction model, introduces neural networks and genetic algorithms to optimize the model, and applies it to the fault prediction of locomotive turbochargers. By fully exploiting the advantages of each method and organically combining them, the operating condition of locomotive turbochargers can be accurately and efficiently predicted, ensuring efficient operation and enhancing the reliability of the locomotive.

Keywords Fault prediction · Multivariable grey prediction model · Neural network · Genetic algorithm · Turbocharger

Y. Xing
Dalian CRRC Diesel Engine Co., LTD., Dalian, China

W. Wei
Institute of Internal Combustion Engine, Dalian University of Technology, Dalian, China

H. Li
Centre for Marine Technology and Ocean Engineering (CENTEC), Instituto Superior Técnico, Universidade de Lisboa, Lisbon, Portugal
e-mail: HeLi@centec.tecnico.pt

J. Guo (✉)
School of Mechatronic Engineering, Southwest Petroleum University, Chengdu, Sichuan, China
e-mail: junyguo@163.com

H. Li et al. (eds.), *Data-Driven Methods for Reliability and Safety Engineering: Applications in Industrial Systems*, Springer Series in Reliability Engineering,
https://doi.org/10.1007/978-3-032-22873-4_47

1 Introduction

With the rapid development of high-speed rail, the performance requirements for various train components have become increasingly stringent. As one of the key components of the locomotive engine, the turbocharger has long been a focal point of research for scholars both domestically and internationally. Turbochargers significantly increase the power of the engine, which greatly improves the performance and fuel efficiency of the locomotive [1–3]. The performance of the turbocharger is closely related to the overall performance of the locomotive, especially in terms of safety. Accurate fault prediction during the operation of the turbocharger can effectively reduce the likelihood of failure, thereby enhancing the safety and efficiency of the locomotive [4]. Therefore, more and scholars and industries are adopting fault prediction techniques to ensure the stable operation of turbochargers [5].

Fault prediction involves analyzing sensor data to identify trends, forecasting future data points, and assessing the predicted results [6]. In the past, equipment maintenance was typically reactive-problems were addressed only after they occurred, with efforts focused on identifying causes and implementing solutions. However, this reactive approach often led to increased maintenance costs, longer downtime, and, in some cases, severe accidents. Such "fix-it-after-it-breaks" methods are no longer suitable for modern operational demands. A Japanese institution conducted a study on fault prediction technology and found that its adoption could reduce the incident rate to one-quarter of the original level, while halving maintenance costs. Similarly, a survey conducted in the UK among large domestic enterprises revealed that the use of fault prediction technology could save £300 million in maintenance costs annually, resulting in a net profit of £250 million. In China, companies such as Liaoyang Petrochemical and China Petrochemical Corporation, which have successfully implemented fault prediction technology, have also seen significant improvements in their economic performance [7].

In the early stages, research on fault prediction for equipment had already begun, employing various methods and achieving promising results. Currently, equipment fault prediction methods can be broadly classified into two categories: physical model-based approaches and data-driven artificial intelligence (AI) techniques [8]. The former involves dynamic modeling to simulate the system and make predictions; however, developing an accurate physical model for complex systems is often challenging. The latter, by contrast, utilizes deep learning techniques to analyze historical data collected from equipment status parameters for learning and prediction. Due to its speed, cost-effectiveness, and wide range of applications, artificial intelligence has become increasingly prevalent in fault prediction systems. As early as 1978, Dr. R. A. Collacott, chairman of the UK's "Mechanical Health Monitoring Centre", began developing fault diagnosis technology. His techniques have been widely applied to the condition monitoring of automobiles and generators, and have gained wide broad acceptance in the engineering community. Between 1961 and 1972, following the successful implementation of manned lunar missions, the U.S. established a Mechanical Fault Prevention Group within NASA, recognizing the need to ensure the reliability of key equipment during the mission. However, at that time, fault prediction technology was primarily focused on aerospace and military equipment and had not yet been applied to common engineering fields. Subsequently, in 1971, Japan introduced the concept of Total Productive Maintenance (TPM). The country dedicated efforts to the development of fault diagnosis technology, which was applied to real-world engineering projects. As a result, equipment maintenance costs were significantly reduced [9]. In France, an institution developed a condition monitoring and fault diagnosis system (PSAD system) for large generator sets. The system has been implemented in major power stations, significantly reduced the failure rate of large generators and yielding substantial economic benefits [7]. Hountals et al. [10] established a fault prediction model for diesel engines, simulating multiple faults and analyzing changes in state parameters based on varying levels of fault severity in the engine [11]. Ugo et al. [12] used MATLAB's Simulink simulation module to predict and diagnose faults in the diesel engine's fuel injection, intake, and exhaust systems [13]. Riccardo et al. [14] conducted mathematical modeling of the entire air pathway of the diesel engine, analyzing common air system faults and turbocharger malfunctions. They applied sensitivity analysis and variable thresholds for fault prediction and diagnosis, and the results showed minimal deviation from actual conditions. Li et al. [15] using transient speed signals from diesel engine tests, employed Independent Component Analysis (ICA) and Empirical Mode Decomposition (EMD) to select characteristic parameters. They then used Support Vector Machines (SVM) to predict and diagnose diesel engine faults, with results indicating that the mathematical model developed was effective for fault prediction and diagnosis.

This paper establishes a highly accurate fault prediction model. First, the operating characteristics of the turbocharger are analyzed, and the relevant literature is reviewed to identify the state parameters that affect the performance of locomotive turbochargers. Then, using the state parameter data provided by the locomotive manufacturer, the correlations between the variables are calculated. A multivariable grey prediction model is then constructed to forecast the variables, and the accuracy of the model is validated through residuals and relative error analysis. Building on the previously developed multivariable grey prediction model, a BP neural network is introduced to fully leverage the strengths of both the multivariable grey prediction model and the BP neural network. This combination results in the development of a more accurate prediction model-the Serial Grey Neural Network (SGNN) model. The SGNN model is then applied to predict turbocharger data, and its accuracy is tested. To further improve the local optimization and convergence issues of the SGNN model, a genetic algorithm is incorporated, resulting in the Genetic Algorithm-Grey Neural Network (GA-SGNN) model. The enhanced model is applied to predict turbocharger data, and the accuracy of the combined prediction model is evaluated using residuals and relative error analysis.

2 Multivariate Grey Prediction Model

In the 1980s, based on the concept of the "black box", Grey System Theory was first proposed and gained widespread attention. Common systems can be categorized into three types: black systems (completely unknown information), white systems (completely known information), and grey systems (partially known information). In practical engineering and daily life, grey systems are the most commonly encountered. A key characteristic of grey systems is their ability to handle "small samples and limited information". Table 1 provides a comparison between black, white, and grey systems.

Grey prediction is widely applied in the field of engineering. Grey models, ranging from one-dimensional to n-dimensional, are often used to predict the evolution of characteristic parameters in engineering systems. The grey prediction model is particularly suitable for addressing three types of uncertainties in engineering. The first is informational ambiguity, where precise mathematical equations cannot be formulated; fuzzy mathematics is ideal for solving such problems by predicting the future values of turbocharger state parameters using fuzzy methods. The second is mechanistic uncertainty, such as the difficulty in accurately forecasting the future operating state of a locomotive turbocharger. The third is informational uncertainty. In practical problems, multiple state parameters that describe a system often interact with each other, and a single state parameter cannot fully capture the operating state of the system. Grey prediction models, however, can analyze the developmental trends among multiple variables, identify patterns in their changes, and predict the overall system behavior. In addition, grey models can assess the correlation between variables, calculate data sequences, and establish corresponding differential equations. These models offer high precision and strong fitting capabilities. Grey prediction demonstrates significant advantages in handling grey data, requiring multiple state parameters to describe a system, which are often interrelated. Therefore, grey prediction models are highly suitable for fault prediction in turbochargers. Based on the characteristic parameters of turbocharger states, a multivariate grey prediction model can be constructed and applied to the fault prediction of locomotive turbochargers.

Table 1 Comparison of three systems of black, white and gray

Concept	White system	Grey system	Black system
Information	Completely known	Partially known	Completely unknown
Appearance	Clear	Grey	Dark
Process	Old	Alternation of new and old	New
Method	Certain	Uncertain	Denied
Nature	Pure	Multiple components	Mixed
Result	Unique solution	Multiple solutions	No solution

2.1 Grey Correlation Matrix

The grey relational matrix is widely applied in practical engineering, serving as a tool to analyze the degree of correlation between various factors. The calculation of the relational degree is based on the comparison between the test data and the reference data, where the relational degree reflects the strength of the relationship between the data sets. The greater the relational degree, the more similar the variation patterns of the two data sets are. By utilizing the grey relational matrix, key factors among the variables can be identified, enabling the construction of grey time series.

2.2 MGM(1, n) Model

In the locomotive turbocharger system, multiple variables are present, and these variables are interrelated. Given these characteristics, this paper employs a multivariate grey prediction model, namely the MGM(1, n) model, to describe the turbocharger system. The GM(1, n) model takes into account several variables within the system, thus overcoming the limitation of the GM(1, 1) model, which cannot accurately capture the characteristics of the system with a single variable. In addition, the prediction accuracy of the GM(1, n) model is superior to that of the GM(1, 1) model.

The specific steps for constructing the MGM(1, n) model are as follows:

(1) Data Accumulation

Data is often influenced by various factors and mixed with "noise", making it challenging to extract valuable information. Data accumulation is an effective approach for denoising, facilitating the revelation of underlying patterns and actual trends in the data.

2.3 Application of the MGM(1, n) Model

Based on the turbocharger fault data provided by a locomotive factory in Dalian, the MGM(1, n) model is constructed with the 265-series diesel engine turbocharger from this factory as the research object. Due to space limitations, this paper selects one of the common turbocharger faults, "overheating", as the subject of study and applies the MGM(1, n) model to predict this fault. The fault data is shown in Table 2.

1. Grey relational degree analysis

For the grey time series in the table above, since the rotor speed is a direct indicator of the turbocharger's performance, while the other parameters indirectly represent the turbocharger's condition, the time series of rotor speed is selected as the reference series. The remaining five time-series are selected as comparison series. Using

Table 2 Time series of turbocharger characteristic variables

Time sequence	Rotational speed (r/min)	Oil temperature (°C)	Intake pressure (KPa)	Exhaust pressure (KPa)	Coolant temperature (°C)	Bearing pressure (MPa)
1	69,280	41.0	397	850	39.2	7.41
2	75,421	42.3	388	831	41.0	7.16
3	79,326	44.2	376	820	42.9	6.82
4	81,874	46.5	349	807	45.6	6.51
5	84,183	48.9	337	769	51.4	6.37
6	87,322	52.8	321	781	55.0	6.19
7	89,727	55.4	305	770	58.9	6.11
8	91,375	58.9	289	761	64.4	6.02
9	93,946	64.1	278	752	68.1	5.87
10	97,315	68.9	267	740	71.5	5.67
11	101,327	71.7	251	733	74.2	5.52
12	105,218	76.9	240	724	76.6	5.31
13	108,025	85.9	228	716	84.0	5.01
14	110,121	88.3	218	697	86.8	4.79
15	115,716	93.9	200	687	89.3	4.78

MATLAB software, the grey relational degrees between rotor speed and oil temperature, intake pressure, coolant temperature, and bearing pressure are calculated. The specific values are shown in Table 3.

As shown in Table 3, the variable with the highest grey relational degree with rotor speed is oil temperature, with a relational degree of 0.7463, followed by coolant temperature. The variable with the lowest relational degree is intake pressure, with a value of 0.5930. The high relational degrees among these six variables indicate that oil temperature, intake pressure, exhaust pressure, coolant temperature, and bearing pressure can be used as state parameters for predicting turbocharger faults.

2. MGM(1, n) modeling

First, based on the data from Table 2, the first 12 sets of 6 variables are extracted to form sequences for establishing the MGM(1, 6) model. The calculated fitted values of the model are compared with the actual values, and the residuals and relative

Table 3 Gray correlation of reference series and comparison series

Variable name	Oil temperature	Intake pressure	Exhaust pressure	Coolant temperature	Bearing pressure
Grey relational degree	0.7463	0.5930	0.6629	0.6779	0.6469

errors are used to evaluate the accuracy of the model. The residuals and relative error results between the fitted values and the actual values obtained from the multivariate grey prediction model are shown in Table 4.

As shown in Table 4, the MGM(1, 6) model constructed using the first 12 sets of 6 variables allows for the calculation of fitted values and measured values. By comparing the fitted values with the actual measurements, the residuals and relative errors are obtained. The maximum relative error is 2.766%, and the minimum is 0%, with all relative errors remaining below 5%. This indicates that the MGM(1, 6) model exhibits excellent data fitting performance and relatively high accuracy [16]. Therefore, this grey prediction model can be utilized for fault prediction in locomotive turbochargers.

Based on the established model using the first 12 sets of data, the MGM(1, 6) model is applied to forecast the variable values for the next three sequences. The residuals and relative errors between the predicted values and measured values are calculated. The error calculation results for the predicted values and actual values are shown in Table 5.

As shown in Table 5, based on the first 12 sets of data, the MGM(1, 6) model has provided predicted values for the next three sequences, and the residuals and relative errors between the predicted and measured values have been calculated. All variables exhibit relative errors within 5%, indicating that the use of the MGM(1, 6) model for fault prediction in locomotive turbochargers meets the accuracy requirements.

From Tables 4 and 5, it is evident that the fitting accuracy of the MGM(1, 6) model is higher than its predictive accuracy. While the predictive precision of model meets the necessary standards, some predicted values show relatively high errors, with the maximum relative error reaching 4.954%. Additionally, oil temperature, intake pressure, coolant temperature, and bearing pressure exhibit relative errors exceeding 4%, raising concerns about the reliability of the model. To ensure broader application of this model in engineering, it is essential to improve its predictive accuracy.

The analysis of the above data indicates that relying solely on a single mathematical model does not guarantee satisfactory predictive accuracy in turbocharger fault prediction. Therefore, it is necessary to leverage the advantages of multiple models, integrating various mathematical approaches to enhance the reliability and generalizability of the model.

3 Grey Neural Network Model

As established in the previous section, a single mathematical model cannot effectively ensure predictive accuracy and reliability. A combined prediction model can effectively address the shortcomings of individual predictive models and provide insights into machine operating conditions from multiple perspectives and various types of sample data. This comprehensive approach enables more thorough system diagnostics and significantly enhances prediction accuracy and stability. Therefore,

Table 4 Point-by-point error of the fitted and actual values of the MGM(1, 6) model

Variable		Sequence 1	Sequence 2	Sequence 3	Sequence 4	Sequence 5	Sequence 6
Rotational Speed (r/min)	Measured Value	69,280	75,421	79,326	81,874	84,183	87,322
	Fitted value	69,280	75,530	78,786	82,116	84,820	86,595
	Residual	0	0.145	− 0.681	0.296	0.757	− 0.832
	Relative error (%)	0	0.091	0.857	0.081	0.752	− 0.746
Oil temperature (°C)	Measured value	41.0	42.3	44.2	46.5	48.9	52.8
	Fitted value	41.0	42.2	44.3	46.5	48.7	51.7
	Residual	0	− 0.1	0.1	0	− 0.2	− 1.1
	Relative error (%)	0	− 0.329	0.228	0	− 0.349	− 2.018
Intake pressure (KPa)	Measured value	397	389	376	349	337	321
	Fitted value	397	389	372	354	335	318
	Residual	0	1	− 4	5	− 2	− 3
	Relative error (%)	0	0.264	− 1.157	1.377	− 0.483	− 0.949
Exhaust pressure (KPa)	Measured value	850	831	820	807	796	781
	Fitted value	850	831	819.6	805.9	792.9	781.8
	Residual	0	0	− 0.047	− 0.142	− 0.391	0.106
	Relative error (%)	0	0	− 0.122	− 0.248	− 0.628	− 0.128
Coolant temperature (°C)	Measured value	39.2	41.0	42.9	45.6	51.4	55.0
	Fitted value	39.2	40.7	43.0	46.1	50.0	55.0
	Residual	0	− 0.3	0.1	0.5	− 1.4	0
	Relative error (%)	0	− 0.674	0.28	1.09	− 2.766	0
Bearing pressure (MPa)	Measured value	7.41	7.16	6.82	6.51	6.37	6.19
	Fitted value	7.41	7.16	6.79	6.52	6.35	6.23
	Residual	0	0	− 0.03	0.01	− 0.02	0.04

(continued)

Table 4 (continued)

Variable		Sequence 1	Sequence 2	Sequence 3	Sequence 4	Sequence 5	Sequence 6
	Relative error (%)	0	0	− 0.471	0.142	− 0.360	0.656
Rotational speed (r/min)	Measured value	89,727	91,375	93,946	97,315	101,327	105,218
	Fitted value	88,320	91,128	94,833	98,084	100,392	103,171
	Residual	− 1407	− 247	887	769	− 935	− 2047
	Relative error (%)	− 1.569	− 0.271	0.944	0.790	− 0.923	− 1.945
Oil temperature (°C)	Measured value	55.4	58.9	64.1	68.9	71.7	76.9
	Fitted value	55.8	60.1	63.8	67.2	71.3	76.6
	Residual	0.4	1.2	− 0.3	− 1.7	− 0.4	− 0.3
	Relative error (%)	0.654	2.021	− 0.444	− 2.514	− 0.533	− 0.331
Intake pressure (KPa)	Measured value	305	289	278	267	251	240
	Fitted value	304	292	280	265	250	237
	Residual	− 1	3	2	− 2	− 1	− 3
	Relative error (%)	− 0.449	1.001	0.565	− 0.839	0.594	− 1.345
Exhaust pressure (KPa)	Measured value	770	761	752	740	733	724
	Fitted value	770.8	758.4	746.5	738.4	733.3	726.7
	Residual	0.8	− 2.6	− 5.5	− 1.6	0.3	2.7
	Relative error (%)	0.104	− 0.346	− 0.730	− 0.220	0.039	0.367
Coolant temperature (°C)	Measured value	58.9	64.4	68.1	71.5	74.2	76.6
	Fitted value	60.4	64.6	67.2	69.6	73.5	78.0
	Residual	1.5	0.2	− 0.9	− 1.9	− 0.7	1.4
	Relative error (%)	2.610	0.380	− 1.350	− 2.656	− 0.982	1.848
Bearing pressure (MPa)	Measured value	6.11	6.02	5.87	5.67	5.52	5.31
	Fitted value	6.11	5.98	5.83	5.700	5.54	5.32

(continued)

Table 4 (continued)

Variable		Sequence 1	Sequence 2	Sequence 3	Sequence 4	Sequence 5	Sequence 6
	Residual	0	– 0.04	– 0.04	0.03	0.02	0.01
	Relative error (%)	0	– 0.736	– 0.658	0.443	0.318	0.168

Table 5 Point-by-point error of predicted and actual values of the MGM(1, 6) model

Variable		Sequence 13	Sequence 14	Sequence 15
Rotational speed (r/min)	Measured value	108,025	110,121	115,716
	Fitted value	107,888	113,670	118,147
	Residual	– 137	3549	2431
	Relative error (%)	– 0.127	3.223	2.101
Oil temperature (°C)	Measured value	85.9	88.3	93.9
	Fitted value	81.9	86.0	89.6
	Residual	– 4	– 2.3	– 4.3
	Relative error (%)	– 4.657	– 2.605	– 4.579
Intake pressure (KPa)	Measured value	228	218	200
	Fitted valuc	225	210	191
	Residual	– 3	– 8	– 9
	Relative error (%)	– 1.042	– 3.670	– 4.500
Exhaust pressure (KPa)	Measured value	716	697	687
	Fitted value	716	706	700
	Residual	0	9	13
	Relative error (%)	0	1.291	1.892
Coolant temperature (°C)	Measured value	84.0	86.8	89.3
	Fitted value	81.1	82.5	85.2
	Residual	– 2.9	– 4.3	– 4.1
	Relative error (%)	– 3.452	– 4.954	– 4.591
Bearing pressure (MPa)	Measured value	5.01	7.79	4.78
	Fitted value	5.05	4.81	4.60
	Residual	0.04	0.02	– 0.18
	Relative error (%)	0.798	0.418	– 3.766

in this section, the MGM(1, 6) model is integrated with the BP neural network model to establish the Grey Neural Network (SGNN) model.

3.1 *Artificial Neural Network*

The human brain is an extraordinarily intelligent machine, with human thought is determined by the brain itself. Neurons are the most fundamental units of the brain, with more than 10 billion neurons in the cerebral cortex. The brain receives and processes information through neurons. Artificial neural networks (ANNs) are abstracted from brain science and consist of numerous interconnected artificial neurons. These networks mimic the function of biological neurons and serve as mathematical models for analyzing and processing information. Artificial neural networks exhibit four key characteristics:

(1) Non-locality: A neural network consists of many interconnected neurons. A single neuron does not influence the entire system; rather, the behaviour of the system is determined by the interactions among many neurons.
(2) Non-linearity: Neurons in a neural network can be either excitatory or inhibitory, corresponding to activation or suppression, respectively.
(3) Non-stationarity: Neural networks possess strong self-learning, self-adapting, and self-organizing capabilities. When the transmitted information changes, the nonlinear dynamic system adapts accordingly.
(4) Non-convexity: The state function determines the trajectory of change of the system, while non-convexity ensures that the state function has multiple stable equilibrium points, guaranteeing diversity during the evolution of the system.

3.2 *Introduction to BP Neural Networks*

In 1985, McClelland and colleagues first proposed the BP (Backpropagation) neural network model. The BP neural network is capable of learning nonlinear relationships between multiple inputs and outputs and has robust memory capabilities. It is commonly applied in fields such as image and text recognition, medical diagnosis, function approximation, and data compression. Currently, it is one of the most widely used artificial neural networks.

The structure of the BP neural network is illustrated in Fig. 1. As shown in the figure, neurons within the same layer are not interconnected, while neurons between different layers are fully connected. The neurons in the input layer are connected to those in the hidden layer through adjustable weights, which represent the degree of signal transmission between the two neurons.

Figure 2 illustrates the flowchart of BP neural network learning. The BP learning algorithm is based on the principle of the steepest gradient descent method, continuously adjusting the weights to minimize the overall system error. The goal is to bring the actual output as close as possible to the desired output. The learning process consists of two phases: forward propagation and backward propagation [17]. As shown in Fig. 2, during forward propagation, when sample data is provided to the neurons in the input layer, the information is sequentially passed through the input

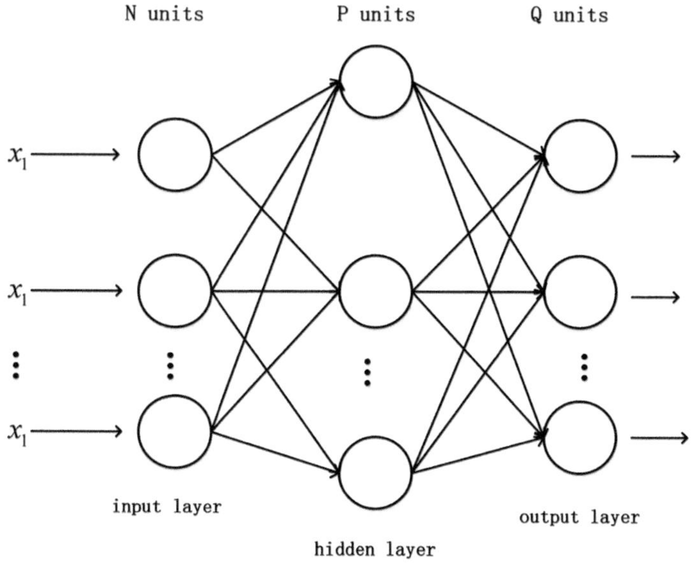

Fig. 1 BP neural network structure

layer, hidden layer, and output layers. Each node transmits the information to varying degrees depending on the weights. The neurons in the output layer integrate the information received from the previous layers, adding a bias during this process. If there is a significant difference between the output and the desired values, the information is transmitted backwards. That is, the neurons in the output layer propagate the error back to the input layer through the hidden layers, following the principle of reducing the error between the actual and expected output. This process continues by iteratively optimizing the weights between neurons until the error between the output and the desired values reaches a minimum.

3.3 Introduction to Grey Neural Network Models

The combination of the Grey Prediction Model and the BP Neural Network Model allows for a complementary integration of their strengths. The advantage of the Grey Prediction Model lies in its ability to handle "poor information" and "small sample" data without requiring knowledge of the underlying trends. It is simple and quick to establish, with fast computation speeds. However, its drawbacks are also apparent, such as the lack of self-learning ability and weak performance in handling nonlinear problems. On the other hand, the BP Neural Network has self-learning capabilities and is well-suited for handling nonlinear problems, compensating for the weaknesses of the Grey Prediction Model. By leveraging the strengths of both models, a more accurate fault prediction model for locomotive turbochargers can be

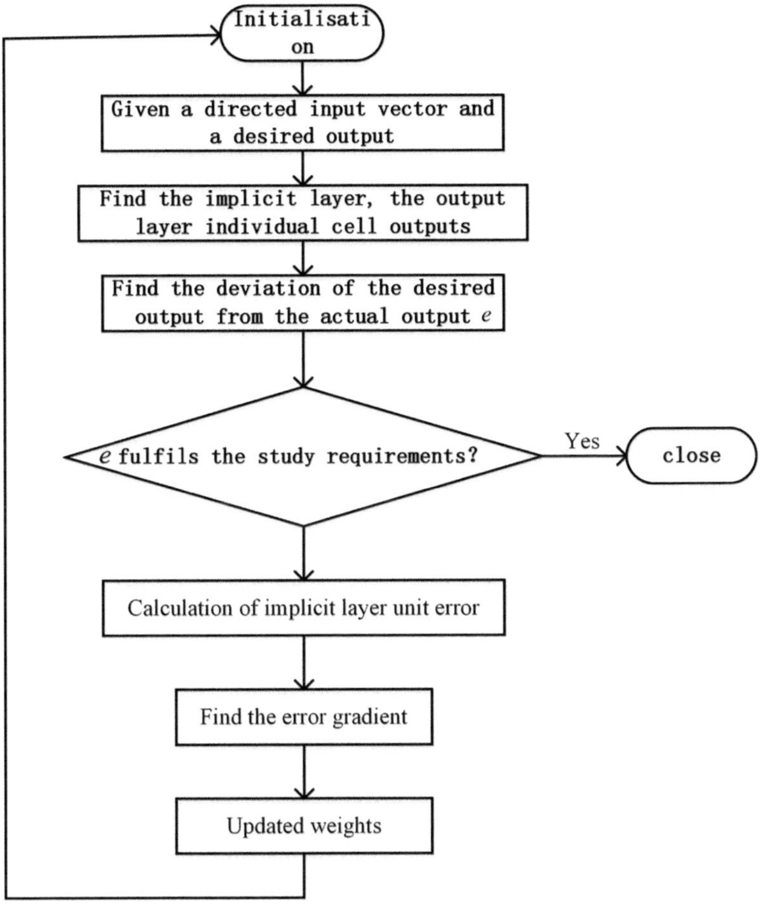

Fig. 2 BP neural network learning process diagram

developed through the organic integration of the Grey Prediction Model and the BP Neural Network.

In the field of engineering, there are various ways to combine the Grey Prediction Model and the BP Neural Network Model, which can generally be categorized into four structures: Serial Grey Neural Network (SGNN), Parallel Grey Neural Network (PGNN), Hybrid Grey Neural Network (IGNN), and Embedded Grey Neural Network (BGNN). This paper focuses on the Serial Grey Neural Network model (SGNN), with the schematic diagram shown in Fig. 3.

The principle of the SGNN is that the fitting results from the multivariable grey prediction model are used as input samples for the BP neural network, while the actual measured values serve as target samples. Through the learning and training of the neural network, a set of corresponding weights and thresholds are derived. Then, the predicted data from the grey prediction model are fed into the BP neural network

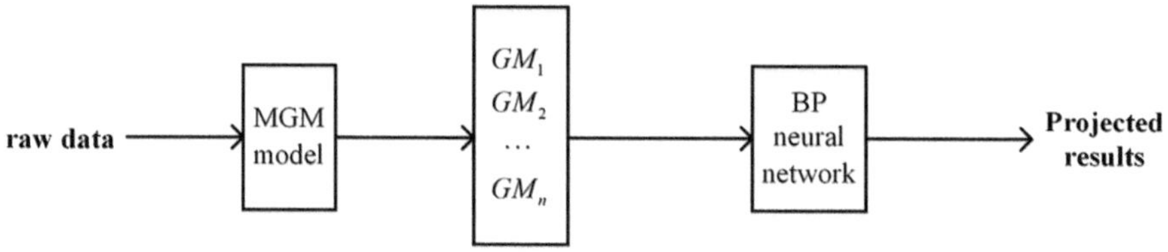

Fig. 3 Tandem gray neural network model

for simulation, and the output is the predicted result of the Grey Neural Network model.

3.4 Application of the Grey Neural Network Model

Through the calculations of the multivariable grey prediction model in the previous section, the fitted values and predicted values of the turbocharger variables were obtained. The actual measured values, along with the model fitted and predicted values, are shown in Table 6.

Based on the data from Table 6, the BP neural network model is applied on top of the multivariable grey prediction model. The specific steps for constructing the SGNN are as follows:

(1) Select six influencing factors, rotational speed, oil temperature, inlet pressure, exhaust pressure, coolant temperature, and bearing pressure, as the study objects. Normalize both the MGM(1, 6) model data sequence and the actual data sequence for these six factors, ensuring that all values fall within the [0, 1] range.
(2) Use the first 10 sets of normalized fitted values from the MGM(1, 6) model as the input samples P for the neural network, and the normalized actual measured values as the output samples T. Train the neural network to determine the weights and biases for each neuron. Set the 11th and 12th data sets from the MGM(1, 6) model as test samples, and validate the trained BP neural network using these test samples.
(3) Input the next three predicted sequences from the MGM(1, 6) model for future time steps into the trained neural network to perform simulation predictions. The resulting data will serve as the final prediction of the Grey Neural Network combined model.

The BP neural network has six inputs and six outputs, with the number of hidden layers set to 3 and the number of hidden layer nodes set to 7. The transfer function for the input and output layers is the *purelin* function, while the hidden layer uses the *tansig* function. The maximum number of training iterations is set to 5000, the learning rate of the network is set to 0.005, and the target error for the training network is set to 0.0001. The training results of the BP neural network are shown in

Table 6 Measured value and calculated value of MGM(1, 6)

Time No.	Rotational speed (r/min)		Oil temperature (°C)		Inlet pressure (KPa)		Exhaust pressure (KPa)		Coolant temperature (°C)		Bearing pressure (MPa)	
	Measured	MGM	Measured	MGM	Measured	MGM	Measured	MGM	Measured	MGM	Measured	MGM
1	69,280	69,280	41.0	41.0	397	397	850	850	39.2	39.2	7.41	7.41
2	75,421	75,530	12.3	42.2	388	389	397	831.0	41.0	40.7	7.16	7.16
3	79,326	78,786	44.2	44.3	376	372	389	819.6	42.9	43.0	6.82	6.79
4	81,874	82,116	46.5	46.5	349	354	372	805.9	45.6	46.1	6.51	6.52
5	84,183	84,820	48.9	48.7	337	335	354	792.9	51.4	50.0	6.37	6.35
6	87,322	86,595	52.8	51.7	321	318	335	781.8	55.0	55.0	6.19	5.23
7	89,727	88,320	55.4	55.8	305	304	318	770.8	58.9	60.4	6.11	6.11
8	91,375	91,128	58.9	60.1	289	292	304	758.4	64.4	64.6	6.02	5.98
9	93,946	94,833	64.16	63.8	278	280	292	746.5	68.1	67.2	5.87	5.83
10	97,315	98,084	68.9	67.2	267	265	280	738.4	71.5	69.6	5.67	5.70
11	101,327	100,392	71.7	71.3	251	250	265	733.3	74.2	73.5	5.52	5.54
12	105,218	103,171	76.9	76.6	240	237	237, 250	726.7	76.6	78.0	5.31	5.32
13	108,025	107,888	85.9	81.9	228	225	225	716	84.0	81.1	5.01	5.05
14	110,121	113,670	88.3	86.0	218	210	210	706	86.8	82.5	4.79	4.81
15	115,716	118,147	93.9	89.6	200	191	687	700	89.3	85.2	4.78	4.60

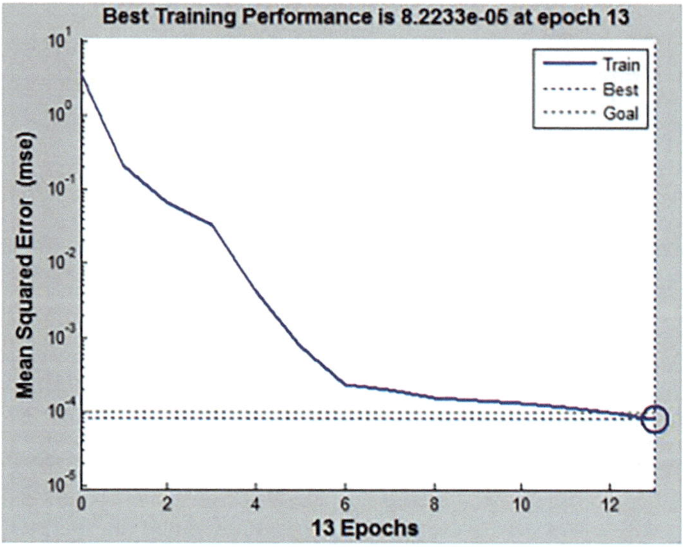

Fig. 4 BP neural network training results

Figs. 4 and 5. As shown in Fig. 4, the network achieved the target error after just 13 iterations, indicating that the network structure is well-designed. Figure 5 shows that the regression coefficient R is 0.9999. The closer R is to 1, the stronger the fitting ability of the BP neural network, demonstrating higher accuracy.

The trained Grey Neural Network model is then validated using the 11th and 12th sets of fitted values from the MGM(1, 6) model as the test samples for the neural network. The relative error between the simulation data obtained from the Grey Neural Network model and the test sample data is plotted as a line graph, as shown in Fig. 6. From Fig. 6, it can be seen that the relative errors are all less than 5%, indicating that the neural network model is suitable for prediction.

As shown in Table 7, for Sample 13's rotational speed, Sample 14's oil temperature, Sample 13's inlet pressure, Sample 13's exhaust pressure, and Sample 14's bearing pressure, the relative error predicted by the SGNN model is higher than that predicted by the MGM(1, 6) model. However, the maximum increase in relative error is only 3.509%. Therefore, after applying the SGNN, the maximum relative error of the prediction sequence is reduced, indicating that the SGNN model provides higher accuracy.

To better compare the accuracy of the MGM(1, 6) model and the SGNN model, the relative errors of the predicted values of each model are presented in the form of a line graph. The relative error comparison chart is shown in Fig. 7. In this figure, the horizontal axis represents the three prediction samples for the six variables, while the vertical axis represents the absolute values of the relative errors for the MGM(1, 6) model and the SGNN model. It can be seen from the graph that the peak relative

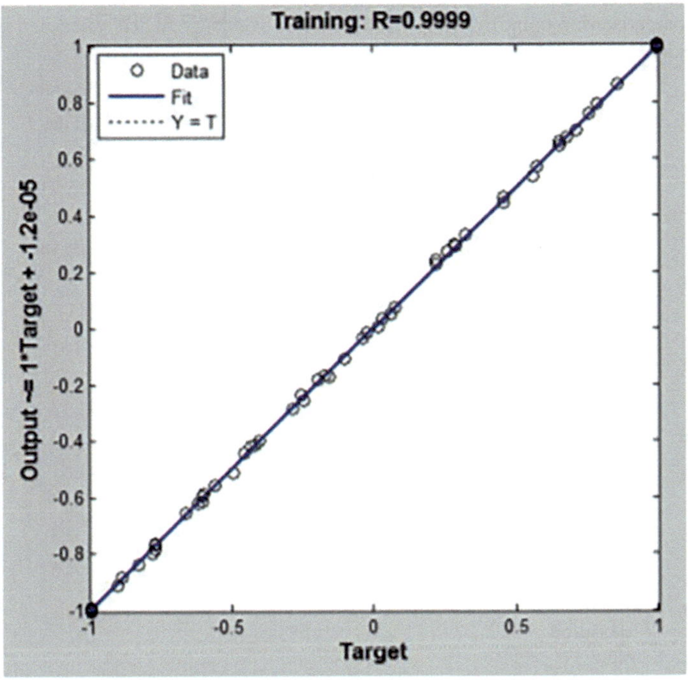

Fig. 5 Regression plot of the BP neural network

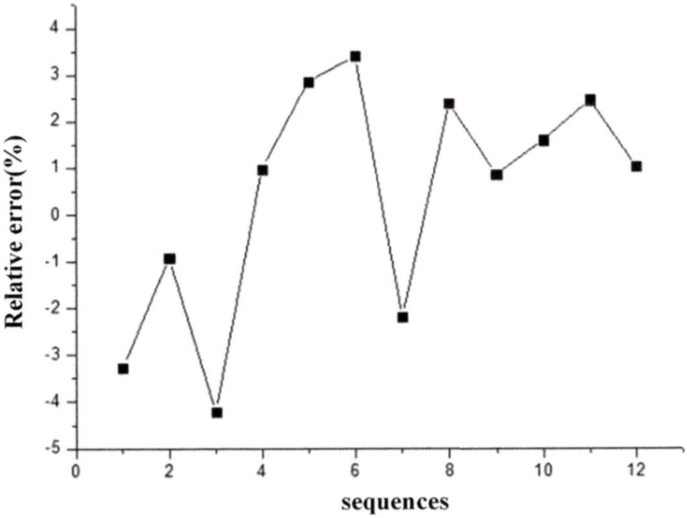

Fig. 6 Relative error between simulation data and test samples

Table 7 Comparison of point-by-point error between SGNN predicted value and MGM(1, 6) measured value and measured value

Variable		Sequence 13		Sequence 14		Sequence 15	
		MGM(1, 6)	SGNN	MGM(1, 6)	SGNN	MGM(1, 6)	SGNN
Rotational speed (r/min)	Measured value	108,025		110,121		115,716	
	Predicted value	107,888	108,767	113,670	113,058	118,147	114,062
	Residual	− 137	742	3549	2937	2431	− 1684
	Measured value	− 0.127	0.687	3.223	2.667	2.101	− 1.455
Oil temperature (°C)	Measured value	85.9		88.3		93.9	
	Predicted value	81.9	85.1	86.0	91.1	89.6	92.5
	Residual	− 4	− 0.8	− 2.3	2.8	− 4.3	− 1.4
	Measured value	− 4.657	0.940	− 2.605	3.171	− 4.579	− 1.491
Intake pressure	Measured value	228		218		200	
	Predicted value	225	236	210	224	191	208
	residual	− 3	8	− 8	6	− 9	8
	Measured value	− 1.042	3.509	− 3.670	2.752	− 4.500	4.000
Exhaust pressure (KPa)	Measured value	716		697		687	
	Predicted value	716	706	706	693	700	690
	Residual	0	− 10	9	− 4	13	3
	Measured Value	0	− 1.397	1.291	0.577	1.892	0.437
Coolant Temperature (°C)	Measured value	84.0		86.8		89.3	
	Predicted value	81.1	81.6	82.5	81.1	81.6	82.5
	Residual	− 2.9	− 2.4	− 4.3	− 2.9	− 2.4	− 4.3
	Measured value	− 3.452	− 2.857	− 4.954	− 3.452	− 2.857	− 4.954
Bearing Pressure (MPa)	Measured value	5.01		4.79		4.78	
	Predicted value	5.05	4.97	4.81	4.70	4.60	4.64

(continued)

Table 7 (continued)

Variable		Sequence 13		Sequence 14		Sequence 15	
		MGM(1, 6)	SGNN	MGM(1, 6)	SGNN	MGM(1, 6)	SGNN
	Residual	0.04	− 0.04	0.02	0.09	− 0.18	− 0.14
	Measured value	0.798	− 0.798	0.418	1.879	− 3.766	− 2.929

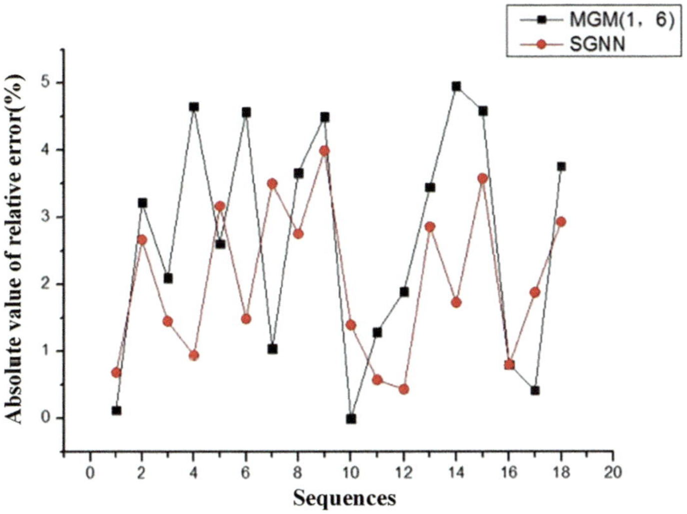

Fig. 7 Comparison of prediction relative error between MGM(1, 6) and SGNN

errors are significantly reduced after the integration of the grey prediction model with the neural network model, indicating that the SGNN model offers higher accuracy.

4 Genetic Algorithm-Based Grey Neural Network Model

As discussed in the previous section, incorporating the BP neural network significantly enhances the prediction accuracy of the model. However, the Grey Neural Network model also has some limitations:

(1) The BP neural network model employs local search for optimization, making it prone to getting trapped in local minima, which can ultimately lead to failure in network learning and training [18].
(2) The results of the BP neural network vary with each run, and there is no established standard for selecting the best result. The choice of network structure is often determined through multiple trials [19].

(3) The BP neural network has a relatively slow convergence speed.

To address these shortcomings, this paper combines the genetic algorithm with the Grey Neural Network, aiming to overcome the local search limitations of the neural network and enhance the stability of the model.

4.1 Introduction to Genetic Algorithms

In the 1970s, Professor J. H. Holland first introduced the concept of "genetic algorithms" in his work Adaptation in Natural and Artificial Systems [18]. Genetic algorithms (GA) are computational models based on the theories of natural selection, survival of the fittest, and genetic variation in biology. Unlike traditional gradient search methods, genetic algorithms are adaptive, global optimization probabilistic search algorithms [19]. In biological evolution, populations are the fundamental unit of evolution, and genetic mutation is the core mechanism. The process of biological evolution can be divided into three basic stages:

(1) Heritable variation: Heritable variation provides the raw material for evolution, including gene mutation and recombination. These variations are nondirectional and do not determine the direction of species evolution [20].
(2) Natural selection: Natural selection directs the evolution of species. It ensures the survival of the fittest, preserving favorable variations and eliminating unfavorable ones, thereby guiding species to evolve in beneficial directions.
(3) Isolation: Isolation results in the formation of new species and is a necessary condition for speciation.

At its core, the genetic algorithm remains an iterative process. Starting from an initial population, it iteratively evolves to find the optimal solution. Unlike traditional iterative methods, genetic algorithms simulate the processes of natural selection, evolution, and genetic mutation. Starting from an initial population, they evolve through genetic variation and natural selection to ultimately converge on an optimal solution (the optimal population) [21–24].

4.2 Basic Principles of Genetic Algorithms

The genetic algorithm consists of three key components: parameter encoding, determination of the fitness function, and selection of genetic operators [25]. The genetic operators include selection, crossover, and mutation. The basic process of the genetic algorithm is as follows:

(1) Select an encoding strategy to convert the problem space parameters into a bit-string structure.
(2) Define the fitness function f(x).

(3) Determine the genetic strategy (population size, methods for selection, crossover, and mutation, as well as the crossover probability and mutation probability).
(4) Randomly generate the initial population.
(5) Compute the fitness function for each individual in the population based on the encoded bit-strings.
(6) Apply the determined genetic strategy to generate the next population.
(7) Evaluate whether the performance of the newly generated population meets the desired criteria. If it does not, return to the previous step or adjust the genetic strategy, refer to Fig. 8.

The first step in the genetic algorithm is genetic encoding. When applying the genetic algorithm to solve a problem, it is essential to link the actual problem parameters to the chromosome bit-string structure used in the algorithm. Genetic encoding must adhere to three basic principles:

(1) Completeness: Every point in the problem space should be representable by a chromosome bit-string phenotype.

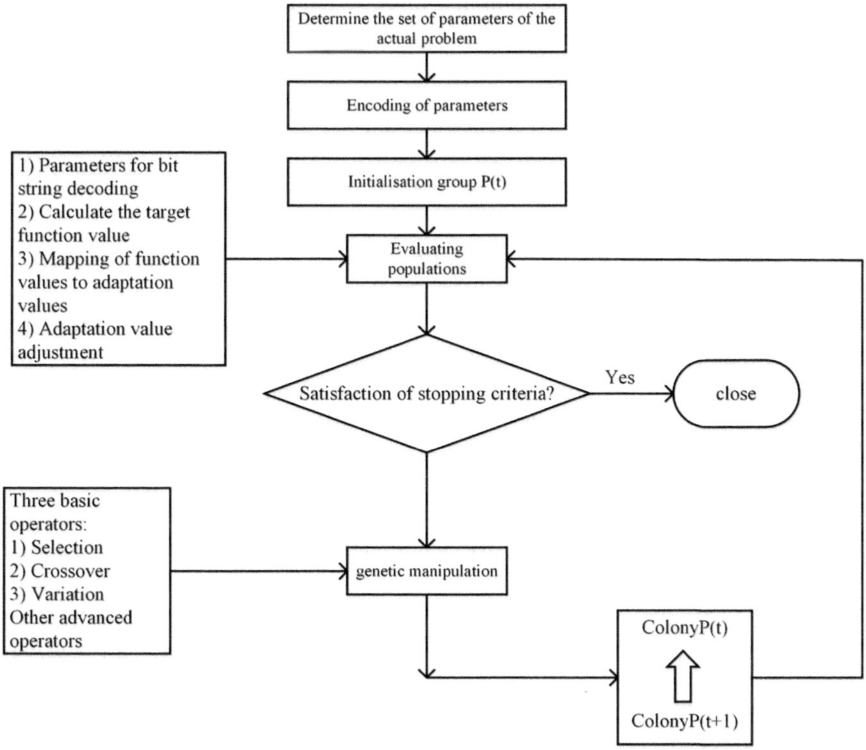

Fig. 8 Basic flow chart of genetic algorithm

(2) Validity: The chromosome encoding must correspond to a feasible solution to the actual problem.
(3) Non-redundancy: Each chromosome must uniquely correspond to a potential solution.

The fitness function is crucial in genetic algorithms, as it determines the speed of population evolution and the quality of the optimal population (optimal solution). It measures the fitness of the individuals in the population, and the search for an optimal solution in the genetic algorithm is driven by the fitness function.

There are three key genetic operations: crossover, recombination, and mutation. These operations increase the diversity in the population, giving the genetic algorithm its powerful search capability. Crossover mimics the biological process of gene recombination, where two individuals exchange parts of their genes according to a crossover probability. Recombination and mutation introduce genetic diversity by randomly flipping alleles on the bit-strings of individuals based on a mutation probability. When generating the next population, crossover is the primary factor, while mutation plays a secondary role. Therefore, the crossover probability is set much higher than the mutation probability. Selection simulates the process of natural selection in biology, generating a mating pool based on the fitness values of individuals in the population. Individuals with higher fitness have a better chance of passing on their genetic code to the next generation, while those with lower fitness are more likely to be eliminated.

4.3 Genetic-Grey Neural Network Model Applications

In the previous section, the Grey Neural Network model effectively captured the relationship between input and output variables, and its prediction accuracy surpassed that of the Grey Prediction model. However, for certain data points, the relative error remained excessively large. To further improve the accuracy of the model, this section utilizes a genetic algorithm to optimize the parameters of the BP neural network.

The genetic algorithm optimizes the BP neural network in two main aspects. The first aspect is to improve the initial weights and biases and optimize the entire topology of the BP neural network, which is particularly useful when the network structure has not yet been determined. The second aspect focuses on optimizing the initial weights and biases of the BP neural network when its topology is already established. In the previous section, the number of samples, input/output parameters, and hidden layers have already been defined. Therefore, this study focuses solely on optimizing the initial weights and biases of the BP neural network using the genetic algorithm.

Based on the Grey Neural Network model, a Genetic Algorithm-Grey Neural Network (GA-SGNN) model is established, and the parameters of the genetic algorithm are configured. The BP network algorithm is set as the objective function of

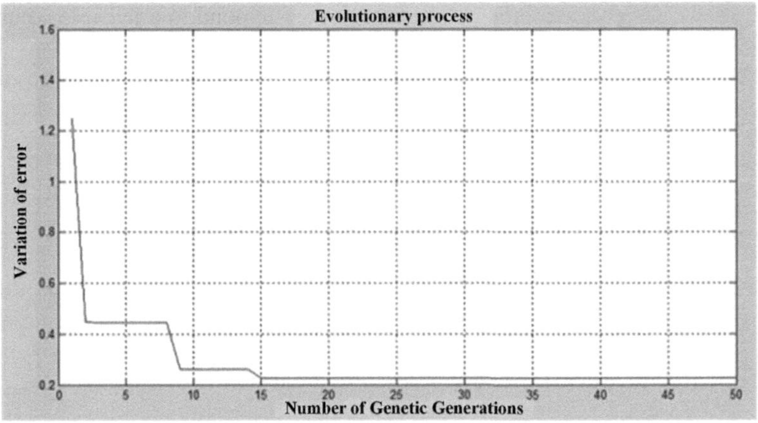

Fig. 9 Change graph of errors

the genetic algorithm, and the output of this function is the norm of the prediction error. The smaller the error of an individual, the higher the prediction accuracy of the network, which increases the fitness value of that individual in the genetic algorithm. The population size is set to 70, the maximum number of generations to 50, the generation gap to 0.8, the crossover probability to 0.8, and the mutation probability to 0.1. The required data for the model is shown in Table 5. The genetic algorithm processes-encoding, initializing the population, evaluating the population, and genetic operations (selection, crossover, and mutation) are applied to the configured parameters. The graph of error variation with genetic generations is shown in Fig. 9. From Fig. 9, it can be observed that after the 15th generation, the model's error stabilizes at 0.225, indicating a relatively small error.

The optimal initial weights and biases obtained from the genetic algorithm are applied to the neural network, yielding the predicted values of the turbocharger's influencing factors and calculating the relative error of the model's predictions. Table 8 presents a comparison of the prediction errors between the GA-SGNN model, the SGNN model, and the MGM(1, 6) model. From Table 8, it is evident that variables with relatively large errors in the SGNN model, such as coolant temperature in Sample 14, oil temperature in Sample 13, inlet pressure in Sample 13, coolant temperature in Sample 13, and bearing pressure in Sample 13, show significantly reduced relative errors after incorporating the genetic algorithm. This indicates that the genetic algorithm effectively optimizes the initial weights and biases, enhancing the accuracy of the GA-SGNN model.

To better compare the prediction accuracy of the GA-SGNN model, the MGM(1, 6) model, and the SGNN model, the relative errors of the predicted values for each model are represented in the form of a line graph. The relative error comparison chart is shown in Fig. 10. In Fig. 10, the horizontal axis represents the three prediction samples for six variables, while the vertical axis represents the absolute values of the relative errors for the GA-SGNN model, MGM(1, 6) model, and SGNN model.

Table 8 Point-by-point error comparison table of GA-SGSN, SGSN and MGM(1, 6) predicted

Variable		Sequence 13			Sequence 14			Sequence 15		
		MGM(1, 6)	SGNN	GA-SGNN	MGM(1, 6)	SGNN	GA-SGNN	MGM(1, 6)	SGNN	GA-SGNN
Rotational speed (r/min)	Measured value	108,025			110,121			115,716		
	Predicted value	107,888.0	108,767.0	104,430.0	113,670.0	113,058.0	107,721.0	118,147.0	114,062.0	112,547.0
	Residual	137.0	742.0	3595.0	3549.0	2937.0	2400.0	2431.0	1684.0	3169.0
	Relative error (%)	0.127	0.687	3.328	3.223	2.667	2.179	2.101	1.455	2.739
Oil temperature (°C)	Measured value	85.9			88.3			93.9		
	Predicted value	81.9	85.1	86.2	86.0	91.1	90.1	89.6	92.5	94.3
	Residual	4.0	0.8	0.3	2.3	2.8	1.8	4.3	1.4	0.4
	Relative error (%)	4.657	0.94	0.349	2.605	3.171	2.039	4.579	1.491	0.426
Inlet pressure (KPa)	Measured value	228			218			200		
	Predicted value	225.0	236.0	227.0	210.0	224.0	219.2	191.0	208.0	206.9
	Residual	3.0	8.0	1.0	8.0	6.0	1.2	9.0	8.0	6.9
	Relative error (%)	1.042	3.509	0.439	3.67	2.752	0.55	4.5	4.0	3.45
Exhaust pressure (KPa)	Measured value	716			697			687		

(continued)

Table 8 (continued)

Variable		Sequence 13			Sequence 14			Sequence 15		
		MGM(1, 6)	SGNN	GA-SGNN	MGM(1, 6)	SGNN	GA-SGNN	MGM(1, 6)	SGNN	GA-SGNN
	Predicted value	716.0	706.0	734.0	706.0	693.0	711.0	700.0	690.0	709.0
	Residual	0.0	10.0	18.0	9.0	4.0	14.0	13.0	3.0	22.0
	Relative error (%)	0.0	1.397	2.514	1.291	0.577	2.009	1.892	0.437	3.202
Coolant temperature (°C)	Measured value	84.0			86.8			89.3		
	Predicted value	81.1	81.6	86.3	82.5	85.3	88.0	85.2	86.1	89.9
	Residual	2.9	2.4	2.3	4.3	1.5	1.2	4.1	3.2	0.6
	Relative error (%)	3.452	2.857	2.738	4.954	1.728	1.382	4.591	3.583	0.672
Bearing pressure (MPa)	Measured value	5.01			4.79			4.78		
	Predicted value	5.05	4.97	5.19	4.81	4.7	4.94	4.6	4.64	4.81
	Residual	0.04	0.04	0.18	0.02	0.09	0.15	0.18	0.14	0.03
	Relative error (%)	0.798	0.798	3.593	0.418	1.879	3.132	3.766	2.929	0.628

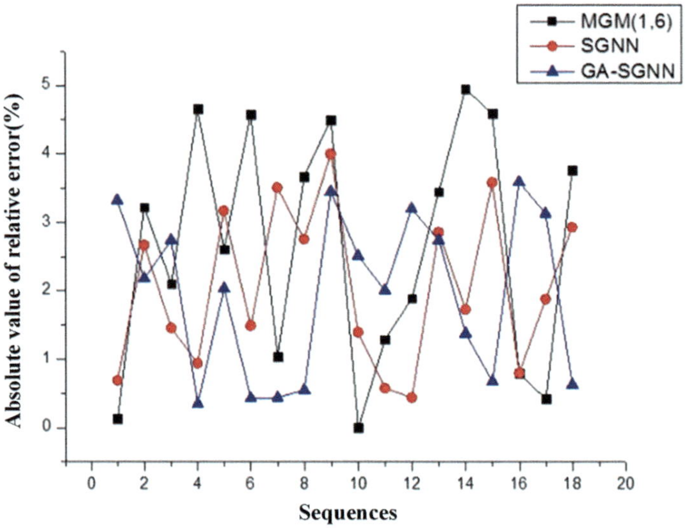

Fig. 10 Comparison of prediction error between GA-SGN and MGM(1, 6) and SGN

As shown in Fig. 10, after incorporating the genetic algorithm to optimize the weights and biases, the maximum peak of the relative error is significantly reduced compared to the previous models. Also, the absolute value of the relative errors for all predicted values in the GA-SGNN model is less than 4%. Compared to the MGM(1, 6) model and the SGNN model, the GA-SGNN model exhibits the lowest relative error. This demonstrates that the combined model offers higher accuracy and reliability than the individual models.

5 Conclusion

This paper first analyzes the operating characteristics of the turbocharger and, by reviewing the relevant literature, identifies the state parameters that influence the performance of locomotive turbochargers. Using data provided by a locomotive factory in Dalian, the research variables are determined, and the correlations between the variables are calculated. A multivariable grey prediction model is then established to predict the variables, and the residuals and relative errors are calculated based on the predicted and actual values. The accuracy of the multivariable grey prediction model is evaluated using the relative error. The results show that the accuracy of the MGM(1, 6) model meets the requirements, although some predicted values have a relative error close to 5%.

To improve the model's accuracy, the BP neural network is integrated with the previously established multivariable grey prediction model, combining the strengths of both approaches to build a more accurate prediction model—the Serial Grey

Neural Network Model (SGNN). The model is then applied to predict turbocharger data, and its accuracy is evaluated using residuals and relative errors. A comparison is conducted with the prediction accuracy of the MGM(1, 6) model, showing that the SGNN model significantly reduces the relative error of the MGM(1, 6) model.

To further enhance the accuracy and reliability of the model, a genetic algorithm is introduced to address the local optimum and convergence issues in the SGNN model. This led to the development of the Genetic Algorithm-Grey Neural Network Model (GA-SGNN). The GA-SGNN model is applied to predict turbocharger data, and its accuracy is evaluated using residuals and relative errors. The prediction accuracy is compared with that of the previous two models. The results indicate that the relative error of the GA-SGNN model is within 4%, ensuring high reliability and accuracy of the model.

References

1. Saito H, Kanzaki D, Yonekura K (2024) Applications of machine learning in surge prediction for vehicle turbochargers. Mach Learn Appl 16:100560
2. Zhao Y, Liu Z, Fu L, Zhao M, Li X (2023) Redesign of a turbocharger compressor based on multi-component full-passage optimization. Aerosp Sci Technol 142:108644
3. Wang J, Shen L, Bi Y, Lei J (2023) Influences of a variable nozzle turbocharger on the combustion and emissions of a light-duty diesel engine at different altitudes. Fuel 349:128626
4. Farhan A, Siraj M T, Nishi A A, Jamil S N S, Ahmed R, Kabir A and Hossen S (2023) Analyzing failure modes of turbochargers for engine-based power plants by combined SWARA-FMEA approach. AIP Conf Proc 2825(1)
5. Dong F, Yang J, Cai Y, Xie L (2023) Transfer learning-based fault diagnosis method for marine turbochargers. Actuators 12(4):146
6. Daniels RK, Kumar V, Prabhakar A (2025) A comparative study of data-driven thermal fault prediction using machine learning algorithms in air-cooled cylindrical Li-ion battery modules. Renew Sustain Energy Rev 207:114925
7. Qiu S, Cui X, Ping Z, Shan N, Li Z, Bao X, Xu X (2023) Deep learning techniques in intelligent fault diagnosis and prognosis for industrial systems: a review. Sensors 23(3):1305
8. Zhang F, Zhang K, Xie G, Ba D and Jiang A (2024) A review of fault prediction methods for high speed elevator brakes for service safety. In: International workshop of advanced manufacturing and automation. Springer, Singapore, pp 522–528
9. Feng Z, Ma H, Zuo MJ (2016) Vibration signal models for fault diagnosis of planet bearings. J Sound Vib 370:372–393
10. Hountalas DT (2000) Prediction of marine diesel engine performance under fault conditions. Appl Therm Eng 20(18):1753–1783
11. Hountalas DT, Kouremenos DA, Sideris M (2004) A diagnostic method for heavy-duty diesel engines used in stationary applications. J Eng Gas Turbines Power 126(4):886–898
12. Benvenuto G, Campora U (2007) Performance prediction of a faulty marine diesel engine under different governor settings. In: International conference on marine research and transportation, pp 35–44
13. Campora U, Figari M (2003) Numerical simulation of ship propulsion transients and full-scale validation. Proc Inst Mech Eng M J Eng Maritime Environ 217(1):41–52
14. Ceccarelli R (2012) Model-based fault detection in diesel engines air-path. Université de Grenoble
15. Li Z, Yan X, Yuan C, Peng Z (2012) Intelligent fault diagnosis method for marine diesel engines using instantaneous angular speed. J Mech Sci Technol 26:2413–2423

16. Chen Y, Qiu J, Wang M, Rao J, Xiao T, Yang Y (2024) Fault prediction of pneumatic valves in an LNG plant by the DGM (1, 1) model. Int J Syst Assurance Eng Manage 15(3):775–785
17. Zhang H, Long H, Chen F, Luo Y, Xiao X, Deng Y, Lu N, Liu Y (2024) Temperature field prediction for a PC beam bridge with corrugated steel webs using BP neural network and measured data. Structures 68:107232
18. Holland JH (1992) Adaptation in natural and artificial systems: an introductory analysis with applications to biology, control, and artificial intelligence. MIT Press
19. Butt MK, Cao L, Wan C, Lei K, Khan I (2024) Integrating the deep learning and multi-objective genetic algorithm to the reloading pattern optimization of HPR1000 reactor core. Nucl Eng Des 428:113531
20. Li H, Yazdi M (2022) What are the critical well-drilling blowouts barriers? A progressive DEMATEL-game theory. Advanced decision-making methods and applications in system safety and reliability problems: approaches, case studies, multi-criteria decision-making, multi-objective decision-making, fuzzy risk-based models. Springer International Publishing, Cham, pp 29–46
21. Li H, Peng W, Adumene S, Yazdi M (2023) Cutting edge research topics on system safety, reliability, maintainability, and resilience of energy-critical infrastructures. Intelligent reliability and maintainability of energy infrastructure assets. Cham, Springer Nature Switzerland, pp 25–38
22. Li H, Peng W, Adumene S, Yazdi M (2023) Advances in failure prediction of subsea components considering complex dependencies. Intelligent reliability and maintainability of energy infrastructure assets. Cham, Springer Nature Switzerland, pp 93–105
23. Li H, Yazdi M (2022) Reliability analysis of correlated failure modes by transforming fault tree model to Bayesian network: a case study of the MDS of a CNC machine tool. Advanced decision-making methods and applications in system safety and reliability problems: approaches, case studies, multi-criteria decision-making, multi-objective decision-making, fuzzy risk-based models. Springer International Publishing, Cham, pp 15–28
24. Guo J, Wang Z, Li H, Yang Y, Huang CG, Yazdi M, Kang HS (2024) A hybrid prognosis scheme for rolling bearings based on a novel health indicator and nonlinear Wiener process. Reliab Eng Syst Saf 245:110014
25. Li H, Yazdi M, Huang HZ, Huang CG, Peng W, Nedjati A, Adesina KA (2023) A fuzzy rough copula Bayesian network model for solving complex hospital service quality assessment. Complex Intell Syst 9(5):5527–5553

GPSR Compliance

The European Union's (EU) General Product Safety Regulation (GPSR) is a set of rules that requires consumer products to be safe and our obligations to ensure this.

If you have any concerns about our products, you can contact us on ProductSafety@springernature.com

In case Publisher is established outside the EU, the EU authorized representative is:

Springer Nature Customer Service Center GmbH
Europaplatz 3
69115 Heidelberg, Germany

Batch number: 10732064

Printed by Printforce, the Netherlands